Handbook of Tunnel Fire Safety

Second edition

ice
Institution of Civil Engineers
publishing

Handbook of Tunnel Fire Safety

Second edition

Edited by

Alan Beard
Civil Engineering Section, School of the Built Environment,
Heriot-Watt University, Edinburgh, UK

Richard Carvel
BRE Centre for Fire Safety Engineering, University of Edinburgh, UK

Published by ICE Publishing, 40 Marsh Wall, London E14 9TP.

Full details of ICE Publishing sales representatives and distributors can be found at: www.icevirtuallibrary.com/info/printbooksales

First published 2005

Also available from ICE Publishing

Specification for Tunnelling, Third edition.
British Tunnelling Society. ISBN 978-0-7277-3477-8
ICE Manual of Health and Safety in Construction.
C. McAleenan. ISBN 978-0-7277-4056-4
Designers' Guide to EN 1991-1-2, EN 1992-1-2, EN 1993-1-2 and EN 1994-1-2.
T. Lennon. ISBN: 978-0-7277-3157-9

Associate Commissioning Editor: Victoria Thompson
Production Editor: Imran Mirza
Market Development Executive: Catherine de Gatacre

www.icevirtuallibrary.com

A catalogue record for this book is available from the British Library

ISBN: 978-0-7277-4153-0

FSC
www.fsc.org
MIX
Paper from
responsible sources
FSC® C013604

Typeset by Academic + Technical, Bristol
Index created by Indexing Specialists (UK) Ltd, Hove, East Sussex
Printed and bound by CPI Group (UK) Ltd, Croydon, CR0 4YY

Contents

Preface

It is several years since the first edition of *The Handbook of Tunnel Fire Safety* appeared in 2005, which followed a series of very serious fires in tunnels. Since then, much has taken place. Firstly, there has been research under the aegis of various programmes, such as UPTUN and SAFE-T. Secondly, there have been more very serious fires in both road and rail tunnels. The fact that such fires continue to take place means that there is no room for complacency. Tunnel construction worldwide has continued at a fast pace in recent years, in both urban (e.g. the Shanghai metro) and non-urban (e.g. the Gotthard Base Tunnel) settings. Rapid changes, both in the technology and methods directly associated with tunnels and in society in general, mean that research and learning from the experience of others becomes ever more important. In addition, every tunnel is unique, and there is no 'one size fits all' solution. Alongside this there is a concern within society about tunnel safety and fire safety in particular. This Handbook is for all those involved, from fire brigade personnel, who are at the sharp end when a tunnel fire occurs, to tunnel designers, operators and regulators, as well as researchers. It is intended that a central theme runs through the book: the need to see fire risk as a product of the working of a system. Therefore, considerations regarding emergency planning and design against fire need to be brought in at the beginning of the design stage; the philosophy of regarding fire safety measures as a 'bolt-on' after a design has largely been completed is now totally unacceptable, especially given the ever longer and more complex tunnels being built or planned.

Within this context, it is hoped that this text will provide a bridge between tunnel fire research and those who need to know the basic results, techniques and current thinking in relation to tunnel fire safety. It is also a vehicle for the transmission of experience gained from real-world tunnels. The Handbook covers a wide spectrum, and the chapters are written by international experts in various fields. Much remains to be done, however. For example, while we know more than we did about human behaviour during tunnel fire incidents, there is still a long way to go. In addition, the prevention of fires in tunnels, as opposed to trying to provide protection once a fire has broken out, requires much more consideration. The issue of which fire protection measures to adopt has come to the fore in recent years, and this is reflected in these pages. A key question is 'What is most appropriate?' rather than, necessarily, 'What is the newest?'. Furthermore, the general move towards performance-based decision-making implies the use of models, and this is problematic, as it relates to the question of what is 'acceptable risk' in relation to tunnel fires. Much consideration and debate needs to take place in this area, including all those involved and affected. This Handbook is intended to represent the broad sweep of knowledge available at the present time, and it is hoped that it will become a valuable resource for all those concerned with tunnel fire safety.

Alan Beard

Richard Carvel

Contributor's list

Alan Beard is Honorary Research Fellow at the Civil Engineering Section, School of the Built Environment, Heriot-Watt University, Edinburgh. His research is in modelling – including deterministic and probabilistic – and qualitative research, especially applying the concepts of systems to safety management. He was commissioned by the European Parliament to conduct a study on tunnel safety risk assessment and make recommendations for future policy in the European Union (published 2008) and was presented with a 'Life-time Achievement Award' in 2010. His work includes helping to develop a framework for the acceptable use of theoretical models in fire safety decision-making.

Richard Carvel is Research Fellow and Assistant Director of the BRE Centre for Fire Safety Engineering at the University of Edinburgh. He has a background in Chemistry and Physics and has worked in the field of tunnel fire phenomena since 1998. He has been involved in post-graduate research at several institutions, including Heriot-Watt University where he completed his PhD on Fire Size in Tunnels in 2004, and has worked for International Fire Investigators and Consultants (IFIC). He currently provides fire safety engineering consultancy services to a number of companies and tunnel operators, primarily on tunnel fire safety.

A. Bendelius	A & G Consultants, Inc., USA
A. Bergqvist	Stockholm Fire Brigade, Sweden
R. Borchiellini	Dipartimento di Energetica, Politecnico di Torino, Italy
K. Both	Efectis Nederland BV, The Netherlands
D. Burns	Formerly of Merseyside Fire Brigade, UK
C. Calisti	Formerly Technical Consultant, Police force of Paris and expert to the French Courts of Appeal
P. Cassini	Institut National de l'Environnement Industriel et des Risques (INERIS), Verneuil-en-Halatte, France
D. Charters	BRE Global, Watford, UK
F. Colella	Dipartimento di Energetica, Politecnico di Torino, Italy; BRE Centre for Fire Safety Engineering, University of Edinburgh, UK
O. Delémont	Institut de Police Scientifique (IPS), Ecole des Sciences Criminelles (ESC), University of Lausanne
A. Dix	University of Western Sydney, Australia
M. Egger	Conference of European Directors of Roads, Paris, France
H. Frantzich	Lund University, Sweden
G. Gash	Kent Fire Brigade, Kent Fire and Rescue Service, UK
J. Gillard	Formerly of Mersey Tunnels, UK
G. Grant	Optimal Energies Ltd, UK
K. Hasselrot	BBm Fire Consulting, Stockholm, Sweden
P. Hedley	Tyne Tunnels, Wallsend, UK

H. Ingason	SP Technical Research Institute of Sweden, Borås, Sweden
S. Jagger	Health and Safety Laboratory, Buxton, UK
H. Knoflacher	Technical University of Vienna, Austria
A. Lönnermark	SP Technical Research Institute of Sweden, Borås, Sweden
S. Maciocia	Formerly of Securiton AG, Switzerland
G. Marlair	Institut National de l'Environnement Industriel et des Risques (INERIS), France
J.-C. Martin	Formerly of the Institut de Police Scientifique (IPS), Ecole des Sciences Criminelles (ESC), University of Lausanne
M. Nielsen	Formerly of Tyne & Wear Fire Brigade, UK
A. Noizet	SONOVISION Ligeron, Saint Aubin, France
J. Olesen	Slagelse Fire Brigade, Denmark
G. Rein	School of Engineering, University of Edinburgh, UK
N. Rhodes	Hatch Mott MacDonald, New York, NY, USA
A. Rogner	Metaphysics SA, Switzerland
J. Santos-Reyes	SARACS Research Group, SEPI-ESIME, Instituto Politecnico Nacional, Mexico City, Mexico
P. Scott	FERMI Ltd, UK
J. Shields	University of Ulster, UK
M. Shipp	Fire Safety Group, Building Research Establishment, UK
B. Truchot	Institut National de l'Environnement Industriel et des Risques (INERIS), Verneuil-en-Halatte, France
V. Verda	Dipartimento di Energetica, Politecnico di Torino, Italy
Y. Wu	Department of Chemical and Process Engineering, Sheffield University, UK

Introduction

The general shift from prescriptive to performance-based decision-making implies the use of models, both experimental and theoretical, and risk assessment. Risk assessment may be qualitative or quantitative, deterministic or non-deterministic; or a mixture of these elements. While, in principle, this shift has advantages, there are many problems as well (Beard and Cope, 2008). As a general rule, the conditions do not yet exist for reliable and acceptable use of complex computer-based models as part of tunnel fire safety decision-making, and these conditions need to be created (see the Chapter 29). Given the degree of flexibility possible in the application of models, and the uncertainty in the conceptual and numerical assumptions of models, it may be quite possible to present an option desired (for whatever reasons) by a client as justifiable, whether using a probabilistic or deterministic model, or a mixture of both. Some countries, most notably Japan, do not use risk assessment (i.e. quantitative risk assessment at least) other than in relatively rare cases.

Prescriptive requirements represent a rich seam of knowledge and experience. There are good reasons why it is sensible to move to a system that includes risk assessment. However, it would be wise not to throw prescriptive regulations overboard *en masse*. Indeed, an element, perhaps a large element, of prescriptive regulation should be retained in a system of design and regulation. The question becomes: what is a 'healthy mixture' of prescriptive requirements, qualitative risk assessment and quantitative risk assessment?

If risk assessment is to be justifiably employed as part of tunnel fire safety decision-making, then a sound knowledge base needs to be created, and this needs to be continually updated because systems change. Research, and the translation of research results into practice, become essential, and experience gained in one quarter needs to be shared. This is where the present Handbook comes in. While a lot has been done over the last few years to add to the knowledge base, there are many questions and issues remaining, and new questions and issues will emerge as time passes. While a significant amount of research has been conducted in recent years, only a relatively small amount has been published in the standard literature and subjected to peer review and critical comment by others. While peer review certainly has problems, it is important for research to be subjected to critical comment. Furthermore, much research and testing is carried out but results are kept confidential, for commercial or other reasons, or only selected results may be released. In addition, the basis of decisions may not be made public. Such secrecy is not acceptable in relation to public safety decision-making.

Questions and issues are raised throughout the chapters of this book. Some basic issues, in no particular order of importance, which exist in relation to tunnel fire safety are given below; there is no doubt that there are many others.

■ Fire risk in tunnels is a result of the working of a system involving design, operation, emergency response and tunnel use. That is, fire risk is a *systemic product*. Furthermore, this 'tunnel system' involves both 'designed parts' and 'non-designed parts' (e.g. traffic volume or the individual behaviour of users). The designed parts need to take account of the non-designed parts as much as possible.

- Tunnels are becoming ever larger and more complex. How do we deal with this?
- The system changes – the tunnel system that exists at the time of opening will be different from the tunnel system that exists a few years later.
- What is to be regarded as acceptable for fire risk with regard to: (1) fatality/injury, (2) property loss and (3) disruption of operation? This applies to both new and existing tunnels.
- What is to be an acceptable methodology for tunnel fire safety decision-making?
- Risk assessment needs to be carried out in as 'independent' a way as possible.
- Given the need for a 'healthy mixture' of prescriptive and non-prescriptive elements in decision-making, a framework needs to be created within which both deterministic and non-deterministic models may play a part. A synthesis of deterministic and non-deterministic elements needs to be achieved.
- Both large- and full-scale experimental tests, as well as small-scale experimental tests, are needed. Tests are needed that simulate, as far as possible, real tunnel incidents, such as collisions. Tests using real vehicles and real (different) cargoes, not just mock-ups, are also required, as are tests on modern vehicles.
- Experimental tests need to be replicated, because of the variability in the results obtained using ostensibly 'identical' tests.
- With regard to operator response, (a) to what extent is automation feasible or desirable, and (b) to what extent can decision-making during an emergency be simplified and yet still be suitable to cope effectively with different emergency situations, in increasingly complex tunnel systems?
- While more is now known about the dynamics of tunnel fires, much more still needs to be known. For example, more knowledge is needed about the conditions for flame impingement, factors affecting flame length, the effect of ventilation on fires, and the production of smoke and toxic gases.
- With regard to fire suppression, what kinds of system are appropriate, and how do such systems interact with ventilation?
- How is real human behaviour (both users and operational/ emergency staff) to be taken into account in tunnel-fire emergencies? While more is now known about this, there is still a very long way to go.
- Is a reliance on computers for the operation of tunnel safety-related equipment introducing a new vulnerability?
- How can we learn as much as possible from real events, near misses, etc.? There needs to be an openness about reporting on such matters, without repercussions.

The research needed is implied by the issues raised above. More specifically, to pinpoint a very few, some key research questions that need to be answered are as follows.

- What are effective ways of preventing fires occurring in tunnels?
- What are the factors affecting tunnel-fire size and spread?
- What are the characteristics of different tunnel-fire suppression systems?
- How do human beings behave in tunnel-fire emergencies – both users and tunnel staff and fire brigade personnel? The behaviour of the elderly, the young and the disabled, as well as 'average' people, needs to be understood.
- What are effective evacuation systems?
- To what extent should the emergency response be 'automated'?
- How can the flexibility and uncertainty inherent in the models used as part of fire safety decision-making be dealt with?

These matters are addressed to some degree in this Handbook, as are other issues and the required research areas implied in the associated chapters (e.g. Chapter 25). International collaboration in research has played an important role in the past, and may be expected to continue to do so. There needs to be a continuing strategy for tunnel-fire research, involving both international collaboration and effort by individual countries. Furthermore, there needs to be an openness about research results and decision-making. However it is done, these issues and the associated implied research areas need to be effectively addressed for the benefit of all countries and their citizens.

REFERENCE

Beard AN and Cope D (2008) *Assessment of the Safety of Tunnels.* Prepared for the European Parliament, and available on their website under the rubric of Science and Technology Options Assessment (STOA).

Alan Beard
Civil Engineering Section, School of the Built Environment, Heriot-Watt University, Edinburgh, UK

Part I

Real tunnel fires

Handbook of Tunnel Fire Safety, 2nd edition
ISBN: 978-0-7277-4153-0

ICE Publishing: All rights reserved
doi: 10.1680/htfs.41530.003

Institution of Civil Engineers

publishing

Chapter 1
A history of fire incidents in tunnels

Richard Carvel
BRE Centre for Fire Safety Engineering, University of Edinburgh, UK
Guy Marlair
Institut National de l'Environnement Industriel et des Risques (INERIS), France

1.1.   Introduction

According to French (Perard, 1996), German (Baubehörde Highways Department, 1992), Swiss (Ruckstuhl, 1990) and Italian (Arditi, 2003) statistics, accidents seem to occur less frequently in road tunnels than on the open road. This is possibly because tunnels are a more 'controlled' environment than the open road; there are generally no complications caused by weather, there are fewer junctions and no sharp bends, and drivers tend to be more attentive when driving in tunnels. However, there is no doubt that the consequences of a fire in a tunnel can be far more serious than the consequences of a fire in the open air. According to the French statistics, there will only be one or two car fires (per kilometre of tunnel) for every hundred million cars that pass through the tunnel. Similarly, out of every hundred million heavy goods vehicles (HGVs) passing through a tunnel, there will be about eight fires (per kilometre of tunnel), only one of which will be serious enough to cause any damage to the tunnel itself. On the basis of the statistics, it has been estimated that there will be between one and three very serious fires (i.e. involving multiple vehicles and fatalities) out of every thousand million HGVs (per kilometre of tunnel).

The chance of a serious accidental fire may sound vanishingly small from these statistics, but when one considers that many road tunnels have very high traffic densities (e.g. about 37 million vehicles per year travelled through the Elb Tunnel in Germany in the mid-1990s (Baubehörde Highways Department, 1992)), there are over 15 000 operational road, rail and underground railway/metro tunnels in Europe alone, and that some of these tunnels are many kilometres long, the chance of a serious fire incident in a tunnel may be greater than is commonly thought. Indeed, significant and fatal accidental fires in tunnels seem to occur on an annual basis.

This is a serious problem, and has the potential to become much worse in the future as more and longer tunnels are constructed and as traffic densities increase. Several of the chapters in this Handbook deal directly with issues relating to the fire safety of tunnels in the future, but this chapter looks to the past to see what can be learned from the fire incidents that have already happened in tunnels. First, a small number of case studies is discussed, and this is followed by a list of serious and significant fire incidents in tunnels. As the issues involved in road-tunnel fires and rail-tunnel fires are significantly different, these are considered separately.

1.2. Fires in road tunnels

On 24 March 1999, a HGV travelling through the Mont Blanc Tunnel from France to Italy caught fire, possibly due to the engine overheating. This HGV stopped 6 km into the tunnel, when the driver became aware of the fire; he was unable to put the fire out, and fled, on foot, towards Italy. Within minutes the tunnel operators were aware of the fire, and prevented further vehicles from entering the tunnel. However, 18 HGVs, nine cars, a van and a motorcycle had already entered the tunnel from France after the first HGV and before the tunnel was closed. Of these 29 vehicles, four HGVs managed to pass the burning HGV and travel on towards Italy in safety, but the other 25 vehicles became trapped in the smoke and eventually became involved in the fire. Nobody travelling in any of these vehicles survived. Due to the prevailing wind direction (from the south) and the different ventilation regimes at either end of the tunnel (all ventilation ducts at the Italian end were set to supply fresh air, whereas at the French end some ducts were set to supply and some were set to exhaust), virtually all the smoke from the fire was carried towards France. As the airflow velocity was more than 1 m/s, the smoke did not remain stratified, and within minutes there was no fresh air in the tunnel downstream of the fire. The fire grew to involve the 25 vehicles behind the first HGV, eight HGVs (which had been abandoned by their drivers travelling from Italy to France, the nearest one being some 290 m from the initial HGV fire), and the first fire-fighting vehicle which entered the tunnel from the French side (which was almost half a kilometre from the nearest vehicle on fire). It is unclear how the fire managed to spread such distances, although explanations such as burning liquid fuels and the involvement of pavement materials in the fire have been proposed. At the height of the fire, the blaze was estimated to have been about 190 MW in size, with temperatures in the tunnel exceeding 1000°C. The fire took 53 hours to extinguish, and hot spots were still being dealt with after 5 days. Thirty-eight tunnel users and one firefighter died as a result of the fire, 27 in their vehicles, two in an emergency shelter (designed to protect life in the event of a fire) and the rest on the roadway trying to reach the French portal (Lacroix, 2001). This was the greatest ever loss of life in any road-tunnel fire (except the Salang fire in 1982, which involved an explosion as well as a fire, see below).

The incident in the Mont Blanc Tunnel was the eighteenth HGV fire recorded in the tunnel since it opened in 1965. Of the other 17 incidents, most had been minor, and only five required the intervention of the fire brigade; none of the other incidents resulted in any fatalities. No single factor was responsible for the severity of the fire; rather the fire became a tragedy due to a combination of factors, including the weather conditions, the different ventilation regimes at each end of the tunnel, and the highly flammable nature of the trailer (insulated with polyurethane foam) and its cargo (margarine and flour) on the initial HGV.

On 7 April 1982, a collision in the Caldecott Tunnel, Oakland, USA, occurred when a passenger car, driven by a drunk driver, collided with the roadside and came to an abrupt halt. The stationary car was struck by a petrol tanker, and subsequently the tanker was struck by a bus, causing the tanker to turn over, partially rupturing and spilling some of its load. The spilt petrol soon ignited, and the blaze grew to involve the tanker, the car and four other vehicles in the tunnel. Seven people were killed (NTSB, 1982). Although some of the petrol spilled out of the tanker, it appears that a large quantity of the fuel remained in the tank. Once the fire had reached a sufficient temperature to melt the aluminium walls of the tanker, the 'top' of the tanker (i.e. the side that was uppermost after the crash) collapsed, creating a large, deep pool of petrol, which ignited and burned fiercely for over 2 hours until firefighters were able to extinguish it (Satoh and Miyazaki, 1989). Once again, a combination of factors brought about a disaster.

On 11 July 1979, a collision in the Nihonzaka Tunnel, near Yaizu City, Japan, resulted in a fire that destroyed 173 vehicles. However, the seven fatalities that occurred during the incident were all a result of the crash itself, not the fire; over 200 people escaped the tunnel on foot before the fire became established. The Nihonzaka Tunnel is unique among those described here, in that it had a water spray system to suppress fires. This automatic system began sprinkling the tunnel only 11 minutes after the crash, and this successfully suppressed the growth of the fire for about half an hour – long enough for all the people in the 2 km long tunnel to walk out. After this time, however, the unburned fuel vapours reignited, and the fire established itself once more. One hour later, the sprinkler reservoirs ran dry, and the fire grew dramatically. It was 2 days before the blaze was 'under control', and a further 5 days before it was extinguished altogether (Oka, 1996).

It seems that human behaviour (including, but not restricted to, human error) is a major factor contributing to fatalities in road tunnel fires. The fires in the Nihonzaka and Caldecott Tunnels both started as a result of a collision, whereas many of the people who died in the Mont Blanc incident may have survived if they had evacuated their vehicles quickly and run away from the smoke. The conditions in the Mont Blanc Tunnel may also have been less severe if the operators in the control centres had adopted a different ventilation strategy; during the fire incident, ventilation duct 5 was set to supply air at a number of locations, including the fire location; if it had been set to extract smoke, there would have been less oxygen to feed the fire (Lacroix, 2001). In the list of tunnel-fire accidents given at the end of this chapter, approximately one-third of all road-tunnel fires started as the result of human behaviour, while just over half of the incidents started due to mechanical or electrical failure. One of the primary lessons to be learned from these incidents is that road-tunnel users (i.e. vehicle drivers) need to be appropriately informed in case of an emergency, and this is particularly true in long road tunnels. It is also incumbent upon tunnel designers and operators to take account of human behaviour in tunnel situations (see Chapters 19–21). However, at present, very little is known about human behaviour in emergency situations in tunnels.

The list of accidents in road tunnels given at the end of this chapter shows that any type of vehicle may be involved in fire incidents, either as the first ignited object, or due to fire spread. Fire incidents in road tunnels have involved private cars, HGVs of all kinds and with various cargoes (sometimes, but not always, including dangerous goods), motorcycles, vans, camper trucks, and urban and tourist buses. Most of the incidents that have led to multiple fatalities have involved one or more HGVs; these have generally contributed greatly to the overall fire load involved in the incident, and these large fires have meant that direct fire-fighting has been extremely difficult and rescue operations have been hindered. These HGVs were not necessarily carrying 'dangerous' or 'hazardous' goods.

It may also be observed from the list of accidents that several fire incidents involving passenger coaches or buses have been recorded. However, most of these incidents involved few, if any, casualties. Considering that the number of potential victims in a single passenger coach is comparable to the overall death toll observed in the Mont Blanc fire disaster (which involved numerous vehicles), this seems quite surprising. The lack of serious incidents involving buses and coaches is possibly due to improved safety features, and to some extent, a greater awareness of safety issues at the design stage, of modern buses and coaches. These features have come to be in place as a consequence of a number of tragic incidents in the past, such as the accident near Beaune, France, in July 1982, which involved 53 fatalities, including 44 children. Hopefully, tunnels can also be made safer environments at the design stage; indeed, it is essential that fire safety be considered at the design stage, and not later as an 'add on'. This will only come about if we can learn the lessons of the past.

1.3.   Fires in rail tunnels

The majority of mass-transit systems used for public transport are railway systems, which generally consist of trains, each with the capacity for carrying several hundred passengers. These systems clearly have a much higher potential for a large number of casualties in the event of a fire, compared with fires in road tunnels. Two incidents in recent years have highlighted the horrendous scale of possible consequences of fires in mass-transit (metro) systems: in 2003, nearly 200 people died following an arson attack on an underground railway/metro train in South Korea; and, in 1995, over 200 people died following an electrical fire on an underground railway/metro train in Azerbaijan (see below).

Fire disasters have also occurred on conventional and funicular railways. On 11 November 2000, a fire started on a funicular railway carrying skiers up to the Kitzsteinhorn glacier, near Kaprun in Austria. Had the train been travelling up the side of the mountain rather than through a tunnel, it is unlikely that the fire would have made the news outside of Austria, but due to the confines of the tunnel the fire was directly responsible for the deaths of 151 people on the train, the driver of a second train in the tunnel and three people near the top portal of the tunnel. The handful of survivors were those who fled down the tunnel (past the fire); those trying to escape the fire by going up the tunnel were all killed by smoke.

The reports on the Kaprun incident indicate that the majority of the passengers on the train did not manage to get off the train before they succumbed to the poisonous smoke (Schupfer, 2001). This was also the case in the fire incident on a Baku underground railway/metro train on 28 October 1995. The bodies of 220 passengers were found on the train itself, while a further 40 passengers succumbed to the fumes while making their way along the tracks toward the station (Hedefalk et al., 1998).

In both these cases the lack of a safety management system was partly responsible for the number of deaths. In the Kaprun incident, the train was held to be 'fire proof' so the consequences of a fire onboard had never been considered, and the possibility that the passengers could be carrying flammable materials, at least in the form of clothing, also appears to have never been considered. In the Baku incident, the lack of communication and the ad hoc operation of the ventilation system also led to fatalities. Some 15 minutes after the fire started, the emergency ventilation system was switched on, and this directed the smoke towards the majority of the passengers – the lack of communication meant that those in the control room had no idea what was going on at the fire (Hedefalk et al., 1998).

Aside from these incidents, large-scale fires rarely happen on passenger trains, as there is comparatively little fuel to burn and, usually, many people are able to extinguish the fire while it is still small. This is not the case for goods trains.

On the evening of 18 November 1996, one of the HGVs onboard a HGV carrier shuttle in the Channel Tunnel caught fire. Upon entering the tunnel, the fire size has been estimated to have been about 1.5 MW; when the fire reached its maximum extent, its size was as much as 350 MW and involved ten HGVs and their carrier wagons. Unlike the Baku underground railway/metro and the Kaprun funicular tunnels, the Channel Tunnel is a very well equipped, modern tunnel with good communications and a carefully planned safety management system, which utilises frequent cross-passages to the service tunnel along the length of the tunnel, a supplementary ventilation system to control smoke movement, and a positive-pressure ventilation system in the service tunnel to prevent smoke entering the cross-passages. Although the supplementary ventilation system was not fully operational until some 30 minutes after the fire was first detected, the safety management system was able to enable the escape of all the HGV drivers and the crew into the service tunnel, with no fatalities and only two escapees requiring significant hospital attention (Allison, 1997). However, this should not allow for complacency. The result may have been very different had the fire started on a HGV close behind the amenity coach, which carried the HGV drivers; instead of starting a considerable distance behind the amenity coach, towards the other end of the train. After all the people had been successfully evacuated, the supplementary ventilation system maintained a ventilation velocity of 2.5 m/s in the tunnel, which may have helped fan the fire and thus enable it to grow to such a size. The main fire was extinguished by 5 a.m. the following day, some 7 hours after the fire was first detected, by firefighters working in relays. Minor smouldering and fires were still in evidence 24 hours later. Two other significant fires have occurred in the Channel Tunnel since 1996. In 2006, a fire completely destroyed one HGV but, remarkably, did not spread to either of the adjacent vehicles, and in 2008 a very large fire developed in much the same manner as the 1996 fire, destroying many vehicles and causing major structural damage to the tunnel (see below).

Possibly the most severe train fire in a tunnel also resulted in no fatalities. On 20 December 1984, a goods train pulling 13 petrol tankers derailed in the Summit Tunnel, near Rochdale, England. The resulting fire burned for 3 days and the flames from the fire reached heights of 120 m above the top of the tunnel ventilation shafts. Remarkably, the train driver escaped unharmed (Jones, 1985).

The majority of train fires in tunnels appear to have started as the result of electrical or mechanical failure, with only a small percentage starting as a direct result of human behaviour. The picture for underground railway/metro fires is slightly different; very few fires started as the direct result of human error, about two-thirds started as the result of electrical or mechanical failure, while more than one in ten of the fires were started deliberately.

An important fire safety issue concerning existing railway tunnels has been raised. While a significant number of safety measures have been implemented in tunnels constructed in the last few decades, there are a vast number of older tunnels, particularly in Europe, many of which are over a hundred years old. It has been questioned whether fire safety requirements can ever be met in these tunnels and, indeed, if it is even safe for fire-fighters to attempt to fight fires or mount rescue attempts in such tunnels (Nothias, 2003). In many cases (e.g. the Mornay Tunnel fire in 2003), there is not even a local water supply for fire-fighters to use. The problems associated with fire-fighting and rescue are discussed in more detail in Chapters 26–28 on emergency procedures and fire and rescue operations.

1.4.   Concluding comments

Although there have been a number of fire incidents on trains in tunnels, it seems that fires in road tunnels are more frequent, although the number of fatalities in road-tunnel incidents generally seems to be far smaller than in train fires. Indeed, the combined death toll of the Baku and Daegu underground railway/metro system fires is greater than the total number of fatalities in all recorded road-tunnel fires.

The main consideration for fire safety management in tunnels is reducing the risk of fatalities. However, there are also life-safety issues involved in freight transport, despite there generally being few crew members on freight trains. These factors cannot be demonstrated by actual historical events, as such fires have not (yet) happened, but it is unwise to wait for a disaster to occur before trying to avoid such events in the future. A study was undertaken to investigate what the consequences might have been if the freight train fire that occurred in Baltimore, USA, in 2001 had involved nuclear waste. Under those circumstances, the fire would have resulted in the exposure of more than 390 000 people to extremely high levels of radiation, and would have had significant consequences over many hundreds of square miles of the surrounding area (Lamb and Resnikoff, 2001). This example may sound rather extreme, but the rail tunnel in which the fire occurred was actually being considered as part of the route for the transport of nuclear waste.

Fires in tunnels are an international concern. The list given in the following section details 75 major fires that have occurred in 23 countries in Europe, Asia, Africa, North America and Oceania. It is apparent from the list that tunnel fire incidents have not only occurred while tunnels are in normal operation, but have also occurred during the excavation and construction phase, during repair and refurbishment works, and even during the pre-commissioning period. Indeed, a minor fire broke out on a bus carrying about 50 attendees through the Laerdal Tunnel in Norway to the opening ceremony of the tunnel. During construction or repair, evacuation routes may be limited or possibly not even available, and some extraordinary fire loads may be present, requiring specific emergency response tactics.

Very often the length of a tunnel is presented as an important factor contributing to risk, in particular to the extent that a long tunnel may have a negative influence on drivers' behaviour. However, the list of tunnel fires presented below demonstrates that serious fire incidents have often occurred in relatively short tunnels, not only long ones. Of particular note is the incident in the Isola delle Femmine tunnel, Italy, in 1996. At only 148 m long, this tunnel is one of the shortest recorded here, yet the incident was one of the most severe.

The analysis provided here also clearly illustrates that valuable lessons may be learnt from past accidents. In addition, cases that may be perceived as 'near misses', a few of which are presented below, may be more instructive than imagined by tunnel designers and operators (Bodart *et al.*, 2004). This is the reason why any initiative leading to more efficient reporting and recording of any significant incidents in tunnels should be encouraged in the future. Easily accessed, online databases were developed as part of recent European Union initiatives. These represent a step in the right direction, making the information available to a wide audience and allowing for systematic updating of data records in the future. Moreover, legislators have the power to render these processes mandatory. The French authorities have already done this by publishing a circular (French Ministry of Equipment, Transportation and Lodging, 2000) which, among other requirements, demands that tunnel operators report all accidents of significance in tunnels according to a prescribed framework. This framework has been used as the basis of the structure of the online tunnel accidents database, which is available on the European Thematic Network of Fires in Tunnels (FIT) website (see reference list).

1.5. A history of tunnel-fire incidents

The list of fire incidents presented in the first edition of this handbook was as comprehensive as was possible at the time of writing, in 2004. That list contained details of all fatal tunnel-fire incidents and significant multi-vehicle fires from 1842 to 2004. It also contained details of many minor fires. The list presented here is shorter and does not aim to be comprehensive, but rather is a list of fatal, or otherwise significant, tunnel-fire incidents from the nineteenth century through to 2010.

1.5.1 Fires since 2000

Eiksund Tunnel, 7.7 km long, Norway, 2009
A lorry and a small van collided in the middle of the Eiksund Tunnel on 28 June 2009. The vehicles caught fire immediately. Five young people died. The emergency services could not reach the incident to fight the fire, due to heat and dense smoke. The Eiksund Tunnel, which opened in 2008, is notable for being the deepest underwater tunnel in the world, reaching 267 m below sea level. (See EuroTest website, in the reference list.)

Channel Tunnel, 51 km long, UK/France, 2008
On 11 September 2008, a fire started in one of the goods vehicles near the front of a Channel Tunnel shuttle travelling from the UK to France. The fire was detected several minutes after the train entered the tunnel, and the train was stopped in the French half of the tunnel, about 11.5 km from the French portal. The fire appears to have grown and spread very rapidly, much like the 1996 fire, but the fire on this occasion involved several more carriages and the damage to the tunnel extended to 650 m. The crew and passengers were safely evacuated to the service tunnel, with only minor injuries (BEA-TT and RAIB, 2010; Carvel, 2010).

Newhall Pass tunnel, 166 m long, Interstate 5, California, USA, 2007
An accident occurred, in rainy conditions, at night on 12 October 2007 in a short tunnel on Interstate 5, the main road between Los Angeles and San Francisco. One truck lost control and struck the barrier at the side of the carriageway. Other trucks collided with the first, and a large fire started, just outside the exit portal of the tunnel. There were three fatalities and 23 people were injured. The fire, driven by the prevailing wind, spread to involve the queue of vehicles that had formed in the tunnel. It took 24 hours to bring the fire under control, and major structural damage occurred (Bajwa *et al.*, 2009; see EuroTest website).

Cabin Creek hydropower plant, Colorado, USA, 2007
A fire broke out during the refurbishment of a water tunnel at a power station on 2 October 2007. The fire, which involved flammable chemicals used for cleaning purposes, blocked the only escape route for the workers, five of whom were trapped and died from smoke inhalation. (See Penn Energy website, in the reference list.)

San Martino Tunnel, 4.8 km long, near Lecco, Italy, 2007
On 10 September 2007, a lorry crashed into the tunnel wall and caught fire. This triggered a pile-up. It took rescue services 45 minutes to arrive at the scene of the accident. Two people died and ten people were taken to hospital suffering from smoke poisoning. (See EuroTest website.)

Burnley Tunnel, 3.5 km long, Melbourne Australia, 2007
On 23 March 2007, a rear-end collision caused a pile-up involving three lorries and four cars, and this resulted in explosions and a fire. Three people died in the accident. The tunnel's deluge system was operated very quickly and contained the fire. About 400 tunnel users were able to safely escape the tunnel on foot. (See the EuroTest website, and Chapter 4 on the Burnley fire.)

Eidsvoll Tunnel on E6, 1.2 km long, near Oslo, Norway, 2006
On 26 October 2006, there was a head-on collision between a car and a fuel tanker. The tanker caught fire immediately. The car driver died in the accident, while the tanker driver was able to escape injured. (See EuroTest website.)

Viamala Tunnel, 0.7 km long, Switzerland, 2006
On 16 September 2006, there was a crash involving a bus and two cars, which resulted in a fire. The fire spread to two further cars. Nine people died in the incident and a further five were injured. (See EuroTest and FIT websites.)

Highway tunnel on B31, 0.2 km long, near Eriskirch, Germany, 2005
On Christmas day 2005, a car skidded and collided with an oncoming car, and hit the tunnel wall. The vehicle caught fire, and four people aged 18–23 years burned to death, and a fifth victim was thrown from the car. (See EuroTest website.)

Channel Tunnel rail link (during construction), Kent, UK, 2005
On 16 August 2005, an explosion on a train led to a fire that killed one worker at the site and resulted in the death of a second worker, due to severe burns, 4 days later (BBC News, 17 and 21 August 2005, 17 March 2011).

Fréjus Tunnel, 12.9 km long, France/Italy, 2005
On 4 June 2005, a HGV with a load of tyres caught fire and stopped in the tunnel. The fire spread to three other HGVs on the opposite carriageway. There were two fatalities in one of the HGVs. It appears that these people remained for too long in the cab of their vehicle and did not attempt to leave it until it was too late. Fire-fighters attending the scene described how the intense heat from the fire melted the roadway under their feet. (See EuroTest website; CNN, 5 June 2005.)

Baregg Tunnel, 1.1 km long, near Baden, Switzerland, 2004
On 14 April 2004, a lorry collided with a car and two other lorries, which had stopped in the tunnel because of an earlier collision. The car was completely crushed and caught fire, and the fire subsequently spread to one of the lorries. The driver of the car died and five other people were injured. (See EuroTest website.)

Dullin Tunnel, 1.5 km long, near Chambéry, France, 2004
On Sunday 18 January 2004, a fire started in the engine compartment at the rear of a coach carrying 37 tourists to the ski resort of Courchevel. Rather than stopping in the tunnel, the driver drove the bus for about 1 km to the tunnel portal, despite flames in the passenger compartment forcing the passengers to the front of the coach. Once outside the tunnel the passengers were able to evacuate the bus safely, but the fire spread rapidly to consume the entire vehicle. The local fire brigade praised the bus driver for his actions (BBC News, 18 January 2004).

Fløyfjell Tunnel, 3.1 km long, Bergen, Norway, 2003
At about 4.20 p.m. on Monday 10 November 2003, a car fire occurred about 1.9 km into the southbound tube. The car initially crashed into the left-hand wall before careering across the carriageway into an emergency telephone box on the right. The car immediately burst into flames, and the fire soon spread to involve the tunnel lining material. Unusually for a tunnel in Europe, the Fløyfjell Tunnel has a sprinkler system. Eleven sprinkler heads activated automatically (within one minute of the crash), and this quickly extinguished the fire involving the

tunnel lining, but not the car fire. Some tunnel users attempted to fight the car fire using portable extinguishers, but were unable to approach the car due to the severity of the fire. The fire brigade arrived after 6 minutes and quickly extinguished the car fire. The driver of the car was trapped in his vehicle because of the crash, and died in the fire (Damsgaard and Svendsen, 2003).

Guadarrama rail tunnel, 30 km long (under construction), Spain, 2003

An accident occurred on 6 August 2003 on a train near the tunnel portal. The train crew escaped the tunnel before the smoke became too thick, but 34 construction workers were trapped about 3 km inside the tunnel by heavy smoke. They took refuge in an air pocket in the tunnel and were rescued after 5 hours. Initial reports suggested, incorrectly, that the trapped workers had been on the train when it caught fire (Tunnels & Tunnelling International, 2003).

Mornay Tunnel, 2.6 km long, between Bourg-en-Bresse and La Cluse, France, 2003

On 2 May 2003, a fire broke out in a passenger carriage on an 'autorail' train. Once the fire was detected the train stopped automatically, about 300 m from the tunnel portal. The tunnel, constructed in 1877, is a single-tube, single-track tunnel, with no lighting and no ventilation system. All the 17 passengers on board were able to self-rescue before the arrival of the fire service. On arrival, however, the fire-fighters had to overcome major problems to fight the fire; there was no local water supply and the railway company could not provide assistance immediately. The fire-fighters blocked a nearby river (using parts of the railway fixtures) to obtain a water supply. The fire took 5 hours to bring under control (Nothias, 2003).

Jungangno underground railway/metro station, Daegu, South Korea, 2003

On the morning of Tuesday 18 February 2003, an arson attack on an underground railway/metro train in Jungangno station, near Daegu city centre, led to the deaths of at least 189 people. The arsonist used a small quantity of petrol and a cigarette lighter to start the fire on a stationary train in the station. The fire quickly spread to engulf the whole six-carriage train. After the underground railway/metro operators became aware of the fire, a second train entered the station and stopped near the burning train. The doors of this second train did not open. The fire spread to the second train, where most of the fatalities occurred. In addition to the fatalities, more than a hundred people, including the arsonist, were treated for smoke inhalation. Following the incident, the head of the underground railway/metro corporation was fired, and six members of the railway staff were arrested and charged with negligence (BBC News, 18, 19, 23 and 24 February 2003; Burns, 2003; Marlair et al., 2004, 2006).

St. Gotthard Tunnel, 16.9 km long, near Airolo, Switzerland, 2001

On Wednesday 24 October 2001, a very large fire resulted from a head-on collision between two HGVs, one carrying a load of rubber tyres. The fire resulted in 11 fatalities, the destruction of 23 vehicles and the collapse of over 250 m of the tunnel lining. Had the tunnel not been equipped with a parallel service tunnel, it is likely that the death toll would have been much higher. As the result of a review following the Mont Blanc Tunnel incident, the St. Gotthard Tunnel had recently had its lighting improved and was scheduled to have its ventilation system upgraded in the summer of 2002. The fire burned for over 2 days (Turner, 2001).

Gleinalm Tunnel, 8 km long, near Graz, Austria, 2001

A fire resulted from the head-on collision of two cars near the middle of the Gleinalm Tunnel on Tuesday 7 August 2001. The fire was successfully extinguished by fire-fighters shortly after their arrival. Five people died and four were injured (Strhhaussl, 2001).

Howard Street Tunnel, Baltimore, OH, USA, 2001

At 3.10 p.m. on Wednesday 18 July 2001, a freight train passing through a tunnel in downtown Baltimore had an 'emergency brake application'. Following standard procedure, the drivers detached the locomotives from the train, and removed them from the tunnel. Of the 60 cars that made up the train, eight were carrying hazardous materials, including hydrochloric acid, chemicals used to make adhesives and various solvents. By mid-morning the following day the fire was mostly extinguished, except for a few hot spots. All the wreckage was removed within 4 days. The fire resulted in gridlock on the roads in Baltimore, as all the major routes were closed to traffic for about 12 hours. Two baseball games at the nearby Camden Yards had to be rescheduled (Lamb and Resnikoff, 2001).

Kitzsteinhorn funicular tunnel, 3.3 km long, near Kaprun, Austria, 2000

At about 9 a.m. on Saturday 11 November 2000, a fire broke out at the rear of the ascending train shortly after leaving the lower terminal. The train stopped automatically 600 m inside the tunnel. The doors of the train failed to open. Twelve passengers escaped by smashing the windows and fleeing down the tunnel. The remaining 150 people on the train died on the train or during their attempt to flee up the smoke-filled tunnel. The smoke also killed two people on the descending train, some 1.5 km further up the tunnel, and three people in the arrival hall of the upper terminal, 2.7 km away. The fire on the supposedly fire-proof train is thought to have been started by hydraulic oil leaking into the heater in the rear driver's cab, and to have spread via the clothes and baggage of the passengers on the train. The ventilation velocity in the tunnel at the time of the incident was approximately 10 m/s (Schupfer, 2001).

Rotsethhorn Tunnel, 1.2 km long, Norway, 2000

On 29 July 2000, a collision and subsequent fire led to the deaths of two people (Nilsen et al., 2001).

Seljestad Tunnel, 1.3 km long, Norway, 2000

Just before 9 p.m. on 14 July 2000, a truck collided with the rear of a line of stationary vehicles in the tunnel, causing an eight-vehicle pile-up. Immediately after the collision, one of the vehicles caught fire, and within minutes the fire spread to involve all the vehicles. The fire quickly destroyed the communications cables in the tunnel. An ambulance arrived within 15 minutes and fire-fighters within half an hour. Twenty people were admitted to hospital, but none had major injuries. Despite the fact that four people were trapped in the smoke for over an hour, there were no fatalities. Due to the prevailing wind, there was breathable air in the tunnel and fire-fighters were able to approach the scene of the fire easily. It has been estimated that, if the wind had not been as strong, there would have been at least four fatalities due to smoke inhalation (Madsen, 2001; Nilsen et al., 2001).

1.5.2 Fires during the 1990s

Tauern Tunnel, 6.4 km long, south-east of Salzburg, Austria, 1999

On Saturday 29 May 1999, a HGV collided with a queue of stationary traffic 800 m from the northern portal of the tunnel. Eight people died as a direct result of the crash. A fire broke out and quickly engulfed the incident HGV, another HGV loaded with a cargo of spray cans, including paints, and the four-car pile-up between them. There were four fire-related fatalities: one HGV driver, who was overcome by fumes in the tunnel; two car passengers, who did not leave their car; and one HGV driver who had initially fled to safety, but returned to his vehicle to collect some documents. The fire destroyed 16 HGVs and 24 cars, and took 15 hours to extinguish. The transverse ventilation system worked very well during the incident (Pucher and Pucher, 1999; Eberl, 2001).

Railway tunnel, 9 km long, near Salerno, Italy, 1999
On Sunday 23 May 1999, a train carrying Italian football fans caught fire as a result of the 'rowdy' behaviour of the passengers. Four people died and at least nine were injured. It was alleged that the fire started as a result of one of the passengers lighting a smoke bomb (Danger Ahead, 1999).

Mont Blanc Tunnel, 11.6 km long, France/Italy, 1999
At 10.46 a.m. on Wednesday 24 March 1999, a HGV carrying a refrigerated cargo of margarine and flour entered the Mont Blanc Tunnel from the French side. At that time, no fire or smoke was observed, but a few kilometres into the tunnel the HGV began emitting 'white smoke'. At 10.53 a.m. the HGV stopped, near lay-by 21, some 6.3 km into the tunnel. Immediately the cab burst into flames, producing 'black smoke', which propagated mostly in the direction of the French portal, and the driver fled towards Italy. One motorcycle, nine cars, 18 HGVs and a van were in the tunnel behind the HGV that was on fire, and eight HGVs and several cars entered the tunnel from the Italian side before the tunnel was closed. Nobody who entered the tunnel from the Italian side was injured. The fire resulted in the death of 39 people (27 of them in their vehicles), the destruction of 34 vehicles (over 1.2 km of tunnel), severe damage to the tunnel lining (over 900 m) and a blaze that took 53 hours to extinguish. The fire is thought to have started because of diesel fuel leaking onto hot surfaces in the HGV's engine compartment. The uncontrolled spread of the toxic smoke (which was responsible for the majority of the deaths) has been blamed on poor operation of the ventilation system and a lack of communication between the French and Italian operators (Lacroix, 2001).

Oslofjord Tunnel, Norway, 1999
An explosion during construction of the tunnel started a fire. Two fire-fighters were killed and several people were injured during the fire-fighting and rescue operations (de Vries, 2002).

Gueizhou Tunnel, 800 m long, between Guiyang and Changsha, China, 1998
On 10 July 1998, gas canisters exploded on a train in south-west China, killing more than 80 people. The tunnel collapsed, and railway workers sent to repair the damage were also killed by a further explosion caused by a build up of gas (Disaster database, 1999).

Exilles rail tunnel, 2.1 km long, near Susa, Italy, 1997
On 1 July 1997, a train transporting 216 cars on 18 wagons caught fire. The fire began because one of the car doors swung open and was dragged along the electrical wiring on the tunnel wall. The fire alarm was raised at 1.30 p.m. Firefighters from Susa arrived 20 minutes later, and were joined by firefighters from Turin after a further 25 minutes. Due to the gradient, the downhill side of the tunnel was smoke free, allowing the firefighters to approach the blaze. Fire-fighting was hazardous, due to the high temperatures and explosive spalling of the concrete lining. The fire took 5 hours to bring under control, and was eventually extinguished by 8 p.m. One locomotive, 13 freight wagons and 156 cars were destroyed. Two train crew members escaped by running uphill out of the tunnel; both were treated for smoke inhalation (Colcerasa, 2001).

Channel Tunnel, 51 km long, France/UK, 1996
On 18 November 1996, a HGV on board a HGV carrier shuttle in the Channel Tunnel caught fire. Upon entering the tunnel, the estimated fire size was about 1.5 MW, but at its maximum the fire

size was as much as 350 MW and involved ten HGVs and their carrier wagons. The train was stopped in the tunnel, and it took about half an hour to get the emergency ventilation system working properly. All the HGV drivers and the train crew were evacuated safely into the service tunnel, with no fatalities, and only two escapees required significant hospital attention. Following the evacuation of all the people, the ventilation system maintained an airflow of 2.5 m/s in the tunnel, which may have helped fan the fire and explain why it grew to such a size. The main fire was extinguished by firefighters working in relays for some 7 hours after the fire was first detected. Minor smouldering fires were still in evidence 24 hours later (Allison, 1997).

Isola delle Femmine motorway tunnel, 148 m long, near Palermo, Italy, 1996

On 18 March 1996, a 16-vehicle pile-up occurred in the two-lane eastbound tube of the Isola delle Femmine motorway tunnel. A tanker carrying liquid petroleum gas safely performed an emergency stop in the tunnel without joining the pile-up. However, a tourist coach following the tanker, and four other vehicles crashed into the back of the tanker. The upper part of the tank ruptured, and a small explosion followed within seconds. This led to only minor injuries and started a fire at the front of the tourist coach. All but five passengers evacuated the bus through a smashed window at the rear. Four of the five remaining passengers were later found dead on the bus, and another was found dead on the roadway. After about 6–7 minutes, by which time all survivors had evacuated the tunnel, there was a massive explosion. Thick smoke and violent flames were observed at both ends of the tunnel and at cross-passages into the other tunnel tube. Eye witnesses outside the tunnel reported experiencing a 'shock wave' from the blast. The second explosion is believed to have been a boiling liquid, expanding vapour explosion (BLEVE), which occurred inside the body of the tanker; of which only four large pieces remained after the incident. A total of 34 people were treated for burns, and of these 16 were admitted to hospital. All those hospitalised were kept in for over 10 days, and five were hospitalised for over a month (Ciambelli et al., 1997; Masellis et al., 1997; PIARC, 1999; Amunsen, 2000).

Baku underground railway/metro, Azerbaijan, 1995

On 28 October 1995, an electrical fault led to a fire breaking out at the rear of the fourth car of a packed five-car underground railway/metro train during rush hour. The train stopped 200 m after departing Uldus station. Initially, the ventilation conditions in the tunnel tended to move the smoke gently towards the rear of the train. There were problems getting the doors open for the passengers to escape. Windows were smashed and some passengers began to evacuate the train in both directions. Soon the fire grew such that it was impossible to pass – passengers in the front three cars would have to evacuate towards Narimanov station, 2 km away. Fifteen minutes after the train stopped, the ventilation conditions in the tunnel changed, drawing all the smoke towards Narimanov and the front three cars. The entire tunnel was filled with smoke in this direction. In total, 220 passengers were killed in the front three cars, 40 being killed by the fumes in the tunnel in the direction of Narimanov station. In addition, 256 people were admitted to hospital following the incident, 100 of whom were still in hospital 5 days later. The tunnel reopened within 24 hours (Wahlstrom, 1996; Hedefalk et al., 1998).

Pfänder Tunnel, 6.7 km long, Austria, 1995

On 10 April 1995, a car driver fell asleep while driving through the Pfänder Tunnel, and his car strayed into the path of an oncoming truck. After colliding with the car, the truck skidded along the wall on the wrong side of the road, finally colliding with a minibus. Three people on the minibus died as a result of the crash. Although emergency ventilation was activated almost immediately, it was not sufficient to control the smoke, which filled the tunnel and hindered the fire-fighting operation. However, no one died as a result of the fire. The fire was extinguished after an hour, by which time it had caused serious damage to the tunnel lining (Pucher, 1996).

Huguenot Tunnel, 4 km long, near Paarl, South Africa, 1994
On Sunday 27 February 1994, a fire broke out in the gearbox of a bus carrying 45 passengers. After a failed attempt by the co-driver to smother the fire, several passengers jumped from the moving vehicle. The driver then lost control of the vehicle, which veered across the oncoming traffic, hitting the tunnel wall. An approaching articulated truck jack-knifed, blocking the tunnel, in an attempt to make an emergency stop. At this stage the fire was relatively small and could have been extinguished by using any of the nearby fire extinguishers. However, this was not done and the fire grew, filling the tunnel with smoke and killing the bus driver. The fire destroyed the bus and many of the tunnel installations nearby, but it did not spread to any other vehicles. Firefighters arrived at the scene within 12 minutes of the first alarm, which was raised by one of the bus passengers. The fire took slightly over an hour to extinguish. The tunnel reopened, following temporary repairs, 4 days later (Gray and Varkevisser, 1995).

Serra a Ripoli Tunnel, 442 m long, Italy, 1993
An out-of-control vehicle crashed, starting a fire that ultimately involved five HGVs, one loaded with rolls of paper, and 11 cars. There were four fatalities and four reported injuries. The fire lasted for 2.5 hours (PIARC, 1999; Amunsen, 2000).

Unnamed tunnel, South China, 1991
In August 1991, a fire broke out on board a train passing through a tunnel. The fire engulfed one of the train cars. Fifteen passengers panicked, and jumped from the train into the path of an oncoming locomotive.

Moscow underground railway/metro, Russia, 1991
On 1 June 1991, an electrical fire under a train in a station resulted in seven fatalities and at least ten people were injured (Atkins, 1996).

New York underground railway/metro, USA, 1990
On 28 December 1990, a cable fire in a tunnel near a station produced dense smoke. A passenger pulled the emergency cord, halting a train near the fire. The fire brigade arrived quickly, but evacuated the wrong train first. Two people died, and 200 were injured (Atkins, 1996).

Mont Blanc Tunnel, 11.6 km long, France/Italy, 1990
On 11 January 1990, a HGV that was on fire stopped nearly 6 km into the tunnel. The driver had first noticed the smoke about 1.5 km into the tunnel, but did not stop until flames entered his cab. He alerted the tunnel operators by telephone, and French firefighters arrived within 10 minutes. Despite the fact that the fire had spread to involve the entire vehicle by this point, the firefighters were able to control the blaze and extinguish it. Two people were treated for smoke inhalation (Ministère de l'Equipement, des Transports et du Logement, 1999; Amunsen, 2000).

1.5.3   Fires during the 1980s
Brenner Tunnel, Austria, 1989
On 18 May 1989, dangerous goods used during construction works exploded in the Brenner Tunnel, resulting in a fire that lasted 7 hours. In total, 155 firefighters and 21 fire-fighting vehicles

attended the scene. Two people were killed and five were injured (Crez, 1990; Le Sapeur Pompier, 2001).

Kings Cross Station, London, UK, 1987
On 18 November 1987, a fire involving the steps and sides of a wooden escalator, and which spread into the ticket hall, was responsible for the deaths of 31 people, and many others were injured. The fire grew and spread very rapidly, probably due to preheating of the escalator floor by a fire which is thought to have started in the grease and fluff under the escalator track. The fire behaved in an unexpected manner: the flames and plume from the fire did not rise from the fire location toward the ceiling of the escalator tube, but rather they tended to hug the steps and sides of the escalator all the way up and into the ticket hall at the top of the escalator. This phenomenon has become known as the 'trench effect'. The fire reached flashover in a matter of minutes, and burned for 6 hours (Fennel, 1988).

Gumefens Tunnel, 343 m long, Switzerland, 1987
On 18 February 1987, a collision resulted in a fire that ultimately involved two HGVs, one van and five cars. There were two fatalities and three reported injuries (possibly all as the result of the crash). The fire burned for 2 hours (Crez, 1990; Le Sapeur Pompier, 2001).

L'Arme Tunnel, 1.1 km long, France, 1986
On 9 September 1986, a collision involving a car started a comparatively small fire, involving the car and a trailer. The incident resulted in three fatalities and five people were injured. The duration of the fire and the damage to the tunnel are not recorded (Day, 1999; Amunsen, 2000).

San Benedetto Tunnel, 18.5 km long, Italy, 1984
On 23 December 1984, a bomb attack was responsible for the deaths of 17 people and 120 injured. The resulting fire was surprisingly small. Firefighters and rescue personnel were unable to reach the scene of the incident for over 2 hours due to the destruction (The World's Longest Tunnels Page).

Summit Tunnel, 2.6 km long, UK, 1984
On 20 December 1984, a train consisting of a diesel locomotive and 13 tankers carrying petroleum spirit derailed approximately a third of the way through the Summit Tunnel. After seeing flames, the driver and the guard evacuated and called the fire brigade. At 9.30 a.m. (3.5 hours after the train derailed), one of the tankers exploded, producing flames 120 m high, which extended up several of the ventilation shafts and into the open air. The fire was contained by introducing high-expansion foam into one nearby ventilation shaft and water into another. By 10.30 a.m. the following day no fire was in evidence, but intense heat prevented anyone from reaching the fire location. A cooling, ventilation and inspection strategy was maintained until 6.30 p.m. on 24 December, by which time the conditions in the tunnel were back to ambient. All train debris had been removed from the tunnel by the 17 January 1985. The tunnel was shut for several months (Howarth, 1996).

Pecorile Tunnel, 662 m long, Italy, 1983
A collision involving a fish lorry started a fire that claimed nine lives and injured a further 20 people (Day, 1999; Amunsen, 2000).

Salang Tunnel, Afghanistan, 1982

On 2 or 3 November 1982, a gas tanker, part of a military convoy, exploded in the Salang road tunnel. The explosion started a large-scale fire. Reports on the number of fatalities vary from 176 to several thousand (Trueheart, 1999; White, 2003).

Caldecott Tunnel, 1 km long, Oakland, CA, USA, 1982

On 7 April 1982, a collision occurred when a passenger car, driven by a drunk driver, collided with the roadside and came to an abrupt halt. The stationary car was struck by a petrol tanker, and subsequently the tanker was struck by a bus, causing the tanker to turn over, partially rupture and spill some of its load. The spilt petrol soon ignited, and the blaze grew to involve the tanker, the car and four other vehicles in the tunnel. Seven people were killed. Although some of the petrol spilled out of the tanker, it appears that a large quantity of the fuel remained in the tank. Once the fire had reached a sufficient temperature to melt the aluminium walls of the tanker, the 'top' of the tanker (i.e. the side that was uppermost after the crash) collapsed, creating a large, deep pool of petrol, which ignited and burned fiercely for over 2 hours until firefighters were able to extinguish it (Hay, 1984; Amunsen, 2000).

London Underground railway/metro, UK, 1981

On 21 June 1981, a fire between two underground stations resulted in one fatality and 15 people injured (Le Sapeur Pompier, 2001).

Sakai Tunnel, 459 m long, Japan, 1980

On 15 July 1980, a collision involving a truck resulted in a fire that burned for 3 hours and ultimately involved ten vehicles, killed five people and injured five others (Amunsen, 2000).

Kajiwara Tunnel, 740 m long, Japan, 1980

On 17 April 1980, a gearbox fire led to a fire that ultimately involved two trucks, one laden with 200 cans of paint, and that burned for 1 hour and 20 minutes, and resulted in one fatality (Amunsen, 2000).

1.5.4   Fires during the 1970s

Nihonzaka Tunnel, 2 km long, Japan, 1979

On 11 July 1979, an accident involving four trucks and two passenger cars led to a fire in the westbound tube of the Nihonzaka Tunnel. Ultimately, the fire killed seven people and spread to involve 189 vehicles. Traffic congestion led to a delay in the arrival of fire-fighters, and their fire-fighting activities were cut off after only half an hour as the water tanks ran dry. The fire was not extinguished for 160 hours (Oka, 1996).

San Francisco underground railway/metro (BART), USA, 1979

On 17 January 1979, a short circuit underneath an underground railway/metro car led to a fire. One person died and 56 were injured. The movements of other underground railway/metro cars spread the smoke rapidly, and over a thousand people had to be evacuated from the underground railway/metro (Demoro, 1979).

Velsen Tunnel, 768 m long, Haarlem, The Netherlands, 1978

On 11 August 1978, a collision killed five people and started a fire that ultimately involved two HGVs and four cars (Amunsen, 2000).

Moorgate underground railway/metro station, London, UK, 1975

On 28 February 1975, an incident involving a train hitting a wall and the subsequent fire resulted in 44 fatalities and 73 people injured (Le Sapeur Pompier, 2001).

Mexico City underground railway/metro, Mexico, 1975

Fifty people died and another 30 were injured due to a train collision and the resulting fire (Le Sapeur Pompier, 2001).

Porte d'Italie underground railway/metro station, Paris, France, 1973

On 27 March 1973, an arson attack on an underground railway/metro carriage in a station led to two fatalities and several injured, even though the response by fire-fighters was very quick (Atkins, 1996).

Hokoriku Tunnel, near Fukui, Japan, 1972

On 6 November 1972, a fire in the restaurant car of a passenger train ultimately led to 30 fatalities, 690 injured people and the destruction of two carriages (Le Sapeur Pompier, 2001).

Vierzy Tunnel, France, 1972

The Vierzy Tunnel collapsed on a passenger train, starting a fire. There were 108 fatalities, mostly due to the tunnel collapse (Le Sapeur Pompier, 2001).

Henri Bourassa underground railway/metro station, Montreal, Canada, 1971

On 12 December 1971, an underground railway/metro train collided with the end of a tunnel, igniting a fire. There was one fatality (Atkins, 1996; Le Sapeur Pompier, 2001).

Crozet Tunnel, France, 1971

On 20 March 1971, a collision between a goods train and a train carrying hydrocarbon fuels caused a derailment and a fire. There were two fatalities (Le Sapeur Pompier, 2001).

Wranduk Tunnel, 1.5 km long, near Zenica, Yugoslavia, 1971

At 5.48 a.m. on 14 February, a train carrying workers to an iron works in Zenica was passing through a 1500 m tunnel. It had to stop 300 m from the exit as a fire had started in the engine. The heat was such that no passengers could pass the engine, and they had to return the 1200 m to the other end of the tunnel. In total, 33 died and 120 were taken to hospital, where 57 were admitted.

Sylmar Tunnel, 8 km long, Los Angeles, CA, USA, 1971

On 24 June 1971, during the tunnel construction, a gas explosion killed 17 of an 18-man team working in the tunnel at the time. The explosion occurred as a result of 'a complete breakdown of corporate safety', and the tunnel project manager, Loren Savage, was sentenced to 20 years and 6 months in prison, on 16 counts of gross negligence and nine labour code violations (Zavattero, 1978).

New York City underground railway/metro, USA, 1970

On 1 August 1970, a tunnel fire near Bowling Green station resulted in 50 injured and one fatality. The woman who died did so because she returned to the train, after evacuation, in an attempt to recover her purse (AP News Agency, 1995).

1.5.5   Fires before 1970

London Underground railway/metro, Holland Park, UK, 1958

On 28 July 1958, a fire occurred in an underground railway/metro train. In total, 48 passengers and three staff were taken to hospital with smoke inhalation. Ten passengers were admitted to hospital, and one subsequently died (Semmens, 1994).

Holland Tunnel, New York, USA, 1949

On 13 May 1949, a fire started as a result of a HGV shedding its load. The fire burned for 4 hours, destroying ten HGVs and 13 cars. Sixty-six people were injured (Amunsen, 2000; Le Sapeur Pompier, 2001).

London Underground railway/metro, UK, 1945

On 31 December 1945, three people died as the result of a collision and fire (Le Sapeur Pompier, 2001).

Torre Tunnel, Spain, 1944

On 3 January 1944, a multi-train collision in a tunnel led to a fire that lasted for at least a day. There were 91 fatalities as a result of the collision and fire (The World's Longest Tunnels Page).

St. Gothard, Giorinco, Switzerland, 1941

A train derailment and the resulting fire led to seven fatalities (Le Sapeur Pompier, 2001).

Gütschtunnel, Lucerne, Switzerland, 1932

Six people died due to a train collision and the fire that started as a result (Le Sapeur Pompier, 2001).

Riekentunnel, Switzerland, 1926

Nine people died from smoke inhalation when a goods train caught fire and stopped in a tunnel (Le Sapeur Pompier, 2001).

Batignolles Tunnel, 1 km long, Paris, France, 1921

On 21 October 1921, a passenger train was allowed into the Batignolles Tunnel while another train was stationary inside. The trains collided, resulting in a very large fire. At least 28 people died, mostly as a result of the fire rather than the crash. The severity of the fire is thought to have been due to the gaslight system installed in the carriages. These gaslight systems were removed from all passenger trains in France following this incident (The World's Longest Tunnels Page).

Couronnes underground railway/metro station, Paris, France, 1903

On 10 August 1903, an underground railway/metro train caught fire as the result of an electrical fault. All the passengers were safely evacuated from the train and from another train in the tunnel. During an attempt to push the first train, still on fire, out of the tunnel, the fire grew dramatically in size, and smoke billowed into the station. There were at least 84 fatalities (some records claim more than 100) (Brader, 1995; Trueheart, 1999).

Mendon, France, 1842

A fire on a train in a tunnel led to 150 fatalities (Le Sapeur Pompier, 2001).

REFERENCES

Allison R (1997) *Inquiry into the Fire on Heavy Goods Vehicle Shuttle 7539 on 18 November 1996.* HMSO, London.

Amunsen FH (2000) *Data on Large Tunnel Fires.* Preliminary Report. Norwegian Public Road Administration, Oslo, 20 April 2000. Report prepared for the OECD study on transport of dangerous goods through tunnels.

AP News Agency (1995) Reported on USENET (misc.transport.urban-transit). 2 November.

Arditi R (2003) Data presented at: Discussion Forum 1. *5th International Conference on Safety in Road and Rail Tunnels, Marseille, 6–8 October 2003.* Data presented showed that, while 6% of Italian roads are in tunnels, only 5% of road accidents and 2% of road fatalities occur in tunnels.

Atkins WS (1996) *Quantified Risk Assessment – Central Line Ventilation.* Report No. M4055.550. London Underground.

Bajwa C, Mintz T, Huczek J, Axler K and Das K (2009) FDS simulation of the Newhall Pass tunnel fire. *NFPA World Safety Conference, Chicago, IL, USA,* 8–11 June 2009.

Baubehörde Highways Department (1992) *Statistics on the Traffic in the Elb Tunnel from the Year 1975 to the Year 1992.* Hamburg.

BBC News (2011) Available at: http://news.bbc.co.uk (accessed 22 June 2011).

BEA-TT and RAIB (2010) *Technical Investigation Report concerning the Fire on Eurotunnel Freight Shuttle 7412 on 11 September 2008.* Report No. BEATT-2008-015. Bureau d'Enquêtes sur les Accidents de Transport Terrestre/Rail Accident Investigation Branch, Paris/London.

Bodart X, Marlair G and Carvel RO (2004) Fire in road tunnels and life safety: lessons to be learned from minor accidents. *Proceedings of the 10th International Fire Science & Engineering Conference (Interflam 2004), Edinburgh, UK,* 5–7 July 2004, pp. 1517–1527.

Brader M (1995) Worst underground railway/metro disasters. Contribution to USENET. 1 November.

Burns D (2003) External observations of the Daegu underground railway/metro fire disaster 18th February 2003. *Proceedings of the 5th International Conference on Safety in Road and Rail Tunnels, Marseille, France,* 6–10 October 2003, pp. 13–26.

Carvel R (2010) Fire dynamics during the Channel Tunnel fires. In: Lonnermark A and Ingason H (eds), *Proceedings of the 4th International Symposium on Tunnel Safety & Security, Frankfurt am Main, Germany,* 17–19 March 2010, pp. 463–470.

Ciambelli P, Bucciero A, Maremonti M, Salzano E and Masellis M (1997) The risk of transportation of dangerous goods: BLEVE in a tunnel. *Annals of Burns and Fire Disasters* **10**(4): 241–245.

CNN (2005) Criminal Probe into Fatal Tunnel Fire. 5 June. Available at: http://edition.cnn.com.

Colcerasa F (2001) Rail-car fire in the Exilles Tunnel (Italy). In: Colombo AG (ed.), *NEDIES Project: Lessons Learnt from Tunnel Accidents.* Report of a NEDIES Meeting, 13–14 November 2000. EUR Report, March 2001, pp. 22–24. Available at: http://nedies.jrc.it (accessed 22 June 2011).

Crez A (1990) *Voies de Circulation Souterraine et Protection Incendie.* Dossier Technique DT 84. Supplement to La Revue Belge du Feu, ANPI, December.

Damsgaard E and Svendsen RH (2003) Omkom i voldsom bilbrann. Sprinkleranlegg hindret større katastrofe i tunnel. Sjanseløs mot flammene. *Bergens Tidende* 11 November: 1–2.

Danger Ahead (1999) Italy: 4 Die in 'Football Special' Fire, 24 May. Available at: http://danger-ahead.railfan.net/reports/rep99/salerno_tunnel_fire.html (accessed 22 June 2011).

Day JR (1999) Active and passive safety systems for road tunnels. *Proceedings of the International Tunnel Fire & Safety Conference, Rotterdam, The Netherlands*, 2–3 December 1999, Paper Number 12.

Demoro HE (1979) Fire blackens BART image. *Mass Transit Journal* **6**(7).

de Vries H (2002) Mining for answers. *Fire Chief Magazine* January.

Disaster database (1999) Disaster database: fires and explosions. *Disaster Prevention and Management* **8**(4).

Eberl G (2001) The Tauern Tunnel incident: what happened and what has to be learned. *Proceedings of the 4th International Conference on Safety in Road and Rail Tunnels, Madrid, Spain*, 2–6 April 2001, pp. 17–30.

EuroTest (2011) Brussels: http://www.eurotestmobility.com (accessed 22 June 2011).

Fennel D (1988) *Investigation into the King's Cross Underground Railway/Metro Fire*. HMSO, London.

FIT (European Thematic Network on Fire in Tunnels) (2011) http://www.etnfit.net (accessed 22 June 2011).

French Ministry of Equipment, Transportation and Lodging (2000) *Circulaire Interministérielle No. 2000-63 Relative à la Sécurité dans les Tunnels du Réseau Routier National* (in French). Official Bulletin of the French Ministry of Equipment, Transportation and Lodging, September.

Gray DJ and Varkevisser J (1995) The Huguenot toll tunnel fire. *Proceedings of the 2nd International Conference on Safety in Road and Rail Tunnels, Granada, Spain*, 3–6 April 1995, pp. 57–66.

Hay RE (1984) *Prevention and Control of Highway Tunnel Fires*. Report No. FHWA-RD-83-032. US Department of Transportation, Federal Highway Administration, McLean, VA. Available at: http://ntl.bts.gov/lib/2000/2400/2416/708.pdf (accessed 22 June 2011).

Hedefalk J, Wahlstrom B and Rohlen P (1998) Lessons from the Baku Underground railway/metro fire. *3rd International Conference on Safety in Road and Rail Tunnels, Nice, France*, 9–11 March 1998, pp. 15–28.

Howarth DJ (1996) Tunnel fires – the operational challenge. *1st International Conference on Tunnel Incident Management, Korsør, Denmark*, 13–15 May 1996, pp. 233–242.

Jones A (1985) *The Summit Tunnel Fire*. Incident Report No. IR/L/FR/85126. Health & Safety Executive Research and Laboratory Services Division, Buxton.

Lacroix D (2001) The Mont Blanc Tunnel Fire: what happened and what has been learned. *Proceedings of the 4th International Conference on Safety in Road and Rail Tunnels, Madrid, Spain*, 2–6 April 2001, pp. 3 16.

Lamb M and Resnikoff M (2001) *Radiological Consequences of Severe Rail Accidents Involving Spent Nuclear Fuel Shipments to Yucca Mountain: Hypothetical Baltimore Rail Tunnel Fire Involving SNF*. Radioactive Waste Management Associates, New York. Available at: http://www.state.nv.us/nucwaste/news2001/nn11458.htm (accessed 22 June 2011).

Le Sapeur Pompier (2001) 35 incendies sous la loupe. *Le Sapeur Pompier* **929**(Nov.): 40–45.

Madsen HK (2001) Fire in the Seljestad Tunnel (Norway). In: Colombo AG (ed.), *NEDIES Project: Lessons Learnt from Tunnel Accidents Report of a NEDIES Meeting*, 13–14 November 2000. EUR Report, March 2001, pp. 15–18. Available at: http://nedies.jrc.it (accessed 22 June 2011).

Marlair G, Le Coze JC, Woon-Hyung K and Galea ER (2004) Human behavior as a key factor in tunnel fire safety issues. In: Kim ES, Kim JD, Park YH, Dlugogorski BZ, Kennedy EM and Hasemi Y (eds), *Proceedings of the 6th Asia–Oceania Symposium on Fire Science & Technology*, 17–20 March 2004.

Marlair G, Le Coze JC and Kim WH (2006) The Daegu metro fire: a review of technical and organisational issues. *Proceedings of the 2nd International Symposium on Tunnel Safety & Security, Madrid, Spain*, 15–17 March 2006, pp. 15–25.

Masellis M, Iaia A, Sferrazza G, Pirillo E, D'Arpa N, Cucchiara P, Sucameli M, Napoli B, Alessandro G and Giairni S (1997) Fire disaster in a motorway tunnel. *Annals of Burns and Fire Disasters* 10(4): 233–240.

Ministère de l'Equipement, des Transports et du Logement (1999) Task Force for Technical Investigation of the 24 March 1999 Fire in the Mont Blanc vehicular Tunnel. Report, 30 June 1999. Original report in French. English translation available at: http://www.firetactics.com/MONTBLANCFIRE1999.htm (accessed 22 June 2011).

Nilsen AR, Lindvik PA and Log T (2001) Full-scale fire testing in sub-sea public road tunnels. *Proceedings of the 9th International Interflam Conference, Edinburgh, UK*, 17–19 September 2001, pp. 913–924.

Nothias JN (2003) Tunnels ferroviaires: Les pompiers tirent le signal d'alarme. *Le Figaro* 29 December.

NTSB (1982) *Highway Accident Report: Multiple Vehicle Collisions and Fire, Caldecott Tunnel, Near Oakland, California, April 7, 1982.* NTSB Report HAR-83/01, NTIS number: PB83-916201.

Oka Y (1996) The present status of safety systems for Japanese road tunnels. *Proceedings of the 1st International Conference on Tunnel Incident Management, Korsør, Denmark*, 13–15 May 1996, pp. 55–66.

Penn Energy (2011) www.pennenergy.com (accessed 22 June 2011).

Perard M (1996) Statistics on breakdowns, accidents and fires in French road tunnels. *Proceedings of the 1st International Conference on Tunnel Incident Management, Korsør, Denmark*, 13–15 May 1996, pp. 347–365.

PIARC (1999) *Fire and Smoke Control in Road Tunnels.* PIARC Report 05.05.B.

Pucher K (1996) Fire in the 7 km long Pfänder Tunnel. *1st International Conference on Tunnel Incident Management, Korsør, Denmark*, 13–15 May 1996, pp. 301–307.

Pucher K and Pucher R (1999) *Fire in the Tauern Tunnel. Proceedings of the International Tunnel Fire & Safety Conference, Rotterdam, The Netherlands*, 2–3 December 1999, Paper Number 8.

Ruckstuhl F (1990) Accident statistics and accident risks in tunnels. *Reports on the OECD Seminar on Road Tunnel Management, Lugano, Switzerland*, November 1990, pp. 346–349.

Satoh K and Miyazaki S (1989) *A Numerical Study of Large Fires in Tunnels.* Report of Fire Research Institute of Japan, No. 68.

Schupfer H (2001) Fire Disaster in the tunnel of the Kitzsteinhorn funicular in Kaprun on 11 Nov. 2000. Presented at: *4th International Conference on Safety in Road and Rail Tunnels, Madrid, Spain*, 2–6 April 2001. Paper not in proceedings, for details contact: info@itc-conferences.com.

Semmens PWB (1994) *Railway Disasters of the World: Principal Passenger Train Accidents of the 20th Century*, p. 165. Patrick Stephens, London.

Strhhaussl E (2001) Osterreich: Erneut Brandfall im Tunnel. *Feuerwehr Fachzeitschrift* November: 665–670.

The World's Longest Tunnels Page (2011) http://www.lotsberg.net (accessed 22 June 2011).

Trueheart C (1999) European tunnel fire toll may hit 40. *Washington Post* March.

Tunnels & Tunnelling International (2003) Fire Drama at Guadarrama 1st Steptember 2003.

Turner S (2001) St. Gotthard Tunnel fire. *New Civil Engineer* 1 November: 5–7.

Wahlstrom B (1996) The Baku underground railway/metro fire. *1st International Conference on Tunnel Incident Management, Korsør, Denmark*, 13–15 May 1996, pp. 291–299.

White M (2003) Technological disasters. In: *Historical Atlas of the 20th Century*. Available at: http://users.erols.com/mwhite28/techfail.htm (accessed 22 June 2011).

Zavattero J (1978) *The Sylmar Tunnel Disaster*. Everest House, New York.

Handbook of Tunnel Fire Safety, 2nd edition
ISBN: 978-0-7277-4153-0

ICE Publishing: All rights reserved
doi: 10.1680/htfs.41530.025

Chapter 2
Tunnel fire investigation I: the Channel Tunnel fire, 18 November 1996

Martin Shipp
Fire Safety Group, Building Research Establishment, UK

2.1.   Introduction

The investigation of a fire in a tunnel is, in principle, no different from any other fire investigation; the processes and methodology of a systematic fire investigation can be found in a number of well-established references (e.g. Cooke and Ide, 1985; De Haan, 1997). However, in practice, the investigation of a tunnel fire can involve particular and unusual problems.

In this chapter, the investigation of fires in rail tunnels is discussed by reference to the fire in the Channel Tunnel, which took place in 1996. There have been a few significant fires in rail tunnels, but this incident achieved worldwide notice. It became a very useful source of information in a number of areas, including fire investigation; in part because of the scale of the inquiry. Since 1996 there have been two more serious fires involving heavy goods vehicles (HGVs) in the Channel Tunnel (see Appendix 2A).

2.2.   The Channel Tunnel fire

On Monday 18 November 1996, at around 10 p.m., a fire occurred on HGV Shuttle 7539 on the French side of the Channel Tunnel. There were no fatalities, but passengers and crew were treated for smoke inhalation. One half of the Tunnel was out of service for about 6 months.

Three inquiries were immediately started: one by the French Judiciary; one by the Channel Tunnel Safety Authority (CTSA); and by Eurotunnel. The report of the CTSA inquiry (Channel Tunnel Safety Authority, 1997) has been called upon extensively for background in this chapter.

2.3.   The tunnel system

The full details of the Channel Tunnel system are widely available elsewhere (e.g. Channel Tunnel Safety Authority, 1997) and are only briefly summarised here. Some changes have since been made to the tunnel and shuttle systems following the recommendations in the CTSA report (and following the 2008 fire), so the description here relates to the situation at the time of the fire.

2.3.1   The tunnels

The system comprises three individual tunnels linking the terminals at Cheriton, Kent, England, and Coquelles, Pas de Calais, France.

The two running tunnels handled rail traffic from the closed-loop terminals or from each national rail network. Running Tunnel North (RTN) normally handled traffic from the UK to France and

Running Tunnel South (RTS) handled traffic from France to the UK. There were four main categories of rail traffic

- tourist shuttles carrying coaches or cars between terminals
- HGV shuttles carrying lorries and trucks between terminals
- passenger trains from the national networks
- freight trains from the national networks.

The central service tunnel ran between the two running tunnels. It was linked to these at approximately 375 m intervals by cross-passages, and carried special 'road' vehicles. The service tunnel had three main roles

- to provide normal ventilation to the running tunnels
- to provide a protected safe haven for passengers and crew in the case of evacuation
- to provide access for emergency vehicles.

It was also used for maintenance.

2.3.2 The HGV shuttles

Each HGV shuttle usually comprised

- a front locomotive
- an amenity coach (club car)
- a front rake of wagons, consisting of
  - a front loader wagon
  - about 15 carrier wagons
- a loader wagon in the middle of the train
- a rear rake of wagons (consisting of about 15 carrier wagons and a loader wagon at the rear)
- a rear locomotive.

The lorries would drive onto the rear loader wagon of the assigned rake, and then through the carrier wagons to the one allocated. The drivers, and any passengers, were then taken to the amenity coach for the journey. On arrival, they returned to their lorries to drive off from the front loader wagon.

The crew of the train comprised a driver, in the front locomotive, a chef de train and a steward, both in the amenity coach.

2.4. The fire safety system
2.4.1 The tunnels

The in-tunnel safety systems were provided for life safety. There was a comprehensive detection system consisting of 66 detection stations, each of which comprised

- four flame detectors with ultraviolet and infrared sensors
- a smoke detector with optical and ionisation sensors
- a carbon monoxide sensor.

The detection system notified the Fire Equipment Management Centre and the Rail Control Centre using a logic system designed to provide redundancy while screening out false alarms.

There was a supplementary ventilation system (SVS) that was used only in emergencies. This was intended to blow smoke away from an incident, to assist the escape of passengers and crew, and to help emergency services.

Within the tunnels was a fire main (connected to hydrants every 125 m), a high-level lighting system and a number of communication systems.

The cross-passages were protected both by heavy fire-resisting doors and by the normal ventilation system, which provided positive pressurisation in the service tunnel and cross-passages.

2.4.2 The HGV shuttle

The four loader wagons were each equipped with two smoke detection units, which signalled to the chef de train's station in the amenity coach. The amenity coach was air-conditioned and constructed of carefully selected materials. There were a number of communication systems between the chef de train and the driver. Air-inlet dampers fitted to the air-conditioning system were designed to close on the actuation of the 'tunnel fire' alarm by either the driver, in the locomotive, or the chef de train, in the amenity coach.

2.4.3 The safety procedures

The procedures laid down to deal with a reported fire on a HGV shuttle were three-fold.

1. The train would first attempt to continue to drive on out of the tunnel to a specially prepared emergency siding.
2. If the train had to stop, the crew would attempt to uncouple the amenity coach and locomotive from the rest of the train, and then leave the tunnel.
3. If neither of these were possible, the crew and passengers would wait for the operation of the SVS and then evacuate into the nearest cross-passage, and thence into the service tunnel for rescue.

2.5. The incident

Very briefly, the incident on 18 November 1996 unfolded as follows:

21.19–21.32	HGV Shuttle 7539 loads with lorries at the French terminal
21.42	The train leaves the platform
21.48	The train enters RTS at 57 km/h; security guards see a fire just prior to tunnel entry, and notify the supervisor who notifies the control centre
21.49	Unconfirmed detection in the tunnel
21.50–21.52	Further unconfirmed alarms
21.51	Control notifies the driver of the fire; unconfirmed alarm in the cab; fire on the rear locomotive; train travelling at 140 km/h
21.52	Train 4899 (single locomotive) enters RTS; encounters smoke
21.53	Confirmed alarm at the chef de train's workstation; fire on the rear locomotive
21.53	Confirmed alarm from in-tunnel detectors; control issues a general message to slow all trains and shut the pressure-relief-duct dampers
21.55	Train 4899 stops; an empty HGV shuttle in RTN encounters smoke around the cross-over
21.56	The French first line of response (FLOR) leaves the emergency centre
21.57	Control shuts the UK cross-over doors

21.57	The chef de train receives a fault warning; the driver receives a stop warning
21.58	The train stops at cross-passage 4131; thick smoke engulfs the locomotive; power is lost from the catenary
21.59	The driver tries to leave the cab, but is prevented by smoke
22.01	The chef de train opens the rear door of the amenity coach; smoke comes into the coach and the door is closed immediately
22.02	The French FLOR enters the tunnel
22.02	Problems with communications
22.03	The UK FLOR enters the tunnel
22.04	Control closes the French cross-over doors
22.11	The driver of 4899 evacuates; the SVS is activated in RTS, but the blades are incorrectly set
22.15	Train 6518 stops in RTN to serve as the evacuation train
22.20	The SVS is correctly set; cross-passage doors 4101 and 4131 are opened remotely; the chef de train evacuates the passengers
22.28	The French FLOR helps passengers
22.29	The driver is rescued
22.30	The UK FLOR arrives
22.42	The driver and some passengers board the evacuation train; smoke enters from RTN
22.53	The UK FLOR enters the tunnel to fight the fire
23.24	The evacuation train arrives at the French terminal
23.39	Casualties are taken out by service tunnel ambulances
05.00	Most of the fire is out

2.6.   The investigation

Within a week of the incident, the CTSA (see Appendix 2B) appointed two of its members (M. Pierre Desfray, Ingénieur Divisionnaire des Travaux Publics de l'Etat in the French Ministry of Transport, and Mr Jeremy Beech, the Chief Fire Officer of Kent) to lead a team to carry out an investigation. They assembled a team that included technical specialists from both countries, and were assisted by the French and UK emergency services and by Eurotunnel and its staff.

The cause of the fire was of great concern, and, as the incident could have been the result of a criminal act (i.e. arson), the French authorities instigated a judicial inquiry. It was therefore agreed that the CTSA inquiry would consider only the events that occurred after the incident train had left the platform at the French terminal, and thus be concerned with assessing the performance of the fire safety systems and learning lessons for the future.

The CTSA investigation comprised a number of parallel and interrelated activities. These included

■ an analysis of the sequence of events up to, during and after the incident, including people on duty and train movements
■ a review of the state of the fixed equipment prior to the fire
■ determining the composition of the incident train
■ determining the composition of the evacuation train
■ a review of the performance of the detection system and fixed equipment during the fire
■ a review of the communications procedures during the incident

- a review of the fire-fighting and rescue operations
- a review of the performance of the rolling stock during the fire
- interviews with victims, staff and firefighters
- an analysis of smoke movement during the incident
- an examination of the damage to the tunnel and the rolling stock
- an estimation of the development and growth of the fire.

2.7.   Method

Many of these studies involved the examination of paper or computer records and log-sheets, and interviews with staff, passengers and other witnesses. Regarding the on-site investigation; the first rule of forensic-scene examination is not to disturb the evidence. However, the nature of this particular incident, being confined within the tunnel and some way from normal facilities, had led to an early decision to clear the site. Within 24 hours of the incident, the front locomotive, the amenity coach, the front rake and most of the rear rake had been driven out from the UK portal, round the terminal loop, and back to France. The severe damage to the tunnel lining was deemed to be dangerous, and so loose material was removed. Figure 2.1 shows the fire damage to the tunnel around one of the piston relief ducts. The undamaged rolling stock (with the exception of the amenity coach) was quickly cleaned and put back ready for service, and undamaged HGVs were returned to their owners.

The badly damaged rolling stock and HGVs were removed from the tunnel as quickly as possible and taken to the French terminal. Figure 2.2 shows one of the fire-damaged carrier wagons still in the tunnel. Many of these vehicles, and in particular the vehicles that were considered to be those where the fire might have started, and the amenity coach, were embargoed by the French police. Some of these vehicles were initially put under cover, but all were later left out on a siding.

Figure 2.1 Fire damage to the tunnel around a piston relief duct. (Photograph courtesy of the author)

This activity resulted in a very unusual fire investigation. The site to be investigated was spread out over about 15 km, from the damaged part of the tunnel to the location of the affected rolling stock and HGVs. The shuttle wagons, amenity coach, rear locomotive and rear loader wagon were spread out over at least 1 km at the French terminal, and not in their correct order.

The separation of the wagons from the tunnel made an assessment of fire severity quite difficult, and the separation and distribution of the vehicles made an assessment of the fire spread quite difficult. In addition, it is usually the case in fire investigations that much can be learned from the less damaged parts, and in this case much of these had been removed.

The delays in gaining access to the tunnel, and then to the wagons, meant that much valuable evidence was lost, in particular from the chemistry of the soot deposits. The vehicles in the open began to 'age' and to become weathered. In addition to the effects of the weather on the evidence, the bad winter conditions (wind, wet and cold) also created difficulties, and some dangers, with regard to carrying out the on-site investigation.

Naturally, some minor difficulties arose from the bilingual nature of the project.

As it had been agreed that the CTSA investigation would not concern itself with the cause of the fire, a number of vehicles could be examined only from outside, but this only marginally limited the work. The rest of the investigation was largely conventional, with the available evidence being assessed for forensic indicators (items that allow a representation of the fire scene to be developed), and included, in particular, the examination of metallic components that provide a ready indication of exposure temperatures. These included aluminium fixtures (such as makers' plates), copper wire, and the stainless steel of the shuttle wagons.

Despite the limitations and difficulties, a wealth of data was gathered from the on-site investigation.

2.8.　Findings from the incident

As the CTSA inquiry developed, so the findings from the on-site investigation were integrated continuously with the information that was being derived from the other aspects. This meant, for example, that the reports from the victims of the conditions in the amenity coach could be compared with the actual soot deposits found.

The observations of the damage to the wagons and HGVs were also brought together, and this enabled a fairly consistent estimate to be made of the severity of the fire during the various stages of the incident. These stages were characterised as follows.

- *Immediate development after ignition.* The fire probably started on wagon 7 or 8 of the rear rake.
- *Early development prior to tunnel entry.* The fire had grown to around 1–1.5 MW when it was seen by the security guards.
- *In-tunnel development prior to confirmed alarm.* If the fire was below detector thresholds, then it was probably no bigger than 4 MW.
- *From confirmed detection to the train stopping.* The fire grew, but without damaging the tunnel. It was probably around 10 MW by the time the train stopped, and was producing large quantities of smoke.
- *Growth on the stationary train.* When the train stopped the smoke moved forward, engulfing the amenity coach. The fire would have grown to cause some fire damage to wagons forward of wagon 7. It would have been limited by the air supply, but may have grown as large as 50 MW.
- *Growth after actuation of the SVS.* After the SVS eventually took effect, the available air increased, and most of the 'fuel' downstream of wagon 7 was consumed. During this stage the fire could have been as large as 350 MW for a short while.

As well as filling the tunnel in its wake, the smoke from the fire spread across the open cross-over and through open pressure-relief ducts to affect the RTN and cause problems during the evacuation. Once the train stopped, the smoke spread forward and engulfed the amenity coach. This smoke entered the coach through the opened door and through other leakage paths. Conditions became untenable, and evacuation into the cross-passage was essential. The SVS should have been holding back the heat and smoke by this stage, but its late operation made it ineffective.

The apparently anomalous performance of the detection system, in particular that on the loader wagons, had to be examined.

2.9.　Issues, problems and lessons for fire investigation

As has been discussed above, there were unusual issues, problems and lessons associated with the fire investigation with this incident, and these fall into three main areas: procedural (relating to the top-level conduct of the investigation), operational (relating to the logistical and practical issues of getting to the investigation site) and technical (relating to the scientific investigation). There was naturally some overlap between these areas.

2.9.1　Procedural

Although the tunnel was a fully bi-national project, the fire had occurred on the French side of the 'border' (the halfway mark at midpoint in the tunnel). Therefore, the French delegation of the CTSA took the lead in the CTSA inquiry. Of immediate importance, however, was the fact that a crime was suspected, and the incident fell within French jurisdiction. The French police took charge of the incident and impounded all vehicles (rail and road) that they considered relevant to their inquires. This initially significantly inhibited the CTSA investigation, and top-level discussions between the French delegation and the judiciary became necessary to agree working arrangements and terms of reference. However, there was no exchange of information between the two bodies, and this did have a number of other implications (see below).

The CTSA inquiry was multi-agency, involving the French delegation of CTSA, the UK delegation of CTSA, Kent Fire Brigade, Kent Police (who were assisting the civil inquiry), the Health and Safety Executive, the Health and Safety Laboratory, FRS (the fire division of the Building Research Establishment; now Fire Safety Group, BRE) and other specialists. However, many of the individuals from these various bodies had worked together previously, either on the Channel Tunnel project, or elsewhere. Consequently, there was very good team-work across the enquiry. In addition, the investigation required the cooperation of Eurotunnel, and their associated organisations, such as WS Atkins. This worked well, and there was a very good working relationship with both Eurotunnel and Atkins. In order to simplify the relation-ship with Eurotunnel, the CTSA investigation team was appointed as notified inspectors under the Channel Tunnel Treaty, as this assigned formal powers of access and inspection.

2.9.2 Operational
The fire in the Channel Tunnel was a big fire which had a major impact on a large transportation system with a very high profile. The commercial losses as a result of the closure of the tunnel took effect at once. Consequently, there were immediate efforts to seek to reopen the tunnel as quickly as possible.

Access to the incident site was an issue; the fire was on 18 November 1996, but the first systematic site investigation in the tunnel did not occur until 22 November. Access to shuttles and HGVs was not possible until 31 January 1997, and the last shuttle (where the fire originated) was examined on 10 June 1998. Given the degree of media interest, the Eurotunnel site had high security, which slowed access; and, as mentioned, there were minor language difficulties, mostly relating to technical terminology.

2.9.3 Technical
The limitations of access to and the tidying and disposal work going on in the tunnel naturally confused the investigation, as, for example, soot deposits were being cleaned. A similar limitation resulted from the disposal of undamaged or lightly damaged vehicles (shuttles and HGVs).

The investigation in the tunnel was difficult due to dust, surface conditions (spalled concrete) and the dark, so that artificial lighting was needed throughout. At the Coquelles site, the investigation had to contend with wind, rain and the cold (but these are typical problems with any investigation). However, the exposure and the time delays resulted in some 'ageing' of soot samples, etc.

The scale and the disordering of the scene was significant in that it was difficult to relate the fire- and smoke-damage patterns to the component elements (shuttle wagons, HGVs and tunnel). Despite these limitations, however, the structural, fire and smoke damage was still visible on the available components.

The investigation generated a large amount of information, and it became necessary to be selective. The investigation identified a lack of appropriate knowledge about the physics and chemistry of wind-blown flames and the effects of the movement of the train on the burning rate.

2.10. Discussion

As well as its naturally high profile, and the degree of media interest, the fire in the Channel Tunnel in November 1996 was of scientific and technical interest, for a number of reasons. The components of the tunnel fire safety system, and the system as a whole, had been fully 'fire safety engineered', with every element being the outcome of a carefully considered analysis of the likely events, and their consequences, and of the potential impact of the various protection methods. There were no predetermined safety 'rules'. The final safety system was the result of a probabilistic risk analysis, in which the probabilities of different events and actions were cascaded to derive the likely loss of life from any particular event. To support these analyses, there had been a large amount of experimental work, both on models and at full scale, that included a fire test on a HGV in a tunnel. The components of the safety system were tested extensively (e.g. in furnaces). All this work was for the protection of life.

The fire in 1996 tested the design of the safety system to its limit, but the fire that developed was one of the predicted scenarios for which the system had been designed to cope.

There was a substantial programme of other work carried out during and following the investigation by Eurotunnel, including computational fluid dynamics (CFD) modelling of both the incident and proposed remedial measures. The recommendations of the CTSA report were all addressed, mostly in the form of operational and procedural changes, although some technical developments were implemented.

In addition, and of specific relevance here, there were a number of lessons to be learned (or reinforced) regarding the conduct of fire investigations in tunnels, in particular bi-national tunnels. These included

- the need to quickly and formally establish responsibilities of the agencies and individuals participating
- formal and agreed liaison and routes for communications between agencies
- inter-agency cooperation
- minimal disturbance of the fire scene, and retention of apparently undamaged material
- early access to the scene
- fully self-contained equipment, including lighting and means of dealing with dust
- means of translation and interpretation
- site maps and design drawings.

2.11. Conclusions

The incident illustrated an issue that was, and still is, of increasing significance to the whole of the fire-safety community; that is, however well designed and effective the engineering of a fire-safety system may be, it will depend, on the day, upon its proper maintenance and operation by the people charged with that responsibility.

The investigation of the incident has given us some opportunity to learn how to deal with the investigation of rail tunnel fires in the future.

Acknowledgements

The CTSA inquiry called upon the help of a number of bodies, from both sides of the Channel, who formed a team to work together with the French and UK delegations of the CTSA on the incident. Eurotunnel and their consultants cooperated fully with the CTSA.

The author wishes to acknowledge in particular the CTSA, the Health and Safety Laboratory (Buxton), Kent Fire Brigade and Kent Police, for their help.

This chapter is contributed by permission of the Chief Executive of the Building Research Establishment Ltd. The views expressed here are those of the author and are not necessarily those of the CTSA.

Appendix 2A: The Channel Tunnel fires of 2006 and 2008

Two serious fires involving HGVs have occurred in the Channel Tunnel since 1996.

On 21 August 2006, a fire broke out in the load compartment of a lorry on a HGV shuttle travelling from the UK to France. The shuttle train was brought to a controlled stop 20.5 km from the UK portal. All persons on board (30 lorry drivers and four Eurotunnel staff) were evacuated into the service tunnel. They were then evacuated out via the service tunnel to the French terminal. No one was injured. The fire was extinguished after about 2.5 hours. The fire was on the penultimate carrier wagon, close to the rear of the train. It destroyed the lorry and damaged adjacent ones. The carrier wagon was structurally damaged but could be moved from the tunnel. The catenary parted and the tunnel lining was damaged to a depth of 30 mm at the top of the tunnel over a length of 10 m. Normal tunnel operation was resumed on 22 August 2006. The official report is available (Rail Accident Investigation Branch, 2007).

On the afternoon of 11 September 2008, a fire broke out on a HGV shuttle going from the UK to France. The train stopped 11.5 km from the French portal (39 km from the UK portal). As soon as it was known that there was a fire on board the train, the train was stopped by a door giving access to the service tunnel. The 29 passengers and three crew evacuated the train and walked to the service tunnel. They were later taken from there to the French terminal. Some were suffering from the effects of smoke inhalation, and some had minor cuts and bruises. The fire spread to involve other HGVs on the train, and was not finally extinguished until mid-morning on Friday 12 September. Full service resumed in February 2009, after repairs costing 60 million Euros. (See BEA-TT and RAIB, 2010.)

(Note: see also Carvel (2010), Chapter 19 and the section on secrecy in Chapter 30.)

Appendix 2B: About the CTSA

The CTSA was established by the French and UK governments in accordance with the provisions of the Treaty of Canterbury, signed on 12 February 1986. The Intergovernmental Commission (IGC), also established under the Treaty, oversees all aspects of construction and operation of the Channel Tunnel. The CTSA advises and assists the IGC on all matters concerning safety.

At the time of the incident, the CTSA comprised delegations from France and the UK as follows:

UK
Edward Ryder CB (Head of Delegation)
Roderick Allison
Sandra Caldwell
Jeremy Beech CBE QFSM
Peter Moss
Victor Coleman

France
Roger Lejuez (Chef de la Délégation)
François Barthelemy
Jean-Pascal Cogez
Claude Charmeil
Pierre Desfray

These members represented different departments from each government, to cover the different areas of interest in tunnel safety. The CTSA called upon a range of specialist technical people to help with its deliberations.

FRS (now Fire Safety Group, BRE) was requested to assist the CTSA in its inquiry, in particular to assist with the on-site investigation. On the afternoon of Friday 22 November 1996, Martin Shipp received a request from Jeremy Beech to join a Kent Fire Brigade team that was going into the tunnel as part of the CTSA investigation of the fire. From that time, Martin Shipp and, shortly after, his colleague Nigel Smithies were fully involved in the inquiry as members of the inquiry team.

REFERENCES

BEA-TT and RAIB (2010) *Technical Investigation Report Concerning the Fire on HGV Freight Shuttle 7412 on 11 September 2008.* Bureau d'enquêtes sur les Accidents de transport terrestre/Rail Accident Investigation Branch, Paris/Derby. English translation of the French original is available at: http://www.raib.gov.uk/cms_resources.cfm?file=/101122_ReportET2010_eurotunnel_eng.pdf (accessed 22 June 2011).

Carvel R (2010) Fire Dynamics during the Channel Tunnel fires. In: Lonnermark A and Ingason H (eds), *4th International Symposium on Tunnel Safety and Security, Frankfurt, Germany,* 17–19 March 2010, SP Report 2010:08, pp. 463–470.

Channel Tunnel Safety Authority (1997) *Inquiry into the Fire on Heavy Goods Vehicle Shuttle 7539 on 18 November 1996.* The Stationery Office, London.

Cooke RA and Ide RW (1985) *Principles of Fire Investigation.* Institution of Fire Engineers, Leicester.

De Haan JD (1997) *Kirk's Fire Investigation,* 4th edn. Brady Prentice Hall, Upper Saddle River, NJ.

Rail Accident Investigation Branch (2007) *Rail Accident Report: Fire on HGV Shuttle in the Channel Tunnel 21 August 2006. RAIB, Derby.* Available at: http://www.raib.gov.uk/cms_resources.cfm?file=/071023_R372007_Channel%20Tunnel.pdf (accessed 22 June 2011).

Handbook of Tunnel Fire Safety, 2nd edition
ISBN: 978-0-7277-4153-0

ICE Publishing: All rights reserved
doi: 10.1680/htfs.41530.037

Chapter 3
Tunnel fire investigation II: the St. Gotthard Tunnel fire, 24 October 2001

Jean-Claude Martin
Formerly of the Institut de Police Scientifique (IPS), Ecole des Sciences Criminelles (ESC), University of Lausanne
Olivier Delémont
Institut de Police Scientifique (IPS), Ecole des Sciences Criminelles (ESC), University of Lausanne
Claude Calisti
Formerly Technical Consultant, Police force of Paris and expert to the French Courts of Appeal
Translated by R. Carvel

3.1.   Introduction

This chapter is an abbreviated translation of part of the official report on the investigation into the fire in the St. Gotthard Tunnel on 24 October 2001. The original report was produced exclusively for Mr Antonio Perugini, the Prosecutor of the Public Ministry of the canton of Tessin, Switzerland, who directed the investigation; it is not publicly available. Publication of this account of the fire investigation has been authorised by Mr Perugini.

The authors would like to thank the Examining magistrate, who not only granted all the authorisations necessary for the full investigation of the accident and the fire which caused the catastrophe, but took part himself in many aspects of the investigation, including interviewing witnesses, laboratory experiments and on-site investigations.

3.2.   Incident summary

At 9.39 a.m. on 24 October 2001, two heavy goods vehicles (HGVs) were travelling through the St. Gotthard Tunnel in the Swiss Alps. The tunnel comprises a single roadway with two lanes, bored on a north–south axis.

A Belgian vehicle, travelling north from Airolo, ran up against the wall on its right-hand side, careered onto the southbound carriageway, collided with the wall on its left-hand side and swerved back towards the middle of the road. An Italian HGV, coming south from Göschenen, slowed down and moved into the adjacent lane, the northbound carriageway, in order to try to avoid collision with the Belgian HGV. The two HGVs collided.

After the collision, the Belgian trailer ended up in the southbound lane, with the cab straddling the centre line of the road. The Italian trailer ended up blocking both lanes, with its back on the southbound carriageway, and the front of the trailer and cab on the northbound carriageway.

Shortly after the collision, a fire broke out on one of the HGVs, and it spread rapidly to the other vehicle. Before long, both trailers were fully involved in the fire. The Italian truck driver raised the alarm and instructed the traffic entering the tunnel from the south to turn back. This enabled the

Figure 3.1 First photograph taken of the disaster; looking north. (Photograph courtesy of Mr C. Grassi.)

fire brigade to reach the incident location rapidly, in less than 7 minutes. The fire scene when the firefighters arrived is shown in Figure 3.1.

The firefighters (approaching from the south) were not able to get within 15–20 m of the vehicles, due to the ferocity of the blaze. Fire-fighting was further hampered by an explosion about half an hour after the fire-fighting began.

To the north of the incident location there was a line of HGVs that had been following the Italian HGV southbound and had stopped in the tunnel when the accident occurred. The fire spread to each of these vehicles in turn.

Moreover, the ventilation in the tunnel was such that the smoke, toxic gases and a considerable amount of the heat produced by the fire were directed northwards along the tunnel. This meant that the firefighters from Göschenen in the north were seriously hampered in their approach.

The fire burned for approximately 24 hours. After the fire was brought under control, the bodies of 11 people were found among the debris to the north of the incident location.

3.3. Aims of the investigation into the fire and explosion

The main aims of the investigation were

- to determine the origin and the cause of the fire which occurred after the collision of the two HGVs on 24 October 2001, at 9.39 a.m., in the St. Gotthard Tunnel

- to carry out all the examinations necessary to establish the dynamics of the disaster according to the nature and quantity of fuels present in the incident zone and the geometry of the site
- to explain the mechanisms of the explosions that were observed by the fire-fighters during their fire-fighting activities in the tunnel.

3.4.　Summary description of the incident zone

The St. Gotthard Tunnel (16.918 km long), bored from Airolo in the South to Göschenen in the North, comprises two lanes. A 'safety gallery' runs parallel to the main tunnel on the east side. Refuges are spaced every 250 m along the length of the tunnel, and these connect the tunnel to the safety gallery.

Figure 3.2 shows a view of the tunnel, and indicates the extent of the fire, the zone subjected to high heat and the extent of the smoke damage.

Figure 3.2 Topographic chart of the St. Gotthard Tunnel, showing fire, heat and smoke damage

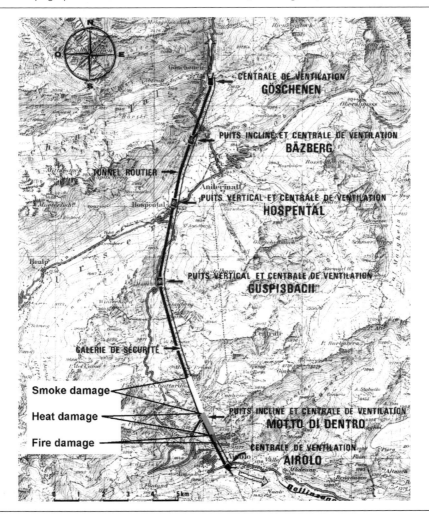

During the investigation, each vehicle involved in the fire or affected by heat or smoke was allocated a number.

1.   The Belgian HGV (carrying a mixed cargo, mainly clothes and textiles), initially travelling north, which was partially burned out after the fire, located on the western carriageway.
2.   The Italian HGV (carrying a cargo of vehicle tyres), initially travelling south, which was totally burned out after the fire, located across both carriageways.
3–7.   Five HGVs, initially travelling south, some partially, some completely, burned out, all located on the western carriageway.
8–13.   Six light vehicles which were behind the HGVs; these were not burned, but did suffer damage through exposure to heat.
14.   A truck, covered in soot, not damaged by heat.
15–18.   Light vehicles, covered in soot, not damaged by heat.
19–22.   Four HGVs, covered in soot, not damaged by heat.
23.   A light vehicle, covered in soot, not damaged by heat.

A body was found on the eastern carriageway just behind the last of these vehicles.

The collision between the two HGVs occurred approximately 1.2 km from the southern portal. This is where the fire started, within a minute of the accident. Flaming began in the lower part of one of the HGV tractors, and quickly spread to involve all of both vehicles. Before long the flames reached the concrete ceiling of the tunnel, which separated the vehicle space from the two ventilation ducts above. The eastern duct supplied fresh air to the tunnel by means of openings, just above the roadway, at 8 m intervals along the length of the tunnel. The western duct was connected to the vehicle space by openings every 8 m, through which the polluted air of the tunnel was extracted (although in the section of tunnel in which the fire occurred, the exhaust openings were 16 m apart). The peak supply/extraction rates of the ventilation system, before the fire started, were about 130 m^3/s. The configuration of the tunnel ventilation is shown in Figure 3.3. (In fact, the ventilation system of the tunnel at that time was complex. It could be longitudinal, i.e. with supply and exhaust systems switched off; semi-transverse, i.e. with only air supply or only exhaust operating; or fully transverse, i.e. with both supply and exhaust operating.) In 'fire mode' the ventilation system was set to fully transverse, having been in longitudinal mode before the fire.

Fresh air was drawn into the air ducts at the portals at both ends of the tunnel, and this air was supplied to the vehicle space through the vents on the eastern side of the tunnel, while the polluted air was extracted at ceiling height on the western side. Although the two HGVs on fire blocked the centre of the tunnel, the flames were pushed to the western side of the tunnel by the ventilation. This was evident from the damage to the upper parts of the tunnel lining (Figure 3.4).

Above the Italian vehicle, on the eastern carriageway, the ceiling was still intact in places, but had partially collapsed and was supported by the concrete walls of the tunnel lining. Above the Belgian truck, on the western carriageway, the ceiling had come away from its side anchoring and collapsed onto the HGV, but was supported in the centre of the tunnel by the concrete wall that separated the air ducts (Figure 3.5).

The fire plume, containing a significant proportion of unburnt gases as well as soot and flame, was forced in the direction of the northern portal. This was evident from the distribution of soot and

Figure 3.3 Cross-sectional view of the St. Gotthard Tunnel, looking north

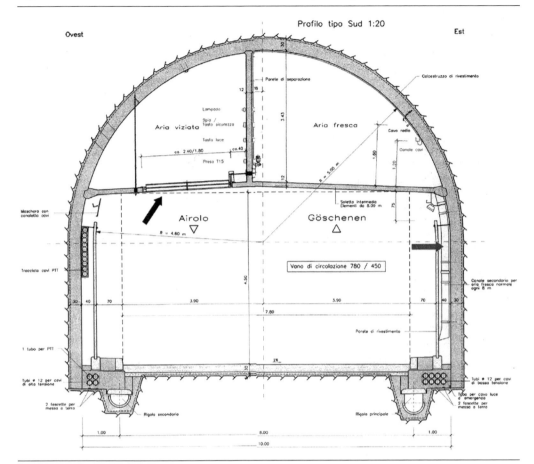

Figure 3.4 Photographs of trucks 1 and 2 taken (a) during and (b) after the fire. (Photograph (a) courtesy of Mr C. Grassi.)

(a)   (b)

Figure 3.5 Photograph of the tunnel ceiling above HGVs 1 and 2

other combustion products in the tunnel. This displacement of the flames, as well as the intense heat release of the blazing inferno, drove the fire northwards, in the upper layer of the tunnel, successively igniting several HGVs behind the Italian HGV on the western carriageway. Seven HGVs were destroyed, the two involved in the initial crash and the five behind the Italian truck (vehicles 3 to 7) (Figure 3.6).

The fire-fighting operations concentrated on the following vehicles.

- From the south the firefighters from Airolo, in the interior of the tunnel, attacked the fire involving the two collided HGVs.
- From the north, the firefighters from Göschenen, in the interior of the tunnel, tried to extinguish all vehicle fires within reach of their hoses.
- Another group of firefighters attacked the fire from the refuge beside HGV 7; by their efforts, the fire involving the tractor of HGV 7 was extinguished. The fire did not spread northwards of this point.
- At the refuge near vehicles 3 and 4, another group of firefighters was able to set up hoses on vehicles 3 and 4 and another in the direction of vehicle 2. The firefighters had to retreat into the safety gallery due to the collapsing ceiling.

Other vehicles, both light and heavy, had been travelling southwards following the HGVs. Several vehicles had been abandoned on the western carriageway, and others had been abandoned

Figure 3.6 Photographs of vehicles 3 to 7

following unsuccessful attempts to turn about, presumably due to lack of visibility in the smoke. All these vehicles were covered with a thick layer of soot. Some plastic elements of the vehicles, exposed to the thermal radiation and, to a lesser extent, the hot gas flow, had melted. The damage due to heat diminished with increasing distance from the two collided HGVs.

The victims of the fire were discovered to the north of the incident location, within a distance of approximately 1250 m. Some were inside their vehicles, and others were on the roadway. Ten died as a result of smoke inhalation. The burned remains of the eleventh victim, discovered in the cabin of HGV 4, did not allow an assumption to be made regarding the cause of death.

3.5. Chronology of the incident

Identifying the chronology of the events that occurred in the early stages of the fire is of great importance, as it makes it possible to estimate not only the evolution of the initial fire which involved the two collided HGVs, but also the propagation velocity of the smoke northwards inside the tunnel. It also allows an estimation to be made of the length of time for which the concrete ceiling was able to resist the high heat release of the fire.

A precise reconstruction of the chronology of events was, therefore, one of the major stages of the investigation. This required the interviewing of a number of witnesses (vehicle drivers, firefighters, maintenance personnel, and tunnel and police control centre operators), an analysis of the alarm records and a reconstruction of the events at full scale.

09.39	Collision of HGV 1 with the eastern wall of the tunnel. Collision with HGV 2. Both lanes are blocked.
09.40	Emission of smoke at the accident location. Discharge of a liquid onto the roadway.
09.41	Ignition of the fire at ground level, underneath the HGVs. Automatic fire alarms are triggered. The fire service in Airolo is alerted automatically. The fire service at Göschenen is informed of the accident by the staff of the Airolo fire service.
09.42	Fire spreads onto the trailer of HGV 2. All the emergency and maintenance services in Airolo are alerted.
09.43	The passenger from HGV 4 leaves the tunnel via the safety gallery. The driver returns to the cabin of HGV 4, where he dies. Monitors in the control room in Airolo show thick smoke in the location of the accident, but no flames are visible; the back of HGV 1 is visible.
09.46	Initial photographs of the fire are taken from the southern side (see Figure 3.1).
09.47	The firefighters from Airolo begin their operations. The first people evacuate the tunnel to the south, using the safety gallery. North of the accident, several people also evacuate the tunnel and find refuge in the safety gallery.
09.49	Some small explosions, attributed to the bursting of tyres, are heard. At the location of the incident, the door between the safety gallery and the tunnel cannot be opened because of very high temperatures.
09.51	Road police begin evacuation of vehicles on the southern side of the incident. HGVs are reversed to the portal.
09.52	The Göschenen fire service arrives at a place approximately 2.5 km north of the accident. A wall of smoke is heading northwards.
09.59	The driver of HGV 21 makes a phone call. He loses consciousness and dies in the cabin.
10.17	The ceiling collapses above HGV 1.
10.28	The ceiling partially collapses on HGV 3.
10.58	The heat in the vicinity of HGVs 1 and 2 means that it is impossible to enter the tunnel from the safety gallery at this location.
11.02	The fire involving HGV 1 is extinguished.
11.03	The ceiling collapses above HGVs 4 to 6.

12.55   Start of fire-fighting procedures at HGV 7.

14.00   (time very approximate) The fires involving HGVs 3 to 7 are extinguished.

3.6.   Discussion of the chronology

The chronology was structured according to testimonies collected from professional people (magistrates and police officers) accustomed to providing this kind of information. It is also based on the computerised record of the events, as recorded by the motorway monitoring stations at Airolo and Göschenen. Consequently, the collected elements represent a very high degree of reliability and relevance, as all the information was carefully checked and cross-referenced. The evidence provided significant clues that made it possible to investigate the fire itself.

- Following the collision between HGVs 1 and 2, a liquid flowed onto the roadway.
- In the first minute after the accident, thick smoke was produced. Shortly after this, flames were visible on the ground
- Three minutes after the incident, the fire spread to the whole of both vehicles; the flames reached the ceiling of the tunnel.
- Within 15 minutes after the collision, the smoke and combustion products had travelled 2.5 km in the direction of the northern portal.
- The smoke was extremely thick and flowed as fast as an avalanche.
- The concrete members of the ceiling resisted the direct influence of the flames for approximately half an hour before breaking down.
- A strong explosion occurred approximately half an hour after the beginning of the firefighters' intervention, i.e. approximately 37 minutes after the accident.
- The strength of the explosion took the firefighters by surprise; its strength was sufficient to shake the fire truck on the Airolo side.

3.7.   The origin of the fire

3.7.1   Summary description of the HGVs involved in the accident

The two HGVs involved in the collision that started the fire had the following characteristics. Vehicle 1 was a heavy articulated vehicle comprising a tractor unit (12 816 cc turbo-diesel engine, 460 horsepower with a 600 litre fuel tank; as the driver died in the accident, there is no way of knowing how much fuel was in the tank at the time of the crash) and a trailer (closed body, with two doors at the rear, approximately 7500 kg of cargo at the time of the crash). Vehicle 2 was also a heavy articulated vehicle comprising a tractor unit (13 798 cc turbo-diesel engine, 470 horsepower with a 600 litre fuel tank and a 300 litre reserve fuel tank; at the time of the crash the reserve tank contained 200–300 litres of diesel and the main tank was virtually empty) and a trailer (open body covered with a tarpaulin, approximately 15 200 kg of cargo).

The trailer of HGV 1 contained approximately 7.5 tonne of various goods, mainly clothes, clothing accessories, textiles and 2230 kg of photographic film. Several hundred kilograms of the latter goods, packed in cardboard and laid out on the floor of the trailer, were discovered intact after the incident. The packing of the materials served to prevent any contact with oxygen in the air. Consequently, only the cardboard exteriors and plastic wrappings were partially burnt; the plastic wrappings had melted and protected the remainder of the goods (Figure 3.7).

The trailer of HGV 2 contained 1099 car tyres, 15.2 tonne in total. This cargo was entirely destroyed; only the metal frame of the trailer remained, covered with a very thick layer of soot, and the steel wire forming the radial structure of the tyres (Figure 3.8).

Figure 3.7 Photograph of the trailer of HGV 1

3.7.2   Fire origin

On their entry into the tunnel, the fire investigators discovered a confined space, entirely covered in a thick layer of soot. There were seven burnt-out HGVs, covered with detached concrete slabs from the ceiling. No investigation was carried out on the vehicles that were not involved in the fire. The first task consisted of accurately locating the origin of the fire.

Several elements can aid in the identification of the original location of a fire: the progression of the fire, traces of combustion, the chronology of the event, and declarations of witnesses and the firefighters who took part in the operations. In this instance, many material clues were destroyed because of the severity and duration of the fire. It was necessary to study the evidence of the photographs taken in the early stages of the disaster.

3.7.2.1 Fire progression

All fires begin at the place where the following three factors were combined: fuel vapour or gas, oxygen (generally in air), and an initial source of heat. Whatever its speed, whether fire or explosion, the combustion reaction always starts at a precise point – the origin – and progresses while consuming fresh fuel. After a fire, it is sometimes possible to follow the spread of the flames and to work back to the original fire seat.

In fact, the examination of the zone of the tunnel reached by the flames and the smoke indicated that the fire progressed from the south towards the north, i.e. from HGVs 1 and 2 in the direction

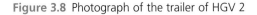
Figure 3.8 Photograph of the trailer of HGV 2

of the other vehicles that had stopped on the western carriageway behind HGV 2. It was clear that the fire began in the proximity of the two HGVs involved in the road accident.

3.7.2.2 Traces of combustion

It is sometimes possible to locate the origin of a fire that has caused only limited damage. When the thermal load of fuel and the contribution of air are considered together, the original fire seat may be localised at the place where there are traces of the most intense burning. It is then assumed that the most significant burning appeared where the flames were active for longest, this generally being at the origin of the fire.

In the case of the St. Gotthard Tunnel fire, it was difficult to use the traces of burning to locate the initial fire seat precisely, due to the extent of the destruction. All the combustible material of HGV 2, in the tractor, trailer and cargo, was destroyed. It was impossible to distinguish a surface or a volume that was more intensely burned than any other. All the combustible materials of the tractor unit of HGV 1 were destroyed by the fire. However, in the trailer some goods resting on the bridge as well as the rear double doors were partially saved from the flames. This situation indicates a spread of the fire from the tractor towards the trailer, assuming that the firefighters used their hoses in a uniform manner across the whole of the vehicle. This assumption is valid for the means used (a water cannon and high lance pressure), which made it possible to deluge simultaneously all the surfaces on fire.

Figure 3.9 Photograph showing HGVs 1 and 2 during the fire. The burning liquid in the gutter is indicated. (Photograph courtesy of Mr C. Grassi.)

3.7.2.3   Chronology of the event and declarations of the witnesses

All the witness statements, including the declarations of the driver of HGV 2, locate the origin of the fire at one of the HGV tractors involved in the collision. The following observations were also made: immediately after the collision, a large quantity of liquid spilled out onto the roadway; the fuel tanks of both HGVs were fixed to the back and side of each tractor unit using metal straps; a few seconds later flames developed on the roadway. A burning liquid can clearly be seen in the eastern gutter in the first photographs taken after the accident (Figure 3.9).

3.7.2.4   Discussion

By using the three parameters of fire progression, traces of burning and the chronology of the event, the origin of the fire can be identified. The fire began somewhere in the region of the tractor of HGV 1, before spreading to the trailer of HGV 2. This is supported by the following observations. On HGV 1, the traces of burning locate the origin of the disaster close to the tractor unit. As the two vehicles collided before the fire, it is logical to suppose that the initial source of heat was generated in the collision zone of the two trucks. If the fire began at the tractor of HGV 1, it is logical to assume that the flames were initially transmitted to the nearest part of the adjacent HGV, its trailer, before reaching the tractor of HGV 2. This assumption is supported by the extent of the visible fire in the first photographs of the accident.

3.8.   Cause of fire

3.8.1   Nature of fuel

Interpretation of several elements of the investigation makes it possible to deduce the nature of the fuel which, once ignited, was at the origin of the fire. These elements are the testimonies of the driver of HGV 3, which stopped behind HGV 2, and the first photographs of the accident.

3.8.1.1   Testimony: flowing liquid

Immediately after the collision, the driver of HGV 3 left his vehicle and, as he approached the scene of the incident, heard the noise of a liquid pouring onto the roadway. On a HGV, there are several different types of liquid that could be released following a crash: engine oil in the lubrication system, hydraulic oil, cooling water or diesel oil fuel. Some HGVs have other hydraulic systems, for example, to control the movements of the trailer, but neither HGV 1 nor HGV 2 had these systems.

3.8.1.2 Testimony: smoke release

Before any flames were observed, the witness describes having seen a smoke-like gas being released from the incident location. The circumstances of the accident and the nature of the cargoes that were transported exclude any massive emission of dust. As the fire did not appear to have broken out by this point, the smoke is assumed not to be the result of an unobserved combustion. It is deduced that the gas release that was observed by the witness could only have originated from the vaporisation of a liquid.

Temperature measurements, carried out on two HGVs identical to those involved in the collision and transporting the same weight of cargo, were taken on the southern slope of the St. Gotthard Tunnel approach on the 18 and 19 February 2002. The results of these experiments showed that the hottest surfaces in the engine, the turbocompressor and the exhaust system, reach temperatures of 400–600°C. These temperatures, functions of the work done by the engine, depend only on the slope of the roadway, as the loadings were kept constant during the tests: 7.5 tons for HGV 1, and 7–8 tons for HGV 2. These temperature values are higher than the boiling points of all the liquids mentioned above (oil, water, fuel, hydraulic oil). Therefore, it is logical to assume that the smoke observed by the driver of HGV 3, soon after the incident, was due to the vaporisation of a liquid coming into contact with a hot surface in one of the tractors.

3.8.1.3 Testimony: ignition of the fuel

After the vapour emission, the witness observed the appearance of 'short flames', which developed at the level of the roadway. Clearly, the conditions at the roadway were such that ignition was favourable. As the liquid fuel poured out onto the roadway, some fuel vapour mixed with the ambient air and ignited. In order for any fuel vapour to ignite and sustain burning, the ratio of the fuel to the air in the mixture must lie within the limits of flammability, which vary from fuel to fuel. Furthermore, the vapour pressure determines the concentration of the combustible vapour above a liquid fuel. This depends on the temperature; the higher the temperature, the higher the fuel vapour concentration.

These considerations indicate that a liquid poured onto a surface burns in the following way: only fuel in the gaseous phase (the concentration being a function of the vapour pressure and the temperature) will burn, and the flame will be above the liquid surface. In the initial phase of a pool fire, the relatively low temperature of the liquid produces only a limited quantity of vapour which, when released, enters the range of flammability and burns; this gives a short flame. As the temperature of the liquid rises (as a result of the heat released by the flames), larger amounts of vapour are released. This has the effect of 'pushing' the flames up to a few centimetres above the surface, and the quantity of oxygen close to the surface is then insufficient to maintain combustion. As the vapour rises, it comes into contact with fresh air and a fuel/air mixture within the flammability limits is established; this mixture burns and, as a result, the height of the flame grows.

3.8.1.4 Photographic evidence: flames on the roadway and in the gutter

On the first photograph, taken approximately 5–6 minutes after the appearance of the 'short flames' described by the driver of HGV 3, the flames covered a significant part of the road surface below HGVs 1 and 2. Moreover, they extended in front of HGV 2, and are also visible at the drains of the gutter on the eastern carriageway.

From this it is clear that a large quantity of liquid fuel flowed onto the ground of the tunnel and into the gutter. It flowed down the slope of the roadway which, at this location, slopes from north to south and west to east. By considering this evidence it is possible to deduce the nature of the poured liquid. Clearly, the material was combustible, and thus could not have been water from the cooling system of either of the tractors. In addition, the viscosity of the liquid was too low to allow a significant spread on the ground, and thus the liquid could not have been engine oil or the gearbox oil, which can therefore be discounted. The volume of the liquid is highly significant. It is estimated that the various oils were present in the following quantities: 20–25 litres of engine oil, about 10 litres of gearbox oil, 5–10 litres of hydraulic oil, and 600 litres of fuel oil for HGV 1 and 900 litres of fuel oil for HGV 2 (these values correspond to the fuel-tank capacities). Even though the quantities of fuel oil contained in the tanks of HGVs 1 and 2 were unknown at the time of the incident, only vehicle fuel oil satisfies all the conditions described above.

This assertion is confirmed by the investigation carried out on the engines of the two tractors after they were dismantled. It was observed that the plastic pipes transporting the hydraulic oil had burned. It is thus impossible to determine whether the rupture of one of the circuits occurred at the time of the accident or following the heat released by the fire. However, even if the liquid flowed at the time of the accident, its volume would be too low to cover the road surface and the parts of the gutter covered by the flames as seen in the photograph. Although exposed to the heat of the fire, the casings of the gearboxes of both HGVs did not exhibit any cracks or traces of bursting which would have allowed the oil to escape. These metal engine parts were not destroyed during the fire.

3.8.1.5 Sampling and analyses

The findings reported in Sections 3.8.1.1. to 3.8.1.4. rely entirely on the testimony of one person and the examination of the incident scene in the initial phase of the fire. From this it was deduced that the fuel from only one of the two HGVs could be responsible for the burning as it appears on the photographs. This conclusion is supported by material evidence from the scene of the accident; traces of fuel oil were discovered in the roadway material under and in the immediate vicinity of HGVs 1 and 2.

When a liquid fuel is poured onto a bituminous surface, part of the fuel is adsorbed by the surface. In the tunnel, the fuel-oil spillage was spread across a large part of the road surface, and some of this fuel vaporised, mixed with the air and burned. However, the adsorbed fraction underwent only a limited amount of heating and, importantly, did not come into contact with oxygen from the air. Because of this lack of exposure to oxygen, the adsorbed fuel did not burn, even under the extreme conditions of the fire.

Eight samples of the bituminous road surface were collected from the part of the road that had been subjected to the fuel spillage. Six other samples, of equivalent weight to the previous samples, were collected at 14, 20, 25, 60, 110 and 210 m from the location of the accident.

All the samples were transported in air-tight containers and were analysed in the laboratories of the Institut de Police Scientifique et de Criminologie (IPSC) using gas chromatography followed by mass spectrometry. Traces of fuel oil were detected in all the samples, but the concentration of the fuel oil was much higher in the samples collected from the fire zone.

From this it is certain that fuel oil was spread over all the surface in the zone in which the flames developed in the initial phase of the fire. It is therefore concluded that a spillage of heated fuel oil was responsible for the ignition of the fire at road level, and that this spillage propagated and maintained the fire at road level.

3.8.1.6 Discussion
The ambient temperature in the St. Gotthard Tunnel remains approximately constant at 20°C. The temperature of fuel oil contained in the tanks of HGVs is slightly higher than this at about 25°C, based on temperature measurements made on HGVs on their entry into the tunnel by the southern portal. This temperature difference is primarily due to fuel return from the fuel injection system. At 25°C, the vapour pressure of fuel oil is very low (about 2.5 mbar). At atmospheric pressure, the amount of gaseous fuel in air is about 0.25%. However, the flammability limits of fuel oil are 0.9–2.2%, which means that the amount of fuel in the air must be between 0.9% and 2.2% for the mixture to burn. Thus, in the incident described here, the fuel must have been heated significantly above ambient temperature for sustained burning to be possible.

This assertion is corroborated by the flashpoint of fuel oil, which is generally given as 80–110°C in tables of constants. The flashpoint of a fuel is defined as the lowest temperature at which there is a flammable vapour/air mixture above the surface. In practical terms, this is the lowest temperature at which the liquid will combust using a small, non-luminous, pilot flame. The flashpoint of this fuel oil was established as being 83.6°C during experiments, carried out on 18 litres of fuel oil, at the laboratories of CSI (a certification and testing organisation), Milan. From this, it is deduced that the fuel oil that leaked from a ruptured tank or pipe during the incident in the St. Gotthard Tunnel came into contact with a hot surface (the temperature of which was higher than the boiling point of the fuel, between 175–360°C) before spilling onto the roadway. In these circumstances, part of the liquid would have vaporised almost instantly, creating a localised atmosphere where the percentage of fuel in the air was above the upper flammability limit, so the vapours could not have ignited. As the liquid fuel spilled onto the roadway, the fuel vapours, which are heavier than air, would have spread out across the ground, cooled down and become diluted by the air, eventually reaching a concentration within the limits of flammability. Once flammable conditions were achieved, either a source of energy ignited the mixture of vapour and air, or spontaneous ignition took place. This flaming would have raised the temperature of the liquid fuel on the roadway, causing more of the fuel to be vaporised. On mixing with the air, this fuel is likely to have ignited, and so the flame would have spread across the surface of the spilt fuel.

3.8.2   Origin of fuel
Because of the quantity of the spilled liquid, it is clear that the fuel oil which spread on the roadway came from at least one fuel tank, which must have ruptured during the collision. Careful examination of the two vehicles allowed the precise origin of the fuel to be identified.

3.8.2.1 HGV 1
In HGV 1, the fuel injector system was to the left of the engine, whereas the turbocompressor, the exhaust and the fuel tank were on the right. During the examination of the vehicle in the tunnel

and the engine in a garage, the following observations were made. The copper pipes and the fuel injection pump were not affected by the shock of the collision or by falling debris. These parts, installed laterally on the left of the engine, were protected by the engine. The turbocompressor and part of the exhaust system were torn off during the collision. The traces of damage show that the mechanical stress was exerted from front to back, i.e. in the direction of the tank and the pipes that brought the fuel oil to the injection pump. The aluminium tank, fixed by two metal straps on a support, on the back right of the tractor, was discovered on the roadway. The bottom of the container was still recognisable, whereas the walls and the upper surface had melted. The remains showed no indication of bursting due to an explosion; this significant observation is discussed below.

3.8.2.2   HGV 2

In HGV 2, the turbocompressor, the exhaust and the secondary fuel tank were on the left of the engine, whereas the fuel injection system and the main fuel tank were on the right. During the examination of the vehicle in the tunnel and the engine in a garage, the following observations were made. The left part of the engine which, after the collision, was close to the east tunnel wall, did not exhibit any trace of shock. The turbocompressor and the exhaust system were still in place. The two fuel tanks were found to be in different conditions: only the bottom of the main tank was found on the roadway, and the parts of the unit closest to the vehicle chassis were still held in place by metal straps. Jagged aluminium fragments were discovered under and near the vehicle, which indicate that this tank exploded. Although it was severely damaged, the secondary tank, discovered hidden under a mass of several hundred kilograms of soot, did not appear to have undergone an explosion.

3.8.2.3   Discussion

From an examination of the fuel tanks and the part of the roadway where the collision occurred, it is possible to deduce the origin of the fuel spillage. The single tank of HGV 1 and the main tank of HGV 2 were both installed in the lower right part of the tractors, i.e. in the collision area. Consequently, the fuel oil spread on the roadway most likely originated from HGV 1 or the main tank of HGV 2. At the time of the accident, the tank of HGV 1 must have contained fuel, as there was no other fuel tank on that vehicle. During the intense heat of the fire, the walls of this tank melted, without any trace of bursting. However, intense heating of an empty or partially filled closed vessel containing combustible vapour generally causes an explosion. In this case, the quantity of fuel oil in the tank at the time of the accident is unknown. However, the testimony of witnesses who followed the HGV on its approach to the tunnel, state that the HGV did not stop to refuel at the station before the tunnel. From this testimony it is clear that the tank was not full. Therefore, it was deduced that this tank must have been opened before being exposed to high heat, otherwise it would have exploded.

According to the testimony of the driver of HGV 2, the main tank was virtually empty at the time of the collision. Indeed, shortly before entering the north end of the tunnel, the driver had switched fuel supply from the main tank to the secondary tank. It is clear that the empty tank (filled with fuel vapour) exploded during the fire; this is consistent with the explanation given above. The secondary tank, filled with fuel, appears to have survived the rise in the ambient temperature during the fire, until the seals on the tank failed, which allowed the fuel to escape onto the roadway without an explosion. It is therefore concluded that, in all probability, it was the fuel oil from the tank of HGV 1 that initially spilled onto the roadway and which, once ignited, was the origin of the fire.

3.8.3   Source of ignition

On a stationary vehicle, there are two sources of energy that could bring about the ignition of a mixture of fuel vapour and air: hot surfaces, or sparks produced by an electrical fault.

3.8.3.1   Heating of the mixture of vapour of fuel oil and air

From the experimental measurements carried out on the turbocompressors and exhausts of HGVs approaching the St. Gotthard Tunnel, it appears that the temperature of these vehicle parts can reach 600°C. In general, fuel oil will spontaneously combust at temperatures close to 300°C, with some variation depending on the quality of the oil. This means that, if the fuel vapour was heated up to 300°C, it could have ignited without the need for a 'pilot' flame or a spark. These factors could explain the ignition of the fuel vapours in this instance; the turbocompressor and the exhaust system could have behaved as powerful radiators, which vaporised the fuel oil. This hot gaseous fuel could then have mixed with the air and spontaneously ignited once the mixture was within the limits of flammability.

3.8.3.2   Ignition by the electric spark

Electrical sparks could have provided a weak source of energy, perhaps a few millijoules, but even sparks with this low energy would have been sufficient to ignite a mixture of gaseous fuel mixed with air.

Production of sparks by a short circuit

When two conductors with different electrical potentials come into contact, the short circuit often produces sparks. In a domestic or industrial situation, this phenomenon can produce significant, but short-lived, periods of heating. In general, this type of circuit failure is stopped by a circuit breaker, but if the safety measures fail, the release of sparks and heat can continue.

Power supply of HGV 1

In general, electrical devices on a vehicle are powered by the potential difference between the positive terminal of the battery and the vehicle chassis, which is connected to the negative terminal of the battery. In this way, it is possible to feed each electrical device on a vehicle with only a single wire. Any defect in the insulation of the supply wires can bring the positive potential into contact with the negatively supplied chassis, causing a short circuit and, possibly, sparks. From the examination of HGV 1, it was apparent that the main electrical circuitry, connecting the battery to the relay, was installed on the right-hand side of the engine, within the collision zone. Significant traces of electrical fusion were found on these wires, and this is direct evidence of electrical sparking between the damaged wires and the main chassis of the vehicle or the engine. This short circuit could only have occurred because of damage to the electrical insulation sustained during the crash between the two HGVs. Photographic evidence of the sparking is shown in Figure 3.10. As there was no circuit breaker on the vehicle, and many wires were damaged in the crash, the effects of the short circuit, i.e. the production of sparks and the production of heat, could have continued for several minutes, until the battery was totally discharged.

3.8.3.3   Discussion

There were two possible heat sources on HGV 1 capable of bringing about the ignition of the gaseous fuel/air mixture: the hot surfaces of the turbocompressor and the exhaust system, or electrical sparks produced by a short circuit of the wires connecting the battery to the starter relay. Physically, both of these possibilities are likely: the first, by raising the temperature of the fuel vapour up to the point of spontaneous combustion; and the second, by producing a

Figure 3.10 Photographs of the traces of electrical sparking observed in the circuitry of HGV 1

stream of sparks, for several minutes, of sufficient energy to ignite a fuel/air mixture, even at ambient temperature. However, it is much easier to ignite a flammable gas/air mixture with sparks than with a hot surface. This is why it was concluded that the fuel spillage was most probably ignited by sparking from the damaged electrical circuit.

3.8.3.4 Explosions perceived by the firemen

The explosions heard by the fire-fighters were most probably due to the tyres of the vehicles and the main fuel tank of HGV 2. These explosions were probably of the same type. At ambient temperature the vehicle tyres would have contained air under pressure, but at the elevated temperatures during the fire the pressure would have been dramatically increased, and this over-pressure would have brought about the explosions. In fact, the bursting of the tyres under the influence of extreme heat was probably due to a combination of the overpressure of the air and the thermal failure of the rubber envelope of the tyres. In a similar manner, the empty tank of HGV 2 would have contained fuel vapour at atmospheric pressure, and under extreme heat this would have created an overpressure that was sufficient to rupture the tank.

3.9.   Propagation of the fire across HGV 1 and HGV 2
3.9.1   Ignition of a tyre

As soon as the first flames appeared on the surface of the spilt fuel, they would have reached the tyres of the tractor units of HGVs 1 and 2. The tyre material, a mixture of rubber, polymers, carbon and sulphur, was moulded on a metal carcass. When tyres like this are subjected to an intense thermal stress, they degrade quickly, collapse, pyrolyse and emit flammable vapours as a result.

Full-scale experiments of HGV tyres subjected to the heat of a burning fuel fire were carried out in the laboratories of CSI, Milan. In the experiments, the liquid was heated until it reached its flash-point (84°C), and it ignited shortly thereafter. All times quoted below are relative to this ignition time. After 90 seconds, the tread of the tyre emitted vapours, which ignited, and the intensity of the fire grew rapidly after this point. Over the following few minutes the flames reached the height of the test chamber (2.5 m), and enormous quantities of soot and smoke were produced. After 5 minutes the temperature reached 1200°C at 40 cm above the tyre.

3.9.1.1 Ignition of the tractors of HGV 1 and HGV 2

The experiments showed that a tyre will burn after only 1.5 minutes of being subjected to the heat from burning fuel. Therefore, it is logical to assume that the fire was transmitted to the tractors of HGVs 1 and 2 in the following way. The tyres were in an upright position compared with the burning fuel surface, and this would have aided the upward spread of the flames to the engine compartments and into the cabs. As the temperature of the burning fuel increased, the dimensions of the flames increased. The flames would have spread into the vehicle cabs through the openings for controls and through the air vents. The high heat release in the cabs would have caused the windscreens and side windows to burst. At this moment, about 4–5 minutes after the first appearance of flame on the ground, the fire developed both inside and outside of the tractor units.

3.9.1.2 Ignition of the trailers of HGVs 1 and 2

The trailer of HGV 1 had a rigid body and the cargo it was carrying was loaded compactly on its floor. These factors would have slowed down the rate of fire spread from the tractor unit to the trailer in the initial stages of the fire. However, after a few minutes of being subjected to the intense heat of the tractor fire, it is likely that the wall and the floor at the front part of the trailer would have been penetrated by the fire, which would then have spread to the cargo.

The configuration of the trailer of HGV 2 was completely different. It did not have a rigid body, but merely had its cargo of more than a thousand tyres, stacked on their sides, covered by a tarpaulin. It is supposed that the fire would have spread to this cargo only a few seconds after the wheels caught fire. The tarpaulin would have burned easily, and the fire would have spread quickly to the cargo of tyres. A wall of flame would have developed from the floor of the trailer to the ceiling of the tunnel. This HGV burned out completely.

3.9.1.3 Discussion

The first photographs of the accident (see Figure 3.1) were taken between 5 and 6 minutes after the appearance of the flames on the roadway. The experiments carried out in Milan showed that a fire involving a HGV tyre can be very intense as soon as 3–4 minutes after the ignition of the fuel on the ground. Although there were a number of differences between the conditions in the tunnel and those in the laboratory, there was sufficient similarity to explain two of the observations. First, the speed with which the burning fuel spillage became an extremely violent fire was mainly due to the ease of ignition of a vehicle tyre by the burning fuel and by the very high fire load of vehicle tyres. In the Milan experiments, the liquid fuel was all consumed within the first 5 minutes, yet the rate of heat release of the tyre fire remained high. Second, the mass of particulate products (e.g. soot) produced by a fire depends on the quantity and the quality of the fuel involved. The speed of soot production is a function of the reaction speed. In this instance, in the open experimental conditions, the liquid fuel and tyres produced extensive quantities of soot and smoke.

3.10. Spread of the fire to HGVs 3 to 7

The back part of the trailer of HGV 1 was partially protected from the fire by the fire-fighting efforts of the firefighters from Airolo. The trailer of HGV 7 was not destroyed by fire because the firefighters from Göschenen succeeded in containing and extinguishing the fire that devastated the front parts of the vehicle.

Although the flames released by the burning of HGVs 1 and 2 were deflected by the tunnel ceiling and pulled northwards by the ventilation in the direction of vehicle 3, the distances between the vehicles were too large for the fire to reach them directly. However, convection and radiation processes were responsible for causing the fire to spread successively to HGVs 3 to 7.

Convective heat transfer, the mechanism of heat transfer from a hot fluid to a solid surface, and radiation, a mechanism of heat transfer between discrete objects, can be used to explain the spread of fire to the other HGVs. It is assumed that these mechanisms of heat transfer are responsible for the spread of the fire across the 30 m distance between the back of HGV 2 and the front of HGV 3. The reasoning used for the fire spread from HGVs 1 and 2 to HGV 3 is obviously applicable also to the spread from HGVs 1, 2 and 3 to HGV 4, and so on. The maximum distance between vehicles across which the fire spread was the 48 m from the back of HGV 3 to the front of HGV 4.

During the incident, HGV 3 would have undergone heating, both by radiation and convection, from the fire involving HGVs 1 and 2. It would also have been subjected to a stream of hot effluents (smoke, soot and firebrands) from the initial fire. The temperature of HGV 3 would have risen progressively. The front parts of the cab, the front tyres and the upper parts of the front of the trailer would have experienced the greatest amount of radiated heat, as they were in direct line-of-sight of the initial fire. These surfaces would have experienced the greatest amount of heating, and therefore would probably have ignited first. The fire spread on each HGV would have been from front to back. This is why the cab of HGV 7 was destroyed but, due to the intervention of the firefighters, the trailer was left intact. Initially, HGV 4 would have been shielded from the heat radiated from the initial fire by HGV 3, although it would still have been exposed to heat convected from the fire. However, once HGV 3 was fully ablaze, HGV 4 would have experienced direct radiated heat from the fire of HGV 3. The line of vehicles would have ignited successively, from south to north, in this manner.

3.11. Thermal degradation of the vehicles beyond HGV 7

In the 110 m beyond HGV 7, a number of vehicles were affected by the heat from the fire, even though none of the vehicles burned. Nine vehicles (numbers 8 to 16) were exposed to the hot gases, smoke, soot, firebrands and radiated heat that propagated northwards, in the direction of Göschenen, from the fires involving vehicles 1 to 7. Some of the south-facing parts of the vehicles show traces of thermal degradation; rubber and plastic parts of vehicles 8 to 16 exhibited signs of melting. It is certain that this damage was due to the intense radiated heat produced by the fires involving the HGVs. The extinguishing of the fire on HGV 7 reduced the potential heat transfer to vehicle 8, and hence reduced the chances of fire spread to vehicle 8 by preventing the heat from HGV 7 being radiated northwards. Even though HGVs 1 to 6 continued to burn for several hours, this fire-fighting success was sufficient to prevent further fire spread to the north.

Finally, for approximately 1750 m beyond vehicle 16 there was a zone containing seven vehicles (numbers 17 to 23), which were all covered in soot but not damaged by the heat of the fire. However, the driver of HGV 21 was found dead in his cab; he had communicated with his company by mobile phone before succumbing to the toxic gases in the smoke.

3.12. General discussion

The investigation carried out at the site of the collision of HGVs 1 and 2 and the examination of the two vehicles led to the identification of the initial fuel of the fire and its source, the identification of the heat source that vaporised the fuel, and the source of ignition of the vapours. This evidence was supplemented by experimental measurements carried out on two HGVs, identical to those involved in the crash, on the southern approach to the St. Gotthard Tunnel. These measurements help explain the speed with which the fire spread to involve both vehicles.

The fuel of the initial fire originated from the fuel tank on the back right of HGV 1. The heat from the turbocompressor and the exhaust system helped vaporise the fuel. Two possible ignition sources were present: the hot turbocompressor and exhaust system, and electrical sparks produced by a short circuit between the battery and the starter motor. Experiments have shown that ignition by sparks is more likely, and is faster than ignition by the hot surfaces, so this is the more likely mechanism in the present case. The fuel spilled onto the roadway, spread under the vehicles and ignited. This led to the fire spreading rapidly to HGV 2.

The mass of burned fuels, in particular the cargo of HGV 2, demonstrates that an enormous quantity of heat was released during the fire of HGVs 1 and 2. The high heat flux that arose from this fire, as well as the large flames produced by the burning of the cargo of tyres being transported by HGV 2, readily explains the ignition of HGV 3, which led in turn to the ignition of HGV 4, and so on, until the fire reached the tractor of HGV 7. The visible degradation to the vehicles beyond HGV 7 was due to the northward transport of heat.

3.13. Conclusions

It is the opinion of the authors that, during the collision of the Belgian registered HGV 1, travelling northwards from Airolo, and the Italian registered HGV 2, coming from Göschenen, the fuel of HGV 1 was spread on the roadway and vaporised due to the heat from various hot parts of the engine. In all probability, this fuel oil was ignited by electrical sparks due to a short circuit caused by the crash.

Within approximately 1 minute the fire spread, by means of the tyres and the vehicle chassis, to involve both HGVs. Their combustion, especially that of the cargo of HGV 2, caused an extremely high heat release which, in spite of the rapid intervention of the firefighters, caused the fire to spread to HGVs 3 to 7, which had stopped in line to the north of the collision.

The small explosions perceived by the firefighters were due to the bursting of the vehicle tyres as they became subjected to the high heat released by the fire.

The large explosion, which occurred approximately 30 minutes after the collision, and which shook the fire truck from Airolo, was due to the rupture of the virtually empty fuel tank of HGV 2, which would have been filled with fuel-oil vapour.

Investigations carried out in the tunnel, the detailed examinations of the tractors involved in the accident and their engines, analysis of various witness statements and photographic evidence, as well as the results of various experiments (recording the temperatures of various parts of HGVs, identical to those involved in the accident, after climbing the ascent to the St. Gotthard Tunnel portal; experimental burning of fuel oil and a tyre) made it possible to specify the origin and the cause of the fire, and demonstrate the mechanisms of the fire propagation.

Appendix 3A: Important factors relating to the investigation of a fire in a road tunnel

1. Do not disturb the site

The site of a tunnel fire will always be profoundly affected by the following.

- The fire itself. In a confined space the duration of the fire will probably be substantially longer than a fire in the open because of the limited access for the firefighters.
- The fire-fighting operations, rescue operations, recovery of the deceased and the analysis of the debris.

It is thus essential, as soon as the investigation begins, to brief all those working at the scene of the incident before the investigators arrive about the issue of conservation of evidence.

2. On-site activities

- Always wear protective clothing and gas masks, as the site will be severely contaminated.
- The vehicles at the fire origin are localised. Identify a sensible security zone around these vehicles, bearing in mind the operations of others in the tunnel (it is unnecessary to prohibit access to a 1 km stretch of tunnel if 50 m would be sufficient for the investigation).
- During the investigation, the use of powerful lighting is essential. It should be easily removable.
- Heavy-lifting apparatus will be required to move the debris. Ensure that the debris removal is directed in consultation with the investigators.

3. Circumstantial and chronological evidence

This evidence should be collected by the investigators as soon as possible after the fire, this information is vital to the investigation.

- Was the fire preceded by an accident or not?
- Was there any dysfunction of the vehicle before the disaster? That is, was there deceleration, smoke emission, appearance of flames, etc.?

4. Work on the vehicles

- Locating the source of ignition is always difficult, and sometimes it is physically impossible due to the deformation, fusion, and sometimes total absence of parts of the engine, the chassis, the electrical circuits, etc.
- Analyse potential sources of fuel. For liquids: hydraulic systems, brake systems, lubrication, fuel, air-conditioning, etc. For solids: electric cabling and insulation, body parts, plastic panelling, tyres, etc.
- Analyse potential sources of ignition. For hot engine parts: turbocompressor, exhaust system, brakes, etc. For sparks: mechanical origin (friction) or electric (alternator, short circuit due to damaged insulation, etc.).

It is important to stress that the fire expert is a general practitioner in the problems of physical and chemical thermodynamics. It will thus be essential to call upon the services of specialists in the fields of vehicle mechanics.

5. Examination of the vehicles in a laboratory

Corrosion is extremely fast. Immediately after the on-site examination, the bodies of the vehicles at the centre of the fire should be transported to the laboratory, the engine examined, the chassis stripped, the electric circuits investigated, etc. From the evidence of traces on the chassis and engine, damaged areas, fuel lines, electrical damage, etc., it is possible to propose one or several physically realistic hypotheses regarding the source of ignition.

Handbook of Tunnel Fire Safety, 2nd edition
ISBN: 978-0-7277-4153-0

ICE Publishing: All rights reserved
doi: 10.1680/htfs.41530.059

publishing

Chapter 4
Tunnel fire investigation III: the Burnley Tunnel fire, 23 March 2007

Arnold Dix
University of Western Sydney, Australia

4.1. Introduction

On 23 March, 2007, an incident occurred in the Burnley Tunnel, Melbourne, Australia. The Burnley Tunnel is one of a non-identical pair of tubes, three lanes wide, servicing Melbourne's busiest toll road networks. Vehicle numbers typically exceed 100 000 vehicles per day, per tube.

The Burnley tunnel is 3.4 km long, and at its deepest point is 60 m below ground level. The gradients within the Burnley Tunnel are very high, exceeding a 6% down-slope after the entrance portal prior to its deepest point, and exceeding a 5% up-slope for nearly 1 km prior to the exit portal.

The Burnley Tunnel has three 3.5 m lanes, a vertical traffic clearance of 4.9 m and two 0.5 m shoulders, as well as an elevated 0.8 m wide walkway. It carries a mixed fleet, with a significant proportion of heavy goods vehicles (HGVs). On the morning of the accident, the traffic mix was 28% HGVs and 72% cars.

Traffic flow in the tunnel is unidirectional. At the time of the collisions there was no prohibition on lane changes or any restriction on vehicle types in lanes. It is not permitted to carry dangerous goods through the tunnel.

4.2. Fire-fighting systems

The Burnley Tunnel is divided into 60 m fire-response segments and has a fixed fire-fighting suppression (FFFS) system. Each segment has

- two independently operable 30 m zones of deluge fire-fighting technology
- a fire cabinet (Figure 4.1) containing
  - a fire hydrant
  - a hose
  - two dry chemical extinguishers
  - a tunnel controller intercom.

Melbourne CityLink uses a deluge system, which comprises a conventional sprinkler system, which has larger droplet sizes, higher volumes and lower pressure than a water mist system.

Figure 4.1 (a) Fire cabinets are situated at every 60 m inside the Burnley Tunnel. (b) Tunnel controller intercom – located at each fire cabinet throughout the Burnley Tunnel

(a)

(b)

The system is arranged in individually operable zones that are 30 m long and 11.5 m wide. It is a dry system – once remotely operated, quick action valves are commanded to open and an otherwise dry set of pipes is filled with water.

The deluge discharge rate is around 2850 l/min, per 30 m zone. The tunnels are equipped with large collection sumps. In the emergency discussed here, these sumps were large enough to contain all the deluge water.

Importantly, the decision to install a deluge system was coupled with a detailed technical analysis of the nozzle performance, including critical factors such as droplet-size distribution, and trajectory modelling of droplet movement over a range of longitudinal ventilation velocities. From this analysis, a speed of 10 m/s was selected as the design longitudinal velocity for achieving the required spread and flow performance for given water pressures. A spiral nozzle located in the area where the fire occurred is shown in Figure 4.2.

4.3.   Smoke and thermal detection

A linear heat-detection system is installed above the traffic lanes to detect temperatures in excess of 68°C or a change in temperature greater than a predetermined rate. These alarms did not play a critical role in the Burnley Tunnel fire.

Smoke detectors are located in all other underground service areas.

4.4.   Communications

4.4.1   Radio re-broadcast

Ten FM and ten AM radio stations have their transmissions re-broadcast within the tunnel. This re-broadcasting enables CityLink to disrupt normal transmissions and superimpose messages

Figure 4.2 A spiral nozzle located in the area where the tunnel fires occurred

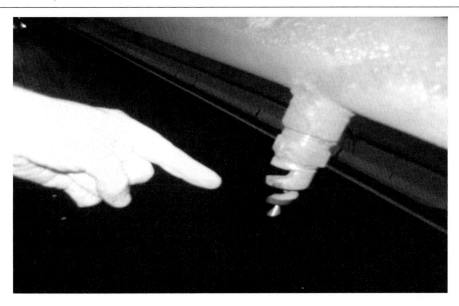

over active radios while vehicles are in the tunnels. Usually these re-broadcasts are coupled with a tunnel public address system.

4.4.2 Public address system

The tunnel control operator can make public announcements using the public address system, which comprises 178 speakers spaced at 30 m intervals. Normally the announcements are the same as those being played through the radio re-broadcast system.

4.4.3 Emergency services radio

A common fire brigade, ambulance and police services communications re-broadcast system is provided in the tunnels. This system replicates the functionality of the emergency services' radios used on the surface.

4.4.4 Fire telephone system

A fire telephone system is installed at 120 m intervals. This was installed especially for fire-fighters. It has a 5-hour battery backup and a high level of reliability.

Despite issues with communications during the incident discussed here, the fire telephone system was not used.

4.5. Signs
4.5.1 Variable message signs

Variable message signs are located at regular intervals throughout the tunnels.

4.5.2 Variable speed limit signs

Variable speed limit signs allow the tunnel operator to vary the allowed speed displayed to motorists in the tunnel.

4.5.3   Closed-circuit television (CCTV)

The 60 CCTV cameras within the Burnley Tunnel, all with pan, tilt and zoom capability, are operated by the tunnel control operator. The cameras were used extensively and effectively during the emergency.

4.5.4   Automatic incident detection (AID) video cameras

Sixty AID cameras provide images of the Burnley Tunnel to the operator. These images are constantly analysed by computers in order to detect incidents. It was the AID image analysis that first alerted the tunnel operator to a truck stopped in the tunnel.

4.6.   Access and egress

Access into and egress from the tunnel by road is normally via one or other of the two tunnels. However, in addition, there are: three passages linking the Burnley Tunnel to the adjoining Domain Tunnel; two safe havens, where a tenable environment is sought to be maintained by active ventilation of underground caverns; and a separately constructed parallel pedestrian emergency tunnel, providing a safe place for pedestrians following an incident.

4.7.   The incident

On 23 March 2007, at 09.52:30 a.m. a truck travelling eastbound in the tunnel made an unscheduled stop. Over the following approximately 2 minutes, 103 vehicles passed the stopped truck without incident. By 09.54:24 several vehicles, including four HGVs and seven light vehicles had crashed, three people were dead, and a fire and a series of explosions had been initiated.

By 09.56:00 a.m. (approximately 2 minutes after ignition), emergency ventilation and the FFFS system had been activated.

At the time of writing, the official coroner's report into this incident was not a public document (although it had been prepared for the coroner by the author). In the absence of legal authority to refer to the coroner's report the following eye-witness evidence from the criminal court case against the truck driver who allegedly caused the deaths is used to illustrate the sequence of events (Victorian Supreme Court Proceedings: *Director of Public Prosecutions* v. *David Lawrence Kalwig 16/07/2009*):

> Essentially I heard the screeching of tyres. I looked in the rear vision mirror; saw the car careering into the back of the truck. The nose of the car went down, the car lifted up like that so … and then there was another smash from behind by a truck.

> I saw the truck hit it and the … I can only assume that it was the gas tank of the vehicle that exploded.

> I saw the explosion.

> Well I continued to drive through the tunnel. There were … there was another explosion shortly after that which was a much bigger explosion. I remember the windows of my car vibrating as a result. There was also another announcement that came over the speaker saying that there had now been an incident in the tunnel and that vehicles were to slow down to 60 kilometres an hour …

This evidence graphically describes the crash, fires and subsequent explosions. The initiating events for this incident were large – large in the context of prior catastrophic events such as the engine-compartment fire in the Mont Blanc Tunnel incident. Yet, despite the severity of these initiating events, the fires were contained, with no generalised spontaneous ignition or other significant fire growth occurring once the deluge FFFS system had been initiated.

However, it was not only the operation of the FFFS system that was critical. The ventilation system was also effective in that it stopped backlayering (up a steep tunnel gradient of 6%) and rapidly reduced the longitudinal airflow (to approximately 2 m/s) in order to optimise smoke extraction and minimise ventilation-induced fire growth. It was the fire brigade that put out the fires, the deluge FFFS system merely keeping the fires small enough to allow effective intervention by the emergency services.

The following is a summary of the key events in the 2007 Burnley Tunnel fires, as revealed in the Supreme Court criminal jury trial against the driver of the truck who initiated the incident (Victorian Supreme Court Proceedings: *Director of Public Prosecutions* v. *David Lawrence Kalwig 16/07/2009*). This evidence is indicative only of the events, and is not from the more detailed Coronial investigation.

At 9.52:30 a truck stopped in the left (slow) of three lanes in the east bound Burnley Tunnel.

Between 9.54:26 and 9.54:30 a series of collisions, explosions and fires occur when a truck crashes into several cars and HGVs in the region immediately behind the stopped truck. Eventually the truck which initiated the series of crashes hit the stopped truck, pushing it many metres forward.

At 9.55:37 the tunnel operator enabled emergency mode in preparation for the smoke extraction, deluge operation and evacuation.

At 9.55:50 the emergency response plan was initiated by the tunnel controller including activation of emergency smoke extraction and the deluge system.

At 9.55:54 the smoke extraction system was activated.

At 9.56 the FFFS system (Deluge) was activated.

If the FFFS system had operated purely automatically, without operator intervention, then the wrong section would have been activated. Three people were killed in three different vehicles. Two of the three deaths were determined to be 'effects of fire'. The fires that killed these people were not, and could not be, extinguished by the FFFS system, because deluge systems cannot penetrate inside the shell of a vehicle. All those killed suffered fatal serious physical injuries in the car crashes, and would have been expected to die even without the 'effects of fire'.

The lack of damage to the outsides of the vehicles involved in the fires, and the absence of fire growth along the tunnel are entirely consistent with the expected impact of a deluge system on the growth and spread of fires in tunnels.

The incident resulted in several hundred people being evacuated from the tunnel – and their vehicles. None of the evacuees or their vehicles was injured or damaged. The tunnel suffered only minor damage, and could have been reopened tens of hours later if the extent of the damage could have been determined more rapidly (Figure 4.3).

4.8.   Discussion

In the Burnley Tunnel incident, the ventilation system rapidly reduced the longitudinal air velocity to approximately 2 m/s. This low ventilation rate was sufficient to stop backlayering, despite the buoyancy effect caused by the tunnel's steep gradient (in excess of 6.2% at the incident location).

Figure 4.3 Burnley Tunnel after the fire. The road surface, tunnel walls and tunnel services are still intact

The rapid activation of the FFFS system and the comparatively low longitudinal air velocities coincided with only minimal growth of the fire following the crash, explosions and subsequent fire. This is entirely consistent with the theoretical results supporting the merits of suppression systems.

The fire burning under the truck, in the engine compartment and tyres, was not extinguished by the FFFS system, but by fire-fighters. This is entirely consistent with the expectations derived from experiments involving shielded fires.

The absence of spontaneous ignition and the lack of accelerated fire growth is consistent with the experimental data on the effects of a FFFS system with a water application rate of the order of 10 mm/min.

4.9. Conclusion

The rapid and accurate use of the Burnley Tunnel FFFS system, coupled with effective control of the longitudinal air velocity coincided with minimal tunnel damage, no non-crash, fire-related injuries and rapid reopening of the facility.

It is likely that the use of an FFFS system coupled with advanced tunnel-ventilation control, rapid incident detection and accurate response positively contributes to tunnel fire safety and asset protection. See also, Dix, 2010.

REFERENCE

Dix A (2010) Tunnel fire safety in Australasia. *4th International Symposium on Tunnel Safety and Security (ISTSS), Frankfurt am Main, Germany*, 17–19 March 2010, pp. 69–79.

Part II

Prevention and protection

Handbook of Tunnel Fire Safety, 2nd edition
ISBN: 978-0-7277-4153-0

ICE Publishing: All rights reserved
doi: 10.1680/htfs.41530.067

publishing

Chapter 5
Prevention and protection: overview

Alan Beard
Civil Engineering Section, School of the Built Environment, Heriot-Watt University, Edinburgh, UK
Paul Scott
FERMI Ltd, UK; Provided additional material

5.1.   Introduction

A number of different kinds of safety system are installed in road, rail and underground railway/ metro tunnels, including fire detection systems, ventilation systems, suppression systems and alarm systems. While these systems are very diverse, they all have the same basic aims, i.e. to reduce the risk of injury or fatality for tunnel users and to reduce the risk of damage to the tunnel structure or operation. Central to the understanding of the expediency of these systems is an understanding of risk itself, and the related concepts of prevention and protection. This chapter gives an introduction to risk and its associated concepts in general terms, before relating these concepts to tunnel fire safety. Related information may be found in other chapters in this Handbook.

5.2.   Risk as a systemic product

If there is a single concept that is crucial to the understanding of risk, it is the concept of the term *system*. However, the word 'system' is used in countless different ways, and there are a very large number of definitions in the literature. For example, the word 'system' applies to the solar system, philosophical systems, social systems, technical systems, etc. Cutting to the essence of the concept, it can be said that: *a system is any entity, conceptual or physical, which consists of interdependent parts*. (For more on the concepts 'system', 'systemic' and 'reductionistic', see Beard (1999).)

Given the general definition above, two specific kinds of system are

■ a human activity system, which comprises people and non-human parts
■ a functional system, which has a function or purpose.

It is important that 'systemic' not be confused with 'systematic'. To be systematic is to be 'methodical' or 'tidy'. It is desirable to be systematic, but to see things in a systemic way is more than this – it is to see events as the result or product of the working of a system. For example, we may talk about a person as being 'happy' or 'sad'. This refers to the entire person, even though the 'person' consists of 'parts', such as neurons and muscles, within a context. It does not make sense to talk about a 'happy muscle'. That is, being happy or sad is an *emergent property* of the whole person, not a property of any single part.

Given this, we may see 'failure' as a product of a system and, within that, see death, injury, property loss, etc., as results of the working of systems. This may be summarised as shown in

Figure 5.1 Risk as a systemic product

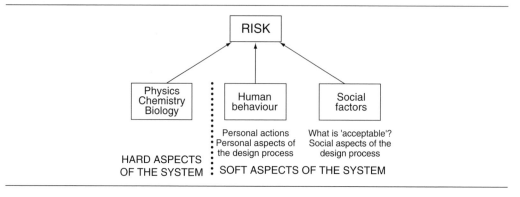

Figure 5.1. In the figure, 'social factors' is intended to be interpreted generally, to include economic or political factors. 'Social aspects of the design process' for example, may be taken to include economic or time pressures. There may also be institutional pressure.

After the start of operation, a similar figure to Figure 5.1 would apply, where the domain would be 'operation' rather than 'design'. Institutional pressure, often related to socio-economic or political influences, may be very strong (see Chapters 23 and 30).

Some specific examples are given below of how risk may be seen as a product of a system. There could be many other examples.

5.2.1 Example 1: interaction between different materials/items

The behaviour of a particular material must not be looked upon purely in isolation. For example, polystyrene packing pellets will not burn very easily on their own (e.g. in a pile on a concrete floor). However, if they are put into a cardboard box and the corner of the box is ignited, the polystyrene would be expected to burn very readily and produce thick black smoke. Thus, if there is sufficient heat flux over a sufficient area, packing pellets will cause a serious fire. One of the heavy goods vehicles (HGVs) involved in the Channel Tunnel fire in November 1996 contained polystyrene.

This example shows the inadequacy of testing materials in isolation. As a material, when in pure isolation, polystyrene packing pellets might be regarded as not representing a particularly great risk. However, as part of a system they may represent a considerable risk.

The question then becomes: 'Where are polystyrene packing pellets often found in the real world?' One answer is that they are very often found packed around electronic equipment or fragile items within cardboard boxes. Thus, in real situations, the system of which this material is often a part is such that there may be a significant risk.

5.2.2 Example 2: geometrical arrangement

Waste paper ignited in a metal bin in the centre of a room will very likely burn out without the fire spreading. If the bin is in a corner, however, the fire may well spread to the walls, up to the ceiling, and lead to flashover.

The question is then: 'Where are waste paper bins often found in the real world?' One answer to this is that they are often found in the corners of rooms. Thus, as with the polystyrene example above, the real-world system of which the waste paper bin is a part is often such as to create a significant risk.

In a tunnel context, the geometrical arrangement of items, in a HGV, for example, may strongly influence the development of a fire. It has been found, for example, that changing the structure of a wooden crib, without changing the fire load, has a 'considerable effect' on combustion (Chang *et al.*, 2006/07).

5.2.3   Example 3: integration of infrastructure and operation
In the Burnley Tunnel fire in Melbourne in 2007, a fixed fire-fighting system played a crucial part. However, according to Dix (2010):

> it was not merely the presence of a fixed fire fighting system which was critical – it was that the ventilation system was effective and that the system was operated in a timely and accurate manner. Furthermore it was the fire brigade which put the fires out – the deluge system merely kept the fires small enough to allow effective emergency services intervention.

It seems that if the fire-fighting system had been purely automatic, it would not have been effective, and Dix (2010) goes on to say:

> provision of a fixed fire fighting system does not, of itself, make a tunnel safe.

5.2.4   Example 4: vehicle design
Vehicle design will affect the chance of ignition and/or fire spread in a tunnel; as well as on the open road.

5.2.5   Summary
Three general points are evident from the above discussion.

- Materials and other parts of a system must not be looked upon in isolation.
- The risk implicit within a system depends on how the system is put together and how it operates.
- How a system is put together and operates depends on decision-making.

As a corollary to the above, it is worth noting that fire testing is often carried out in isolation; although testing of assemblies takes place to some extent. While individual properties are important, it is not efficacious to take a purely reductionist approach. A coherent understanding of risk must employ the concept of 'system' in a meaningful way.

5.3.   Incompleteness of assessment: allow for the unanticipated
The complexity and multi-faceted nature of the entire system of which tunnel risk is a product means that an assessment should be expected to be incomplete, especially as analysts are often working under restraints of time and money. It follows from this that it would be prudent to err on the side of caution and not to assume that we can know all possible sequences (Beard, 2002, 2004). That is, it is wise to try to allow for the unanticipated, by building flexibility and redundancy into the system.

5.4. The system changes

It is necessary to realise that, in the real world, the system changes. It is not adequate to carry out an assessment for a current system and then, effectively, assume that the system remains unchanged. For example, in recent decades there has been a considerable increase in the number of HGVs on the roads; that is, traffic patterns and vehicles change over time. In general, it is essential to detect significant changes in the system as soon as possible, and to take account of these. In addition, relying to a large extent on historical statistics (e.g. the average number of tunnel fires or fatalities over two or three decades) may be misleading, as recent changes in the system may not be reflected in these data to any significant degree (see also Chapter 27).

The specific issues of global warming and rising sea levels, and the possible effects on tunnel safety, have been raised by Dix (2008), and this certainly requires further work.

5.4.1 Degradation of the system: increase in 'entropy'

It has been argued that human–technical systems have a general tendency for their 'entropy' (i.e. their 'degree of disorder') to increase over time (Critchley, 1988). Among other things, this links to the crucial importance of maintenance. It also links to the 'boiled frog syndrome', whereby risk has a tendency to creep up on us (Senge, 1993; Beard and Cope, 2008). Seriously attempting to address this issue is very important. The concept of 'degradation' also relates to the concepts of 'safety-critical' element and system (Redmill and Anderson, 1993) and 'vulnerability' of systems (Agarwal *et al.*, 2001).

5.4.2 Information

Information needs to be up to date and communicated to the people who need to know. For example, in 2010 a nuclear submarine ran aground off the Isle of Skye, Scotland. It was later discovered that the charts of the area that were being used were out of date (Paterson, 2010).

5.4.3 The terms 'hazard' and 'risk'

Unfortunately, there is considerable lack of consistency in the use of the terms 'hazard' and 'risk', both in the literature and in society in general. Some people tend to use the word 'hazard' alone, and refrain from using the word 'risk', while for other people the opposite is true. Etymologically, the origin of the word 'hazard' is not known. A common interpretation is that it derives from the Arabic *al zar*, meaning 'the die' (i.e. as used in the game of dice). The word 'risk' seems to come from Italian, but is of uncertain origin. Conceptually, it is convenient to distinguish between hazard and risk and, historically, the essence of the usage of these term seems to be

- the word *hazard* tends to be associated with the *factors* that may contribute to some kind of 'harm' (e.g. loss of life, injury or property damage)
- the word *risk* tends to be associated with *consequences*, or the chance of a particular kind of harm.

It is important to be aware that these words are often used in quite different ways. If either of these words is used, either verbally or in writing, then the question should be asked: 'What is meant by the use of this word here by this writer or speaker?'. With this caveat in mind, the following definitions may be employed.

- *Hazard* pertains to the potential, within a system, for harm. A system may, therefore, contain different kinds of hazard, for example, fire hazard or electrical hazard.

Figure 5.2 A crucial event

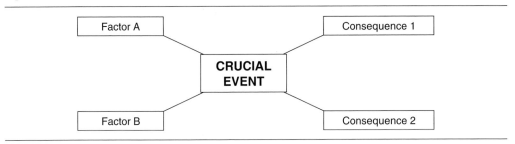

Beyond this, the following concepts may be defined.

■ A *crucial event* is an event that may lead to harm.
■ A *hazardous system* is a system that has the potential for a crucial event.
■ A *hazardous factor* is a factor within a hazardous system.

The term 'risk' may be thought of as relating to the chance or probability of a particular kind of harm occurring (e.g. loss of life) following a crucial event. The word 'crucial' comes from the Latin *crucis*, meaning a 'cross', and the relevance of this can be seen in Figure 5.2 (in this case resembling a Saint Andrew's cross). Factors A and B come together within a system to produce a crucial event, and may lead to consequence 1 or consequence 2. A diagram of this kind can be generalised to include more incoming factors and more outgoing consequences. The word 'crucial' is particularly apt (rather than, for example, a term such as 'initiating event'), because it emphasises the 'crossing' nature of the event, i.e. considering what comes 'before' is just as important as what comes 'after'.

5.5. Prevention and protection as basic concepts

As with the terms 'hazard' and 'risk', there is often lack of clarity in the use of the terms *prevention* and *protection*. A measure that one person refers to as 'prevention' may be referred to by another person as 'protection'. There is no absolute sense in which these terms may be understood. However, these concepts may be seen in a straightforward manner in relation to the concept of a crucial event. Measures relating to prevention may be seen as those that reduce the probability of a crucial event occurring, while measures relating to protection may be seen as those aimed at reducing the consequences after a crucial event. (The word 'protection' here may be regarded as including 'total protection' (i.e. no harm occurs), or 'mitigation' (i.e. 'partial protection'), in which some harm would come about.) This concept is illustrated in Figure 5.3, which is a more generalised form of Figure 5.2.

As a specific example, in relation to 'fire hazard', one might consider the diagram shown in Figure 5.4, where ignition is the crucial event. Bear in mind that ignition might be preceded by other events that may also be regarded as crucial events (e.g. a petrol spillage). In Figure 5.4, the major, medium and minor consequences may be thought of as:

■ *Major consequence*: a consequence involving fatalities or severe injuries, or major property damage, or major disruption of operation.
■ *Medium consequence*: a consequence involving medium level injuries, or a medium level of property damage, or a medium level of disruption of operation.

Figure 5.3 Prevention and protection

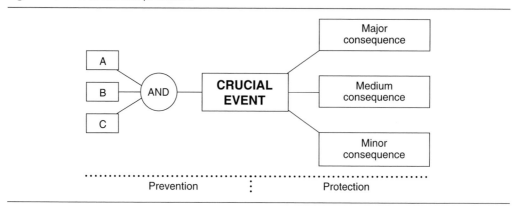

- *Minor consequence*: a consequence involving minor or no injuries, or minor property damage, or minor disruption of operation.

It follows that a preventive measure is preventive in relation to a particular crucial event; in relation to another crucial event it may be regarded as protective. For example, if ignition is regarded as a crucial event, then first-aid fire-fighting may be regarded as a protective measure. However, if major fire is taken as a crucial event, first-aid fire-fighting immediately after ignition may be regarded as preventive.

A more specific case, where fire in a vehicle is the crucial event, is represented in Figure 5.5.

5.6.   Context and causation

Whether or not an event may be regarded as a crucial event depends on context. For example, the event 'hydrocarbon release' could only be regarded as a crucial event with respect to fire hazard in the context of a system in which there is the possibility of ignition. Furthermore, a factor (i.e. condition or event) that plays a part in bringing about a crucial event is only a 'cause' within a given context. That is, a fire is only produced by a 'cause' where that cause is within a particular system. It is more clear to refer to 'causal factors' bringing about a fire; and, even then, those causal factors are within the context of a system.

Figure 5.4 Ignition as a crucial event

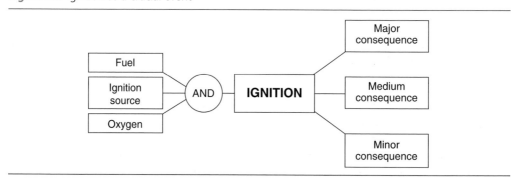

Figure 5.5 Fire in a vehicle as a crucial event

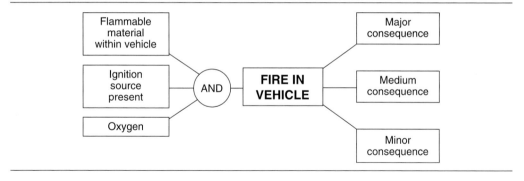

5.7.   Prevention and protection in tunnels

Fires in tunnels are generally different to 'similar' fires in the open (e.g. in a car or HGV). The tunnel environment creates effects such as those associated with confinement of heat and smoke and difficulties of fire suppression. Because the nature of a fire in a tunnel depends considerably on the tunnel system it is in, including the basic characteristics of the tunnel, it is vital that fire safety be considered from the beginning as an intrinsic part of the design process.

Given this, whether or not a measure is to be regarded as part of 'prevention' or 'protection' depends on the particular crucial event that these terms are defined in relation to. If 'fire spreads beyond the first vehicle' is taken as a crucial event, then confining a fire to the first vehicle, by manual or automatic fire suppression, would be part of 'prevention'. However, if 'fire starts in first vehicle' is taken as a crucial event, then stopping the fire spreading beyond the first vehicle is part of 'protection'. In this way, almost every measure could be regarded as prevention, if the defining crucial event were taken as, say, 'extended conflagration in tunnel'. Conversely, almost every measure could be regarded as protection if, for example, the defining crucial event were 'fire starts in vehicle engine 100 m before enters tunnel'. Prevention and protection are not defined in absolute terms, but only in relation to a crucial event and the system that the crucial event pertains to.

If prevention and protection are to be distinguished clearly in a valuable way in relation to tunnel fire safety, then it becomes necessary to define a *fundamental crucial event* (FCE), which would apply to all tunnels. Furthermore, the FCE should allow definitions of prevention and protection that are valuable and seem to contain the essence of what prevention and protection have implied as they have been used in the past. Historically, 'prevention' has implied ideas such as 'locking the stable door *before* the horse has bolted' and, in terms of healthcare, 'prevention is better than cure' (i.e. stopping a disease coming about rather than trying to treat people after they are already ill).

With this in mind, the following crucial event is put forward as a FCE in the context of tunnel fire safety: 'fire exists in tunnel'. It is proposed that this crucial event be used as a defining crucial event for the concepts of prevention and protection in a 'fundamental' sense in relation to tunnel fires. (It should be emphasised that the FCE is intended to include non-flaming combustion as well as flaming combustion.) With this definition, therefore, avoiding a major tunnel fire would involve both preventive and protective measures. This is illustrated in Figure 5.6, where event B consists of

Figure 5.6 A fundamental crucial event (FCE) for tunnel fire prevention and protection

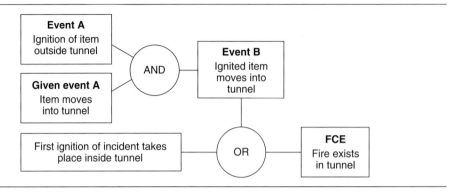

the events that lead to it (i.e. an item is ignited and that item moves into the tunnel). (Note that the two events preceding event B are conditionally related, and are not independent.) Events combined through AND and OR gates in this way constitute a logic tree or 'fault tree'. Fault trees were developed in the nuclear industry, and the faults consisted of events such as the sticking of a switch or failure of a power supply leading to, for example, failure of control rods to drop. The fault-tree concept is valid for events in general, however, not just 'faults'.

Figure 5.6 may be used as a basis for defining 'fire prevention' and, by implication, 'fire protection' in relation to tunnels. The FCE may be prevented by preventing the two crucial events that precede it, i.e. by preventing: 'Ignited item moves into tunnel' and 'First ignition of the incident takes place inside tunnel'. Fire protection consists of trying to reduce the unwanted impact of the consequences of these two events. These events would result from *constitutive crucial events* (CCEs), such as those in Figure 5.7. In this figure, the OR gate would probably be regarded as a mutually exclusive OR gate (indicated as XOR), unless it is assumed that the first ignition of an incident could take place at precisely the same time on two different items (in this case, a HGV and a vehicle other than a HGV). While this may seem unlikely, in principle it could occur in any system as a result of a failure due to a common cause (e.g. related to electricity supply). If the possibility of common-cause failure is assumed to exist, then very great care must be taken when constructing trees in order to avoid false logic, which will lead to erroneous results and conclusions.

CCEs, such as 'First ignition of the incident takes place inside tunnel, in a HGV' may be used as a more specific basis for managing prevention and protection. The possible consequences

Figure 5.7 Constitutive crucial events

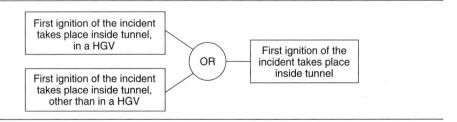

Figure 5.8 An event tree

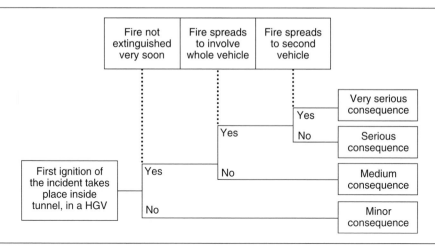

after such a crucial event may be examined more closely using an 'event tree', an example of which is given in Figure 5.8. Such event trees may be used as a basis for examining 'protection'. In Figure 5.8, it has been assumed, for simplicity, that the first burning vehicle needs to become fully involved before the fire spreads to a second vehicle, although in reality this may well not be the case. While the tree in Figure 5.8 is simple, for illustrative purposes, such a tree may be used as a basis for considering 'protection' with respect to that particular crucial event.

Effectively, therefore, there is a 'fault tree' before a crucial event and an 'event tree' after a crucial event. An example of a more specific event tree is given in Appendix 5A. A fundamental task, in a specific case, involves the identification of possible crucial events both before and after the FCE. Crucial events that may come about after the FCE may be termed constitutive, not just those crucial events that occur before the FCE. It is proposed here that the FCE be used as a basis for defining prevention and protection in a 'fundamental' sense. That is, measures aimed at reducing the probability of the FCE occurring are 'fundamentally preventive'. It remains the case that all crucial events, both before and after the FCE, carry implicit definitions of prevention and protection in relation to those crucial events. However, all measures aimed at reducing the probability of the FCE occurring, whether protective in relation to a CCE or not, are defined as 'fundamentally preventive'. Likewise, all measures aimed at reducing the magnitude of the consequences after the FCE has occurred are 'fundamentally protective', whether a measure is preventive in relation to a constitutive crucial event or not. In this sense, a measure may be constitutively protective, but fundamentally preventive (e.g. using an extinguisher to put out a small fire on a HGV before it has entered a tunnel). In a corresponding manner, a measure may be fundamentally protective but constitutively preventive (e.g. stopping a fire on a vehicle in a tunnel from spreading to a second vehicle).

There are, however, problems with constructing fault trees and event trees for fire safety purposes. In particular, it is difficult to represent dynamic change, i.e. to take account of developments over time. This may be overcome through the use of stochastic models, which may be constructed to both lead to a crucial event and away from a crucial event (see Chapter 29).

5.8.   Fire Safety Management

Fire safety management may be seen in terms of the two key concepts of prevention and protection, where protection is assumed to include mitigation. This is a special case of the management of risk in general. In broad terms, measures aimed at reducing the probability of ignition may be regarded as preventive, and measures aimed at eliminating or reducing harm after ignition may be regarded as protective. As part of overall fire safety management, it is essential to have a coherent fire safety management system, and this is considered in more detail in Chapter 23.

Very considerable effort should be put into prevention as part of managing the fire safety for tunnels. Furthermore, it is proposed that prevention be defined in relation to the FCE and the CCEs, leading to the concepts of fundamentally preventive and constitutively preventive. However, while great effort should be made with regard to prevention, there must be adequate protection, should prevention fail, in order to maintain the risk within an acceptable range. As a counterpart to prevention, the concepts of fundamentally protective and constitutively protective emerge as part of the structure created above. This conceptual division is intended to introduce greater clarity and structure to the assessment and management of fire risk in tunnels, and to help to 'focus the mind'.

To be as effective as possible, it is essential that both prevention and protection be considered at the start of the design stage. With ever more complex tunnels, it is no longer acceptable to consider fire prevention and protection as a 'bolt-on' at the end of the design stage.

Categories of measures that fall within the ambit of fire safety management are summarised in Figure 5.9. (Useful general material on fire prevention and protection is contained in DiNenno *et al.* (2008) and NFPA (2008).) Sometimes a measure may be regarded as both preventive and protective. For example, maintaining spacing between vehicles may be regarded as part of prevention, in that it may help to reduce the chance of ignition via a collision, and it may also be regarded as protective, in that it may help to reduce the chance of the spread of fire after ignition.

5.9.   Fire prevention

In general, the essence of fire prevention is to try to ensure that ignition does not take place. In the tunnel context, this has been interpreted in terms of the FCE (i.e. 'Fire exists in tunnel'). Four general methods for achieving this are

- Category 1: eliminate or reduce the number of ignition sources and hot surfaces
- Category 2: have materials or items of 'low ignitability' wherever possible (e.g. fire retarded)
- Category 3: keep ignition sources and potential fuels well separated
- Category 4: eliminate or reduce the chance of spontaneous ignition.

As shown in Figure 5.6, two strategic methods are: to prevent an ignited item moving into a tunnel; and to prevent ignition of an item inside a tunnel. The word 'item' here is used for simplicity; it could just as easily be a fuel, such as a combustible liquid pool.

5.9.1   Prevention: some specific topics

A few specific topics are discussed below as examples of the kinds of practical measures that might be taken to try to reduce the probability of a fire occurring in a tunnel. The examples are meant to be illustrative, and are certainly not exhaustive. It is not intended that topics not mentioned here be regarded as being of lesser importance, and no significance should be attached to the order of the items below. What might be appropriate would depend on the particular tunnel. (For other topics, see Box 22.1 in Chapter 22, and Beard and Cope (2008) and UPTUN (2008a).)

5.9.1.1 Detection of overheated vehicles
An inspection facility at the entrance to a tunnel may be put in place, as at the Mont Blanc Tunnel. If a vehicle (especially a HGV) were found to be overheated, then it would not be allowed to enter the tunnel until it had cooled sufficiently. This may be regarded as being in category 1 above.

5.9.1.2 Use of white surfaces
Increasing the illumination level in a tunnel may help to reduce the chance of an accident. A measure that can contribute to this is to use a white or light-coloured road and/or wall surface. Such an option has been adopted in the Baregg Tunnel, Switzerland (Day, 2007), where a light-coloured tarmac has been used for the road surface. This measure may be regarded as belonging to category 3 above, i.e. keeping potential fuels and ignition sources well separated by reducing the chance of a collision. It may also be regarded as protective to some extent, in that a white or light-coloured road surface may possibly assist escape.

5.9.1.3 Speed control
Controlling the speed of vehicles in tunnels is extremely important. This may be regarded as being in system risk category 3 above, i.e. reducing the chance of a collision.

5.9.1.4 Spacing between vehicles
Maintaining sufficient space between vehicles is also extremely important. This measure may also be regarded as part of fire protection (see Section 5.10). As part of fire prevention, it may be seen as in category 3 above.

5.9.1.5 Shuttle transport and road-ferries
A report on tunnel safety, prepared for the National Assembly in France, was published in 2000 (OPECST, 2000). In that report, it is pointed out that the concept of carrying vehicles on carriers through tunnels (the 'piggyback' method) had been used in Switzerland and Austria. A version of this approach is used in the Channel Tunnel. While the Channel Tunnel is a rail tunnel, the idea may also be applied to road tunnels. The approach has also been proposed for use on congested bridges, where it has been called a 'road-ferry' (Salter, 2004). It has been estimated that the system proposed would considerably increase, not decrease, the carrying capacity of the bridge. In relation to tunnels, it would provide one possible way of controlling the passage of HGVs and other vehicles through a tunnel. Such a measure may be regarded as in being in category 1 above.

5.9.1.6 Fire-resistant materials and fire retardants
Using fire-resistant materials in vehicles, rail rolling stock and other parts of a tunnel system may help to reduce combustibility, and this measure would come under category 2 above. In the sense that fire-resistant materials may not stop ignition but may impede fire growth, they may also be regarded as part of protection.

An example is seen in the case of the train in the Kaprun disaster, where the train has been described as being 'fire-proof'. However, the clothing of the passengers and the goods they were carrying were not necessarily fire resistant, and these seem to have played a major part in the fire (see Chapter 1). While some materials may be regarded as 'naturally' fire resistant, relative to other materials, other materials may have retardants added to them to increase their fire resistance. However, it needs to be borne in mind that, while fire-retarded materials may have a lower ignitability than similar non-fire-retarded materials, once burning they may produce more smoke or toxic gases than similar materials that are not fire-retarded (see e.g.: Cisek and Piechoki, 1985; Blomqvist *et al.*, 2004, 2007; Stec *et al.*, 2009). The degree to which smoke and toxic products will be produced in a fire-retarded material appears to be related to whether or not the material will form a char (Hull, 2010a). Also, a distinction has been made between 'fire retardants' and 'flame retardants' (Hull, 2010b). For a text on fire toxicity, see Stec and Hull (2010). In their review of the hazards associated with fire emissions, Blomqvist *et al.* (2007) argue that 'faithful use of full protective gear' should provide adequate protection for firefighters on active duty (although this review is not related specifically to fires in tunnels).

5.9.1.7 Vehicle design

Changing vehicle design may have a very significant effect on the chance of a fire starting or spreading in a tunnel. Changes may have the effect of increasing or decreasing the chance of harmful effects. For example, increasing the fuel tank size in HGVs is an undesirable change. Amending vehicle design to reduce fire risk has taken place to some degree (e.g. in France). This issue illustrates how tunnel safety results from the action of the broader system, not just the immediate tunnel system itself.

5.10.   Fire protection

As indicated in Figure 5.9, fire protection consists of passive fire protection and active fire protection.

5.10.1   Passive fire protection

Passive measures of fire protection are those related to the features of the tunnel structure itself, its subdivisions and the envelope of the structure. They are properties of the tunnel's construction

Figure 5.9 Categories of fire safety management

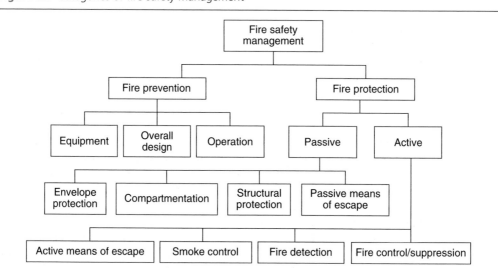

that serve to limit the spread of fire and smoke, should a fire occur. Essentially, they are there for the lifetime of the tunnel.

Such measures may be considered under four headings.

- *Structural protection*, i.e. protection against the effects of heat passed to the structural elements of the building.
- *Compartmentation*, i.e. the division of the tunnel into different spaces and the fire and smoke resistance afforded by that subdivision. While this may not have been carried out in the past, it should be considered as a possibility.
- *Passive means of escape*, i.e. those 'fixed' aspects of the tunnel that are intended to aid escape in the case of fire; the features of intended escape routes. The boundaries of escape routes should have a relatively high degree of fire resistance.
- *Envelope protection*, i.e. the elimination or reduction of the threat posed by a fire to other properties and people outside a tunnel containing a fire. Limitation of the possibility of ignition of another property. For example, flaming particles (brands) may be carried upwards by convection currents and land on other buildings.

5.10.2 Active fire protection

Active measures of fire protection include: measures that operate in the event of a fire; and measures (e.g. operational procedures) that exist perhaps all the time, and are aimed at protection should a fire occur in the tunnel. Measures in the first of these categories require some form of communication or activation to occur, by informing people or equipment of the presence of the fire, and thus enabling action to be taken to contain the spread of the fire or to initiate escape. Several methods exist for fire detection, and it has been recommended that the fire-detection system incorporate redundancy by consisting of several different systems using different technologies (Rogner, 2009). It has become evident that, as a general rule, very rapid action is required to extinguish or substantially control a tunnel fire, especially fires that involve HGVs (Beard, 2009, 2011) (see also Chapter 4). Because of this, very rapid detection is needed, and detection systems are required that can do this reliably in an operational setting. Rapid action by operational staff is also a necessary part of the system.

Many active measures of containment are concerned with the control of smoke spread rather than the control of the spread of the fire itself. Measures of fire or smoke control generally depend on detection of the fire triggering some kind of countermeasure, which may be manual or automatic. Fire brigade action is effectively part of active fire protection. In recent years, much consideration has been given to fixed fire-suppression systems, and the World Road Association (PIARC) now recognises that such systems may be appropriate, even if it does not actually recommend these systems in all new tunnels. The first such system was installed in a Japanese road tunnel in around 1950, and today Japan has about 80 road tunnels fitted with these systems (Brinson, 2010). Worldwide there are more than 100 tunnels with fixed fire-suppression systems (Brinson, 2010). More detail on active fire protection can be found in Chapters 6 and 8. See also Brugger (2003), Collins (2003) and Jacques (2003), on automatic incident detection.

5.10.3 Protection: some specific topics

Some information on a few specific topics is given below. The topics covered should not be regarded as exhaustive, and it should not be assumed that topics that are not mentioned are less important. (For other topics see Beard and Cope (2008) and Box 22.1 in Chapter 22.)

5.10.3.1 SCADA systems

Supervisory, control and data acquisition (SCADA) systems have come into use in tunnels and are intended to provide a comprehensive system that enables an operator to receive information from equipment located in the tunnel (e.g. detectors) and to control equipment (e.g. fans). The data acquisition and control are routed through a computer system. SCADA systems have been used in different types of facilities, not just tunnels. It needs to be borne in mind that any particular SCADA system may be unsuitable for its intended purpose, and may cause serious problems. As an example, in the case of a prominent European tunnel, a report (Prime Time, 2008) found that the specific SCADA system in place 'generates dangerous situations' and was found to be 'unsuitable to ensure the tunnel's safe operation'. It is essential, therefore, to ensure that, if a SCADA system is to be used, it is reliable and adequate for the intended purpose. This would, of course, apply to any system involving electronic or computer-based elements, whether a SCADA system or not. As general related points: it is essential to avoid common-cause failure, and sufficient redundancy must be built into any system; and systems need to be as robust as possible.

5.10.3.2 Fire detection in tunnels: some specific documents

Many documents exist on fire detection in tunnels, and a very few are mentioned here (Directorate General for Public Works & Water Management, 2002; Comeau and Flynn, 2003; Liu *et al.*, 2008; Aralt and Nilsen, 2009; Day, 2009). Some key points taken from these documents are given below.

Road Tunnel Fire Detection Research Project

This project (Liu *et al.*, 2008) was commissioned by the Fire Protection Research Foundation, USA, and carried out by the National Research Council of Canada and Hughes Associates Inc. It included tests conducted in laboratory tunnel facilities, the Carre-Viger Tunnel in Montreal and the Lincoln Tunnel, New York. The detector systems tested were

- linear heat detection systems (two types)
- optical flame detector (one type)
- video-imaging detection (VID) systems (three types)
- spot heat detectors (two types)
- air-sampling smoke-detection system (one type).

The fire scenarios simulated were

- small unobstructed pool fires
- pool fires beneath a simulated vehicle
- pool fires behind a large vehicle
- engine compartment fire
- passenger compartment fire
- moving vehicle fire.

Tests were carried out with minimal air flow and with longitudinal air flow up to 3 m/s. The report should be consulted for full details.

Some key general results taken from the report include the following.

- Longitudinal air flow had a significant effect on fire behaviour and detector performance. Response times to a fire could be increased or shortened with longitudinal ventilation, depending on fuel type, fire size, fire location and growth rate, ventilation velocity and detection type.

■ Small moving vehicle fires (conducted for minimal air flow only) were difficult to detect.
■ Overall: the air-sampling smoke-detection system performed well; the linear heat detectors were able to detect the fires for most scenarios; systems that rely on field of view had problems detecting fires that were concealed by obstructions, but multiple detectors could be used to address this; for VID systems there was a variation in performance depending on the method used to determine the presence of a fire; spot heat detectors were not able to detect fires smaller than 1.5 MW.

It needs to be borne in mind that the results obtained are relevant only to the conditions of the tests (e.g. the direction of the forced ventilation flow and the fuels used).

Automatic fire detection in road traffic tunnels
In this study (Aralt and Nilsen, 2009), fire experiments were carried out in the Runehamar Tunnel in 2007. One of the conclusions of the study is: 'earliest detection of a car fire, fire starts inside, was by smoke detection given fixed limits (3000 µg/m^3)'.

Second Benelux Tunnel tests, 2001
This was a major test series covering many aspects of tunnel fire safety (Directorate General for Public Works & Water Management, 2002). For linear detection systems, it was found that 'for a rapidly developing fire, an alarm will usually be activated within 3 minutes'. However, the important caveat is included that 'conclusions are based on the sensitivities set'.

The Big Dig
It is reported (Comeau and Flynn, 2003) that during a fire which occurred in the Ted Williams Tunnel, Boston, MA, USA, in 2002, 'the first alarm to the operations control centre (OCC) came from the carbon monoxide detectors in the ceiling. The linear heat detectors didn't activate until after the fire department was on the scene'. The point is also made that a fire may well start inside a vehicle, and the amount of heat released into the tunnel may be relatively low.

Linear temperature/heat detectors
It has been reported (Day, 2009) that: 'The most common detection technology is the linear temperature/heat detection system. These systems claim reliable, fast and accurate detection of the fire and its location. Numerous examples have demonstrated that those claims are not always forthcoming in practice, particularly the reliable detection within the required period of time'.

5.10.4 Protection: some specific measures
Given below are a few examples of the kinds of measures that might be taken to reduce the severity of consequences should a fire occur in a tunnel. The list of topics covered is meant to be illustrative, and certainly is not exhaustive. As with the specific topics on prevention above, it should not be assumed that measures not mentioned here are of lesser importance, and no significance should be attached to the order of the items. (Also, see the UPTUN work, especially that of UPTUN, 2008b.)

5.10.4.1 Spacing between vehicles
Maintaining sufficient spacing between moving vehicles and between stationary vehicles is extremely important. Measures aimed at this may be regarded as part of protection, in that it

may help to reduce the chance of fire spread and is part of fire control (see e.g. Perard, 2001). It may also be regarded as part of fire prevention in that it may help to prevent a collision and a fire starting. Preventing a collision not only reduces the chance of ignition, but also reduces the chance of loss of life or injury resulting directly from a crash, whether a fire results or not.

It is now known that fire can spread over very great distances between vehicles. In the Runehamar test series (Lonnermark and Ingason, 2006), it was found that, with a ventilation velocity of about 2.5 m/s, flame length extended approximately 95 m downstream of a simulated HGV trailer fire. It should be assumed, therefore, that ignition of a second vehicle by flame impingement would be possible at about that distance or even greater. Furthermore, spontaneous ignition (via hot smoke) is possible beyond the distance at which flame impingement would be possible, and there is also the chance of ignition via burning brands or transfer of liquid. In the Mont Blanc Tunnel fire of 1999, distances between ignited vehicles were sometimes considerable; a fire engine even caught fire while stationary at 450 m behind the last HGV to have entered at the French portal (Lacroix, 2001).

5.10.4.2 Interspersing more and less flammable loads
While vehicles carrying loads designated as 'dangerous goods' are subject to special regulations, other vehicles also carry loads that are 'dangerous' but which are not designated as such. Goods vehicles, especially HGVs, carry loads of different flammability. For example, one HGV may be carrying plastics and another may be carrying fruit. If vehicles entering a tunnel are controlled such that those with more flammable loads are interspersed amongst those carrying less flammable loads, then the chance of fire spread may be reduced. In the Channel Tunnel fire of 1996, fire spread from one HGV to another, and at least ten wagons were severely damaged. It appears that the fire essentially stopped spreading when it reached two HGVs carrying pineapples, which seem to have effectively acted as a 'fire break'. The HGV beyond the pineapples did not sustain significant damage (Liew and Deaves, 1998).

The interspersing of more flammable loads between less flammable loads could be implemented by design, rather than chance. While the Channel Tunnel operator employs a shuttle system, similar to a road-ferry, and control is thus easier than in other tunnels, the basic principle of interspersing more flammable loads amongst less flammable ones may be implemented in other tunnels, including road tunnels.

5.10.4.3 Preventing vehicles entering a tunnel: use of a water screen
Tunnel operators may find it very difficult to prevent vehicles entering a tunnel when an incident has occurred inside the tunnel. Red lights and similar measures may not be adequate. A novel system for preventing vehicles entering a tunnel has been developed and implemented in the Sydney Harbour Tunnel. This consists of a water screen at the entrance, directly facing drivers, with a very large 'Stop' sign projected onto it (Allen, 2009). Overall, it has been found to be effective, although a few vehicles have actually passed through the screen.

5.10.4.4 Reducing the number of HGVs and dangerous-goods vehicles
Reducing the number of HGVs and dangerous-goods vehicles passing through tunnels would be expected to reduce fire severity. In Europe, the overwhelming majority of goods is transported by road, and it appears that the market share of rail freight (about 8%) is set to fall further (Beard and Cope, 2008). This share is much lower than in, for example, the USA.

Increasing the proportion of goods transported by rail would be expected to have environmental benefits, as well as reducing the risk of major tunnel fires.

5.10.4.5 Audio beacons
An audio beacon may be placed by an exit, sounding a loud auditory message such as 'exit here' (Arch, 2007), in order to help guide people to the exit. Systems of this kind have been considered as part of research on human behaviour and evacuation. Tesson (2010) outlines the main results of a research project on human factors in tunnels, conducted by the Centre d'Etudes des Tunnels (CETU), France, between 2004 and 2008. In that study, it was found that the beacons used remained audible even with a forced ventilation system in operation. Tesson (2010) makes the general, but very important, point that measures need to be evaluated carefully before being deployed in a widespread manner.

5.11. Summary
Fire risk must be seen as a product of the working of a system. That is, it is essential to try to adopt a *systemic* approach to understanding fire risk. Given this, fire prevention and protection for tunnels may be considered in relation to the concept of a *crucial event*. Measures aimed at reducing the probability of a crucial event are preventive, and measures aimed at reducing the magnitude of the consequences after a crucial event are protective. In the tunnel context, this leads to the concepts of fundamental crucial event (FCE) and constitutive crucial events (CCEs), as described earlier. Corollaries of this are the ideas of fundamentally preventive and fundamentally protective, and constitutively preventive and constitutively protective. Given this context, it is desirable for great effort to be placed on bringing about prevention in relation to the FCE. If this fails, however, it is vital that there be sufficient fire protection (including measures for detection, suppression and evacuation) to ensure that the risk implicit in the system as a whole lies within an acceptable range. It is intended that this structure help to provide clarity and focus to assessing and managing fire risk in tunnels.

Appendix 5A: Thoughts on avoiding major tunnel fires
Paul Scott, FERMI Ltd, UK
It is vital that design and planning aimed at avoiding major fires starts early in the concept design stage. Designers have traditionally responded to this challenge by adopting one of two design routes:

- application of a recognised prescriptive code of practice
- the use of 'risk-informed' design, using engineering principles to determine appropriate measures and their specification.

Each design route has its own advantages and proponents, and several codes use both methods. Each method calls on the other to a greater or lesser extent. The practicalities of tunnelling do not allow the strict achievement of prescriptive design measures, and risk-informed designs are always compared with current best practice and code-based design measures, to ensure that designers do not deviate too radically from accepted good practice.

5A.1 Prescriptive codes
It is accepted that prescriptive codes may not always be appropriate, but they can represent a good solution to maintaining risk within an acceptable range.

The danger of prescriptive codes is that they do not, in general, require a detailed investigation of hazards, no requirement to assess the 'risk profile' of the tunnel is laid down, and there is no requirement to justify the operational and other parameters required to achieve an acceptable risk. As tunnel use changes and operational practices change or develop away from that originally understood by the designers, the risk profile of the tunnel may change, and this may go unrecognised. However, the tunnel may well conform to all audit requirements and be considered to represent a well-operated facility.

This effect is noted in most major incidents, as the tunnels in which they have occurred have all been operating legally, within the framework of national safety legislation; yet many have started with minor incidents that were entirely preventable.

5A.2   'Risk-informed' methods

Risk-informed design practices can, at best, provide an excellent solution to tunnel design when the parameters, operating practices and life cycle are well understood.

However, the studies that support safety design proposals may be based on incomplete information, and the requirement to use 'too much' engineering judgement, too many unfounded assumptions, and data that are not applicable to the system under consideration. Often, in the early stages

Figure 5A.1 Event tree for a rail tunnel fire

Frequency of fire in sector which could result in a train stop in tunnel	Does the train proceed out of the tunnel	Is the stop controlled or uncontrolled?	Is the fire extinguished locally?	Does the fire separation maintain its integrity during evacuation	Type of evacuation required	Fault sequence number
Incident/train km	Y/N	C/UC	Y/N	Y/N		

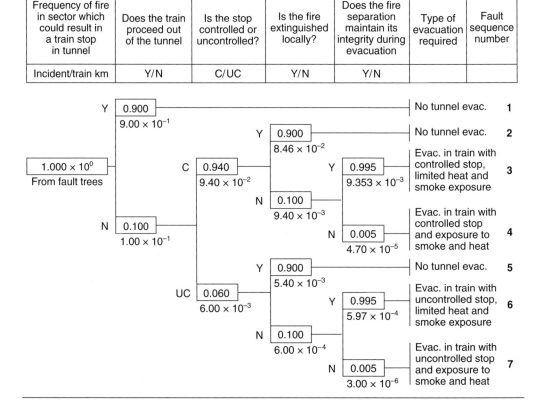

of a design, the final operator has not yet been identified, and operational procedures and safety management systems can only be assumed.

Finally, risk-informed methods may suffer from an over-optimistic interpretation of 'reasonableness'; i.e. what it is reasonable for the designer and operator to mitigate, and what scenarios need not be considered as the basis of design parameters. What is considered reasonable may change, as the public may become sensitised to tunnel fires due to repeated major accidents. Thus a tunnel design may ultimately be considered to be inadequate at a public inquiry.

5A.3   Major fire development

There appear to be no simple design methods that achieve the aim of preventing major tunnel fires. As with all design concepts, there is a requirement for compromise.

Figure 5A.2 A major tunnel fire sequence and prevention/protection

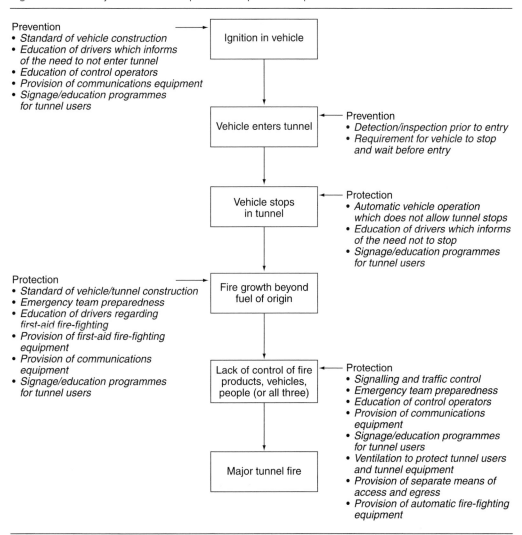

In order to understand the sequence of events that lead to a tunnel fire, we can utilise a traditional risk model, which combines base events that could cause ignition into a fault tree, and the resulting fire can then be analysed using a conditional probability tree to predict event sequences (i.e. an event tree). A similar procedure may be followed for other fires.

Of the seven scenarios in the example given in Figure 5.A1, only two could lead to major tunnel fires. The remainder would not be expected to produce serious consequences. What are the sequences, therefore, that produce major tunnel fires?

A simple sequence that applies to both road and rail tunnels is shown in the flow chart in Figure 5.A2. In this chart the terms 'prevention' and 'protection' are used in relation to the FCE as defined in the main part of this chapter. The safety measures in italics are placed in relation to where they disrupt the sequence of major fire development.

It can be seen that there are recurring themes, and a general progression of safety features, from vehicle construction standards, management actions and education through to technical measures (i.e. mechanical and electrical equipment that mitigates the effects of fire, or tunnel design features that enhance the means of escape when a fire has occurred).

REFERENCES

Agarwal J, Blockley D and Woodman N (2001) Vulnerability of systems. *Civil Engineering & Environmental Systems* **18**: 141–165.

Allen B (2009) Operational experience with water screen technology. Presented at: *2nd International Tunnel Safety Forum for Road and Rail, Lyon, France, 20–22 April 2009*. (NB: not in published proceedings.)

Aralt TT and Nilsen AR (2009) Automatic fire detection in road traffic tunnels. *Tunnelling and Underground Space Technology* **24**: 75–83.

Arch P (2007) Provision of emergency refuge facilities for the Mersey Queensway Tunnel – human factors. *1st International Tunnel Safety Forum for Road and Rail, Nice, France, 23–25 April 2007*.

Beard AN (1999) Some ideas on a systemic approach. *Civil Engineering and Environmental Systems* **16**: 197–209.

Beard AN (2002) We don't know what we don't know. *7th International Symposium on Fire Safety Science, Worcester, MA, USA*, pp. 765–775.

Beard AN (2004) Risk assessment assumptions. *Civil Engineering and Environmental Systems* **21**: 19–31.

Beard AN (2009) HGV fires in tunnels: fire size and spread. *2nd International Tunnel Safety Forum for Road and Rail, Lyon, France, 20–22 April 2009*, pp. 103–111.

Beard AN (2011) HGV fires in tunnels: spread from front unit to load unit and second vehicle. *3rd International Tunnel Safety Forum for Road and Rail, Nice, France, 4–6 April 2011*, pp. 195–204.

Beard AN and Cope D (2008) *Assessment of the Safety of Tunnels*. Report to the European Parliament. Available on the website of the European Parliament under the rubric of 'Science and Technology Options Assessment (STOA)'.

Blomqvist P, Rosell L and Simonson M (2004) Emission from fires. Part I: fire retarded and non-fire retarded TV sets. *Fire Technology* **40**: 39–58.

Blomqvist P, Andersson P and Simonson M (2007) *Review of Fire Emissions from Products with and without BFRs and the Hazard of Exposure for Fire-Fighters and Clean-Up Crews*. Report 2007:74. SP Fire Technology, Boras, Sweden.

Brinson A (2010) Active fire protection in tunnels. *4th International Symposium on Tunnel Safety and Security, Frankfurt, Germany*, 17–19 March 2010, pp. 47–58.

Brugger S (2003) Rapid fire detection concept for road tunnels. In: *Proceedings of the 5th International Conference on Safety in Road and Rail Tunnels, Marseille, France*, 6–10 October 2003, pp. 191–200.

Chang SW, Lin CY, Tsai MJ and Hsieh WD (2006/07) The effects of the type of wood crib arrangement on burning behaviour. *Journal of Applied Fire Science* **16**: 71–82.

Cisek T and Piechoki J (1985) Influence of fire retardants on smoke generation from wood and wood derived materials. *Fire Technology* **21**: 122–133.

Collins S (2003) *Automatic incident detection, linking traffic and tunnel safety – experiences from operational sites. 5th International Conf on Safety in Road and Rail Tunnels*, Marseille, France, 6–10 October 2003, pp. 171–179.

Comeau E and Flynn W (2003) The Big Day: NFPA 502 drives safety of world's biggest tunnel system. *NFPA Journal* May/June.

Critchley OH (1988) Reliability degradation – the problem of an insidious hazard. *International Conference on Radiation Protection in Nuclear Energy, Sydney, Australia*, 18–22 April 1988, Vol. 1. International Atomic Energy Agency.

Day J (2007) Prevention, Detection, Suppression, Protection: Are We Going in the Right Direction? *1st International Tunnel Safety Forum for Road and Rail, Nice, France*, 23–25 April 2007.

Day J (2009) Do smoke exhaust systems really increase safety in road tunnels? *2nd International Tunnel Safety Forum for Road and Rail, Lyon, France*, 20–22 April 2009.

DiNenno PJ *et al.* (eds) (2008) *The SFPE Handbook of Fire Protection Engineering*, 4th edn. Society of Fire Protection Engineers, Boston, MA.

Directorate General for Public Works and Water Management (2002) *Project 'Safety Test': Report on Fire Tests*, Second Benelux Tunnel Test Series, Directorate General for Public Works and Water Management, Utrecht.

Dix A (2008) Referred to in Beard and Cope (2008).

Dix A (2010) Tunnel fire safety in Australasia. *4th International Symposium on Tunnel Safety and Security, Frankfurt am Main, Germany*, 17–19 March 2010.

Hull TR (2010a), Private communication, 2010

Hull TR (2010b) *Fire Retardants – Classification*. Centre for Fire and Hazards Science, School of Forensic and Investigative Sciences, University of Central Lancashire, Preston.

Jacques E (2003) Safer traffic on the Cointe Link with AID. In: *Proceedings of the 5th International Conference on Safety in Road and Rail Tunnels, Marseille, France*, 6–10 October 2003, pp. 181 189.

Lacroix D (2001) The Mont Blanc Tunnel fire: what happened and what has been learned. *4th International Conference on Safety in Road and Rail Tunnels (SIRRT4), Madrid, Spain*, 2–6 April 2001.

Liew SK and Deaves DM (1998) Eurotunnel HGV Fire on 18th November 1996 – Fire Development and Effects. *3rd International Conference on Safety in Road and Rail Tunnels (SIRRT3), Nice, France*, 9–11 March 1998.

Liu ZG, Kashef A, Lougheed G, Crampton G and Gottuck D (2008) *Summary of International Road Tunnel Fire Detection Research Project – Phase II*. National Research Council of Canada and Hughes Associates, Inc., for National Fire Protection Research Foundation, Quincy, MA.

Lonnermark A and Ingason H (2006) Fire spread and flame length in large scale tunnel fires. *Fire Technology* **42**: 283–302.

NFPA (2008) *Fire Protection Handbook*, 20th edn. National Fire Protection Association, Quincy, MA.

OPECST (2000) *Rapport sur Securite des Tunnels Routieres et Ferroviaires Francais* (in French). Office Parliamentaire d'Evaluation des Choix Scientifiques et Technologiques (OPECST), National Assembly, Paris.

Paterson K (2010) That wasn't astute! The charts were out of date. *Metro* (newspaper, UK) 25 October.

Perard M (2001) Spacing and speed of vehicles in road tunnels. *4th International Conference on Safety in Road and Rail Tunnels (SIRRT4), Madrid, Spain*, 2–6 April 2001.

Prime Time (2008) current affairs programme; RTE1 (Irish television channel); transmitted 23 September, 2008. Accessible via www.rte.ie; then search for 'News', 'Prime Time' and 'Archive'.

Redmill F and Anderson T (eds) (1993) Directions in Safety-Critical Systems. *Proceedings of the Safety-Critical Systems Symposium*, 9–11 February, Bristol, UK. Springer Verlag, London. Organised by the Safety-Critical Systems Club, UK. See also later conferences in this series organised by the Safety-Critical Systems Club; www.scsc.org.uk

Rogner A (2009) Fire detection in tunnels: an actual overview on technologies and systems. *2nd International Tunnel Safety Forum for Road and Rail, Lyon, France*, 20–22 April 2009, pp. 273–283.

Salter S (2004) *Road-Ferries for Congested Bridges*. School of Engineering and Electronics, Edinburgh University, UK.

Senge PM (1993) *The Fifth Discipline*. Century Business, London.

Stec AA and Hull TR (2010) *Fire Toxicity*, 1st edn. CRC Press, Cambridge, UK.

Stec AA, Hull TR *et al.* (2009) Effects of fire retardants and nano-fillers on the fire toxicity. In: *Fire and Polymers V – Materials and Concepts for Fire Retardancy*. ACS Symposium Series 1013. Oxford University Press, Oxford, Ch. 21, pp. 342–366.

Tesson M (2010) Recent research results on human factors and organizational aspects for road tunnels. *4th International Symposium on Tunnel Safety and Security, Frankfurt, Germany*, 17–19 March 2010, pp. 191–202.

UPTUN (2008a) *Workpackage 1: Prevention, Detection and Monitoring*. 5th Framework Programme of the European Commission, Contract G1RD-CT-2002–766. See especially Part 12b: Recommendations for prevention solutions. Available at: http://www.uptun.net (accessed 22 June 2011).

UPTUN (2008b) *Workpackage 2: Fire Development and Mitigation Measures*. 5th Framework Programme of the European Commission, Contract G1RD-CT-2002–766. Available at: http://www.uptun.net (accessed 22 June 2011).

Handbook of Tunnel Fire Safety, 2nd edition
ISBN: 978-0-7277-4153-0

ICE Publishing: All rights reserved
doi: 10.1680/htfs.41530.089

publishing

Chapter 6
Fire detection systems

Sandro Maciocia
Formerly of Securiton AG, Switzerland
Arnd Rogner
Metaphysics SA, Switzerland; Provided updated chapter

6.1. Introduction

This chapter describes the problems of detecting fires in tunnels and performance requirements. Different approaches to alerting tunnel users in case of fire are briefly discussed. A review of state-of-the-art fire detection systems currently in use and the different approaches in different countries and cultures leads to the assessment of the technologies in respect of the outlined performance requirements. Finally, this chapter highlights future trends and emerging new technologies that aim to increase further the efficiency of fire detection and alarm activation. In this chapter 'alarm' is taken to mean alerting the tunnel operator via a fire-detection control panel; not information for the public about a fire (see also Chapter 5).

6.2. Problems in detecting fires
6.2.1 Smoke and heat development in a tunnel fire

The development of fires mostly depends on the goods that are burned, although ventilation has a major effect. Common commuter cars (typical heat-release rate (HRR) in tunnels 3–5 MW) are the cause of most tunnel fires (see Chapter 1). However, when heavy goods vehicles (HGVs) are involved the HRR may be considerably greater: 20–30 MW (World Road Association, 1999) or possibly much more; about 200 MW has been found experimentally (see Chapter 14). Tanker fires may be even larger. With such high HRR values, the consequences can be extremely serious. In fact, a very large amount of smoke, often very toxic, is released, and may fill an entire tunnel. Smoke and gases from burnt vehicles in tunnels are very poisonous to tunnel users, and most casualties in recent major tunnel fires have resulted from inhaled smoke.

The first signs of a fire are smoke or heat. Not all car or HGV fires start with a flaming fire (which generally occurs after an accident where liquid fuel spills out). Often, fires start with a smouldering phase, and tunnel users are able to extinguish the smouldering fire without further consequences (Figure 6.1). But analysing some recent tunnel-fire catastrophes leads to the following observations.

■ Mont-Blanc Tunnel fire, 24 March 1999: a HGV travels inside the tunnel emitting smoke (smouldering fire), stops and flames burst into the HGV cabin (Lacroix, 2001).
■ Tauern Tunnel fire, 29 May 1999: a HGV crashes into stopped cars, fuel pours out and a flaming fire starts immediately (Eberl, 2001).
■ St. Gotthard Tunnel fire, 24 October 2001: a HGV crashes into the tunnel wall and another HGV, and a flaming fire starts immediately (see Chapter 3).

Figure 6.1 The usual course of fire development

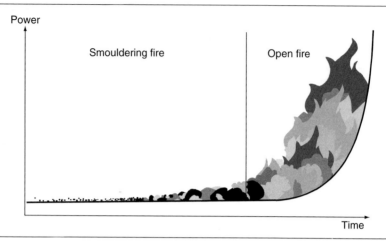

During these incidents, the fire grew very quickly and smoke emission was very strong. The analysis of the data of the St. Gotthard Tunnel fire has shown that the visibility monitors (see below) were the first instruments to detect the fire, but at that time this signal was not used for fire detection.

The smoke and heat spread inside a tunnel can be controlled and limited by maintaining smoke stratification for a limited period of time. By this means the possibility for endangered people to escape and survive are significantly increased. Therefore, the most important 'duty' of the fire-detection system is to activate the smoke-extraction system in the quickest and most reliable way possible (Figure 6.2).

Figure 6.2 The relationship between intervention time and damage

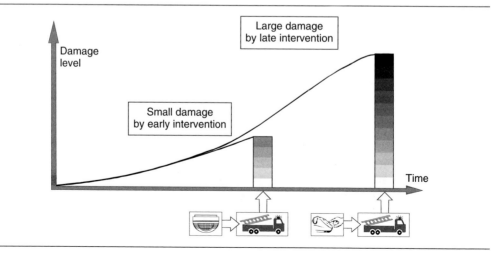

Figure 6.3 Fire-detector types

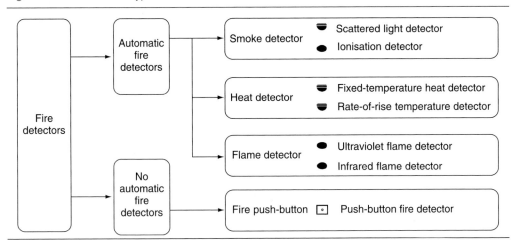

6.2.2   Detection principles

A fire gives rise to material and energy conversions, the end products of which are referred to as 'fire phenomena'. Automatic fire detectors convert fire phenomena into electrical signals. The fire phenomena of smoke, heat and flame radiation are suitable for early detection of fire. The different detector types suited to each specific fire phenomenon under normal ambient conditions are described below (Figure 6.3).

6.2.2.1 Smoke

As a consequence of the disastrous fires mentioned above, in some countries, such as Germany and Switzerland, visibility monitors or special tunnel smoke detectors installed at intervals of 100–300 m are used for early fire detection in road tunnels (FGSV, 2006; Swiss Confederation, 2007). Visibility or smoke-opacity monitors are usually installed in a tunnel to measure the smoke density or air pollution (as the number of particles) for ventilation control. Basically, these systems use light emitters and light receivers to measure scattered light (Figure 6.4) or

Figure 6.4 The principle of operation of smoke detectors

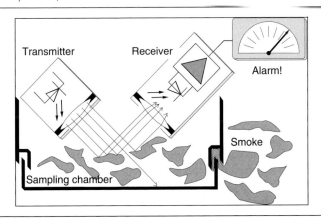

Figure 6.5 The principle of operation of beam detectors

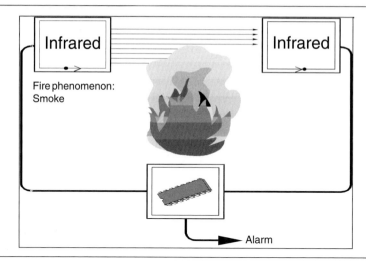

light absorption. The best-known devices are optical smoke detectors, which are based on the reflection principle. When smoke enters the chamber, the light is reflected to the receiving photo cell (see Figure 6.4). When a specific smoke density is reached (in tunnels this is equivalent to an extinction coefficient of, for example, 30×10^{-3} per metre), a fire alarm is activated (see Chapter 13). In recent years, a new generation of tunnel smoke detectors has been developed, which are based on visibility monitors. These smoke detectors are of simplified construction for the specific task of smoke detection in tunnels.

Another well-known system is the light barrier, or beam detector, which is based on the measurement of light absorption. A light emitter sends, over a distance of several metres, a light beam, which is received by a light-sensitive device. Depending on the loss of light intensity due to absorption by the smoke within the measurement length, the system activates an alarm (Figure 6.5). The reason why this technology cannot be used in tunnels is discussed later.

Smoke can also be detected using video-image-processing software. These systems generally detect smoke and fire with a high reliability, but they still show too much cross-sensitivity with other phenomena, leading to false alarms. This problem is discussed further in Section 6.6.

6.2.2.2 Flame
A flame is a radiation-emitting fire phenomenon. Different wavelengths of electromagnetic radiation, ranging from infrared to ultraviolet, can be detected by flame detectors. Flame detectors are light-sensitive sensors (calibrated for a certain wavelength) that are activated when the specific radiation is received (Figure 6.6). This technology is used mainly in some Asian countries, such as Japan or China. The flame detectors are installed at intervals of approximately 25 m. However, the rapidly developing smoke will impede the sensors' 'view' of the flame.

In addition, flames can be detected by video-image-processing systems; but again these systems suffer from a certain false alarm rate due to interference with other phenomena.

Figure 6.6 The principle of operation of flame detectors

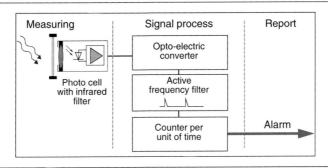

6.2.2.3 Heat

Reliable fire detection can also be achieved by measuring the ambient temperature. When a pre-set maximum temperature is detected, the system activates an alarm. In addition to this, European Standards (EN 54-5) require evaluation of the rate of temperature rise in order to activate an alarm, even before the maximum temperature is reached (Figure 6.7).

There are two types of heat-detection systems: point detectors and line-type detectors. Point heat detectors incorporate a heat sensor (heat-sensitive resistance) connected to an evaluation unit. Line-type heat detectors have no longitudinal interruption between sensors, and measure, and in some cases even mathematically integrate, the temperature increase. They thus provide higher accuracy of detection. Further details of line-type heat detection are given in Section 6.5.

6.2.3 Fire-detection principles appropriate for use in tunnels

Tunnel ventilation keeps down carbon monoxide levels in the air and is intended to ensure good visibility. Under normal ventilation conditions (0–3 m/s), smoke and heat can rise to the tunnel

Figure 6.7 The response behaviour of a heat detector

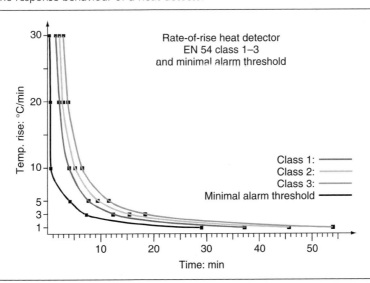

ceiling within a short time, and can be detected very quickly and reliably. Fire tests show an increasing difficulty in detecting the rise of heat when the speed of longitudinal ventilation is in the range 6–8 m/s, due to heavy air pollution. Even with a high air speed and a small increase in temperature of 2–3°C/min over a 50 m section of tunnel ceiling (typically 5 m high), a fire-detection system should be able to activate a fire alarm.

Because of the critical ambient conditions in tunnels, fire detection is very demanding. Exhaust fumes from fuel-driven engines are dirty and partly corrosive, and elevated temperature and humid conditions are present because of the special geological situations of tunnels.

A fail-safe and false-alarm-safe fire-detection system must take into account disturbing factors such as

- strong air fluctuations (ventilation)
- corrosive air, heat, dirt and dust
- electrical interference from electronic devices, cables and boards
- temperature changes due to external conditions affecting the portal area
- resistance to tunnel-washing machines
- hot exhaust fumes from a truck with a high exhaust pipe held up in a traffic jam
- mechanical forces from goods shed from a HGV, or a sailing boat mast touching the tunnel ceiling.

For fire detection in tunnels, conventional point-type fire detectors are not adequate. The ambient conditions lead these devices to become dirty very quickly, and the corresponding electronic circuits will be rapidly destroyed by corrosion. The consequence of this would be high rates of false alarms and faults in the detection system. Therefore, fire detectors used in tunnels must be adapted for this specific environment.

Today, line-type heat detectors for temperature-based fire detection, and specially constructed smoke, flame and beam detectors have been developed to be used in the aggressive tunnel environment.

6.3. Performance requirements for fire detection systems

The different principles on which detection systems are based, and the requirement for early warning systems for tunnel fires, give rise to the questions of what type of system engineering is appropriate and what operational performance is required. There are, in fact, no worldwide standards for fire-detection systems in tunnels. Many countries do equip new tunnels with fire detection, but no further technical requirement is expressed. Quite often, only a detection sensor is specified, which does not have to meet any standards. Usually, very little is specified with regard to the system architecture.

6.3.1 Fire-alarm system engineering

The design and installation of fire-detection systems in tunnels requires a high level of attention and technical competence. As the effectiveness of a fire alarm system is responsible for the very fast alarm activation required, no element in the chain

can be neglected in the system design, installation and certified performance. An efficient fire-alarm system should operate over several years, and be both fail-safe and false-alarm-safe. State-of-the-art fire alarm systems are, therefore, designed according to various standards; including EN 54 and NFPA 502. These standards require

■ overall autonomy of the fire-alarm system
■ self-monitoring of all devices (detectors, alarm activation, alarm transmission)
■ battery back-up for operation in the case of a power supply failure
■ a fault-tolerant system architecture for control, remote and repeater panels
■ fault-tolerant installation of the cables that collect the field data
■ a defined response behaviour for all detection devices
■ a defined automatic alarm activation and alarm transmission
■ product and system certification.

With the new means of detection now available, such as smoke detectors and video detection, the situation becomes more complicated. On the one hand, detection should be as rapid as possible, and this was the reason for integrating smoke detectors into fire-alarm systems. However, smoke detectors, and certainly video-based systems, suffer from a certain false-alarm rate, and thus the alarm generated by such a system should generally be given only in the control room. On the other hand, the final alarm initiated by a rise in temperature (which is in fact the only clear indication of a fire) should be generated automatically, and not depend on human reliability of interpreting data. Figure 6.8 shows an example of the architecture of such an optimised reaction system.

6.3.2 Smoke and heat control to reach full operation within 3 minutes

State-of-the-art detection systems for tunnels activate an alarm in an emergency situation (flaming fire) and limit false-alarms caused by the difficult ambient conditions, such as humidity, dust or corrosion, and thus guarantee fail-safe operation. Present-day fire-detection systems are able to detect rapid increases in temperature resulting from open fires of 1 MW ($1 m^2$ open fire, ethanol) and activate an alarm within 30–60 s under different ventilation conditions (0–8 m/s wind speed).

Figure 6.8 System architecture of a state of-the-art fire-alarm system (Copyright Securiton)

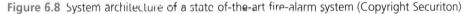

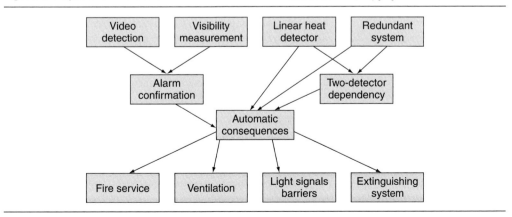

In the event of a fire, one of the main tasks of a fire alarm system is to enable activation (perhaps automatically) of the ventilation system (or, perhaps, extinguishing systems) in order to control the smoke layer in the upper part of the affected tunnel segment. As a general rule, the full operation of smoke control (open smoke extraction, and power up of the ventilation system to maximum speed) should be achieved within 3 minutes of the onset of flaming combustion (Swiss Confederation, 2008). Most current directives for tunnel-fire detection require the detection of a 5 MW fire within 60 s (FGSV, 2006; Swiss Confederation, 2007). The same fire alarm can be used for other automatic actions, such as switching on the emergency lights, turning traffic-control signals to 'stop' (red), activating CCTV cameras for viewing on a monitor, triggering a deluge system, etc.

6.3.3   Other operational performance requirements

In addition to the basic system requirements, there are requirements for the operational performance of fire-alarm systems (Table 6.1). By meeting these requirements, operators limit the operational costs and operational problems of the system. In other words, operators should receive a guarantee that the fire-alarm system will detect a fire within the required time and activate an alarm, as well as eliminate false alarms and faults.

Table 6.1 Operational performance requirements

Item	Desirable performance requirement
Detection principle	Line-type heat-detection systems, including rate-of-rise detection combined with smoke or video detection for early fire warning
Detection accuracy (line-type heat-detection system)	1–2°C for absolute temperature measurement 0.5°C for rate-of-rise temperature measurement ±10 m of a 50 m tunnel segment
Alarm activation/detection time	Within 60 s from the start of an open fire
System approval	EN 54-5 Class A1 (VdS, AFNOR, LPCB[a])
System interface to other systems	Monitored hardware contacts (relay) to different other safety relevant systems Serial interface for temperature and data transmission (information level)
Fail-safe and false-alarm-safe operation	Defined false alarm rate, for example, one false alarm per year and 2 km (Swiss Confederation, 2008), automatic system self-testing
Tunnel-washing machine	Immunity to mechanical forces and water
System life time	>15 years, in order to protect the operator's investment
Operation costs	Zero operation costs for the fire-detection system
Maintenance costs	Maintenance-free operation through system self-test procedures and long operation life of materials Nevertheless, European standards do require a system test once a year

[a] Certification bodies for fire-alarm systems: AFNOR, Association Française de Normalisation (France); LPCB, Loss Prevention Certification Board (UK); VdS, Verein deutscher Sachversicherer (Germany).

6.4.   Different approaches to alerting tunnel users

In the case of a major fire breaking out, the external emergency services will likely arrive too late to extinguish the fire immediately. Considering the worst-case scenario of a HGV in full open fire, smoke control should be effective after a maximum of 120 seconds. Therefore, authorities often choose the conceptual basis of 'enable the people to help themselves'. Since the tunnel disasters in 1999 (e.g. in the Grand St. Bernhard Tunnel between Switzerland and Italy), as a preventive measure some tunnel operators in central Europe have started to distribute information leaflets to users entering a tunnel. In other cases, the signage indicating the route to emergency niches has been improved, so that even under the most difficult conditions (intensive smoke) tunnel users are enabled, in principle, to find their way and to rescue themselves. All these measures are described in the European Directive 2004/54/EC (2004) and the corresponding national guidelines.

However, all these initiatives do not change the fact that tunnel users are usually informed about an incident only when they find themselves within the incident area. Often, tunnel users are not informed at all about an incident of any kind until the emergency services arrive. All in all, there is no common way between different countries and cultures of alerting tunnel users to a fire within a tunnel.

Looking at an emergency scenario from a systemic point of view, tunnel users may find themselves inside a tunnel at a certain moment during an incident. Considering this scenario and the unpredictable behaviour of humans in such a situation, users should be informed about the danger facing them and how to reach a safe place (which is preferably outside the tunnel). Prior to receiving any emergency information, users must know what possible information may be given and what it means. In the case of a tunnel incident, users should be informed both visually and audibly about the type of danger and its location, escape routes and what they are supposed to do. There is a wide variety of possible means of doing this, such as: horns, sirens, bells, loudspeakers, paging systems, mobile phones, FM radio broadcast, all-round signal lamps, flashing lamps, indicator panels and layout-plan panels. The activation of the means of providing information should be triggered automatically by the fire-alarm system. Manual activation is not suitable, as such events happen rarely. The risk of an operator acting in an unsuitable way due to a lack of experience or instruction could result in severe negative consequences. At the very least, visual and audible fire-alarm indication devices should be installed inside the tunnel.

6.5.   Currently available tunnel fire detectors
6.5.1   Line-type, digital heat-sensing cables

A line-type, digital heat-sensing cable consists of a twisted pair of electrical conductors with temperature-sensitive insulation. When the ambient temperature reaches the temperature rating of the cable, the thermoplastic insulation melts and short circuits the conductors (Figure 6.9). The short circuit is electrically processed and an alarm is activated. Depending on the application, different melting temperatures can be chosen. Temperature-sensing cables were the first linear fire-detection systems, and they have been used for years, initially in the UK and the USA. They were generally used for tunnel-fire detection until the late 1980s. However, as the detection speed using these cables is rather slow, and the exact location of the fire in the tunnel cannot be pinpointed, such systems have been superseded by more advanced technologies. In general, systems using line-type, digital heat-sensing cables cannot detect small car fires (or do so only at a very late stage), and they therefore do not meet current requirements.

Figure 6.9 A heat-rated cable

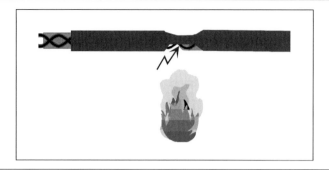

6.5.2   Line-type heat-detection cable with semiconductor temperature sensors (multipoint system)

Semiconductor temperature sensors, or temperature-sensitive resistors, are connected to a bus for data transmission. The bus consists of several conductors and uses special transmission technologies (Figure 6.10). The sensors and transmission conductors are encased in a plastic coating, protecting them from aggressive ambient and mechanical influence. Data are processed in the central evaluation unit at both ends of the detection cable. The connection of the second cable end to a second evaluation unit results in a high system-fault tolerance. The distance between the sensors can vary from 1 m to 20 m (usually 7–10 m in tunnel applications) according to requirements, and they can have an accuracy of up to 0.5°C. The maximum length of the cable is usually 2 km, and the sensitivity (detection speed) is independent of cable length. These systems activate an alarm when a pre-programmed rate-of-rise in temperature and maximum temperature are reached. Determining the exact location of the fire within the tunnel is theoretically possible (depending on ventilation speed). These systems have been in use since the early 1990s.

6.5.3   Line-type heat detector with fibre-optic cable

A line-type heat detector with fibre-optic cable consists of an optical fibre encased within a metal tube (Figure 6.11). Connected to one end of the optical fibre is the evaluation unit, which continuously sends laser light pulse signals along the fibre and receives a specific response signal resulting from reflection inside the fibre (using the principles of time domain

Figure 6.10 Semiconductor heat sensors

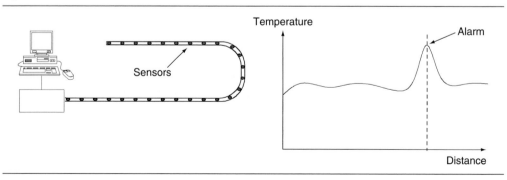

Figure 6.11 A fibre-optic cable

Fibreglass   Metal   Plastic

reflectometry and frequency domain reflectometry). A change in temperature changes the reflection and scattering characteristics of the fibre, and thus reflected laser light detected is changed. The maximum length of the cable is 8 km, and the speed of detection is strongly dependent on cable length (the longer the cable the slower the detection speed). Determining the exact location of the fire within the tunnel is possible, depending on the ventilation air speed and the measuring point distances. This detection principle was first presented in the early 1990s.

6.5.4   Line-type heat detector with hollow copper-tube sensor

The copper-tube sensor system was used widely in European tunnels in the 1980s and 1990s. This fire detection system is a pneumatic system (Figure 6.12). A hollow copper tube (1), with a typical length of 100 m, contains an hermetically enclosed atmosphere (normal air). One side of the tube is connected to a pressure sensor (3) in the evaluation unit (2). As the tube becomes warmer, the pressure inside the tube rises, and this can be detected by electronic means (4). When a pre-programmed rate-of-rise of temperature or maximum temperature is reached, the system activates an alarm. Determining the exact location of the fire in the tunnel is theoretically possible, but is limited to the length of one stretch of tube (typically 100 m). The first pneumatic systems were installed in Swiss tunnels in the early 1970s.

6.5.5   Visibility monitors and smoke detectors

Visibility monitors measure the smoke (particle) concentration by measurement of scattered light or light absorption (see also Section 6.2.2). Recently, a new generation of smoke detectors, based

Figure 6.12 A copper-tube sensor and evaluation unit

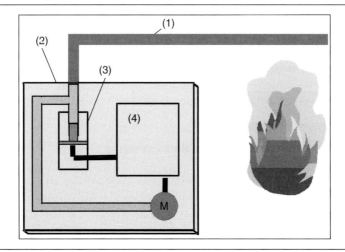

on visibility monitors, for use in tunnels has been developed. These smoke detectors use the principle of scattered light, and have a simplified construction for the specific task of smoke detection in a tunnel.

As can be seen in Figure 6.1, smoke detection can help to detect a fire at a very early stage, and the reaction speed is high. However there is a risk of interference from the high air pollution from vehicles or from fog. While the latter can be suppressed using detectors that have air-heating systems, a false-alarm triggered by high air pollution (e.g. in the case of a traffic jam where a truck is halted at the position of the detector) cannot be excluded. Most currently used systems give a pre-alarm to the control room to initiate further investigation and activities. Experiments done in the Gotthard road tunnel indicate that a smoke-detector system can be optimised by recording and investigating the behaviour of the system under different conditions, including fire (Grässlin, 2007). Thereby an optimised set of parameters can be established so that certain actions trigger the system with sufficient security.

6.5.6 Video-based systems

During the last few years, analysis of video images as a means of fire detection has been widely discussed, and several public studies have been performed (e.g. Haack *et al.*, 2005). The idea is to use the CCTV cameras that are already installed in the tunnel for monitoring purposes and for detecting events such as 'wrong way' vehicle movement, stopped vehicles, people in the tunnel, etc., and for fire detection. Special algorithms have been developed to detect smoke or flames. These algorithms use, for example, contrast, variation of background, and growth of objects for smoke detection, and brightness, colour, oscillations, or shape variations for flame detection. The detection of such events usually works well, and the reaction speed is high. The problem is the false alarm rate, due to interference from lights, light reflection in wet conditions, fog, etc. Even if the rate of false alarms is reduced significantly by optimising the algorithm (e.g. one false alarm per camera per month), this can still give several alarms a day if there are 100 cameras.

Because of this, a fire alarm raised on the basis of CCTV-image analysis needs to be confirmed by the control room personnel, who need to look at the images from the relevant camera to make a decision on the situation. CCTV-image analysis can certainly help to provide very early detection, and also gives a direct view of the event, and so it may make sense as an additional tool for early detection in, for example, tunnels with a high risk potential.

6.5.7 Flame detectors

Flame detectors measure infrared and ultraviolet radiation, and can detect a flame by identifying specific wavelengths and intensity oscillations (see Section 6.2.2). To achieve sufficient sensitivity and local resolution, the detectors must be installed every 20–25 m in the tunnel. The reaction speed is high. At present, this type of detector is used only in some Asian countries, such as China or Japan.

6.5.8 Assessing state-of-the-art fire alarm systems

6.5.8.1 Definition of 'state of the art'

In several countries, such as Germany and Switzerland, there are already very detailed specifications on how fire detection in road tunnels should be realised (FGSV, 2006; Swiss Confederation, 2007). However, in many countries there may be only a general indication that fire detection systems should be employed (2004/54/ EC, 2004), and it is left to the owner of the tunnel to decide on the details of a state-of-the-art system installed. The definition of 'state of the art' is very often the subject of individual interpretation by tunnel designers, although from a legal point of view the meaning of a 'state-of-the-art system' is based on all relevant publications and experience that are publicly known and accessible (Dix, 1999).

For example, consider the case where a failure of a fire detection and alarm system is evident and has led to damage or casualties during a fire incident in a tunnel. In this case the appropriate legal body will undertake an investigation to determine whether all known technical measures were taken to ensure system performance at the time when the system design was approved. In doing so it will take into account the available expertise and knowledge, and identify the publicly accessible knowledge on system-performance requirements available at the time when the system was installed. If it is found that technical measures that would have ensured system performance had been neglected, the responsible system design engineer might be considered liable.

6.5.8.2 Assessment criteria

Bearing this in mind, the assessment criteria for state-of-the-art performance can be summarised as follows.

- Detection speed
  - 30–60 s detection time for a 1 MW, 1 m^2 ethanol fire with an air speed up to 3 m/s and a tunnel height of 5 m
  - 30–60 s detection time for a 5 MW, 2 m^2 ethanol fire with an air speed up to 5 m/s and a tunnel height of 5 m.
- Fire-alarm system design
  - 100% system autonomy (detection, signal processing, alarm activation, direct automatic activation)
  - redundant system architecture for control, remote and repeater panels
  - self-monitoring of all connected devices.
- Operation
  - product and system certification
  - recognises and excludes disturbing signals (no false alarms)
  - 100% system availability for fail-safe operation (one technical fault will not affect system operation)
  - a detector resistant to a corrosive, dirty and wet environment.

6.5.8.3 Assessing line-type heat detectors

This assessment is limited to line-type heat detectors, as they are the only ones used for automatic fire detection.

Three detection technologies have been proven in the field and will be considered further here

- line-type heat detector cable with temperature sensors (multi-point)
- line-type heat detector cable with optical fibre
- line-type heat detector with pneumatic evaluation.

All these detectors are theoretically able to fulfil the required detection speed. However, depending on the design of the installation, the detection characteristics will vary in such an important way that detection sensitivity may be lost (detection times up to 180 s).

The latest generation line-type heat detector cable with temperature sensors based on semi-conductor technology (multipoint measurement) is able to fulfil the following requirements for an installation with increased availability

- detection cable length up to 2000 m
- 30–60 s alarm activation time
- guaranteed operation in the event of cable breakage or short circuit, by means of fail-safe functions (loop installation, separation of defect segments)
- possible monitored integration into a fire alarm system (50 m segments)
- serial interface for temperature and data interchange to supervisory control and data acquisition (SCADA) systems
- needs two independent evaluation units at each end of the cable for coverage of 2000 m tunnel.

The latest generation line-type heat detector cable with a fibreglass conductor fulfils the following requirements for an installation with increased availability

- detection cable length up to 8000 m
- 30–180 s alarm activation time (the longer the detection cable, the longer the alarm activation time)
- guaranteed operation in the event of cable breakage, by means of redundancy functions
- limited possibility of monitored integration into a fire-alarm system (50 m segments)
- serial interface for temperature and data interchange to SCADA systems
- needs two independent evaluation units at each end of the cable for coverage of 4000 m tunnel.

A line-type heat detector with a copper tube and pneumatic evaluation fulfils the following requirements

- detection tube length typically 100 m, maximum 130 m
- 30–40 s alarm activation time
- in the event of tube breakage, no detection within that 100 m tunnel segment
- possible monitored integration into a fire alarm system (50 m segments)
- serial interface for temperature and data interchange to SCADA systems
- needs 20 evaluation units for coverage of 2000 m tunnel.

A detailed comparison of these systems is given in Haack *et al.* (2005).

6.6.   Future trends and emerging new technologies

The latest developments in fire detection for tunnel safety feature CCTV-image-processing systems as the most important technical innovation. As part of a professional fire-alarm system, image-processing systems enable early detection of smoke and fire. Whereas monitoring the rate of rise of heat allows for detection of a fast flaming fire (which corresponds to detection of emergency situations), the new fire-detection technologies including image processing allow fast detection of dangerous and emergency situations. Integrating all relevant fire-detection criteria, such as smoke, flame and heat, under the control of a state-of-the-art fire-alarm system, new technologies will effectively be able to provide greater safety in tunnels.

Different manufacturers are currently undertaking intensive research to develop the right algorithms for good image-processing software. Meanwhile, several test installations have been studied (Haack *et al.*, 2005). In addition, several installations of video-based fire detection under commercial conditions are now in place.

6.6.1 Smoke and flame detection with video-image processing

Equipped with the latest image-processing technologies incorporated in state-of-the art fire detection systems, tunnel operators are, in principle, able to detect dangerous situations very early. Smoke-emitting smouldering fires can be detected automatically by the fire-detection system, long before an open fire breaks out. Image-processing technology for smoke and fire detection was first used in 1994 by British companies. Other systems, in Germany, use infrared cameras equipped with special filters. These systems have proven efficiency in monitoring outdoor open areas, but are not suitable for use in tunnels due to the presence of disturbance factors such as moving vehicles and people, changing lights and reflexes, etc.

Video-image processing is at the core of the technology for the detection of smoke and flame in tunnel applications. The technology is based on the existing CCTV system within a tunnel, installed to monitor traffic flow. The video signal is captured and digitised by a grabber, and then evaluated via algorithms to identify flame and smoke (Figures 6.13 to 6.15). The processed image results are made available as a pre-alarm or alarm to a fire alarm system, and control signals are sent to the CCTV system for automatic activation of the specific cameras and monitors. A PC-based man–machine interface is responsible for parameter configuration, alarm picture recording and history functions.

The false-alarm and fail-safe operation of an image-processing system are of crucial importance for overall system efficiency, and are as important as the system's reliable detection of smoke

Figure 6.13 Digital image-processing algorithms

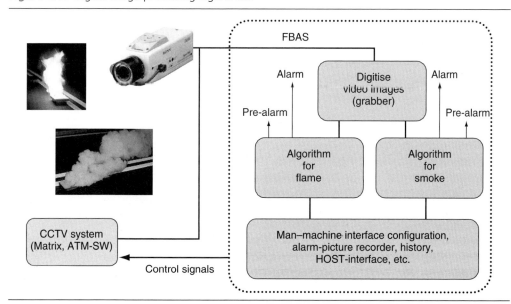

Figure 6.14 Smoke test and detection in a tunnel

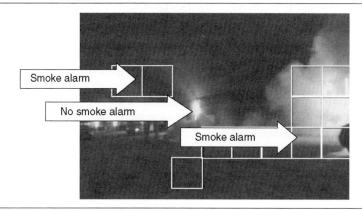

and flames. The following disturbing parameters, among others, should be excluded by the system

- car lights, constant or intermittent
- heavy traffic (movement)
- people (movement)
- day, night, variation of illumination
- fog, rain, snow (portal area)
- light reflexes, snow
- large white or grey areas, as produced by large trucks
- high levels of pollution of tunnel air.

By monitoring image segments with image-processing algorithms, an alarm is activated only when predefined threshold levels are reached. Both flame and smoke can activate an alarm, and the system distinguishes between pre-alarm and alarm levels of both phenomena, and the alarm thresholds are adjustable.

As yet there are no international standards for fire tests or system approval for early fire detection. It is not clear what physical values should be applied for testing and certification: it might be

Figure 6.15 A flaming fire test within a tunnel, and fire detection

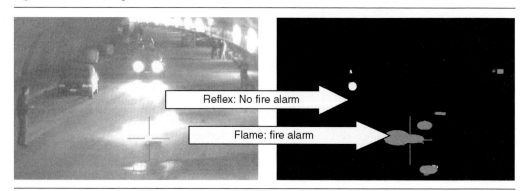

defined as a number of cubic metres of smoke, or as a defined opacity per metre. Therefore, CCTV-image processing can, at present, only be regarded as a complementary detection mode to a state-of-the-art fire-alarm system that complies with international standards. Recent published drafts of the American standard NFPA 502 mention image-processing systems as an addition to heat-type fire detection (e.g. for use as early fire detection). Meanwhile, in Europe similar discussions are currently at only a very early stage.

6.6.2 System architecture of fire-alarm systems with video-image processing

Image-processing systems will never be a substitute for fire-alarm systems as a whole because of the important limitations imposed by poor visibility. Fires may be hidden and not seen by CCTV cameras, or the cameras may have no clear view because of poor illumination, dirt, high opacity air and fog. Therefore, events can only be detected if clearly visible. Because of the lack of international standards at present, it is impossible to certify or achieve approval for fire-alarm systems that are based on image processing only.

It is suggested that the fire-alarm system installed in a tunnel should consist of a line-type heat-detection cable connected to a fire-alarm control panel in the technical operation building. The pre-alarm, alarm and fault outputs generated by the video-image-processing control unit are connected to the same fire-alarm control panel. Furthermore, the following inputs and outputs should be monitored via dedicated and redundant data lines: manual call points, alarm inputs from fire extinguishers/doors/etc., point-to-point detectors and alarm indication devices. Again all these devices should be connected to the fire-alarm control panel. Long tunnels require several control panels connected via a dedicated control-panel network (Figure 6.16).

Figure 6.16 Fire-alarm system architecture, with early fire detection (image processing technology)

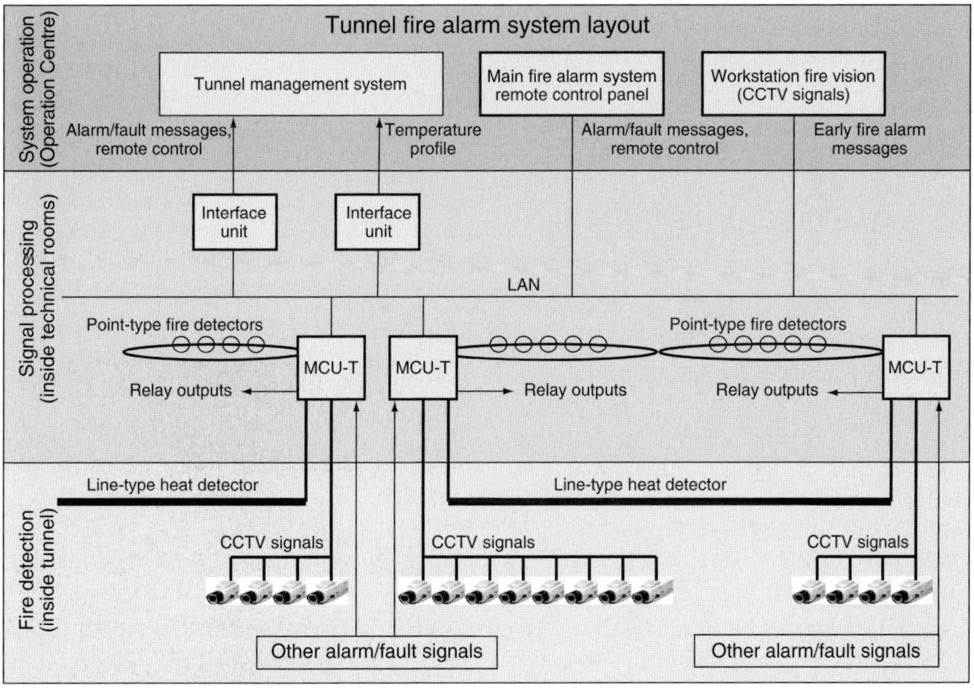

The CCTV system forms an autonomous system that makes duplicate video signals available to the video-image-processing control unit. This image-processing control unit is responsible for the early detection of fire and smoke, while the linear heat detector is responsible for monitoring the rate of rise of heat.

6.7.   Conclusions

Line-type heat detection has become established as the most reliable detection technology for fast fire detection, and hundreds of kilometres of tunnels worldwide are equipped with this technology. However, systems that can react on the basis of a change in the temperature gradient must be used, as the maximum temperature of a car fire might not reach more than 50°C. Systems of this type fulfil all the necessary requirements, including a detection speed of 60 s for a 5 MW fire even at a high air speed of 10 m/s. Modern systems provide full redundancy of operation, even in the event of a short circuit, and can be combined into networks for use in large or complex tunnel systems.

In recent years, smoke detection for early fire warning by visibility monitors or video-image processing has been introduced. It is recommended that these facilities be used in addition to heat detection. Alarm signals produced by these detectors have to be confirmed by control room personnel, although in some installations they automatically switch on the fire ventilation. In the future, with more experience in their application and the definition of possible thresholds and interference, such detectors might be used for automatic triggering of the complete fire-alarm scenario.

Up to now, the system engineering of fire-alarm systems has not been based on proven design criteria (e.g. system design and product standards for fire-alarm systems in buildings), and therefore a wide variety of system architecture is to be expected. EN 54 and NFPA 502 are proven bases for the design of professional fire-alarm systems. Even though the ambient conditions in tunnels are very different from those in buildings, the specific operational performance requirements for a tunnel fire-alarm system are the same as those for building fire alarms.

Video-image processing for fire-alarm applications is a major technical innovation, and it is suitable for use in tunnels. However, the software currently available is still under development (see Chapter 5 for more on detection).

REFERENCES

2004/54/EC (2004) *Directive on Minimum Safety Requirements for Tunnels in the Trans-European Road Network*. European Union, Brussels.

Association Française de Normalisation (2011) Available at: http://www.afnor.fr (accessed 22 June 2011).

Dix A (1999) Beyond deals – professional liability in a world of blame. *Proceedings 1st International Conference on Long Road and Rail Tunnels, Basel, Switzerland*, December 1999. Independent Technical Conferences Ltd, pp. 113–124.

Eberl G (2001) The Tauern Tunnel incident – What happened and what has to be learned. *Proceedings 4th International Conference on Safety in Road and Rail Tunnels, Madrid, Spain*, April 2001, pp. 17–28.

European Standard EN 54 (2011) *Fire Detection and Fire Alarm Systems*. Including: EN 54-1:1996, *Part 1: Introduction*; EN 54-2:1997, *Part 2: Control and Indicating Equipment*; EN 54-3:2001, *Part 3: Fire Alarm Devices – Sounders*; EN 54-4:1997, *Part 4: Power Supply*

Equipment; EN 54-5:2000, *Part 5: Heat Detectors – Point Detectors*; EN 54-7:2000, *Part 7: Smoke Detectors – Point Detectors using Scattered Light, Transmitted Light or Ionization*; and other parts. European Committee for Standardization. Available at: http://www.cenorm.be (accessed 22 June 2011).

FGSV (2006) *Richtlinie für die Ausstattung und den Betrieb von Strassentunneln (RABT)*, (*Guidelines for the Equipment and Operation of Road Tunnels*) (in German). FGSV-Verlag, Cologne.

Grässlin U (2007) Erfahrung aus dem Gotthardstrassentunnel durch Einsatz der Sichttrübung und Rauch-, Branddetektion. *Proceedings of 6th Symposium: Sicherheit im Strassentunnel durch Einsatz moderner Messtechnik*. Sigrist-Photometer, Brunnen.

Haack A, Schreyer J, Grünewald M, Steinauer B, Brake M and Mayer G (2005) *Brand- und Störfalldetektion in Straßentunnel – Vergleichende Untersuchung*. Wirtschaftsverlag NW Verlag für neue Wissenschaft, Bremerhaven.

Lacroix D (2001) The Mont Blanc Tunnel fire – What happened and what has been learned. *Proceedings 4th International Conference on Safety in Road and Rail Tunnels, Madrid, Spain*, April 2001, pp. 3–16.

Loss Prevention Certification Board (2011) Available at: http://www.brecertification.com (accessed 22 June 2011).

NFPA 502 (2008) *Standard for Road Tunnels, Bridges, and Other Limited Access Highways*. National Fire Protection Association Standard. Available at: http://www.nfpa.org (accessed 22 June 2011).

Swiss Confederation (2007) Branddetektion in Strassentunneln. ASTRA 13004. *Richtlinie* **V2.10**.

Swiss Confederation (2008) Lüftung der Strassentunnel. Systemwahl, Dimensionierung und Ausstattung. ASTRA 13001. *Richtlinie* **V2.01**.

Verein deutscher Sachversicherer (2011) Available at: http://www.vds.de (accessed 22 June 2011).

World Road Association (1999) *Fire and Smoke Control in Road Tunnels*. PIARC Committee on Road Tunnels, Publication 05.05.B. Available at: http://www.piarc.org (accessed 22 June 2011).

Handbook of Tunnel Fire Safety, 2nd edition
ISBN: 978-0-7277-4153-0

ICE Publishing: All rights reserved
doi: 10.1680/htfs.41530.109

Chapter 7
Passive fire protection in concrete tunnels

Richard Carvel
BRE Centre for Fire Safety Engineering, University of Edinburgh, UK
Kees Both
Efectis Nederland BV, The Netherlands

NOTE: References to commercial products in this chapter are not to be taken to imply endorsement by the authors or editors, they only serve purposes of illustration.

7.1.   Introduction

A number of incidents in the past have shown that fires can pose a serious threat to the structural integrity of a tunnel, both during and after the fire. Fires can occur during tunnel operation (e.g. the fires in the Channel Tunnel (UK/France) in 1996 and 2008, and the Mont Blanc Tunnel (France/Italy) in 1999), as well as during the construction phase (e.g. the fire in the Great Belt Tunnel (Denmark) in 1994).

Modern national and international economies rely heavily on sustainable transport systems. In many of these systems, tunnels are key elements. Thus, continuity of tunnel operation is crucial, and various fire-safety standards are required for tunnels. Proper assessment of safety in the case of fire addresses all the links in the so-called 'safety chain': prevention, protection (passive and active), after-care and evaluation.

■ Prevention: adequate design and operational safety measures are adopted to avoid incidents that lead to fire (e.g. one-way traffic avoids head-on collisions).
■ Protection: safety measures that limit fire damage and maintain structural integrity during and after fire. These systems fall into two broad categories:
  – passive fire protection (e.g. non- or less-flammable tunnel linings)
  – active fire protection (e.g. sprinklers, traffic control, emergency response).
■ After-care and repair: actions and measures to make the tunnel fully operational (retrofitting and repair).
■ Evaluation: the usefulness of the adopted safety strategy and safety measures is reassessed in the light of lessons learned.

In this chapter, the focus is on the requirements with respect to structural integrity, which is part of passive fire protection. The issue of structural integrity is an important one. If a tunnel is not adequately protected:

■ There may be loss of structural integrity during a fire, which may cause severe leakage and even collapse, possibly before evacuation is complete. This is an especially important consideration in underwater tunnels and in tunnels in soft ground.

- A fire may cause significant damage. This may lead to large direct repair and retrofitting costs, as well as large indirect economic damage due to long non-operational times, etc.
- The emergency service response teams may encounter dangerous situations (of unknown risk) in which they have to work (assisting evacuation and fire-fighting).
- Other fire-safety measures, including ventilation channels, evacuation routes (above false ceilings in some Alpine tunnels), anchorage systems for ventilation systems, cable trays, etc., may fail, possibly causing casualties and perhaps leading to increased severity of the incident itself.

7.2.   Types of tunnel

When classifying tunnels, distinctions are usually made on the basis of traffic type; i.e. road, rail and mass-transit tunnels. As the fire risk in each type of tunnel is quite different, in terms of probability as well as consequences, this distinction is also commonly adopted in fire-safety analyses. When assessing the structural integrity of a tunnel during a fire, however, it is also convenient to make a distinction on the basis of material use and construction method.

Most modern tunnels are constructed using one of four methods.

- Cut and cover: the tunnel is formed by excavating an open trench. The walls and floor are constructed, and then the ceiling is cast or placed using prefabricated elements such as prestressed concrete beams. Finally, the ground is replaced over the tunnel. These tunnels generally have a rectangular profile.
- Immersed tube: the tunnel is constructed of large prefabricated tunnel segments. These are normally assembled in a trench excavated in the sea floor or river bed. These tunnels often have a rectangular profile, but other profiles are also used.
- Bored tube: the tunnel is bored through rock or other strata and lined with prefabricated steel and/or prefabricated concrete sections inside the bore. These tunnels generally have an oval or circular profile.
- New Austrian Tunnel Method (NATM): this is a bored construction, but the tunnel lining is formed using shotcrete (a rough aggregate concrete that is sprayed onto the surface of the excavated tunnel bore) rather than prefabricated sections. These tunnels generally have a horseshoe-shaped profile.

While there are unlined tunnels cut out of solid rock, and tunnels lined with iron or steel sections, stone or brick, the majority of modern tunnels are constructed of concrete. Therefore the discussion in this chapter is focused on fire protection in concrete tunnels. A discussion of fire protection in cast-iron-lined tunnels is given in Pope and Barrow (1999).

7.3.   The behaviour of concrete subject to fire

High temperatures affect the material properties of concrete in several ways. The strength, stiffness, and deformation behaviour in general are influenced through a combination of chemical and physical phenomena.

Figures 7.1 to 7.3 give indications of the temperature influence on some of the most important structural parameters of concrete: strength, stiffness and thermal elongation. For more elaborate descriptions and background, refer to a textbook such as the one by Purkiss (1996). As can be seen from Figures 7.1 to 7.3, the relationships are highly non-linear. The influence of temperature on

Figure 7.1 Typical temperature influence on the (normalised or relative) stress–strain relationship of concrete (normal weight, siliceous aggregates)

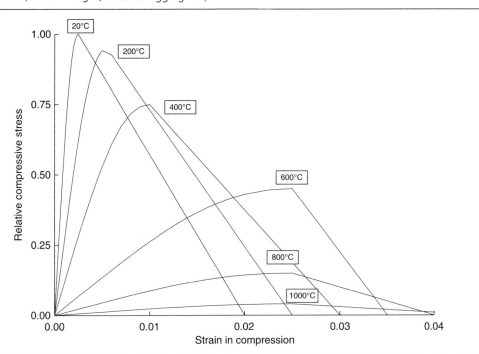

the strength and stiffness of concrete and structural steel is compared in Figure 7.2. See Buchanan (2001) for more details.

Not all concrete mixes are the same, and neither is the manufacturing process used (casting, curing) or the environment the concrete is placed in (climate, physical or chemical attack). However, the general trend in tunnel construction appears to be towards more durable concrete, which generally implies lower porosity and permeability, and (consequently) higher strength. Concretes used in bored-tube tunnels tend to be high-strength, high-density, low-porosity mixes, while those used in immersed-tube tunnels are generally of lower density and higher porosity (Both, 1999). These differences greatly influence the behaviour of the concrete when subjected to fire conditions.

In tunnels constructed from high-strength, low-porosity concrete mixes, the dominant failure process is that of 'spalling'. Under high-temperature conditions, chunks of concrete explode away from the surface at high velocities. The exact mechanism of spalling is not yet fully understood, but it is thought to be governed by (a combination of) two processes.

- The pressure build-up within the concrete, which is in turn due to the formation of water vapour (van der Graaf *et al.*, 1999). In practice, porous concrete will contain a certain amount of liquid water. This will obviously become water vapour if the temperature of the concrete exceeds 100°C. If the build up of vapour pressure exceeds the capabilities of the concrete pores to release the pressure, the concrete will spall.

Figure 7.2 Comparison of the strength and stiffness of concrete and steel as a function of temperature

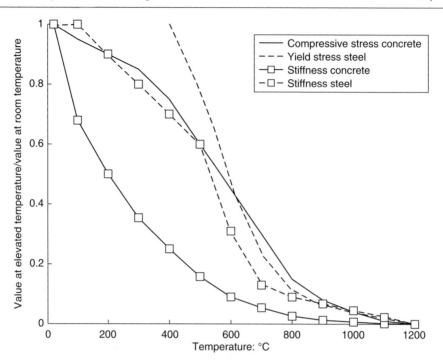

- Thermal stresses in the concrete, resulting from the restraint on thermal expansion (and also, for example, differences in thermal expansion between aggregates in the concrete and the matrix), and the thermal gradient over the thickness/cross-section of the concrete lining.

Secondary processes include physical and chemical processes, such as dehydration and material phase changes. These processes also influence the material properties governing pressure build up, the development of thermal stresses and the formation of (micro) cracks in the concrete, and, consequently, eventually loss of strength and stiffness.

At temperatures above 400°C the calcium hydroxide in the concrete cement will dehydrate and produce water vapour:

$$Ca(OH)_2 \rightarrow CaO + H_2O$$

Not only does this process accelerate the spalling, but it also greatly decreases the strength of the concrete (Wetzig, 2001). Other chemical processes may occur in the concrete aggregates at elevated temperatures. For example, quartz undergoes a mineral transformation at 575°C, which brings about an increase in volume, and limestone aggregates will decompose at temperatures above 800°C (Wetzig, 2001):

$$CaCO_3 \rightarrow CaO + CO_2$$

If the porosity of the concrete is not sufficient to allow the carbon dioxide gas to escape, the pressure will build up and further spalling will occur.

Figure 7.3 The thermal expansion of concrete and steel as a function of temperature

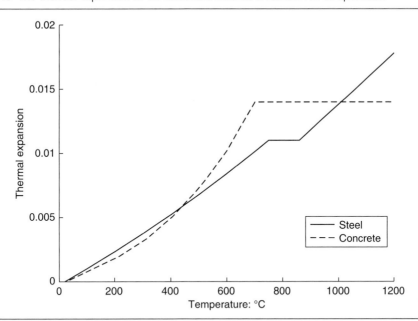

Spalling has been observed in certain types of concrete, even at temperatures below 200°C (Both *et al.*, 1999). Also, once the spalling process has started, it is not likely to stop until structural failure (Both *et al.*, 1999). At present, there is no generally applicable model to predict or simulate explosive spalling in different compositions of concrete (Both and Molag, 1999), so experimental testing of tunnel lining materials is essential before they are used in practice. Various research programmes are currently being carried out to develop such a model. For example, Delft University of Technology and Eindhoven University of Technology (2011) are collaborating on a project focusing on the development of a model incorporating temperature and moisture transport, as well as mechanical behaviour. It is expected that it will take many more years before such a model can be applied to the design of new concrete mixes and structures.

As well as having an influence on the concrete mix, high temperatures will also have an influence on the steel reinforcements within the concrete. In general, all metals will tend to expand with increasing temperature, and will exhibit a marked reduction in their load-bearing capacity. For example, at 700°C the load-bearing capacity of common hot-rolled steel reinforcement will be reduced to as little as 20% of its value at ambient temperature (see Figure 7.2). Steels with a low carbon content are also known to show irregularity (blue brittleness) between 200°C and 300°C. As a consequence, it is usually recommended that steel reinforcement is protected from temperatures above about 250–300°C (Wetzig, 2001).

Like most materials, concrete will also exhibit thermal expansion. This may lead to bending of the concrete members, and possibly to collapse of the structure. This effect may be significant in, for example, intermediate ceilings within tunnels and partitions between ventilation ducts. The thermal expansion of steel and concrete is similar in the temperature range 0–400°C, but above this temperature their expansion behaviours are significantly different, which will lead to damaging stresses within the mix (Wetzig, 2001). This is shown in Figure 7.3, which gives the

Figure 7.4 Schematic illustration of the cracking problem

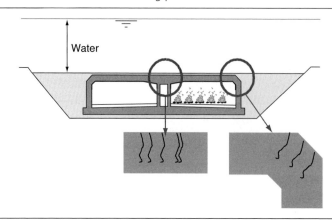

thermal expansion of concrete and steel as a function of temperature. Notice the plateaus at around 700°C, indicating phase changes in the materials. For more information, see standard textbooks such as those by Purkiss (1996) and Buchanan (2001).

Although the concrete used in bored tunnels generally contains steel reinforcement, this reinforcement is largely redundant after completion of the tunnel. The reinforcement is required to endure the pressures and stresses of construction (e.g. jack pressures), but once construction has been completed, the tunnel lining is in compression and the reinforcement is not required to prevent sagging, etc. (van der Graaf *et al.*, 1999). Thus, in bored tunnels, fire-protection measures are utilised exclusively to prevent explosive spalling.

This is not the case in immersed and cut-and-cover tunnels. These tunnels are generally constructed out of less dense, more porous concrete mixes, and the steel reinforcement in the mix is required to prevent shear failure in and sagging of the ceiling (Both *et al.*, 1999). The greater levels of porosity and the lower strength of the concrete, combined with a construction that is not in compression, mean that explosive spalling of the concrete is not necessarily the main failure mechanism of these tunnels under fire conditions. Of greater concern, especially in underwater tunnels, is the prevention of sagging in the reinforcement of the roof. If not prevented, this would lead to leakage, and possibly collapse, of the tunnel ceiling.

Another point of concern in immersed and cut-and-cover underwater passages is the development of cracks in hogging regions at the unexposed side. A recent study (Both *et al.*, 2010) has revealed this phenomenon, which may result in durability problems of the reinforcement and premature shear failure of the ceiling (Figure 7.4). In that study, various model scale (1:10) tests were performed. It was shown that, under fire conditions, cracks up to several millimetres long may be formed on the unexposed side of the concrete (Figure 7.5). These hidden cracks may pose a serious threat to the durability and shear resistance of the structure. As a consequence of these tests, the requirements for passive fire protection have had to be reassessed. The tests have also been used to develop and validate computer models of structural behaviour.

Although the mechanisms of failure in bored and immersed tunnels are different, the fire-protection requirements are the same in both types of tunnel: either to prevent or delay the

Figure 7.5 Model-scale fire tests and simulations

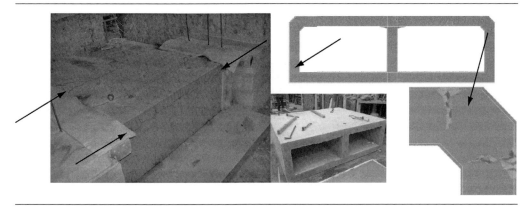

build up of temperature in the concrete of the tunnel lining; or to prevent or mitigate the effects of excessive heat flow in the lining material.

Fire-protection methods fall into two broad categories

■ passive fire protection (e.g. insulating materials restricting the flow of heat from the tunnel tube into the concrete lining)
■ active fire protection (e.g. water sprinkler systems removing heat from the tunnel void and cooling the walls).

Active fire protection is considered in Chapters 8 to 11. Only passive protection is considered here.

7.4.   Passive fire protection

Passive fire-protection measures generally take one of three forms

■ a secondary layer of a concrete or cementitious material applied to the inner surface of the tunnel.
■ cladding – panels of protective material, fixed to the tunnel walls and ceiling.
■ addition of certain fibres, etc., to the main concrete mix, to make the concrete more 'fireproof'.

These measures are illustrated in Figure 7.6.

Another means of fire protection is a variation of cladding, where the material applied is not so much an insulator, but a shield, protecting the concrete structure from radiated heat transfer. For example, thin, perforated metal sheets have been shown to perform well in tests (Haack, 1999).

Finally, fire protection could be achieved by simply overdimensioning the components; if the concrete members are sufficiently large, then, although some surface spalling may take place, the structure will be able to endure extremes of temperature for an extended period of time (Wetzig, 2001). This is not 'added' fire protection, and so will not be discussed further in this chapter.

Figure 7.6 Types of passive fire protection (the bars within the concrete represent steel reinforcement rods). (a) A secondary layer of insulating material applied directly to the tunnel lining; (b) A panel cladding system. Certain types of cladding may be bolted directly to the tunnel lining, not separated by an air gap as shown. (c) The addition of fibres to the concrete mixture

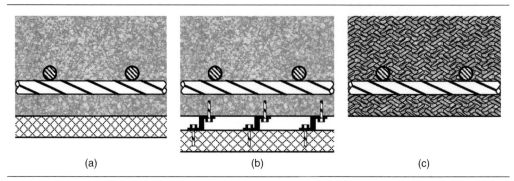

(a)   (b)   (c)

7.5. Requirements

At present there are no internationally accepted standards or legislation specifying the level of fire protection to be used in tunnels (Beeston, 2002). One country that has set prescriptive standards is The Netherlands. This is not surprising as, due to the fact that much of the country is below sea level, the collapse of a tunnel in The Netherlands may not merely lead to the flooding of the tunnel but possibly the flooding of the surrounding areas (Peherstorfer *et al.*, 2002). (Other countries also have prescriptive standards, for example, the ZTV-RABT standard used in Germany.) In The Netherlands, the Rijkswaterstaat (Ministry for Public Works and Water Management, hereafter referred to as RWS) has responsibility for safety in tunnels. Their tests (Dekker, 1986; van Olst, 1998; Bjegovic *et al.*, 2001; Breunese *et al.*, 2008) are used to assess

- the spalling behaviour (in principle only for bored tunnels, but also for immersed tunnels using higher grade concretes)
- the temperature development within the structure in the case of application of insulation.

When assessing the propensity for spalling in tests, the 'pass' criterion is simply that the structural integrity should be sufficient to withstand the required fire design temperature–time curve. Spalling is allowed, therefore, to the extent that it does not lead to premature failure.

The following criteria hold when testing an insulation system

- that the temperature of steel reinforcement within concrete shall not exceed 250°C (to prevent sagging and collapse)
- that the temperature at the surface of the concrete shall not exceed 380°C (to prevent spalling).

RWS requires that these criteria are met in the event of all fire incidents up to a fuel tanker fire (taken to be the most severe tunnel fire scenario, having very rapid growth and resulting in very high temperatures) with a duration of up to 2 hours.

Samples of concrete linings and protection cladding or coatings are tested in laboratory experiments to determine whether or not these, or other, requirements are met. While the RWS

Figure 7.7 Time–temperature curves used to test tunnel-lining systems: (a) ISO 834, (b) hydrocarbon fire, (c) RWS curve, (d) RABT and EBA curves (approximate representations)

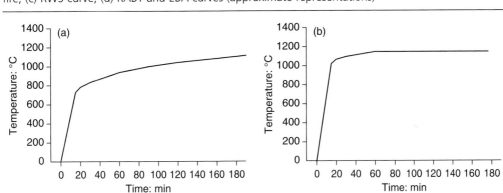

standard is required in The Netherlands, it is not necessarily required in other countries. A number of different time–temperature curves are, and have been, used in laboratory experiments to test the capabilities of fire protection systems.

The international standard time–temperature curve ISO 834-1:1999 is routinely used around the world to test the fire-resisting properties of building components. This curve, represented in Figure 7.7(a), is based on a 'cellulose fire', and is more typical of the temperature development in a compartment fire than in a tunnel fire.

Vehicle fires have a time–temperature curve more similar to the curve for hydrocarbon fires used to test building components for the petrochemical industry (Norwegian Petroleum Directorate (NPD) hydrocarbon curve) (see Figure 7.7(b)). The NPD curve has a much faster rate of growth than the ISO 834 curve, exceeding 1000°C within 15 minutes.

Fires involving fuel tankers or other dangerous goods may grow more rapidly and attain higher temperatures than those attained in a hydrocarbon fire. The RWS time–temperature curve exceeds 1100°C within 5 minutes and peaks at 1350°C after an hour (see Figure 7.7(c)). This curve is based on experimental testing carried out by RWS together with TNO (The Netherlands Organisation for Applied Scientific Research) in the early 1990s (van de Leur, 1991). The fire curve was confirmed later in numerous research projects, including the Runehamar fire tests in the UPTUN project (see Chapter 31).

While several other time–temperature curves may be used to test tunnel fire-protection systems, two others are worthy of note. In Germany, tunnel-lining protection systems are tested according to the RABT (*Richtlinien für die Ausstattung und den Betrieb von Straßentunnel ı*, also known as the ZTV tunnel curve (*Zusätzlichen Technische Vertragsbedingungen*)) (German Federal Ministry of Traffic, 1995, 1999, 2002) and the German Federal Railway Authority (Eisenbahn-Bundesamt (EBA)) curves. These curves differ from those described above in that they feature a cooling-off phase after 30 minutes (RABT) or 60 minutes (EBA) (see Figure 7.7(d)). The importance of a cooling-off phase in the assessment of a sample's resistance to heat was demonstrated during a test of some concrete tunnel elements at the Hagerbach test gallery, Switzerland. During the test, a concrete sample resisted temperatures of up to 1600°C for 2 hours without collapsing, but half an hour into the cooling-off phase the sample collapsed explosively (Wetzig, 2001).

Tunnel fire-lining systems are tested according to one or more of the above time–temperature curves, as described in the literature (Barry, 1998; Both *et al.*, 1999; Bjegovic *et al.*, 2001; Wetzig, 2001).

7.6.   Secondary tunnel-lining systems

The simplest, and often cheapest, form of tunnel-lining system is the application of a layer of insulating material to the interior surfaces of the primary tunnel lining. These are generally known as 'passive thermal barriers', and are often of the form of vermiculite cements, which can be sprayed on to the tunnel lining to the desired thickness. Vermiculite cements are inorganic materials that do not burn, produce smoke or release toxic gases under high temperature conditions. Tests (Barry, 1998) have shown that layers of vermiculite cement, 20–60 mm thick, may suffice, depending on the tunnel structure and concrete mix, to protect a concrete lining material to the requirements of the RWS, when tested using the RWS time–temperature curve. Other experiments have demonstrated that this form of lining material can protect concrete structures for up to 24 hours under extreme fire conditions (Barry, 1998). Coatings of this type are easy to repair if damaged. Promat International Ltd is one of the leading producers of passive thermal barriers.

As an alternative to spray mortars, boards, generally containing calcium silicate, may be used. In this case, board thicknesses of 20–60 mm may also suffice to protect against severe hydrocarbon fires.

Of course, fire resistance is not the only requirement of such systems, and the following other factors must also be taken into consideration

- frost–thaw resistance
- resistance to vibrations and air-pressure fluctuations, resulting from (air flow caused by) traffic/rolling stock
- repair options after damage (e.g. due to collisions)
- inspection possibilities (e.g. after leakages)
- influence on acoustics (e.g. when tunnel operators need to address occupants in the tunnel using loudspeakers).

Last, but not least, the passive thermal barriers must be designed in such a way that additional equipment (ventilators, traffic lights, etc.) can be safely suspended from ceilings and walls. It has occasionally been observed that the anchors used in suspension systems for additional equipment have provided a pathway for heat leakage into the concrete, resulting in unexpected spalling (Figure 7.8).

Figure 7.8 An example of spalling due to poorly designed equipment fixings

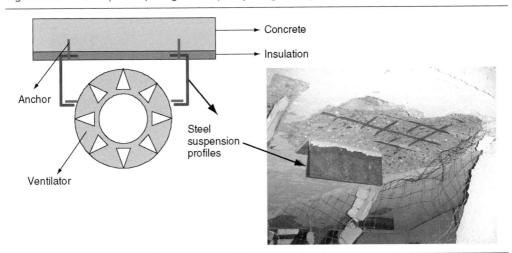

Caution must be expressed in the use of such sprayed materials, however, as illustrated by the collapse of the World Trade Centre towers in New York on 11 September 2001 after an aeroplane impact. There is evidence that some of the sprayed insulation was blown off by the blast, exposing the structural members to extremely high temperatures. In addition, where the insulation was not blown off, it was sometimes found to have been of insufficient thickness to effectively resist the ensuing fire (Quintiere *et al.*, 2002). The application of these materials must be done professionally.

A relatively new form of tunnel-lining material was used to protect the concrete lining of the refuges in the Mont Blanc Tunnel during its refurbishment following the 1999 fire (Figure 7.9). This was a 'refractory' ceramic cementitious material called FireBarrier. The new safety guidelines for the Mont Blanc Tunnel require that the temperature within the refuges should remain below 60°C for 2 hours when subjected to a hydrocarbon fire, and below 60°C for 4 hours assuming an ISO 834 fire. FireBarrier was deemed to have a better fire resistance than traditional vermiculite cements (Beeston, 2002).

The word 'refractory' denotes a material that is highly resistant to heat and pressure. Refractory ceramics are more hard wearing than traditional vermiculite cements, i.e. they are less prone to erosion by weather and high pressure cleaning, etc. The surface of the coating is also suitable for painting or the addition of pigment for decoration. Its high compressive strength also makes it structurally suitable for the support of lighting fixtures, etc. (Beeston, 2002). Refractory products need not be sprayed onto tunnel linings directly, and are sometimes applied as panelling or cladding systems.

7.7. Tunnel cladding and panelling systems
Many tunnel systems, particularly road tunnels, have a secondary lining composed of panels. These were originally employed for mainly cosmetic purposes, but in recent years these cladding systems have been developed to provide structural fire protection as well.

Figure 7.9 The refurbishment of the refuges in the Mont Blanc Tunnel (Photograph courtesy of Morgan Thermal Ceramics)

Tunnel claddings may consist of monolithic panels of materials such as calcium silicates, vermiculite cement, fibre cement or mineral wool, or may be composite panels with rigid outer surfaces (e.g. steel) and an insulating core of a material such as mineral or glass wool. These products are said to be easy to install, and have the advantage of being prefabricated rather than having to be constructed on site. However, the disadvantage of these systems is that their installation can be time consuming (and therefore more expensive) compared with some coating systems (Barry, 1998).

In The Netherlands (Huijben, 1999), it is deemed necessary to clad the ceiling and upper parts of the walls (1 m) of immersed-tube and cut-and-cover tunnels with prefabricated tunnel-lining systems like Promatect (produced by Promat International Ltd) (Figure 7.10).

Until recently, cladding systems generally required panel thicknesses of about 70 mm (Peherstorfer *et al.*, 2002), but it is claimed that recent developments have enabled much thinner panels to provide the same degree of fire protection as older, thicker panels. It has been claimed that panels only 27 mm thick are required in immersed-tube tunnels to provide 2 hours of protection, compared with 45 mm of sprayed fire-resistant lining in bored tunnels (Huijben, 1999).

Recent innovations have led to the development of other types of tunnel lining for fire protection. Of particular note is the ceramic–steel panel system currently in development in Germany (Rauch, 2001). This system consists of two layers of enamel-coated steel, separated by an air gap of about 20 mm. The enamel protects the steel at high temperatures, so that the structure does not fail, even at fire temperatures above the melting point of the steel. The air gap has an insulating effect, and

Figure 7.10 The Promatect tunnel-lining system (Photograph courtesy of Promat International Ltd – Tunnel Fire Safety)

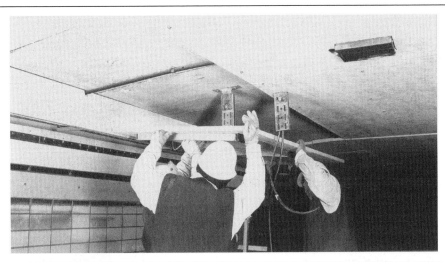

allows transport of the heat along the longitudinal direction of the tunnel. In full-scale experimental fire testing of the double-shell lining system, the temperature of the structural wall did not exceed 200°C, even though the recorded temperature of the outer shell exceeded 650°C (Rauch, 2001). However, directly above the fire location, the structural members of the ceiling were exposed to temperatures of over 400°C, which would not satisfy the RWS requirements.

It is said that this tunnel-lining system does not merely slow down the transport of the heat into the structural concrete like most conventional linings, but also allows transport of the heat away from the concrete, by reflection (back into the tunnel void), by conduction (longitudinally, through the steel panelling) and by convection (longitudinally, within the air gap). While this is clearly beneficial from a structural protection point of view, it should be noted that this may have effects on the development of the fire itself and on life safety; if more heat is radiated back into the tunnel void, the burning of the fire may be enhanced and lead to a more severe blaze.

7.8. Concrete additives

Another method of protecting the structural concrete of a tunnel is to make the concrete itself more fire resisting. The spalling failure mechanism is thought to occur due to a build up of vapour pressure within the concrete. If a concrete mix is used that contains materials which can release the pressure more effectively, the concrete will be able to withstand fire conditions for longer before failure. Recent investigations have focused on two different (combinations of) additives in the mix

- polypropylene fibres
- steel fibres.

The aim of introducing polypropylene fibres into the concrete mix is to provide expansion channels for the water vapour that is produced under fire conditions (Wetzig, 2001). The aim of including steel fibres in the mix is to increase the ductility of the concrete, and to enable the concrete to withstand higher internal pressures without spalling (van der Graaf *et al.*, 1999).

Results from a series of experiments testing a variety of different concrete mixes under fire conditions (ISO 834 and RWS curves) have been presented (Shuttleworth, 2001). Two of the main findings of the study were

- the inclusion of monofilament polypropylene fibres (1 kg/m^3) in high-strength, low-porosity concrete significantly reduces the risk of explosive spalling when exposed to severe hydrocarbon fires (RWS curve)
- steel fibres apparently do not contribute to the ability of concrete mixes to resist explosive spalling when used without polypropylene fibres.

Steel fibres do increase the fracture energy of the concrete, and as such limit the risk of propagating spalling. However, the quantities necessary to bring about this effect generally introduce an unacceptable influence on the casting properties of the concrete. In general, using only steel fibres to increase spalling resistance is not very popular.

The study also observed that monofilament polypropylene fibres give better fire protection than fibrillated polypropylene fibres. The first finding above directly contradicts an earlier study, which reported that the inclusion of polypropylene fibres did not have an influence on the spalling behaviour of concrete (Steinert, 1997). It would appear that the proper mixing of the fibres is key to good performance. The contradictory results could be partly attributable to inadequate mixing of the fibres in the concrete. Another factor that may play an important, although not governing, role is that, at the onset of micro-cracking in the concrete as a result of the extreme heating, the fibres bridge these micro-cracks. It has been shown that such micro-cracks may develop below 100°C, a temperature at which the fibres still have some tensile strength properties (Breunese *et al.*, 2006). Figure 7.11 shows the results of a detailed simulation, using a finite-element model, of micro-crack development in heated concrete. The

Figure 7.11 Computer simulation of micro-cracks between aggregates

model includes aggregates and cement paste, as well as the interface zone between aggregates and paste.

7.9.   Other passive fire protection systems

In addition to the three main methods of passive fire protection described above, a number of other methods of fire protection have been described. These include

- a method of prefabricating structural tunnel-lining elements with composite layers of fire protection built in (Bjegovic *et al.*, 2001; Planinc *et al.*, 2002)
- the use of fibre-reinforced composites (to protect cabling, etc., rather than structural members) (Gibson and Dodds, 1999)
- the use of organic coatings, including: intumescent products, which foam up on heating to form a thick insulating layer; ablative products, which use up large quantities of heat as they erode under high-temperature conditions; and subliming products, which also absorb heat as they vaporise (Barry, 1998).

At first glance, the use of organic coatings may seem to solve many fire-protection problems, but it should be noted that these coatings decompose to form many organic products, some of which are toxic, and so their use in passenger tunnels is not recommended, and needs further investigation.

7.10.   Active fire protection

Active systems for fire protection are also used in tunnels. By definition, active systems need to be switched on, either by manual or automatic means, to be effective. The two most common forms of active fire protection used in tunnels are ventilation systems and water suppression systems. The capabilities and effects of suppression and ventilation systems are described in detail in Chapters 8 and 9. See also Chapters 4, 5 and 32.

7.11.   Concluding comments

Fire protection is a necessity in most tunnel systems. It has a vital role in maintaining the structural integrity of the tunnel in the event of a fire, and thus in reducing repair and refit time and costs. It also has an influence on life safety. However, to date, fire-protection systems have not been considered primarily as life safety devices. When selecting a fire-protection product for use in a tunnel, care must be taken to ensure that the proposed system does not make any concessions on life safety in order to increase the protection of the concrete lining.

REFERENCES

Barry I (1998) Fire resistance of tunnel linings. *Proceedings of the International Conference on Reducing Risk in Tunnel Design and Construction, Basel, Switzerland*, 7–8 December 1998, pp. 99–111.

Beeston A (2002) Refractory solutions for fire protection of tunnel structures. *Proceedings 4th International Conference on Tunnel Fires, Basel, Switzerland*, 2–4 December 2002, pp. 63–72.

Bjegovic D, Planinc R, Carevic M, Planinc M, Betonidd T and Zuljevic M (2001) Composite fire-resistant tunnel segments. *Proceedings 3rd International Conference on Tunnel Fires and Escape From Tunnels, Washington, DC, USA*, 9–11 October, 2001, pp. 177–185.

Both C and Molag M (1999) Safety aspects of tunnels. *Proceedings International Tunnel Fire and Safety Conference, Rotterdam, The Netherlands*, 2–3 December 1999, Paper No. 3.

Both C, van de Haar P, Tan G and Wolsink G (1999) Evaluation of passive fire protection measures for concrete tunnel linings. *Proceedings International Conference on Tunnel Fires and Escape From Tunnels, Lyon, France*, 5–7 May 1999, pp. 95–104.

Both C, Breunese AJ and Scholten PG (2010) Emerging problem for immersed tunnels: fire induced concrete cracking. *Proceedings 4th International Symposium on Tunnel Safety and Security, Frankfurt am Main, Germany*, 17–19 March 2010.

Breunese AJ, Fellinger JJH and Walraven JC (2006) Constitutive model for fire exposed concrete in FE analyses. *Proceedings of the 2nd International fib congress, Naples, Italy*, 5–8 June 2006, pp. 1–12.

Breunese AJ, Both C and Wolsink GM (2008) *Fire Testing Procedure for Concrete Tunnel Linings*. Report 2008-R0695. Efectis Nederland BV, Rijswijk.

Buchanan AH (2001) *Structural Design for Fire Safety*. Wiley-Blackwell, Oxford.

Dekker J (1986) *Tunnel Protection Fire Test Procedure*. Institute TNO for Building Materials and Structures (IBBC-TNO), Rijswijk.

Delft University of Technology/Eindhoven University of Technology (2011) Explosive spalling of concrete: towards a model for fire resistant design of concrete elements. *PhD Project*. Available at: http://www.citg.tudelft.nl/live/pagina.jsp?id=3123a9be-7f85-4b61-9bfd-a571de87dc77&lang=en (accessed 22 June 2011).

Eisenbahn-Bundesamt (EBA) (1997) The German Federal Railway Authority. Available at: http://www.eisenbahn-bundesamt.de (accessed 22 June 2011).

German Federal Ministry of Traffic (1995) *Zusätzlichen Technische Vertragsbedingungen und Richtlinien für den Bau von Strassentunneln* (*Additional Technical Conditions for the Construction of Road Tunnels*), Part 1. Research Society for Roads and Transport, Working Committee for the Equipment and Operation of Road Tunnels.

German Federal Ministry of Traffic (1999) *Zusätzlichen Technische Vertragsbedingungen und Richtlinien für den Bau von Strassentunneln* (*Additional Technical Conditions for the Construction of Road Tunnels*), Part 2. Research Society for Roads and Transport, Working Committee for the Equipment and Operation of Road Tunnels.

German Federal Ministry of Traffic (2002) *Richtlinien für die Ausstattung und den Betrieb von Straßentunneln* (*Guidelines for Equipment and Operation of Road Tunnels*). Research Society for Roads and Transport, Working Committee for the Equipment and Operation of Road Tunnels.

Gibson G and Dodds N (1999) Fire resistance of composite materials. *Proceedings of the International Tunnel Fire and Safety Conference, Rotterdam, The Netherlands*, 2–3 December 1999, Paper No. 16.

Haack A (1999) Fire protection system made of perforated steel plate lining coated with insulating material. *Tunnel* **18**(7): 31–37.

Huijben JW (1999) Designing safe tunnels in The Netherlands. *Proceedings of the International Tunnel Fire and Safety Conference, Rotterdam, The Netherlands*, 2–3 December 1999, Paper No. 5.

ISO 834-1:1999 (1999) *Fire-resistance Tests – Elements of Building Construction – Part 1: General Requirements*. Available at: http://www.iso.ch (accessed 22 June 2011).

Norwegian Petroleum Directorate (NPD) (1987) Hydrocarbon curve. Given in Appendix D of: BS 476: Part 20: 1987 Method for the Determination of the Fire Resistance of Elements of Construction (General Principles); and BS 476: Part 21: 1987 Method for the Determination of Fire Resistance of Load Bearing Elements of Construction. Available at: http://www.bsigroup.com/en/Standards-and-Publications (accessed 22 June 2011).

Peherstorfer H, Eisenbeiss AA and Trauner T (2002) Provisions and measures to improve fire resistance of tunnel structures from the point of view of an accredited test institute. *Proceedings International Conference Tunnel Safety and Ventilation – New Developments in Fire Safety, Graz, Austria*, 8–10 April 2002, pp. 235–238.

Planinc R, Pavetić J, Planinc M, Tisovec I, Bejgovic D, Stipanovic I, Drakulic M and Carevic M (2002) Tunnel for safe traffic. *Proceedings of the International Conference Tunnel Safety and Ventilation – New Developments in Fire Safety, Graz, Austria*, 8–10 April 2002, pp. 303–310.

Pope CW and Barrow H (1999) Protection of cast iron tunnel linings against damage caused by a serious fire. *Proceedings International Conference Long Road and Rail Tunnels, Basel, Switzerland*, 29 November–1 December 1999, pp. 327–336.

Promat International Ltd. (2011) Available at: http://www.promat-spray.com (accessed 22 June 2011).

Purkiss JA (1996) *Fire Safety Engineering, Design of Structures*. Butterworth-Heinemann, Oxford.

Quintiere JG, di Marzo M and Becker R (2002) A suggested cause of the fire-induced collapse of the World Trade Towers. *Fire Safety Journal* **37**: 707–716.

Rauch J (2001) Fire protection in road tunnels: development and testing process of a tunnel lining system for more efficiency and safety. *Proceedings 4th International Conference on Safety in Road and Rail Tunnels (SIRRT), Madrid, Spain*, 2–6 April 2001, pp. 271–280.

Shuttleworth P (2001) Fire protection of concrete tunnel linings. *Proceedings 3rd International Conference on Tunnel Fires and Escape From Tunnels, Washington, DC, USA*, 9–11 October, 2001, pp. 157–165.

Steinert C (1997) *Brandverhalten von Tunnelauskleidungen aus Spritzbeton mit Faserzusatz (Fire Behaviour in Tunnel Linings made of Shotcrete with added Fibres)*. MFPA, Leipzig.

UPTUN (2006) Available at: http://www.uptun.net (accessed 22 June 2011).

van de Leur PHE (1991) *Tunnel Fire Simulations for the Ministry of Public Works* (in Dutch). TNO Report B-91-0043. Institute TNO for Building Materials and Structures (IBBC-TNO), Rijswijk.

van der Graaf JG, Both C and Wolsink GM (1999) Fibrous concrete, safe and cost-effective tunnels. *Proceedings International Tunnel Fire and Safety Conference, Rotterdam, The Netherlands*, 2–3 December 1999, Paper No. 17.

van Olst D (1998) Fire protection in tunnels, a necessity or an exaggeration? *Proceedings 4th International Conference on Safety in Road and Rail Tunnels (SIRRT), Nice, France*, 9–11 March 1998, pp. 337–345.

Wetzig V (2001) Destruction mechanisms in concrete material in case of fire, and protection systems. *Proceedings 4th International Conference on Safety in Road and Rail Tunnels (SIRRT), Madrid, Spain*, 2–6 April 2001, pp. 281–290.

Handbook of Tunnel Fire Safety, 2nd edition
ISBN: 978-0-7277-4153-0

ICE Publishing: All rights reserved
doi: 10.1680/htfs.41530.127

publishing

Chapter 8
Water-based fire-suppression systems for tunnels

Yajue Wu
Department of Chemical and Process Engineering, Sheffield University, UK
Richard Carvel
BRE Centre for Fire Safety Engineering, University of Edinburgh, UK

8.1.   Introduction

Water sprinkler systems have been used in fire suppression and loss control in buildings and warehouses for over 100 years. The first standard sprinkler was developed and installed in 1953 (Thomas, 1984). At the present time, automatic water sprinkler systems are widely used as fixed fire-protection systems, sometimes called fixed fire-fighting systems (FFFSs), in both industrial and residential buildings. The design of the sprinkler system is guided by regularly updated codes and standards, such as NFPA 13 (NFPA, 2007a) and BS 9251 (BSI, 2005). The latest development in the technology includes early suppression–fast response (ESFR) sprinkler systems, which are often used in warehouses, and water-mist systems, which are now regularly used in marine fire-protection applications and some residential and industrial applications, and are becoming popular for tunnels.

ESFR systems are designed to suppress a fire at an early stage of its development, and thus to minimise losses. They use an aggressive large-droplet spray and a flame-penetrating central core of water. The protection performance of ESFR systems depends on rapid fire detection and response, and they are designed to use fewer sprinkler heads than the normal control-mode sprinkler systems. The concept of 'fast response' promoted in the ESFR literature has since remained as one of the main features of most modern fire sprinkler systems. However, such systems rely on rapid fire detection and accurate localisation of the fire. This is relatively easy to ensure in situations such as warehouses, where the commodities generally remain at fixed locations during a fire incident. However, it is much harder to achieve in vehicle tunnels, where the objects on fire (generally vehicles) may move after fire detection, and where there are often significant ventilation flows, which may mean that the first detector to activate during an incident may not be the detector nearest to the incident. Thus, ESFR systems are not generally considered for use in tunnels.

The most recent development of sprinkler technology focuses on water-mist systems. Instead of large, fast droplets, water-mist systems employ very small water droplets. The systems were originally developed as an alternative to gas-extinguishing systems. The term 'water mist' is defined as a fine water spray in which 99% of the volume of the spray comprises water droplets with a mean diameter of less than 1000 μm. Water-mist systems use much less water than traditional sprinkler systems, and consequently produce considerably less water damage. Systems of this type have been shown to be highly efficient in suppressing certain kinds of fire in certain environments. However, until recently, water-mist systems have not generally been used as a substitute for conventional fire sprinkler systems, but rather as a supplementary measure in areas where a more specialist system is required.

Although the benefits of fire sprinkler systems in buildings have been widely recognised for many years, the acceptance of, and attitude towards, implementing sprinkler systems in tunnel infrastructures varies around the world. For example, sprinkler systems have been routinely used in Japanese tunnels since the 1960s, and in Australian tunnels since the early 1990s, but have generally not been installed in European or North American tunnels until recently. (In 1999, the World Road Association (PIARC) reported that no road tunnels in Belgium, Denmark, France, Italy, The Netherlands, the UK or the USA were equipped with water-based sprinkler systems, while Sweden had only one and Norway had two (PIARC, 1999). One tunnel system in the USA at that time was protected with a foam based-system.)

The tunnel environment is a confined space with complex ventilation systems, electrical systems and limited water drainage capacity. Until recently, the main organisations making recommendations for fire safety in tunnels, the World Road Association (PIARC) and the US National Fire Protection Association (NFPA), have taken a stance against the installation of FFFSs in tunnels for life-safety purposes. The 1999 PIARC report on fire and smoke control in road tunnels (PIARC, 1999), which summarised previous work by PIARC, stated:

> ... the use of sprinklers raises a number of problems which are summarised in the following points:
>
> ■ water can cause explosion in petrol and other chemical substances if not combined with appropriate additives,
> ■ there is a risk that the fire is extinguished but flammable gases are still produced and may cause an explosion,
> ■ vaporised steam can hurt people,
> ■ the efficiency is low for fires inside vehicles,
> ■ the smoke layer is cooled and de-stratified, so that it will cover the whole tunnel,
> ■ maintenance can be costly,
> ■ sprinklers are difficult to handle manually,
> ■ visibility is reduced.
>
> As a consequence, sprinklers must not be started before all people have evacuated.
>
> Based on these facts, sprinklers cannot be considered as an equipment useful to save lives. They can only be used to protect the tunnel once evacuation is completed. Taking into account this exclusively economic aim (protection of property and not life safety) sprinklers are generally not considered as cost-effective and are not recommended in usual road tunnels.

Following the catastrophic fire events of 1999 and the early 2000s (see Chapter 1), and the experience of initiatives such as the UPTUN project (see Chapter 31), PIARC produced a further report in 2007 on systems and equipment for fire and smoke control in road tunnels (PIARC, 2007), which acknowledged some advances in suppression systems and, in principle, permitted the installation of sprinklers in road tunnels, with the following recommendation:

> At the moment, an owner/operator who wants to install new detection and new fire-fighting measures must properly verify that the conditions for installing, using and maintaining contribute to the overall safety and are compatible with the framework of the entire safety concept for that specific tunnel. He must also ensure the effectiveness of the proposed measures.

Thus, before a suppression system can be installed, it is important that the operator can demonstrate the effectiveness of the system and that it works in conjunction with the other components of the overall fire-safety strategy.

In 2008, PIARC issued a report on FFFSs in road tunnels (PIARC, 2008), which is, essentially, an elaboration of the recommendations made in 2007. The report summarises its recommendations as follows:

In most cases, FFFS are not capable of extinguishing vehicle fires. The aims are to: slow down fire development, reduce or completely prevent fire from spreading to other vehicles, provide for safe evacuation, maintain tenability for fire-fighting operations, protect the tunnel structure and limit environmental pollution.

To fulfil these purposes, the FFFS must:

- be supported by effective and rapid fire detection and location systems that are optimized to ensure proper functioning of the FFFS, resulting in a highly reliable integrated system,
- be designed to handle air velocities in the range of 10 m/s that can result from ventilation system operation or natural effects,
- be able to mitigate fire development and infrastructure damage by utilizing an agent with good cooling effect,
- have an acceptable influence on visibility, especially during the self-rescue phase,
- be able to reduce radiant heat.

The report acknowledges that the current understanding of fire suppression in tunnel environments is incomplete, and it goes on to state that:

It is recommended that FFFS applications in road tunnels be researched further in order to better determine:

- the effect on tunnel fires,
- the effect on visibility and air quality,
- the effect on the rescue process (e.g. possible destruction of smoke stratification, deterioration of visibility, cleaning of the smoke, improvement of fire-fighters' approach to the fire source),
- the best activation procedure (e.g. automatic; by the tunnel operator; by fire-fighters only; before, during or after evacuation),
- possible interaction with dangerous goods (e.g. gasoline spills),
- installation and maintenance costs,
- cost savings related to minimization of infrastructure damage.

Water-based fire-suppression technologies have been the subject of fundamental research and practical engineering development for many years. There are many publications relating to sprinkler systems in the literature. Detailed technical information can be found elsewhere, including comprehensive reviews (Liu and Kim, 1999; Grant et al., 2000; BRE, 2005; Williams and Jackman, 2006) and NFPA handbooks (Cote and Linville, 2008; NFPA, 2007a,b, 2010). This chapter does not provide a literature review of the technologies, but instead focuses on the application of water-based fire-suppression systems in tunnel infrastructures. Although sprinkler systems for fire protection in tunnels operate on the same basic principles as those used in buildings and other contained spaces, tunnels involve unique problems, including ventilation and traffic.

The interactions between the sprinkler and the self-rescue of tunnel users, and with the ventilation system, have always been issues of concern for fire protection in tunnels. This chapter presents a review of several series of full-scale tests of fire-protection systems in tunnels identified in the literature. The next two sections provide some basic technical information on water-based fire-suppression systems, and the principles and dynamics involved in the interaction of water spray with the fire plume and extant ventilation. This information is used in the sections that follow to aid coherent discussion on the outcomes of the large-scale tests and experience gained in implementing the systems in tunnels.

8.2. The principles of fire sprinklers

A water spray from a conventional sprinkler head can generally be divided into three regions (Figure 8.1)

- the spray-formation region, where the water jets break into water droplets
- the vaporisation region, where the water droplets interact with flames and fire plumes, reducing the plume temperature by evaporation
- the fire-suppression region, where the larger water droplets, having passed through the plume, cool the fuel bed directly.

The performance of the spray is generally characterised by the atomisation cone angle, the droplet-size distribution and the water-volume discharge rate.

8.2.1 Spray formation

Many different types of spray nozzle have been developed over the years. However, the atomising mechanisms employed in the nozzles to produce fine water droplets can be classified into three categories

- atomisation by mechanical means (deflection, impingement or rotary), to break the water jet
- atomisation by pressurised jet at low, medium or high pressure
- atomisation by utilising an additional pressurised atomising fluid (e.g. air or an inert gas), sometimes called an 'inert gas propellant'.

Some spray nozzles may use more than one of these mechanisms in the spray-formation process.

Figure 8.1 Regions of the spray from a fire sprinkler head.

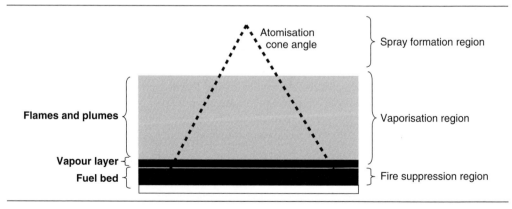

The process of water-jet break-up follows the mechanism suggested by Rayleigh's instability theory, i.e. that a column of liquid becomes unstable and breaks into droplets if its length is greater than its circumference (Rayleigh, 1878). The column breaks down into a series of similarly sized droplets, separated by smaller satellite droplets. Because of the irregular character of the atomisation process, non-uniform threads are produced, and this results in a wide range of droplet sizes.

Spray nozzles using a mechanical mechanism usually have a deflector or a rotary cup. The discharged vertical water jets are deflected mechanically in all directions in a random fashion, and naturally form a cone distribution. These types of nozzle produce relatively large water droplets, and are often used in deluge systems. The disadvantage is that the droplets are generated randomly and have low spray momentum.

Pressurised jet nozzles operate at low, medium or high pressure, and are most commonly used to produce water sprays. The pressurised jets can produce water spray in a much more controlled way, and generate finer droplets. Water-mist systems mainly use pressure jet technologies. The operating pressure is split into three ranges in NFPA 750 (NFPA, 2010), and the definition of these ranges is accepted internationally.

- Low-pressure water-mist system: operational pressure $\leqslant 12.5$ bar.
- Medium-pressure water-mist system: operational pressure 12.5–35 bar.
- High-pressure water-mist system: operational pressure $\geqslant 35$ bar.

Higher pressure systems result in smaller droplet sizes.

Systems in which the water spray is produced using pressurised air or an inert gas as the atomising medium are also known as 'twin-fluid water-mist systems'. Both the water and the atomising-medium lines operate in the low-pressure region of 3–12 bar. In this type of system, the air or inert gas is the driving force that maintains the spray momentum. The water mist produced by a twin-fluid system behaves as a gas, and has been considered as a substitute for gaseous Halon fire-suppression systems. However, the spray momentum is relatively low compared with the water mist generated by high-pressure systems. It is also significantly affected by ventilation flow and, it is claimed, may not be as effective as high-pressure water mist in extinguishing fires (Liu et al., 2001). The twin-fluids water mist is sometimes used in mobile fire fighting devices. As an FFFS, twin-fluids water mist is used in some specialised areas such as electrical cable tunnels.

8.2.2   Spray classification

Water-droplet distribution can be measured experimentally (Lawson et al., 1988; Widmann et al., 2001; Wang et al., 2002). The droplet size is affected by the angle and the distance away from the spray head. The standard practice is to measure the droplet population on a horizontal plane 1 m away from the spray head.

One of the commonly used representative diameters for a spray as a whole is the volume median diameter, denoted by D_{V50} (Grant et al., 2000). Here, half of a volume of water is contained in droplets greater than this diameter and half in droplets smaller than this diameter. Percentages other than 50% may be considered and the droplet distribution defined by giving a percentage of the droplet population smaller or greater than a particular diameter. For example, water

Table 8.1 Water spray classifications according to droplet volume and diameter

Class 1	Class 2	Class 3
$D_{V99} \leqslant 200\ \mu m$	$D_{V99} \leqslant 400\ \mu m$	$D_{V99} \geqslant 400\ \mu m$
$D_{V10} \leqslant 100\ \mu m$	$D_{V10} \leqslant 200\ \mu m$	$D_{V10} \geqslant 200\ \mu m$

mist is defined in NFPA 750 (NFPA, 2010) as $D_{V99} < 1\ mm$, where D_{V99} is the 99% volume diameter. That is, for a given volume of water mist, 99% of its droplets (by volume) have diameters smaller than $1000\ \mu m$. Using the volume diameter, water sprays are often classified into three categories, as shown in Table 8.1.

The performance of nozzles is often described using the parameters of spray classification, atomising cone angle and water discharge rate. One of the most desirable benefits of a small droplet size is the significant increase in the water surface area available for heat absorption and water evaporation. If the droplets are treated as being uniform in size, each having the mean diameter, the surface area of water in a $0.001\ m^3$ volume is given in Table 8.2. The surface area increases significantly with decreasing droplet size, and therefore significantly increases the cooling effect on gases. Smaller droplets do not necessarily have a greater cooling effect on solid or liquid surfaces, as these are cooled by the formation of a water layer on the surface.

8.2.3 Fire-fighting modes
The trajectory of a droplet depends on its size and momentum. Large, fast droplets and fine droplets interact with the fire plume in very different ways and result in very different trajectory patterns. Given this, systems tend to be used in different fire-fighting modes, depending on the size and momentum of the droplets that they produce.

8.2.3.1 Large, fast droplets
Generally, large (fast) droplets have a short residence time in the fire-plume zone. Large droplets have a relatively small surface area per unit mass, and therefore the heat-transfer rate is generally not significant enough to vaporise the droplet. During the interaction with the fire plume, large droplets may break into smaller droplets, but the high momentum is generally sustained and the droplets reach the fire seat and wet the hot burning fuel surfaces (and the surrounding unburned fuel). The direct contact with the fire surfaces and fuel bed leads to direct suppression of the fire. In some circumstances, the vapour formed as the water evaporates may form a layer near the fire surface, and droplets arriving later may have to penetrate this vapour layer, which may influence the cooling effect at the fuel surface. Conventional sprinkler systems produce

Table 8.2 Water surface area per $0.001\ m^3$ of water for different droplet sizes

	Droplet size: mm		
	6	1	0.1
Total number of droplets	8.8×10^3	1.9×10^6	1.9×10^9
Total surface area: m^2	1	6	60

large droplets. This fire-fighting mode, where the target is the fuel bed itself, is commonly called the 'fire-suppression mode'.

8.2.3.2 Fine droplets

The trajectories of fine droplets, such as those in a water mist, are heavily dependent on the interaction of the droplets with ventilation flows, flames and fire plumes. When it interacts with the fire plume, water mist can be considered as behaving like a gas. The small droplets in water mist have a relatively much larger surface area per unit mass. This leads to a greater heat-transfer rate and, therefore, to a high evaporation rate. Evaporating water provides effective cooling in the fire-plume region. Water-mist sprays have relatively low momentum compared with fast, large droplet sprays. Most of the droplets in water mist are likely to evaporate in the fire plume region, and therefore very little water is expected to reach the fuel bed.

One of the well-known advantages of water-mist systems is that they produce considerably less water damage than traditional sprinkler systems, because they use less water. However, a shortcoming is that water mist has a very limited effect on the fuel bed, and therefore cannot extinguish deep-seated fires. With water-mist systems the fire-fighting strategy is to generate a sufficient volume of mist to cool the plume and flame envelope. This cooling, by evaporating water, should bring the plume temperature down to a sufficient level to extinguish the flames. Thus, fine droplets operate in 'flame-extinguishing mode'.

8.3. The dynamics of fire suppression by water sprays

Fire suppression using water sprays operates on three principles: cooling of the fire plume by the evaporating water droplets; cooling of the fire surfaces and the fuel bed by the droplets that penetrate the fire plume; and cooling of the environment around the flame.

The following mechanisms are identified as being responsible for the extinguishing of flames by water sprays

- heat extraction (cooling of the fire plume; wetting/cooling of the fuel surface)
- displacement (displacement of oxygen; dilution of the fuel vapour)
- radiation attenuation
- kinetic effects and other factors.

8.3.1 Heat extraction

Water is a highly effective fire-protection agent. The average specific heat of water before reaching boiling point is 4.2 kJ/kg K, and its latent heat of vaporisation at atmospheric pressure is 2458 kJ/kg. Evaporating water can absorb a significant quantity of heat from flames and fuels. The cooling effect of water sprays has been considered as one of the most dominant fire-suppression mechanisms. The thermal and dynamic interaction of the water spray and the fire plume provide a cooling effect on the fire plume, both by the evaporation of the water droplets and by convective cooling (inducing a flow of cool air into the plume). The cooling effect is influenced by the spray pattern, water-droplet size and water discharge rate. A numerical study has suggested that a finer water spray produces a higher water evaporation rate and higher entrained air velocity, which in turn will deliver more cool air to the fire plume and hence provide more effective fire suppression (Hua *et al.*, 2002). The finer spray effectively cools the fire plume and reduces the buoyancy effect of the fire plume. Consequently, a fine water spray is able to penetrate and suppress the fire plume more effectively than a less fine spray. However, if the droplets are too small, the spray may not possess the momentum required to penetrate the flame, and fire-fighting efficiency may be compromised.

8.3.2 Displacement

The rapid volume expansion of evaporating water reduces the oxygen concentration in the fire-plume area. The specific volume of water at standard atmospheric pressure and 20°C is 0.001 m^3/kg, and the specific volume of steam at the boiling point of water is 1.694 m^3/kg. Assuming that the ideal gas law can be applied, the specific volume could reach 4.87 m^3/kg at a typical plume temperature of 800°C. Therefore, the volume of the water has the potential to expand 1700 times during the evaporation process, and by 5000 times if its temperature reaches that of the plume when interacting with the fire plumes. The expansion of the water volume displaces the oxygen locally, and also leads to a local dilution of fuel vapour. To extinguish a flame, the aim is to bring the fuel/air ratio below its lower flammability limit. For ordinary hydrocarbon fuels, a flame would not be sustained when the oxygen level is below 15% (Drysdale, 1985).

8.3.3 Radiation attenuation

The attenuation of thermal radiation by water is attributed to two mechanisms. Firstly, the water spray can act as a radiation shield, inhibiting the spread of fire from one burning object to another. The transmission of thermal radiation is reduced by absorption and scattering of radiation by the droplets in the spray. In addition, thermal radiation can also be attenuated in the gaseous phase. The high water vapour concentration within the gaseous species reduces the emission and transmission of radiation in the gaseous species of the flames and combustion products (Baillis and Sacadura, 2000; Sacadura, 2005). Various studies have been carried out on radiation absorption by water droplets and the spectral attenuation factors of water sprays (Ravigururajan and Beltram, 1989; Tseng and Viskanta, 2007). The radiation attenuation factor is influenced by droplet diameter and number density. Experimental tests have shown that the thermal shielding provided by a water spray around a storage tank fire, with a typical water mass flow rate per curtain-metre of 2 kg/m^2, gives an attenuation factor of 50–75% (Buchlin, 2005), and a water-spray curtain impinging on the wall could lead to higher shielding performance. Attenuation values as high as 90% can be achieved with the falling water film continuously refreshed by the spray. Water mist can act as an effective radiation shield to impede radiated energy transfer from hot surfaces. The attenuation of the thermal radiation depends weakly on the droplet size, but more strongly on the water load (water/air mass ratio). Yang et al. (2004) examined that the penetration depth of radiant energy emitted from 1300–2100 K black-body sources into a uniform water mist, with water loads of >10% (by mass), and found that the radiation was reduced to 20% of its initial intensity over a path of approximately 3 m.

8.3.4 Kinetic effects and other factors

The kinetic effects of water spray are mainly secondary effects, such as the results of water-spray-induced turbulence, the drop in flame temperature and the direct interaction of the spray with fuel surfaces.

The rates of chemical reactions are normally influenced by turbulence intensity and flame temperature. Some experimental studies of the interaction between water spray and flames have shown that the entrainment of water droplets increases the intensity of the turbulence and enhances the combustion rate (van Wingerden, 2000). The effect is normally strongly influenced by the size of the droplets and the water discharge rate. Water-spray-induced turbulence could enhance the reaction rate in fast combustion flames, such as turbulent premixed flames and gas explosions. However, it is rare that water-droplet-induced turbulence would have any noticeable effect on diffusion flames, such as fire plumes.

It is also possible that the application of a water spray could momentarily increase the fire size of a liquid pool fire (Kim and Ryou, 2004, 2008; Liu *et al.*, 2006). The rate of heat release from a liquid pool fire is strongly associated with the combustion process in the gaseous phase immediately above the liquid surface. The momentary fire flare-up effect in a pool fire is attributed to the enlarged flame surface caused by the impingement of a water spray on the surface and the resulting increase in the mixing area between the oxidant and the fuel.

Water impingement on liquid or solid burning surfaces has various effects on the chemical reactions and reaction rates in the flames. In some instances, the boiling of water droplets can enhance the evaporation of a liquid fuel or the rate of release of volatile matter from solid fuels. Experimental tests have been carried out to investigate the effect of water mist on the burning of solid materials such as poly(vinyl) chloride (PVC) (Zhang *et al.*, 2007), various plastics (Chow *et al.*, 2005) and poly(methyl methacrylate) (PMMA) (Jiang *et al.*, 2004). However, the risk of these effects is not deemed to be significant if the water spray is applied at an early stage with a sufficient water discharge rate.

Water vapour may also directly influence the kinetics of chemical reactions. The chemical reactions of hydrocarbons involve a series of reaction mechanisms, including initiation, propagation and termination. Free radicals such as H$^{\cdot}$ and OH$^{\cdot}$ play an important role in the breakdown of heavy hydrocarbon molecules to lighter molecules. The presence of water vapour could directly influence the production of free radicals, and therefore influence the reaction rate in the flames, and the production of carbon dioxide, carbon monoxide and soot. Water spray can also bring some other effects and benefits, such as the dispersion of unignited fuel.

8.4.   The principle of water-based fire protection for tunnels

Many publications on tunnels use the terms 'deluge' and 'sprinkler' interchangeably, while others distinguish between 'sprinkler nozzles' (spray heads that are actuated by the failure of a fused link, for example, a glass bulb, at a critical temperature) and 'deluge nozzles' (open spray heads that are activated by the supply of water to the nozzle). In the discussion that follows, the term 'sprinkler' is used to denote suppression systems that have relatively large water droplets, irrespective of the nozzle type, while 'water mist' is used to denote systems with smaller droplets. The terms 'deluge' and 'temperature-actuated nozzle' will be used to distinguish between nozzle types.

Both conventional sprinkler and water-mist systems are currently used for fire protection in tunnels. Some important parameters relating to the tunnel structure, facilities and rescue strategies should be considered in the design of FFFSs for use in tunnels. These parameters include

- tunnel geometry and cross-sectional area
- traffic flow (unidirectional or bidirectional)
- potential fire size expected if unsuppressed (e.g. are 'dangerous goods' permitted)
- escape-route provisions
- ventilation system type (natural, transverse, semi-transverse, longitudinal or hybrid)
- surveillance methods within the tunnel and other fire-detection methods used.

Possible accident scenarios form the basis of a fire strategy, and thus also are the basis of the design specification of tunnel safety systems, including FFFSs. The accident scenarios are strongly influenced by traffic flow. For the case of unidirectional traffic and longitudinal ventilation, the considered fire scenarios commonly assume that all traffic in front of the incident will be able to exit the tunnel freely, while a queue of traffic might build up behind the incident. Thus, part of the fire-safety strategy might be to blow all smoke away from the queuing traffic. In the case of bidirectional traffic and transverse ventilation, stationary traffic could build up on both sides of the fire, and the fire development could threaten vehicles in both directions. Considerations such as these must be taken into account when defining sprinkler zones inside a tunnel.

The zone method has commonly been used as the principle for the design of a water-based fire-protection system for use in a tunnel. The tunnel is divided into zones, and independent controlling systems are installed in each zone. Some systems may also have an independent water supply for each zone. Both the conventional sprinkler systems used in Japan and the water-mist systems used in Europe operate on this principle. A key part of the design is to dimension the zones appropriately, and determine the number of zones to be activated simultaneously in each considered fire scenario. It is common for two or three adjacent zones to be activated simultaneously to give sufficient coverage. Zone size is calculated according to the maximum vehicle length allowed, the expected fire load, the accuracy of the fire-detection system and the safety margins associated with each of these parameters. Zone lengths typically vary between about 24 and 50 m, and the whole coverage length typically varies between 50 and 100 m.

The determination of which sprinkler zones should be activated is dependent on the fire-detection system in the tunnel. The fire-detection sensors should be distributed frequently enough to locate the fire seat with an acceptable accuracy. Figure 8.2 illustrates an example of the selection of sprinkler zones in a unidirectional tunnel with a longitudinal ventilation system. It is based on the practice in Japan (Stroeks, 2001). Here, the zone length is 50 m and the detection sensors are located at 25 m intervals. In this example, fire alarms are triggered in zones 3, 4 and 5. The system decides the fire is likely to be in zone 3, and the sprinkler heads in zones 3 and 4 are opened to prevent fire spread. It is important that the management of the sprinkler system is integrated with the tunnel operation.

It is currently common practice that sprinkler systems in tunnels are not activated before the tunnel operator has assessed the situation brought to his attention by alarms. While, the activation of the controlling valves to the sprinkler zones (priming of the system) is frequently

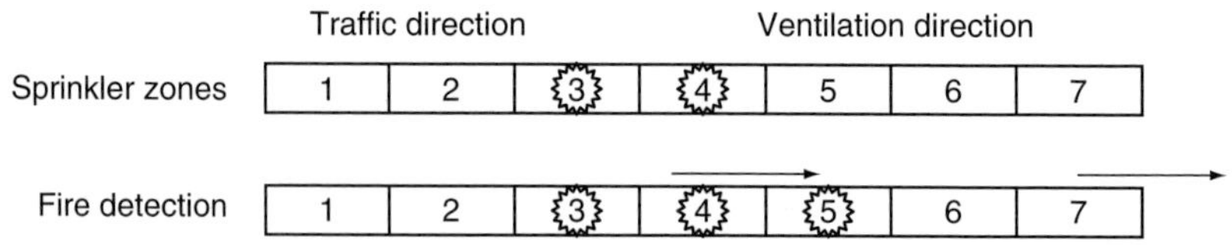

Figure 8.2 Example of zones for a sprinkler system and fire detection in a unidirectional tunnel. The alarm is triggered by the sensors in zones 3, 4 and 5. The system decides the fire is likely to be in zone 3, and sprinklers are activated in zones 3 and 4 to prevent fire spread

done automatically, the system is not fully automatic and the water sprays are not activated until the tunnel operator verifies the situation and confirms the location of the fire, using CCTV or other independent monitoring systems. For example, many systems use dry pipes from the sprinkler head to the main water supply, and it is common practice that the caps or valves on the sprinklers heads in the activated zones are opened as soon as the detection system determines a fire. Thus, the activation zones are selected and the sprinkler system is ready for water discharge, but the final authorisation to open the valves and allow water discharge is controlled manually by the tunnel operator.

This delay in sprinkler activation can minimise false alarms in the fire-detection system and prevent unnecessary activation of the sprinkler systems. It has also been claimed that the delay is necessary to allow the self-rescue of the tunnel users before the sprinkler system is activated, although this remains a contentious issue. Some of the main concerns in the ongoing debate over sprinklers during the self-rescue phase include the reduction in visibility and destratification of the smoke layer during sprinkler operation. One opinion is that the water discharge should be delayed until all tunnel users have moved out of the sprinkler zones to be activated. However, the duration of the self-rescue phase is determined by factors such as the proximity of escape routes and human behaviour. The danger is that the effectiveness of the fire-suppression system could be compromised by a delay in sprinkler activation.

Another issue concerning the application of sprinkler systems in tunnels is the interaction of the water sprays with the tunnel ventilation systems. This will be discussed below.

In the following section, the results of some large-scale tests on sprinkler and water-mist systems in tunnels are discussed, with particular reference to the following aspects

- effectiveness in reducing the rate of heat release
- effectiveness on different types of fuel package
- assessment of fire spread
- interaction with the ventilation
- effect of delay of activation of the water discharge
- effect on the tunnel users' self-rescue (visibility, smoke stratification layer, air quality and flow temperature).

8.5.   Large-scale trials
A number of large-scale tests and test series on the effects of FFFSs on fires in tunnels have been carried out. This section examines nine series of tests that are reasonably well documented and are available in the public domain. These test series are summarised in Table 8.3.

Two of the series are tests carried out in Japanese tunnels (Stroeks, 2001), using a low-pressure sprinkler system installed at the top of the tunnel wall on one side of the roadway. One other series were carried out using a conventional, ceiling-mounted sprinkler system in the second Benelux Tunnel in The Netherlands (DGPWWM, 2002).

The rest of the tests were carried out using ceiling-mounted water-mist systems. Some of these tests formed part of the UPTUN project (see Chapter 31) and the SOLIT project (Starke, 2010), while others were carried out for specific instillations including the A86 near Paris (Guigas et al., 2005) and the M30 Madrid ring road (Tuomisaari, 2008).

Table 8.3 Large-scale water fire-protection system tests in tunnels

Year	Test site	Tunnel dimensions	Sprinkler type and layout	Fuel package	Ventilation: m/s	Designed unsuppressed HRR: MW
Sprinkler systems						
1969	Futatsugoya Tunnel in the old Kuriko National Expressway (Stroeks, 2001)	Horseshoe shaped W: 6 m L: 384 m	Corner installation 36 m spray L: 12 m upstream of the fire	Three light vans and fire pans (2, 4 and 6 m²) with 50–300 litres of petrol	0–3	
2000 2001	Second Benelux tunnel near Rotterdam (DGPWWM, 2002)	Box shaped H: 5 m W: 10 m L: 900 m	Ceiling installation Deluge heads	Van loaded with wood Aluminium-covered truck with wood Wood cribs	3–5	15 20 36
2001	New Tomei Expressway; three traffic lanes in one direction (Stroeks, 2001)	Horseshoe shaped W: 8 m	Corner installation 36 m spray zone length	9 m² fire pan	5	23
Water-mist systems						
2002	IF Assurance fire and safety test centre, Norway (Kratzmeir, 2006)	Horseshoe shaped H: 8 m W: 6 m L: 100 m	Three-row ceiling installation Water mist	Stable, concealed Spray	1.1	8.1
2004	IF Assurance fire and safety test centre, Norway (Kratzmeir, 2006)	Horseshoe shaped H: 8 m W: 6 m L: 100 m	Three-row ceiling installation Water mist	Pools, wood pallets (UPTUN type)	3	<27
2005	Virgl/Virgolo tunnel of the Brenner motorway, Italy (UPTUN, 2005)	Horseshoe shaped H: 6.5 m W: 10 m L: 860 m	Ceiling installation 80–120 bar Water mist	Four pools, 70 litres of burning diesel oil in each pool	2.25	20

Year	Reference	Tunnel	System	Fuel		
2005	VSH Hagerbach, Switzerland (Guigas et al., 2005; Tuomisaari, 2008)	Box shaped H: 2.55 m W: 9.0 m L: 230 m	Three-row ceiling installation Water mist	Passenger vehicles 10 tests	3	5–30
2006	San Pedro de Anes test tunnel, Spain (Tunnel Safety Testing SA; Kratzmeir and Lakkonen, 2008; Mawhinney and Trelles, 2008; Tuomisaari, 2008)	Box shaped H: 5.2 m W: 9.5 m L: 630 m	Three-row ceiling installation Water mist	Simulated HGVs with wooden and plastic pallets	3.5	95
2007 2008	Runehamar Tunnel in Åndalsnes, Norway (Lemaire and Meeussen, 2008, 2010)	Horseshoe shaped H: 6.5 m W: 9.5 m	Ceiling water mist 75 m spray length 1% AFFF 1% AFFF No additive 1% Bioversal No additive	Pallets Diesel pool Diesel pool Diesel pool Pallets	3.7 4 4 4 4	50 200 200 200 200

In the discussion that follows, tests with sprinkler systems are discussed separately from the tests with water-mist systems.

8.5.1 Sprinkler systems

In Japan, it is recommended that all tunnels longer than 1000 m have a sprinkler system. Some tunnels between 300 and 1000 m long are also recommended to have sprinkler systems, depending on the traffic volume and bidirectional traffic flow. In Japan, guidelines on sprinkler systems are issued by two regulatory bodies: the Ministry of Land Infrastructure and Transport and the Japan Highway Public Corporation. Their main requirements are

- the total operation length should be at least 50 m
- the standard water-supply rate is 6 l/min per m^2
- the water-discharge time should be at least 40 minutes
- the sprinkler-control method should be chosen taking into consideration the tunnel length and structure, and the ventilation system.

For tunnels with a horseshoe-shaped cross-section, a single row of sprinkler heads is installed along the top of one wall at about 6 m height. This single row and corner installation method is also used in tunnels with a rectangular cross-section. Ceiling installation is sometimes used in rectangular tunnels in cases when the corner installation would not be efficient (e.g. if there are three lanes of traffic). Side installation is preferred as it facilitates easy inspection and maintenance. The location and type of nozzles are determined based on the tunnel layout (ventilation duct ceiling, lay-bys, jet fans).

8.5.1.1 Futatsugoya Tunnel on the old Kuriko National Expressway, 1969

A series of tests was carried out in this 384 m long tunnel, involving three light vans and three sizes of fuel pool (2, 4 and 6 m^2), with 50–300 litres of petrol (Stroeks, 2001). Jet fans were temporarily installed in the tunnel, and the tests were carried out with natural ventilation and forced ventilation at 3 m/s. Tests were carried out with and without sprinklers, which were temporarily installed at 4 m intervals along a length of 36 m, of which 12 m was upstream of the fire location. Water pressure was 2.9 bar (3 kg/cm^2) and the water discharge rate was 95 l/min.

The main results regarding the performance of the sprinkler system were as follows.

- There was a rapid reduction in temperature at all measured locations. No zones of very high temperature were measured after system activation.
- Vertical extension of the flames was hindered by the sprinkler action, and the flames tended to elongate in the horizontal direction.
- The sprinklers extinguished open fires, but were unable to extinguish fires inside or under vehicles.
- Sprinklers reduced the size of petrol fuel fires, but did not manage to extinguish them.

8.5.1.2 The New Tomei Expressway, March 2001

The New Tomei Expressway tunnel is three lanes wide. However, the sprinkler system was installed in only one corner, contrary to common Japanese practice for three-lane tunnels. This had been fitted with custom-built nozzles to generate a uniform spray across all three lanes. The spray zone was 36 m long, and fire detectors were installed at 12 m intervals. The road was 8 m wide, resulting in a spray surface of 288 m^2. Fire tests were carried out with fuel pans and passenger cars (Stroeks, 2001).

Table 8.4 The performance of sprinkler heads under different pressures and at different installation heights in the New Tomei Expressway tunnel

Water pressure: bar	Spray volume: l/min	Spray angle: °	Installation height: m	Spray diameter: m	Water-droplet diameter: mm	Spray velocity: m/s
1.9	14.7	90	0.5	0.9		11.3
			1.0	1.6	0.8–1.5	10.1
			1.5	2.1		
2.9	18.0	90	0.5	0.9		13.9
			1.0	1.6	0.5–1.0	12.1
			1.5	2.1		
3.9	20.8	90	0.5	0.9		15.5
			1.0	1.6	0.3–0.7	
			1.5	2.0		

The main results were as follows:

- The sprinkler system was able to suppress the 23 MW fire produced by a $9 \, m^2$ fire pan.
- In a test with multiple cars and a ventilation velocity of 5 m/s, the fire damage was confined to three passenger cars, one on each side of the burning vehicle.

The visibility during operation of the Japanese tunnel sprinkler systems is reported to be about 10 m. The droplet-size distribution and water flow rates under different pressures (2–4 bar) are shown in Table 8.4. The droplets were mainly in class 3 (see Table 8.1), and provided good coverage and uniformity in cross-sections.

8.5.1.3 The second Benelux tunnel near Rotterdam, 2002

A series of tests was carried out in the newly built second Benelux tunnel near Rotterdam (DGPWWM, 2002). The tunnel is an immersed-tube tunnel consisting of three tubes for traffic, one tube for pedestrians and cyclists, and two tubes for the Rotterdam metro. The tests were carried out in one of the traffic tubes, which has a rectangular cross-section and is 5 m high and about 10 m wide. The tunnel length is about 900 m, and the test site was located 265 m from the exit portal. The tunnel is unidirectional and is equipped with longitudinal ventilation and escape doors every 100 m. A section of the tunnel was protected for the tests, and the tunnel was refurbished afterwards, before reopening the tunnel to traffic. Four tests with sprinklers were carried out with heat-release rates ranging from about 15 to 35 MW. In one test the fire load was a transit van loaded with wooden pallets, in two tests it was wooden pallets under an aluminium covering (taken to represent a HGV trailer), and in the final test it was uncovered wooden pallets. Two sprinkler sections were temporarily fitted, one above the zone of the seat of the fire (17.5 m long) and one above an adjacent zone (20 m long) downstream. Three rows of traditional sprinkler heads arranged at 2.5 m intervals were fitted on the ceiling. The spray density on the floor was designed as $12.5 \, l/m^2$ per minute (i.e. mm/min). In three of the four tests, the activation of the deluge system was significantly delayed (10, 14 and 22 minutes) after ignition of the fuel. Each of these tests was designed to investigate the effect of the deluge system on the vehicles and the environment around the fire, and not to investigate the suppression effect of the deluge. Different ventilation conditions were used in each of the four tests.

The cooling effects of the deluge system were clearly demonstrated in these tests. The fire-suppression effects were not discussed in the report. It was observed that the enclosed fires were not extinguished by the deluge systems. The tests also made observations on visibility and smoke layer destratification.

8.5.2   Water-mist systems

Over the past decade a number of tunnel fire test series have been carried out to investigate the capabilities of water-mist systems in suppressing tunnel fires. Many of these tests have been carried out confidentially by private companies, with only a small percentage of the test data being presented in the public domain, primarily as marketing material. Thus it is to be expected that the publicly released data will generally present the success stories and may mask those occasions when water-mist systems were not able to suppress fires.

8.5.2.1   IF Assurance tunnel, Norway, 2002/2004

The IF Assurance tunnel, near Hobøl, south of Oslo, is primarily a training tunnel used for training of fire-fighters and similar activities. It is of conventional shape, 100 m long, 8 m wide and 6 m high at the apex. It has a cross-sectional area of about 40 m². The tests carried out are not described in detail in the public literature. They involved three different types of fire: (i) uncovered liquid pool fires, (ii) partially covered liquid pool fires, and (iii) partially covered wooden pallet fires. Nineteen fire tests were carried out with a low-pressure water-mist system, and 56 fire tests with a high-pressure water-mist system. Eight tests were carried out with no suppression. Airflow speeds of up to 3 m/s were used in the tests (Kratzmeir, 2006).

In general, the water-mist systems had a better suppressing effect on pool fires than on pallet loads. For the pallet-load fires, it is claimed that the low-pressure water mist reduced the heat-release rate by 40%, while the high-pressure water mist performed better, reducing the heat-release rate by 50–80%.

The principal observations from this study appear to be that water mists are not very efficient at extinguishing small fires (i.e. less than 5 MW) but are able to suppress larger fires down to lower burning rates. It was reported that visibility was drastically reduced (down to less than 1 m) in the early stages of activation, but was better after a few minutes. Downstream temperatures were greatly reduced by the activation of the water-mist systems.

8.5.2.2   Virgil/Virgolo tunnel of the Brenner motorway, Italy, 2005

The tunnel is an operational two-lane tunnel on the Brenner motorway in northern Italy. The tunnel is of standard shape, has an internal diameter of 5 m, is 6.5 m high at its apex and is 860 m long. Three full-scale fire tests were carried out in the tunnel in February 2005 as part of the UPTUN project (UPTUN, 2005). The tests were designed to investigate multiple research topics and technologies. Only Test 2 was designed to test the suppression capabilities of a water-mist system.

The tunnel was fitted with 28 thermocouples distributed from 90 m upstream of the fire to 90 m downstream. All the thermocouples were mounted near to the tunnel walls or ceiling. The temperature within the concrete was measured. Gas properties were analysed using a single CO gas analyser, mounted on the tunnel wall 40 m downstream of the fire location. Two anemometers were used to record the airflow. The water-mist system was installed by Fogtec/Semco across a 30 m section of tunnel, straddling the fire location. The working pressure of the water-mist system was 80–120 bar.

In Test 2 (nominally 20 MW), the fire load was four 1.5×1.0 m fuel pools, each containing 70 litres of diesel fuel. The fire was allowed to burn with no suppression for 3 minutes, and then the water-mist system was activated. This resulted in an immediate reduction in the gas temperature above the fire location from about 180°C to below 50°C. However, within a further 3 minutes the gas temperature had increased to over 200°C. The water-mist system was active for 5 minutes, after which time it was switched off and the fire burned itself out within another 5 minutes. The airflow in the tunnel was forced using jet fans and remained at about 2–2.3 m/s throughout the test.

From the details presented in the report it is not apparent that the water-mist system had any significant influence on suppressing the fire itself. It appears to have only had a transient beneficial effect on the temperature distribution in the tunnel, and this coincided with an increase in carbon monoxide levels within the tunnel. Throughout the test, the temperatures within the concrete did not reach 100°C, so the capabilities of the water-mist system as a structural protection system were demonstrated.

8.5.2.3 VSH Hagerbach, Switzerland, 2005

The Hagerbach facility in Switzerland consists of a number of tunnels, galleries and caverns of varying shape and size cut into rock. For the tests carried out in 2005, a 150 m length of one of the galleries was fitted with a false ceiling and a smooth concrete floor, to reproduce the geometry of the car-only section of the A86 ring road tunnel, near Paris (2.55 m high). The rough rock walls of the tunnel were excavated and smoothed to give the same approximate width as the A86 tunnel (9.3 m). The tunnel was fitted with a large number of sensors, primarily for airflow measurement upstream of the fire zone, temperature and heat flux within the fire zone, and opacity and gas analysis downstream. The heat-release rate was calculated by enthalpy flow and from the gas concentration (Tuomisaari, 2008).

Nineteen fire tests were carried out (Guigas et al., 2005), ten of them with Marioff's HI-FOG system (Marioff and HI-FOG are registered trademarks of Marioff Corporation Oy) (Tuomisaari, 2008). HI-FOG water mist is generated by a high pressure of up to 140 bar if the system is powered by a constant-pressure electric or diesel pump, or up to 200 bar if powered by a pressurised gas cylinder. Most of the droplet sizes were below 200 μm. The fuel loads used in the tests were various configurations of cars, with a range of water-mist activation times used. A full description of the test configurations is not in the public domain.

For most tests, the simulated fire scenario was a collision of two or more cars, which resulted in either a fire in a single vehicle, or a fire involving several vehicles in either a two-lane or three-lane arrangement. Other cars were positioned around the simulated crashed vehicles. The fire in each case was started in the engine compartment of one of the cars. Real cars in working condition were used, and had warm engines and fuel tanks partially filled with petrol. The water-mist system was generally activated early in the test. During all the tests, the tunnel wind speed was 6 m/s before ignition, and then reduced to 3 m/s. In most instances the water-mist system appears to have had a positive influence on the tenability in the tunnel, and prevented the fire from spreading between vehicles.

Test C is summarised in one paper (Guigas et al., 2005). In this test, three cars were arranged in a collision configuration, and the engine compartment of one of the cars was set on fire. Various other cars were positioned close to the incident cars. The water mist was operated early, and no fire spread to the adjoining vehicles (i.e. those supposedly involved in the collision) was observed. At 22 minutes after ignition, the water mist system was switched off. This was followed almost immediately by the fire spreading to both the adjoining vehicles, and the combined heat-release rate of the vehicle fires grew sharply from about 2 MW to over 15 MW within 4 minutes. The water mist was switched on again and, after briefly peaking at about 21 MW, the heat-release rate dropped to about 12 MW. However, while the water mist was active, the fire was observed to spread to two of the adjacent (but non-incident) vehicles, causing the total heat-release rate to grow to over 15 MW once again. The water mist was maintained, and the fires eventually burned out.

8.5.2.4   San Pedro de Anes test tunnel, Spain, 2005–2006

The San Pedro de Anes test tunnel in northern Spain was used for a series of tests in the SOLIT project, and also for various tests done for the Madrid M30 ring road tunnel project. The tunnel itself is 600 m long, 9.5 m wide and 8.12 m high. A false, flat ceiling can be installed at a height of 5.17 m above the roadway if required.

According to the website of Tunnel Safety Testing SA (the company that runs the tunnel), Marioff carried out 'more than 35' full-scale fire tests in the facility between December 2005 and February 2006, and FOGTEC carried out a 'large testing campaign' from March to May 2006. The FOGTEC tests were part of the SOLIT project (Starke, 2010).

Marioff tests

Full details of the tests carried out by Marioff are not in the public domain, and are likely to remain confidential. What little public information there is on these tests comes from Marioff themselves and their collaborators, and is therefore unlikely to report any negative findings from the tests, if there were any.

Marioff report carrying out 'about 40 tests in total', with the fuel load in most of the tests being constructed of stacks of wooden pallets. In some tests, some plastic pallets were included also. Marioff estimate the total *potential* heat-release rate of these fire loads to be over 75 MW for the arrangements of wooden pallet stacks, and over 95 MW for the wooden and plastic pallet stacks. In some tests a (non-fire-retardant) polypropylene tarpaulin was used to cover the load. The ventilation for these tests was longitudinal, and was between 2 and 3 m/s (Tuomisaari, 2008).

In the majority of tests, the fires were allowed to grow to between 15 and 20 MW before the water-mist system was activated. The water mist was then active for half an hour before it was switched off and the fire service entered the fire zone to extinguish the fire. The published paper only details two of the tests, one with an uncovered fuel load of wooden pallets, suppressed with a 'hybrid' water-mist system, and one with a mixed load covered with a tarpaulin, with a 'zoned' water-mist system.

TEST WITH UNCOVERED WOODEN PALLETS

In this test, the water-mist system was activated at 5 minutes 40 seconds after ignition, at which time the heat-release rate was about 20 MW. In the 'hybrid' water-mist system used in this test, half the sprinkler heads produced mist immediately on system activation, and the other half were only activated if a certain (unspecified) temperature was reached at the sprinkler head. This configuration meant that all the unburned fuel at the time of activation was soaked with water. The water mist halted the growth of the fire, holding it at about 20 MW for about

20 minutes, after which time the heat-release rate began to slowly diminish. The fire did not spread to a target object, 5 m downwind of the fire location. It would appear that, in this instance, the water mist halted the fire growth, but did not reduce the fire size in any way (there was no reduction in the heat-release rate for 20 minutes, and there was even a slight increase in the heat-release rate after about 10 minutes).

Another paper (Mawhinney and Trelles, 2008) refers to a different test with a similar fuel load but only a 'minimal application of water mist'. In this test, the fire grew to a peak heat-release rate of over 55 MW.

TEST WITH A COVERED, MIXED LOAD

The water-mist system in this test used purely temperature-activated nozzles. The first nozzle activated after 2 minutes, when the heat-release rate was a little below 10 MW. Within the next 4 minutes the fire grew to over 30 MW, and another five nozzles were activated. A further ten nozzles activated shortly after this. The activation of the first six nozzles does not seem to have influenced the development of the fire significantly, but the growth rate of the fire was slowed slightly when all 16 nozzles were active. However, the fire continued to grow, attaining a peak heat-release rate of over 60 MW some 12 minutes or so after the first nozzle was activated. After this time the fire diminished in size over a period of 20 minutes. The fire did not spread to the target.

It would appear that the water mist did no more than slightly slow the growth of the fire. If, as several dictionaries define it, 'suppression' is taken to mean 'halt the growth of' or 'reduce the size of', then no suppressing effect was observed in this test.

Summary of the Marioff tests

Despite the fact that both the above tests failed to meet the success criteria (the first test failed because the downstream temperatures exceeded 50°C at the specified location, the second test failed because the fire consumed more than 90% of the fuel load and the downstream temperatures exceeded 50°C), the Marioff paper (Tuomisaari, 2008) still presents these tests as successful, as the fire did not spread to the target object in either test. Indeed, the paper states: 'in the two large fire test programs [referring also to the Hagerbach tests for the A86 tunnel] the fire propagation was stopped and the fire suppressed during system operation'.

FOGTEC tests for the SOLIT project

More than 50 full-scale tests were carried out at the San Pedro de Anes test tunnel for the SOLIT project (Kratzmeir and Lakkonen, 2008). Few of the results are available in the public literature, but the estimated heat-release rate graphs for two of the tests have been published. These tests both involved a large fuel load of wooden pallets, similar to the tests carried out by Marioff. The exact number and configuration of wooden pallets in the fire load is not stated. The two tests differ because one was carried out with a tarpaulin over the fire load, and the other was carried out without a tarpaulin.

In both instances the water-mist system was activated about 4 minutes after ignition, when the heat-release rate was less than 10 MW. In the next 8 minutes of fire growth, the difference between the two tests is not significant; in both tests the fire grew from about 10 MW to about 35–40 MW with the water-mist system active. Beyond this point, the water-mist system appears to have controlled the fire load without the tarpaulin, causing a slow but steady reduction in the heat-release rate from 35 MW to about 20 MW over 25 minutes. The fire in the case of the fire load with a tarpaulin continued to grow in severity up to about 55 MW over 10 minutes, although the rate of growth to this point was slower than it was up to 35 MW.

A reference test done without sprinklers was also carried out, and in this case the fire was extinguished after 8 minutes, by which time it had grown to about 25 MW. By comparison with this reference test, it is clear that the water-mist system did slow down the fire development, but the fire continued to grow in size considerably (to almost four times the heat-release rate) after activation of the water-mist system.

Once again, these results are presented as being positive, as the fire did not spread to the adjacent target object and because the conditions were such that the fire brigade could get close enough to the fire to successfully fight the fire with the water-mist system in operation.

8.5.2.5 Runehamar Tunnel, Åndalsnes, Norway, 2007–2008

A series of five large-scale tests was carried out in December 2007 and January 2008 in the Runehamar Tunnel near Åndalsnes, Norway (Lemaire and Meeussen, 2008, 2010). The tests were commissioned by Rijkswaterstaat, the department responsible for tunnel safety within the Ministry of Public Works of The Netherlands. The series consisted of three pool fire tests and two tests with wooden pallets, with four of the tests having an estimated unsuppressed potential heat-release rate of 200 MW. The two main objectives of the tests were to determine the suppression and extinguishing effect of a water-mist system on fully developed fires, and to investigate the risk of a fuel tanker undergoing a BLEVE (boiling liquid, expanding vapour explosion) in the locality immediately downwind of the fire. The tests were carried out by SINTEF NBL, Aquasys and Efectis. The water-mist nozzles were arranged in three zones, with a total spray length of 75 m. In some tests, foaming agents were added to the water supplied to the nozzles. The experimental data were also used to assess the tenability conditions downwind of the fire.

As in the water-mist test series described above, the ability of the water-mist system to reduce the temperatures in the locality of the fire and downwind was clearly demonstrated. However, one of the negative aspects of water-mist systems is clearly highlighted by this set of test results. While in the three tests involving pool fires the fire was rapidly suppressed and extinguished by the water-mist system, in the two tests with wooden pallets it was not, and this resulted in very high ('lethal') levels of carbon monoxide for hundreds of metres downstream of the fire for an extended period of time (up to the end of the test period).

The tests demonstrate the ability of water-mist systems to protect fuel tankers from the risk of BLEVE, assuming that the system can be operated early. The only risk of BLEVE in any of the tests was identified when the activation of the water-mist system was delayed until 7 minutes after ignition.

8.6. Evaluation of fixed fire-fighting systems for tunnels

Some consistent outcomes can be drawn from the large-scale tests described above. The cooling effect of the water spray on the tunnel environment in the presence of a fire was confirmed in all the large-scale tests, for both traditional sprinkler and water-mist systems. The cooling effect appears to be achieved by the water both blocking the heat radiating from the fire and by the water directly cooling the air. Thus, these suppression systems are able to offer a degree of protection to the tunnel structure and also to people and vehicles in the vicinity of the fire. The ability of

FFFSs to protect adjacent fuel tankers from the risk of BLEVE has been demonstrated, provided the system is activated early enough.

It is generally observed that sprinkler and water-mist systems are able to extinguish or effectively suppress open fires, such as pools of fuel (e.g. fuel spillages).

In Japan, the objectives of deploying sprinklers in tunnels are stated clearly as: to cool the environment; to prevent fire spread; to protect the tunnel structure, and to protect tunnel equipment. While the achievement of these objectives is generally confirmed by the large-scale tests, there remain uncertainties and unknowns with FFFSs, particularly with regard to life safety.

Although there are as yet no common standards for the deployment of FFFSs in tunnels, it is commonly understood that extinguishing the fire itself may not be achievable with such systems, and the role of the FFFS is thermal management in the tunnel. The final extinction of the fire should be provided by mobile fire-fighting. Therefore, the role of FFFSs in tunnels has more to do with fire control than fire suppression.

From a thermal management point of view, both traditional deluge systems and water-mist systems are effective in cooling the air temperature, and sometimes in reducing the fire size. While both water-mist and traditional sprinkler systems are effective in providing fire protection for tunnels, the water-mist system seems, currently, to be preferred on the grounds that it uses significantly less water, and thus smaller pipes and reservoirs are required.

Water-mist has better suppression effects on pool fires, and can extinguish pool fires and oil flow fires. This can be attributed to the direct cooling effect of the water mist on the fire plume above the pool, thus suppressing the fire by reducing the radiant heat transfer to the pool, thus reducing the rate of production of gaseous fuel.

However, water mist appears to be less effective on solid fuels than are traditional sprinkler systems. This is mainly due to the ability of the large drops of water from a traditional sprinkler to penetrate the fire plume and provide direct wetting and cooling of the solid fuel surfaces.

8.7.   Questions remaining

With the exception of the test series in the Hagerbach tunnel, and one or two tests done using conventional sprinklers (in Japan and The Netherlands), the vast majority of suppression tests in tunnels have used open fuel pools or piles of wooden pallets. One question that must be raised is whether these tested fuel loads are in any way representative of real tunnel fire scenarios.

For example, in the Mont Blanc Tunnel fire incident in 1999 the initial truck load on fire was a refrigerated trailer containing margarine and flour, and in the St. Gotthard Tunnel incident in 2001 one of the initial cargo loads involved was a trailer load of rubber tyres. To what extent are such fuel loads represented by piles of wooden pallets? It is currently unknown if either water-mist systems or traditional sprinklers could have dealt with these incidents, yet it is exactly these types of incident that FFFSs are intended to prevent.

Much of the research on suppression systems has focused on open pools and wooden pallets, under ventilation conditions with air speeds generally not higher than 3 m/s, so it may be reasonable to say that we understand the capabilities of FFFSs under these conditions and for these kinds of fuel loads. What is of greater interest for future research is to identify the conditions under which FFFSs could 'fail' in some way. For example, how large does an initial fire have to be before it will spread to an adjacent vehicle, even with a suppression system active? Or, what kind of HGV trailer constructions reduce the effectiveness of suppression systems?

One of the main areas of research still needing investigation is the influence of longitudinal ventilation on the effectiveness of fire suppression. Are water-mist systems less effective with a longitudinal air flow of 6 m/s than they are with a longitudinal flow of 3 m/s? Are traditional sprinkler systems better than water-mist systems at higher ventilation velocities? Is there an optimal ventilation flow for effective fire suppression?

Very small mist droplets are highly susceptible to longitudinal ventilation. Simple modelling studies have shown that under 'emergency' ventilation conditions, a large proportion of water mist droplets could be carried hundreds of metres downstream in the tunnel before they reach the road level (Rein *et al.*, 2008). This suggests that there may be a fundamental incompatibility between higher ventilation rates and fine-water-mist suppression systems.

Another of the biggest unanswered (and rarely asked) questions is, for a given tunnel configuration (length, slope, geographical location, number of lanes, ventilation conditions, etc.), which is better, a traditional sprinkler or a water-mist system? At the time of writing, the preferred option seems to be water-mist systems, but this appears to be based largely on economics and on the marketing strategies of companies promoting water mist, not on any scientific basis. Research should be aimed at investigating the most effective suppression systems for tunnel applications.

Finally, the question of whether or not FFFSs should be considered as life safety devices still remains to be adequately addressed. It is clear that water-mist systems (and, to a lesser degree, traditional sprinkler systems) can substantially increase the levels of carbon monoxide produced by a fire (especially if the fire is not extinguished), while effectively reducing visibility to only a few metres in the operational zones. This will clearly hinder egress, and may endanger people trapped downwind of the fire location. Therefore, from a life-safety perspective, is it desirable to delay the activation of a FFFS until all (or most) of the tunnel users have evacuated the locality around the fire? This question has been raised at international safety forums on a number of occasions, without having ever been adequately answered. The current thinking seems to be that early activation of a FFFS should control or minimise a fire, and thus minimise the risk to escaping tunnel users. This strategy was clearly successful in the 2007 incident in the Burnley Tunnel, but there is little other historical evidence to support or refute this position. As far as the authors are aware, no water-mist system in a tunnel has yet been deployed during a fire incident.

8.8. Outlook

The benefits of FFFSs in tunnels have been clearly demonstrated, under a narrow range of conditions, in large-scale tests. However, the FFFS technologies for tunnel safety are still in their infancy, and there are not yet any generally accepted guidelines or standards for system design.

Society seems to have decided that FFFSs are an acceptable and, indeed, desirable fire-protection measure in tunnels, and many of the old justifications for not using sprinklers have been debunked or discarded. It is clear that water-based FFFSs will play a much greater role in tunnel fire safety in the future. The challenge for the future is not whether FFFSs should be used, but rather which types of FFFS should be used and in which tunnels. It is hoped that future research will push towards a better understanding of fire suppression, and towards better systems that are designed to work on a wider range of possible fire types.

REFERENCES

Baillis D and Sacadura J-F (2000) Thermal radiation properties of dispersed media: theoretical prediction and experimental characterization. *Journal of Quantitative Spectroscopy and Radiative Transfer* **67**(5): 327–363.

BRE (2005) *Fire Suppression in Buildings using Water Mist, Fog and Similar Systems.* Project Report 213293V3. Building Research Establishment, Watford.

BSI (2005) BS 9251: *Sprinkler Systems for Residential and Domestic Occupancies – Code of Practice.* British Standards Institution, London.

Buchlin J-M (2005) Thermal shielding by water spray curtain. *Journal of Loss Prevention in the Process Industries* **18**: 423–432.

Chow WK, Qin J and Han SS (2005) Bench scale tests on controlling plastic fires with water mists. *Chemical Engineering and Technology* **28**(9): 1041–1047.

Cote AE and Linville JL (eds) (2008) *Fire Protection Handbook*, 20th edn. National Fire Protection Association, Quincy, MA.

DGPWWM (2002) *Project 'Safety Test' – Report on Fire Tests.* Directorate-General for Public Works and Water Management, Civil Engineering Division, Utrecht.

Drysdale DD (2011) *An Introduction to Fire Dynamics*, Third edition. Wiley, New York.

Grant G, Brenton J and Drysdale D (2000) Fire suppression by water sprays. *Progress in Engineering and Combustion Science* **26**: 79–130.

Guigas X, Weatherill A, Bouteloup C and Wetzig V (2005) Water mist tests for the A86 east tunnel. *International Congress on Safety Innovation Criteria Inside Tunnels, Gijón, Spain*, 29 June–1 July 2005, pp. 163–173.

Hua J, Kumar K, Khoo BC and Xue H (2002) A numerical study of the interaction of water spray with a fire plume. *Fire Safety Journal* **37**: 631–657.

Jiang Z, Chow WK, Tang J and Li SF (2004) Preliminary study on the suppression chemistry of water mists on poly(methyl methacrylate) flames. *Polymer Degradation and Stability* **86**(2): 293–300.

Kim SC and Ryou HS (2004) The effect of water mist on burning rates of pool fire. *Journal of Fire Sciences* **22**(4): 305–323.

Kratzmeir S (2006) Wassernebelanlagen – Innovativer Brandschutz für Tunnel (Water mist systems – innovative fire protection for tunnels). *Technische Überwachung* Jan./Feb.: 10–13.

Kratzmeir S and Lakkonen M (2008) Road tunnel protection by water mist systems – implementation of full scale fire test results into a real project. *Proceedings of the 3rd International Symposium on Tunnel Safety and Security, Stockholm, Sweden*, 12–14 March 2008, pp. 195–203.

Lawson JR, Walton WD and Evans DD (1988) *Measurement of Droplet Size in Sprinkler Sprays.* NBSIR 88-3715. National Institute of Standards and Technology, Gaithersburg, MD.

Lemaire A and Meeussen V (2008) *Effects of Water Mist on Real Large Tunnel Fires: Experimental Determination of BLEVE Risk and Tenability during Growth and Suppression.* Report 2008-Efectis-R0425. Efectis Nederland BV, Rijswijk.

Lemaire A and Meeussen V (2010) Experimental determination of BLEVE risk near very large fires in a tunnel with a sprinkler/water mist system. *Proceedings of the 4th International Symposium on Tunnel Safety and Security, Frankfurt am Main, Germany*, 17–19 March 2010, pp. 143–152.

Liu Z and Kim AK (1999) A review of water mist fire suppression system – fundamental studies. *Journal of Fire Protection Engineering* **10**: 32.

Liu Z, Kim AK and Su JZ (2001) Examination of performance of water mist fire suppression system under ventilation conditions. *Journal of Fire Protection Engineering* **11**: 164.

Liu Z, Carpenter D and Kim AK (2006) Characteristics of large cooking oil pool fires and their extinguishment by water mist. *Journal of Loss Prevention in the Process Industries* **19**(6): 516–526.

Liu Z, Carpenter D and Kim AK (2008) Cooling characteristics of hot oil pool by water mist during fire suppression. *Fire Safety Journal* **43**(4): 269–281.

Mawhinney JR and Trelles J (2008) The use of CFD-FDS modeling for establishing performance criteria for water mist systems in very large fires in tunnels. *Proceedings of the 3rd International Symposium on Tunnel Safety and Security, Stockholm, Sweden*, 12–14 March 2008, pp. 29–42.

NFPA (2007a) NFPA 13: *Standard for the Installation of Sprinkler Systems*. National Fire Protection Association, Quincy, MA.

NFPA (2007b) NPFA 15: *Standard for Water Spray Fixed Systems for Fire Protection*. National Fire Protection Association, Quincy, MA.

NFPA (2010) NFPA 750: *Standard on Water Mist Fire Protection Systems*. National Fire Protection Association, Quincy, MA.

PIARC (1999) *Technical Report on Fire and Smoke Control in Road Tunnels*. Report 05.05.B. PIARC (World Road Association), La Defense, France.

PIARC (2007) *Technical Report on Systems and Equipment for Fire and Smoke Control in Road Tunnels*. Report 05.16.B. PIARC (World Road Association), La Defense, France.

PIARC (2008) *Technical Report on Road Tunnels: Assessment of Fixed Fire Fighting Systems*. Report 2008R07. PIARC (World Road Association), La Defense, France.

Ravigururajan TS and Beltram MR (1989) A model for attenuation of fire radiation through water droplets. *Fire Safety Journal* **15**: 171–181.

Rayleigh JWS (1878) On the instability of jets. *Proceedings of the London Mathematical Society* **x**: 4–13.

Rein G, Carvel RO and Torero JL (2008) Study of the approximate trajectories of droplets from water suppression systems in tunnels. *Proceedings of the 3rd International Symposium on Tunnel Safety and Security, Stockholm, Sweden*, March 2008, pp. 163–171.

Sacadura JF (2005) Radiative heat transfer in fire safety science. *Journal of Quantitative Spectroscopy and Radiative Transfer* **93**(1–3): 5–24.

Starke H (2010) Fire suppression in road tunnel fires by a water mist system – results of the SOLIT project. *Proceedings of the 4th International Symposium on Tunnel Safety and Security, Frankfurt am Main, Germany*, 17–19 March 2010, pp. 311–321.

Stroeks R (2001) *Sprinklers in Japanese Road Tunnels*. Project Report BFA-10012. Bouwdienst Rijkswaterstaat, Directorat-Generaal Rijkswaterstaat, Ministry of Transport, The Netherlands.

Thomas S (1984) Revolution in sprinkler technology. *Building Standards* **53**(3): 6–8.

Tseng CC and Viskanta R (2007) Absorptance and transmittance of water spray/mist curtains. *Fire Safety Journal* **42**: 106–114.

Tunnel Safety Testing SA (2011) San Pedro de Anes, Spain. Available at: www.tunneltest.com (accessed 22 June 2011).

Tuomisaari M (2008) Full scale fire testing for road tunnel applications: evaluation of acceptable fire protection performance. *Proceedings of the 3rd International Symposium on Tunnel Safety and Security, Stockholm, Sweden*, 12–14 March 2008, pp. 181–193.

UPTUN (2005) *Test Report – Virgl/Virgolo Tunnel*. Report No. 875-05-004. Department of Structural Engineering and Natural Hazards, University of Natural Resources and Applied Life Sciences, Vienna.

van Wingerden K (2000) Mitigation of gas explosions using water deluge. *Process Safety Progress* **19**(3): 173–178.

Wang XS, Wu XP, Liao GX, Wei YX and Qin J (2002) Characterization of a water mist based on digital particle images. *Experiments in Fluids* **33**: 587–593.

Widmann JF, Sheppard DT and Lueptow RM (2001) Non-intrusive measurements in fire sprinkler sprays. *Fire Technology* **37**: 297–315.

Williams C and Jackman L (2006) *An Independent Guide on Water Mist Systems for Residential Buildings*. Building Research Establishment, Watford.

Yang W, Parker T, Ladouceur HD and Kee RJ (2004) The interaction of thermal radiation and water mist in fire suppression. *Fire Safety Journal* **39**: 41–66.

Zhang Y, Zhu W, Zhou X, Qin J and Liao G (2007) Experimental study of suppressing PVC fire with water mist using a cone calorimeter. *Journal of Fire Sciences* **25**(1): 45–63.

Handbook of Tunnel Fire Safety, 2nd edition
ISBN: 978-0-7277-4153-0

ICE Publishing: All rights reserved
doi: 10.1680/htfs.41530.153

publishing

Chapter 9
Tunnel ventilation: state of the art

Art Bendelius
A & G Consultants, Inc., USA

9.1.   Introduction

Webster's dictionary defines ventilation simply as 'circulation of air'. Ventilation does not necessarily mean the use of mechanical devices such as fans being employed: non-fan or natural ventilation is still considered to be ventilation. From that simple definition of ventilation, we move forward to the ventilation of tunnels. The use of tunnels dates back to early civilisations, and so too does ventilation in the form of natural ventilation. However, the ventilation of tunnels has taken on greater significance within the past century, due to the invention and application of steam engines and internal combustion engines, which are prevalent as motive power in the transport industry. This all emerged as increasing quantities of combustion products and heat became more troublesome to the travelling public.

Exposure to the products of combustion generated by vehicles travelling through a tunnel can cause discomfort and illness to vehicle occupants. Ventilation became the solution, by providing a means to dilute the contaminants and to provide a respirable environment for the vehicle occupants. Visibility within the tunnel will also be aided by the dilution effect of the ventilation air.

In the past quarter century, great concern has arisen regarding the fire life safety of vehicle occupants in all transport tunnels. Much effort has been made to improve the fire life safety within tunnels, thus focusing more attention on the emergency ventilation systems installed within tunnels

The use of the term 'tunnel' in this chapter refers to all transportation-related tunnels, including road tunnels, transit (metro, underground or subway) tunnels and railway tunnels.

Road tunnels, from a ventilation viewpoint, are defined as any enclosure through which road vehicles travel. This definition includes not only those facilities that are built as tunnels but those that result from other construction such as the development of air rights over roads. All road tunnels require ventilation, which can be provided by natural means, traffic-induced piston effects and mechanical ventilation equipment. Ventilation is required to limit the concentration of obnoxious or dangerous contaminants to acceptable levels during normal operation and to remove and control smoke and hot gases during fire-based emergencies. The ventilation system selected must meet the specified criteria for both normal and emergency operations, and should be the most economical solution considering both construction and operating costs.

The portions of transit (metro) systems located below the surface in underground structures most likely will require control of the environment. In transit (metro) systems, there are two types of tunnel: the standard underground transit (metro) tunnel, which is usually located between stations and normally constructed beneath surface developments with numerous ventilation shafts and exits communicating with the surface; and the long tunnel, usually crossing under a body of water, or through a mountain. The ventilation concepts for these two types will be different, since in the long tunnel there is usually limited ability to locate a shaft at any intermediate point, as can be accomplished in the standard underground transit (metro) tunnel. The characteristics for a long transit tunnel will be similar to the ventilation requirements for a railway tunnel.

Ventilation is required in many railway tunnels to remove the heat generated by the locomotive units and to change the air within the tunnel, thus flushing the tunnel of pollutants. Ventilation can take the form of natural, piston effect or mechanical ventilation. While the train is in the tunnel, the heat is removed by an adequate flow of air with respect to the train, whereas the air contaminants are best removed when there is a positive airflow out of the tunnel portal.

9.1.2 Early ventilation concepts

The earliest evidence of the serious consideration of ventilation appeared in transit (metro) tunnels, where the ventilation of transit (metro) tunnels was accomplished by utilising the piston effect generated by moving trains and by constructing large grating-covered openings in the surface, sometimes called 'blow-holes', thus permitting a continuous exchange of air (when trains were running) with the outside and subsequently lowering the tunnel air temperature. However, in the early part of the twentieth century, when the air temperatures in the tunnels began to rise in both London and New York, mechanical means of ventilation (fans) began to be employed.

One of the first formal ventilation systems in a road tunnel was in the Holland Tunnel (New Jersey–New York) in the 1920s (Singstad, 1929). As a part of the planning and design process for the Holland Tunnel, a series of innovative tests were conducted by the US Bureau of Mines (Fieldner et al., 1927). The tests conducted included

- full-scale tests to determine the carbon monoxide generation rate
- tests on people to determine carbon monoxide criteria
- scale-model tests to develop the coefficients for the pressure-loss equations.

The Holland Tunnel opened to traffic in 1927. The use of mechanical ventilation in road tunnels coincided with the growing concern for the impact of the exhaust gases from internal-combustion-engine-propelled vehicles in road tunnels.

9.2. Types of ventilation systems

There are two basic types of ventilation airflow systems used in transport tunnels: longitudinal and transverse. In the longitudinal systems the air moves longitudinally through the tunnel; while in the transverse systems the air moves both transversely and longitudinally, depending on the type of transverse system employed.

- *Longitudinal.* The airflow is longitudinal through the tunnel and essentially moves the pollutants and/or heated gases along with the incoming fresh air, and provides fresh air at the beginning of the tunnel or tunnel section and discharges heated or polluted air at the

Figure 9.1 Longitudinal ventilation configurations

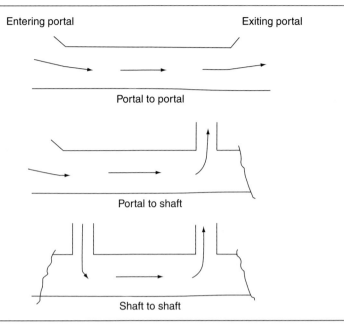

tunnel portal or at the end of the tunnel section (Figure 9.1). Longitudinal ventilation can be configured either portal to portal, portal to shaft or shaft to shaft, as shown in Figure 9.1. The air entering the tunnel is at ambient conditions, and is impacted by the pollution contaminants and the heated gases from vehicles moving through the tunnel, as clearly seen in Figure 9.2. It is longitudinal airflow that is applied most often in transit (metro) and railway tunnels, as the moving trains themselves create longitudinal air flow via what is known as the 'piston effect'.

■ *Transverse*. Transverse flow is created by the uniform distribution of fresh air and/or the uniform collection of vitiated air along the length of the tunnel. This airflow format is used

Figure 9.2 Longitudinal ventilation airflow characteristics

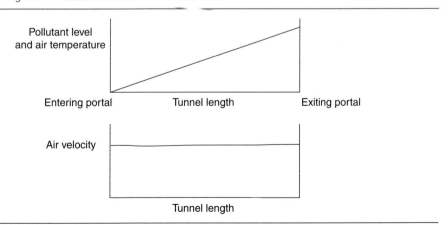

mostly in road tunnels, although it is occasionally used for unique circumstances in transit (metro) tunnels. The uniform distribution and collection of air throughout the length of a tunnel will provide a consistent level of temperature and pollutants throughout the tunnel. The transverse ventilation system can be configured as fully transverse or semi-transverse.

9.2.1   Mechanical versus natural ventilation systems

An evaluation of the natural ventilation effects in a tunnel must determine whether a sufficient amount of the heat and/or pollutants emitted from the vehicles is being removed from the tunnel during normal operations. Mechanical ventilation (or, possibly, cooling) is required if the natural ventilation does not adequately remove the heat. However, the primary thrust of current tunnel ventilation design is tied to the requirement for ventilation during fire-based emergencies.

9.2.2   Natural ventilation

Tunnels that are naturally ventilated rely primarily on meteorological conditions and the piston effect of moving vehicles to maintain satisfactory environmental conditions within the tunnel. The chief meteorological conditions affecting tunnels is the pressure differential between the two tunnel portals created by differences in elevation, ambient temperatures or wind. Unfortunately, none of these factors can be relied upon for continued consistent results. A sudden change in wind direction or velocity can rapidly negate all of these natural effects, including, to some extent, the vehicle-generated piston effect. The natural effects defined above are usually, in the majority of cases, not sufficiently reliable to be considered when addressing emergency ventilation during a fire except in relatively short tunnels or in tunnels with unique potential smoke storage configurations.

Naturally ventilated tunnels can be configured with airflow from portal to portal (Figure 9.3), from portal to shaft or from shaft to shaft. As can be seen in Figure 9.4, the air velocity within the roadway is uniform, and the temperature and pollutant level increase to a maximum at the exit portal or at the end of the section. If adverse meteorological conditions occur, the velocity is reduced, and the temperature and pollutant level are increased.

The benefit of the 'chimney effect' ('stack effect') of the ventilation shaft in a naturally ventilated tunnel is dependent on air temperature differentials, rock temperatures, wind direction and velocity, and shaft height.

A mechanical ventilation system has been installed in many originally naturally ventilated urban road tunnels. Such a system is often required to purge the smoke and hot gases generated during a fire emergency, and may also be required to remove stagnant polluted and heated gases or haze during severe adverse meteorological or stalled traffic conditions.

Figure 9.3 Natural ventilation – portal to portal configuration

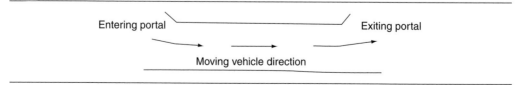

Figure 9.4 Natural ventilation airflow characteristics

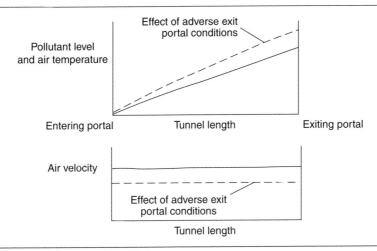

The reliance on natural ventilation for tunnels should be carefully and thoroughly evaluated, specifically the effect of adverse meteorological and operating conditions. If the natural mode of ventilation is not adequate, a mechanical system with fans must be considered. There are several types of mechanical ventilation systems, which are outlined below.

9.3. Mechanical ventilation
9.3.1 Longitudinal ventilation systems

A longitudinal ventilation system is defined as any system where the air is introduced to or removed from the tunnel roadway at a limited number of points, thus creating a longitudinal airflow within the tunnel. There are three distinct types of tunnel longitudinal ventilation systems: those that employ an injection of air into the tunnel from centrally located fans, those that use jet fans mounted within the tunnel cross-section and those that employ a push–pull concept.

The injection-type longitudinal system (Figure 9.5) has been used extensively in railway tunnels; however, it has also found application in road tunnels. Air is injected into the tunnel through a Saccardo nozzle at one end of the tunnel, where it mixes with the air brought in by the piston effect of the incoming traffic and induces additional longitudinal airflow. The air velocity within the tunnel is uniform throughout the tunnel length; and the level of pollutants and/or temperature increase from ambient at the entering portal to a maximum at the exiting

Figure 9.5 Saccardo nozzle longitudinal ventilation system

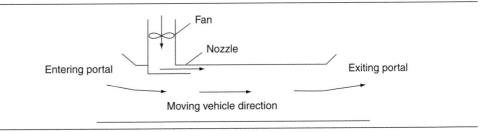

Figure 9.6 Jet fan longitudinal ventilation system

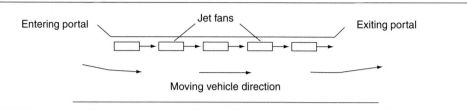

portal. Adverse external atmospheric conditions can reduce the effectiveness of this system. The levels of pollutants and temperature increase as the airflow decreases or the tunnel length increases.

The jet fan longitudinal ventilation concept is based on the installation of a series of axial flow fans (jet or booster fans) in series, within the tunnel, usually mounted at the tunnel ceiling or roof (Figure 9.6). These fans have a high discharge thrust and velocity, which in turn induce additional longitudinal airflow within the tunnel. Although these systems are primarily employed in road tunnels, they have been used in special transit (metro) tunnel situations.

The longitudinal system with a shaft is similar to the naturally ventilated system with a shaft, except that it provides a positive fan-induced stack effect.

The push–pull type of longitudinal ventilation system is employed primarily in transit (metro) applications, where a series of ventilation shafts are constructed connecting the tunnel environment with the ambient environment (Figure 9.7). Reversible fans are installed in these shafts, and ultimately operated to create a longitudinal airflow in the tunnel sections between the shafts. The primary purpose for this mode of operation is usually to control smoke in the transit (metro) tunnel during a fire emergency.

An alternative longitudinal system uses two shafts located near the centre of the tunnel or tunnel section, one for exhausting and one for supplying (Figure 9.8). This configuration will provide a

Figure 9.7 Push–pull longitudinal ventilation system

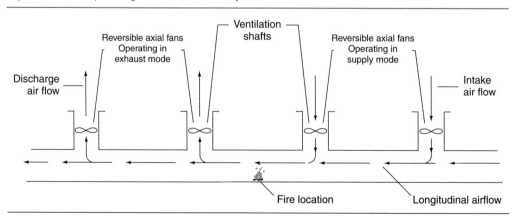

Figure 9.8 Two-shaft longitudinal ventilation system

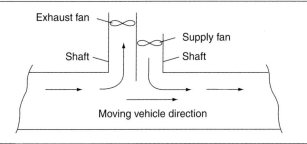

reduction in temperature and the pollutant level in the second half, because a portion of the tunnel airflow is exchanged with ambient air at the shaft. Adverse wind conditions can cause a reduction in airflow and a rise in the pollutant level and temperature in the second half of the tunnel, and 'short-circuiting' of the fan airflows.

9.3.2 Transverse ventilation systems

A transverse ventilation system is defined by the uniform distribution of fresh air and/or uniform collection of vitiated air along the length of the tunnel. There are three system configurations in use: fully transverse, semi-transverse – exhaust and semi-transverse – supply, as described below.

■ *Fully transverse ventilation systems.* A fully transverse ventilation system incorporates a full-length supply duct and a full-length exhaust duct, which achieves a uniform distribution of supply air and a uniform collection of vitiated air (Figure 9.9). This system was originally developed for the Holland Tunnel (New York). A pressure differential between the ducts and the roadway is required to assure proper distribution of air under all ventilation operating conditions. This system has been used primarily in long road tunnels, and has limited application in transit (metro) and railway tunnels.
During a fire, it has been demonstrated, in the Memorial Tunnel Fire Ventilation Test Program (MTFVTP) (Massachusetts Highway Department, 1995), that the dilution provided by the supply element and the extraction provided by the exhaust element of a

Figure 9.9 Fully transverse ventilation system

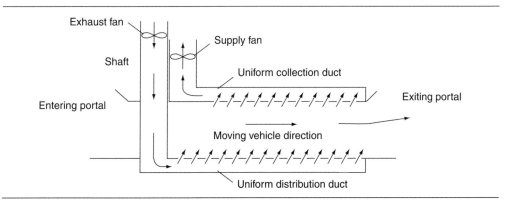

Figure 9.10 Supply semi-transverse ventilation system

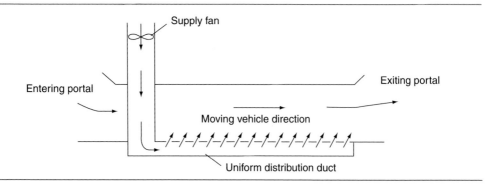

fully transverse ventilation system will not control smoke and heated gases from a large fire. Most long tunnel systems employ the multiple-zone concept, where the ventilation system in the fire zone is placed in exhaust mode and adjacent zones in supply mode, to maximise the longitudinal airflow to limit the movement of smoke.

- *Semi-transverse ventilation systems.* Uniform distribution or collection of air throughout the length of a tunnel is the chief characteristic of a semi-transverse system.
  - A *supply air semi-transverse system* (Figure 9.10) produces a uniform level of pollutants and temperature throughout the tunnel due to the fact that the air and the vehicle-generated pollutants and heat enter the roadway area at the same relative rate. Supply air is transported to the roadway in a duct, and uniformly distributed. During a fire within the tunnel, the air supplied from a semi-transverse system will provide dilution of the smoke. However, to aid in fire-fighting efforts and in emergency egress, the fresh air should enter the tunnel through the portals to create a respirable environment for these activities. For these reasons, the fans in a supply semi-transverse system should be reversible.
  - The *exhaust semi-transverse system* (Figure 9.11) will produce a maximum level of pollutants and temperature at the exiting portal in a tunnel with unidirectional traffic. In a tunnel with bi-directional traffic, the peak level of pollutants will occur somewhere in the middle portion of the tunnel; the exact location will depend on directional traffic volumes and meteorological conditions. In the event of a fire, the exhaust semi-transverse system will extract smoke into the exhaust duct, but it will most likely not control the smoke and heated gases from a large fire.

Figure 9.11 Exhaust semi-transverse ventilation system

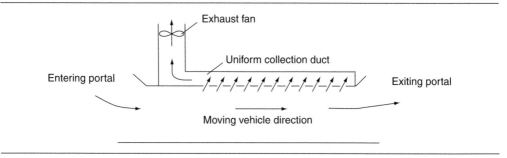

Figure 9.12 Schematic of Sydney Harbour Tunnel ventilation system (Bendelius and Hettinger, 1988)

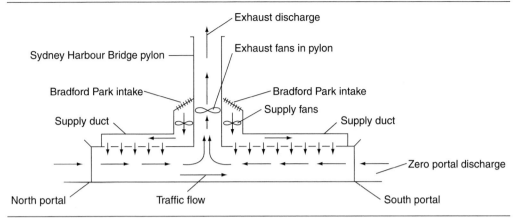

9.3.3   System enhancements

While most of the tunnel ventilation systems developed over the last 50 years fit one of the system descriptions noted above, there have been continuing efforts to examine enhancements, which include the combination of several systems in the same tunnel in addition to some unique concepts. Tunnel ventilation engineers have not been hesitant to seek appropriate solutions for the specific tunnel situation. This is nowhere more evident than in Australia, where stringent criteria based on the concern for the environment near the tunnel portals have forced the tunnel ventilation engineers to use combined systems to minimise the portal emissions. This began in Australia with the design and construction of the Sydney Harbour Tunnel in the late 1980s, where both transverse and longitudinal ventilation systems are employed, as seen in Figure 9.12.

The application of single-point extraction (SPE) to road tunnels has increased since the successful testing noted below. Austria is one of the countries that has used the concept extensively. Figure 9.13 clearly shows the installation of an SPE damper in the ceiling of the Katschberg Tunnel. The Katschberg Tunnel is a 6 km tunnel with two unidirectional bores, and is on the A10 Tauren Route in Austria. A view of the SPE damper from inside the tunnel overhead exhaust air duct is show in Figure 9.14. The damper operator is located in the adjacent parallel overhead supply air duct.

The increased use of extraction systems has been encouraged by the results from the several extensive test programmes conducted, such as the EUREKA Programme (Studiengesellschaft Stahlanwendung, 1995) and the MTFVTP (Massachusetts Highway Department, 1995). Results of the MTFVTP clearly showed that the use of longitudinal ventilation to control the movement of smoke and heated gases was extremely successful; that dilution, as provided by transverse ventilation, had limited success (small fires only); and that extraction was very effective. In the MTFVTP, SPE was tested using both mechanical dampering devices and meltable panels. The tests involving meltable panels were not successful, whereas the tests using mechanical dampering methods were highly successful in demonstrating the effectiveness of SPE, a concept now being employed in many tunnel retrofits. In addition, oversized exhaust ports (expansion of existing normal exhaust ports) were also tested in the MTFVTP, with limited success.

Figure 9.13 View of single point extraction damper in tunnel roadway (Katschberg Tunnel, Austria) (Reproduced by permission of Peter Sturm of the Technische Universität Graz)

Figure 9.14 View of single point extraction damper in tunnel exhaust air duct (Katschberg Tunnel, Austria) (Reproduced by permission of Peter Sturm of the Technische Universität Graz)

9.4.   Ventilation system components

Each tunnel ventilation system is composed of many components, such as fans, dampers, motors and controls. There is also a history of development evident here. The fans considered for use in tunnels have changed: originally, many were centrifugal; however, in the last 30 years the axial flow fan has gained more prominence. This was spearheaded by the increased use of axial flow fans as jet fans in longitudinal ventilation systems and the development of the 100% reversible fan for transit (metro) applications. In addition, the ability to electrically reverse the flow direction of an axial flow fan became of greater significance as the interest in fire and smoke control grew more intense.

9.4.1   Fans

A ventilation fan is a rotary, bladed machine that maintains a continuous airflow created by aerodynamic action. A fan has a rotating impeller carrying a set of blades that exert a force on the air thereby maintaining the airflow and increasing the pressure. A fan is a constant-volume device, since it delivers the same air volume regardless of the air density.

Two basic types of fans, axial and centrifugal, are used predominantly in tunnel ventilation systems. The fan type selected is determined by the required airflow and pressure, the available space, and the tunnel and ventilation building (structure) configuration.

9.4.1.1   Axial flow fan

The flow of air through this fan is virtually parallel to the impeller shaft. The radial component of velocity is nearly zero. The axial fan impeller with blades rotates in a cylindrical housing (Figure 9.15).

Figure 9.15 Axial fan configuration (Bickel *et al.*, 1996)

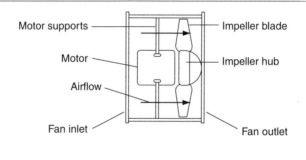

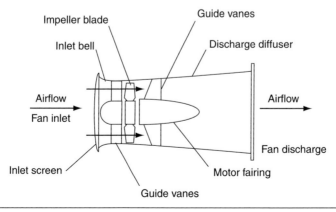

Figure 9.16 Centrifugal fan configuration (Bickel *et al.*, 1996)

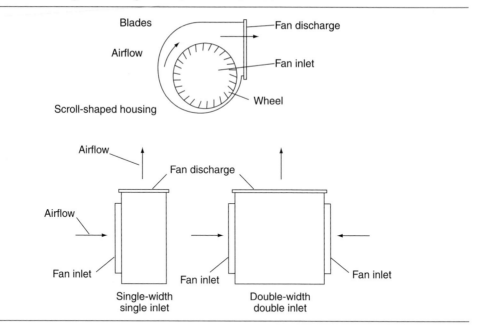

Fans used in tunnel ventilation systems should be constructed to withstand the maximum pressure and temperature anticipated. Flow reversibility is frequently required in a tunnel ventilation system. An axial flow fan can be reversed electrically by reversing the rotation of the motor.

9.4.1.2 Centrifugal fan
A centrifugal fan has a wheel rotating within a scroll-shaped housing or casing (Figure 9.16). The air enters the centrifugal fan parallel to the wheel shaft, and is discharged at 90° to the shaft. Centrifugal fans can be single width or double width.

9.4.2 Dampers
A damper is a device used to control the flow of air in a ventilation system. The control of airflow is accomplished by varying the resistance to flow created by the damper, much as a valve does in a water system. There are two primary damper applications, flow shut-off dampers and smoke dampers, utilised in tunnel ventilation systems. There are two general damper types, those having sliding blades and those having rotating blades. The rotating type can be furnished either with a single blade or with multiple blades. Dampers used in tunnel ventilation systems should be constructed to withstand the maximum pressure and temperature anticipated. Dampers and their operators are particularly critical in SPE applications, as described above, as they are the device that must function if the system is to operate successfully in extracting smoke.

9.4.3 Motors
Most tunnel ventilation fans are driven by electric motors. The selection of a motor for a fan is based on the full-load horsepower requirements, the fan speed and the starting characteristics.

9.4.4   System control
Tunnel ventilation systems can be controlled manually or automatically, and operated remotely or locally. A basic operational philosophy is involved regarding whether there will be a human operator continuously present at the tunnel control centre. Manually operated tunnels require an operator to be present at all times, whereas fully automatic systems can function without the attendance of an operator. However, fully automatic systems are not completely without some human participation, in that a number of system operating conditions must be monitored to prevent serious equipment breakdown.

9.5.   Facilities
There are requirements for facilities to house fans and ancillary equipment and to transport the air in all tunnel ventilation systems. These include ventilation buildings, ventilation structures, ventilation shafts, fan rooms and ducts. The ventilation shafts can be with or without fans. The extent of the facility needed will be determined by the type of ventilation employed and the tunnel configuration. In tunnels employing jet fan longitudinal ventilation systems, there will be no requirement for ventilation structures to house fans, but there may be a requirement to house ancillary systems and equipment.

All of the facilities must be designed to provide ease of inspection and maintenance of the fans and related equipment as well as sufficient access for replacement. Also, moisture on the emergency control dampers may ice up the linkages or controllers during cold weather. The design of these dampers must also include consideration of material strength and resistance to corrosion. Dampers subjected to repeated stresses may be the most sensitive to this problem because of corrosion fatigue.

9.6.   Technology
Much has been technologically accomplished over the past 50 years regarding tunnel ventilation. This technological growth has provided the ventilation engineer with new tools, guidelines and standards to apply to the tunnel ventilation business.

In 1973, the British Hydromechanics Research Association (BHRA), which is now the BHR Group Limited (BHRG), organised and held the First International Symposium on Aerodynamics and Ventilation of Vehicle Tunnels (ISAVVT) at the University of Kent at Canterbury (BIIRA, 1974). This was the first such symposium permitting tunnel ventilation engineers around the world to meet and discuss pertinent topics. The event has continued and has been held every 3 years and is now being held every two years, and has become the primary venue for the presentation, documentation and discussion of all new developments in the tunnel ventilation industry (BHRA, 1974, 1976, 1981, 1982, 1985, 1988; BHRG, 1991, 1994, 1997, 2000, 2003, 2006, 2009, 2011). The most recent BHRG-sponsored International Symposium on Aerodynamics and Ventilation of Tunnels was held in May 2011 (BHRG, 2011).

9.6.1   Analysis
The initial computerised analytical tool to be developed for use in the tunnel ventilation field was the application of one-dimensional network models, which had been used by civil engineers in the water flow field for many years, to airflow in tunnels. A one-dimensional network model is relatively simple, and when configured to address the airflow in a tunnel-shaft configuration, much could be learned by its application. The most significant of the one-dimensional network models for tunnel application is the Subway Environment Simulation Model (SES), developed

as part of a research grant made available by the US Department of Transportation (Associated Engineers, 1975). Up to this point in the history of tunnel ventilation, all calculations had been performed by hand.

The analytical scene for tunnel ventilation changed dramatically with the application of computational fluid dynamics (CFD) to tunnel ventilation. Although CFD had been used in the aircraft industry for years, it has only penetrated the tunnel ventilation industry within the last 15–20 years. The application of the many CFD models has changed the way the tunnel ventilation problem and solution, particularly in the case of a tunnel fire, is viewed. The animation possibilities permit the understanding of the various phenomena by most, including the untrained. They clearly and visibly evaluate and show the impact of ventilation on the control of smoke and heated gases during a fire in a tunnel. It is necessary, however, for such models to be employed by a knowledgeable user, to reduce the chance of misinterpretation of the results (see Chapter 15).

9.6.2   Handbooks, regulations, standards and guidelines

There are now numerous handbooks published both privately and by professional associations that provide material addressing the issue of tunnel ventilation. A brief list of these documents is given in Table 9.1.

The *Subway Environment Design Handbook* (*SEDH*) (Associated Engineers, 1975) is a product of the research grant that funded the development of the SES. The *SEDH* was the first comprehensive such document addressing the ventilation and environmental control of transit (metro) tunnels and underground stations. The *SEDH* provides a detailed description of the evaluation and design process.

At about the time that the *SEDH* was under development, the National Fire Protection Association (NFPA) in the USA was initiating an effort to develop a fire protection standard for transit

Table 9.1 Handbooks addressing tunnel ventilation and tunnel fire safety

Handbook title	Edition	Pertinent chapters	Reference
Subway Environmental Design Handbook, vol. 1. *Principles and Applications*	1st	Entire handbook	Associated Engineers (1975)
Tunnel Engineering Handbook	1st	Chapter 19, 'Tunnel Ventilation' Chapter 20, 'Fire Protection'	Bickel and Kuesel (1982)
Tunnel Engineering Handbook	2nd	Chapter 19, 'Fire Life Safety' Chapter 20, 'Tunnel Ventilation'	Bickel et al. (1996)
2011 ASHRAE Handbook – HVAC APPLICATIONS	2011	Chapter 15, 'Enclosed Vehicular Facilities'	Owen (2011)
Fire Protection Handbook	20th	Section 21.7, 'Fixed Guideway Transit and Light Rail Systems' Section 21.11, 'Road Tunnels and Bridges'	NFPA (2008)
The Handbook of Tunnel Fire Safety	1st	Entire handbook	Beard and Carvel (2005)
Fire Protection in Vehicles & Tunnels for Public Transport	1st	Entire handbook	Alba Fachverlag (2005)

(metro) tunnels. In 1983, the first NFPA standard for transit systems was published as *NFPA 130, Standard for Fixed Guideway Transit Systems* (NFPA, 1983). This standard has been updated on a regular basis, and is in widespread use today, and is currently known as *NFPA 130, Standard for Fixed Guideway Transit and Passenger Rail Systems* (NFPA, 2010), with the addition of passenger rail systems including tunnels.

In the area of road tunnels, the NFPA published for several years, beginning in 1972, a tentative standard, *NFPA 502T, Tentative Standard for Limited Access Highways, Tunnels, Bridges and Elevated Structures* (NFPA, 1972), that was ultimately withdrawn in 1975. In 1981, it was converted and republished as a recommended practice, entitled *NFPA 502, Recommended Practice on Fire Protection for Limited Access Highways, Tunnels, Bridges, Elevated Roadways and Air Right Structures* (NFPA, 1981). This document was completely revised in the 1990s, and ultimately emerged as an NFPA standard: *NFPA 502, Standard for Road Tunnels, Bridges and other Limited Access Highways*, which was first published in 1998 (NFPA, 1998). The latest version of this standard has recently been published (NFPA, 2011).

The World Road Association (formerly the Permanent International Association of Road Congresses), now known simply as PIARC, has for over 40 years, been publishing technical reports on tunnels and tunnel ventilation in conjunction with its quadrennial World Road Congresses. The PIARC Technical Committee on Road Tunnel Operations and its working groups have published several important specific documents on tunnel fire safety and ventilation, as listed in Table 9.2.

Many countries publish tunnel regulations, standards and guidelines, primarily for use in their country; in addition, a number of international agencies also produce such documents. However, many of these documents do provide an insight into numerous unique tunnel applications. A partial list of such documents is provided in Table 9.3.

9.6.3 Testing

Within the past 20 years, there have been a number of significant full-scale fire tests involving tunnels. Both the EUREKA test programme (Studiengesellschaft Stahlanwendung, 1995), conducted in Europe, and the MTFVTP (Massachusetts Highway Department, 1995), conducted in the USA, have returned tremendous benefits relating to the knowledge of fire behaviour in tunnels. They have demonstrated clearly the ability of the tunnel ventilation system, when properly designed and configured, to control the smoke and heated gases resulting from a tunnel fire (see also Chapter 10).

9.6.4 The future

The future for tunnel ventilation appears to be bright, as we continue to develop more advanced analytical methods and tools. This is being accomplished along with the acquisition of more actual test data to assist in testing the various models developed. It is the integrated evaluation of safety elements that will continue to be critical. The industry can no longer simply evaluate the emergency ventilation system independently based solely on the size of the design fire. The tunnel safety industry must find the tools to be able to evaluate the combined integrated impact of each system on its companion systems. The capability must exist to permit evaluation of the impact of each of any proposed fire safety systems on the remaining systems. As an example, these proposed fire safety systems could include emergency ventilation, egress route and a fixed fire suppression system. In addition, there are several specific system improvements

Table 9.2 World Road Association (PIARC) publications on tunnel ventilation and tunnel fire safety

PIARC publications	PIARC report No.	Reference
Road Tunnel Committee Reports to World Road Congresses		
Report to the XVth World Road Congress, Mexico City	–	Road Tunnels Committee (1975)
Report to the XVIth World Road Congress, Vienna	–	Technical Committee on Road Tunnels (1979)
Report to the XVIIth World Road Congress, Sydney	–	Technical Committee on Road Tunnels (1983)
Report to the XVIIIth World Road Congress, Brussels	–	Technical Committee on Road Tunnels (1987)
Report to the XIXth World Road Congress, Marrakech	19.05.B	Technical Committee on Road Tunnels (1991)
Report to the XXth World Road Congress, Montreal	20.05.B	Technical Committee on Road Tunnels (1995)
Report to the XXIst World Road Congress, Kuala Lumpur	21.05.B	PIARC (1999)
Road Tunnel Committee technical reports		
Classification of Tunnels, Existing Guidelines and Experiences, Recommendations	05.03.B	PIARC (1995)
Road Tunnels: Emissions, Environment, Ventilation	05.02.B	PIARC (1996a)
Road Safety in Tunnels	05.04.B	PIARC (1996b)
Fire and Smoke Control in Road Tunnels	05.05.B	PIARC (1999)
Pollution by Nitrogen Dioxide in Road Tunnels	05.09.B	PIARC (2000)
Road Tunnels: Vehicle Emissions and Air Demand for Ventilation	05.14.B	PIARC (2004)
Systems and Equipment for Fire and Smoke Control in Road Tunnels	05.16.B	PIARC (2007a)
Integrated Approach to Road Tunnel Safety	2007R07	PIARC (2007b)
Road Tunnels: A Guide to Optimising the Air Quality Impact upon the Environment	2008R04	PIARC (2008a)
Road Tunnels: An Assessment of Fixed Fire Fighting Systems	2008R07	PIARC (2008b)
Human Factors and Road Tunnel Safety Regarding Users	2008R17	PIARC (2008c)
Tools for Road Tunnel Safety Management	2009R08	PIARC (2009)
Road Tunnels: Operational Strategies for Emergency Ventilation	2011R02	PIARC (2011)

that would enhance the ability to deal with tunnel fires, including developing better fire detection systems that will properly function within the tunnel environment. The recent research conducted on tunnel detection systems has clearly shown this need, particularly for road tunnels. Timely detection ties directly into the appropriate timing of safety system deployment that needs to be better understood and defined. When should the operation of the tunnel ventilation system and the other safety systems be initiated in the event of a fire within the tunnel? There also needs to be improved understanding of the appropriate design fire heat release rate for a specific tunnel.

Table 9.3 Global regulations, standards and guidelines

Country	Document title	Reference
National regulations, standards and guidelines		
Australia	*Fire Safety Guidelines for Road Tunnels*	AFAC (2001)
	Road Tunnel Design Guidance – Fire Safety Design	RTA (NSW) (2006)
Austria	*Design Guidelines Tunnel Ventilation*	TRRA (2008)
Croatia	*Rules on Technical Standards for Design Preparation and Construction of Road Tunnels*	Croatia (1991)
Czech Republic	*Design of the Road Tunnels*	Czech Republic (2005)
	Road Tunnel Equipment	Czech Republic (2004)
France	*Safety in the Tunnels of the National Highways Network*	Ministry of Public Works (2000)
Germany	*Guidelines for Equipment and Operation of Road Tunnels*	Federal Ministry of Traffic (2002)
Greece	*Design Guidelines for Road Works – Tunnels (Electromechanical Works)*	Greece (2003)
Italy	*Safety of Traffic in Road Tunnels with Particular Reference to Vehicles Transporting Dangerous Materials*	Italy (1999)
Japan	*Tunnel Ventilation Design Guidelines*	Japan Road Association (1985)
	National Safety Standard of Emergency Facilities on Road Tunnels	Japan Road Association (2001)
The Netherlands	*Ventilation of Road Tunnels*	KIVI (1993)
	Recommendations on Ventilation of Road Tunnels	RWS Bouwdienst (2005)
New Zealand	*Fire Safety Guidelines for Road Tunnels*	AFAC (2001)
Norway	*Norwegian Design Guide – Road Tunnels*	Public Roads Administration (1992/2006)
Nordic Countries	*Ventilation of Road Tunnels*	NVF (1993)
	Safety Concept 2004 for Road Tunnels	NVF (2004)
Slovakia	*Technical Standard for Design of Road Tunnels*	Slovenia (2008)
	Guideline on Road Tunnel Safety	Slovakia (2006)
Slovenia	*Technical Standards and Requirements for Road Tunnel Design in Slovenia*	Slovakia (2006)
Spain	*Manual for the Design, Construction and Operation of Tunnels*	Spain (1998)
Sweden	*Tunnel 2004 – General Technical Specification for New and Upgrading of Old Tunnels*	SNRA (2004)
	Safety in Road Tunnels	Sweden (2005)
	Regulations on Road Tunnel Safety	Sweden (2007)
Switzerland	*Ventilation of Road Tunnels*	ASTRA (2001)

Table 9.3 Continued

Country	Document title	Reference
United Kingdom	*Design Manual for Roads and Bridges*	Highways Authority (1999)
United States	*Road Tunnel Design Guidelines*	FHWA (2004)

Agency	Document title	Reference
International regulations, standards and guidelines		
European Union	*Directive 2004/54/EC of the European Parliament and of the Council on Minimum Safety Requirements for Tunnels in the Trans-European Road Network*	European Union (2004)
NFPA	*Standard for Road Tunnels, Bridges and Other Limited Access Highways*	NFPA (2011)
NFPA	*Standard for Fixed Guideway Transit and Passenger Rail Systems*	NFPA (2010)
PIARC	*Fire and Smoke Control in Road Tunnels*	PIARC (1999)
PIARC	*Systems and Equipment for Fire and Smoke Control in Road Tunnels*	PIARC (2007a)
United Nations	*Recommendations of the Group of Experts on Safety in Road Tunnels*	UN Economic Council (2001)

In addition, the fire smoke release rate and the fire growth rate must also be included in this improved understanding in order to permit the design of a realistic integrated set of safety systems. It is clear that with the continued development of analytical tools and methods the tunnel ventilation engineer will ultimately be able to predict more accurately the interactions between the tunnel fire and the tunnel ventilation system. Lastly, maybe the industry needs to take a harder look at minimising the hazard rather than maximising the protection.

REFERENCES

AFAC (2001) *Fire Safety Guidelines for Road Tunnels*. Australasian Fire Authorities Council, East Melbourne.

Alba Fachverlag (2005) *Fire Protection in Vehicles & Tunnels for Public Transport*, 1st edn. Alba Fachverlag, Dusseldorf.

Associated Engineers (1975) *Subway Environmental Design Handbook*, vol. 1. *Principles and Applications*. Report No. UMTA-DC-06-0010-74-1. Transit Development Corporation, US Department of Transportation, Washington, DC.

ASTRA (2001) *Ventilation of Road Tunnels*. Selection of System, Design and Operation, Swiss Federal Roads Office (ASTRA), Bern.

Beard A and Carvel R (2005) *The Handbook of Tunnel Fire Safety*, 1st edn. Thomas Telford, London.

Bendelius AG and Hettinger JC (1988) Ventilation of the Sydney Harbour Tunnel. In: *Proceedings of the 6th International Symposium on the Aerodynamics and Ventilation of Vehicle Tunnels*. BHRA Fluid Engineering, Durham.

BHRA (1974) *Proceedings of the 1st International Symposium on the Aerodynamics and Ventilation of Vehicle Tunnels (ISAVVT), Canterbury*, 10–12 April 1973. British Hydromechanics Research Association, Fluid Engineering Centre, Cranfield.

BHRA (1976) *Proceedings of the 2nd International Symposium on the Aerodynamics and Ventilation of Vehicle Tunnels (ISAVVT), Cambridge*, 23–25 March 1976. British Hydromechanics Research Association, Fluid Engineering Centre, Cranfield.

BHRA (1981) *Proceedings of the 3rd International Symposium on the Aerodynamics and Ventilation of Vehicle Tunnels (ISAVVT), Sheffield*, 19–21 March 1979. British Hydromechanics Research Association, Fluid Engineering Centre, Cranfield.

BHRA (1982) *Proceedings of the 4th International Symposium on the Aerodynamics and Ventilation of Vehicle Tunnels (ISAVVT), York*, 23–25 March 1982. British Hydromechanics Research Association, Fluid Engineering Centre, Cranfield.

BHRA (1985) *Proceedings of the 5th International Symposium on the Aerodynamics and Ventilation of Vehicle Tunnels (ISAVVT), Lille*, 20–22 May 1985. British Hydromechanics Research Association, Fluid Engineering Centre, Cranfield.

BHRA (1988) *Proceedings of the 6th International Symposium on the Aerodynamics and Ventilation of Vehicle Tunnels, Durham*, 27–29 September 1988. British Hydromechanics Research Association, Fluid Engineering Centre, Cranfield.

BHRG (1991) *Proceedings of the 7th International Symposium on the Aerodynamics and Ventilation of Vehicle Tunnels Brighton*, 27–29 November 1991. BHR Group, Fluid Engineering Centre, Cranfield.

BHRG (1994) *Proceedings of the 8th International Conference on the Aerodynamics and Ventilation of Vehicle Tunnels (ISAVVT), Liverpool*, 6–8 July 1994. BHR Group, Fluid Engineering Centre, Cranfield.

BHRG (1997) *Proceedings of the 9th International Symposium on the Aerodynamics and Ventilation of Vehicle Tunnels (ISAVVT), Aosta Valley*, 6–8 October 1997. BHR Group,

Fluid Engineering Centre, Cranfield.

BHRG (2000) *Proceedings of the 10th International Symposium on the Aerodynamics and Ventilation of Vehicle Tunnels (ISAVVT), Boston, 1–3* November 2000. BHR Group, Cranfield.

BHRG (2003) *Proceedings of the 11th International Symposium on the Aerodynamics and Ventilation of Vehicle Tunnels (ISAVVT), Luzern, 7–9* July 2003. BHR Group, Cranfield.

BHRG (2006) *Proceedings of the 12th International Symposium on the Aerodynamics and Ventilation of Vehicle Tunnels (ISAVVT), Portoroz, 11–13* July 2006. BHR Group, Cranfield.

BHRG (2009) *Proceedings of the 13th International Symposium on the Aerodynamics and Ventilation of Vehicle Tunnels (ISAVVT), New Brunswick, NJ, 14–16* May 2009. BHR Group, Cranfield.

BHRG (2011) *Proceedings of the 14th International Symposium on the Aerodynamics and Ventilation of Tunnels (ISAVT), Dundee, Scotland, UK, 11–13* May 2011. BHR Group, Cranfield.

Bickel JO and Kuesel TR (1982) *Tunnel Engineering Handbook*, 1st edn. Van Nostrand Reinhold, New York.

Bickel JO, Kuesel TR and King EH (1996) *Tunnel Engineering Handbook*, 2nd edn. Chapman and Hall, New York.

Croatia (1991) *Book of Rules on Technical Standards for Design Preparation and Construction of Road Tunnels*. Zagreb.

Czech Republic (2004) *Road Tunnel Equipment*. Prague.

Czech Republic (2005) *Design of the Road Tunnels*. Prague.

European Union (2004) *Directive 2004/54/EC of the European Parliament and of the Council on Minimum Safety Requirements for Tunnels in the Trans-European Road Network*. European Union, Brussels.

Federal Ministry of Traffic (2002) *Guidelines for Equipment and Operation of Road Tunnels*. Road and Transportation Research Association (RABT), Federal Ministry of Traffic, Berlin.

FHWA (2004) *Road Tunnel Design Guidelines*. Federal Highway Administration (FHWA), US Department of Transportation, Washington, DC.

Fieldner AC, Henderson Y, Paul S, Sayers RR *et al.* (1927) *Ventilation of Vehicular Tunnels*. Report of the US Bureau of Mines to New York State Bridge and Tunnel Commission, and New Jersey Interstate Bridge and Tunnel Commission, American Society of Heating and Ventilating Engineers (ASHVE), New York.

Greece (2003) *Design Guidelines for Road Works – Tunnels (Electromechanical Works)*. Athens.

Highways Authority (1999) *Design Manual for Roads and Bridges*, vol. 2. *Highway Structure Design Materials*, section 2. *Special Structures*, part 9, BD 78/79. *Design of Road Tunnels*. Stationery Office, London.

Italy (1999) *Safety of Traffic in Road Tunnels with Particular Reference to Vehicles Transporting Dangerous Materials*. Rome.

Japan Road Association (1985) *Tunnel Ventilation Design Guidelines*. Japan Road Association, Tokyo.

Japan Road Association (2001) *National Safety Standard of Emergency Facilities on Road Tunnels*. Japan Road Association, Tokyo.

KIVI (1993) *Ventilation of Road Tunnels*. Royal Institute of Engineers (KIVI), Amsterdam.

Massachusetts Highway Department (1995) *Test Report – Memorial Tunnel Fire Ventilation Test Program (MTFVTP)*. Massachusetts Highway Department, Boston, MA.

Ministry of Public Works (2000) *Safety in the Tunnels of the National Highways Network*. Inter-Ministerial Circular No 2000-63. Ministry of Public Works, Paris.

NFPA (1972) *NFPA 502T: Tentative Standard for Limited Access Highways, Tunnels, Bridges and Elevated Structures*. National Fire Protection Association, Quincy, MA.

NFPA (1981) *NFPA 502: Recommended Practice on Fire Protection for Limited Access Highways, Tunnels, Bridges, Elevated Roadways and Air Right Structures*. National Fire Protection Association, Quincy, MA.

NFPA (1983) *NFPA 130: Standard for Fixed Guideway Transit Systems*. National Fire Protection Association, Quincy, MA.

NFPA (1998) *NFPA 502: Standard for Road Tunnels, Bridges and Other Limited Access Highways*. National Fire Protection Association, Quincy, MA.

NFPA (2008) *Fire Protection Handbook*, 20th edn. National Fire Protection Association, Quincy, MA.

NFPA (2010) *NFPA 130: Standard for Fixed Guideway Transit and Passenger Rail Systems*. National Fire Protection Association, Quincy, MA.

NFPA (2011) *NFPA 502: Standard for Road Tunnels, Bridges and Other Limited Access Highways*. National Fire Protection Association, Quincy, MA.

NVF (2004) *Safety Concept 2004 for Road Tunnels*. Nordic Road Association (NVF), Oslo.

NVF Sub-committee 61 (1993) *Ventilation of Road Tunnels*. Report No. 6. Sub-committee 61, Nordisk Vejteknisk Forbund, Helsinki.

Owen MS (ed.) (2011) *2011 ASHRAE Handbook – Heating Ventilating, and Air-conditioning (HVAC) Applications*. American Society of Heating, Refrigerating and Air-Conditioning Engineers (ASHRAE), Atlanta, GA.

PIARC (1995) *Classification of Tunnels, Existing Guidelines and Experiences*. World Road Association (PIARC), Paris.

PIARC (1996a) *Road Tunnels: Emissions, Ventilation Environment*. World Road Association (PIARC), Paris.

PIARC (1996b) *Road Safety in Tunnels*. World Road Association (PIARC), Paris.

PIARC (1999) *Fire and Smoke Control in Road Tunnels*. World Road Association (PIARC), Paris.

PIARC (1999) *Report to the XXIst World Road Congress, Kuala Lumpur, 3–9 October 1999*. World Road Association (PIARC), Paris.

PIARC (2000) *Pollution by Nitrogen Dioxide in Road Tunnels*. World Road Association (PIARC), Paris.

PIARC (2004) *Road Tunnels: Vehicle Emissions and Air Demand for Ventilation*. World Road Association (PIARC), Paris.

PIARC (2007a) *Systems and Equipment for Fire and Smoke Control in Road Tunnels*. World Road Association (PIARC), Paris.

PIARC (2007b) *Integrated Approach to Road Tunnel Safety*. World Road Association (PIARC), Paris.

PIARC (2008a) *Road Tunnels: A Guide to Optimizing the Air Quality Impact upon the Environment*. World Road Association (PIARC), Paris.

PIARC (2008b) *Road Tunnels: An Assessment of Fixed Fire Fighting Systems*. World Road Association (PIARC), Paris.

PIARC (2008c) *Human Factors and Road Tunnel Safety Regarding Users*. World Road Association (PIARC), Paris.

PIARC (2009) *Tools for Road Tunnel Safety Management*. World Road Association (PIARC), Paris.

PIARC (2011) *Road Tunnels: Operational Strategies for Emergency Ventilation.* World Road Association (PIARC), Paris.

Public Roads Administration (1992/2006) *Norwegian Design Guide – Road Tunnels.* Public Roads Administration, Directorate of Public Roads, Oslo.

Road Tunnels Committee (1975) *Report to the XVth World Road Congress, Mexico City,* 12–19 October 1975. World Road Association (PIARC), Paris.

RTA (NSW) (2006) *Road Tunnel Design Guidance – Fire Safety Design.* Roads and Traffic Authority, New South Wales, Sydney.

RWS Bouwdienst (2005) *Recommendations on Ventilation of Road Tunnels.* Rijkswaterstaat RWS Bouwdienst, Utrecht.

Singstad O (1929) *Ventilation of Vehicular Tunnels.* World Engineering Congress, Tokyo.

Slovakia (2006) *Guideline on Road Tunnel Safety.* Bratislava.

Slovakia (2006) *Technical Standards and Requirements for Road Tunnel Design in Slovenia.* Bratislava.

Slovenia (2008) *Technical Standard for Design of Road Tunnels.* Ljubljana.

SNRA (2004) *Tunnel 2004 – General Technical Specification for New and Upgrading of Old Tunnels.* Swedish National Road Association (SNRA), Stockholm.

Spain (1998) *Manual for the Design, Construction and Operation of Tunnels,* IOS-98. Madrid.

Studiengesellschaft Stahlanwendung (1995) *Fires in Transport Tunnels. Report on Full Scale Tests. EUREKA Project EU 499: FIRETUN.* Studiengesellschaft Stahlanwendung, Düsseldorf.

Sweden (2005) *Safety in Road Tunnels.* Stockholm.

Sweden (2007) *Regulations on Road Tunnel Safety.* Stockholm.

Technical Committee on Road Tunnels (1979) *Report to the XVIth World Road Congress, Vienna,* 16–21 September 1979. World Road Association (PIARC), Paris.

Technical Committee on Road Tunnels (1983) *Report to the XVIIth World Road Congress, Sydney,* 8–15 October 1983. World Road Association (PIARC), Paris.

Technical Committee on Road Tunnels (1987) *Report to the XVIIIth World Road Congress, Brussels,* 13–19 September 1987. World Road Association (PIARC)), Paris.

Technical Committee on Road Tunnels (1991) *Report to the XIXth World Road Congress, Marrakech,* 22–28 September 1991. World Road Association (PIARC)), Paris.

Technical Committee on Road Tunnels (1995) *Report to the XXth World Road Congress, Montreal,* 3–9 September 1995. World Road Association (PIARC)), Paris.

TRRA (2008) *Design Guidelines Tunnel Ventilation.* Transportation and Road Research Association (TRRA), National Roads Administration, Vienna.

UN Economic Council (2001) *Recommendations of the Group of Experts on Safety in Road Tunnels.* United Nations, UN Economic Council, Economic Commission for Europe, Inland Transport Committee, Geneva.

FURTHER READING

Bendelius AG and Caserta AS (1999) The Memorial Tunnel Fire Ventilation Test Program. Phase IV – development of a tunnel specific CFD code (05–01). *XXIst World Road Congress,* Kuala Lumpur, 3–9 October 1999. World Road Association (PIARC), Paris.

Caserta A and Bendelius A (1995) The Memorial Tunnel Fire Ventilation Test Program. *XXth World Road Congress,* Montreal, 3–9 September 1995. World Road Association (PIARC), Paris.

PIARC (1967) Road Tunnels Committee Documentation and Studies Report. *XIIIth World Road Congress, Tokyo,* 5–11 November 1967. World Road Association (PIARC), Paris.

PIARC (1995) *The First Road Tunnel – A Planner Guide for Countries Without Previous Experience of New Road Tunnels*. World Road Association (PIARC), Paris.

PIARC (1999) Road tunnels. *Introductory Report of the Technical Committee to the XXIst World Road Congress*, Kuala Lumpur, 3–9 October 1999. World Road Association (PIARC), Paris.

PIARC (1999) *Road Tunnels: Reduction of Operating Costs*. World Road Association (PIARC), Paris.

PIARC (1999) Summaries of individual papers. *XXIst World Road Congress*, Kuala Lumpur, 3–9 October 1999. World Road Association (PIARC), Paris.

PIARC (1999) Transport of Dangerous Goods Through Road Tunnels: Introductory Report of the Technical Committee (C5). *XXIst World Road Congress*, Kuala Lumpur, 3–9 October 1999. World Road Association (PIARC), Paris.

PIARC (2002) *Cross Section Geometry in Uni-directional Tunnels*. World Road Association (PIARC), Paris.

PIARC (2003) *Proceedings of the XXIInd World Road Congress, Durban,* 19–25 October 2003. World Road Association (PIARC), Paris.

PIARC (2003) *Traffic Incident Management Systems Used in Road Tunnels*. World Road Association (PIARC), Paris.

PIARC (2004) *Cross Section Design of Bidirectional Road Tunnels*. World Road Association (PIARC), Paris.

PIARC (2004) *Good Practice for the Operation and Maintenance of Road Tunnels*. World Road Association (PIARC), Paris.

PIARC (2007) *Proceedings of the XXIIIrd World Road Congress, Paris,* 17–21 September 2007. World Road Association (PIARC), Paris.

PIARC (2008) *Management of the Operator – Emergency Teams Interface in Road Tunnels*. World Road Association (PIARC), Paris.

PIARC (2008) *Risk Analysis for Road Tunnels*. World Road Association (PIARC), Paris.

PIARC (2008) *Urban Road Tunnel – Recommendations to Managers and Operating Bodies for Design, Management, Operation and Maintenance*. World Road Association (PIARC), Paris.

TRB (2006) *Making Transportation Tunnels Safe and Secure*, Transportation Research Board (TRB), Transit Cooperative Research Program (TCRP Report 86)/National Cooperative Highway Research Program (NCHRP Report 525), Washington, DC.

Handbook of Tunnel Fire Safety, 2nd edition
ISBN: 978-0-7277-4153-0

ICE Publishing: All rights reserved
doi: 10.1680/htfs.41530.177

Chapter 10
The use of tunnel ventilation for fire safety

George Grant
Optimal Energies Ltd, UK
Stuart Jagger
Health and Safety Laboratory, Buxton, UK; Provided updated chapter

10.1. Introduction

A series of major tunnel fires in Europe during the last 10–20 years showed the potentially serious consequences of such incidents and highlighted the need for a reappraisal of the fire safety philosophy for these structures. Although the majority of these incidents occurred in 'older' tunnels (the Mont Blanc and St Gotthard Tunnels, for example, were opened in 1965 and 1980, respectively), the Channel Tunnel fires in 1996 and later showed that state-of-the-art transport systems are not immune from such events. Consequently, tunnel fire safety is currently high on the European political agenda, as both trans-Alpine traffic and public apprehension continue to increase. For example, the 11.6 km Mont Blanc Tunnel, originally designed for 450 000 vehicles per year, was used by some 1.1 million vehicles in 1997; similarly, annual traffic levels in the 16 km St Gotthard Tunnel have more than doubled to 6.5 million per year since 1981 (Edwards, 1999). Neither is the problem of reduced public confidence limited to the familiar trans-Alpine routes, since several major new tunnel projects are currently underway, some of great length, and all the indications are that tunnel construction is set to increase worldwide in the coming decades.

Due to the confined conditions, such fires can be more severe, in terms of rate of growth and temperature, than equivalent fires in the open. This is clearly indicated by reference to the proposed tunnel infrastructure fire resistance test protocols such as the Rijkwaterstaat (RWS) curve (SINTEF, 1999), which involve more severe temperature–time exposures than the usual hydrocarbon or cellulosic fire curves. In addition, the consequences may be more severe due to difficulties in evacuating from the fire zone to a place of safety or due to transport of toxic combustion products to locations remote from the fire.

In the early days of vehicle tunnel construction, few tunnels were equipped with any system of forced mechanical ventilation. Reliance was placed on natural air movements induced by differences in pressure due to elevation, temperature and meteorological effects at the tunnel portals and/or the circulation produced by vehicle movements to refresh the tunnel with clean cool air and to remove smoke and other pollutants.

However, as traffic levels grew, it became apparent, particularly for road tunnels in which usage could be continuous and heavy, that such mechanisms could not always provide sufficiently frequent air changes to maintain satisfactory air quality levels and ensure safe levels of irritant and toxic pollutants, for example engine particulate emissions and oxides of carbon, sulphur and nitrogen. Consequently, for long road tunnels (though not at this stage for rail tunnels), particularly those in urban areas, the removal of such emissions became a major design consideration, and mechanical ventilation systems optimised to this requirement were developed. These included systems such as fully or semi-transverse ventilation, in which ventilating air was delivered to or removed from various locations along the tunnel length by dedicated ducts incorporated in the tunnel fabric. Guidance in designing systems against such requirements is summarised in publications such as Bickel and Kuesel (1991) or those prepared, for example, by the World Road Association (formerly known as the Permanent International Association of Road Congresses and still known by the abbreviation PIARC) (World Road Association, 1996).

As a result of incidents in long single-bore, bi-directional road tunnels, some have advocated the addition of a second and a third bore to serve as an additional uni-directional traffic tunnel and a dedicated evacuation tunnel, respectively (i.e. a configuration similar to the Channel Tunnel). Others support the introduction of additional mitigation systems in the form of active fire protection (e.g. sprinklers or water mist systems), combined with improved fire detection and traffic management, particularly for heavy goods vehicles (HGVs), which have been implicated in the most serious of recent incidents. Proponents of rail freight suggest removing the HGV fire hazard from road tunnels completely, by transferring all freight to the railways. Although the last of these might be shown to be safer statistically, the loss of ten HGVs and their loads during the 1996 Channel Tunnel fire shows that this is not a complete answer in itself. However, it should be stressed that despite the seriousness of the incident, there were no fatalities in this case, although the result might well have been different if the fire had started in a HGV close to the amenity carriage, instead of towards the other end of the train. The Channel Tunnel fire is also relevant to the mitigation argument, since it prompted Eurotunnel to explore the use of high-pressure water mist as an on-board fire suppression system for the protection of its HGV shuttle fleet (Grant and Southwood, 1999). This on-board system was not installed, however (see Chapter 30).

Drawing on knowledge obtained from the general fire engineering field (Heselden, 1976), much attention has been paid to the ventilation system and, particularly, its use to control smoke movement to maintain a smoke free evacuation route for tunnel occupants. Regardless of future developments in active and/or passive fire protection systems for tunnels, it is inevitable that ventilation systems will continue to play a key role during 'abnormal' tunnel operations, particularly during a fire, because they have a critical impact on incident management with respect to both evacuation and subsequent fire-fighting/rescue operations. Thus, their use to mitigate fire incidents is now one of the main considerations, if not the main one, in the design of a tunnel ventilation system (World Road Association, 1999). Thus, the Channel Tunnel ventilation system was developed with this in mind (Bradbury, 1998), and the revamped Mont Blanc system has given smoke control detailed attention (Brichet et al., 2002). The following summarises the current state of knowledge regarding the response of fires in tunnels, paying particular attention to the critical issue of smoke control.

10.2. Modes of operation of tunnel ventilation systems during a fire

The current state-of-the-art in tunnel ventilation has been described in Chapter 9. However a brief resumé of the various design options is useful here as an aid to the subsequent discussion. In broad terms, tunnel ventilation systems fall into one of two categories: either 'natural' or 'mechanical'. The first of these encompasses air movements induced by temperature or pressure gradients (i.e. meteorological effects) and those induced by the tunnel traffic itself. Mechanical ventilation configurations include: longitudinal, fully transverse, semi-transverse (and reversible semi-transverse) and partial- (or pseudo-) transverse, and comprise a combination of fans to move air and ducts/dampers to deliver it to the required location.

10.2.1   Natural ventilation

Natural ventilation is common in 'short' tunnels, whereas mechanical ventilation becomes a requirement for longer structures. Although the boundary between 'short tunnels' and 'long tunnels' is somewhat fuzzy, some national guidelines have been established as an aid to designers (World Road Association, 1999). In Germany, for example, tunnels between 350 and 700 m in length do not require mechanical ventilation, whereas in the UK an upper limit of only 400 m is allowable, provided there is sufficient technical justification. Under some conditions, 'natural movements' may be significant. For example, in a 1 km tunnel with a gradient of 10%, a flow of ~0.5–1 m/s may be set up due to static pressure differences at the portals. The flows produced by vehicle movement will depend on a number of factors such as the vehicle speed, frequency and, especially, the fit of the vehicle to the tunnel cross-section. Typically, for example, Torbergsen *et al.* (2002) reported measurements in a single-bore railway tunnel, and found that, during passage of a train with a cross-section ~30% that of the tunnel at 44 m/s, the flow velocity reached ~6.5 m/s and decayed to background levels over the next 5 minutes or so.

The case for natural ventilation is based on the observation that hot smoke rises from a fire towards the ceiling, where it 'stratifies' (i.e. forms a discrete layer under the tunnel ceiling) and subsequently propagates longitudinally away from the fire. It is argued that this natural phenomenon can maintain a tenable environment for tunnel occupants at a lower level, thus facilitating their escape from the incident site. This situation is shown in Figure 10.1. However, the spatial and temporal extent of this phenomenon is highly variable, depending on such factors as fire size and growth rate, tunnel dimensions and slope, boundary temperatures and, critically, on ambient air movements (either natural or vehicle-induced). The main mechanisms by which the buoyancy of the hot layer is eroded are heat transfer to the cooler walls coupled with increased turbulent mixing at the interface with the fresh air layer beneath. Since both of these effects are proportional to the length of travel of the hot layer, it is clear that the natural ventilation strategy becomes increasingly risky in longer tunnels. Indeed, past experience shows that stable stratification of smoke layers in tunnels may break down at relatively short distances from the fire, with the smoke being subsequently recirculated to the fire zone after mixing with the colder air beneath. In a relatively short time, conditions within a few hundred metres of the ignition zone may have deteriorated to the point that neither evacuation nor fire-fighting are possible, and the only recourse is to let the fire burn out – and bear the loss of both life and asset. In the case of the Japanese experiments reported by Haerter (1994) and depicted in Figure 10.1, it can be seen that stable stratification was maintained initially for a distance of 400–600 m from the fire, but mixing and recirculation became problematic thereafter. For this particular tunnel/fire combination, then, it seems that natural ventilation could be a viable option provided the fire zone is never more than 400–600 m from a portal. Alternatively, where practical, vent shafts may be sited at appropriate intervals down the tunnel to remove smoke before it fills the entire tunnel height (Viot and Vauquelin, 2002). For greater smoke travel distances, the approach becomes unreliable, and stable stratification may collapse before the smoke layer has a chance to vent to atmosphere.

10.2.2   Mechanical systems

Where the use of mechanical ventilation is indicated, local smoke extraction from the incident site is the ideal solution, since smoke obscuration is effectively eliminated during evacuation, and the

Figure 10.1 Smoke stratification in a tunnel containing a 4 m² petrol pool fire with ambient air velocity <0.5 m/s (Based on Haerter (1994) Copyright SP Technical Research Institute of Sweden)

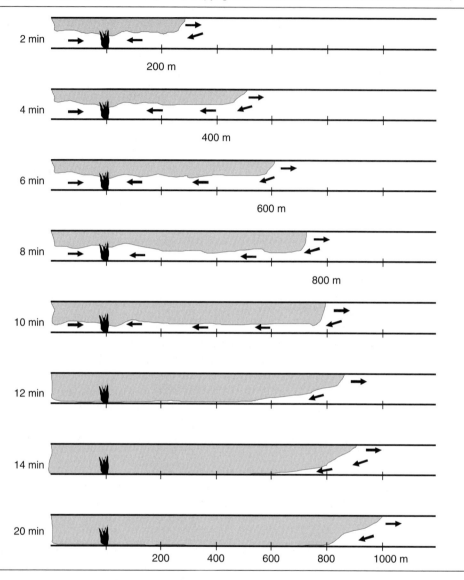

fire-fighters' approach to the scene is then limited only by radiant heat levels. In the case of transverse ventilation systems, this goal is achievable, at least in principle. However, for longitudinal systems, smoke extraction is possible only where the incident occurs near the base of a fan shaft (assuming the fan is capable of being run in exhaust mode); for ceiling-mounted jet fans, 'local extraction' can only be achieved for the trivial case of a fire near a portal.

The advantage and disadvantage of the longitudinal system is that the traffic space also acts as the ventilation duct, thus obviating the need for either a separate ventilation bore or a significant increase in the size of the running bore (fans are usually mounted in shafts off the tunnel or are of the jet fan type with a relatively small diameter, which can be accommodated at intervals along the tunnel ceiling). In fully transverse systems, the supply of fresh air and the extraction of vitiated air takes place along the tunnel length via a system of adjustable louvres servicing two separate plenums on opposite sides of the tunnel (usually either above a false ceiling or below floor level). In this arrangement, the inflow and outflow rates of air per unit length of tunnel are identical, and the two air streams create a flow in the traffic space that is transverse to the longitudinal axis of the tunnel. Semi-transverse systems are similar to the fully transverse case but with no dedicated exhaust plenum; fresh air is added to the traffic space along its length, and the vitiated air flows longitudinally towards the tunnel portals, where it is vented to the atmosphere. Where reversed flow is possible, semi-transverse systems may draw fresh air into the tunnel via the portals, while the polluted air is extracted along the length of the tunnel; usually, however, the reverse-mode operation is used only in the case of a tunnel fire. Partial transverse systems have characteristics that are somewhere between those of the fully transverse and semi-transverse designs, depending on the ratio of the supply-to-exhaust rate along the tunnel axis. Transverse systems are common in road tunnels but, for rail tunnels, they become prohibitively expensive, and longitudinal systems dominate this area.

With a transverse system it is possible to extract smoke from the vicinity of a blaze into the exhaust plenum (Chan *et al.*, 1988) (Figure 10.2), but the operator of a longitudinal system does not have the option of local pollutant extraction. Longitudinal ventilation can, however, be employed to provide clear air upstream of the fire (Figure 10.3), creating a smoke-free escape route and allowing the emergency services to attend the fire in the manner proposed by Fuller (1985/ 1986) and Eisner and Smith (1954). The obvious drawback of this strategy is the possibility that it may lead to a deterioration of the environment downstream of the fire where people may still be trapped. The question then arises, 'Which fan configuration is best?', and indeed the whole issue of fire emergency procedures is complex (Vardy, 1988). Notwithstanding this important consideration, the ability to provide *positive smoke control* remains a key element in the design of tunnel ventilation systems, the need for which is confirmed by experience. Many examples of severe tunnel fires have been documented over the last half-century (e.g. Donato, 1972, 1975; Egilsrud, 1984; Chan *et al.*, 1988; Channel Tunnel Safety Authority, 1997; SINTEF, 1999; Lacroix, 2001; see also Chapter 1); the full range of forced-ventilation configurations, including none at all, is represented in these incidents. An analysis of these events reveals a consistent need for positive smoke control to enable the fire to be attacked quickly and effectively; many of the incidents where this was not possible resulted in the abandonment of fire-fighting efforts, with the fire being left to burn out.

Despite the apparent benefits associated with providing positive intervention, caution is needed in using forced ventilation because high air velocities may 'fan the flames' (Parsons Brinckerhoff, 1996). In the case of the recent disaster in the Mont Blanc road tunnel, it does seem that the fire was exacerbated by the inappropriate use of the tunnel's semi-transverse ventilation system (Lacroix, 2001). In this particular case, the disproportionately high rate of air supply through the roadside ventilation slots created an excessive longitudinal airflow, which may have intensified the heat release rate (HRR) of the fire. In addition, some of the reversible high-level exhaust openings near the fire were left in supply mode, increasing the air supply and accelerating the vertical mixing of the hot smoke layer into the traffic space below. Although there were additional factors that may have led to the rapid growth of the fire and the high rate of smoke production, it is clear that the Mont Blanc Tunnel ventilation system was not operating in the optimum 'fire emergency' mode. The current PIARC recommendations (World Road Association, 1999) for transverse

Figure 10.2 Common tunnel ventilation configurations: (a) longitudinal ventilation (jet fans); (b) fully transverse ventilation (normal operation); (c) fully transverse ventilation (emergency operation in case of fire); (d) semi-transverse ventilation (normal operation); (e) semi-transverse ventilation (emergency operation in case of fire)

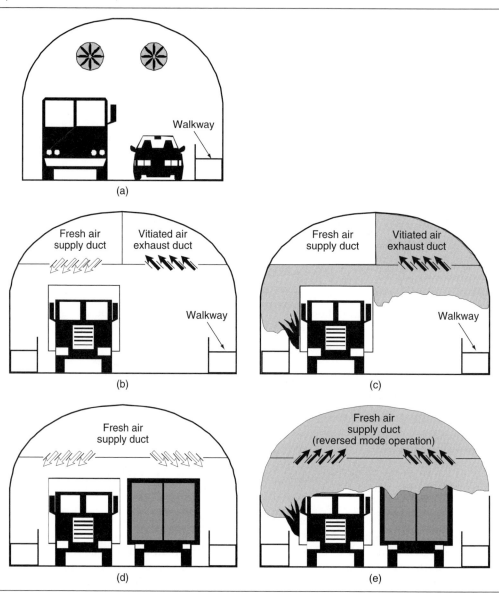

systems are to set the extraction rate to maximum around the fire zone and simultaneously to stop altogether any local fresh air supply. For all other zones remote from the fire, it is recommended that the rate of supply of fresh air should be throttled back to between one-half and one-third of the full capacity, to ensure that the induced longitudinal air velocities remain below 2 m/s (in order to preserve the two-layer flow, which is of critical importance both during the evacuation phase and in the subsequent fire-fighting operations). In the context of fanning the fire, the data from

Figure 10.3 Typical longitudinal ventilation arrangements in tunnels: (a) longitudinal ventilation (vertical fan shafts); (b) longitudinal ventilation (jet fans); (c) longitudinal ventilation (vertical fan shafts), emergency operation in case of fire; (d) longitudinal ventilation (jet fans), emergency operation in case of fire

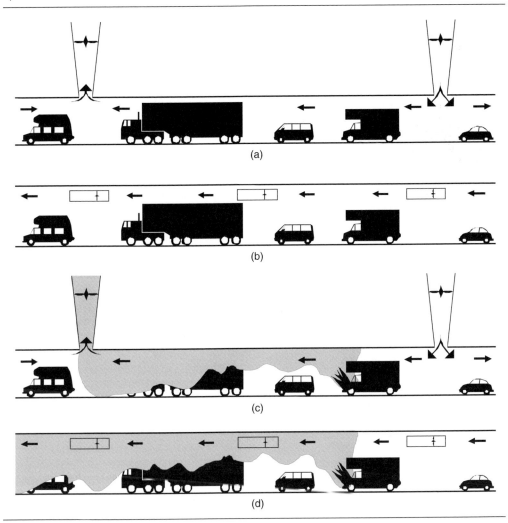

the full-scale Memorial Tunnel Fire Ventilation Test Program (MTFVTP) (see Section 10.3) showed that there was a generally sufficient quantity of combustion air within the tunnel itself and that: 'The possible increase in fire intensity resulting from the initiation of ventilation did not outweigh the benefits' (Parsons Brinckerhoff, 1996). It is the case, though, that the MTFVTP series only involved pool fires, and it has been demonstrated by Carvel *et al.* (2001b) that forced ventilation generally has a far more dramatic enhancing effect on the HRR of HGV fires than it does on the HRR of pool fires in tunnels (see Chapter 11).

Some tunnels make use of both longitudinal and transverse ventilation systems. In the Channel Tunnel, for example (Figure 10.4), the normal ventilation system is of the non-reversible

Figure 10.4 The Channel Tunnel: (a) general arrangement; (b) schematic of ventilation systems

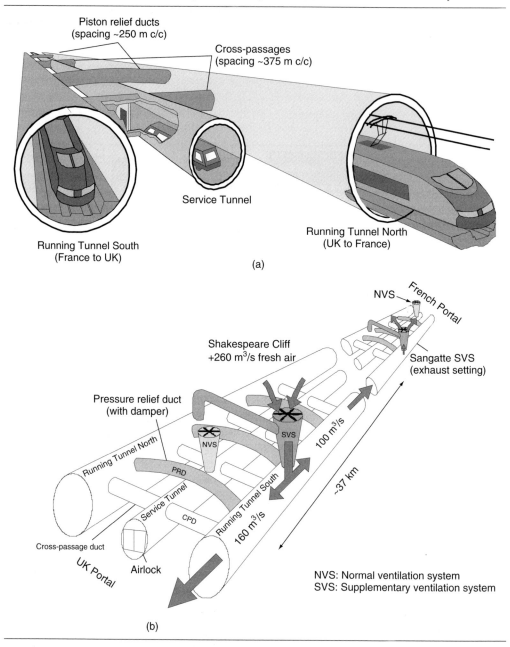

semi-transverse type. The fresh air intakes are sited at either end of the 37 km-long undersea section of the tunnel, and these fans maintain the service tunnel (ST) at an ambient pressure that is somewhat greater than the two running tunnels (RTs). The fresh air is subsequently fed into the RTs via a number of air distribution units (ADUs), which are located within 38 of the cross passages that connect the RTs to the ST. In normal operation, this is all that is required, since the 'piston effect' of the trains ensures that pollutants in the running tunnels are initially well mixed with the fresh air and subsequently vented to atmosphere at the tunnel portals. Under normal service conditions, there is also free circulation of air between the two RTs, via the piston relief ducts (PRDs), located along the tunnels at ~250 m intervals and whose primary function is to alleviate extreme air pressures in the RTs. In the event of a fire, the priority is to bring the affected train to a controlled halt adjacent to the cross passages (which are located throughout the entire length of the tunnels at ~375 m spacings). Thereafter, the PRDs are closed remotely, to prevent the recirculation of smoke into the non-incident tunnel. Next, the supplementary ventilation system (SVS) in the incident tunnel is activated, which induces a longitudinal airflow over the affected train and maintains smoke-free conditions upstream of the fire. Once these conditions are confirmed, it is the task of the train crew to open the cross passage doors leading to the ST and then assist the passengers to disembark towards the nearest cross passages and the 'safe haven' of the ST. In the fire that occurred on a Eurotunnel HGV shuttle train on 18 November 1996, it was necessary to open only one cross passage door, upstream of the fire near the front of the train, since the HGV drivers were all located in the amenity coach just behind the forward locomotive. This was fortunate, as there was a delay in establishing the correct longitudinal SVS airflow in the incident tunnel (Channel Tunnel Safety Authority, 1997). In the event, the train crew managed to evacuate everyone to the safety of the ST, and there were no fatalities, but they were exposed to smoke.

Similarly, the revamped Mont Blanc system employs a combination of transverse and longitudinal systems, to extract smoke at or near the fire, to ensure that both up- and downstream of the fire remains smoke-free and to try to maintain near-zero velocities at the site of the fire (Brichet *et al.*, 2002).

10.2.3 Design objectives for smoke control

When a fire breaks out in a vehicle tunnel, the ventilation system must maintain tenable conditions along the escape routes and any additional access routes used by the emergency services for firefighting and rescue operations. They may also be used to maintain tenable conditions in any temporary safe refuge. In the absence of adequate ventilation, the confinement afforded by a tunnel increases the severity of problems faced by those evacuating and the emergency services. These issues have been discussed by Fuller (1985/1986), following the Summit Tunnel fire in the UK in 1984 (Pucher, 1994), and include the provision of adequate communications, access to the site, the optimum duration for breathing apparatus, and the physiological and psychological stresses associated with working in hot, dense smoke. Fuller considered that a reliable ventilation flow provided a significant tactical advantage when tackling such fires, by increasing the visibility and diluting toxic and flammable fumes. Similarly, Eisner and Smith (1954) concluded that it was essential for fire-fighters to be able to approach to within 11–14 m of an underground mine fire in order to conduct an effective attack. Evidently, the ability to fight an underground fire at relatively close range is assisted by the provision of forced ventilation, which improves the visibility of the seat of the fire from the upstream side. However, progress can be impeded if smoke moves against the ventilation stream due to buoyancy effects – a phenomenon known as *back-layering* or *smoke backflow*. The design of a ventilation system to counteract back-layering has traditionally been based on the provision of a minimum mean longitudinal air velocity in the traffic space – the so-called *critical velocity*.

The last two decades have seen some fundamental changes in the design process for tunnel ventilation systems that can be explained by several factors.

- The occurrence of numerous high-profile tunnel fires, many of which have resulted in multiple fatalities.
- An increased awareness, by society in general, of life safety issues during fire emergencies.
- The worldwide increase in underground construction in general and the completion of several major tunnelling projects in particular (e.g. the Channel Tunnel and trans-Alpine links in Europe and the Boston Central Artery/Tunnel Project in the USA).
- The maturation of fire safety engineering as a viable engineering subject and the concomitant requirement for 'performance-based' fire safety solutions in modern construction projects.

Prior to the 1980s, there existed scant full-scale experimental data concerning the interaction between tunnel fires and ventilation systems; the two best known contemporary examples of such studies are those performed in the Ofenegg and Zwenberg Tunnels in 1965 and 1975, respectively. Both test facilities were disused single-track railway tunnels, and the tunnel lengths and cross-sectional areas were 190 m and 23 m^2 (Ofenegg) and 390 m and ~20 m^2 (Zwenberg). Both programmes employed mainly liquid fuels (petrol and diesel) during the tests, which numbered 11 for the Ofenegg experiments (Haerter, 1994) and 30 in the case of the Zwenberg trials (Pucher, 1994). The stated aims of the Ofenegg and Zwenberg trials were also broadly similar.

- To determine the dangers facing tunnel occupants during a fire (in terms of temperature, noxious gases, visibility and fire duration).
- To determine how various ventilation strategies affect the tunnel environment when a fire is present, and hence to determine how best to improve life safety under such conditions.
- To examine the effects of a fire on the tunnel structure and ventilation plant.
- To investigate the influence of sprinklers in mitigating the adverse effects of a tunnel fire (Ofenegg tests only).

Although providing useful data for the time, by today's standards these experiments might be considered somewhat rudimentary in nature, mainly due to the limitations of the instrumentation employed and, in particular, the inability to measure the dynamic variation of the fire HRR during the tests; it is now accepted that the HRR is the single most important parameter in any fire safety problem. In addition, the cross-sectional areas of the test tunnels were considerably less than those for typical two-lane road tunnels (45–60 m^2). Notwithstanding these limitations, the Ofenegg and Zwenberg experiments (plus a few other studies of a similar nature) effectively comprised the entire 'full-scale database' on tunnel fire/ventilation interaction that was available to designers, operators and researchers in this field. In addition, the limited full-scale data were augmented by various small-scale investigations, driven primarily by the mining sector (e.g. Eisner and Smith, 1954; Bakke and Leach, 1960; Leach and Barbero, 1964; Chaiken et al., 1979; Lee et al., 1979), and some elementary theoretical modelling studies (e.g. Thomas, 1970; Hwang et al., 1977; Kumar and Cox, 1986); this situation persisted well into the 1980s.

During this period, there existed limited 'official' guidance on defining and meeting tunnel ventilation design objectives, with the few quantitative 'design rules' being based necessarily on this disparate data set. Notionally, road tunnel systems were covered by the various PIARC publications, while the *Subway Environmental Design Handbook* (US Department of Commerce, 1976) was the primary reference for railway tunnel applications. In both cases, however, the design objectives for ventilation systems are conventionally based on two operating conditions, 'normal' and 'abnormal'. Under normal operation, the ventilation requirements for long road and rail tunnels are similar, i.e. to reduce the concentration of tunnel pollutants and to provide a constant supply of fresh air to the tunnel occupants. Normally, the main pollutant in long railway tunnels is heat, while the dilution of gaseous vehicle emissions (carbon monoxide, carbon dioxide, nitrous oxides, etc.) is the prime concern for road tunnels. In the context of abnormal tunnel operation, PIARC has developed a hierarchy of objectives for fire and smoke control (Lacroix, 1998)

- to save lives by facilitating occupant evacuation
- to facilitate rescue and fire-fighting operations
- to avoid explosions
- to limit damage to the tunnel structure, equipment and surrounding buildings.

Up until the late 1980s and early 1990s, the ventilation objectives for a new tunnel project would be based on the PIARC recommendations and/or *Subway Environmental Design Handbook*. The methods for achieving these broad objectives would be refined using the findings of contemporary research such as the BHRA's International Symposia on the Aerodynamics and Ventilation of Vehicle Tunnels series (e.g. Heselden, 1976; Bendelius and Hettinger, 1988; Desrosiers, 1988) and incidental publications from research institutions such as the Fire Research Station in the UK (e.g. Spratt and Heselden, 1974). The details of the ventilation system would then be developed in the context of the *overall tunnel system*, in order to meet the specific *system requirements*.

Against this background, during the 1970s, the American Society of Heating, Refrigerating and Air-Conditioning Engineers (ASHRAE) tasked one of its technical committees with the production of guidance on the subject of smoke ventilation for 'enclosed vehicular facilities'. Unfortunately, the conclusion of this initiative was that some of the 'rules of thumb' that had been adopted throughout the tunnel ventilation industry had limited scientific basis and that this was particularly true for the ventilation criteria relevant to fire-based emergencies. In contrast to the relative wealth of full-scale vehicle emissions data pertaining to pollutant control under normal operation, it was realised that the accepted design 'rules' for emergency ventilation had not been rigorously tested against comprehensive full-scale data. Despite an increased awareness by society in general of life safety issues during fire emergencies, it was concluded that there was 'no definitive and universally accepted consensus regarding the design and operation of ventilation systems for road tunnel fire emergencies'. The ASHRAE technical committee also found that 'The standards that have been applied to existing facilities have been based upon theoretical analyses, empirical values and individual judgement and/or experience – not drawn from comprehensive testing.' Consequently, opinions differed regarding the capabilities of various types of ventilation system to effectively manage heat and smoke and whether sprinklers could be a useful addition to tunnel fire safety (Parsons Brinckerhoff, 1996).

This state of affairs led to the realisation in the USA that there was an important gap in the knowledge with respect to the design of emergency ventilation for tunnels. It was concluded that there was a pressing need to improve this situation, particularly given the reduction in vehicle emissions which had taken place over the past three decades; hence the trend would be for emergency smoke control to replace 'normal running' emissions dilution as the primary design driver. The commitment to a full-scale test programme in the USA was finally undertaken during the late 1980s as a result of the proposed Boston Central Artery/Tunnel multi-billion dollar road tunnel construction project (Luchian, 1992); it was considered that the latter project would benefit greatly from the acquisition of new (full-scale) data on the interaction between tunnel fires and various ventilation configurations. The MTFVTP (Parsons Brinckerhoff, 1996) comprised a total of 98 full-scale fire tests, conducted between September 1993 and March 1995 in a disused two-lane highway tunnel in Virginia. The facility comprised a tunnel 850 m long with a 3.2% slope; the cross-sectional areas of the two-lane roadway and the overhead ventilation duct were approximately 36 and 24 m^2, respectively. The overhead air duct was divided into two by a centrally located concrete dividing wall running the full length of the tunnel. The comprehensive test matrix included transverse, semi-transverse and longitudinal ventilation régimes (including jet fans which, though common in Europe, were virtually non-existent in the USA at the time) with fires of nominally 20, 50 and 100 MW provided by liquid pools; the fire mitigation effects of foam and sprinklers were also examined. The tunnel was comprehensively instrumented in both the up- and downstream directions, with 15 instrument trees located at various cross-sections throughout the length of the tunnel. Measured quantities included air velocity, gas temperature and gas concentrations (carbon monoxide, carbon dioxide and total hydrocarbons); in addition, the rate of fuel supply to the fuel pans was automatically adjusted to maintain a constant total fuel mass in the pans so that the HRR of the fire was essentially constant throughout the test and could be easily estimated.

Among the most important conclusions arising from the MTFVTP test series were as follows.

- A recognition of the need to determine emergency ventilation criteria based on the individual tunnel physical characteristics and the tunnel ventilation system.
- The confirmation that jet fans are an acceptable smoke and heat management solution for fires up to 100 MW (requiring a mean longitudinal velocity of 3 m/s for the Memorial Tunnel geometry).
- A confirmation of the need to throttle the rate of fresh air supply when transverse systems are used to extract smoke from the traffic space at a high level.
- The realisation that the traditional ASHRAE recommendation on the minimum longitudinal air velocity (critical velocity) required to prevent smoke backflow was not valid as a general design criterion. While the 100 cubic feet per minute (cfm) per lane-foot ($\sim 1.55 \times 10^{-3} \, m^3/s$ per lane-metre) was acceptable for some situations, it was found to be clearly inadequate in other circumstances.

A further outcome of the MTFVTP tests was reported by PIARC (World Road Association, 1999): it was noted that above a certain fire size (i.e. heat release rate), and probably depending also on the tunnel cross-section and slope, the dependence of the critical velocity on the HRR was less than that predicted by the standard empirical design equation. This important discrepancy was also discussed by Grant et al. (1998) in a review of smoke control practice during tunnel fires; this review considered experimental data obtained over a range of scales, and concluded that 'current design methods may be conservative, resulting in forced ventilation capacities significantly greater than those required to control upstream smoke movement'.

The impact of this finding on current tunnel ventilation guidelines has now been evaluated by the relevant PIARC committees and also by a Task Group of the National Fire Protection Association on Ventilation and Tenability in order to develop suitable guidelines and standards such as NFPA 130 (NFPA, 2003). This is important because conventional design practice for the tunnel emergency ventilation requirement is largely based on the answers to two questions: 'What is the maximum size of any fire which might reasonably be expected to occur (i.e. the *design fire*) given the use of the tunnel?' and 'What corresponding magnitude of ventilation velocity is required to prevent smoke backflow?' Notwithstanding the progress that has been made in recent years, the situation is somewhat more complex than these two simple questions might imply.

Thus, information on design fire sizes relating to road tunnels was first suggested by Heselden (1976), but, since then, the EUREKA-EU 499 'Firetun' project (Studiengesellschaft Stahlanwendung, 1995) has yielded much more refined quantitative data on the heat release rates typical of large fires in ventilated tunnels (e.g. French, 1994). Other information is also slowly becoming available (Directorate-General for Public Works and Water Management, 2002; Huijben, 2002), particularly from a number of projects funded by the European Community – for example, UPTUN. However, there is still uncertainty as to how the tunnel ventilation affects certain fire characteristics, such as flame length, HRR, fire spread, smoke evolution and the ratio of carbon monoxide to carbon dioxide production. These issues are explored in more depth in the following, particularly in the light of the objectives set for the refurbished Mont Blanc Tunnel ventilation system (Brichet et al., 2002), which define a more complete set of criteria. These are

- to limit fresh air supply to the fire so as to limit fire growth
- to confine smoke to the smallest distance surrounding the fire location
- to provide tunnel occupants initially with escape routes with tenable levels of temperature, toxicity and visibility from both up- and downstream of the fire, by allowing development of 'stratified' hot layers
- at later stages to allow emergency personnel to approach and effectively fight the fire in safety from close in, thereby minimising damage to the tunnel infrastructure.

At Mont Blanc, these demanding objectives will be achieved by computer control of an existing transverse system configured to extract smoke with additional jet fans to provide additional control. Figure 10.5 gives a visual representation of four ventilation configurations for this system. In particular, Figure 10.5d shows how the system is used to confine smoke in a short length of tunnel centred around the smoke extraction point.

Recently, the International FORUM of Fire Research Directors has recognised the complexity of modern tunnel systems, and has set out a series of potential design requirements for tunnel fire safety systems. In particular, for ventilation, it suggests that systems must be designed to handle a wide variety of fire scenarios and have the additional capability to extract smoke from a number of locations in the tunnel system (Ingason and Wikstrom, 2006). Any balancing longitudinal flow must be sufficient to maintain tenable conditions for stranded tunnel occupants and fire-fighters. However it has also recognised the hazards of rapid fire growth, and additionally suggests that designers should also consider the influence of the tunnel ventilation on fire growth.

10.3. Influence of ventilation on tunnel fire characteristics

The first requirement in the design of any tunnel fire emergency system is to specify the design fire; this might be the maximum credible incident or the representative set of typical incidents that may occur in the facility given its use, size and structure. For a fire, the specification should include, at least, the variation of the HRR over time together with the rates of production of smoke and toxic species. Ideally, there is now also a need to know how such a fire responds to changes in ventilation, including the amount of smoke to be extracted and the way smoke may be controlled.

10.3.1 Variation of heat release with ventilation velocity

Surprisingly little information is available to add to the initial estimates produced by Heselden in 1976 to assist tunnel designers on the type and characteristics of fires that may occur in tunnels.

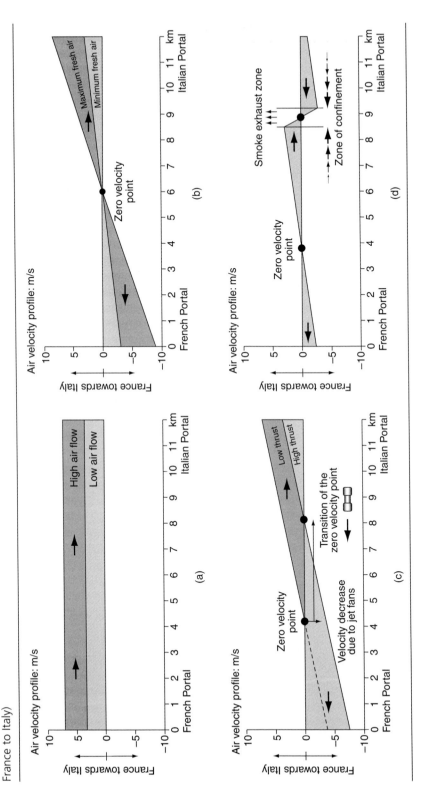

Figure 10.5 Typical air flow velocities in the Mont Blanc Tunnel equipped with both semi- transverse and longitudinal systems: (a) pure longitudinal ventilation under the influence of an ambient pressure difference; (b) semi-transverse with fresh air blowing only; (c) as (b) but with jet fans operating, increasing or decreasing the velocity by a constant value over the entire tunnel (the zero-velocity point moves in a direction opposite to the direction of jet fan flow); (d) the capability to confine the smoke in a restricted length of the tunnel – a convergence zone is set up around the smoke exhaust point, which pulls back smoke that would have otherwise have filled the tunnel without control of the longitudinal velocity (From Brichet et al., 2002; BG Consulting Engineers, Switzerland). (Note that the positive sign has been defined for the air velocity from France to Italy)

From a consideration of the available fuel, Heselden suggested that a HGV and passenger car might produce heat outputs in the region of 20 and 5 MW, respectively. As a result of confinement, it is thought that fires in tunnels are more severe than equivalent fires in the open. Due to the difficulties and expense of conducting large tests in such an environment, there are few experiments to confirm this. Of the large test programmes carried out, many, such as the MTFVTP and the UK Health and Safety Executive large experimental-scale test programme (Bettis *et al.*, 1994a), concentrated on smoke movement and control using well-characterised pool and crib fire sources.

This situation was somewhat eased by the EUREKA 499 'Firetun' project (Haack, 1994; Studien-gesellschaft Stahlanwendung, 1995). This produced valuable practical data from a series of fires on real vehicles in a disused mine tunnel in northern Norway. These tests produced maximum HRR, smoke and toxic species production rates for a variety of vehicles, including a passenger train coach (13.5 MW), an underground railway/metro coach (35 MW) and a school bus (29 MW), and culminated in a fire on a HGV laden with a cargo of furniture. This latter fire produced a maximum heat output of at least 120 MW. These tests also produced data on the rates of smoke production and the resulting tunnel optical densities from these vehicles.

However, despite the urgent need to examine the wide range of cargoes and investigate the influence of a range of ventilation conditions, tunnel aspect ratios, etc., there was little follow-on effort. In particular, since the increase in the forced ventilation flow was now one potential response to mitigate the early stages of a tunnel fire incident, it was important to know how a fire responded to an increase in longitudinal ventilation. In the absence of a large amount of experimental data, Carvel *et al.* (2001a,b,c, 2004) employed a Bayesian statistical approach combining expert judgement with the little existing experimental data in an attempt to answer this question. They recognised that data were available from a number of different fire tests in a variety of different tunnels under a range of different ventilation conditions, but there was no coherent data set and no way of using such mixed data sets using traditional statistical analysis. The methodology, assumptions and results of this study are described in Chapter 11.

Results from fires in a road tunnel in which some repeat fires were conducted at different ventilation velocities have been presented in the Second Benelux Tunnel tests (Directorate-General for Public Works and Water Management, 2002; see also Huijben, 2002). These fires involved simulated truck loads, passenger cars and pool fires. For small fires, it was reported that ventilation had little effect on heat output after about 20 minutes. Indeed, car and pool fire tests with forced ventilation gave a lower heat output during that period. For truck loads there was a marked observed increase in the rate of heat release. However, the increase of about 35% was not as severe as predicted by the model of Carvel *et al.*. These results have been supported by model-scale experiments carried out by Ingason and Li (2010). They observed only a weak dependence of the HRR on ventilation in a long-itudinally ventilated tunnel. They noted an increase in the maximum heat release of about 50% for forced ventilation compared with a naturally ventilated tunnel. However, it should be noted that the Second Benelux Tunnel tests were carried out in a two-lane tunnel, while the results of the study on HGV fires by Carvel *et al.* are for single-lane tunnels. These new experimental data (and others) have been used by Carvel *et al.* to update the results of their study (Carvel *et al.*, 2004). Also, it should be noted that the tests made by Ingason and Li (2010) were at model scale.

Forced ventilation systems can affect fires in tunnels in many ways. In some instances, increasing the ventilation velocity may cause a reduction in the severity of the fire; however, in other instances, increasing the ventilation velocity will cause the fire to engulf a HGV load more rapidly or cause a substantial increase in the HRR of the fire. Adopting a fixed forced ventilation rate (e.g. 3 m/s, as recommended in some countries; Nordmark, 1998) in all tunnels may not be the most sensible option: different tunnel fires respond to forced ventilation in different ways, and the emergency ventilation system used in the event of a tunnel fire should be appropriate to the specific fire scenario, wherever possible.

10.3.2 The influence of ventilation on fire spread

The development of fire depends on how rapidly flame can spread from the point of ignition to involve an increasingly large area of combustible material. It is a very complex process even for the simplest arrangements of fuels and environments. Drysdale (1998) has reviewed the main factors controlling these processes. These fall into two main areas: those related to the fuel–surface orientation, fuel thickness, density, thermal capacity and conductivity, and fuel geometry; and those related to the environment, such as imposed radiative flux, fuel temperature, imposed air movement and atmospheric composition. These factors are also likely to be important in a tunnel environment but, due to the confinement, some, such as ventilation and imposed heat flux, may be relatively more significant.

Historically, the problem of fire spread in a tunnel has received even less attention than fire size. Before the Runehamar tests, the only known experimental study had been carried out in Finland, and involved the spread of fire between a series of wooden cribs (Keski-Rahkonen et al., 1986). Subsequently, experiments conducted in the Netherlands (Directorate-General for Public Works and Water Management, 2002; Huijben, 2002) indicated that increased ventilation resulted in little change in the development of smaller fires such as in passenger cars but substantially increased the rate of development in larger loads such as HGV cargoes. Ingason et al. have attempted to rectify this situation. They have carried out tests with wooden cribs in both full-scale (Runehamar) and laboratory-scale tunnel fire test programmes (see Chapter 14 and Ingason and Li (2010)). Their laboratory data show an important and clear linear increase in the fire growth rate with increasing ventilation velocity. They found that doubling the ventilation velocity produced a similar increase in the fire growth rate.

Due to this lack of data, Carvel et al. also applied their Bayesian techniques to the spread of fire through a HGV (see Chapter 11). Their results suggest that the velocity of forced ventilation needs to be kept as low as possible to minimise the rate of fire spread across a HGV.

Where a deterministic treatment of fire development is required, it is generally considered in very simple terms. The analysis of the Channel Tunnel fire of 1996 by Liew et al. (1998) is a good illustration, and demonstrates that little specific information is available for tunnels, with reliance being put on a general fire engineering practice. They examined a number of modes for fire spread along a rake of HGVs carrying a variety of loads including, among others, flame impingement on downwind vehicles, propagation of burning brands or burning liquid, ignition of downstream vehicles through exposure to some critical thermal dose, and a 'flashover'-type mechanism due to radiation from flames at the tunnel ceiling. (Strictly speaking, the concept of 'flashover' does not apply to tunnels, as it is defined only in relation to compartments. For tunnels, it may be better to refer to 'major fire spread' or 'rapid major fire spread'.) By working from the damage produced and timescales for fire development, they concluded that the major mode of fire spread between vehicles was flame impingement. However, they also suggested that this might only occur over a restricted window of ventilation velocities. Thus, at low velocities, flame impingement on the ceiling is likely, and propagation by a remote-ignition-type mechanism may be more appropriate, whereas, at high velocities, flame lengths may be shortened and attached to burning vehicles. Thus, a thermal dose mechanism might be more important. The flame length used by Liew *et al.* was derived by reference to the little data available at the time from the Hammerfest fire and the MTFVTP, and also drew on open-air crib and pool fires carried out by Thomas (1963) and Pritchard and Binding (1992). This indicated a linear dependence on the HRR and a decreasing flame length with increasing ventilation velocity. The flames were shorter and were tilted in the direction of flow to a significantly greater extent than in the open. The model took the form

$$L = L_c(Q/Q_c)(u/u_c)^{-0.4}$$

where L_c, Q_c and u_c are a reference length, the HRR and the velocity, with, respectively, values of 22 m, 120 MW and 10 m/s.

Further data on flame lengths and tilt are given by Oka *et al.* (1998) and Kuwana *et al.* (1998), who examined the variation of flame length and tilt for flames that both did not touch and touched the ceiling. They obtained a similar dependence on the ventilation rate, though the dependence on the HRR was somewhat weaker. Li *et al.* (2010) and Wang *et al.* (2009) have also carried out similar measurements to examine flame length and tilt in test facilities from laboratory to full scale and covering a range of fire sizes. They have also examined the variation of the total heat flux at the floor and the maximum ceiling temperature with ventilation velocity.

Ignition and fire growth criteria used by Liew *et al.* were based on standard fire engineering practice and were independent of the tunnel ventilation rate. Thus, Following Heselden (1976), ignition was assumed when a vehicle had been exposed to temperatures close to 600°C, identified with impingement of the flame tip, for a prescribed time, and fire growth followed a quadratic time dependence of the form αt^2, with values of the growth constant α taken from standard literature values with no allowance for possible enhanced spread due to the tunnel environment.

The only comprehensive treatment of fire spread in a tunnel has been that due to Beard (Beard, 1997, 1998, 2001, 2003; Beard *et al.*, 1995). He has developed a suite of non-linear models of thermal instability in fires to examine the ignition of objects downwind of an existing fire in a tunnel. The series of models has been devised under the generic title 'FIRE-SPRINT', which is an acronym for 'fire spread in tunnels'. Longitudinal forced ventilation is assumed, and illustrative simulations have been presented for a tunnel similar in size to the Channel Tunnel. The 'A' series of models assumes that the flame from an initial fire does not impinge on a target object downstream whereas the 'B' series assumes flame impingement (see Chapter 16). For the 'A' series, a control volume partially surrounding the target object is assumed downstream of the fire, and an energy conservation equation is used.

The 'A' series models determine how the temperature T of the gases in a control volume, partially surrounding the target object, varies with the unenhanced source fire mass loss rate M_{fun}. Thus, fire spread occurs by a process of remote ignition through a build-up of heat around an object due to radiation from the fire and hot combustion products. Under certain circumstances, Beard has suggested that the variation of T with M_{fun} will produce the type of curve shown in Figure 10.6, a so-called 'S curve' with two stable (solid) branches and an unstable (dashed) branch. In real terms, the fire may be regarded as developing along the lower stable branch. When the unenhanced mass

Figure 10.6 A typical (FIRE-SPRINT A2) 'S' curve for a HGV in a tunnel subject to fire attack for a vehicle separation of 8.75 m. The upper curve shows how the equilibrium temperatures of the system vary for each assumed value of the unenhanced fuel mass loss rate at the upstream fire obtained from solution of the energy balance equation. The solid lines show stable solutions; the dashed lines unstable solutions. The lower curve shows the eigenvalue, which provides a numerical measure associated with the nature and stability of the system. The negative parts of the curve show stable solutions. Spread of fire is identified with transition from one stable part of the curve to the other (From Beard (1998), with acknowledgements to ITC Ltd and the University of Dundee)

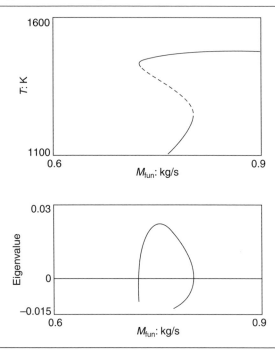

loss rate reaches the lower fold point of the S curve, the system would jump to the upper stable branch, this transition being interpreted as the spread of fire to the target object. This novel way of investigating fire spread has several difficulties, most notably the requirement to significantly simplify the equations for energy gain and loss within the control volume to make the system tractable. This shows that care must be taken when choosing these. However, Beard has codified the models and used them (see Chapter 16). Beard also found that the results were sensitive to the assumptions used in the model. For example, the results are dependent on the length of the control volume considered. In general, he found that for a tunnel similar in size to the Channel Tunnel (near-circular cross-section with a diameter of about 7.6 m), then the critical rate of heat release (Q_{fc}) for spread to a HGV-like object increases with increasing longitudinal ventilation velocity. That is, for spontaneous ignition.

The 'B' series models assume that the flame from the initial fire does impinge on the target object. So far, there has been one model in the 'B' series: FIRE-SPRINT B1, which has largely the same assumptions as FIRE-SPRINT A3, but additionally assumes a small but persistent flame impingement from the initial fire onto the target object (Beard, 2003). The 'B' series models

Table 10.1 Values of the critical HRR, Q_{fc}, found using FIRE-SPRINT A3 and FIRE-SPRINT B1, compared with Runehamar experimental test results, assuming the ventilation velocity $= 2.5$ m/s

Test	Q_{fc}: MW	Q_{fc}: MW	Q_{fc}: MW
	FIRE-SPRINT A3	FIRE-SPRINT B1	Experiment
T3	46	15	43 (observed directly)
T2	43	14	16 (inferred by experimenters)
T1	43	14	29 (inferred by experimenters)

have two control volumes: a large control volume similar to the one considered in the 'A' series of models and a small control volume at the point of impingement. In reality, there would be cases where flame impingement occurs but would not be persistent and other instances where the degree of impingement would be much greater than currently assumed in the B1 model. It has been found that the critical HRR (Q_{fc}) for spread in a tunnel similar in size to the Channel Tunnel increases with increasing ventilation velocity, but Q_{fc} is considerably less when assuming flame impingement. However, it has also been found from a study that increasing velocity increases the probability of flame impingement (Carvel *et al.*, 2005).

Until recently there had been no experimental estimates of the critical HRR for fire spread in a tunnel, from an initial fire to a target object, with which to compare FIRE-SPRINT calculations. The Runehamar test series (Lonnermark and Ingason, 2006) contained such estimates, and simulations conducted on an *a priori* basis have been carried out. An *a priori* basis means (see Beard (2000) and Chapter 29) that results from the tests being used for the comparison were not used in constructing input to the simulations, and adjustments to the input to obtain a 'good fit' were not made. All FIRE-SPRINT models were run for each of three tests in the Runehamar series (T1–T3), and approximate correspondences looked for (Beard, 2007). Table 10.1 provides some of the results. For two tests (T2, T3), approximate correspondence was found, enabling inferences to be drawn about whether ignition had been spontaneous or via flame impingement. For the third test (T1), the experimental result was midway between predicted values for 'spontaneous' or 'flame impingement'. In this case, it might be inferred that ignition had been via a burning brand or intermittent flame impingement.

The Runehamar test series gives one experimental set; it would be desirable to have replications. In general, it is very important to have replications of tests because ostensibly 'identical' tests may well produce quite different results. (See also Chapter 29.) Beyond the Runehamar tests, many more experimental observations of critical HRRs are needed, covering different conditions. Also, large-scale tests are needed, not just small-scale investigations.

10.3.3 Ventilation control of tunnel fires

It is known that under certain circumstances, fires in compartments undergo a transition to ventilation-controlled burning. Thus, for a fully developed fire, the rate of burning is determined by the maximum rate at which air can flow into the compartment. Such conditions are often accompanied by the onset of external flaming, and if the supply of oxygen falls below that required for stoichiometric burning, then the production rates of smoke and carbon monoxide may increase significantly.

The possible occurrence of ventilation-controlled burning conditions with a forced ventilated tunnel fire is not settled. There are suggestions that periods during the Hammerfest HGV fire exhibited some characteristics of ventilation-controlled burning, and some of the tests reported by Bettis *et al.* (1994b) produced very low oxygen concentrations in the exhaust, and might, at least, have produced some local ventilation control. Its importance has been demonstrated by the one-dimensional energy balance model of Atkinson (1997), who showed that increasing the ventilation to a ventilation-controlled fire could increase temperatures downstream and thus increase the exposure of people there and/or any temporary safe refuge.

The importance of local ventilation control in ducts and tunnels has been highlighted in work carried out for the mining industry (Hwang *et al.*, 1991). Experiments were carried out in an experimental duct and very high flame spread rates on conveyor belting were noticed when it was raised towards the ceiling. In these circumstances, the flow in the belt–ceiling gap was much reduced, and this in turn produced high fuel–air ratios, high levels of soot and radiant fluxes, and enhanced rates of flame spread, often by a factor of up to five. Thus, there may be occasions when such conditions occur within the confines of a transport tunnel, and further work is necessary to fully understand the conditions that promote such rapid rates of flame spread.

10.3.4 Control of smoke by ventilation

Following on from knowledge of the type and size of fire that can occur in tunnels, the other most important piece of information required by tunnel designers and safety engineers is the way in which a buoyant fire plume moves within the tunnel environment and how it interacts with a forced ventilating flow. Since the realisation that a tunnel ventilation system could be used to control smoke movement, this is a topic that has received increasing attention.

The issue of smoke control and, particularly, the ventilating flow required to prevent the backflow of smoke from a fire has received great attention, especially the way in which it varies with the HRR. Indeed, the topic was considered to be so important as to merit a seminar dedicated solely to its discussion (ITC, 1996). In general, workers have concentrated on the critical velocity, defined as that flow which just prevents any flow upstream of the fire. Much of the work has been carried out at laboratory scale (Oka and Atkinson, 1995; Megret, 1999) since this is a phenomenon where extrapolation to full scale can be easily justified by the use of Froude number scaling. The use of small-scale experimental rigs has allowed a large number of situations to be investigated, including different tunnel cross-sectional shapes, aspect ratios (Wu *et al.*, 1997; Wu, 2003), fire shapes and heights within the tunnel and slopes (Atkinson and Wu, 1996). The main result from this work was that, at low HRRs, the critical velocity increased with the one-third power of the HRR, but, at a certain fire size, it remained constant for further increase in the fire size – the so-called 'super-critical velocity'. The explanation for this effect is that the transition in behaviours occurs when the fire is of such a size that the intermittent flaming region impinges on the ceiling. At this stage the flame structure changes, not allowing smoke to back up along the ceiling. Figure 10.7 clearly shows the form of this relation. The equations relating the critical velocity and fire size are given by

$$V^*(0) = 0.40(0.20)^{-1/3}Q^{*1/3} \qquad \text{for } Q^* < 0.20$$

$$V^*(0) = 0.40 \qquad \text{for } Q^* > 0.20$$

$$V^*(\theta) = V^*(0)\,(1 + 0.014\theta)$$

Figure 10.7 Variation in the critical velocity with the HRR for a variety of experimental and large-scale experiments. The behaviour at high HRRs is apparent (From Wu (2003) with permission)

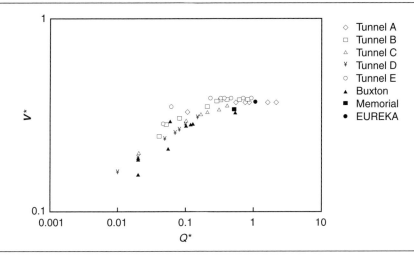

where $V^*(0)$ is a non-dimensionalised critical velocity given by $V/(gH_D)^{1/2}$ for a horizontal tunnel, Q^* is a non-dimensionalised HRR given by $Q/[\rho_0 C_p T_0 (gH_D^5)^{1/2}]$, H_D is a tunnel hydraulic diameter given by $A/4P$, A is the tunnel cross-sectional area, P is the tunnel perimeter, θ is the tunnel slope in degrees, g is the acceleration due to gravity, ρ_0 is the density of incoming air, C_p is the specific heat of air and T_0 is the temperature of the inflow.

Figure 10.7 also includes some larger-scale data obtained from large experimental facilities (Hwang *et al.*, 1991) and from the MTFVTP (Parsons Brinckerhoff, 1996). These are in reasonable agreement and give supporting evidence for the constancy of critical velocity at high HRRs.

There is also work by Lee and Ryou (2004) that suggests that the critical velocity depends on the aspect ratio of the tunnel, for a given value of the mean hydraulic diameter H_D (defined later in this section). Kunsch (2002) has also produced simple correlations incorporating the tunnel aspect ratio using a one-dimensional modelling approach.

There have been a number of further experimental tunnel fire programmes, at both full and laboratory scale, that have further examined the variation in the critical velocity with fire size. These have largely confirmed the relationship expressed above. Examples include work reported by Li *et al.* (2010) and Wang *et al.* (2009).

A related but different quantity has been investigated by Telle and Vauquelin (2002), again using a Froude-number-scaled laboratory scale rig. They have investigated the control of smoke using transverse smoke extraction at the ceiling, and defined a velocity analogous to the critical velocity – 'the confinement velocity'. In this arrangement, backflowing smoke is extracted upstream of the fire, and any flow beyond the vent is prevented by an induced longitudinal flow into the tunnel through the portal. By reducing the extraction velocity, the length of the smoke layer upstream of the extraction point is seen to increase. The advantage of this arrangement is that the overall flow velocity at the fire is zero. Telle and Vauquelin have used helium–air mixtures to simulate the buoyant flow from the fire, and carried out experiments in a 1/20th-scale rig modelling scaled fire sizes up to 20 MW. The functional form of the confinement velocity was identical to that given by Wu *et al.* for low HRRs, with magnitudes agreeing to within 10%. The transition to a constant confinement velocity at high HRRs was not observed, though this is possibly due to the range of fire powers used in the experiments and differences between the helium–air and real fire plumes.

Vantelon *et al.* (1991) employed yet another approach. Again using a scale model, they attempted to obtain a relation between the length of the backlayer L and the longitudinal ventilation velocity. This was further investigated by Saito *et al.* (1995). They found that the length of the layer, as expected, varied as the one-third power of the convective HRR (Q), and was inversely proportional to the ventilation velocity u. The relationship took the form

$$L/R \sim (gQ/\rho C_p Tu^3 R)^{0.33}$$

where R is the tunnel radius and ρ, C_p and T are the density, specific heat and temperature of the inflowing air, respectively.

Since only low heat outputs were used in this study, any transition to a constant value at high HRRs was not observed. Li *et al.* (2010) have also examined the behaviour of backlayer length using a laboratory experimental facility, and observed similar functional dependencies. They reported the backlayer length as a function of the ratio of the longitudinal ventilation velocity to the critical velocity. This work also indicated that at higher HRRs the behaviour of backlayer length mirrored that of critical velocity in that there was a fire size above which the layer length was constant. Their relationships took the form

$$L^* = 18.5 \ln(0.81 Q^{*1/3}/V^*) \qquad \text{for } Q^* < 0.15$$

$$L^* = 18.5 \ln(0.43/V^*) \qquad \text{for } Q^* > 0.15$$

where L^* is the non-dimensionalised backlayer length (L/H_D), and the non-dimensional HRR and velocity are as previously defined.

Viot and Vauquelin (2002) have considered the use of smoke extraction in further detail, using two vertical extracts of different widths, shapes and locations, positioned symmetrically around a fire. They have examined the use of such extraction ducts in the natural ventilation mode. They found that such a system had the advantage of preserving any stratification in the smoke layer, and that it was possible to remove smoke from substantial 25–30 MW fires using such techniques, preventing any smoke transport beyond the vent. The capability of the vents was largely dependent on their length and height. Thus, the higher and longer the vent, the more efficient they were in capturing smoke. However, Viot and Vauquelin indicated that further investigation was required since vent performance was very sensitive to external influences such as the tunnel slope and longitudinal movement due to pressure and temperature differences at the portals.

The use of forced ventilation for smoke control is currently being examined in greater detail by various other workers and designers of specific new systems (e.g. Brichet *et al.*, 2002; Weatherill *et al.*, 2002). Also, Ingason (2002) has investigated the use of a mobile fan to clear smoke and to allow fire-fighting. In a 1.1 km tunnel of 50 m^2 cross-section, he used a fan of 30 m^3/s capacity, and found the velocity varied between 1.5 and 2.0 m/s. This cleared smoke from a fire of 2.6 MW output in some 6–7 minutes. Elsewhere, the subject of jet fan performance has been widely addressed, especially the siting of fans for optimum performance in a fire relative to the tunnel portal.

10.4. Modelling tunnel flows

To assess the hazards arising from tunnel fires and to plan effective mitigation strategies, it is necessary to have a prior understanding of the behaviour of fire and smoke movement in tunnel environments. It is not practicable to investigate every individual construction experimentally, hence predictive models are required. These models range from simple empirical relationships, through phenomenological or integral/zone models, to complex, three-dimensional, computational fluid dynamics (CFD) simulations. All model types are potentially useful. Thus, the simpler models are ideal for repeated applications, such as would be required in a risk assessment, where cost and practicality are also important. CFD is expensive to apply repeatedly, and so is used primarily as a research tool. Ideally, the requirement is for a well-tested, yet accessible, suite of models for use in ventilation system design and emergency procedure development.

The simulation of fires in longitudinally ventilated tunnels has received the greatest attention in the literature; however, applications to transverse or semi-transverse ventilation systems have been made. Close to the fire, the primary aim has been to predict the critical longitudinal ventilation velocity required to halt the upstream movement of combustion products. The more sophisticated models, typified by CFD analyses, have also captured details of the complex flows in the immediate vicinity of the fire. Further downstream, where conditions are homogeneous, simple models have been devised to simulate the movement and dilution of combustion products through a complete network of tunnels. Here, the physical basis of these diverse models is described, their capabilities and inherent limitations highlighted, and applications surveyed. Experimental testing of model predictions is required periodically to ensure the reliability of the method.

10.4.1 Network ventilation modelling

Models such as SES, VENDIS-FS and MFIRE (Chang *et al.*, 1990), developed by the US Bureau of Mines, are more commonly used in the design of complete ventilation systems for tunnel networks and for modelling the overall response of these networks to fire. Typically, these models calculate the disturbed network ventilation in the event of fire and allow the user to follow the downstream time development of smoke movement and its relative concentration. For simulation purposes, the network of tunnels is considered to be composed of closed circuits of airways intersecting at junctions. The equations for the conservation of energy for each circuit, and mass conservation at each junction, are applied to give a system of algebraic equations that are solved iteratively. Circulation fans can be included.

Network-based approaches generally provide no detail of the local fire-generated flows, nor their interactions with ventilation (but see Chapter 17). This is because they assume that heat and products of combustion are uniformly distributed over the cross-section and length of a tunnel. Nevertheless, these techniques are useful for giving a global view of the consequences of fire in a tunnel network, allowing evacuation strategies to be planned and assessed. In addition, they can provide boundary conditions for more complex modelling methods. Network models appear to have received limited experimental testing in the event of a fire, although there is little reason to suspect that they would give misleading results – at least when combustion products have cooled to near-ambient conditions. The only recent testing carried out has been that by Cheng *et al.* (2001), who used a scale model comprising a system of stainless steel pipes 0.1 m in diameter providing 15 nodes and 27 tunnels, an axial fan and a 1 kW heater to assess the model MFIRE. Cheng *et al.* used the flow rate and temperature to compare model performance with experiment. They found good correlation for the former but poor agreement for simulated and measured temperatures, probably due to scaling effects.

The Subway Environmental Simulation (SES) code, developed under the lead of Parsons Brinckerhoff for predicting ventilation flows in tunnel networks, also includes a simple model to calculate the critical velocity U, devised by Danziger and Kennedy (1982). Once again, this is based on a Froude number modelling approach. The relevant relations are

$$U = K_g k (gQH/C_p \rho_0 A T)^{1/3}$$

$$T = (Q/C_p \rho_0 A u) + T_0$$

In the above, Q should again be the convective HRR from the fire, although this is not stated explicitly by Danziger and Kennedy. The cross-sectional area of the tunnel appears as A, and K_g is a 'grade correction factor', to be applied for fires in sloping tunnels, and derived from the work of Bakke and Leach (1960), who studied methane layer propagation in sloping tunnels. However, the magnitude of the density difference, and the nature of the source of the buoyant flow, are very different in these two cases. It is by no means certain that gradient effects can be represented in such a simple manner, nor even that Bakke and Leach's data are relevant to ventilated tunnel fires. The SES fire model predicts that the critical velocity rises as the fire heat output increases. However, the appearance of T in the denominator ensures that at large heat outputs the critical velocity tends to an asymptote of near-constant velocity.

This behaviour initially appears to show some similarity to that observed in the Buxton fire trials (Wu, 2003). For the model results to remain physical, though, T must be set to an upper limit concomitant with some maximum potential flame temperature. For very large heat outputs, T is thus fixed at this upper limit, which for further increases in fire size means that the critical velocity again rises with the one-third power of heat output. The value of the constant k was set to 0.61 by Danziger and Kennedy. This figure is based on experimental work carried out by the US Bureau of Mines on fires in small-scale wood-lined ducts, reported by Lee *et al.* (1979). Only four tests were carried out, and yet again the applicability of these limited data to full-scale fire trials must be questioned. For instance, the duct is just 0.27 m square and lined with fuel on all exposed surfaces, over much of its length. In addition, there are only ten tunnel heights upstream from the fire, and over this length the duct area at its open end decreases to just 55% of its nominal area. Under these conditions, the flow will be far from fully developed, and thus far removed from conditions encountered in reality for fires occurring well inside long vehicle (or mine) tunnels.

Danziger and Kennedy's value for k, of 0.61, contrasts to that of Thomas (1970), who recommended a value of unity, and Heselden (1976), who advocated a value of 0.8. These different k values arise because the experimental measurements, on which they are based, derive from a wide variety of tunnel shapes, sizes and fire scenarios. The differences in the Froude number at which backing-up just occurs, $Fr_{m,cr}$, are even more marked. This is due to the cube-root form of these relations. Thus, $Fr_{m,cr}$ ranges from unity in Thomas's model to 4.5 in the SES model. Notwithstanding the above, the SES fire model has often been used in tunnel ventilation system design. The model is also incorporated in D'Albrand and Bessiere's VENDIS-FS tunnel network code (D'Albrand and Bessiere, 1992), developed at the Institut National de l'Environ-ment Industriel et des Risques (INERIS), as well as Te Velde's zero-dimensional code for simulating flow in a vehicle tunnel incorporating slip roads (Te Velde, 1988).

10.4.2 Phenomenological models

Phenomenological models attempt to predict the main features of fire-generated flows in unobstructed tunnels, both upstream and downstream, and their interaction with longitudinal ventilation. They thus provide more detail than the simple empirical approaches, but less than the multi-dimensional information available from CFD techniques. Solutions are typically achieved by solution of a set of ordinary differential equations. There are relatively few published models that fall into this category. The examples given below are by Hwang *et al.* (1977), Daish and Linden (1994), Kunsch (2002) and Charters *et al.* (1994). The first of these employs an integral approach, while Charters presents a zone model of tunnel fires. Daish and Linden's model is a hybrid of the two, using an integral method only for the hot stratified layer, coupled with an algebraic relation for the thickness of the mixing layer that exists below the hot layer. In each approach, the domain of interest has typically been split into several distinct regions, for instance a plume impinging on the tunnel crown and downstream hot and cold layers. In this way, attempts are made to address the key phenomena that occur in identifiable regions of interest.

In the integral approach, one-dimensional integral equations for mass, momentum and energy conservation are approximated over each of these regions. The resulting ordinary differential equations are then solved using commonly available techniques. With a zone modelling approach, the fire environment is split into discrete volumes, and zero-dimensional relations are approxi-mated over each of these regions, or zones. These relations express the exchange of mass, momentum and energy at zone boundaries, assuming uniform conditions within each zone. Consequently, the domain may be split into many zones in an attempt to retain some dimensional detail; in the case of a compartment fire, suitable zones would be the flame plume, the hot smoke layer and the cold air layer below. The main advantage of this class of model is that, in principle, the major physical phenomena of importance are represented, and so gross features such as the existence of a backed-up smoke layer, or evolution of ceiling layer temperatures, should be reproduced. In addition, the models run quickly on a personal computer. This approach, does, however, demand that the important flow and heat transfer processes can be identified and understood *a priori*.

A major criticism of phenomenological approaches is that they are of necessity based on gross simplifications of a complex combustion and fluid flow scenario. This is acceptable if there is a strong experimental foundation for such simplifications, as exists with zone models of compart-ment fires. Unfortunately, it is debatable if the same can be said to be true of these approaches as applied to fires in tunnels.

Hwang *et al.* (1977) assumed that three distinct regions exist: a plume whose angle of inclination is affected by the imposed longitudinal ventilation, a plume–ceiling impingement zone referred to as a 'turning region' and a hot ceiling layer. For each of these regions, assumptions are required to yield closed forms of the governing equations. For instance, the plume is initially assumed to be two-dimensional, vertical and with the fuel being instantaneously burned at its base. Entrainment of ambient air is estimated from plume theory, and properties are assumed to be uniform across the plume. Within the turning region, the fluid density is assumed to be constant. In the ceiling layer, all properties are assumed to be uniform over the cross-section, and entrainment rates are again estimated. In addition, there is no feedback from the computed upstream conditions to the plume region. The modelling of the plume is carried out simply to supply initial conditions for calculating the developing upstream and downstream hot layers. In reality, any changes in the upstream flow will affect the fire, with the two regions being coupled in a complex non-linear manner. Each of these assumptions is open to question, especially those concerning the two-dimensionality of the plume. The only comparison that Hwang et al. make between model predictions and measurements is with a single data point due to Eisner and Smith (1954). In this instance, the backed-up layer length is grossly underestimated.

Daish and Linden's (1994) model shows some similarity to that of Hwang et al., but is simpler and based on additional assumptions. The authors assume that the flow is composed of a hot plume and ceiling layers travelling both up- and downstream. In the region downstream from the fire, it is assumed that a layer of air at ambient temperature exists at low level. Such a layer proved difficult to identify in the experiments reported by Bettis et al. (1994b), but has previously been identified in other trials, probably with much lower heat outputs. This fire plume is considered to behave according to classical theory, and its trajectory is assumed to be unaffected by imposed ventilation. The effect of neglecting the inclination of the fire plume is, as yet, uncertain. However, several workers have observed that the plume can be inclined at substantial angles to the vertical. The simple model of Hwang et al. for the plume–ceiling impingement region indicates that, at an inclination of between 45° and 60° to the vertical, the mass fraction of plume products travelling upstream could be as small as 15% and 7%, respectively.

The performance of Daish and Linden's (1994) model was evaluated by comparison with the data obtained from the comprehensive Buxton fire trials (Bettis et al., 1994a). The model allows predictions to be made of the evolution of the hot layers – their temperature, depth and propagation distance. Daish and Linden concluded that the model is able to give a fair representation of the existence of any backed-up flow, the evolution of temperature in the hot layer, and the downstream distance at which the hot layer breaks down to well-mixed conditions, providing that 'the fire is not too large and the ventilation not too strong'. The position at which breakdown of the hot layer is deemed to have occurred in the experimental trials is, however, somewhat arbitrary, because in reality this layer is not clearly defined in experimental data. The theoretical model does display the same behaviour of continued rise in the critical velocity with fire size as the simpler models based on Thomas's work. The authors speculate that the resulting disagreement with the observations of recent experiments may be related to the assumptions made for entrainment into the fire plume or non-Boussinesq effects.

Kunsch (2002) has used a similar approach to that above. He derived simplified mass and momentum conservation equations for a control volume, including the front of a ceiling backlayer, and derived analytical expressions for the variation of critical velocity with the HRR. Interestingly, he found asymptotic behaviour, so that the critical velocity was independent of the HRR for large fires.

The model of Charters et al. (1994), referred to as FASIT (Fire growth And Smoke movement In Tunnels), was designed to simulate either natural ventilation conditions or longitudinal ventilation in which all combustion products travel downstream. It assumes that the near-fire region can be represented by a Gaussian plume and that downstream of the fire three distinct layers can be identified: a hot layer, a mixing layer and a cool layer (but still above ambient temperature). Each of these layers is subdivided into multiple zones. The model output consists of a time evolution of each layer's temperature and depth. Once again, the model is based on a number of assumptions. It has already been mentioned that the fire plume is assumed to be Gaussian, but the conditions for this assumption to hold true are unlikely to be met for fires with flaming over a substantial part of the tunnel height. The initial temperature of the hot layer is taken from the Gaussian plume model, and its initial velocity is calculated using a Thomas-type relation. Charters *et al.* (1994) claim that the length of any backed-up layer can be calculated using Vantelon's model, and compare their results to temperature profiles measured downstream of a wood fire in a tunnel. The FASIT model captured the qualitative flow behaviour, but generally overpredicted the measured temperatures by about 100°C. It should be noted that in this experiment the fire source was a very large wooden crib, almost one-half the tunnel height, guaranteeing flame impingement on the tunnel crown. In these circumstances, a Gaussian plume model is not tenable. Similar work has been carried out by Schwenzfier *et al.* (2002) using a modified zone model that splits the tunnel volume into two gaseous volumes and wall zones. This has been used to predict zone temperatures both down- and upstream of the fire and to compare these with CFD simulations and experiment.

10.4.3 Methods based on computational fluid dynamics

The increasing availability of powerful computers, combined with the development of efficient numerical techniques, has spawned many complex mathematical models capable of analysing general problems in fluid mechanics through the solution of the fundamental conservation equations. These models are referred to as CFD models, or more generically as 'field models', in contrast to the 'zone models' considered above. In a field model, the region of interest is divided into small volumes, and the equations representing the conservation of momentum, energy and species concentration, etc., are solved at a point within each volume: this approach permits, in theory, a very fine resolution of the problem in terms of both space and time for all the parameters of interest. The need for powerful computers and efficient solution algorithms stems from the non-linear nature of the problem. In a zone model, perhaps two or three regions are prescribed empirically. A field model comprises several tens of thousands of 'zones', coupled together through the fundamental equations of fluid motion. This approach allows a more mechanistic representation of the processes that occur to be inserted in the model. Consequently, there is less reliance on gross empirical approximations, and representations of phenomena such as combustion or turbulence can be incorporated at a more fundamental level. Sophisticated graphical post-processing of the raw numerical output data enables the most important features of the flow field to be visualised; for example, as carried out by Gobeau and Zhou (2002).

The flow behaviour associated with fires is usually three-dimensional, turbulent and strongly influenced by buoyancy forces. Rhodes (1989) noted that the simplest field models needed to solve equations for velocity components in the coordinate directions (e.g. u, v and w in the Cartesian system), enthalpy h and pressure p, and would employ a fixed value of turbulent viscosity μ_t to represent the effects of turbulent mixing. In these simple models, a fire was represented by a prescribed heat source within the computational domain. The general form of the field model equations were given by Rhodes as

$$\frac{\partial}{\partial t}(r\rho\phi) + \nabla \times (r\rho V_\phi - r\Gamma_\phi\nabla\phi) = rS_\phi$$

where r is the 'phase volume fraction', ρ is the density, ϕ is the dependent variable, V_ϕ is the velocity vector, Γ_ϕ is the 'exchange coefficient' (laminar or turbulent) and S_ϕ represents the source or sink terms (e.g. heat or mass). The numerical form of this equation is obtained by integration over a control volume. The resulting set of partial differential equations (PDEs) are solved for appropriate boundary conditions. The inclusion of turbulence in the models requires further assumptions, since an exact description of turbulence effects would require an infinite set of PDEs. This is known as the 'closure problem', and is conventionally overcome through the incorporation of an approximate turbulence model, the best known of these being the 'k–ε' version.

Rhodes also considered various potential enhancements to the simple CFD fire model. In general terms, these refinements involve the additional cell-wise solution of new equations for phenomena previously described by fixed values of global parameters. Such refinements include the following.

- *A turbulence model.* The most widely used turbulence model is the two-equation k–ε approximation, in which the kinetic energy of turbulent fluctuations k and the rate of dissipation of turbulence energy ε are solved for each computational volume using additional PDEs. Although not a 'universal' model of turbulence, it is widely used, and has been developed and tested for many flows of engineering interest. A modified k–ε model has been described by Woodburn and Britter (1995); it includes the effects of buoyancy and wall damping on the turbulence in a more realistic manner, and is thus of particular relevance to tunnel fire modelling.

  More recently, the more sophisticated large eddy simulation (LES) method is being employed. In this method, an attempt is made to directly model the effects of the largest eddies involving the main part of the turbulent energy but using a sub-grid scale model for the small-scale turbulent viscosity. In general, these follow the widely used treatment of Smagorinsky (1963) for the small-scale turbulent viscosity. This has been stimulated through the public availability of the National Institute of Standards and Technology (NIST) Fluid Dynamics Simulator (FDS) software package for general fire problems (McGrattan *et al.*, 2001).

  Thus, Hwang and Edwards (2005) have used the LES turbulence model in FDS to examine the occurrence of the maximum critical velocity with fire size. They concluded that this maximum was caused by temperature stratification in the tunnel, and corresponded to a maximum temperature at the ceiling above the fire. There has been further work using both Reynolds averaged stress models (RANS) and LES – for example, see Van Maele and Merci (2008).
- *A combustion model.* Combustion models require the solution of further transport equations, specifically for the concentrations of reacting and inert gaseous species. Source terms are also required for the prescription of any kinetically controlled reaction rates. The most basic implementation of a combustion model assumes a diffusion-controlled, 'single-step' reaction of the following form: fuel + oxidant → product. This dictates an instantaneous reaction between any fuel and oxidant contained within a cell. More elaborate schemes can accommodate chemical kinetic effects by employing a combination of an Arrhenius expression (Drysdale, 1998) and some formulation describing turbulent interactions. While these approaches have their uses, they cannot be considered 'fundamental', as they do not address the numerous competing chemical reactions occurring in reality. More recent developments in the field modelling of fires have been reported by Moss (1995); in particular, the 'laminar flamelet' representation of combustion chemistry has been successfully applied to simple gaseous hydrocarbon flames. The emerging technology of direct numerical simulation (DNS) has the potential to model the

fundamental chemical and physical phenomena associated with such complex combustion systems. The advantage of DNS is that the governing equations of turbulent combustion are solved without the approximations used in conventional CFD models. The drawback of the technique is that it relies on the availability of significant computing resources. In the future, DNS may provide the means of simulating turbulent diffusion flames while retaining all of the important intermediate reaction steps.

■ *A radiation model.* Radiation can account for a large percentage of the heat release from a fire source (40–50% is not uncommon). Flux models utilise further transport equation solutions, and therefore fit conveniently into the overall solution scheme. However, either discrete transfer or Monte Carlo methods are thought to be more accurate, and are therefore preferred. The accuracy of radiation models is further dependent on a knowledge of the absorption and scattering coefficients of the fluid medium and the emissivities of the various solid surfaces.

The boundary conditions also require specification, and some are applied very simply during the set-up of a model, for example, no-slip velocity conditions at walls. Heat losses through the walls can be calculated from the wall conductivity and the local temperature gradients. Provided that the boundary conditions and turbulence models are adequate, field models can be used to investigate flow configurations where a zone model would be worthless in the absence of established empirical data for a similar configuration. Rhodes (1989, 1998) has examined the application of CFD modelling to tunnel fire problems in some detail. He has drawn attention to the source of errors in such modelling and indicated the need for greater awareness of model limitations and capabilities of turbulence and other component models, numerical solution techniques and more robust codes. The accuracy of the models is affected by both numerical errors and physical approximations in the model. Numerical errors might arise from too coarse a computational grid, and may be reduced by successively refining the grid. Errors in the physical models are less easy to identify, and underline the continuing need for model testing by comparison with experimental results. Much of this experience will be derived by application of field models by both fire researchers testing models and architects involved in building design. These comparison studies have included, amongst others, room and compartment fires, a simulated six-bed hospital ward and a number of tunnel fire studies (Rhodes, 1998). The latter include the reduced-scale experiments of Ledin *et al.* (2002) and the application of CFD modelling to the MTFVTP and EUREKA 'Firetun' test data.

The early published literature on CFD studies of tunnel fire and ventilation problems include an unsteady two-dimensional model of a fire in a corridor (Ku *et al.*, 1976), a similar model but with the addition of a combustion model (Brandeis and Bergmann, 1983) and the work in the UK to develop the JASMINE (Smoke Movement in Enclosures) code (Kumar and Cox, 1986, 1988). More recent contributions include those of Lea (1995), Woodburn and Britter (1995), work at CETU (Chasse and Biollay, 1998) and work using LES (Cochard, 2002; Kunikane *et al.*, 2002a,b).

Ku *et al.* (1976) developed a two-dimensional computer model for solving the transient flows induced by a fire in a room/corridor geometry. Combustion effects were not modelled, and the fire was represented by a volumetric heat source only. The results were compared with small-scale experimental configurations, and gave fairly close agreement except in the region close to the fire source. The discrepancies were thought to be due to the coarseness of the computational grid, inaccuracies in the turbulence model and the omission of a numerical model of radiation effects. Some simulations were also compared with analytical solutions, and once again the agreement was very good.

Brandeis and Bergmann (1983) described another two-dimensional numerical model, designed to investigate the problem of accidental highway tunnel fires. This work was motivated by the Caldecott Tunnel fire in 1982, and examined the effect of forced ventilation on the transient development of a hydrocarbon spill fire. In order to achieve this, the model included a mathematical representation of the idealised ('one-step') combustion reaction between hexane and air, although this was not used for all runs in order to simplify the model and to give computational economy. The study concluded that combustion and fluid mechanics were closely coupled, and found that combustion was the primary factor affecting the flow field. The external wind and ventilation configuration were of secondary importance.

The work of Kumar and Cox (1986) represents some of the most relevant early research into the use of CFD for tunnel fire ventilation problems. Their JASMINE model stemmed from the Fire Research Station's interest in developing field models to analyse fire and smoke spread in buildings. The model is similar in principle to that of Brandeis and Bergmann (1983), but fully three-dimensional. In the approach by Kumar and Cox (1986), a series of steady state and transient runs is described, these runs being performed to provide data for comparison with the Zwenberg Tunnel experiments (Feizlmayr, 1976). The effects of ventilation and tunnel gradient were investigated, and, as in Brandeis and Bergmann's (1983) study, the combustion was represented by a simple hexane–air reaction with a constant HRR (and therefore reaction rate) for a given ventilation rate. Therefore, in this case the combustion model was used merely to predict the concentrations of various combustion products. The loss of heat due to convection and radiation were lumped together in a local empirical heat transfer coefficient varying linearly between 5 and 40 $W/m^2 K$ (from ambient to 100°C above ambient). The approximation allows for heat transfer from the smoke to its surroundings by radiation but does not model radiative heat exchange between neighbouring cells in the smoke layer. This refinement is included in a later study by Kumar and Cox (1988). A number of simulations were carried out with variations in ventilation configuration with and without the effects of gradient. The model output was in terms of carbon dioxide and oxygen distributions and temperature throughout the tunnel. The former only posed a threat to life at very close proximity to the fire, in agreement with the Zwenberg tests. In the natural ventilation run, a well-defined interface was set up between the out-flowing hot gas layer in the upper region of the tunnel and the inflowing fresh air towards the fire region. This stratification was maintained in the steady state solution. For the forced ventilation cases, fresh air was maintained behind the fire source, but the stratification was eroded downstream with increasing mean inlet velocity. The study concluded that the predictions in the 'far field' (i.e. at fairly large distances from the fire) were in close agreement with the experiments, but that in the 'near field' (close to the fire) the errors were increased. The discrepancies were ascribed to grid coarseness and inaccuracies in the simple turbulence–chemistry interaction theory. However, the model was thought to be of sufficient accuracy to be useful for the study of the tunnel environment in the case of ventilated fires. Gas composition predictions were reasonable in the forced-ventilation cases but poor for the naturally ventilated run. The recommendations of the study were that an improved radiation treatment needed to be incorporated and that a time-dependent conduction equation was necessary at the solid boundaries.

The inclusion of a more elaborate radiation model and improved simulation of convective heat transfer at rough surfaces into the JASMINE code is described by Kumar and Cox (1988).

This study concluded that the radiation model improved the realism of the predictions and enabled a more accurate assessment of the thermal radiation hazard; however, the numerical predictions still showed the same variance with experimental data. The work also highlighted the inadequacy of the present knowledge of convective heat transfer coefficients for modern building materials.

Bettis *et al.* (1994b) compared the results of the one-third-scale HSL 'Phase 2' tests (with obstructed geometry) (Bettis *et al.*, 1994a) with FLOW3D simulations for two of the nine experiments. Crucially, the CFD simulations were performed 'blind', so that the modellers were unaware of the experimental behaviour in advance: these circumstances were contrary to previous 'tests' of CFD, and thus represented a pragmatic trial of the worth of the technique as a design tool. The CFD results were found to be qualitatively correct but quantitatively inaccurate. The latter inaccuracy was characterised by an underprediction of the extent of backflow L for a given ventilation velocity U and fire size Q. An alternative interpretation is that the critical velocity U_{cr} was underpredicted for a given fire size Q. It was concluded that a ventilation system based on these 'design' calculations would suffer from a reduced safety margin and would not be capable of positive smoke control for fire sizes approaching the maximum design case. Overall, it was felt that CFD was 'still an immature technology, very far from the design tool it is sometimes purported to be', and that the further elucidation of tunnel fire problems would require a combination of better-quality CFD simulation and experiment.

Lea (1995) conducted a series of three-dimensional FLOW3D simulations of the HSL 'Phase 1' (unobstructed geometry) tests using a buoyancy modified k–ε turbulence model and omitting radiative heat transfer effects. The CFD predictions were in qualitative agreement with the HSL trials, revealing the same weak dependence of U_{cr}. It was suggested that the significant blockage presented by the fire plume produces local flow accelerations that tilt the plume markedly in the downstream direction, reducing the tendency for upstream smoke propagation. Additional simulations were designed to investigate how the flow behaviour, in particular the value of U_{cr}, was influenced by the tunnel aspect ratio (H/W) and the fire size Q. It was found that as the tunnel width was increased, U_{cr} became more dependent on the heat output of the fire and that U_{cr} increased with W for a given value of Q. It was stressed that this behaviour ran contrary to the expectations associated with simpler modelling approaches, the observed realignment of the plume being identified as a key factor.

Woodburn and Britter (1995) employed both standard and buoyancy-modified k–ε implementations of FLOW3D to model one of the HSL experiments; the physical fire size was 2.3 MW, with an associated mean ventilation velocity of 1.7 m/s, which led to a backlayer length of 11 m. It was found that the modified turbulence model accurately predicted the extent of the upstream layer whereas the standard k–ε model predicted zero backflow. The value of L was found, unexpectedly, to be sensitive to variations in Q; however, this was attributed to the characteristics of the combustion model employed and differences between the theoretical and experimental methods for reducing the heat input rate. Predictions of the downstream flow régime were found to be independent of the turbulence model selected. Sensitivity studies indicated that the downstream region was most sensitive to variations in the prescription of natural convective and radiative heat transfer and also to the wall roughness. It was concluded that the upper limit of accuracy of the simulations was ±10%, due to uncertainties in the empirical wall roughness data and ventilation velocity profile.

Typical of more recent research applications of CFD is the work of Cochard (2002), who used MTFVTP data to test the FDS V2.0 model for tunnel fires work. In particular, he attempted to reproduce Test 321A, which comprised a 40 MW pool fire with a single-point ceiling ventilation supply. Detailed measurements of velocity, temperature and gas concentration profiles were available. This model involves solution of the Navier–Stokes equations simplified for low Mach number flows together with a mixture fraction combustion, subgrid-scale Smagorinsky turbulent viscosity and finite-volume grey gas radiation models. The equations were solved over a grid of 480 000 cells. Cochard considered the model gave good predictions of the overall phenomena and good estimates of temperature on a par with other 'engineering codes'. He, however, identified the modelling of the flame sheet as an important area in determining the accuracy of predicted temperatures. For this, a fine grid was required.

Kunikane *et al.* (2002a,b) have employed a similar approach using an LES for the velocity field only together with a Smagorinsky turbulent viscosity model for the turbulent plume. Combustion and radiation were ignored while little detail of the numerical approach has been given. They applied this code to experimental pool fires of 1, 4 and 9 m^2 carried out in a three-lane tunnel of 115 m^2 cross-sectional area some 1.1 km long, to examine the behaviour of the smoke plume. They compared the predictions with measurements of smoke density, temperature, and the time and position when the ceiling smoke layer descended to the tunnel floor, examining how agreement varied with different values of cell size and Smagorinsky constant. They concluded that the smoke descent point was well predicted, but it was very dependent on how the wall effects were modelled. This in turn was dependent on the grid resolution. In such a large tunnel they also found that the ventilation had little effect on the fire heat output.

Experience continues to be gained with commercially available fluid flow codes. Thus, Yau *et al.* (2001) used MTFVTP data to test the STAR-CD code with k–ε turbulence and eddy break-up combustion models, and concluded that 'CFD codes could be used for fire safety engineering design in tunnels'. Torsbergen *et al.* (2002) and Shahcheraghi *et al.* (2002) have used CFX software to examine transient smoke movement and the use of longitudinal ventilation to clear underground stations of smoke. Similarly, Chasse and Biollay (1998) studied the behaviour of the critical velocity under longitudinal ventilation, and compared predictions with the EUREKA 'Firetun' H32 test. They found that the shape, size and slope had little influence on the critical velocity, as did the location of the fire. They concluded that the critical velocity was bounded at heat outputs up to 100 MW, and varied little if the tunnel contained a substantial blockage. Ledin *et al.* (2002) explored the application of a commercial code in a little more detail, as encouraged by Rhodes (1998) for situations for which they had collected some detailed laboratory-scale data. These included situations representative of station platforms, booking halls and sloping access tunnels. Thus, they examined the influence of different numerical differencing schemes, wall heat transfer and component models. They concluded that there were problems even with the simple situations considered, and any application of CFD should be carried out with caution at all stages in order to make correct choices where available. Thus

- higher-order differencing should be used in complex situations
- finer meshes generally produce more accurate solutions, by comparison with experimental results (this may not be the case with LES sub-models)
- there are problems with currently employed turbulence models and wall heat transfer formulations for some situations examined

■ for studies of ventilation, it is important to use a combustion model that calculates the flame shape and volume rather than one which involves its prescription.

10.5.   Conclusions

Fire safety is now one of the principle considerations in the design of tunnel ventilation systems that are used to control the movement of smoke and provide tunnel occupants with a tenable atmosphere for evacuation from the fire scene. In designing a ventilation system to mitigate the effects of a fire, two main items of information are required.

■ Specification of the fires likely to occur in the facility – the design basis fires. This ideally should include the time variable heat output and the rate of production of smoke and other toxic products.
■ The response of the fire and combustion products to changes in the applied ventilation.

The most accessible form for this information would be as a set of well-tested and widely accepted predictive modelling tools.

The environment in which designers operate has significantly improved over the past 20 years. There is much better availability of data as a result of the large-scale MTFVTP, the EUREKA 'Firetun' and other projects and smaller-scale work carried out on the basis of Froude number scaling, and the data collected are now more relevant to the quantification of the true fire hazard. Thus, a small number of tests, restricted by cost and the availability of facilities, have identified the heat output from a number of different vehicle types in a limited range of ventilation situations. However, a need remains to examine the influence of different cargoes carried by HGVs. Significant effort has been devoted to the response of the fire plume to changes in ventilation, largely as a result of the ease of scaling this interaction. As a result, the control and movement of smoke is reasonably well understood.

The information from these experiments has also provided a fuller background for the development of modelling tools. Many workers now use these tools as a matter of course in developing tunnel ventilation strategies as a result of the various published studies that have compared model predictions with experimental results. Advances in computational hardware have also made the widespread application of models based on CFD a more practical and accessible option. More advanced models, such as those based on LES, until recently applied by the research community only, are also being used more widely.

However, there is a requirement for further experiment and model development to give greater certainty to the design process, particularly where safety critical systems and decisions are involved. In particular, since modification to the ventilation is now a standard response to a fire incident, there is a need to examine the influence of ventilation on the fire itself. Thus, the issue of the ventilation control of a tunnel fire must be addressed, initially through experiment, since this is one of the situations in which confidence in modelling is low, to determine how the rate of burning, fire development/flame spread and production of smoke/toxic species vary with imposed flow. Implicit in this is an understanding of how fire spreads within a tunnel. Such detailed studies should then allow more informed deployment of predictive models and provide the basis for detailed in-depth model evaluation of the type encouraged by Rhodes (1998) and conducted by Ledin et al. (2002).

The state of knowledge does not therefore permit the identification of any standard approach to system design against fire but requires that all situations must be considered individually and approached through the deployment of both experiment and modelling tools. The latter can still only be applied through close reference to relevant data.

REFERENCES

Atkinson GT (1997) A model to simulate conditions around a HGV shuttle amenity coach. Unpublished manuscript. HSL, Buxton.

Atkinson GT and Wu Y (1996) Smoke control in sloping tunnels. *Fire Safety Journal* **27**(4): 335–341.

Bakke P and Leach SJ (1960) *Methane Roof Layers*. Research Report No. 195. SMRE, Sheffield.

Beard AN (1997) A model for predicting fire spread in tunnels. *Journal of Fire Sciences* **15**: 277–307.

Beard AN (1998) Modelling major fire spread in a tunnel. In: Vardy AE (ed.), *3rd International Conference on Safety in Road and Rail Tunnels, Nice*. ITC, Tenbury Wells.

Beard AN (2000) On *a priori*, blind and open comparisons between theory and experiment. *Fire Safety Journal* **35**: 63–66.

Beard AN (2001) Major fire spread in a tunnel: a non-linear model. In: Vardy AE (ed.), *Proceedings of the 4th International Conference on Safety in Road and Rail Tunnels, Madrid, 2–6 April 2001*. ITC, Tenbury Wells, pp. 467–476.

Beard AN (2003) Major fire spread in a tunnel: a non-linear model with flame impingement. In: Vardy AE (ed.), *Proceedings of the 5th International Conference on Safety in Road and Rail Tunnels, Marseille, 6–10 October 2003*. ITC, Tenbury Wells, pp. 511–519.

Beard AN (2007) Major fire spread in tunnels: comparison between theory and experiment. In: *Proceedings of the 1st International Tunnel Safety Forum for Road and Rail, Nice, 23–25 April 2007*. Tunnel Management International, Tenbury Wells.

Beard AN, Drysdale DD and Bishop SR (1995) A non-linear model of major fire spread in a tunnel. *Fire Safety Journal* **24**: 333–357.

Bendelius AG and Hettinger JC (1988) Ventilation of the Sydney Harbour Tunnel. In: *Proceedings of the 6th International Symposium on the Aerodynamics and Ventilation of Vehicle Tunnels, Durham, 27–29 September 1988*. BHRA, Fluid Engineering Centre, Cranfield, pp. 321–347.

Bettis RJ, Jagger SF and Hambleton RJ (1994a) *Interim Validation of Tunnel Fire Consequences Models: Summary of Phase 1 Trials*. RLSD Report IR/FR/94/02. Health and Safety Executive, Buxton.

Bettis RJ, Jagger SF and Moodie K (1994b) Reduced scale simulations in partially-blocked tunnels. In: *Proceedings of the International Conference on Fires in Tunnels, Borås, 10–11 October 1994*, SP Report 1994:54, pp. 162–186.

Bickel JO and Kuesel TR (1991) *Tunnel Engineering Handbook*, Krieger Publishing Company, Florida.

Bradbury W (1998) Smoke control during the Channel Tunnel fire of 18 November 1996. In: *Proceedings of the 3rd International Conference on Safety in Road and Rail Tunnels, Nice, 9–11 March 1998*. ITC, Tenbury Wells, pp. 623–640.

Brandeis J and Bergmann DJ (1983) A numerical study of tunnel fires. *Combustion Science and Technology* **35**: 133–155.

Brichet N, Weatherill A, Crausaz B and Casale E (2002) The new ventilation systems of the Mont Blanc Tunnel active smoke control: from simulation to successful operation. In: *Proceedings of the 4th International Conference on Tunnel Fires, Basel, 2–4 December 2002*. ITC, Tenbury Wells, pp. 95–104.

Carvel RO, Beard AN, Jowitt PW and Drysdale DD (2001a) Variation of heat release rate with forced longitudinal ventilation for vehicle fires in tunnels. *Fire Safety Journal* **36**: 569–596.

Carvel RO, Beard AN and Jowitt PW (2001b) The influence of longitudinal ventilation systems on fires in tunnels. *Tunnelling and Underground Space Technology* **16**: 3–21.

Carvel RO, Beard AN and Jowitt PW (2001c) A Bayesian estimation of the effect of forced ventilation on a pool fire in a tunnel. *Civil Engineering and Environmental Systems* **18**: 279–302.

Carvel RO, Beard AN and Jowitt PW (2004) How longitudinal ventilation influences fire size in tunnels. *Tunnel Management International* **7**(1): 46–54.

Carvel RO, Beard AN and Jowitt PW (2005) Fire spread between vehicles in tunnels: effects of tunnel size, longitudinal ventilation and vehicle spacing. *Fire Technology* **41**: 271–304.

Chaiken RF, Singer JM and Lee CK (1979) *Model Coal Tunnel Fires in Ventilation Flow*. US Bureau of Mines Report of Investigations No. 1979–603–002/54. US Department of Interior, Washington, DC.

Chan R, Murphy RE and Sheehy JC (1988) Post Transbay tube fire tests. In: *Proceedings of the 6th International Symposium on the Aerodynamics and Ventilation of Vehicle Tunnels, Durham, 27–29 September 1988*. BHRA, Fluid Engineering Centre, Cranfield, pp. 593–606.

Chang X, Laage LW and Greuer RE (1990) *User's Manual for MFIRE: A Computer Simulation Programme for Mine Ventilation and Fire Modelling*. US Bureau of Mines, Washington, DC.

Channel Tunnel Safety Authority (1997) *Inquiry Into The Fire On Heavy Goods Vehicle Shuttle 7539 On 18 November 1996*. Stationery Office, London.

Charters DA, Gray WA and McIntosh AC (1994) A computer model to assess fire hazards in tunnels (FASIT). *Fire Technology* **30**: 134–154.

Chasse P and Biollay H (1998) CFD validation for tunnel fires under longitudinal ventilation and application to the study of the critical velocity. In: *Proceedings of the 3rd International Conference on Safety in Road and Rail Tunnels, Nice, 9–11 March 1998*, pp. 169–184.

Cheng LH, Ueng TH and Liu CW (2001) Simulation of ventilation and fire in the underground facilities. *Fire Safety Journal* **36**(6): 597–619.

Cochard S (2002) Validation of the freeware 'Fire Dynamics Simulator Version 2.0' for simulating tunnel fires. In: *Proceedings of the 4th International Conference on Tunnel Fires, Basel, 2–4 December 2002*. ITC, Tenbury Wells, pp. 377–386.

Daish NC and Linden PF (1994) *Interim Validation of Tunnel Fire Consequence Models: Comparison between the Phase 1 Trials Data and the CREC/HSE Near-fire Model*. Report No. FM88/92/6. Cambridge Environmental Research Consultants, Cambridge.

D'Albrand N and Bessiere C (1992) Fire in a road tunnel: comparison of tunnel sections In: *Proceedings of the 1st International Conference on Safety in Road and Rail Tunnels, Basel*, pp. 439–449.

Danziger NH and Kennedy WD (1982) Longitudinal ventilation analysis for the Glenwood Canyon Tunnels. In: *Proceedings of the 4th International Symposium on the Aerodynamics and Ventilation of Vehicle Tunnels*. BHRA, Fluid Engineering, Cranfield, pp. 169–186.

Desrosiers C (1988) Improving emergency-ventilation procedures in subways thru' hot-smoke tests. In: *Proceedings of the 6th International Symposium on the Aerodynamics and Ventilation of Vehicle Tunnels, Durham, 27–29 September 1988*. BHRA, Fluid Engineering Centre, Cranfield, pp. 549–566.

Directorate-General for Public Works and Water Management (2002) *Project 'Safety Test' – Report on Fire Tests*. Directorate-General for Public Works and Water Management, Civil Engineering Division, Utrecht.

Donato G (1972) *Montreal Metro Fire December 9th 1971.* International Union of Public Transport International Metropolitan Railway Committee, Working Group for Protection of Electrical Cables used in Underground Metro and Subways, New York.

Donato G (1975) *Report on Fire in the Montreal Metro on January 23rd 1974.* International Union of Public Transport International Subway Committee, Working Group on Electric Cables, New York.

Drysdale DD (1998) *An Introduction to Fire Dynamics,* 2nd edn. Wiley, Chichester.

Edwards R (1999) Where there's smoke .... *New Scientist* **2191**: 22–23.

Egilsrud PE (1984) *Prevention and Control of Highway Tunnel Fires.* Report No. FHWA/-RD-83/032. US Department of Transportation, Federal Highway Administration, Washington, DC.

Eisner HS and Smith PB (1954) *Convection Effects from Underground Fires: The Backing of Smoke Against the Ventilation.* Research Report No. 96. SMRE, Sheffield.

Feizlmayr AH (1976) Research in Austria on tunnel fire. In: *Proceedings of the 2nd International Symposium on the Aerodynamics and Ventilation of Vehicle Tunnels, Cambridge, 23–25 March 1976.* BHRA, Fluid Engineering Centre, Cranfield, paper J2.

French SE (1994) EUREKA 499-HGV fire test (Nov. 1992). In: *Proceedings of the International Conference on Fires in Tunnels, Borås, 10–11 October 1994,* SP Report 1994:54, pp. 63–85.

Fuller B (1985/1986) Problems of safety for firemen underground. *Fire International* Dec./Jan.: 37–39.

Gobeau N and Zhou XX (2002) *Evaluation of CFD to Predict Smoke Movement in Complex Enclosed Spaces.* Project Report CM/02/12. Health and Safety Executive, London.

Grant G and Southwood P (1999) Development of an onboard fire suppression system for Eurotunnel HGV Shuttle Trains. In: *Proceedings of Interflam '99, 8th International Fire Science and Engineering Conference,* Edinburgh, 29 June–1 July 1999. Interscience, Chichester, pp. 651–662.

Grant GB, Jagger SF and Lea CJ (1998) Fires in tunnels. *Philosophical Transactions of the Royal Society of London* A **356**: 2873–2906.

Haack A (1994) Introduction to the EUREKA-EU499 Firetun project. In: *Proceedings of the International Conference on Fires in Tunnels, Borås, 10–11 October 1994,* SP Report 1994:54, pp. 3–19.

Haerter A (1994) Fire tests in the Ofenegg Tunnel in 1965. In: *Proceedings of the International Conference Fires in Tunnels, Borås, 10–11 October 1994,* pp. 195–214.

Heselden AJM (1976) Studies of fire and smoke behaviour relevant to tunnels. In: *Proceedings of the 2nd International Symposium on the Aerodynamics and Ventilation of Vehicle Tunnels,* Cambridge, 23–25 March 1976. BHRA, Fluid Engineering Centre, Cranfield, pp. J1.1–J1.18.

Huijben J (2002) The influence of longitudinal ventilation on fire size and development. In: *Proceedings of the 4th International Conference on Tunnel Fires,* Basel, 2–4 December 2002. ITC, Tenbury Wells, pp. 115–124.

Hwang CC and Edwards JC (2005) The critical ventilation velocity in tunnel fires – a computer simulation. *Fire Safety Journal* **40**: 213–244.

Hwang CC, Chaiken RF, Singer JM and Chi DNH (1977) Reverse stratified flow in duct fires: a two-dimensional approach. In: *16th Symposium (International) on Combustion.* Combustion Institute, Pittsburgh, PA, pp. 1385–1395.

Hwang CC, Litton CD, Perzak FJ and Lazarra CP (1991) Modelling the fire-assisted flame spread along conveyor belt surfaces. In: *Proceedings of the 5th US Mine Ventilation Symposium.* Society for Mining, Metallurgy and Exploration (SME), Littleton, CO, pp. 39–44.

Ingason H (2002) Trial of a mobile fan for smoke ventilation in a railway tunnel. In: *Proceedings of the 4th International Conference on Tunnel Fires, Basel*, 2–4 December 2002. ITC, Tenbury Wells, pp. 151–160.

Ingason H and Li YZ (2010) Model scale tunnel fire tests with longitudinal ventilation. *Fire Safety Journal* **45**: 371–384.

Ingason H and Wikstrom U (2006) The international FORUM of fire research directors: a position paper on future actions for improving road tunnel fire safety. *Fire Safety Journal* **41**: 111–114.

ITC (1996) *Proceedings of the one-day seminar on Smoke and Critical Velocity in Tunnels, London Heathrow Airport*, 2 April 1996. ITC, Tenbury Wells.

Jones A (1985) *The Summit Tunnel Fire*. Incident Report No. IR/L/FR/85126. Health and Safety Executive, Research and Laboratory Services Division, London.

Keski-Rahkonen O, Holmlund C, Loikkanen P, Ludvigsen H and Mikkola E (1986) *Two Full Scale Pilot Fire Experiments in a Tunnel*. VTT Report VTT/RR-453. Palotekniikan Laboratory, Espoo.

Ku AC, Doria ML and Lloyd JR (1976) Numerical modelling of unsteady buoyant flows generated by a fire in a corridor. In: *16th Symposium on Combustion*. Combustion Institute, Pittsburgh, PA, pp. 1373–1384.

Kumar S and Cox G (1986) *Mathematical Modelling of Fire in Road Tunnels – Validation of JASMINE*. Contractor Report 28. Transport and Road Research Laboratory, Department of Transport, London.

Kumar S and Cox G (1988) *Mathematical Modelling of Fire in Road Tunnels – Effects of Radiant Heat and Surface Roughness*. Contractor Report 101. Transport and Road Research Laboratory, Wokingham.

Kunikane Y, Kawabata N, Takekuni K and Shimoda A (2002a) Heat release rate induced by gasoline pool fire in a large cross section tunnel. In: *Proceedings of the 4th International Conference on Tunnel Fires, Basel*, 2–4 December 2002. ITC, Tenbury Wells, pp. 387–396.

Kunikane Y, Kawabata N, Yamamoto N, Takekuni K and Shimoda A (2002b) Numerical simulation of smoke descent in a tunnel fire accident. In: *Proceedings of the 4th International Conference on Tunnel Fires, Basel*, 2–4 December 2002. ITC, Tenbury Wells, pp. 357–366.

Kunsch JP (2002) Simple model for control of fire gases in a ventilated tunnel. *Fire Safety Journal* **37**(1): 67–81.

Kuwana H, Satoh H and Kurioka H (1998) Scale effect on temperature properties in tunnel fire. In: *Proceedings of the of 3rd International Conference on Safety in Road and Rail Tunnels*, Nice, 9–11 March 1998, pp. 87–96.

Lacroix D (1998) Fire and smoke control in road tunnels. *Routes/Roads*, No. 300(IV). World Road Association (PIARC), Paris, pp. 33–43.

Lacroix D (2001) The Mont Blanc Tunnel fire – what happened and what has been learned. In: *4th International Conference on Safety in Road and Rail Tunnels*, Madrid, 2–6 April 2001, pp. 3–15.

Lea CJ (1995) CFD modelling of the control of smoke movement from tunnel fires using longitudinal ventilation. In: *Abstracts of the 1st European Symposium on Fire Safety Science, Zurich*, 21–23 August 1995.

Leach SJ and Barbero LP (1964) *Experiments on Methane Roof Layers: Single Sources in Rough and Smooth Tunnels with Uphill and Downhill Ventilation, with an Appendix on Experimental Techniques*. Research Report No. 222. SMRE, Sheffield.

Ledin S, Bettis R, Allen JT and Ivings M (2002) Validation of CFD approaches for modelling smoke movement from fires in tunnels. In: *Proceedings of the 4th International Conference on Tunnel Fires, Basel*, 2–4 December 2002. ITC, Tenbury Wells, pp. 347–356.

Lee CK, Chaiken RF and Singer JM (1979) Interaction between duct fires and ventilation flow: an experimental study. *Combustion Science and Technology* **20**: 59–72.

Lee SR and Ryou HS (2004) *An Experimental Study of the Effect of the Aspect Ratio on the Critical Velocity in Longitudinal Ventilation Tunnel Fires*. Department of Mechanical Engineering, Chung-Ang University, Seoul.

Li YZ, Lei B and Ingason H (2010) Study of critical velocity and backlayering length in longitudinally-ventilated tunnel fires. *Fire Safety Journal* **45**(6–8): 361–370.

Liew SK, Deaves D and Blyth A (1998) Eurotunnel HGV fire on 18 November 1996 – fire development and effects. In: *Proceedings of the of 3rd International Conference on Safety in Road and Rail Tunnels, Nice*, 9–11 March 1998, pp. 29–40.

Lonnermark A and Ingason H (2006) Fire spread and flame length in large-scale tunnel fires. *Fire Technology* **42**: 283–302.

Luchian SF (1992) The Central Artery/Tunnel Project: Memorial Tunnel fire test program. In: *1st International Conference on Safety in Road and Rail Tunnels, Basel*, 23–25 November 1992, pp. 363–377.

McGrattan KB, Baum HR, Rehm RG, Forney GP, Floyd JE and Hostikka S (2001) *Fire Dynamics Simulator (Version 2)*. Technical Reference Guide Technical Report NISTIR 6783. National Institute of Standards and Technology, Gaithersburg, MD.

Megret O (1999) Experimental study of smoke propagation in a tunnel with different ventilation systems. PhD thesis. University of Valenciennes, Valenciennes.

Moss JB (1995) Turbulent diffusion flames. In: Cox G (ed.), *Combustion Fundamentals of Fire*. Academic Press, London, pp. 221–272.

NFPA (2003) *NFPA 130: Standard for Fixed Guideway Transit and Passenger Rail Systems*. National Fire Protection Association, Quincy.

Nordmark A (1998) Fire and life safety for underground facilities: present status of fire and life safety principles related to underground facilities. *Tunnelling and Underground Space Technology* **13**(3): 217–269.

Oka Y and Atkinson GT (1995) Control of smoke flow in tunnel fires. *Fire Safety Journal* **25**: 305–322.

Oka Y, Yamada T, Kurioka H, Sato H and Kuwana H (1998) Flame behaviour in a tunnel. In: *Proceedings of the of 3rd International Conference on Safety in Road and Rail Tunnels, Nice*, 9–11 March 1998, pp. 159–168.

Parsons Brinckerhoff (1996) *Memorial Tunnel Fire Ventilation Test Program*. Interactive CD and comprehensive test report. Parsons Brinckerhoff 4D Imaging, Boston.

Pritchard MJ and Binding TM (1992) FIRE2: a new approach for predicting thermal radiation levels from hydrocarbon pool fires. *IChemE Symposium Series No. 130*, pp. 491–505.

Pucher K (1994) Fire tests in the Zwenberg Tunnel (Austria). In: *Proceedings of the International Conference on Fires in Tunnels, Borås*, 10–11 October 1994, pp. 187–194.

Rhodes N (1989) *Prediction of Smoke Movement: An Overview of Field Models*, CH-89-12-4. ASHRAE, Atlanta, GA.

Rhodes N (1998) The accuracy of CFD modelling techniques for fire prediction. In: *Proceedings of the of 3rd International Conference on Safety in Road and Rail Tunnels, Nice*, 9–11 March 1998, pp. 109–115.

Saito N, Yamada T, Sekizawa A, Yanai E, Watanabe Y and Miyazaki S (1995) Experimental study on fire behaviour in a wind tunnel with a reduced scale model. In: *2nd International*

Conference Safety in Road and Rail Tunnels, Granada, pp. 303–310.

Schwenzfier L, Bodart X and Sanquer S (2002) Comparison of zone and field modelling in tunnel fire simulations. In: *Proceedings of the 4th International Conference on Tunnel Fires, Basel*, 2–4 December 2002. ITC, Tenbury Wells, pp. 367–376.

Shahcheraghi N, McKinney D and Miclea P (2002) The effects of emergency fan start time on controlling the heat and smoke from a growing station fire – a transient CFD study. In: *Proceedings of the 4th International Conference on Tunnel Fires, Basel*, 2–4 December 2002. ITC, Tenbury Wells, pp. 233–246.

SINTEF (1999) *Norwegian Fire Research Laboratory, Annual Report 1999*. SINTEF, Trondheim.

Smagorinsky J (1963) General circulation experiments with primitive equations. I: the basic experiment. *Monthly Weather Review* **91**: 99–164.

Spratt D and Heselden AJM (1974) *Efficient Extraction of Smoke from a Thin Layer Under a Ceiling*. Fire Research Note 1001. Fire Research Station, Borehamwood.

Studiengesellschaft Stahlanwendung (1995) *Fires in Transport Tunnels. Report on Full Scale Tests. EUREKA-Project EU499: FIRETUN*. Studiengesellschaft Stahlanwendung, Dusseldorf.

Telle D and Vauquelin O (2002) An experimental evaluation of the confinement velocity. In: *Proceedings of the 4th International Conference on Tunnel Fires, Basel*, 2–4 December 2002. ITC, Tenbury Wells, pp. 161–170.

Te Velde K (1988) A computer simulation for longitudinal ventilation of a road tunnel with incoming and outgoing slip roads. In: *Proceedings of the 6th International Symposium on the Aerodynamics and Ventilation of Vehicle Tunnels, Durham*, 27–29 September 1988. BHRA Fluid Engineering, Cranfield, pp. 179–201.

Thomas PH (1963) The size of flames from natural fires. In: *Proceedings of the 9th International Combustion Symposium of the Combustion Institute*, Pittsburgh, pp. 844–859.

Thomas PH (1970) Movement of smoke in horizontal corridors against an air flow. *Institution of Fire Engineers Quarterly* **30**(77): 45–53.

Torbergsen LE, Paaske P and Schive C (2002) Numerical simulation of fire and smoke propagation in long railway tunnels. In: *Proceedings of the 4th International Conference on Tunnel Fires, Basel*, 2–4 December 2002. ITC, Tenbury Wells, pp. 339–346.

US Department of Commerce (1976) *Subway Environmental Design Handbook*, vol. 1. *Principles and Applications*, 2nd edn. US Department of Commerce, National Technical Information Service, Washington, DC.

Van Maele K and Merci B (2008) Application of RANS and LES field simulations to predict the critical ventilation velocity in longitudinally-ventilated horizontal tunnels. *Fire Safety Journal* **43**: 598–609.

Vantelon JP, Guelzim A, Quach D, Son DK, Gabay D and Dallest D (1991) Investigation of fire-induced smoke movement in tunnels and stations: an application to the Paris metro. In: *Proceedings of the 3rd International Symposium on Fire Safety Science*. IAFSS, Boston, pp. 907–918.

Vardy AE (1988) A safe ventilation procedure for single-track tunnels. In: *Proceedings of the 6th International Symposium on the Aerodynamics and Ventilation of Vehicle Tunnels*, Durham, 27–29 September 1988. BHRA, Fluid Engineering Centre, Cranfield, pp. 567–574.

Viot J and Vauquelin O (2002) Investigation of natural ventilation for smoke extraction. In: *Proceedings of the 4th International Conference on Tunnel Fires, Basel*, 2–4 December 2002. ITC, Tenbury Wells, pp. 141–150.

Wang Y, Jiang J and Dezhi Z (2009) Full-scale experiment research and theoretical study for fires in tunnels with roof openings. *Fire Safety Journal* **44**: 339–348.

Weatherill A, Guigas X and Trottet Y (2002) The new ventilation systems of the Mont Blanc Tunnel: specificities and performance of the fire ventilation. In: *Proceedings of the 4th International Conference on Tunnel Fires, Basel*, 2–4 December 2002. ITC, Tenbury Wells, pp. 105–114.

Woodburn PJ and Britter RE (1995) The sensitivity of CFD simulations of fires in tunnels. In: *Abstracts of the 1st European Symposium on Fire Safety Science, Zurich*, 21–23 August 1995.

World Road Association (1996) *Road Tunnels: Emissions, Environment, Ventilation.* PIARC Committee on Road Tunnels (C5), Paris.

World Road Association (1999) *Fire and Smoke Control in Tunnels.* PIARC Committee on Road Tunnels (C5), Paris.

Wu Y (2003) Smoke control in tunnels with slope using longitudinal ventilation: effect of slope on critical velocity. In: *Proceedings of the 11th International Symposium on Aerodynamics and Ventilation of Vehicle Tunnels, Luzern*, 7–9 July 2003, pp. 77–86.

Wu Y, Bakar MZ, Atkinson GT and Jagger SF (1997) A study of the effect of tunnel aspect ratio on control of smoke flow in tunnels. In: *Proceedings of the 9th International Conference on Aerodynamics and Ventilation of Vehicle Tunnels, Aosta Valley*, 6–8 October 1997, pp. 573–587.

Yau R, Cheng V, Lee S, Mingchau L and Zhao L (2001) Validation of CFD models for room and tunnel fires. In: *Proceedings of Interflam 2001. 9th International Fire Science and Engineering Conference, Edinburgh*, 17–19 September 2001, pp. 807–815.

Handbook of Tunnel Fire Safety, 2nd edition
ISBN: 978-0-7277-4153-0

ICE Publishing: All rights reserved
doi: 10.1680/htfs.41530.217

Chapter 11
The influence of tunnel ventilation on fire behaviour

Richard Carvel
BRE Centre for Fire Safety Engineering, University of Edinburgh, UK
Alan Beard
Civil Engineering Section, School of the Built Environment, Heriot-Watt University, Edinburgh, UK

11.1. Introduction
The last two chapters have discussed what is currently possible with tunnel ventilation systems, and the use of ventilation to control smoke. There is, however, a downside when using ventilation to control smoke: the ventilation may also enhance the burning rate of the fire, which would be expected to have implications for life safety and fire-fighting. It would also be expected to pose more of a threat to the integrity of the tunnel structure as well as increase the risk of spread. In addition, forced ventilation may cause significant flame deflection and increase the probability of spread through closer proximity of flame and smoke to a second object.

11.2. Basic fire science
For centuries, people have known that blowing on a fire helps to get it going. However, many people will blow on a lit match to put it out. These two examples highlight the two main conflicting influences that ventilation has on fire: that of enhancing burning by providing additional oxygen at the fire location, but also reducing burning by cooling the fuel. (Flame deflection may also play a significant part in both cases. That is, enhancement may occur by flames being deflected by the wind towards neighbouring fuel; reduction may also occur by deflection reducing radiation feedback to the fire base.)

In order for there to be a fire, three things are required: fuel, oxygen and heat. If any of these are removed, or reduced by a significant margin, then the fire will be extinguished or diminished in size. Any airflow at the fire location will have some cooling effect, but will also provide more oxygen for the fire.

By their very nature, tunnels are confined environments. A sizeable fire in a tunnel will quickly consume much of the oxygen in the locality of the fire. The burning rate of the fire will then be constrained by the amount of oxygen supplied to the fire, either by natural convection and entrainment processes, or due to the ventilation system in the tunnel. This type of fire is known as a 'ventilation-controlled' (V-C) fire.

Rather than discuss fires in terms of their 'burning rate', which may mean different things to different people, this chapter will concern itself with the 'heat release rate' (HRR) of a fire. The

HRR of a fire is considered by some to be the single most important factor contributing to the severity of a fire (Babrauskas and Peacock, 1992).

For a given fire scenario, it is a relatively simple matter in principle to calculate approximately the maximum HRR allowed by the available airflow:

$$\mathrm{HRR} = \dot{Q}_{max} = V \eta_{ox} \rho_{ox} \Delta H_{Cox} \times 10^3 \ (\mathrm{kW})$$

where V is the volumetric flow of air (m^3/s), η_{ox} is the mole fraction of oxygen in the airflow (generally 0.21), ρ_{ox} is the density of oxygen (1 kg/m^3 at standard temperature and pressure) and ΔH_{Cox} is the 'heat of combustion for oxygen', which can be taken to have a value of about 13 kJ/g in most cases. Thus, the maximum HRR of a fire in an enclosed space can generally be approximated by

$$\dot{Q}_{max} = 2.73V \ (\mathrm{MW})$$

For example, the Channel Tunnel is circular with a radius of approximately 3.8 m. If there were a longitudinal airflow in the tunnel with a velocity v of 2.9 m/s (the ventilation rate used during the fire on 11 September 2008), then the maximum heat release of a fire in the tunnel would be about: $2.73 \times (2.9 \times \pi \times 3.8^2) = 359$ MW.

In instances where the HRR of a fire is not ventilation controlled, the fire is considered to be 'fuel controlled' (F-C), that is, the HRR of the fire is limited by the arrangement and the chemical nature of the fuel, not by the amount of available oxygen.

However, applying a forced ventilation to a fuel-controlled fire may still have a small enhancing effect. If burning in naturally ventilated conditions, a fire will entrain air from all sides (unless they are blocked). If a forced airflow is imposed on the fire, this may provide a more efficient way of supplying oxygen to the fire source than the natural burning processes, and thus the HRR may be enhanced.

The interaction between forced airflow and fire size is not a simple one, largely because the burning process is not a simple process. The HRR of a fire is dependent on the amount of energy radiated from the flames to the fuel. If an airflow is imposed on a fire, the flames will be pushed to one side, significantly reducing the amount of radiation transferred back to the surface (Figure 11.1).

A similar effect means that forced ventilation will tend to increase the rate of fire spread across a surface or an object. In many fire situations, the rate of fire spread is dependent on the amount of

Figure 11.1 Diagram of heat transfer from flames to a burning object, subject to natural and forced ventilation

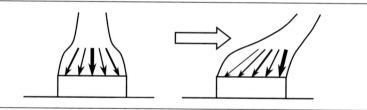

Figure 11.2 Diagram of heat transfer from flames to the unburned fuel ahead of the flames on a burning object, subject to natural and forced ventilation

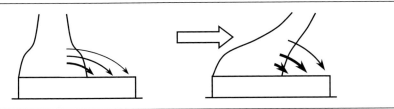

radiation from the flames heating up the unburned fuel ahead of the flame front. If the flames are pushed to one side by an imposed airflow, then the unburned fuel on the downwind side of the fire will tend to receive a much greater amount of radiation from the flames than in the case of natural ventilation (Figure 11.2).

Thus, if there were a fire at the upwind end of a heavy goods vehicle (HGV) trailer in a tunnel, one would expect the fire to spread across the load substantially faster with forced ventilation than with natural ventilation. Conversely, if the fire were at the downwind end of a HGV trailer, the fire would be expected to spread across the load more slowly with forced ventilation. However, in the latter case the enhancing effect of the ventilation should not be forgotten: if the HRR of the fire is enhanced, then the amount of radiation transferred back to the unburned fuel upwind of the fire might also be significantly enhanced.

The influence of forced ventilation on fire size and spread is not well understood. It is therefore desirable that a mathematical model of the interaction between forced ventilation and fire size be developed. As we have seen, there are a number of different, conflicting, influences on fire growth, fire spread and peak fire size for fires in tunnels. Due to the complexity of the relationships and their dependence on factors such as the arrangement of the fuel, it is impractical to try to represent all these processes in a deterministic mathematical model. A probabilistic Bayesian approach is more practical at present.

A research project was undertaken at Heriot-Watt University to address this question. A Bayesian model, based on experimental observations, was developed that describes the interaction of forced ventilation and fire size/fire development. This was first published in 2001 (Carvel *et al.*, 2001a). This model has subsequently been updated as and when new experimental data have become available.

11.3. Definitions

A fire in a tunnel could involve a car, a small truck, a HGV, a bus, a train, a pool of spilled fuel, the tunnel fixings (cables, etc.) or any one of a number of different things that are commonly found in tunnels. Even within these categories there is substantial variation; a 'mini' car will surely burn with different fire characteristics to a 'people carrier'. Previous experience (Beard, 1983) has shown that some of these problems can be overcome by considering the *change* to the fire behaviour in a given scenario rather than considering the absolute HRR.

For example, a small car may burn with a HRR of about 1.5 MW in natural ventilation conditions (Mangs and Keski-Rahkonen, 1994); if this HRR increases to 2.0 MW when a longitudinal ventilation is applied, then there is a 33% change in the HRR. It is reasonable to assume that a larger car, which might have a HRR of 3.0 MW in naturally ventilated conditions, will also experience a 33% enhancement in HRR under the same longitudinal ventilation conditions, this time reaching a peak HRR of about 4.0 MW.

In this chapter, the change in the HRR between natural and forced ventilation will be described in terms of the variable k, defined by

$$\dot{Q}_{vent} = k\dot{Q}_{nat}$$

where $\dot{Q}_{vent}$ is the HRR of a given fire in a tunnel with forced ventilation and $\dot{Q}_{nat}$ is the HRR of a similar fire in a naturally ventilated tunnel.

k can be expected to vary with ventilation rate, i.e. $k = f(v)$. It may prove useful to consider values of k for specific ventilation rates, therefore the nomenclature $k_{2.0}$, $k_{3.5}$, etc., can be used to denote the values of k at 2.0 and 3.5 m/s, and so on.

The assumption that the change with ventilation in the burning properties of a small car fire is similar to the change with ventilation for a larger fire is only valid for comparatively small variations in vehicle size. This kind of assumption is not valid when considering the difference between a small car fire and a HGV fire, or between a HGV fire and a fuel spillage.

In light of this, the project addressed specific fire types separately. Due to the lack of useful fire data from different vehicle types, this chapter will be restricted to considering the following five archetypes

1. a HGV fire, to represent medium/large trucks, HGVs, etc.
2. a small pool fire
3. a medium-sized pool fire
4. a large-scale pool fire
5. a car fire, to represent all sizes of cars, small vans, etc.

The definitions of small, medium and large pool fires are as follows.

■ A small pool fire is one that is less than one-sixth as long, in the direction of the airflow, as the tunnel is wide.
■ A medium-sized pool fire is one that is one that is about half as long, in the direction of airflow, as the tunnel is wide.
■ A large pool fire is one that is as long, in the direction of airflow, as the tunnel is wide.

Thus, it is to be expected that only a large spillage of fuel from a tanker will result in either a medium or large pool, and that any pool fires formed as a consequence of an incident involving vehicles other than tankers are likely to be considered as being small pool fires.

11.4. Methodology

Having defined the specific cases to be considered and having defined, in simple terms, the problem (to find $k = f(v)$ for each case), there still remained the challenge of finding sufficient experimental data upon which to base these mathematical descriptions.

In reality, at the time of the project, there had been only a small number of experimental fire tests carried out in tunnels, and only a small number of these tests had been carried out in tunnels with longitudinal ventilation.

Yet the task remained: how to use a handful of data to estimate a mathematical relationship?

Bayesian methods (Ang and Tang, 1975; Lee, 1989) have been used by statisticians for many years as a way to adjust estimates of probability in the light of new evidence. Bayes' theorem concerns conditional probability and, in its discrete form, may be expressed in this way:

$$P(k = k_i \,|\, I) = \frac{P(I \,|\, k = k_i)P(k = k_i)}{\sum_{i=1}^{n} P(I \,|\, k = k_i)P(k = k_i)}$$

where $P(k = k_i)$ is a 'prior' estimate of probability in the absence of evidence I, $P(I \,|\, k = k_i)$ is the likelihood of evidence I given $k = k_i$ and $P(k = k_i \,|\, I)$ is the 'posterior' probability, that is, the probability that k is k_i, updated in the light of evidence I. In this way, estimates may be updated by consideration of each piece of new evidence as it becomes available. That is, the Bayesian process is a continual one. As Lindley (1970) says, 'today's posterior distribution is tomorrow's prior'. The 'expectation' or mean value of k may then be determined using the Bayesian estimator:

$$E(k) = \sum k_i P(k = k_i \,|\, I)$$

Using Bayes's theorem, it is possible to estimate values of k for specific scenarios, given one or more relevant pieces of 'evidence', assuming that some form of prior probability distribution is available or can be estimated.

Often in probability problems of this type, the prior distribution is produced from expert estimates. In this study, a panel of experts in tunnel fire safety was organised, and each expert independently made estimates, which were then combined and used to generate the prior probability distributions needed for the calculations.

These prior probability distributions were then adjusted (or 'updated') using Bayes' theorem, together with all the experimental evidence that was available for each case.

This process has been described in detail elsewhere (Carvel et al., 2001a, b, c, 2003, 2004a, 2005). This chapter will not discuss the process but will concentrate on the results of the study

11.5. A note on naturally ventilated tunnel fires

This study presents the influence of longitudinal ventilation on the HRR relative to the case of a fire in a tunnel subject to natural ventilation. It should be noted that there are significant differences between fire behaviour in naturally ventilated tunnels and fires in the open air.

As with the influence of ventilation on fire behaviour, there are two main conflicting influences that tunnel geometries have on fire behaviour; there is an enhancing effect due to the confined nature of the tunnel – heat is not lost to the atmosphere but tends to remain in the location of the fire, resulting in enhanced heat feedback to the fire, and there is a diminishing effect due to the lack of inflowing oxygen in the smoke filled tunnel. As with most tunnel fire phenomena, these influences are still not adequately understood.

Figure 11.3 The influence of tunnel geometry on fire size

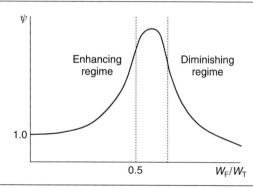

Here we consider the influence of tunnel geometry on fire behaviour in terms of another HRR coefficient, in this instance designated ψ, where

$$\dot{Q}_{tun} = \psi \dot{Q}_{open}$$

where $\dot{Q}_{tun}$ is the HRR of a fire in a naturally ventilated tunnel and $\dot{Q}_{open}$ is the HRR of a similar fire in the open air. It is assumed that the relationship between ψ and tunnel geometry behaves as shown in Figure 11.3. Fires which are small relative to the size of a tunnel will not be significantly influenced by the tunnel geometry (i.e. $\psi \approx 1$). Fires up to about half the width of a tunnel will be enhanced by the tunnel geometry (i.e. $\psi > 1$) whereas fires with dimensions close to the width of the tunnel will be reduced (i.e. $\psi < 1$). ψ values as large as 4.0 and as small as 0.2 have been observed experimentally (Carvel *et al.*, 2001d).

Research (Carvel *et al.*, 2001d, e, 2004) has shown that the relationship in the 'enhancing' regime is

$$\psi = 24 \left(\frac{W_F}{W_T} \right)^3 + 1$$

where W_T is the width of the tunnel and W_F is the width of the fire object. The relationship between tunnel geometry and fire size has yet to be established in the 'diminishing' regime.

11.6. Results for HGV fires
11.6.1 Single-lane tunnels
In most road tunnels and a number of rail tunnels, the greatest fire load regularly passing through is the HGV. Approximately two-thirds of all fatal tunnel fires have involved HGVs in their initial stages (see Chapter 1). Thus, it is not surprising that the HGV is perceived by many to be the greatest fire risk in tunnels.

The initial 'prior' estimates of probability for HGV fires were made by the expert group, considering the case of a single-lane tunnel equivalent to the dimensions of the Channel Tunnel. These estimates were 'updated' using data from the following experiments

■ HGV, 'truck load' and wooden crib fire tests carried out in a disused tunnel near Hammerfest, Norway, in 1992 (EUREKA, 1995)

- reduced-scale fire tests carried out in the Health and Safety Laboratory fire gallery, Buxton, UK, in 1993 (Bettis *et al.*, 1994)
- 'truck load' fire tests carried out in the Second Benelux Tunnel, Rotterdam, the Netherlands, in 2002 (DGPWWM, 2002)
- 'truck load' fire tests carried out in the Runehamar Tunnel, Norway, in 2003 (SP Technical Research Institute, 2003).

It was noted above that forced ventilation may have a considerable influence on the development of a fire on a HGV and on the maximum fire size. As a result, this project considered the growth phase and the fully involved phase of a HGV fire separately.

11.6.1.1 The growth phase

During the growth phase of a fire, k may be considered to be an indication of the rate of growth. If, at a given time during the growth phase of a fire subject to natural ventilation, the HRR of that fire is 3 MW and, at the same time during the growth phase of a fire subject to forced ventilation, the HRR of that fire is 6 MW, then k is taken to have a value of about 2. That is, the fire subject to forced ventilation has attained twice the HRR of the fire subject to natural ventilation in the same time period.

The probabilities of k for a HGV fire in the growth phase are represented by the percentile graphs shown in Figure 11.4. The variation of the expectation (i.e. the mean value) of k is also shown.

Figure 11.4 Graphical representation of the posterior probability data of k for the growth phase of a HGV fire in a single-lane tunnel with forced longitudinal ventilation. Note that the vertical axis of the graph is presented on a logarithmic scale. Based on discrete data, continuous lines represent trends only

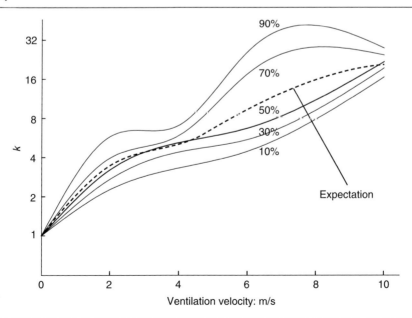

The meaning of the plot lines in Figures 11.4, 11.6–11.8 and 11.10–11.12 are as follows. k must have a value of 1.0 with no forced ventilation by definition, hence all the plots originate at 1 on the k axis. The study calculated discrete probability distributions for k at 2, 4, 6 and 10 m/s for each case. There is no clear way of representing these discrete probability values on a graph of k against ventilation velocity, so the data have been converted into percentile graphs to aid visualisation. The meaning of the tenth percentile is that, at any given ventilation velocity, there is a 10% probability that k will fall below the line and a 90% probability that k will be above the line. Similarly, there is a 30% probability that k will fall below the 30th percentile, and so on. The k values for 80% of HGV fires will fall between the tenth and 90th percentiles, while the average HGV fire will have the k value of the expectation, which is the mean value of k for the probability distribution. An example of the derivation of the percentile values is given in Figure 11.5.

Figure 11.5a shows the posterior probability distribution for a HGV fire in a single lane tunnel in the growth phase at 4 m/s. These discrete data are converted into a cumulative probability distribution as shown in Figure 11.5b. From this, the values of k which correspond to cumulative probabilities of 0.1, 0.3, 0.5, 0.7 and 0.9 can be found. These values are used as the basis for the probability percentiles shown in Figure 11.5c. Graphing software is used to generate a smooth line through the points at 0, 2, 4, 6 and 10 m/s, to aid visualisation.

In essence, these results indicate that any forced ventilation will increase the fire size in the growth phase, that is to say, that any forced ventilation will cause the fire to grow significantly faster than it would under conditions of natural ventilation. From the plot in Figure 11.4, it is clear that 70% of HGV fires (i.e. the portion of the graph above the 30th percentile), subjected to forced ventilation rates of 4 m/s or higher, will be at least four times greater in magnitude, at a given time in their development, than similar HGV fires in naturally ventilated tunnels. (It should be noted that the narrowness of the probability distribution at 4 m/s is not considered to be due to an actual physical process, but rather because the probability distribution at 4 m/s has been updated more than those at 2 and 6 m/s, and hence is narrower than those distributions due to decreased uncertainty.)

Figure 11.5 An example of the derivation of the percentile values for a HGV fire in the growth phase at 4 m/s

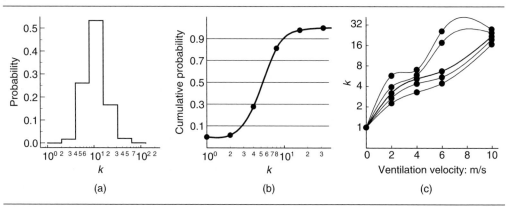

Figure 11.6 Graphical representation of the posterior probability data of k for a fully involved HGV fire in a single-lane tunnel with forced longitudinal ventilation. Note that the vertical axis of the graph is presented on a logarithmic scale. Based on discrete data, continuous lines represent trends only

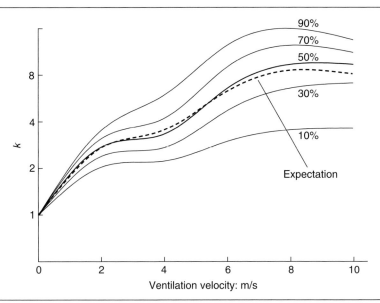

11.6.1.2 The fully involved fire

Once the HGV fire has reached full involvement, the ventilation no longer has the effect of enhancing the growth rate of the fire, but it still may enhance the burning efficiency and hence increase the HRR. The probability data of k for a fully involved HGV fire are represented by the percentile plots shown in Figure 11.6. The variation in expectation value is also shown.

These results indicate that forced ventilation will have a substantial enflaming influence on a fully involved HGV fire. From the graph in Figure 11.6 it is clear that 70% of HGV fires (i.e. the portion of the graph above the 30th percentile) will be at least three times greater in magnitude when subject to a forced airflow of 4 m/s (or above) compared with natural ventilation, in a single-lane tunnel.

11.6.2 Two-lane tunnels

In 2004, the work was extended to consider the influence of ventilation on HGV fire behaviour in two-lane tunnels. The results for the growth phase and the fully involved fire are summarised in Figures 11.7 and 11.8. Full details of the analysis are presented elsewhere (Carvel *et al.*, 2004c).

From these results it is clear that while the same overall trends are observed in two-lane tunnels as in smaller single-lane tunnels, the magnitude of the HRR coefficient k varies significantly with tunnel size as well as with ventilation velocity. Thus, while the expectation of $k_{4.0}$ in the growth phase has a value of about 5 in a single-lane tunnel, this is reduced to about 3.5 in a two-lane tunnel. Similarly, the expectation of $k_{4.0}$ for the fully involved fire is about 3 in a single-lane tunnel, reducing to about 2 in a two-lane tunnel.

Figure 11.7 Graphical representation of the posterior probability data of k for the growth phase of a HGV fire in a two-lane tunnel with forced longitudinal ventilation. Note that the vertical axis of the graph is presented on a logarithmic scale. Based on discrete data, continuous lines represent trends only

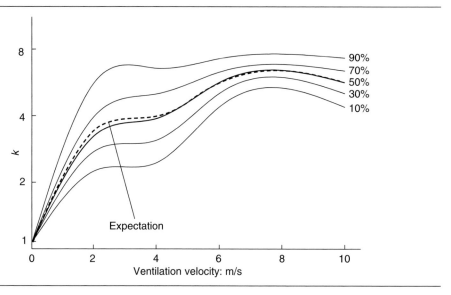

Figure 11.8 Graphical representation of the posterior probability data of k for a fully involved HGV fire in a two-lane tunnel with forced longitudinal ventilation. Note that the vertical axis of the graph is presented on a logarithmic scale. Based on discrete data, continuous lines represent trends only

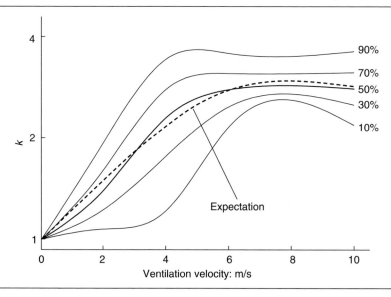

In general, it is to be expected that ventilation will have a diminishing enflaming influence on a HGV fire as the size of the tunnel increases. However, in all sizes of tunnel it is still observed that increasing ventilation velocity will result in an enhancement of the HRR.

11.7. Further observations on the growth rate of HGV fires: aggregated data

An analysis of the growth rates of experimental tunnel fires in their initial stages, carried out in 2008 (Carvel, 2008; Carvel *et al.*, 2009), addressed the problem in a different manner. In that study, it was observed that the growth rate of various 'HGV cargo' fire tests (generally HGV-trailer-sized loads comprised mainly of wooden pallets and some plastic materials, sometimes covered by a tarpaulin) could be approximated by a simple two-stage linear fire growth curve, rather than the 't^2' fire curve that is frequently assumed in fire safety engineering calculations.

The first stage of fire growth is often characterised by a period of relatively slow growth over a period of several minutes after ignition. However, this period is very variable, and may be very short. An analysis of experimental data (Carvel, 2008) suggests that there is a relationship between the length of this delay phase and the applied ventilation velocity (Figure 11.9a). From the limited data available, it appears that both low and high ventilation rates may tend to extend the delay phase, while ventilation rates of about 2–3 m/s might tend to result in the shortest delay. The data have been aggregated from one-lane (Runehamar and Hammerfest simulated truck load) and two-lane (Second Benelux tunnel) test results.

The second stage of fire growth is characterised by a period of rapid fire growth, with some fires growing at more than 20 MW/min. Again, an analysis of results from the same experiments

Figure 11.9 The influence of ventilation velocity on fire growth. (a) Observed variation of duration of the 'delay' phase with ventilation velocity. (b) Observed variation of fire growth rate with ventilation velocity (Aggregated experimental data from one and two lane tunnels.) Note: the graphs shown are polynomial fits to the available data, and should not be taken as anything other than simple trend lines

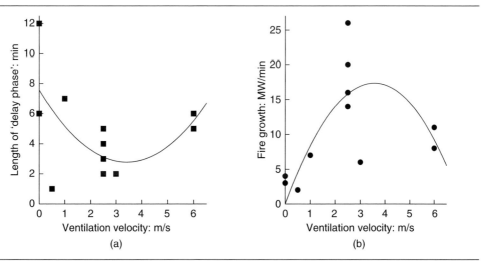

(Carvel, 2008) suggests that there is a relationship between the rate of fire growth and the applied ventilation velocity (Figure 11.9b). In this instance, it is observed that the fastest fire growth rates are observed for ventilation velocities around 3 m/s, whereas velocities significantly lower or significantly higher than this appear to result in lower fire growth rates.

However, it is important to see these results in context.

- The data are all from tests in which a HGV has been represented as a 'mock-up' of the load, and no account has been taken of the 'front unit' (i.e. cab/engine, etc.). For the Hammerfest HGV test, ignition was in the cab, and it took about 11 minutes for the load to ignite, after which fire growth was very rapid. If the data from this test are included, then the maximum in Figure 11.9b is not there, although the minimum in Figure 11.9a is. (For further considerations on early fire growth in HGVs, see Beard (2009, 2011).)
- The data have been aggregated from one- and two-lane tunnels. (The Runehamar Tunnel test rig had a cross-section of about 32 m^2, the Hammerfest Tunnel was about the same, and the Second Benelux Tunnel was about 50 m^2.) Considering one-lane data only, there is no maximum in Figure 11.9b and no minimum in Figure 11.9a. For two-lane data only, the maximum is not there in Figure 11.9b, although the minimum in Figure 11.9a is. Fire behaviour is different in one- and two-lane tunnels, and it is necessary to distinguish. In the Bayesian work, care was taken to distinguish between such tunnels, and different graphs have been constructed for each. In that work, while two-lane results were employed in updating for a one-lane tunnel, the way such data were used, as part of a Bayesian approach, was totally different. In a Bayesian process, data are not simply aggregated (i.e. pooled), they are used in a very different way.

11.8. Results for pool fires
11.8.1 Small pool fires
Pool fires can occur in any road tunnel fire incident. Vehicle crashes frequently result in the rupture of at least one fuel tank, leading to a spillage of fuel on the roadway. Under certain circumstances this can lead to the formation of an open pool of fuel, which may then ignite. Understanding the fire behaviour of fuel pools is therefore of importance. Fire spread across liquid fuels is generally substantially faster than across solid fire loads, so understanding the influence of ventilation on fire development is not as important an issue as it is for vehicles and solid fuels. Because of this, the study only concerned itself with fully involved pool fires.

The initial 'prior' estimates of probability for small pool fires, made by the expert group, were 'updated' using data from fire tests carried out in

- the Ofenegg Tunnel, Switzerland, in 1965 (Haerter, 1994)
- a disused tunnel near Hammerfest, Norway, in 1992 (EUREKA, 1995)
- the Memorial Tunnel, USA, in 1995 (Parsons Brinckerhoff, 1996)
- a motorway tunnel, between Toumei and Meishin, Japan, in 2001 (Kunikane *et al.*, 2002)
- the Second Benelux Tunnel, Rotterdam, the Netherlands, in 2002 (DGPWWM, 2002).

The results of the part of the study concerned with small pool fires (approximately the size of a fuel spillage due to a single HGV fuel tank rupture) are represented by the percentile plots shown in Figure 11.10.

Figure 11.10 Graphical representation of the posterior probability data of *k* for a small pool fire in a tunnel with forced longitudinal ventilation. Based on discrete data, continuous lines represent trends only

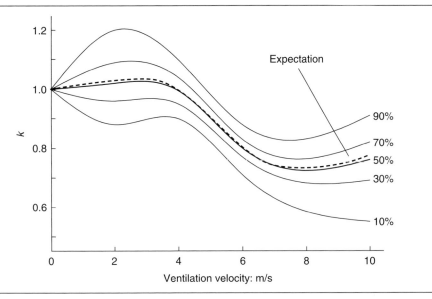

It can be seen that the enhancing effect of ventilation for pool fires is much less significant than with HGVs. Indeed, the probabilities of a small pool fire being enhanced by longitudinal airflow are only about 50% at 2 and 4 m/s velocities and less than 10% at 6 m/s. In general, it is expected that a small pool fire will exhibit a significantly lower HRR when subjected to forced ventilation than it would with natural ventilation.

11.8.2 Medium pool fires
Fuel spillages or pools that form in tunnels may be of any size. If a fuel tanker were to rupture, the resulting spillage/pool could be very large indeed. For this reason, larger fuel pool fire tests have been considered in this study.

The initial 'prior' estimates of probability for medium-sized pool fires, made by the expert group, were 'updated' using data from fire tests carried out in

- a mine tunnel, Londonderry, Australia, in 1991 (Apte *et al.*, 1991)
- a disused tunnel near Hammerfest, Norway, in 1992 (EUREKA, 1995)
- a laboratory-scale tunnel, Japan, in 1995 (Saito *et al.*, 1995).

The data from the part of the study concerned with medium-sized fuel pools are represented by the percentile plots shown in Figure 11.11.

In this instance, it appears that forced ventilation at any velocity will diminish the HRR of a medium-sized pool fire in a tunnel. Conceptually, it is not clear why the reduction in the HRR should be much more significant for a medium-sized pool fire than for a small pool fire. The reason for this may be due to the experiments upon which the study was based: the majority of

Figure 11.11 Graphical representation of the posterior probability data of *k* for a medium pool fire in a tunnel with forced longitudinal ventilation. Based on discrete data, continuous lines represent trends only

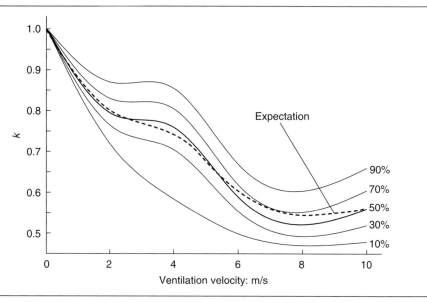

fire tests used as evidence in this instance were small-scale tests in model tunnels. The fire behaviour of small-scale fires may not be the same as their full-scale counterparts. Also, some of the experiments considered here involved methanol fuel, which burns in a significantly different manner to the majority of hydrocarbon fuels. The results from this case should therefore be used with caution.

11.8.3   Large pool fires

A large pool fire could only really occur if a fuel tanker were involved in an accident in a tunnel. This is, however, not without historical precedent. In 1982, a petrol tanker became involved in an incident in the Caldecott Tunnel (Oakland, USA). On that occasion, the tanker collided with a passenger car and was struck by a bus, causing the tanker to turn over, rupture and ignite. This incident not only created a pool fire on the roadway but the uppermost side of the tanker soon melted, to create a large, deep, open pool in the remains of the tank. This fire burned severely for over 2 hours (see Chapter 1). Understanding the fire behaviour of large fuel pools in tunnels is therefore of great importance.

The initial 'prior' estimates of probability for large pool fires, made by the expert group, were 'updated' using data from fire tests carried out in

- the Ofenegg Tunnel, Switzerland, in 1965 (Haerter, 1994)
- the Memorial Tunnel, USA, in 1995 (Parsons Brinckerhoff, 1996).

The data from the part of the study concerned with large fuel pool fires are represented by the percentile plots shown in Figure 11.12.

Figure 11.12 Graphical representation of the posterior probability data of k for a large pool fire in a tunnel with forced longitudinal ventilation. Based on discrete data, continuous lines represent trends only

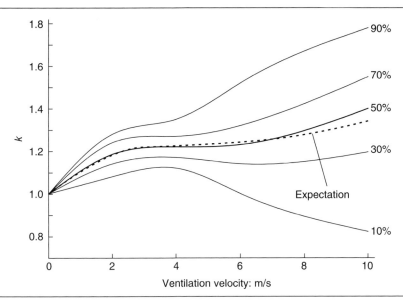

Unlike the smaller fuel pools, it appears that forced ventilation of any velocity will tend to enhance the burning of large fuel pools in tunnels. Indeed, there is less than 30% probability that the fire will be reduced in severity at any ventilation velocity, and there is a significant chance that the fire may be enhanced by as much as 50% at higher ventilation rates. The primary reason for this finding may be that it is likely that a large pool fire in a naturally ventilated tunnel would be significantly under-ventilated, so any application of ventilation would result in an increase in the HRR.

Subsidiary note on pool fires: Chen *et al.* (2008–2009) have conducted experiments in a one-eighth scale model of a tunnel, using heptane as fuel, and found a significant enhancing effect on the HRR with an increase in velocity. These results have not been used as part of the Bayesian study.

11.9. Results for car fires
The initial 'prior' estimates of probability for car fires, made by the expert group considering a single-lane tunnel, were 'updated' using data from

- wooden crib fire tests in a mine tunnel near Lappeenranta, Finland, in 1985 (Keski-Rahkonen *et al.*, 1986)
- car fire test in the Des Monts Tunnel, France, in 1992 (Perard, 1992)
- car and wooden crib fire tests in a disused tunnel near Hammerfest, Norway, in 1992 (EUREKA, 1995)
- car and wooden crib fire tests in a 'blasted rock tunnel', Sweden, in 1997 (Ingason *et al.*, 1997)
- car fire tests in the Second Benelux Tunnel, Rotterdam, the Netherlands, in 2002 (DGPWWM, 2002).

Although there were hardly any experimental data relating to car fires in tunnels at the time of the initial study, there were sufficient to estimate the influence of ventilation velocities of about 1.5–2 m/s. The Second Benelux Tunnel experiments were carried out after the study was concluded, and the results presented here have been updated to include the Benelux data. Although the Benelux experiments were carried out at airflow rates above 2 m/s, there are still not enough data to expand the study on car fires to 2, 4, 6 and 10 m/s as with the other cases considered here.

The results estimate that a car fire in a tunnel subjected to a longitudinal ventilation rate of about 1.5 m/s would not exhibit a HRR substantially different from a similar car fire in a naturally ventilated tunnel. Specifically, the study predicted that $k_{1.5}$ would have a value of 1.0 ($\pm$0.25) in almost 70% of car fires. The study also indicated that the fire growth rate would not be significantly influenced by forced ventilation rates of this magnitude.

It should, however, be noted that in the car fire tests carried out in the Benelux Tunnel, ventilation appeared to have an enhancing influence on fire size, but only after the failure of the car windscreen (DGPWWM, 2002).

11.10. Discussion

Although fire safety in tunnels has become an important issue in the past decade, there are still many large gaps in our understanding of fire behaviour in tunnels: the effect of ventilation systems on fire development and size is only one of many poorly understood aspects of fires in tunnels. Clearly, the best way to understand influences such as these fully would be to perform more fire tests in tunnels, especially full-scale tests, and it is hoped that such work will be carried out in the future. In the meantime, it is important to try to learn as much as possible from all the experimental results that are available, and to implement changes in practice where applicable.

At present, there is no generally accepted consensus regarding the design or operation of ventilation systems for vehicle tunnels in the event of a fire. Where standards or recommendations do exist (Nordmark, 1998), they are largely based on individual judgement or experience and not on experimental results (Luchian, 1996). In this study, the opinions of a panel of experts were sought at the outset, and these estimates were updated with as much experimental data as was available at the time. The Bayesian process is continual, results being updated as new data become available. One thing that became clear from this procedure was that there was no consensus of opinion amongst even a small group of experts, and that even where there was some consensus in part of the group, this did not necessarily agree with the evidence of the experimental data. Very few of the experts expected the ventilation to have as dramatic an effect on the HRR of a HGV fire as the experimental evidence suggests, especially in the growth phase. There was also some diversity of opinion in the case of the large pool fire: half of the group believed that forced ventilation, at any velocity, would enhance the HRR of the fuel pool whereas the other half of the group believed the opposite. It is clear that, if an engineering decision has to be made based on estimates alone, the opinion of one or two experts may not be enough. A method of combining expert judgement with what little experimental evidence is available is not merely desirable but essential if the level of safety in vehicle tunnels is to be increased.

Large fires in tunnels are relatively rare. The majority of fire-fighters who work in areas near vehicle tunnels may have little or no experience of fighting such fires, and may not know what to expect in the event of such fires. Yet these fire-fighters, in conjunction with the tunnel operators, will probably have access to the controls of the tunnel ventilation systems in the event of a fire in a tunnel. Decisions on how to use the tunnel ventilation system will be made. It is hoped that the results presented in this chapter, together with results from other studies on tunnel fires, will enable the decision to be grounded in knowledge of fire behaviour, rather than just a 'best guess'.

The primary concern of a fire brigade is to help any people escape a fire to a place of safety and to control and extinguish the fire. Ventilation systems may be controlled in such a way as to blow all the smoke produced by the fire to one side of the fire location. This will provide a smoke-free escape route upwind of the fire location, but may also significantly affect the size of the fire. This study has demonstrated that, for a HGV fire in a tunnel, forced ventilation causes an increase in fire size. Increasing the ventilation velocity will probably cause the fire to grow significantly. Thus, in the light of the results presented here, it would appear that the best solution would be to use the minimum ventilation velocity that is sufficient to control the smoke.

It has been shown that the sizes of fires on heavy goods vehicles can be greatly increased by application of forced ventilation. It has also been shown that the HRR of large pool fires may be increased, while that of the HRR of smaller pool fires may be reduced significantly, if forced ventilation is used. It has also been shown that fires involving passenger cars are not significantly affected by forced ventilation; at least at the ventilation velocity considered. From this, it is clear that different ventilation strategies should be used when tackling different types of fires – ventilation should be kept to an absolute minimum if the fire involves a HGV or a large fuel spillage, whereas ventilation may be increased if the fire involves a passenger car or a smaller pool fire. Indeed, it has been shown that a high ventilation velocity may be the best response for smaller pool fires as, in some cases, this will cause a decrease in fire size. Adopting a single airflow velocity in the event of any fire in a tunnel does not seem to be the most sensible option – if the fire type can be identified by the tunnel operators or fire brigade, then the tunnel ventilation should be adjusted accordingly.

These results have been both criticised and commended in the literature over the past decade. In 2002, Huijben (2002) was the first to observe that, while the trends predicted by the original (single-lane tunnel) study were consistent with his experimental observations (the two-lane Second Benelux Tunnel tests), the numbers predicted were too large. This is indeed the case, as the effect of ventilation is reduced in larger tunnels, as calculated in 2004 and presented above.

Ingason has also claimed that the numbers predicted in this study are too large, and has explained that the reason for the enhancing effect, observed for ventilation rates between 2 and 4 m/s, compared with naturally ventilated fires, is likely due to the fact that the naturally ventilated fires are underventilated (Lönnermark and Ingason, 2008; Ingason and Li, 2010). The addition of ventilation, up to a certain limit, results in an enhancement of the HRR, as the fire tends towards being well ventilated. However, beyond this limit, there is no reason why a further increase in ventilation velocity should lead to an enhancement in the HRR. Thus, Ingason would expect the variation of k with ventilation to level off beyond a certain velocity (much like it does in Figure 11.8), and possibly even diminish beyond this limit.

Other studies, for example the analysis by Melvin and Gonzalez (2009), have found that not only the trends but also the predicted values are consistent with their analyses.

Irrespective of the accuracy of the numbers predicted in this study – which vary with tunnel dimensions anyway – it is hoped that the trends in the results presented here will be easily remembered and therefore may be recalled when decisions need to be made when fighting real fires in tunnels.

11.11. Conclusions

Vehicle and pool fires in tunnels respond to forced ventilation in different ways. Adopting the same emergency ventilation response to all fires in tunnels does not seem sensible.

In summary, from the study, the following can be determined.

- The size of HGV fires will generally be increased by forced longitudinal ventilation. This influence is even more significant in the growth phase of the fire.
- It is important to distinguish between one- and two-lane tunnels.
- Maxima may exist in HGV fire growth rates and peak HRRs, with variation in ventilation: see the k-value graphs.
- Increasing the ventilation rate will increase the HRR of large pool fires. Increasing the ventilation rate appears to tend to reduce the size of smaller pool fires in tunnels.
- A forced ventilation velocity of 1.5 m/s does not seem to significantly affect the HRR of a car fire in a tunnel, compared with natural ventilation.

Acknowledgements

This project would not have been possible without the expert estimates and advice of the late H. L. Malhotra, R. Chitty and M. Shipp (BRE), S. Miles (currently with International Fire Consultants), G. Atkinson and R. Bettis (HSE, Buxton), J. Beech (formerly Chief of Kent Fire and Rescue Service), G. Grant (currently with Optimal Energies) and D. Drysdale (Professor Emeritus, University of Edinburgh). The original project was funded by the Engineering and Physical Sciences Research Council under grant GR/L69732.

REFERENCES

Ang AH-S and Tang WH (1975) *Probability Concepts in Engineering Planning and Design.* Wiley, New York, vol. 1.

Apte VB, Green AR and Kent JH (1991) Pool fire plume flow in a large-scale wind tunnel. In: *Fire Safety Science – Proceedings of the 3rd International Symposium, University of Edinburgh,* 8–12 July 1991, pp. 425–434.

Babrauskas V and Peacock RD (1992) Heat release rate: the single most important variable in fire hazard. *Fire Safety Journal* **18**: 255–272.

Beard AN (1983) A logic tree approach to the St. Crispin hospital fire. *Fire Technology* **19**: 90–102.

Beard AN (2009) HGV Fires in tunnels: fire size and spread. In: *2nd International Tunnel Safety Forum for Road and Rail, Lyon,* 20–22 April 2009. Organised by Tunnel Management International, pp. 103–111.

Beard AN (2011) HGV fire in a tunnel: spread from front unit to load. In: *3rd International Tunnel Safety Forum for Road and Rail, Nice,* 4–6 April 2011. Organised jointly by TMI, FERMI and BREGlobal, pp. 195–204.

Bettis RJ, Jagger SF and Moodie K (1994) Reduced scale simulations of fires in partially blocked tunnels. In: *Proceedings of the International Conference on Fires in Tunnels, Borås,* 10–11 October 1994, pp. 163–186.

234

Carvel R (2008) Design fires for water mist systems. In: *Proceedings of the 3rd International Symposium on Tunnel Safety and Security (ISTSS), Stockholm*, 12–14 March 2008, pp. 163–171.

Carvel R, Rein G and Torero JL (2009) Ventilation and suppression systems in road tunnels: Some issues regarding their appropriate use in a fire emergency. In: *Proceedings of the 2nd International Tunnel Safety Forum for Road and Rail, Lyon*, 20–22 April, pp. 375–382.

Carvel RO, Beard AN and Jowitt PW (2001a) The influence of longitudinal ventilation systems on fires in tunnels. *Tunnelling and Underground Space Technology* **16**: 3–21.

Carvel RO, Beard AN, Jowitt, PW and Drysdale DD (2001b) Variation of heat release rate with forced longitudinal ventilation for vehicle fires in tunnels. *Fire Safety Journal* **36**: 569–596.

Carvel RO, Beard AN and Jowitt PW (2001c) A Bayesian estimation of the effect of forced ventilation on a pool fire in a tunnel. *Civil Engineering and Environmental Systems* **18**: 279–302.

Carvel RO, Beard AN and Jowitt PW (2001d) How much do tunnels enhance the heat release rate of fires? In: *Proceedings of the 4th International Conference on Safety in Road and Rail Tunnels, Madrid*, 2–6 April 2001, pp. 457–466.

Carvel RO, Beard AN and Jowitt PW (2001e) A method for estimating the heat release rate of a fire in a tunnel. In: *Proceedings of the 3rd International Conference on Tunnel Fires, Gaithersburg, MD*, 9–11 October 2001, pp. 137–144.

Carvel RO, Beard AN and Jowitt PW (2003) The influence of longitudinal ventilation on fire size in tunnels: update. In: *Proceedings of the 5th International Symposium on Safety in Road and Rail Tunnels, Marseille*, 6–8 October 2003, pp. 431–440.

Carvel RO, Beard AN and Jowitt PW (2004a) How longitudinal ventilation influences fire size in tunnels. *Tunnel Management International* **7**(1): 46–54.

Carvel RO, Beard AN, Jowitt PW and Drysdale DD (2004b) The influence of tunnel geometry and ventilation on the heat release rate of a fire. *Fire Technology* **40**: 5–26.

Carvel RO, Beard AN and Jowitt PW (2004c) The influence of longitudinal ventilation and tunnel size on HGV fires in tunnels. In: *Proceedings of the 10th International Fire Science and Engineering Conference (Interflam 2004), Edinburgh*, 5–7 July 2004, pp. 815–820.

Carvel RO, Beard AN, Jowitt PW and Drysdale DD (2005) Fire size and fire spread in tunnels with longitudinal ventilation systems. *Journal of Fire Sciences* **23**: 485–518.

Chen C, Qu L, Yang YX, Kang GQ and Chow WK (2008–2009) Scale modelling on the effect of air velocity on heat release rate in tunnel fire. *Journal of Applied Fire Science* **18**: 111–124.

DGPWWM (2002) *Project 'Safety Test' – Report on Fire Tests*. Directorate-General for Public Works and Water Management, Civil Engineering Division, Utrecht.

EUREKA (1995) *Fires in Transport Tunnels: Report on Full-scale Tests*. EUREKA-Project EU499:FIRETUN Studiengesellschaft Stahlanwendung elV. D-40213 Dusseldorf.

Haerter A (1994) Fire tests in the Ofenegg Tunnel in 1965. In: *Proceedings of the International Conference Fires in Tunnels, Borås*, 10–11 October 1994, pp. 195–214

Huijben JW (2002) The influence of longitudinal ventilation on fire size and development. In: *4th International Conference on Tunnel Fires, Basel*, 2–4 December 2002, pp. 115–124.

Ingason H and Li YZ (2010) Model scale tunnel fire tests with longitudinal ventilation. *Fire Safety Journal* **45**: 371–384.

Ingason H, Nireus K and Werling P (1997) *Fire Tests in a Blasted Rock Tunnel*. Report FOA-R-97-00581-990-SE. FOA, Linköping, Sweden.

Keski-Rahkonen O, Holmjund C, Loikkanen P, Ludvigsen H and Mikkola E (1986) *Two Full Scale Pilot Experiments in a Tunnel*. VTT Research Report 453. VTT, Oulu, Finland.

Kunikane Y, Kawabata N, Takekuni K and Shimoda A (2002) Heat release rate induced by gasoline pool fire in a large-cross-section tunnel. In: *Proceedings of the 4th International Conference on Tunnel Fires, Basel*, 2–4 December 2002, pp. 387–396.

Lee PM (1989) *Bayesian Statistics: An Introduction*. Edward Arnold, London.

Lindley DV (1970) Bayesian analysis in regression problems. In: Meyer DL and Collier RO (eds), *Bayesian Statistics*. FE Peacock, Itasca, IL, p. 38.

Lönnermark A and Ingason H (2008) The effect of air velocity on heat release rate and fire development during fires in tunnels. In: *Proceedings of the 9th International Symposium on Fire Safety Science, Karlsruhe*, 21–26 September 2008, pp. 701–712.

Luchian S (1996) *Memorial Tunnel Fire Ventilation Test Programme*. Interactive CD-ROM and Comprehensive Test Report. Parsons Brinckerhoff 4D Imaging, Boston, vol. 1, p. 1.1.

Mangs J and Keski-Rahkonen O (1994) Characterization of the fire behaviour of a burning passenger car. Part 1: car fire experiments. *Fire Safety Journal* **23**: 17–35.

Melvin B and Gonzalez J (2009) Considering tunnel geometry when selecting a design fire heat release rate for road tunnel safety systems. In: *Proceedings of the 13th International Symposium on Aerodynamics and Ventilation of Vehicle Tunnels, New Brunswick, NJ*, 3–15 May 2009, pp. 225–236.

Nordmark A (1998) Fire and life safety for underground facilities: present status of fire and life safety principles related to underground facilities. *Tunnelling and Underground Space Technology* **13**(3): 217–269.

Parsons Brinckerhoff (1996) *Memorial Tunnel Fire Ventilation Test Program*. Interactive CD-ROM and Comprehensive Test Report. Parsons Brinckerhoff 4D Imaging, Boston.

Perard M (1992) Organization of fire trials in an operated road tunnel. In: *1st International Conference Safety in Road and Rail Tunnels, Basel*, 23–25 November 1992, pp. 161–170.

Saito N, Yamada T, Sekizawa A, Yanai E, Watanabe Y and Miyazaki S (1995) Experimental study on fire behaviour in a wind tunnel with a reduced scale model. In: *2nd International Conference Safety in Road and Rail Tunnels, Granada*, 3–6 April 1995, pp. 303–310.

SP Technical Research Institute (2003) *Proceedings of the International Conference on Catastrophic Tunnel Fires, Borås*, 20–21 November 2003. SP Technical Research Institute, Sweden.

Part III

Tunnel fire dynamics

Handbook of Tunnel Fire Safety, 2nd edition
ISBN: 978-0-7277-4153-0

ICE Publishing: All rights reserved
doi: 10.1680/htfs.41530.239

ice
Institution of Civil Engineers

publishing

Chapter 12
A history of experimental tunnel fires

Richard Carvel
BRE Centre for Fire Safety Engineering, University of Edinburgh, UK
Guy Marlair
Institut National de l'Environnement Industriel et des Risques (INERIS), France

12.1. Introduction

Until the 1960s, fire research in tunnels had been largely concerned with fire safety in mine tunnels, and so the main fire loads that had been considered were coal (on the coal face, in sacks, coal-dust explosions, etc.), wooden structures (e.g. wooden scaffolding supporting a mine-tunnel ceiling) and conveyor belts (seemingly the source of several fires in mines). The consequences of vehicle fires had never really been considered. In the early 1960s, a lot of transport tunnels were being constructed, particularly in the Alps, so in order to better understand what might happen if there were a fire, experimental testing of fires in vehicle tunnels began.

Fire tests have been carried out for a number of different reasons over the years, the two main reasons being

- to gain an understanding of the fire dynamics and related phenomena in tunnels
- to test or commission tunnel installations, including ventilation systems, sprinkler systems and tunnel linings.

As these two reasons have very different agendas, they are considered separately here. This chapter focuses primarily on the former type of tunnel-fire experiment. The conclusions or findings given in the following accounts are those of the people who conducted the tests.

12.2. Fire experiments to gain an understanding of fire phenomena
12.2.1 The Ofenegg tunnel fire experiments, 1965

The primary tunnel fire safety concern in Switzerland in the early 1960s was what would happen if there were an accident involving a fuel tanker in one of the new tunnels. To address this question, a series of fire tests was carried out in an abandoned railway tunnel near Ofenegg, Switzerland, in 1965 (Haerter, 1994). The tunnel was converted to allow for the testing of natural, longitudinal and semi-transverse ventilation systems by building a duct along one side of the tunnel. One end of the tunnel was blocked, so that semi-transverse ventilation ($25 \, \text{m}^3/\text{s}$ per metre) could be simulated by having vents at 5 m intervals along the length of the duct, and longitudinal ventilation (1.7 m/s) could be simulated by sealing these vents and having an opening between the duct and the main tunnel near to the blocked end of the tunnel. Three different sized pools of 'aircraft-quality petrol' were used in the tests (6.6, 47.5 and 95 m^2 containing 100, 500 and 1000 litres of fuel, respectively); each pool filled the width of the tunnel (3.8 m). Data recorded included:

visibility measurements, air temperatures, carbon monoxide and oxygen concentrations, and air velocities, Pieces of hair, meat and wood were hung at various points along the tunnel to determine the effect of the heat from the fires. In all, 12 tests were carried out, the main observations being the following.

- Natural and semi-transversely ventilated fires burned slower than equivalent fires in the open air, due to oxygen depletion. This effect was greater with larger fires.
- Longitudinal ventilation can cause an increase in burning rate (compared with other fires in tunnels, not compared with burning in the open air).
- The velocity and thickness of the smoke layer was greater for larger fires (up to 11 m/s and 4 m for semi-transverse ventilation).
- Longitudinal ventilation can cause the smoke layer to fill the whole tunnel (loss of stratification).
- Maximum temperatures (of pool fires) were achieved within 1–2 minutes from ignition.
- Survival is not possible within 30–40 m of a large pool fire (with any ventilation configuration), and the chances of survival downstream of the fire are substantially reduced with longitudinal ventilation.
- Sprinklers can extinguish the fire, but fuel vapours will remain and may re-ignite, with devastating effects (airflow above 30 m/s and damage to the tunnel facilities).

12.2.2 The West Meon tunnel fire experiments

In the early 1970s, some fire tests were carried out by the Fire Research Station, UK, in a disused railway tunnel near West Meon, Hampshire (Heselden, 1976). The tunnel was 480 m long, 8 m wide and 6 m high. Few details of the tests are reported in the literature, except that a number of cars were burned and the natural ventilation was a wind of about 2 m/s. The main findings were

- the smoke layer from the fire was up to 3 m thick (i.e. the air was breathable at head height)
- observers were able to remain near the fire without any ill effects, except headaches, afterwards.

12.2.3 The Glasgow tunnel fire experiments, 1970

Five fires were carried out in a 620 m long, 7.6 m wide, 5.2 m high disused railway tunnel in Glasgow (Figure 12.1) (Heselden and Hinkley, 1970). The fire load in each case was made up of one, two or four trays of kerosene fuel, each tray (1.2 m × 1.2 m) having an approximate thermal output of 2 MW. Smoke movements were recorded, as were some temperature measurements.

The main findings were as follows.

- The smoke layer in most tests started out at 1–2 m thick, but thickened with time. For the largest fire, the smoke layer reached a thickness of 3–4 m after 10 minutes.
- The smoke layer advanced at about 1–1.5 m/s.
- A 'plug' of smoke was formed at the end of the tunnel as it encountered the crosswind outside. This filled the entire height of the tunnel, and tended to be drawn back into the tunnel by the fire-induced airflow.
- The air below the main smoke layer did not remain smoke free – some mixing occurred shortly after the fire started.

Figure 12.1 The Glasgow tunnel fire experiments, 1970 (Photograph courtesy of BRE Global)

12.2.4　The Zwenberg tunnel fire experiments, 1976

In the early 1970s, the Austrian road network was being improved, and this involved the construction of many road tunnels of between 5 and 13 km long. As the Austrian authorities had no experience of fires in tunnels, in order to gain a better understanding of such fires they carried out 30 fire tests (25 of a 6.8 m^2 petrol pool, three of a 13.6 m^2 petrol pool, one of a 6.8 m^2 diesel pool, and one with a mixed load of wood, car tyres and sawdust) in a 390 m long abandoned railway tunnel (Pucher, 1994). This tunnel was modified in order to simulate natural, longitudinal, transverse and semi-transverse ventilation systems. The main findings were as follows.

- Temperatures at the ceiling were very high (sometimes over 1200°C).
- 'Higher' longitudinal air velocities destroyed thermal layering (stratification).
- With transverse ventilation, fans in the fire section should be set to maximum exhaust, while the power of fans supplying fresh air should be reduced by 20–30% to maintain stratification.
- The position of supply vents (above or below) was not crucial.
- For the smaller fires it was possible to extract all the plume gases over a tunnel length of approximately 260 m with maximum extraction and minimum supply of fresh air.
- The tunnel lining and intermediate ceiling were not destroyed by the fires.

12.2.5　PWRI tunnel fire experiments, 1980

The Japanese Public Works Research Institute (PWRI) carried out 16 full-scale fire tests in a 700 m long fire gallery, and eight full-scale fire tests in a 3.3 km road tunnel. Fire loads tested included 12 petrol pools (ten 4 m^2 pools and two 6 m^2 pools), six passenger cars and six buses. The conditions (temperature, gas concentrations, smoke, etc.) in the gallery/tunnel were

241

monitored during tests, which were done with natural and longitudinal ventilation (Mizutani *et al.*, 1982; PWRI, 1993). The main observations were as follows.

- Stratification of smoke was partially destroyed by longitudinal ventilation at 1 m/s, and totally destroyed by longitudinal ventilation at 2 m/s.
- The 'heat-generation speed' (heat release rate) of a fire increased at higher longitudinal ventilation velocities; for a petrol pool fire the heat release rate was 4 MW at 1–2 m/s and 6 MW at 4 m/s. This effect was 'more evident' for car fires.
- The temperature rise in the tunnel was only significant near to the fire source.
- None of the car, bus or pool fires were totally extinguished by the sprinklers, but the 'heat-generation speed' was reduced in each case.
- The sprinklers had an 'adverse effect on the environment' by causing a reduction in smoke density near the ceiling and an increase in smoke density in the lower part of the tunnel.

12.2.6 Tunnel fire tests at VTT, 1985

As a precursor to the EUREKA project (see Section 12.2.7), two 'pilot' fire tests of wooden cribs were carried out in a 'small' tunnel in Lappeenranta, south-eastern Finland (Keski-Rahkonen *et al.*, 1986). The purpose of the tests was to investigate the effect of the fires on the tunnel lining and also to investigate fire spread across and between objects. The tunnel is 183 m long, between 5.5 and 6.1 m wide, and between 4.3 and 5.0 m high. Ventilation was forced by two fans at one end of the tunnel. The main observations were as follows.

- In test 1 (which involved a 'mock-up' of a subway train – wooden cribs totalling 3.2 m wide and 48.0 m long), a constant rate of fire spread (6.6×10^{-4} m/s, ~1.8 MW) was established after a brief ignition phase.
- In test 2 (which involved a line of eight 'mock-ups' of passenger cars – wooden cribs $1.6 \times 1.6 \times 0.8$ m high, spaced 1.6 m apart), no spread was observed between cribs.
- Identical cribs at opposing ends of the line (in test 2) exhibited very different burning characteristics – the crib at the windward end of the line burned with almost twice the heat release rate of one at the other end.
- As experiments of this scale are never fully reproducible, more than one experiment is needed to make far-reaching conclusions regarding vehicle fires (Keski-Rahkonen, 1994).
- Considerable spalling of rock occurred during the fires.
- Using the theoretical models available at the time, both tests were expected to produce an equivalent of flashover. Neither did. (Strictly speaking, 'flashover' is defined in relation to compartments and therefore cannot apply to tunnels.)

12.2.7 The EUREKA EU-499 'FIRETUN' test series, 1990–1992

The tunnel-fire test series with the largest scope was undoubtedly the EUREKA 'FIRETUN' test series, carried out between 1990 and 1992 by teams of fire researchers representing Austria, Finland, France, Germany, Italy, Norway, Sweden, Switzerland and the UK (EUREKA, 1995). The majority of the fire tests (21 tests) were carried out in an abandoned tunnel near Hammerfest, Norway, in 1992, and involved fire loads such as cars, train carriages, wooden cribs, heptane pools, a 'simulated truck load' and a heavy goods vehicle (HGV) fully laden with a cargo of furniture. Other fire tests carried out as part of the series were wooden crib fire tests in a disused railway tunnel in Germany, wooden crib fire tests at VTT in Finland (see Section 12.2.6), heptane pools at INERIS (see Section 12.6.3) and a laboratory test of samples of tunnel lining. The objectives of the project were to provide information on

- fire phenomena
- escape, rescue and fire-fighting possibilities
- the effect of the surrounding structure on the fire
- reusing the structure (damage done, time required for redevelopment, etc.)
- accumulation of theory (improving the understanding of fire, modifying models, etc.)
- the formation, distribution and precipitation of contaminants.

The Hammerfest tunnel is a 2.3 km long mine tunnel with an irregular cross-section (very approximately square, varying from 30 to 40 m²). During the fire tests various data types were recorded, including mass loss, temperature, gas concentrations, smoke density, airflow velocity, etc. All these were all recorded at many points near the fire load and distant from it (except the mass loss, which was measured directly by load cells under the fire location). Unfortunately, not all these data types were recorded in each experiment. For example, in several of the wooden crib tests and one of the car fire tests, mass loss and gas concentration data were not recorded, so it is not possible to estimate the heat release rate of these fires. Many conclusions have been drawn from this test series.

- The influence of damage, both to the vehicles and the tunnel lining, especially in the crown area, depended on the type of vehicle. The roofs of those vehicles constructed of steel resisted the heat, whereas the roofs of the vehicles made of aluminium were completely destroyed at an early stage of the fire (Haack, 1995; Richter, 1995).
- The temperatures during most of the vehicle fires reached maximum values of 800–900°C. The temperature during the HGV test reached 1300°C. Temperatures decreased substantially within a short distance from each fire location. Temperatures were greater downwind than upwind (Haack, 1995).
- The railway carriages burned between 15 and 20 MW. The HGV burned at over 100 MW (Haack, 1995).
- A 'small fire load', such as a single railway carriage, may exhibit a heat release rate of 45 MW (Steinert, 1994).
- All road and rail vehicles showed a fast fire development in the first 10–15 minutes (Haack, 1995).
- Growth rates of vehicle fires vary from 'medium' to 'ultra fast' (see NFPA 72) (Ingason, 1995b).
- Modern rail cars are more resistant to ignition than older carriages (Haack, 1995).
- In naturally ventilated fire tests of railway carriages, the smoke density and carbon monoxide concentration exceeded 'acceptable' limits over 300 m away from the fire location (Blume, 1995).
- To enable fire-fighters to get to the fire location, a 'sheltered route' needs to be provided (e.g. the service tunnel in the Channel Tunnel) (Blume, 1995).
- The maximum concentration of polycyclic aromatic hydrocarbons (PAHs) and other pollutants was found at about 70–80 m downwind of each fire location (Bahadir et al., 1995).
- The fire growth and burning pattern was strongly influenced by ventilation conditions (Malhotra, 1995).
- If a container or enclosure is not well sealed the prevailing ventilation can introduce sufficient leakage air for combustion to continue at a slow rate (Malhotra, 1995).
- The rate of burning can be significantly accelerated by a free supply of air (Malhotra, 1995).

- Controlled ventilation can be useful in smoke management (Malhotra, 1995).
- Longitudinal ventilation destroyed stratification downwind of the HGV fire (Malhotra, 1995).

12.2.8   The Memorial Tunnel Fire Ventilation Test Program (MTFVTP), 1993–1995

The scope of the EUREKA test series was larger than that of any other series. However, the MTFVTP is the largest tunnel fire test series to date in terms of actual scale, although only pool fire tests were carried out. In all, 98 pool fire tests ranging in size from about 10 MW to about 100 MW were carried out in a disused 850 m long two-lane road tunnel near Charlestown, WV, USA (Parsons Brinckerhoff, 1996). The tunnel was modified to allow for the testing and comparison of natural, semi-transverse, fully transverse and longitudinal ventilation systems, and their ability to control and extract smoke. Extensive arrays of temperature sensors, gas analysers, airflow sensors and smoke-density measuring equipment were located throughout the tunnel, so there are extensive data from each test. The experiments were carried out by the Federal Highway Administration for the Boston Central Artery Tunnel project, and the stated objectives of the test series were as follows.

- 'To develop a comprehensive database regarding temperature and smoke movement from full-scale fire ventilation tests which would permit a definitive comparative evaluation of the capabilities of transverse and longitudinal ventilation systems to manage smoke and heat in a fire emergency.'
- 'To determine, under full-scale fire test conditions, the relative effectiveness of various ventilation system configurations, ventilation rates, and operating modes in the management of the spread of smoke and heat for tunnel fires of varying intensities.'

The influence of the different ventilation strategies on the heat release rate of the fire was not considered. A systematic series of tests, each with a single (different) airflow velocity, was not carried out. Instead, the configurations of active ventilation fans were changed several times (in some instances, many times) during a test, and so changes in the behaviour of the fires due to different ventilation strategies are not clear. The fires were all carried out using diesel fuel, which produces a lot of smoke and reduces the risk of explosion, but unfortunately has not been used in many other fire experiments, so comparison with other pool fire tests (both in and out of tunnels) is not easy. The main conclusions from the test series include:

- Longitudinal ventilation:
  - Longitudinal ventilation using jet fans was highly effective in controlling smoke spread for fires up to 100 MW. However, it is appropriate only for unidirectional tunnels.
  - Longitudinal air velocity was dependent on the number of active fans and the thrust, not on the configuration of the fans.
  - A 10 MW fire tended to reduce the longitudinal airflow by 10%, and a 100 MW fire reduced it by 50–60%.
  - Airflow velocities of 500–580 fpm (2.5–3 m/s) were sufficient to prevent backlayering of smoke from 100 MW pool fires in the Memorial Tunnel.
- Transverse ventilation:
  - It was not sufficient only to supply air in a tunnel fire situation, extraction is also necessary.
  - Longitudinal airflow is a major factor in smoke control for transversely ventilated tunnels.
  - Multiple-zone ventilation systems are better than single-zone ventilation systems at controlling smoke.

- Single-point extraction openings and oversized exhaust ports significantly enhance the ability of a ventilation system to control and extract smoke.
■ Smoke and heat movement:
  - The time taken for smoke to enter the 'occupied zone' at positions distant from the fire location was dependent on the height and geometry of the tunnel ceiling.
  - A significant reduction in visibility was reached more quickly than was debilitating heat.

12.2.9 Cable-tunnel fire testing programme, New Zealand, 1997

A series of tests was carried out in a 50 m long tunnel (part of a sports stadium in Auckland) to investigate the use of water-mist systems for suppression of cable fires, to set performance criteria for such systems and to test the performance of various commercial systems (Mawhinney *et al.*, 1999). Four manufacturers participated in the programme. The fire scenario involved an arrangement of simulated cables above a 0.25 m × 0.25 m square pan containing 1 litre of heptane as an ignition source (the heptane burned out within 3 minutes).

The conclusions from these tests include the following.

■ The fire scenario was reproducible and was suitable for the evaluation of commercial water-mist systems.
■ The tests allowed the establishment of satisfactory performance criteria, in terms of temperature levels at the underside of the cable bundles immediately above the fire.
■ Of the four water-mist systems tested, two met the performance criteria.

12.2.10 Fire suppression tests for Eurotunnel, 1998–1999

A series of tests was carried out in a specially constructed 'half-tunnel' fire gallery at Darchem Flare, UK (Grant and Southwood, 1999; Grant, 2001). The fire gallery was based on the dimensions of the Channel Tunnel, having a radius of 3.8 m, but the gallery was equivalent to only half the width of the Channel Tunnel, having a D-shaped profile. The concept behind this design was that HGV fires in the Channel Tunnel would most likely be approximately symmetrical, and thus the fire behaviour of a HGV in the Channel Tunnel may be modelled by the fire behaviour of a half-HGV in a half-tunnel. The conclusions depend on this basic assumption. The gallery was 38 m long. The test series was designed to test the capabilities of the onboard fire-suppression system (OFSS) intended for use on HGV-carrying trains in the Channel Tunnel.

The conclusions from the test series include the following.

■ The OFSS would be efficient at mitigating the consequences of a fire on a HGV carrier in a tunnel, and thereby increase life safety and reduce damage to facilities and rolling stock.
■ High-pressure water-mist systems would be better at mitigating HGV fires on carrier wagons in tunnels than would deluge sprinkler systems.

In addition to these full-scale tests, other smaller tests were carried out to support the use of the OFSS (Dufresne *et al.*, 1998). Despite positive findings in these tests, the OFSS was not eventually put into service in the Channel Tunnel.

12.2.11 Fire tests as part of the Mont Blanc Tunnel fire investigation, 2000

One HGV and three pool fire experiments were carried out in the Mont Blanc Tunnel almost a year after the disaster of March 1999 (Brousse *et al.*, 2001). A number of non-fire ventilation tests were also carried out. The overall aims of the tests were to evaluate the performance of the ventilation system and to test theories about the fire and smoke behaviour during the disaster. The aim of the first pool fire test was to reproduce the smoke behaviour that occurred during the disaster. The aim of the second pool fire test was to test the limits of the ventilation system. The aim of the third pool fire test was to investigate smoke movements under 'optimal' conditions. The aim of the HGV test was to produce fire conditions similar to the HGV fire in the actual incident. The data produced by these experiments formed part of the body of evidence that helped guide the fire investigation and legal enquiries carried out in the aftermath of the catastrophe.

Some of the experimental conclusions are listed below.

- Pool test 1 accurately modelled the observed smoke movements of the disaster.
- Pool fires 2 and 3 showed that the smoke during the disaster could have been limited to a distance of 700–900 m around the fire zone, with good stratification, if the ventilation had been controlled appropriately. However, it should be noted that the tunnel was free of all vehicles under test conditions, so such optimal smoke control may not have been possible under operational conditions.
- The HGV test reached a maximum heat release rate of about 22 MW after about 45 minutes. However, the fire load of the experimental vehicle was substantially less than that of the actual vehicle involved in the incident, and the fire growth rate of the experiment was also substantially slower than that during the incident. Thus, only data from the first few minutes of the fire test are useful for understanding the fire development that occurred during the Mont Blanc catastrophe.

12.2.12 Project safety test: tests in the second Benelux Tunnel, The Netherlands, 2001

Following the high-profile accidental fires in several Alpine road tunnels in 1999 to 2001, a series of 26 fire tests was commissioned to investigate (a) the spread of heat and smoke from a fire, (b) the influence of ventilation on fire size, (c) the influence of sprinklers, and (d) the capabilities of fire detectors during tunnel-fire incidents. These tests were carried out in the second Benelux Tunnel, Rotterdam, The Netherlands, shortly before the tunnel was made operational (DGPWWM, 2002). The test series included fires involving six fuel pools, three cars, one van, six stacked loads (to represent HGV fires; Figure 12.2) and ten small fuel basins (to test the fire detectors). The test series included tests with natural ventilation and forced longitudinal ventilation. The main conclusions from the test series include the following.

- Due to high levels of heat radiation, conditions were lethal within 6 m of a fully developed passenger-vehicle fire. For a small HGV fire this distance increased to 12 m. There was no threat to life from carbon monoxide at such locations close to a fire due to convection and stratification of the fumes.
- Both with and without longitudinal ventilation there was poor visibility due to smoke at 100–200 m from the fire location. Toxicity limits were not be exceeded at these locations.
- High ventilation rates retarded the development of a fire involving a passenger car by up to 30 minutes, with the fire starting at the front of the vehicle and the ventilation blowing from the rear. High ventilation rates tended to enhance the burning of the 'HGV' fires by up to 20 MW, but the temperatures in the locality of the vehicles were reduced due to the ventilation.

Figure 12.2 Fire experiment with wooden pallets and an aluminium covering carried out in the second Benelux Tunnel, The Netherlands (Photograph courtesy of Steunpunt Tunnelveiligheid, Rijkswaterstaat, The Netherlands)

- Sprinklers substantially reduced the air temperature and the temperature of vehicles in the vicinity of the fires. For the range of vehicles tested, lethal temperatures were not observed and fire spread was controlled. The formation of steam was not observed.
- Linear fire detectors, in general, activated an alarm more than 5 minutes after the start of the fire. For rapidly growing fires this time was reduced to 3 minutes. Under natural ventilation conditions, a fire would generally be detected by a sensor less than 5 m away; with forced ventilation the activated detector may be as much as 20 m away.
- Escape-route signage became invisible very quickly in smoke. It was recommended that the signs be situated at low level.

12.2.13 Fire tests carried out in the Toumei–Meishin expressway tunnel, Japan, 2001

A series of tests were carried out during construction of the No. 3 Shimizu Tunnel on the new expressway between Toumei and Meishin in Japan in 2001. The aims of the experiments were to understand better fire behaviour and smoke control in a tunnel having a very large cross-section and provide data for comparison with simulations done using computational fluid dynamics (CFD). The tunnel is almost semicircular in aspect (8.5 m high, 9.0 m radius, 115 m² cross-sectional area) and has a 1119 m long three-lane roadway. Nine pool fire experiments were carried out as well as a bus fire test (Kunikane *et al.*, 2002a, b). The results of the fire tests were compared with data from previous experimental fire tests in smaller tunnels. The conclusions include the following.

- The evacuation environment in a three-lane tunnel was safer than would be expected in a two-lane tunnel, due to the size of the tunnel.
- Backlayering distances were slightly longer in a three-lane tunnel than would be expected in a two-lane tunnel.

- Longitudinal ventilation did not have as great an effect on the heat release rate in a large tunnel as would be expected in a smaller tunnel.
- The maximum convective heat release rate of a fire in a large tunnel was greater than would be expected in a two-lane tunnel.

12.2.14 The Runehamar Tunnel fire test series, 2003

A series of four fire tests was carried out in a disused two-lane road tunnel in Norway in September 2003 (Ingason and Lönnermark, 2004; Lönnermark and Ingason, 2004). The tunnel is 1.6 km long and has a rough rock cross-sectional area of about 47–50 m². At the location of the fire experiments (approximately 1 km into the tunnel), a 75 m length of the tunnel was lined with fire-protective panels, and this reduced the cross-sectional area of the tunnel to 32 m² in the vicinity of the fire (Figure 12.3). The objectives of the test series were to investigate: (a) fire development in HGV cargo loads, (b) the influence of longitudinal ventilation on fire heat release rate and growth rate, (c) the production of toxic gases, (d) fire spread between vehicles, (e) fire-fighting possibilities, and (f) temperature development at the tunnel ceiling.

Each of the tests comprised a fire load of equivalent size and shape to a standard HGV trailer (10.45 m long, 2.9 m wide, 4.5 m high). In test 1, the fire load consisted of 10.9 tonnes of wooden pallets and plastic materials, and a 'target' object positioned 15 m downstream of the fire. In test 2, the fire load consisted of 6.8 tonnes of wooden pallets, mattresses and plastic materials. In test 3, the fire load consisted of 7.7 tonnes of furniture on wooden pallets, and ten tyres (800 kg) were positioned around the frame at the locations where they would be on a real HGV trailer. In test 4, the fire load consisted of 3.1 tonnes of plastic cups in cardboard boxes on wooden pallets. In each test the amount of plastic materials was estimated to be

Figure 12.3 Fire experiment (test 3) in the Runehamar Tunnel, Norway (Photograph courtesy of Haukur Ingason)

about 18–19% (by weight). Similar tarpaulin coverings were used in all four tests. In each test, two fans positioned near the tunnel portal were used to generate a longitudinal airflow; this was about 3 m/s at the start of each test, but reduced to about 2–2.5 m/s once the fires became fully active. The peak heat release rates of the fires in tests 1 to 4 were 203, 158, 125 and 70 MW, respectively. The conclusions from the test series include the following.

- 'Ordinary' HGV cargoes could give rise to heat release rates comparable to fuel tanker fire scenarios (Ingason and Lönnermark, 2004).
- Temperatures above 1300°C could be achieved in HGV tunnel fires (Lönnermark and Ingason, 2004).
- When the fires reached full involvement, the ventilation flow was reduced and backlayering occurred (Lönnermark and Ingason, 2004).
- A 'pulsing' phenomenon was observed in tests 1 and 2 (Ingason and Lönnermark, 2004).
- The 'RWS fire curve' (see Chapter 7) is the best standard fire curve to represent a HGV fire in a tunnel (Lönnermark and Ingason, 2004).

12.2.15 La Ribera del Folgoso Tunnel fire test series, 2009

A series of five fire tests was carried out in a 300 m long tunnel in Spain with a 50 m^2 cross-sectional area (Vigne and Jönsson, 2009). The fuel source in each case was a 1 m × 2 m rectangular pool of heptane, with a peak heat release rate of approximately 5 MW. The test conditions were ostensibly identical, the idea being to investigate the variability of data within a single experimental configuration and to provide a robust set of experimental data for comparison with (i.e. for validation of) numerical models. The tests show that there is an approximately 10% variation in heat release rate and temperatures in the vicinity of the fire for experimental pool fire tests (Figure 12.4).

Figure 12.4 Pool fire test in La Ribera del Folgoso Tunnel (Photograph courtesy of Jimmy Jönsson, ArupFire Madrid)

12.3.   Fire experiments to evaluate sprinkler performance

In the aftermath of the major tunnel fire disasters that occurred in the late 1990s and early 2000s (see Chapter 1), the issue of suppressions systems became the subject of much debate. Opinion was (and to some extent still is) divided as to whether the benefits of such systems (e.g. reduction of fire size) outweighed the negative aspects (e.g. decreased visibility). The use of water mists, sprinklers and deluge systems in tunnel applications have been the focus of several recent and ongoing testing programmes in Europe, including the UPTUN project (see Chapter 31). The tests used to investigate suppression systems and demonstrate the capabilities of these systems in tunnel environments are discussed in Chapter 8, and are not detailed here.

12.4.   Fire experiments to test or commission tunnel installations

Since the fires in the Mont Blanc and Tauern Tunnels in 1999, there has been renewed interest in tunnel fire safety and in developing new means of fire prevention, suppression and mitigation (Figure 12.5). Several companies are developing new deluge, sprinkler or mist fire-suppression systems, other companies are developing fire-resistant tunnel linings, while others are developing new ventilation systems, etc. Some of these products have been tested in realistic fire situations in real and experimental test tunnels. However, few of the test results have been published in the public domain.

As noted in Chapter 11, there have been no well-documented tests of car fires in longitudinally ventilated tunnels, except for the fire tests in the second Benelux Tunnel in 2001 (DGPWWM, 2002). This does not mean that these are the only fire tests of cars that have been carried out in longitudinally ventilated tunnels. For example, a car fire was used to test a new tunnel-lining

Figure 12.5 Fire experiment carried out to test/commission the linear fire-detection system in the Baregg Tunnel, near Baden, Switzerland (Photograph courtesy of John Day)

system in a tunnel at DMT, Dortmund (Rauch, 2001). This test was carried out with longitudinal ventilation. However, the details of the test have not been released into the public domain, and important measurements such as the mass loss and heat release rate of the burning car (which could have been measured at minimal additional cost, and would have greatly increased our knowledge of the burning behaviour of a car subject to longitudinal ventilation) were not made.

In the early days of experimental tunnel fires, experiments were carried out with specific individual goals, for example, to determine smoke behaviour under certain conditions. More recently, there has been a trend to learn as much as possible from fire tests in tunnels, and to this end there has been collaboration between several groups of people with different interests, best demonstrated by the EUREKA EU499 FIRETUN test series. However, with the majority of tunnel-fire experiments now being carried out by private companies, things appear to be reverting to the old ways again. This is a backward step, which will slow the rate of advance of our understanding of the fire behaviour in tunnels. It is only through collaboration and publishing in the public domain that our knowledge of fire phenomena will be increased, and it is only through increased understanding that our tunnels can be made safer places.

12.5. Fire experiments to investigate detector performance

A 2-year international research project (2006–2008) was carried out to investigate the capabilities of various kinds of fire-detection system for tunnel applications. Various experiments were carried out in an experimental tunnel-like facility (37.5 m × 10 m × 5.5 m high) and in the Carré-Viger Tunnel in Montreal, Canada (Liu et al., 2011a, b). The study demonstrated that longitudinal ventilation has a significant influence on the time to activation of various detection systems used in tunnels, and also investigated the capabilities of various systems (linear detectors, flame detectors, smoke detectors and video fire detectors) in a range of different scenarios.

12.5.1 Fire tests in operational tunnels

The majority of the tunnel-fire tests described above were carried out in abandoned or disused tunnels. This has the advantage of allowing the testing of very large-scale fires, which may destroy the tunnel lining and fittings (e.g. the 100 MW pool fire tests done in the Memorial Tunnel), but the majority of these fire tests (except those done in the Memorial Tunnel) have the disadvantage that the test tunnel was significantly smaller than most operational tunnels, especially two- or three-lane road tunnels. It would be particularly valuable if tunnel-fire experiments could be carried out in real vehicle tunnels, but this is generally unrealistic for large-scale fires. However, some small-scale fires (car and pool fires) have been evaluated in new and operational road tunnels, generally in order to test the capabilities of the ventilation system or as a fire-fighting exercise (Perard, 1992; Perard and Brousse, 1994; Nilsen et al., 2001). Some details of tests carried out in French and Norwegian road tunnels are in the public domain, and these include those described below.

12.5.1.1 Des Monts Tunnel

The tunnel is an 850 m long, twin-tube, two-lane tunnel, with a cross-section of about 80 m^2 and a longitudinal ventilation system. The fire load tested in 1988 was a van, with hay bundles, tyres and petrol ignited to start the fire. Conclusions include

- the fire size was approximately 2 MW (at 1.3 m/s ventilation velocity)
- temperatures in the vicinity of the fire did not exceed 80°C
- the temperature quickly decreased when the ventilation was started
- other vehicles can easily overtake a burning car (Perard, 1992).

12.5.1.2 Nogent-Sur-Marne covered-trench tunnel

The tunnel is a 1.1 km long, twin-tube, three-lane tunnel, with a partially transverse ventilation system. Smoke tests and a car fire test were carried out in 1988–1989. Conclusions include the following.

- The smoke remained stratified with natural ventilation. With the ventilation set to exhaust, part of the tunnel was kept smoke free, but 100 m from the fire location the whole tunnel was filled with smoke.
- The wind-induced airflow in the tunnel dominated the exhaust ventilation in one test, and the partially transverse system was unable to extract the smoke.
- When the wind-induced airflow was blocked (one end of the tunnel was physically blocked), the partially transverse ventilation system was better able to remove smoke.
- An air or water curtain in the tunnel may be able to minimise wind induced airflow (Perard, 1992).

12.5.1.3 Frejus Tunnel

Many tests have been carried out in this 13 km long, twin-lane, tunnel on the border of Italy and France, generally for training emergency personnel and for testing ventilation and communications systems (Perard, 1992).

12.5.1.4 FFF tunnel

The tunnel is a 750 m long, twin-tube, two-lane motorway tunnel. The ventilation system is longitudinal. In 1990, shortly before the tunnel was due to be opened, a passenger car was burned in the tunnel to test the ventilation system. Conclusions include the following.

- Maximum fan operation resulted in a ventilation velocity of 8 m/s, and this was sufficient to prevent backlayering.
- Smoke was transferred from the tube containing the fire to the other tube. Steps were taken to prevent this in future incidents (Perard, 1992).

12.5.1.5 Monaco branch tunnel

The tunnel is a 1.5 km long, single-tube, bidirectional, three-lane tunnel. The tunnel has a slope of 5.5%, and the ventilation system is partially transverse. The tests were carried out in 1992, shortly before the tunnel was due to open. One wood-fire test and six car-fire tests were carried out, primarily to investigate the capabilities of the ventilation system. Conclusions include the following.

- Smoke-removing equipment is useful, as conditions were substantially worse when it was not used.
- The ventilation system was able to control the smoke when the wind-induced airflow was low; however, when the wind-induced airflow was about 1–2 m/s the smoke reached the tunnel portals.
- In every test one side of the fire location was always kept smoke free, this was to help fire-fighting activities.
- Supply of fresh air (via wall jets) caused a loss of stratification (Perard and Brousse, 1994).

12.5.1.6 Grand-Mare Tunnel, Rouen

The tunnel is a 1.5 km long, twin-tube, two-lane tunnel. The ventilation is longitudinal and, in the tube where the tests were carried out, opposes the natural airflow induced by the chimney effect of the sloping tunnel. Three pool-fire tests were carried out, corresponding to approximate sizes of 5, 10 and 20 MW. Conclusions include the following.

- The smoke from a 20 MW fire was controlled by longitudinal ventilation (4 m/s).
- The flames flattened by the airflow caused some damage to the carriageway downstream of the fire.
- The heat release rate of the '10 MW' fire was slightly higher than expected, and the heat release rate of the other two tests was slightly lower than expected (Perard and Brousse, 1994).

SUMMARY OF ABOVE TESTS
The overall lessons learned from all the French fire tests include the following.

- A car fire in a tunnel may not be very big. It might be possible to pass the fire location on foot or by car.
- Increased knowledge of smoke behaviour in various configurations was gained.
- Some quantitative data on temperature and smoke movement was obtained.
- Cold smoke tests are not representative of real fires.
- Tunnel operators and emergency services gained experience (Perard, 1992).

12.5.1.7 Byfjord Tunnel

In August 1998, a fire test of two cars was carried out. The fire was located at the deepest point of the 5.8 km long tunnel, which is near the midpoint of the tunnel and 230 m below sea level. The tunnel is longitudinally ventilated, is 11.4 m wide and has a cross-sectional area of 76 m^2. Measurements of temperature and gas concentrations (carbon dioxide, nitrogen oxides and oxygen) were recorded in the vicinity of the fire location. A 'cold-smoke' test was also carried out as an exercise for fire-fighters. Conclusions from these tests include the following (Nilsen *et al.*, 2001).

- Concentrations of carbon monoxide and nitrogen oxides in the smoke downstream of the fire were not life-threatening.
- Even close to the fire, smoke-layer temperatures were low ($\sim$40°C) due to the large cross-sectional area and efficient ventilation.
- The main hazard was the lack of visibility due to smoke; drivers would have little warning of the smoke before becoming trapped in it. In addition, the possibility of cars colliding with other vehicles or evacuating passengers in the smoke may be a greater hazard than the fire itself.
- Assisting the evacuation of passengers caught in the smoke is difficult. Measures such as early fire-detection systems and adequate warning signs are required to reduce the risk of becoming trapped in the smoke plume.

12.5.1.8 Bømlafjord Tunnel

In December 2000, a fire test, similar to the one done in the Byfjord Tunnel, was carried out. The fire was located on a sloping section of road (8%) about 2.5 km into the 7.8 km long tunnel, which is 260 m below sea level at its midpoint. Like the Byfjord Tunnel, the Bømlafjord Tunnel is longitudinally ventilated, is 11.4 m wide and has a cross-sectional area of 76 m^2. Conclusions from this test include the following (Nilsen *et al.*, 2001).

- Detector cables based on thermocouples may be a low-cost alternative for early fire detection.

- The ventilation strategy adopted (i.e. to force smoke away from the nearest fire-fighters) would be expected to be successful when combined with early automatic fire detection and clear warnings to the public.
- An airflow velocity of 4 m/s was sufficient to force the smoke plume from a 5–10 MW fire down a tunnel incline of 8%.
- The gas temperatures and toxic gas concentrations were low downstream of the fire.

12.5.1.9 Tunnels in NunYan province, China

In recent years, there has been a considerable investment in experimental tunnel-fire research in China, much of it involving small-scale testing (see below) but a number of experiments have been done in operational tunnels.

A series of ten fire experiments was carried out in three tunnels in NunYan province in 2005 (Hu *et al.*, 2007). The tunnels were the three-lane YangZong Tunnel and the two-lane DaFengYaKou and Yuanjiang Tunnels. Experiments were also carried out using a laboratory tunnel. The fires tested were small relative to the tunnel sizes (the largest reached a peak of about 1.8 MW, smaller than most experimental car fires), but conclusions were drawn relating to smoke movements (both under natural ventilation conditions and under forced longitudinal airflow) and smoke temperatures.

12.6.   Experimental testing on a smaller scale

Fire tests involving full-size fires in full-size tunnels are very expensive. The EUREKA FIRETUN test series cost about US$10 million and the MTFVTP series cost in excess of US$40 million (PIARC, 1999). The high cost seriously limits the number of full-size experimental tests that are conducted. To date, many fundamental questions about the behaviour of fire and smoke in tunnels have not been answered by full-scale tests. In order to be able to answer these questions at a lower price, and sometimes to model a specific tunnel design, reduced scale and small-scale experiments have been carried out.

In order for reduced or small-scale experiments to be useful, there must be a well-defined similarity between the scale model and the full-scale case of interest (Quintiere, 1989). If there is a strong similarity then a scale model can be used to investigate specific aspects of the behaviour of the fire or smoke. If the similarity is not so strong, then the behaviour of the fire can only provide information in general terms. In order to scale between a real situation and a scale model with a strong similarity, it is necessary to consider the gas flow at and around the fire.

Gas flow can be classified using certain non-dimensional numbers. The *Froude number* may be used to classify fire type:

$$Fr = \frac{U^2}{gL}$$

where U is the velocity of the gases, g is the acceleration due to gravity, and L is a characteristic dimension of the system. Other non-dimensional numbers are also used to classify flow behaviour, including: the *Reynolds number*

$$Re = \frac{UL}{v}$$

where v is the viscosity; the *Richardson number*

$$Ri = \frac{gL}{U^2} \frac{\Delta\rho}{\rho}$$

where ρ is the density of the plume gases, and $\Delta\rho$ is the difference between this and the density of the surrounding air; and the *Grashof number*

$$Gr = g \frac{L^3}{v^2} \frac{\Delta\rho}{\rho}$$

which is essentially a combination of the Reynolds and Richardson numbers.

Ideally, each of these numbers should be the same in the scale model as in the real situation. However, this is not possible, and models are usually scaled assuming conservation of the Froude number only. This is a reasonable assumption to make for all but the smallest scale models, as variations in the Reynolds number are not particularly significant for turbulent flows.

The Froude number is proportional to q^2/L^5, where q is the heat release rate of the fire and L is a characteristic dimension of the system (often taken to be the tunnel height). This means that, for example, a model fire that is half the size of a full-sized tunnel will have a heat release rate 0.177 times that of the full-scale fire ($0.5^{5/2} = 0.177$). In a similar way, the ventilation velocity in the scale model scales as $L^{1/2}$, so a full-scale airflow velocity of 2 m/s will be modelled by a half-scale airflow velocity of 1.4 m/s. The temperature of the fire will be the same at test scale and full scale (Bettis *et al.*, 1993).

Reduced-scale fire-test series tend not to be as high profile as full-scale test series, and the data and results from these series are not always in the public domain. Some reduced-scale test series include the ones described below.

12.6.1   Small-scale fire experiments, Japan, 1960–1981

In addition to the large-scale fire experiments described previously (see Sections 12.2.5 and 12.2.14), a number of small-scale pool-fire experiments have been carried out in Japan in reduced-scale apparatus (Stroeks, 2001). The majority of these experiments were for the purposes of testing sprinkler systems (petrol pool fire tests in 1960, 1961, 1968, 1973 and 1981). Other experiments included a petrol pool fire experiment to test a foam suppression system (1970) and a petrol pool fire experiment to test a fire-detection system (1980). The results obtained from these tests are only available in Japanese.

12.6.2   Pool fire tests at the Londonderry Occupational Safety Centre, Australia, 1990

A series of five kerosene pool fire tests was carried out in a 130 m long, 5.4 m wide, 2.4 m high 'mine roadway' tunnel near Londonderry, NSW, Australia (Apte *et al.*, 1991). The longitudinal ventilation in the tunnel was maintained by two exhaust fans at one end of the tunnel, with a rectangular grid for 'flow straightening' near the other end of the tunnel. Three tests were carried out using a 1 m diameter pool and ventilation velocities ranging from minimal (~0.5 m/s) to 2 m/s. Tests were also carried out using 0.57 and 2 m diameter pools. An extensive array of thermocouples and airflow probes was arranged around the fire, and mass loss was measured by load cells under the pool tray. Two video cameras were also used to record the experiments. The experiments were carried out to test a numerical model. Some of the observations from the study include the following.

- Increasing the ventilation from 0.5 to 2.0 m/s caused a 25% decrease in the heat release rate of the 1 m diameter fire.
- The observed decrease in the heat release rate is probably due to the fact that less of the plume is above the fuel surface at higher air velocities.
- The rate of mass loss rate in larger fires is proportionately higher than that in smaller fires (subject to ventilation velocities of about 0.9 m/s).
- There is significant backlayering at 0.85 m/s, but this is 'arrested' by a 2 m/s airflow.

12.6.3   Pool fire tests carried out by INERIS, France, 1991–1992

In addition to the 1 m^2 and 3 m^2 heptane pool fire tests carried out in the Hammerfest tunnel (see Section 12.2.7), Institut National de l'Environnement Industriel et des Risques (INERIS) carried out some pool-fire tests in their own fire gallery (Marlair and Cwiklinski, 1993) at Verneuil-en-Halatte, France, and in the open air as part of the EUREKA 499 test series (Casalé and Marlair, 1994; Casalé, 1995). The tests were primarily for comparison with the Hammerfest tunnel tests and for calibration of the heat release rate measuring apparatus. Another objective was the development of the test protocol for remote control of the fire tests carried out in the Hammerfest tunnel (Casalé et al., 1993). The INERIS fire gallery is 50 m long and has an approximately square (10 m^2) cross-section with a concave ceiling. The gallery is naturally ventilated, or use may be made of an exhaust fan to produce forced-ventilation conditions.

Some of the test conclusions were as follows.

- The heat release rate and duration of burning of a liquid hydrocarbon pool fire can be remotely controlled reliably (from a distance of several hundred metres from the pan) by using an appropriate test procedure.
- In order to control the heat release rate and duration of a pool fire, a system where the fuel level is kept constant is required. Control of the system (the external fuel tank and fuel supply line) is straightforward, requiring only simple thermocouples.
- The heat release rate of the pool fire in the fire gallery was substantially greater than that of the open air fire. This enhancement was attributed to 'significant re-radiation from the tunnel walls'.

The INERIS fire gallery was modified to investigate conditions within tunnel refuges in tunnel-fire situations (Casalé and Broz, 1992). Of particular interest was the design and capabilities of the fresh air supply system within the refuge. The aim was to determine the values of air pressure and air flow rate in the refuges necessary to maintain tenable conditions for tunnel users escaping near the fire location. The refuge mock-up was able to undergo different sequences of door operation to simulate various scenarios of people entering the refuge. The fire source in each case was a 0.41 m^2 heptane pool fire (with constant heptane level in the pan), and the temperature and opacity conditions were monitored within the refuge. The refuge tested was modelled on the proposed emergency refuges for the 5 km long Puymorens Tunnel, France.

Conclusions from the experiments include the following.

- Satisfactory conditions were maintained within the refuge when the door was closed.

- Unsatisfactory conditions arose temporarily when the door was open.
- The results were used to finalise the design and size of the air ducts installed in the refuges in the Puymorens Tunnel.
- Other conclusions relating to the geometric dimensions of tunnel refuges were also drawn.

12.6.4 Tunnel fire experiments for the EGSISTES project, INERIS, France, 2006–2009

The INERIS fire galley was again modified to reproduce a realistic tunnel section at one-third scale. A roof was installed to obtain a modified section of $5.4\,m^2$ with a smoke duct positioned above. The resulting section was 3 m wide and 1.8 m high, and experiments were carried out with both longitudinal and semi-transverse ventilation systems. A large series of experiments was carried out to investigate smoke behaviour both in terms of backlayering and in the downstream smoke layer, and also the smoke behaviour in the presence of perturbations (Boehm *et al.*, 2008).

The fires tested were different sized heptane pool fires, with a fuel-control system to obtain constant burning. The fire properties were studied, with a particular focus on the radiated heat fraction, and two fire positions were used to study the upstream and downstream smoke layers. The stability of those two layers was studied with regard to the influence of jet fans located upstream and the influence of vehicles located upstream or downstream of the fire. In both cases, different ventilation regimes were used to study various configurations corresponding to the French ventilation practice. The main conclusions of this project were as listed below.

- Upstream jet fans affect only the fastest part of the backlayering. The flow close to the fire layer is not significantly affected, and the temperature in the plume is independent of the velocity gradient upstream (Truchot *et al.*, 2009).
- Vehicle blockages upstream of the fire influence backlayering, primarily due to the increased velocities around them, due to blockage effects.
- At low ventilation velocities, vehicle blockages downstream of the fire do not reduce the smoke stratification.

12.6.5 Reduced-scale fire tests in the Health and Safety Establishment test tunnel, Buxton, UK, 1993

A series of nine fire tests involving a wooden crib and kerosene pool fires within a mock-up of a Eurotunnel HGV shuttle carriage were carried out in the test tunnel at the Health and Safety Laboratory near Buxton (Bettis *et al.*, 1993, 1994). The concrete, arch-shaped tunnel is 366 m long and has a cross-section of $5.6\,m^2$. It has a maximum height of 2.44 m at the centre of the tunnel and a maximum width of 2.75 m near the floor. The tunnel walls and ceiling were fire-proofed in the vicinity of the fire location with mineral-fibre sheeting. Longitudinal ventilation in the tunnel was supplied by up to three fans housed at one end of the tunnel. Measurements of the temperature, the airflow velocity, the smoke density, the heat flux, the gas concentrations and the mass loss of the fire load were recorded during each test (except test 7, due to instrument failure). These data were gathered primarily for testing of CFD model results (see Section 12.7.3). The conclusions from the study include the following.

- The fires produced small, but measurable, effects on the ventilation rate.
- Many of the fires, particularly the larger ones, showed elements of ventilation control.
- Temperatures near the ceiling reached 1000°C in some tests.

- Backlayering of smoke was prevented at ventilation rates above 1.25 m/s, was controlled (i.e. a stationary hot gas layer was formed upstream of the fire location) at ventilation rates between 0.75 and 1.25 m/s, and was not controlled at lower ventilation rates.
- There was no clear relationship between the critical velocity required to prevent backlayering and the heat output of the fire.
- The air velocity required to prevent any backlayering was less than predicted by existing theories.
- The reduced-scale data can be extrapolated to tunnels of different scales using Froude scaling.

12.6.6 Cable-tunnel fire tests in the HSE test tunnel, Buxton, UK, 2006

The same tunnel as above was used to investigate the behaviour of fires in cable tunnels. A section of the tunnel was fitted with cable racks, and tests were carried out using various cable loadings. The results show that these fires produce a great deal of smoke, but that the heat produced is not sufficient to hamper fire-fighting operations (Bettis *et al.*, 2007).

12.6.7 Test series carried out by the Technical Research Institute of Sweden (SP)

Various series of fire tests have been carried out at the Technical Research Institute of Sweden (SP) over the past two decades. The details presented here do not include all test series to date.

A series of 18 fire tests were carried out in an 11 m long fire gallery at SP in 1995. The gallery was 1.08 m wide and 1.2 m high, and was equipped with extensive measuring equipment and natural or longitudinal ventilation (Ingason, 1995a). The pool fire tests were carried out using heptane (11 tests), methanol (two tests) and xylene (five tests) as fuels. Two sizes of square fuel pan were used (0.09 and 0.16 m^2). The tests were primarily carried out to investigate the effects of under-ventilation on the heat release rate of a fire. Comparable fire tests were also carried out in the open air. The following observations were made.

- Restricting the airflow caused a decrease in the heat release rate but not a significant increase in carbon monoxide production.
- Despite attempts to make the fires underventilated, it was not possible to do so.
- The mass burning rate (and hence the heat release rate) was much higher in the tunnel than in the open for heptane and xylene fuel pools.
- The mass burning rate for methanol fires was slightly lower in the tunnel than in the open air.
- The burning rate decreased with increasing wind velocity for heptane and xylene fuel pools.
- With restricted airflow, the flames (from heptane pools) tended to be deflected by about 30–45°, while with unrestricted airflow the flames were deflected almost horizontally at the fuel pan and did not reach the ceiling until about 1.2 m from the fire location (corresponding to approximate airflow velocities of 0.4 and 0.7 m/s, respectively).

A series of fire tests using heptane fuel pans and wooden cribs as fuel were carried out in a similar tunnel construction in 2005. The tunnel in this instance was 10 m long, but six different configurations were used in the study, with widths of 0.30, 0.45 and 0.60 m and heights of 0.25 and 0.4 m (Ingason, 2005; Lönnermark and Ingason, 2008). Twelve tests were carried out, with one heptane test and one crib test being carried out in each tunnel configuration. The longitudinal airflow was 0.67 m/s for ten of the tests, and 0.5 m/s for the two tests in the largest tunnel configuration. The main findings from this study include the following.

- The gas temperatures in the tunnel were influenced by the tunnel dimensions. In general, temperature decreases with increasing dimensions, but this was not evident near the fire source, where the opposite relationship was observed.
- The near-field conditions (temperature, heat flux) were significantly different between pool fires and wooden crib fires, but the far-field conditions were similar.
- In smaller tunnels, the conditions tended to be more uniform, whereas there tended to be more stratification in larger tunnels.

Another series of 12 tests was carried out in 2006 using the same tunnel (0.4 m wide, with a false 0.2 m high ceiling), wooden cribs as fuel, and a water fire-suppression system installed (Ingason, 2006, 2008). In some tests, there was a second crib positioned downstream of the fire crib. In the tests with the fire-suppression system, the system was activated 1 minute after ignition, which was about halfway through the growth phase of the test with no fire suppression. In the majority of tests, the water-spray system was able to prevent fire spread to the second crib. In most tests, the water spray was also observed to reduce the growth rate of the fire, and to reduce the peak heat release rate. The flow rates used, however, were not able to halt fire growth or extinguish the fire.

A further series of 12 experiments was carried out in 2008 using the same tunnel, with up to three wooden cribs positioned in the tunnel (Ingason and Li, 2010). The primary objective of the study was to investigate the influence of longitudinal ventilation on the fire growth rate. It was found that there was a clear correlation between longitudinal airflow and fire growth rate, but the influence of ventilation on the peak heat release rate was not as great as expected. Other conclusions relating to temperatures and flame lengths were also drawn.

12.6.8   Test series carried out by FOA, the Swedish Defence Agency, 1997

A series of 24 fire tests was carried out in a 100 m long 'blasted rock tunnel', approximately 3 m wide and 3 m high. The tunnel was open to the air at one end and had a large chimney at the other. There was no mechanical ventilation system installed in the tunnel, but different ventilation rates were achieved by restricting the inflow of air in some tests and using two different fire locations within the tunnel. The experiments included fire tests of heptane pools (12 tests), methanol pools (two tests), kerosene pools (occasionally incorrectly referred to in the report as 'diesel' pools) (two tests), polystyrene cups in cardboard boxes (two tests), wooden cribs (three tests), and heptane pools contained within a dummy vehicle (two tests) and a car (Ingason et al., 1997). Although the tests were primarily carried out to provide experimental data for testing of CFD codes, the following conclusions were made.

- All fire tests showed some correlation between the degree of ventilation and the heat release rate of the fire. This effect was more apparent for solid fire loads than for liquid pools.
- Measurements of optical density and gas concentrations indicated that there was a correlation between these two parameters. However, a parameter for the fuel type must be included.
- In a confined space the heat release rate of a small car reached 4 MW for a short time (Ingason, 1998).

12.6.9   One-third scale fire experiments in France, 2009

A series of fire tests was carried out in a custom-built, reduced-scale tunnel facility in France (Blanchard et al., 2011). The tunnel is 43 m long and of conventional arched shape, with a hydraulic diameter of 2.16 m. Tests were carried out using a 0.5 m² heptane fuel pool, and the fire reached a peak of about 3.4 MW in one test. Two different longitudinal ventilation velocities were tested, one of about 1.0 m/s, which is below the critical ventilation velocity for the fire, and one of about 2.2 m/s, which is above the critical ventilation velocity. The ventilation flow was generated by an extraction fan. The results of the experiments were intended for use as a comparison with (i.e. for the validation of) those obtained using numerical fire models.

12.7. Laboratory-scale experiments

In addition to reduced-scale tests, there have also been a number of laboratory-scale tests of fires in 'tunnels'. Ideally, the use of laboratory-scale mock-ups of structures such as tunnels should be considered as complementary to the use of full-scale tests and computer models (Marlair, 1999). Most research institutes, including INERIS in France, the Health and Safety Laboratory in the UK, and SP in Sweden, make use of all these techniques. While the results from laboratory-scale tests can give some information on the behaviour of larger scale fires, the similarity of laboratory experiments to full-scale fires is not necessarily good.

Froude scaling criteria may not apply to fires in tunnels smaller than about 1 m in diameter (Bettis et al., 1993), and small-scale pool fires may behave substantially differently from their full-scale counterparts. For example, a pool of a hydrocarbon fuel will exhibit 'transitional' (laminar turbulent) flame behaviour if its diameter is less than about 1 m, whereas larger pools will exhibit fully turbulent flame behaviour.

Some laboratory-scale models use non-fire processes to model the tunnel-fire and smoke phenomena. For example, a modular apparatus made of Plexiglas is used by INERIS (Ruffin, 1997), and a similar apparatus is used at the University of Valenciennes (Vauquelin and Telle, 2005). These make use of cold helium–air mixtures to simulate the flow of hot gases from fire sources, and thereby study fire smoke patterns, including smoke stratification and backlayering. This type of apparatus is intended to produce additional data for comparison with the results obtained using computational models.

By using gas mixtures of different densities, laboratory-scale apparatus can be used to model the gas/smoke flows produced by tunnel fires more accurately than can be achieved with some small-scale fire experiments where, as noted above, the fire behaviour might be very different to that in a full-scale fire. In addition to gas mixtures of different densities, small-scale experiments using fresh water/salt water mixtures can also be used to model smoke flow from fires (Steckler et al., 1986).

As long as the limitations of these small-scale experiments are borne in mind, laboratory scale experiments can yield useful results and help improve the understanding of some features of tunnel-fire dynamics. Some small-scale test series include those described below.

12.7.1 A scale model to investigate smoke movement in the Paris Metro, 1991

A laboratory-scale 3 m long Pyrex 'tunnel' was used to investigate some aspects of smoke movement and backlayering for the Paris Metro tunnel system (Vantelon et al., 1991). The apparatus was semi-circular, with a diameter of 30 cm. The fire source was a porous burner, 2 cm in diameter, with a variable rotameter regulating the fuel flow. Ventilation was provided by an extraction fan at one end of the apparatus, and ventilation flows of up to 25 cm/s could be attained. Smoke movements were recorded using lasers, video cameras and photography.

Temperatures and airflow were recorded inside the tunnel. Conclusions from the test series include the following.

■ Some backlayering would be expected to occur in the Paris Metro tunnels in the event of a fire if the usual ventilation velocity of ~1.5 m/s is used.
■ In the absence of forced ventilation (natural airflow at about 0.2–0.4 m/s), the smoke from a fire is likely to remain stratified, and so there will most likely be a smoke-free layer near the floor.
■ The behaviour of smoke entering a station from a tunnel was also modelled.

12.7.2   A wind-tunnel study to investigate the influence of ventilation on pool fires, 1995

Many pool-fire experiments were carried out in a $0.3 \times 0.3 \times 21.6$ m long wind tunnel at the Japanese fire research institute, to investigate burning rate, ceiling temperature and backlayering in a number of different experimental configurations (Saito et al., 1995). To investigate the influence of longitudinal ventilation on the burning rate of pool fires, 31 methanol pools (of 10, 15, 20 or 25 cm diameter) were tested at a range of ventilation rates from 0.08 to 1.03 m/s; and five pools of n-heptane (all 15 cm diameter) were also tested, with ventilation rates ranging from 0.43 to 1.3 m/s. To investigate the phenomenon of backlayering, many tests were carried out using different sized methanol pools, different ventilation rates and different tunnel slopes. Conclusions from this study include the following.

■ Burning rates of liquid pools in tunnels depend on the airflow velocity.
■ The burning rate of a pool fire in a tunnel was higher than the burning rate of a similar fire in the open air; the burning rate tended towards the open-air burning rate as the ventilation rate was increased.
■ The ceiling temperature of the tunnel fire was inversely proportional to the wind velocity.
■ Some pool fires reached a steady rate of burning, while in others the rate increased with time. The airflow velocity was the deciding factor in this.
■ The relationship between the critical velocity and the heat output parameter was almost linear.
■ When backlayering is long (20 times the tunnel height), the backlayering length is only weakly dependent on the slope of the tunnel (up to 10°) or the fire size; it can be expressed by a non-dimensional parameter. When backlayering is short, the backlayering length is dependent on both fire size and tunnel angle.

12.7.3   Critical-velocity experiments in a laboratory-scale model of the HSE tunnel, 1995

A series of fire tests was carried out in a one-tenth scale model of the HSE tunnel described in Section 12.6.5. The apparatus was mounted in such a way that different tunnel slopes (up to 10°) could be tested. The fire source used in the test series was a propane burner capable of producing heat release rates of 2.8–14.1 kW. When scaled to 'full size' dimensions of a 5 m diameter tunnel, this corresponds to heat release rates of 5–30 MW. Longitudinal airflow was produced using compressed air at one end of the apparatus (Wu et al., 1997a). The main conclusion from these tests was:

■ The effect of slope on the critical velocity required to control smoke is modest.

The same apparatus was modified to study the influence of tunnel geometry on the critical velocity (Wu et al., 1997b). Burner fire experiments were carried out in the apparatus as described, and then the tunnel section was replaced with a new section of different aspect (250 mm × 250 mm) and the test series repeated. The tunnel section was then again replaced with another new section of different aspect (500 mm × 250 mm) and the test series repeated. The results from the three series were compared. Conclusions include

- the critical velocity required to control smoke varies with the mean hydraulic diameter of a tunnel (a ratio of four times the tunnel cross-sectional area to the tunnel perimeter).
- a universal formula for determining the critical velocity may be possible (subject to further work).

12.7.4   A small-scale model for CFD model testing, 1997

Tests have been carried out in a laboratory-scale 'tunnel' at Turin Polytechnic University, primarily for the testing of CFD models (Bennardo et al., 1997). The apparatus is modular in design, so different lengths of tunnel may be tested, but the configuration described in the literature was 1.0 m long, 0.1 m high and 0.2 m wide. The tunnel is constructed of concrete, but has a ceramic–glass wall on one side for observation and recording purposes. Ventilation ducts along the top of the tunnel can simulate transverse ventilation, and fans at one end can drive longitudinal ventilation. The fire source can be either a fuel pool or a gas burner. A naturally ventilated petrol-pool fire test is presented in the literature. No conclusions are presented from the experimental work, but the advantages and limitations of the CFD model are presented.

12.7.5   Model tunnel fire tests, Korea, 2005–2009

A number of experiments have been carried out in various types of tunnel apparatus at Chung-Ang University, Korea. One study used five different tunnels and ethanol-pool fires to investigate the influence of tunnel aspect on the critical ventilation velocity (Lee and Ryou, 2005). Other studies involved an arched tunnel, 0.4 m high by 0.4 m wide, and a range of fuel pool fires (Roh et al., 2007; Ko et al., 2010). Some conclusions from these studies are listed below.

- The height of the tunnel influences the fire growth rates more than the width of the tunnel does (Lee and Ryou, 2005).
- A new equation for critical velocity is proposed. It agrees well with previous studies.
- The tests (Roh et al., 2007) confirm the relationship between the heat release rate and the critical ventilation velocity proposed by Wu et al. (1997b).
- Longitudinal ventilation has an enhancing effect on heptane-pool fires in tunnels.
- Calculations of the critical ventilation velocity must take into account the enhancing influence of the ventilation on the fire size.
- A new equation for critical velocity, taking into account the slope of the tunnel, is proposed (Ko et al., 2010).

12.7.6   Model tunnel backdraught investigation, China, 2008

Some experiments were carried out in a 15 m long × 0.6 m × 0.6 m tunnel apparatus at Beijing Jiaotong University in China (Mao et al., 2011). The experiments were used to investigate the possibility and consequences of backdraught-like conditions in the Beijing subway system. The relationships between airflow, air humidity and fuel were investigated. It was concluded that, in order to reduce the likelihood of backdraught, airflow rates or air humidity should be increased.

12.7.7   Model tunnel fire tests, China, 2009

A number of different studies have been carried out in a laboratory tunnel apparatus at Southwest Jiaotong University in China (Li et al., 2009a, b). The apparatus is 12 m long, and two different tunnel configurations are tested: one square tunnel, 0.25×0.25 m; and one arched tunnel, 0.39 m high by 0.45 m wide. Some conclusions drawn from these series include the below.

- New equations for critical velocity were derived (Li *et al.*, 2009b). It was claimed that the equations proposed by Wu *et al.* (1997b) underestimate the critical velocity, especially at low heat release rates.
- The height of cross-passage doors is the main parameter controlling the critical velocity for smoke control in cross-passages (Li *et al.*, 2009a).

12.7.8 Summary of laboratory-scale tests

There have been other tunnel-fire test series carried out at laboratory scale. Several research centres are undoubtedly continuing to investigate tunnel-fire phenomena at this scale, and will do so in the future. The few laboratory-scale studies presented here demonstrate the sorts of conclusions that can and cannot be drawn from very small-scale experimental tests.

12.8. Non-tunnel fire experiments

In addition to the experiments described above, a number of other experimental studies have been carried out to investigate issues relating to fire safety in tunnels. However, the experiments described below were not carried out in tunnels.

12.8.1 Fire-safety experiments for the Channel Tunnel, 1987–1991

A number of experimental programmes were carried out to investigate various aspects of fire safety in the Channel Tunnel rail link between the UK and France (Marlière *et al.*, 1995). Among these was a test series, carried out by INERIS (known as the 'Cerchar' tests), to investigate the efficiency of detection and extinguishing systems, the possibilities of various evacuation strategies, and the so-called 'non-segregation' principle used on car-transporting trains in the Channel Tunnel. (In the Channel Tunnel, car drivers and passengers remain with their vehicles during the journey. With the 'segregation' principle, such as used for HGVs in the Channel Tunnel and for all vehicles on ferries, drivers and passengers must leave their vehicles and travel in a separate compartment.) The tests were performed on full-scale mock-ups of the car-transporting wagons containing real cars (Figure 12.6), and the fires were located in a number of different locations

Figure 12.6 Inner view of the full-scale double-deck shuttle wagon mock-up used to investigate fire safety concepts and installations for the Channel Tunnel

within the rig (e.g. engine fires, passenger compartment fires, boot compartment fires, and fires involving fuel leakages under the cars). Simulation of the carriage movement was achieved in the mock-up by use of ventilation systems at the front and rear to reproduce pressure profiles similar to those on an operational train.

In addition to providing data used to aid the design of the rolling stock (Malhotra, 1993; French, 1994) and for comparison with numerical simulations (Kumar, 1994), results from the test series include the following.

- The temperature and toxic gas measurements made during the tests were used as the basis for the design, calibration and testing of the fire detection and suppression system.
- The programme was a clear demonstration that innovative design in tunnel construction requires large-scale experimental testing for important safety issues.
- As a result of the experiments, the 'non-segregation principle' was approved for use by the Safety Committee of the Inter-Governmental Commission (Cwiklinski, 1990).

12.8.2   Tests to investigate the fire safety of tunnel drainage systems, 1993

Some of the aspects of tunnel fire safety not adequately investigated as part of any of the experimental fire tests in tunnels are the consequences of fires involving spillages of fuel in the drainage system. During the planning stage for the construction of the second bore of the Chamoise tunnel, France, a series of full-scale fire tests was carried out to investigate the possibilities of mitigating the consequences of fires involving fuel spillages in tunnels (Lacroix et al., 1995). The study involved several of the stakeholders in the tunnel project, including the motorway operator, Autoroutes Paris–Rhin–Rhône (SAPRR), the tunnel contractor, Scetauroute, the Centre d'Etudes des Tunnels (CETU) and INERIS. A full-scale concrete construction of an 80 m long prototype system (including slot gutters and a twin duct network composed of siphons and fire and explosion 'barriers') was tested (Figure 12.7). The study led to a great increase in knowledge and understanding of fuel/fire behaviour in drainage systems under both normal and abnormal conditions. The work also identified key maintenance considerations for such drainage systems if they are to function as fire-protection devices.

12.8.3   Rail-carriage fire tests, Australia, 2003

Some fire experiments involving rail carriages were carried out by the Commonwealth Scientific and Industrial Research Organisation (CSIRO) in conjunction with Queensland Fire and Rescue Service (QFRS) in Australia (IAFSS, 2004; White and Dowling, 2004). The experimental programme included nine ignition experiments on various rail-car fixtures, and a full-scale fire test of a suburban rail carriage, performed in the open air. Conclusions were drawn concerning the fire behaviour of rail carriages, but it was noted that only a single test was carried out and more detailed research is still needed. The study also included an assessment of CFD modelling capabilities for simulation of railway fires.

12.9.   Concluding comments

Experimental fire testing in tunnels has greatly increased our understanding of tunnel-fire phenomena (see Chapters 11, 13 and 14) and has led to the development of better and more sustainable tunnel-fire prevention, protection and mitigation technologies (see Chapters 5 to 11). If the lessons of the earlier tunnel-fire experiments had not been learned, catastrophes like the fires in the Mont Blanc Tunnel (1999) and the St. Gotthard Tunnel (2001) may have been

Figure 12.7 Full-scale apparatus used to investigate the mitigation of fires and explosions in tunnel drainage systems

far more frequent than they have been. It is only through experimental testing and learning from experience that incidents on this scale will be prevented in the future.

Fire tests in operational tunnels, like those described in this chapter, will undoubtedly become more common in the future. Indeed, there is now a regulation in France (Ministry of the Interior, 2000) which states that tunnel operators must organise full-scale exercises to test emergency procedures at least once a year. These exercises include fire scenarios, and therefore it has induced a significant increase in the number of fire tests in French tunnels since the last decade. It is expected that other countries will implement similar requirements in the future.

Many of the numerous initiatives established in Europe after the Alpine tunnel fire disasters have highlighted new ideas about tunnel-fire testing. Although the capabilities of computer models are increasing rapidly, experimental fire testing remains an essential tool to assess tunnel fire safety. Indeed, the Swiss Office Fédéral des Routes Tunnel Task Force considered 'testing' as a key tool for the improvement of tunnel fire safety (OFROU, 2000). They propose (for implementation at a national level) the following.

- It is desirable to establish definitions, in law, of what constitute acceptable exercises and fire tests for training personnel and testing equipment (from Measure 2.03).
- It is desirable to build or make available a tunnel facility dedicated to training and testing of fire-safety issues outside of the national road network (from Measure 2.04).
- It is highly desirable to establish a directive for the preparation, execution and evaluation of fire tests in tunnel structures at an international level (from Measure 3.07).

Large-scale fire tests will always be expensive and involve highly qualified specialists, but they remain of prime importance to provide essential information for understanding fire behaviour, investigating fire suppression and protection possibilities (ventilation design, water-based suppression systems, etc.), training fire-fighters and tunnel operation personnel, and to provide data for testing computer models.

Finally, the booming interest in the use of alternative energy sources in the transport sector might trigger the need for testing new vehicle types and fuel sources. Design fire scenarios, including vehicle fires involving biofuels (Rivière and Marlair, 2010), or featuring vehicles with fuel cells, as well as fire scenarios relating to electro-mobility, will need to be considered. Indeed, fire development in 'plug-in' electric vehicles and fully electric vehicles, as well as the behaviour of a fire on a HGV carrying a cargo of rechargeable lithium-based batteries in tunnel configurations, need to be assessed, due to the special thermal and chemical risk profiles of such new electric storage systems (Marlair and Torcheux, 2009). It must be noted that lithium batteries are classified as dangerous goods (UN 3090, UN 3480) according to international TDG regulations (see Chapter 22). A facility at INERIS in France is currently being developed to test such systems.

REFERENCES

Apte VB, Green AR and Kent JH (1991) Pool fire plume flow in a large-scale wind tunnel. *Proceedings of the 3rd International Symposium on Fire Safety Science, Boston, MA, USA.* IAFSS, Boston, MA, pp. 425–434.

Bahadir M, Wichmann H, Zelinski V and Lorenz W (1995) Organic pollutants during fire accidents in traffic tunnels. *International Conference on Fire Protection in Traffic Tunnels, Dresden, Germany*, 12–13 September 1995, pp. 46–54.

Bennardo V, Cafaro E, Ferro V and Saluzzi A (1997) Physical modelling and numerical simulation of fire events in road tunnels. *9th International Conference on Aerodynamics and Ventilation of Vehicle Tunnels, Aosta Valley, Italy*, 6–8 October 1997, pp. 649–667.

Bettis RJ, Wu Y and Hambleton RT (1993) *Interim Validation of Tunnel Fire Consequence Models: Data from Phase 2, Test 5.* Section Paper IR/L/93/09 (Part 5). Health and Safety Executive, Buxton.

Bettis RJ, Jagger SF and Moodie K (1994) Reduced scale simulations of fires in partially blocked tunnels. *Proceedings of the International Conference Fires in Tunnels, Borås, Sweden*, 10–11 October 1994, pp. 162–186.

Bettis RJ, Carvel RO and Jagger SF (2007) Fires in cable tunnels: characteristics and development of an experimental facility. *Proceedings of the 5th International Seminar on Fire and Explosion Hazards, Edinburgh, UK*, 23–27 April 2007, pp. 778–787.

Blanchard E, Boulet P, Desanghere S, Cesmat E, Meyrand R, Garo JP and Vantelon JP (2011) Experimental and numerical study of fire in a midscale test tunnel. *Fire Safety Journal* (submitted).

Blume G (1995) Temperature distribution and spread of toxic gases: effects on escape and rescue procedures. *International Conference on Fire Protection in Traffic Tunnels, Dresden, Germany*, 12–13 September 1995, pp. 56–65.

Boehm M, Truchot B and Fournier L (2008) Smoke stratification stability: presentation of experiments. *4th International Conference on Tunnel Safety and Ventilation – New Developments in Tunnel Safety, Graz, Austria*, 21–22 April 2008.

Brousse B, Voeltzel A, Le Botlan Y and Ruffin E (2001) Ventilation and fire tests in the Mont Blanc tunnel to better understand the catastrophic fire of the 24 March 1999. *Proceedings of the 3rd International Conference on Tunnel Fires and Escape from Tunnels, Washington, DC, USA,* 9–11 October 2001, pp. 211–222.

Casalé E (1995) Heptane fire tests in EUREKA 499. *2nd International Conference on Safety in Road and Rail Tunnels, Granada, Spain,* April 1995.

Casalé E and Broz J (1992) Puymorens Tunnel refuges fire tests. *Proceedings of the 1st International Conference on Safety in Road and Rail Tunnels, Basel, Switzerland,* 23–25 November 1992, pp. 451–465.

Casalé E and Marlair G (1994) Heptane fire tests with forced ventilation. *Proceedings of the International Conference on Fires in Tunnels, Borås, Sweden,* 10–11 October 1994, pp. 36–50.

Casalé E, Chassé P, Legrand S, Marlair G and Bonnardel X (1993) Projet EUREKA 499 'FIRETUN' (in French with English summary). In: Reid (ed.), *Underground Transport Infrastructures.* Balkema, Rotterdam, pp. 389–397.

Cwiklinski C (1990) Fire safety inside the Shuttle wagons transporting coaches and passenger cars. Experiments at real scale carried out at Cerchar (in French). *International Workshop on Underground Works for the Transport of Energy, Lille, France,* 16–18 October 1990.

DGPWWM (2002) *Project 'Safety Test' – Report on Fire Tests.* Directorate-General for Public Works and Water Management, Civil Engineering Division, Utrecht.

Dufresne M *et al.* (1998) *Wind Tunnel Experiments for Testing Water Mist Nozzles for Fire Protection of the Eurotunnel Shuttle Wagons carrying HGVs* (in French). CSTB Report EN-AEC98-59C.

EUREKA (1995) *Fires in Transport Tunnels: Report on Full-scale Tests.* EUREKA-Project EU499:FIRETUN Studiengesellschaft Stahlanwendung elV. D-40213. Dusseldorf.

ETN and FIT (2011) http://www.cstc.be/homepage/index.cfm?cat = services&sub = standards_regulations&pag = fire&art = library&niv01 = fit. Available at: http://www.etnfit.net (accessed 24 July 2011).

French SE (1994) Fire safety in the Channel Tunnel, an overview. *Proceedings of the International Conference on Fires in Tunnels, Borås, Sweden,* 1994, pp. 253–275.

Grant G (2001) Fire suppression systems for tunnels. *Brandposten* **25**: 4–5.

Grant G and Southwood P (1999) Development of an onboard fire suppression system for Eurotunnel HGV Shuttle trains. *Proceedings Interflam '99, Edinburgh, UK,* pp. 651–662.

Haack A (1995) Introduction to the EUREKA project/BMBF research project 'Fire protection in underground transportation facilities'. *International Conference on Fire Protection in Traffic Tunnels. Dresden, Germany,* 12–13 September 1995, pp. 6–18.

Haerter A (1994) Fire tests in the Ofenegg tunnel in 1965. *Proceedings of the International Conference Fires in Tunnels, Borås, Sweden,* 10–11 October 1994, pp. 195–214.

Heselden AJM (1976) Studies of fire and smoke behaviour relevant to tunnels. *Proceedings of the 2nd International Symposium on Aerodynamics and Ventilation of Vehicle Tunnels, Cambridge, UK,* 23–25 March 1976. BHRA Fluid Engineering. Also published as BRE Paper CP66/78 (1978).

Heselden AJM and Hinkley PL (1970) *Smoke Travel in Shopping Malls: Experiments in Co-operation with Glasgow Fire Brigade. Part 1.* Fire Research Note No. 832, Joint Fire Research Organization, Borehamwood.

Hu LH, Huo R, Wang HB, Li YZ and Yang RX (2007) Experimental studies on fire-induced buoyant smoke temperature distribution along tunnel ceiling. *Building and Environment* **42**: 3905–3915.

IAFSS (2004) CSIRO: full-scale fire experiments on a passenger rail carriage. *International Association for Fire Safety Science (IAFSS) Newsletter* **16**: 4.

Ingason H (1995a) *Effects of Ventilation on Heat Release Rate of Pool Fires in a Model Tunnel.* SP Report 1995:55. Swedish National Testing and Research Institute, Borås, 1995.

Ingason H (1995b) Findings concerning the rate of heat release. *International Conference on Fire Protection in Traffic Tunnels, Dresden, Germany*, 12–13 September 1995, pp. 94–103.

Ingason H (1998) Unpublished results – personal communication.

Ingason H (2005) *Model Scale Tunnel Fire Tests.* SP Report 2005:49, 2005. Swedish National Testing and Research Institute, Borås.

Ingason H (2006) *Model Scale Tunnel Fire Tests – Sprinkler.* SP Report 2006:56. Technical Research Institute of Sweden, Borås.

Ingason H (2008) Model scale tunnel tests with water spray. *Fire Safety Journal* **43**: 512–528.

Ingason H and Li YZ (2010) Model scale tunnel fire tests with longitudinal ventilation. *Fire Safety Journal* **45**: 371–384.

Ingason H and Lönnermark A (2004) Large scale fire tests in the Runehamar Tunnel: heat release rate. *Proceedings of the International Conference on Catastrophic Tunnel Fires, Borås, Sweden*, 20–21 November 2003. SP Report 2004:05, pp. 81–92. Swedish National Testing and Research Institute, Borås.

Ingason H, Nireus K and Werling P (1997) *Fire Tests in a Blasted Rock Tunnel.* FOA Report FOA-R-97-00581-990-SE. Totalförsvarets forskningsinstitut (FOA) (Swedish Defense Research Agency), Stockholm.

Keski-Rahkonen O (1994) Tunnel fire tests in Finland. *Proceedings of the International Conference Fires in Tunnels, Borås, Sweden*, 10–11 October 1994, pp. 222–237.

Keski-Rahkonen O, Holmjund C, Loikkanen P, Ludvigsen H and Mikkola E (1986) *Two Full Scale Pilot Fire Experiments in a Tunnel.* Research Report 453. Valtion Teknillinen Tutki-muskeskus (VTT), Technical Research Centre of Finland.

Ko GH, Kim SR and Ryou HS (2010) An experimental study on the effect of slope on the critical velocity in tunnel fires. *Journal of Fire Sciences* **28**: 27–47.

Kumar S (1994) Field tunnel simulations of vehicle fires in a channel tunnel shuttle wagon. *Proceedings of the 4th International Symposium on Fire Safety Science, Ottawa, Canada*, June 1994, pp. 995–1006.

Kunikane Y, Kawabata N, Takekuni K and Shimoda A (2002a) Heat release rate induced by gasoline pool fire in a large-cross-section tunnel. *Proceedings of the 4th International Conference on Tunnel Fires, Basel, Switzerland*, 2–4 December 2002, pp. 387–396.

Kunikane Y, Kawabata N, Ishikawa T, Takekuni K and Shimoda A (2002b) Thermal fumes and smoke induced by bus fire accident in large cross sectional tunnel. *Proceedings of the 5th JSME-KSME Fluids Engineering Conference, Nagoya, Japan*, 17–21 November 2002.

Lacroix D, Casalé E, Cwiklinsi C and Thiboud A (1995) Full-scale testing of drainage systems for burning liquids in road tunnels. *Proceedings of the World Congress on Tunnels and STUVA Workshop, Stuttgart, Germany*, 6–11 May 1995, pp. 230–236.

Lee SR and Ryou HS (2005) An experimental study of the effect of the aspect ratio on the critical velocity in longitudinal ventilation tunnel fires. *Journal of Fire Sciences* **23**: 119–138.

Li YZ, Lei B, Xu ZH and Deng ZH (2009a) Small-scale experiments on critical velocity in a tunnel cross passage. *Proceedings of the 13th International Symposium on Aerodynamics and Ventilation of Vehicle Tunnels, New Brunswick, NJ, USA*, 13–15 May 2009, pp. 355–363.

Li YZ, Lei B, Xu ZH and Deng ZH (2009b) Critical velocity and backlayering length in tunnel fires using longitudinal ventilation system. *Proceedings of the 13th International Symposium on Aerodynamics and Ventilation of Vehicle Tunnels, New Brunswick, NJ, USA*, 13–15 May 2009, pp. 365–378.

Liu ZG, Kashef AH, Lougheed GD and Crampton GP (2011a) Investigation on the performance of fire detection systems for tunnel applications – Part 1: Full-scale experiments at a laboratory tunnel. *Fire Technology* **47**: 163–189.

Liu ZG, Kashef AH, Lougheed GD and Crampton GP (2011b) Investigation on the performance of fire detection systems for tunnel applications – Part 2: Full-scale experiments under longitudinal airflow conditions. *Fire Technology* **47**: 191–220.

Lönnermark A and Ingason H (2004) Large scale fire tests in the Runehamar Tunnel: gas temperature and radiation. *Proceedings of the International Conference on Catastrophic Tunnel Fires, Borås, Sweden*, 20–21 November 2003, SP Report 2004:05, pp. 93–104. Swedish National Testing and Research Institute, Borås.

Lönnermark A and Ingason H (2008) The influence of tunnel cross section on temperatures and fire development. *Proceedings of the 3rd International Symposium on Tunnel Safety and Security, Stockholm, Sweden*, 12–14 March 2008, pp. 149–161.

Malhotra HL (1993) Fire protection in channel tunnel car carrying shuttles. *Proceedings Interflam '93*, 30 March to 1 April 1993, pp. 365–374.

Malhotra HL (1995) Goods vehicle fire test in a tunnel. *2nd International Conference on Safety in Road and Rail Tunnels, Granada, Spain*, 1995, pp. 237–244.

Mao J, Xi YH, Bai G, Fan HM and Ji HZ (2011) A model experimental study on backdraught in tunnel fires. *Fire Safety Journal* **46**: 164–177.

Marlair G (1999) Experimental approach of the fire hazard in closed spaces: laboratory and full-scale tests. *International Congress on Fire Safety in Hazardous Enclosed Spaces – Tunnels, Underground Spaces, Parkings, Storages, CNPP-INERIS, Vernon, France*, 8–9 November 1999.

Marlair G and Cwiklinski C (1993) Large-scale testing in the INERIS fire gallery: a major tool for both assessment and scaling-up of industrial fires involving chemicals. In: Cole ST and Wicks P (eds), *Industrial Fires – Workshop, Apeldoorn, The Netherlands*. Proceedings EUR 15340 EN, pp. 303–309.

Marlair G and Torcheux L (2009) Lithium ion safety issues for electric vehicles applications. *Proceedings of the International Power Supply Conference and Exhibition 'Batteries 2009', Cannes–Mandelieu, France*, 30 September 2009 to 2 October 2009.

Marlière F, Cwiklinski C, Houeix JP and Marlair G (1995) Full-scale testing approach to fire safety issues: case study of the tourism TransManche shuttle (in French). *Recherche, Transport, Sécurité* **49**: 138–144.

Mawhinney JR, Soja E and Gillespie R (1999) Mercury Energy CBD tunnel project, New Zealand – performance based fire testing of a water mist suppression system. *Proceedings of Interflam '99, Edinburgh, UK*, pp. 663–673.

Ministry of the Interior (2000) Concerning Safety in the Tunnels of the National Highway Network. Inter-Ministry Circular No. 2000-63, Section 5.3.1. Ministry of the Interior and Ministry of the Establishment, Transport and Housing, Paris.

Mizutani T, Horiuchi K and Akiyama K (1982) Experimental study of tunnel fires. *Journal of the Japan Road Association* 24–28.

Nilsen AR, Lindvik PA and Log T (2001) Full-scale fire testing in subsea public road tunnels. *Proceedings of the 9th International Interflam Conference, Edinburgh, UK*, 17–19 September 2001, pp. 913–924.

OFROU (2000) *Tunnel Task Force, Final Report* (in French). Office Fédéral des Routes, Geneva.

Parsons Brinckerhoff (1996) *Memorial Tunnel Fire Ventilation Test Programme*. Interactive CD-ROM and Comprehensive Test Report. Parsons Brinckerhoff 4D Imaging, Denver, CO.

Perard M (1992) Organization of fire trials in an operated road tunnel. *1st International Conference on Safety in Road and Rail Tunnels, Basel, Switzerland*, 23–25 November 1992, pp. 161–170.

Perard M and Brousse B (1994) Full size tests before opening two French tunnels. *8th International Symposium on Aerodynamics and Ventilation of Vehicle Tunnels, Liverpool, UK*, 6–8 July 1994, pp. 383–408.

PIARC (1999) *Fire and Smoke Control in Tunnels*. Report No. 05.05.B. World Road Association (PIARC/AIPCR), La Defense.

Pucher K (1994) Fire tests in the Zwenberg tunnel (Austria). *Proceedings of the International Conference on Fires in Tunnels, Borås, Sweden*, 10–11 October 1994, pp. 187–194.

PWRI (1993) *State of the Road Tunnel Equipment Technology in Japan – Ventilation, Lighting, Safety Equipment*. Technical Note. Japanese Public Works Research Institute (PWRI), Ministry of Construction, Japan, pp. 35–47.

Quintiere JG (1989) Scaling applications in fire research. *Fire Safety Journal* **15**: 3–29.

Rauch J (2001) Fire protection in road tunnels: development and testing of a tunnel lining system for more efficiency and safety. *Proceedings of the 4th International Conference on Safety in Road and Rail Tunnels, Madrid, Spain*, 2–6 April 2001, pp. 271–280.

Richter E (1995) Smoke and temperature development in tunnels – experimental results of full scale fire tests. *2nd International Conference Safety in Road and Rail Tunnels, Granada, Spain*, pp. 295–302.

Rivière C and Marlair G (2010) The use of multiple correspondence analysis and hierarchical clustering to identify incident typologies pertaining to the biofuel industry. *Biofuels, Bioproducts and Biorefining* **4**(1): 53–65.

Roh JS, Ryou HS, Kim DH, Jung WS and Jang YJ (2007) Critical velocity and burning rate in pool fire during longitudinal ventilation. *Tunnelling and Underground Space Technology* **22**: 262–271.

Ruffin E (1997) *Perfecting a Tool for Simulating Accident Situations in Underground Networks*. BCRD Final Summary Report, Subsidy No. 97-028. French Ministry of Territorial Development and the Environment, Accidental Risks Division.

Saito N, Yamada T, Sekizawa A, Yanai E, Watanabe Y and Miyazaki S (1995) Experimental study on fire behaviour in a wind tunnel with a reduced scale model. *2nd International Conference Safety in Road and Rail Tunnels, Granada, Spain*, 3–6 April 1995, pp. 303–310.

Steckler et al. (1986) Salt water modelling of fire induced flows in multicompartment enclosures. *21st International Symposium on Combustion*, 1986, pp. 143–149.

Steinert C (1994) Smoke and heat production in tunnel fires. *Proceedings of the International Conference Fires in Tunnels, Borås, Sweden*, 10–11 October 1994, pp. 123–137.

Stroeks R (2001) *Sprinklers in Japanese Road Tunnels*. Bouwdienst Rijkswaterstaat, Utrecht.

Truchot B, Boehm M and Waymel F (2009) Numerical analysis of smoke layer stability. *13th International Symposium on Aerodynamics and Ventilation of Vehicle Tunnels, New Brunswick, NJ, USA*, 13–15 May 2009, Vol. 1, pp. 281–295.

Vantelon JP, Guelzim A, Quach D, Son DK, Gabay D and Dallest D (1991) Investigation of fire-induced smoke movement in tunnels and stations: an application to the Paris Metro. *Proceedings of the 3rd International Symposium on Fire Safety Science, Boston, MA*, pp. 907–918. IAFSS, Boston.

Vauquelin O and Telle D (2005) Definition and experimental evaluation of the smoke 'confinement velocity' in tunnel fires. *Fire Safety Journal* **40**: 320–330.

Vigne G and Jönsson J (2009) Experimental research – large scale tunnel fire tests and the use of CFD modelling to predict thermal behaviour. *Advanced Research Workshop on Fire*

Protection and Life Safety in Buildings and Transportation Systems, Santander, Spain, October 2009, pp. 255–272.

White N and Dowling VP (2004) Conducting a full scale experiment on a rail passenger car. *Proceedings of the 6th Asia–Oceania Symposium on Fire Safety Science and Technology, Daegu, Korea,* 17–20 March 2004, Paper 6B-2.

Wu Y, Stoddard JP, James P and Atkinson GT (1997a) Effect of slope on control of smoke flow in tunnel fires. *Proceedings of the 5th International Symposium on Fire Safety Science, Melbourne, Australia,* 3–7 March 1997, pp. 1225–1236.

Wu Y, Bakar MZA, Atkinson GT and Jagger S (1997b) A study of the effect of tunnel aspect ratio on control of smoke flow in tunnel fires. *9th International Conference on Aerodynamics and Ventilation of Vehicle Tunnels, Aosta Valley, Italy,* 6–8 October 1997, pp. 573–587.

Handbook of Tunnel Fire Safety, 2nd edition
ISBN: 978-0-7277-4153-0

ICE Publishing: All rights reserved
doi: 10.1680/htfs.41530.273

publishing

Chapter 13
Fire dynamics in tunnels

Haukur Ingason
SP Technical Research Institute of Sweden, Borås, Sweden

Nomenclature

A	cross-sectional area of the tunnel (m^2)
A_0	area of the openings (m^2)
c_p	heat capacity (kJ/kg K)
C_s	extinction coefficient (m^{-1})
D_h	hydraulic diameter; $D_h = 4A/P$
f_D	friction factor
F_w	view factor
g	acceleration due to gravity (9.81 m/s^2)
h	lumped heat-transfer coefficient (kW/m^2 K)
h_c	average convected-heat transfer coefficient to the walls (kW/m^2 K)
h_r	radiated-heat transfer coefficient (kW/m^2 K)
h_0	height of the opening (m)
h_{cut}	cut-off height of the flame (m)
h_{free}	flame height in the open (m)
h_{hor}	horizontal flame height along the ceiling (m)
Δh	hot smoke stack height (m)
H	tunnel height (m)
H_c	chemical heat of combustion (kJ/kg)
H_{ec}	effective chemical heat of combustion (kJ/kg)
H_T	net heat of complete combustion (kJ/kg)
K_1	the pressure-loss coefficient at the entrance of the tunnel
k_{CO_2},	maximum possible yield of carbon dioxide (CO$_2$) and carbon monoxide (CO)
k_{CO}	(g/g), respectively
L	tunnel length (m)
L_b	backlayering distance of smoke (m)
L_c	combustion length (m)
L_f	flame length (m)
L_s	length of the tunnel, which is exposed to the hot smoke (m)
M_i	molecular weight of gas species i (g/mol)
$\dot{m}_a$	air mass flow rate (kg/s)
$\dot{m}_f$	mass fuel rate (fuel supply) (kg/s)
$\dot{m}_i$	mass flow rate of gas species i (kg/s)
OD/L	the optical density per path length of light through smoke (m^{-1})
Q	heat-release rate (kW)
Q_{max}	maximum heat-release rate (kW)

q''	heat flux (kW/m^2)
$\dot{q}''_{cr}$	critical heat flux (kW/m^2)
P	perimeter of the tunnel (m)
r	air/fuel mass ratio required for stoichiometric combustion
t	time (s)
T_a	ambient air temperature (K)
T_{avg}	average gas temperature over the entire cross-section at actual position (K)
T_{cr}	critical ignition temperature (K)
$\bar{T}$	average gas temperature of the stack height (K)
T_0	ambient temperature within the tunnel (K)
T_w	wall temperature (K)
$T_{avg,\,x=0}(\tau)$	average gas temperature at the fire source ($x=0$) at the time $\tau = t - (x/u)$
ΔT_{avg}	$T_{avg} - T_a$ (K)
u	air velocity (m/s)
u_{avg}	uT_{avg}/T_a (m/s)
x	distance from the centre of the fire source (m)
X_i	concentration of gas species i (or mole fraction)
V	visibility in smoke (m)
θ	tunnel slope (%)
δ	proportionality constant that is a weak function of temperature
ε_{g-w}	effective emissivity between gas and walls, respectively
ρ_a	ambient air density (kg/m^3)
τ	dummy variable (s)
ϕ	air/fuel mass ratio
χ	ratio of chemical heat of combustion to net heat of complete combustion
σ	Stefan–Boltzmann constant ($5.6 \times 10^{-11}\ kW/m^2\,K^4$)

13.1. Introduction

The occurrence of some catastrophic fires has put the focus on fire dynamics in tunnels. In particular there is a need for a better understanding of the influence of ventilation on the heat-release rate, and how fire and the smoke spread within a tunnel.

The knowledge about the dynamics of fire in enclosed spaces is usually related to fire behaviour in ordinary sized compartments. Moreover, much of the chemistry of fires and knowledge about fire plumes is treated without taking into account the direct interaction with the type of environment that the fire is in. Most of the fundamental research on fire dynamics in tunnels (see, for example: Hwang and Wargo, 1986; Grant, 1988; Ingason, 1995b; Atkinson and Wu, 1996; Grant and Drysdale, 1997; Wu et al., 1997; Megret and Vauquelin, 2000; Wu and Bakar, 2000; Carvel et al., 2001), is focused on smoke spread in tunnels, and the local effects of a single burning vehicle. With regard to the characteristics of large fires in Europe, it is surprising that there are only a limited number of studies reported in the literature that deal with the fire spread between vehicles in tunnels (Beard, 1995, 1997; Beard et al., 1995). However, numerous fundamental research studies have been performed on the spread of fire in mines and ducts (de Ris, 1970; Lee et al., 1979; Tewarson, 1982a; Newman and Tewarson, 1983; Newman, 1984; Comitis et al., 1994, 1996). The knowledge and experience from these research projects provides a sound basis for future studies on the spread of fire and smoke in tunnels. Catastrophic fires worldwide have shown us that there are more points of coincidence between duct and mine fires than between compartment fires and tunnel fires.

The principal phenomena of tunnel fires and fire dynamics are discussed in this chapter. Descriptions are given of the development of fires in tunnels and building compartments, and the influence of ventilation on a fire (fuel control versus ventilation control). In addition, methods are given for the calculation of gas temperatures, the stratification of smoke, flame lengths, the incident radiation, and the spread of fire between vehicles.

13.2. Tunnel fires and open fires

Tunnel fires differ from open fires in at least two important ways. First, the feedback of heat to burning vehicles in a tunnel tends to be more effective than that in an open environment, because of the confined space. This effective heat feedback often causes vehicles that do not burn intensely in an open fire to burn vigorously in a tunnel fire. For example, Carvel et al. (2001) concluded that the heat-release rate of a fire within a tunnel could increase by a factor of four compared with the heat-release rate of the same material burning in the open. Furthermore, the oxygen required for combustion is not always as readily available in tunnels as in the open. The fire may develop either as a fuel-controlled fire, where unreacted air bypasses the burning vehicles, or a ventilation-controlled fire. In a ventilation-controlled fire, all the oxygen is essentially consumed within the combustion zone, and this gives rise to large amounts of toxic fumes and products of incomplete combustion, and fuel-rich gases leave the exit of the tunnel.

Second, as a fire develops in a tunnel, it interacts with the ventilation airflow and generates aerodynamic disturbance in the tunnel. This interaction and disturbance may lead to drastic changes in the ventilation flow pattern, such as throttling of airflow (buoyancy effects). and reverse flow of hot gases and smoke from the fire into the ventilation airstream (backlayering). Such effects on the ventilation not only complicate fire-fighting procedures, but also present extreme hazards by pushing toxic fumes and gases far away from the fire.

13.3. Tunnel fires and compartment fires

Tunnel fires differ from building compartment fires in at least three important ways. First, the maximum heat-release rate in compartment fires is usually dictated by the natural ventilation, i.e. the ventilation factor $A_0\sqrt{h_0}$ $(m^{5/2})$, where A_0 is the area of the openings (m^2) and h_0 is the height of the opening (m) (Drysdale, 1992). In tunnels, however, the fire size, tunnel slope, cross-sectional area, length, type of tunnel (concrete, blasted rock) and meteorological conditions at the entrance all govern the natural ventilation. Tunnels work more or less as communicating vessels, and the excess air available for combustion is an order of magnitude higher than that available in compartment fires. Tunnels are also often equipped with mechanical ventilation (forced ventilation), such as exhaust fans and/or jet fans in the ceiling. Therefore, the influence of the ventilation on the combustion efficiency and heat-release rate of fires in tunnels is significantly different from that in compartments.

Second, compartment fires can easily grow to 'flashover' within a few minutes (flashover is defined as the rapid transition to a state of total surface involvement in a fire of combustible material within an enclosure). Indeed, flashover can hardly take place outside of a 'compartment'. Tunnel fires are, therefore, not likely to grow to conventional 'flashover', due to the large heat losses from the fire to the surrounding walls and the lack of containment of hot fire gases. Flashover can, however, easily occur in a train compartment or a truck cabin located inside a tunnel. Furthermore, the risk of secondary fire-gas explosions due to underventilated fires is much lower in tunnels than in building compartment fires. This is mainly due to the difference in the ventilation conditions (as explained above), the geometry and the heat losses to surrounding walls. However, when a powerful ventilation system is suddenly activated during an under-ventilated fire situation, the effects may be dramatic. The flames may suddenly increase in size and length, and the fire may easily spread forward due to the preheated vehicles downstream of the fire, although this phenomenon cannot be defined as 'flashover'. This situation may become very hazardous for fire-fighters and those who are still trapped inside the tunnel. Starting a ventilation system in a tunnel when a fire has been ongoing for some time and there is a high vehicle density is always very risky and should be avoided.

The third difference is in the formation of the stratified smoke layer. In the early stages of compartment fires, an upper quiescent buoyant smoke layer is formed, with a cold smoke-free layer below. A similar smoke layer may be created in tunnels during the early stage of a fire when there is essentially no longitudinal ventilation. The smoke layer will, however, gradually move further away from the fire. If the tunnel is very long, the smoke layer may move to the tunnel surface. The distance from the fire that this may occur depends on the fire size, the tunnel type, and the perimeter and height of the tunnel cross-section. When the longitudinal ventilation is gradually increased, this stratified layer gradually dissolves. A backlayering of smoke is created on the upstream side of the fire, and the degree of stratification of the smoke downstream from the fire is governed by the heat losses to the surrounding walls and by the turbulent mixing between the buoyant smoke layer and the cold layer moving in the opposite direction. The particular dimensionless parameter that determines whether one gas will stratify above another is the Richardson number (Ri). The Richardson number is similar to the inverse of the Froude number (Fr); however, the Richardson number is thought of as controlling mass transfer between layers, whereas the Froude number gives the general shape of a layer in an airstream.

13.3.1 Fire development

The development of a fire within a compartment is usually divided into two separate stages: the *pre-flashover* stage and *post-flashover* stage. In textbooks on fire dynamics in enclosures (Karlsson and Quintiere, 2000; Drysdale, 1992), three distinct time periods of fire development are usually given within the pre- and post-flashover stages. The first is the *growth period*, the second is the *fully developed fire*, and the last stage is the *decay period* (Figure 13.1). The development of a fire in a tunnel can be described in the same way, although the interaction with the enclosure differs considerably.

Figure 13.1 The different phases of a typical compartment fire

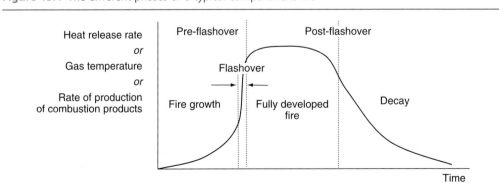

Figure 13.2 Principal sketch of fuel-controlled and ventilation-controlled fire in a compartment, and in a tunnel with a natural draught and forced ventilation

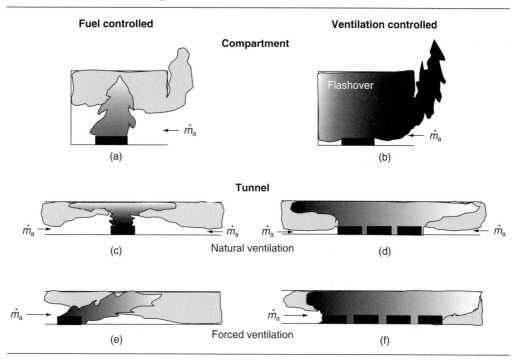

Compartment fires are usually defined as either *fuel-controlled* or *ventilation-controlled*. In the growth period in compartment fires (the pre-flashover stage), when there is sufficient oxygen available for combustion and the growth of the fire depends entirely on the flammability of the fuel and its geometry, the fire is defined as fuel controlled. During the growth period, the fire will either continue to develop up to and beyond a point at which interaction with the compartment boundaries becomes significant (flashover), or it will start to decay. The fire will either start to decay due to lack of fuel, or become ventilation controlled, whereby the fire size is dictated by the inflow of fresh air $\dot{m}_a$ towards the base of the fire source (Figure 13.2).

There are numerous synonyms in the literature for fuel-controlled and ventilation-controlled fires. A fuel-controlled fire may also be described as well-ventilated, overventilated, oxygen rich or fuel lean. A ventilation-controlled fire may be described as fuel rich, oxygen starved or underventilated. This can cause confusion for the reader, but as authors use different words to describe the same physical phenomena it is unavoidable. Due to this confusion it is very important to understand the basic difference between these two combustion modes. For compartment fires, the transition period between these two physical conditions is usually defined as the 'flashover'. In tunnels, fires are usually fuel-controlled, but in severe fires where multiple vehicles are involved (e.g. those that occurred in the Mont Blanc, Tauern and St. Gotthard tunnels), the fires are ventilation controlled.

If the base of the fire source is completely surrounded by vitiated air, which is a mixture of air and combustion products, it may self-extinguish when the oxygen content of the air reaches about 13% (i.e. the flammability limits are exceeded) (Beyler, 1995). However, this value is to some extent temperature dependent (Quintiere and Rangwala, 2003), with increasing temperature tending to lower the flammability limits.

If a door is opened just before a fire has self-extinguished, the entire compartment may erupt into flames if fresh air is entrained into the hot fire source and mixes with the unburned volatile components of the fuel. This phenomenon is better known as 'backdraught', and is very hazardous for fire-fighters. In tunnels, this type of fire development is most likely to occur inside vehicle compartments (a train or truck cabin) and not in the tunnel itself. As mentioned previously, switching on a longitudinal ventilation system in a tunnel when the fire has been ongoing for some time may create a very dangerous situation, because large amounts of the volatile components of fuel may be produced under conditions where the fuel bed and the surrounding walls have become very hot. When the oxygen (air) is suddenly transported towards an area where there is a high level of vaporised fuel, better mixing of the fuel and oxygen results, and the flames may suddenly increase dramatically in size. Although, this phenomenon cannot be defined as 'backdraught', it is in some ways analogous.

In tunnel fires with a high degree of backlayering (e.g. tunnels with natural ventilation), the air that backflows towards the fire may be highly vitiated (being a mixture of combustion products transported backward from the fire and fresh air transported from the entrance towards the fire), which may affect the combustion efficiency. Currents of fresh air along the tunnel floor will, however, usually provide the fire with sufficiently high oxygen levels. Thus the chance of a fire self-extinguishing due to the backflow of vitiated air in a tunnel is low. This phenomenon has been observed in experiments done on a model-scale tunnel, but the experimental conditions in these cases were quite special (Ingason, 1995a; Ingason et al., 1997). The fresh air was choked upstream of the fire by reducing the inlet area. At a certain inlet area, the fire self-extinguished due to the vitiated air (<13% oxygen) created by a mixture of the backlayering combustion products and the inflowing fresh air.

13.3.2 Flashover

The parameters that govern whether a fire will spread dramatically include the fire load, the dimensions of the compartment and the ventilation openings, and the thermal properties of the surrounding walls. Flashover in a compartment has been explained as thermal instability caused by the rate of energy generation increasing faster with temperature than the rate of aggregated energy losses (Thomas et al., 1980). Usually this phenomenon occurs over a short period, and results in a rapid increase in the heat-release rate, gas temperatures and production of combustion products. After a flashover has occurred in a compartment, the rate of heat release will develop to produce temperatures of 900–1100°C. As mentioned above, the period after flashover is called the 'post-flashover stage' or the 'fully developed fire period' (see Figure 13.1). During this period the heat-release rate is dictated by the oxygen flow through the openings, and the fire is said to be 'ventilation controlled' (see Figure 13.2). The amount of heat released depends on the relative quantities of air available within the compartment. The mass flow rate of air through the opening $\dot{m}_a$ can be expressed in general terms (Babrauskas, 1981; Tewarson, 1984) as:

$$\dot{m}_a = \rho_a \delta \sqrt{g} A_0 \sqrt{h_0} \tag{13.1}$$

where δ is a proportionality constant, which is a weak function of temperature, A_0 is the area of the opening (m^2) and h_0 is the height of the opening (m). The value of δ has been estimated to be 0.08 for pre-flashover fires and 0.13 for post-flashover fires (Babrauskas, 1981; Tewarson, 1984).

Therefore, the value of $\rho_a\delta\sqrt{g}$ in the pre-flashover case (fuel controlled) is $0.3\ \mathrm{kg/s\,m^{-5/2}}$ and in the post-flashover case (ventilation controlled) $0.5\ \mathrm{kg/s\,m^{-5/2}}$, assuming that the density ρ_a is $1.22\ \mathrm{kg/m^3}$ and g is $9.81\ \mathrm{m/s^2}$. The term $A_0\sqrt{h_0}$ is better known as the 'ventilation factor', and originates from Bernoulli's equation applied to density flow through a single opening (Drysdale, 1992). It is not suitable to use the ventilation factor $A_0\sqrt{h_0}$ when calculating the airflow in a tunnel, as the derivation of equation (13.1) does not justify this use, and other methods are thus required to estimate $\dot{m}_a$ in tunnels. The simplest method is to use Bernoulli's equations for pipe/duct flow (see equation (13.16)).

Assuming that each kilogram of oxygen used for combustion produces about $13.1\times10^3\ \mathrm{kJ}$ of energy (Huggett, 1980; Parker, 1984) and that the mass fraction of oxygen in air is 0.231, we can approximate the maximum heat-release rate that is possible *within* a compartment during the ventilation-controlled stage. Using the values given earlier for equation (13.1), i.e. $13.1\times10^3\times0.231\times\dot{m}_a$, where $\dot{m}_a=\rho_a\delta\sqrt{g}A_0\sqrt{h_0}$, we obtain the maximum heat-release rate Q_{max} (kW) within the compartment ($\rho_a\delta\sqrt{g}=0.5\ \mathrm{kg/s\,m^{-5/2}}$) as:

$$Q_{max}\approx1500A_0\sqrt{h_0} \tag{13.2}$$

Here we assume that all the oxygen entering the compartment is consumed within the compartment. In many cases the rate at which air enters the compartment is insufficient to burn all the volatile compounds vaporising within the compartment, and the excess volatiles will be carried through the compartment opening with the outflowing combustion products. This is normally accompanied by external flaming outside the opening (Drysdale, 1992), as shown in Figure 13.3.

This phenomenon becomes important when one wishes to estimate the maximum heat-release rate in a 'post-flashover', steel-bodied train coach located *inside* a tunnel. Equation (13.2) may underestimate the maximum heat-release rate within the tunnel, as excess volatile products are

Figure 13.3 A fully developed fire in a train coach (Photograph courtesy of Tomas Karlsson)

burned outside the train coach. Model-scale tests (1 : 10) of a fully developed fire in a train coach showed that the maximum heat release when all windows were open was, on average, 72% higher than the value obtained according to equation (13.2) (Ingason, 2007). This means that 42% of the total fuel vaporised within the coach (assuming that all the oxygen in the entrained air is consumed within the coach) is burned outside the openings. Bullen and Thomas (1979) showed that the amount of excess fuel burning outside the openings is mainly dependent on the surface area of the fuel and the ventilation factor $A_0\sqrt{h_0}$. Thus, assuming that this factor is relatively constant for this type of geometry (a train coach), the maximum heat-release rate according to equation (13.2) can be multiplied by a factor of 1.72. The use of equation (13.2) is demonstrated in the following example.

Example 1

What is the maximum heat-release rate after flashover of the burning train coach shown in Figure 13.3? The coach windows are 1.2 m wide and 1.2 m high, and there are seven windows on each side. The door opening is 1 m wide and 2 m high. The total opening area times the square root of the opening heights is:

$$\sum A_{0,\,window}\sqrt{h_{0,\,window}} + A_{0,\,door}\sqrt{h_{0,\,door}} = 24.9\,\mathrm{m}^{5/2}$$

Solution. According to equation (13.2), the maximum heat-release rate is 37 MW. This means that, if the coach were burning in a tunnel, the total heat-release rate inside the tunnel would be higher than 37 MW, as volatile components of the fuel are burning outside the openings. This can be estimated by multiplying the value obtained from equation (13.2) by a constant 1.72. Thus, the maximum heat-release rate from the burning coach in Figure 13.3 is estimated to be 64 MW.

If the fire were to spread between the train coaches, the total heat-release rate would be much higher than 64 MW. However, it is not possible to merely sum the heat-release rate for each coach, because the first coach would not necessarily reach the peak heat-release rate at the same time as the later ones. There are many other parameters that will affect the fire development in a train coach. These include, for example, the body type (steel, aluminium, fibreglass, etc.), the quality of the glazed windows, the geometry of the openings, the amount and type of combustible interior and its initial moisture content, the construction of the wagon joins, the air velocity within the tunnel, and the geometry of the tunnel cross-section.

13.4. Fuel control and ventilation control

Ventilation is one of the key elements influencing the safety of tunnel fires. Therefore it is of great importance to understand the difference between the two combustion modes, i.e. fuel controlled and ventilation controlled, and the severe fire accidents that have occurred in tunnels have laid focus on the phenomenon of ventilation-controlled fires in tunnels. These two combustion modes are discussed fully in Chapter 14, for both compartment fires and tunnel fires.

A compartment fire is generally fuel-controlled during the pre-flashover stage, and even in the decay period and, in rare cases, the post-flashover period. The fire is generally ventilation controlled in the post-flashover stage, and in rare cases even in the pre-flashover stage. (Note that backdraught in a compartment, which initially corresponds to a ventilation-controlled state, may also be effectively a flashover, given the definition of flashover as a sudden transition from localised to generalised burning.) 'Fuel control' implies that oxygen (i.e. the oxidant) is in unlimited supply and the rate of combustion is independent of the oxygen supply rate (or mass flow rate of air), but is determined by the fuel supply rate (mass flow rate of vaporised fuel). The combustion behaviour is then similar to that of combustion in the open. A ventilation-controlled fire, in contrast, has a limited oxygen supply, and the rate of combustion is dependent on both air and fuel supply rates. The efficiency of combustion depends on the fuel supply rate relative to the oxygen supply rate. When precisely the necessary amount of oxygen is available to enable complete combustion, the mixture is said to be 'stoichiometric'. In the case of the combustion of a generic fuel ($C_aH_bO_c$) in air, the stoichiometric chemical equation is:

$$C_aH_bO_c + \left(a + \frac{b}{4} - \frac{c}{2} \right)(O_2 + 3.76N_2) \rightarrow aCO_2 + \frac{b}{2}H_2O + \left(a + \frac{b}{4} - \frac{c}{2} \right)3.76N_2 \tag{13.3}$$

where the ratio of moles of nitrogen (N_2) to moles of oxygen (O_2) in air is 3.76. The stoichiometric coefficient r, which gives the air/fuel mass ratio required for stoichiometric combustion of fuel to produce carbon dioxide (CO_2) and water (H_2O), can be obtained using the following equation:

$$r = \frac{137.8[a + (b/4) - (c/2)]}{12a + b + 16c} \tag{13.4}$$

where $(1 + 3.76) \times 28.95 = 137.8$, and 28.95 is the molecular weight of air (g/mol). The use of equations (13.3) and (13.4) is demonstrated in the following example.

Example 2
How much air is required to completely burn 1 kg of propane (C_3H_8)?

Solution. Equation (13.3) yields:

$$C_3H_8 + 5(O_2 + 3.76N_2) > 3CO_2 + 4H_2O + 18.8N_2$$

Here $a = 3$, $b = 8$ and $c = 0$. Using equation (13.4), the stoichiometric ratio is obtained as $r = 15.7$. This means that 15.7 kg of air is required to completely burn 1 kg of propane.

There are many different ways of characterising the relationship between the air (oxygen) supply and the fuel supply. One way is with the aid of the air/fuel equivalence ratio ϕ (Tewarson, 1988):

$$\phi = \frac{\dot{m}_a}{r\dot{m}_f} \tag{13.5}$$

where $\dot{m}_a$ is the mass flow rate of the air (oxygen) supply, $\dot{m}_f$ is the rate of loss of fuel mass (fuel supply) and r is the stoichiometric coefficient for complete combustion, as obtained from equation (13.4). (Note: Beyler (1985) uses the fuel/air equivalence ratio, i.e. $1/\phi$.) The air/fuel equivalence ratio ϕ can be used to determine whether a fire is ventilation controlled or fuel controlled.

13.4.1 Fuel control
If the air/fuel mass ratio is greater than or equal to the stoichiometric value r, (i.e. $\dot{m}_a/\dot{m}_f \geqslant r$) or $\phi \geqslant 1$, then the fire is assumed to be fuel controlled, and the heat-release rate is directly proportional to the rate of loss of fuel mass $\dot{m}_f$. To illustrate this, consider the case where the oxygen concentration in the gases flowing out of the compartment or the tunnel exit is greater than zero. The chemical heat-release rate Q (kW), which is directly proportional to the rate of loss

of fuel mass loss $\dot{m}_f$ (kg/s), can then be calculated using the equation:

$$Q = \dot{m}_f \chi H_T \tag{13.6}$$

where H_T is the net heat of complete combustion (kJ/kg), i.e. combustion in which the water produced is in the form of a vapour. The combustion of fuel vapours in a fire is never complete, and thus the effective heat of combustion H_{ec} is always less than the net heat of complete combustion H_T. Furthermore, χ is the ratio of the effective heat of combustion to the net heat of complete combustion, i.e. $\chi = H_{ec}/H_T$ (Tewarson, 1988) (Note: Tewarson calls the 'effective heat of combustion' the 'chemical heat of combustion' (Persson, 2000a).) The rate of loss of fuel mass is sometimes expressed as $\dot{m}_f''$, the rate of loss of fuel mass per unit area of fuel A_f, which means that in equation (13.6) $\dot{m}_f$ can be replaced by $\dot{m}_f'' A_f$, then giving the heat-release rate per unit area.

13.4.2 Ventilation control

If the air/fuel mass ratio is less than the stoichiometric value, $\dot{m}_a/\dot{m}_f < r$, or $\phi < 1$, the fire is defined as ventilation controlled and the heat-release rate Q is directly proportional to the mass flow rate of air $\dot{m}_a$ (i.e. it is proportional to the oxygen supply) available for combustion. Sometimes, but not always, we can simplify this by saying that the oxygen concentration in the gases flowing out of the compartment or the tunnel exit is essentially zero. One notable exception is ventilation-controlled fires in large compartments with small openings (i.e. no flashover). There are numerous options available for calculating the heat-release rate for these conditions. The simplest is the following equation, which assumes complete combustion and that all the air $\dot{m}_a$ is consumed:

$$Q = \dot{m}_a \frac{H_T}{r} \tag{13.7}$$

where the ratio H_T/r is found to be about 3000 kJ/kg of air consumed for most carbon-based materials (Huggett, 1980). It has been argued that, for incomplete combustion, there are counteracting effects, which means that equation (13.7) is approximately valid in this case also, as well as for the case of complete combustion. If all the oxygen were consumed in the fire, the energy developed would correspond to 13×10^3 kJ/kg of oxygen consumed (i.e. 3000/0.231, assuming that 23.1% (mass) of the air is oxygen). This number is well known from calorimeter measurements made in fire laboratories (Janssens and Parker, 1992), which use the average value of 13.1×10^3 kJ/kg.

13.4.3 Determination of the combustion mode

In order to make a simple estimation of the combustion mode, i.e. to determine whether the fire is fuel controlled or ventilation controlled, equations (13.5) and (13.6) can be combined to give:

$$\phi = 3000 \frac{\dot{m}_a}{Q} \tag{13.8}$$

Equation (13.8) assumes that not all the air is necessarily consumed. If all the air is consumed, then $\dot{m}_a = Q/3000$ (from equation (13.7)) and ϕ becomes 1. It is sometimes of interest to know how much oxygen is used in the combustion process of the fire (assuming stoichiometric combustion, i.e. $\phi = 1$). The mass fraction of unreacted oxygen (or air) passing the fire can be estimated using the equation:

$$\beta = \frac{\dot{m}_a - (Q/3000)}{\dot{m}_a} \tag{13.9}$$

(here, not all the air $\dot{m}_a$ is consumed). The following two examples demonstrate the use of equations (13.8) and (13.9).

Example 3
Assume a fire in a heavy goods vehicle (HGV) or semi-trailer in a tunnel. The tunnel is 6 m high and 9 m wide, with a longitudinal ventilation of 2 m/s. The fire is estimated to grow linearly and reach a peak heat-release rate of 120 MW after 10 minutes. When the fire reaches 120 MW, is it ventilation controlled or fuel controlled? The air density within the tunnel is $\rho_a = 1.2 \text{ kg/m}^3$.

Solution. First, calculate the mass flow rate of air: $\dot{m}_a = 1.2 \times 2 \times 6 \times 9 = 130 \text{ kg/s}$. Equation (13.8) gives $\phi = 3000/(130/120\,000) = 3.25$. This value is larger than 1, and therefore the fire is fuel controlled. This also means that unreacted air is passing the combustion zone. The amount of unreacted air can be determined using equation (13.9); $\beta = [130 - (120\,000/3000)]/130 = 0.69$. This means that 69% of the available oxygen in the airflow remains unreacted and 31% has been consumed in the fire.

Example 4
What is the necessary heat-release rate Q to obtain a ventilation-controlled fire in Example 3?

Solution. Assuming the same ventilation rate, i.e. $\dot{m}_a = 130 \text{ kg/s}$, then using equation (13.8) and setting $\phi = 1$ we find that the fire first becomes ventilation controlled when $Q > 390 \text{ MW}$. Furthermore, putting the value 390 000 kW into equation (13.9) gives $\beta = 0$, which implies that all the oxygen has been consumed in the fire.

When the oxygen concentration at a given distance downstream of the fire in a forced ventilation flow is essentially zero, the fire gradually changes from fuel controlled ($\phi > 1$) to ventilation controlled ($\phi < 1$). A good indication of when a fire has become ventilation controlled is when the ratio $\dot{m}_{CO}/\dot{m}_{CO_2}$, where $\dot{m}_{CO}$ is the mass flow rate of carbon monoxide (CO) and $\dot{m}_{CO_2}$ the mass flow rate of CO_2, begins to increase considerably. Tests show that the ratio $\dot{m}_{CO}/\dot{m}_{CO_2}$ increases exponentially as the fire becomes ventilation controlled, for both diffusion flames of propane and propylene and for wood crib fires (Tewarson, 1988). Tewarson (1988) presents a correlation between the ratio $\dot{m}_{CO}/\dot{m}_{CO_2}$ and ϕ which shows that wood crib fires become ventilation controlled when the ratio $\dot{m}_{CO}/\dot{m}_{CO_2} > 0.036$, and gas diffusion flames become ventilation controlled when $\dot{m}_{CO}/\dot{m}_{CO_2} > 0.1$. The value of $\dot{m}_{CO}/\dot{m}_{CO_2}$ can therefore be used as an indicator of the combustion conditions measured in tunnel fires. This ratio can also be expressed as:

$$\frac{\dot{m}_{CO}}{\dot{m}_{CO_2}} = \frac{M_{CO}X_{CO}}{M_{CO_2}X_{CO_2}} = 0.636\frac{X_{CO}}{X_{CO_2}} \tag{13.10}$$

where X is the concentration (or mole fraction) and M is the molecular weight (M is 28 g/mol for CO and 44 g/mol for CO_2). Thus, the limits for ventilation control when using this equation are $X_{CO}/X_{CO_2} > 0.057$ for wood cribs and 0.157 for gas diffusion flames.

13.5. Stratification of smoke in tunnels
In this section we discuss fuel-controlled fires where the smoke will stratify. The stratification of smoke has important implications for those who have to escape from the tunnel. The characteristics of the smoke spread are highly dependent on the air velocity inside the tunnel. In order to

illustrate this, we can identify three typical ranges (groups) of air velocity

■ low or no forced air velocity (0–1 m/s)
■ moderate forced air velocity (1–3 m/s)
■ high forced air velocity (>3 m/s).

In the low forced air velocity range, the stratification of the smoke is usually high in the vicinity of the fire source. This group normally includes tunnels with natural ventilation. The backlayering distance of the smoke is relatively long, and in some cases the smoke travels nearly uniformly in both directions (Figure 13.4a). When the velocity increases and is close to about 1 m/s, back-layering of the smoke occurs upstream of the fire source, at a distance which can be of the order of 17 times the height of the tunnel (see Figure 13.4b). A correlation for the backlayering distance in tunnels L_b can be obtained according to the original work carried out by Thomas (1958, 1968) and Vantelon *et al.* (1991):

$$\frac{L_b}{H} \propto \left(\frac{gQ}{\rho_a c_p T_a u^3 H} \right)^{0.3} \tag{13.11}$$

where H is the tunnel height (m), Q is the heat-release rate (kW), ρ_a is the ambient density (kg/m^3), T_a is the ambient temperature (K), u is the air velocity, g is the acceleration due to gravity, and c_p is the specific heat of air at constant pressure (kJ/kg K). The only 'tunnel' dimension in equation (13.11) is the height H; no other geometrical parameters, such as the tunnel width or radius, are considered. The proportionality constant has to be determined from experiments, but there are limited experimental data available in the literature, especially for large-scale tests. The model-scale tests presented by Vantelon *et al.* (1991), Guelzim *et al.* (1994), Oka and Atkinson (1995), and Saito *et al.* (1995) can be used to estimate this proportionality constant. The proportionality constant is found to vary between 0.6 and 2.2 for model-scale tests. Due to the large variation in the proportionality constant for model-scale tests and the lack of experimental data from large-scale tests, it is difficult to know the validity of the constant. An average value (based on model-scale tests) that can be used until further large-scale data are available is 1.4. Equation (13.11) should not be used for velocities below 0.5 m/s or higher than 3 m/s (the 'critical velocity'). The effect of the geometrical shape of the cross-section and the tunnel slope on the backlayering distance have not been investigated for this correlation.

In the moderate forced air velocity group, the stratification in the vicinity of the fire is strongly affected by the air velocity, particularly at the higher velocities. This group normally includes tunnels with natural ventilation or forced ventilation. The backlayering distance can vary between zero up to 17 times the tunnel height (see Figure 13.4c).

In the high forced air velocity group, there is usually low stratification of the smoke downstream of the fire, and usually no backlayering (see Figure 13.4d). This group normally includes tunnels with forced ventilation. The actual velocity, which is required to prevent any backlayering, is sometimes called the 'critical velocity'. The concept of critical velocity is discussed more thoroughly in Chapter 9.

Example 5
What is the backlayering distance after 10 minutes in the case of the HGV (semi-trailer) described in Example 3? The tunnel is 6 m high and 9 m wide, and the longitudinal ventilation inside the

Figure 13.4 Schematic sketches showing the smoke stratification in tunnels having different forced air velocities (u): (a) u typically 0–0.3 m/s (very low); (b) $u \approx 1$ m/s (low); (c) $u = 1$ to 3 m/s (moderate); (d) typically $u > 3$ m/s (i.e. a wind velocity greater than the critical wind velocity u_c)

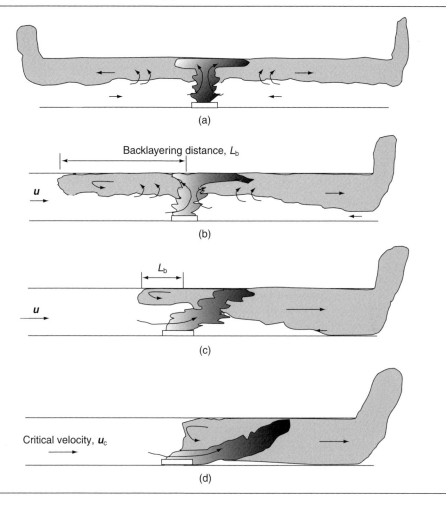

tunnel is 2 m/s. The fire was estimated to grow linearly, and to reach a peak heat-release rate of 120 MW after 10 minutes.

Solution. Putting $H = 6$ m, $\rho_a = 1.2$ kg/m³, $T_a = 293$ K, $c_p = 1$ kJ/kg K, $Q = 120\,000$ kW and $u = 2$ m/s into equation (13.11), and assuming a proportionality constant of 1.4, the backlayering distance is:

$$L_b = 1.4 \times 6 \times \left(\frac{9.81 \times 120\,000}{1.2 \times 1 \times 293 \times 2^3 \times 6} \right)^{0.3} = 30 \text{ m}$$

The stratification downstream of the fire is a result of the mixing process between the cold airstream and the hot plume flow created by the fire. The phenomenon is three-dimensional in

Figure 13.5 The principal flow pathways in the vicinity of a fire plume

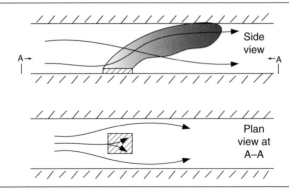

the region close to the fire plume. The principal pathways of the two density flows are illustrated in Figure 13.5. The gravitational forces tend to suppress the turbulent mixing between the two different density flows.

This explains why it is possible for cold unreacted air to bypass or pass beneath the fire plume without mixing, even though the flow is turbulent. The longitudinal aspect of the fuel involved in the fire may, therefore, play an important role in the mixing process between the longitudinal flow and the fuel vapours generated by the fire.

The type of characterisation of the smoke stratification given earlier can only provide limited information. Newman (1984) has shown for duct fires that there is a correlation between the local temperature stratification and the local mass concentration of chemical compounds. Furthermore, Ingason and Persson (1999) have shown that, in tunnels, there is a correlation between the local optical density of the smoke (or visibility) and its local density (or temperature), and the oxygen concentration. Therefore, it is reasonable to assume that there is a correlation between the local temperature stratification, as given by Newman (1984), and the composition of the gases (CO, CO_2, O_2 etc) and the smoke stratification in tunnels. The temperature stratification is, however, not only related to the air velocity, but also to the heat-release rate and the height of the tunnel. These parameters can actually be related through the local Froude number (Fr) or Richardson number (Ri). Newman (1984) presented a very simple method for assessing the local gas-stream stratification in a duct fire using the local Froude number.

Newman (1984) was able to define three distinct regions of temperature stratification.

- Region I, Fr $\leqslant$ 0.9: in this region the stratification is extremely marked. Hot combustion products travel along the ceiling, while the gas temperature near the floor is essentially ambient. This region consists of a buoyancy-dominated temperature stratification.
- Region II, 0.9 $\leqslant$ Fr $\leqslant$ 10: this region is dominated by a strong interaction between the imposed horizontal flow and the buoyancy forces. Although there is no strong stratification or layering, there are vertical temperature gradients, and this region is mixture controlled. In other words, there is significant interaction between the ventilation velocity and the fire-induced buoyancy.
- Region III, Fr $>$ 10: in this region the vertical temperature gradients are insignificant, and consequently there is insignificant stratification.

Figure 13.6 Illustration of the three different temperature stratification regions according to the definition given by Newman (1984)

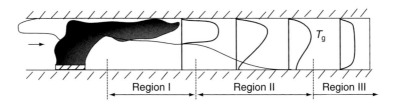

The three temperature-stratification regions are illustrated in Figure 13.6.

Newman (1984) also established correlations for the excess temperatures near the ceiling ($0.88H$) and floor ($0.12H$) for the different regions of stratification, based on weighted average of the gas temperature T_{avg} and the gas velocity u_{avg} over the entire cross-section. He postulated that these correlations could be used for various applications, such as in assessing flame spread and detector response. These correlations have not been tested for tunnel fires, and are therefore not presented here. The majority of the tests used to identify the stratification regions were performed in a large (rectangular) duct measuring 2.4 m wide (B), 2.4 m high (H) and 47.6 m long (L), and with $D_h/L = 19.8$ m, where D_h is the hydraulic diameter. The hydraulic diameter is defined as $D_h = 4A/P$, where A is the cross-sectional area (m^2) and P is the perimeter (m). Newman's method for identifying the temperature-stratification regions appears to be very rational and simple, but it has not been tested for tunnel fires. However, assuming that Newman's correlations for identifying the stratification regions are appropriate for tunnel fires, Fr can be calculated from the equation:

$$\text{Fr} = \frac{u_{avg}^2}{1.5\left(\Delta T_{avg}/T_{avg}\right)gH} \tag{13.12}$$

where H is the ceiling height, T_{avg} is the average gas temperature (K) over the entire cross-section at a given position, $\Delta T_{avg} = T_{avg} - T_a$ (the average gas temperature rise (K) above ambient over the entire cross-section at a given position), and $u_{avg} = uT_{avg}/T_a$ (m/s). The values of T_{avg} and u_{avg} and equation (13.12) can be used to identify the stratification regions I, II and III. The equations for calculating T_{avg} and ΔT_{avg} are presented in the next section.

13.6. Average flow conditions in longitudinal flow

This section gives the equations for calculating the average gas temperature at different locations downstream of the fire location, the average gas velocity due to buoyancy effects (hot smoke stack), gas concentrations, visibility and heat-release rates in one-dimensional tunnel flow. If, due to the longitudinal gas velocity, the fire gases (bulk flow) are assumed to be completely mixed, the average gas temperature, gas velocity and gas concentration, as a function of the distance x from the fire, may be calculated.

13.6.1 Gas temperature and velocity

The governing energy equation for one-dimensional bulk flow in a tunnel can be expressed as:

$$\dot{m}_a c_p \frac{dT_{avg}}{dx} = h_c P(T_{avg} - T_w) + \varepsilon_{g-w} F_w P\sigma(T_{avg}^4 - T_w^4) \tag{13.13}$$

where h_c is the average convected-heat transfer coefficient to the walls (kW/m^2 K), P is the perimeter of the tunnel (m), T_w is the wall temperature (K), F_w is a view factor, ε_{g-w} is the effective emissivity between the gas and the walls, and σ is the Stefan–Boltzmann constant (kW/m^2 K^4). This equation can only be solved numerically. It is possible to obtain a simple analytical solution by assuming a lumped heat-transfer coefficient h for both the convected and radiated heat losses. This can be done by replacing h_c with $h = h_c + h_r$ and ignoring the radiated heat term in equation (13.13). Thus, assuming that the wall temperature is the same as the ambient temperature T_a, equation (13.13) can be solved analytically (Persson, 2000a):

$$T_{avg}(x,\ t) = T_a + \left[T_{avg,\ x=0}(\tau) - T_a\right] e^{-(hPx/\dot{m}_a c_p)} \tag{13.14}$$

where $T_{avg}(x,t)$ is the average temperature across the whole cross-section of the tunnel at a distance x from the fire and at time t, and $T_{avg,\ x=0}(\tau)$ is the average gas temperature at the fire source ($x=0$) at time $\tau = t - (x/u)$. This means that it will take a time x/u to transport the heat generated at the fire location to the distance x. This may become important for growing fires and at long distances from the fire. Here we have not considered the effects of increased velocity on the transport time x/u due to increased gas temperature (especially close to the fire), i.e. the velocity varies along x. A suitable value for the lumped heat-transfer coefficient h for tunnels varies in the range 0.02–0.04 kW/m^2 K. The average gas temperature over the entire cross-section at the fire location can be obtained according to the equation:

$$T_{avg,\ x=0}(\tau) = T_a + \frac{2}{3} \frac{Q(\tau)}{\dot{m}_a c_p} \tag{13.15}$$

where $T_{avg,\ x=0}(\tau)$ is the average gas temperature over the entire cross-section at $x=0$ and at time τ. Here, τ is the time at which the gas, which reaches position x at time t, starts to flow from position $x=0$. This assumes that the fire is centred on position $x=0$ and starts at time $t=0$. This gas temperature is usually much lower than the ceiling temperature in the vicinity of the fire, due to the turbulent mixing effects of the buoyant flow (see Figure 13.5). In large tunnel fires (>50 MW), the ceiling gas temperatures may become as high as 900–1200°C, depending on the ventilation conditions and fuel load. Here we assume that one-third of the total heat-release rate is radiated away from the fire. Now, using equations (13.14) and (13.15), we can calculate the average temperature over the entire cross-section T_{avg} (K) and the average rise in the gas temperature above ambient $\Delta T_{avg} = T_{avg} - T_a$ (K) at different distances from the fire. Consequently, Fr can be calculated according to equation (13.12), assuming that $u_{avg} = uT_{avg}/T_a$ at x, in order to establish the stratification regions I, II and III.

Example 6

What is the stratification region after 10 minutes (600 s) at $x = 150$ m from the fire location for a growing fire like the one considered in Example 3? The longitudinal ventilation is 2 m/s, the ambient temperature T_a is 20°C, the tunnel geometry is $H = 6$ m and $B = 9$, and the fire grows linearly and reaches a peak heat-release rate of 120 MW after 10 minutes. (Note: be careful to distinguish between degrees celsius (°C) and degrees kelvin (K).)

Solution. First we calculate $\tau = t - (x/u) = 600 - (150/2) = 525$ s, which means that $Q(\tau) = (120\,000/600) \times 525 = 105\,000$ kW, as the linear fire-growth constant is $120\,000/600 = 200$ kW/s. Thus, the average temperature at $x=0$ becomes

$$T_{avg,\ x=0}(525) = 20 + \frac{2}{3} \frac{105\,000}{130 \times 1} = 560°C$$

where $\dot{m}_a = 1.2 \times 2 \times 6 \times 9 = 130$ kg/s. The reader must be aware that the corresponding ceiling temperature may be much higher than 560°C ($\sim$1000°C) but, as this is an average bulk temperature used as input to determine the Froude number, it is acceptable. The average temperature 150 m from the fire at time $t = 600$ s can be calculated from equation (13.14):

$$T_{avg}(150,600) = 20 + [560 - 20]\, e^{-((0.025 \times 30 \times 150)/(130 \times 1))} = 247°C$$

where $h = 0.025$ kW/m^2 K, $P = 30$ m, $x = 150$ m, $\dot{m}_a = 130$ kg/s, $T_a = 20°C$ and $c_p = 1$ kJ/kg K. Thus, the Froude number can be determined from equation (13.12):

$$Fr = \frac{\{2 \times [(247 + 273)/(20 + 273)]\}^2}{1.5[(247 - 20)/(247 + 273)] \times 9.81 \times 6} = 0.33 < 0.9$$

This corresponds to region I, i.e. the smoke is still severely stratified 150 m downstream of the fire. If the distance is increased to $x = 593$ m, the average temperature drops to 37°C and $Fr = 0.9$, i.e. the stratification corresponds to region II.

A fire in a sloped tunnel creates a natural draught (natural ventilation) due to the buoyancy forces. If there is no external wind or forced longitudinal ventilation in the tunnel, and it is assumed that the ambient temperature outside (T_a) and inside the tunnel (T_0) is the same, the average 'cold' gas velocity u (bulk) can be obtained from Bernoulli's equation:

$$u = \sqrt{\frac{2g\Delta h_s[1 - (T_a/\bar{T})]}{(\bar{T}/T_a) + K_1 + (f_D/D_h)\{L + L_s[(\bar{T}/T_a) - 1]\}}} \qquad (13.16)$$

where Δh_s is the hot smoke-stack height (m) (Figure 13.7), $\bar{T}$ is the average temperature (K) over the hot smoke stack height Δh_s, T_a is the ambient temperature (K), f_D is the friction factor, L is the tunnel length (m), L_s is the length of the tunnel exposed to the hot smoke (m), D_h is the hydraulic diameter of the tunnel (m), and K_1 is the pressure loss coefficient at the entrance of the tunnel ($K_1 = 0.5$). If the entire tunnel is filled with hot gases, the hot smoke-stack height can be obtained from $\Delta h = \Delta h_s = \theta L/100$, where θ is the tunnel slope, expressed as a percentage, and L is the

Figure 13.7 Definition of the hot smoke-stack height Δh_s and the smoke stack length L_s

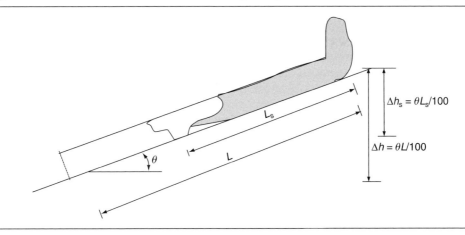

tunnel length (m). If only part of the tunnel is filled with smoke, $\Delta h_s = \theta L/100$, and θ is the corresponding slope of the smoke-filled area, expressed as a percentage.

The friction factor f_D is dependent on the Reynolds number, and for most tunnels varies between 0.015 and 0.03. The average temperature $\bar{T}$ over the hot smoke-stack length L_s can be estimated using the equation (Persson, 2000b):

$$\bar{T} = T_0 + \frac{2}{3}\frac{Q}{PhL_s} \qquad (13.17)$$

where $\bar{T}$ is the average gas temperature (°C) over the stack height and T_0 is the ambient temperature (°C) within the tunnel. Equation (13.16) requires that $T_a/\bar{T} \leqslant 1$, i.e. the average temperature inside the tunnel $\bar{T}$ is greater or equal to ambient temperature outside the tunnel T_a.

Example 7

What is the average 'cold' gas velocity u in a sloped tunnel with a 50 MW fire in the middle of the tunnel. The tunnel slope is 2%, the tunnel length is 2 km, the tunnel geometry is $H = 6$ m, $B = 9$ m and $T_a = T_0 = 20°C$.

Solution. The length of the hot smoke stack L_s is 1000 m, as the fire is located in the middle of the tunnel (see Figure 13.7). Thus $\Delta h_s = (2 \times 1000)/100 = 20$ m and the average temperature $\bar{T}$ is determined using equation (13.17):

$$\bar{T} = 20 + \frac{2}{3}\frac{50\ 000}{30 \times 0.025 \times 1000} = 64.4°C$$

assuming $h = 0.025\ \text{kW/m}^2\ \text{K}$, $f_D = 0.02$ and $K_1 = 0.5$. Other geometrical parameters are $D_h = 7.2$ m, $P = 30$ m and $L_s = 1000$ m. Thus, using equation (13.16) the average 'cold' gas velocity u is determined as:

$$u = \sqrt{\frac{2 \times 9.81 \times 20\{1 - [(273 + 20)/(273 + 64.4)]\}}{[(273 + 64.4)/(273 + 20)] + 0.5 + (0.02/7.2) \times (2000 + 1000\{[(273 + 64.4)/(273 + 20) - 1]\})}} = 2.6\,\text{m/s}$$

The average gas velocity in the hot smoke stack will be $[(273 + 64.4)/(273 + 20)] \times 2.6 = 3.0$ m/s.

13.6.2 Gas concentrations

The average mole fraction $X_{i,\,\text{avg}}$ of CO_2, CO or HCN over the cross-section of the tunnel and at a certain position downstream of the fire can be obtained using the general equation:

$$X_{i,\,\text{avg}} = Y_i\frac{M_a}{M_i}\frac{Q(\tau)}{\dot{m}_a\chi H_T} \qquad (13.18)$$

assuming $\dot{m}_a \approx \dot{m}_g$ where $\dot{m}_g$ is the mass flow rate of combustion gases. Here M_a is the molecular weight of air, M_i is the molecular weight of chemical species i, and Y_i is the mass yield of species i for well-ventilated fires. The value of $X_{i,\,\text{avg}}$ can be converted into a percentage by multiplying by 100. The yields of Y_{CO_2}, Y_{CO} and Y_{HCN} for well-ventilated conditions can be obtained as, for example, in Tewarson (1988, 1995), for different fuels. Some values for different fuels in well-ventilated conditions are given in Table 13.1.

Table 13.1 Yields Y of CO_2, CO, HCN and smoke, the mass optical density D_{mass}, and the effective heat of combustion H_{ec} $(=\chi H_T)$ for well-ventilated fires (Tewarson, 1988, 1995)

Material	Yield, Y: kg/kg					
	CO_2	CO	HCN	Smoke[e]	D_{mass}[e]: m²/kg	H_{ec}: MJ/kg
Wood[a]	1.27	0.004	–	0.015	37	12.4
Rigid polyurethane foam[b]	1.50	0.027	0.01	0.131	304[a]	16.4
Polystyrene[c]	2.33	0.06	–	0.164	335	27
Mineral oil[d]	2.37	0.041	–	0.097	–	31.7

[a] Table 1-13.7 in Tewarson (1988) and 3-4.11 in Tewarson (1995).
[b] Table 1-13.11 in Tewarson (1995).
[c] Table 1-13.7 in Tewarson (1988) and 3-4.11 in Tewarson (1995).
[d] Table 3-4.11 in Tewarson (1995).
[e] Parameters defined in Section 13.6.3.

The average mole fraction of oxygen in the combustion gases at a certain position x from the fire can be approximated using the equation:

$$X_{O_2, \text{avg}} = 0.2095 - \frac{M_a}{M_{O_2}} \frac{Q(\tau)}{\dot{m}_a 13\,100} \tag{13.19}$$

assuming $\dot{m}_a \approx \dot{m}_g$ and 0.2095 is the mole fraction of oxygen in ambient air, M_a is the gram molecular weight of air (28.95 g) and M_{O_2} is the gram molecular weight of oxygen (32 g) and 13 100 kJ/kg is the energy released per unit mass of oxygen.

Example 8
What is the average mole fraction of CO, CO_2 and O_2 at $x = 520$ m for a linearly growing fire $(Q(t) = 200 \times t)$ of a HGV fully loaded with piled wood pallets at time $t = 600$ s (10 minutes) with a longitudinal ventilation of 2 m/s?

The tunnel geometry is $H = 6$ m, $B = 9$ m and $T_a = 20°C$.

Solution. First calculate $\tau = t - (x/u) = 600 - (520/2) - 340$ s, which means that $Q(\tau) = 200 \times 340 = 68\,000$ kW. According to Table 13.1, for wood the yields of CO_2 and CO are 1.27 and 0.004 kg/kg, respectively.

Thus, the mole fractions of CO_2 and CO can be calculated using equation (13.18) and the effective heat of combustion ($H_{ec} = 12.4$ MJ/kg) from Table 13.1:

$$X_{CO_2, \text{avg}} = 1.27 \frac{28.95}{44} \frac{68\,000}{130 \times 12\,400} = 0.035 \text{ or } 3.5\% \text{ } CO_2$$

where $M_{CO_2} = 44$ g and $\dot{m}_a = 1.2 \times 2 \times 6 \times 9 = 130$ kg/s; and

$$X_{CO, \text{avg}} = 0.004 \frac{28.95}{28} \frac{68\,000}{130 \times 12\,400} = 0.00017 \text{ or } 1.7\,\text{ppm} \text{ } (0.017\%) \text{ CO}$$

where $M_{CO} = 28$ g. The mole fraction of oxygen can be calculated using equation (13.19):

$$X_{O_2, avg} = 0.2095 - \frac{28.95}{32} \frac{68\,000}{130 \times 13\,100} = 0.173 \text{ or } 17.3\% \text{ } O_2$$

where $M_{O_2} = 32$ g.

This means that 82.6% ($17.3/20.95 \times 100$) of the available oxygen remains unreacted and 17.4% has been consumed. If we use equation (13.9) we obtain identical results, or $\beta = [130 - (68\,000/3000)]/130 = 0.826$, which means that 82.6% of the oxygen remains unreacted.

13.6.3 Visibility

The practical use of the models for calculating visibility is to determine the visibility along escape routes and thus to assist in estimating escape times. This may be compared with the time to 'untenable conditions'. Equations used to calculate visibility are given in this section. The theoretical approach given here is based on the work presented in Jin and Yamada (1985), Tewarson (1988) and Östman (1992).

The visibility in smoke can be related to the extinction coefficient C_s (Jin and Yamada, 1985) using the equation (Östman, 1992):

$$C_s = \frac{OD}{L} \log_e(10) \tag{13.20}$$

where OD is the optical density (defined as $\log_{10}(I_o/I)$, where I_o is the intensity of incident light and I is the intensity of light transmitted through the mass of smoke), and L is the path length of light through the smoke. Note that C_s as defined here (taken from Jin and Yamada (1985)), is equivalent to the extinction coefficient as defined in the Lambert–Beer law (Drysdale, 1992) multiplied by the mass concentration of smoke particles. The optical density per unit optical path length can also be expressed as (Tewarson, 1988):

$$\frac{OD}{L} = \xi Y_s \dot{m}_f / \dot{V}_T \tag{13.21}$$

where ξ is the specific extinction coefficient of smoke or particle optical density (m^2/kg), Y_s is the yield of smoke (kg/kg) (see e.g. Table 13.1), $\dot{m}_f$ denotes the rate of loss of the mass of the fuel (kg/s) and $\dot{V}_T$ is the total volumetric flow rate ($\dot{V}_T = uA$ in a tunnel) of the mixture of fire products and air (m^3/s) at the actual location x. The parameter ξY_s is defined as mass optical density D_{mass} (m^2/kg), and is compiled by Tewarson (1988) for a number of different materials (see also Table 13.1). The $\dot{m}_f$ can also be written as $\dot{m}_f = Q(\tau)/H_{ec}$, where $Q(\tau)$ is the heat-release rate (kW) at the actual location x, and $H_{ec} = \chi H_T$ is the effective heat of combustion (kJ/kg). Thus, equation (13.21) becomes:

$$\frac{OD}{L} = D_{mass} \frac{Q(\tau)}{uAH_{ec}} \tag{13.22}$$

where u is the longitudinal gas velocity (m/s) and A is the area of the cross-section (m^2). We know also from Jin and Yamada (1985) that, for objects such as walls, floors and doors in an underground arcade or long corridor, the relationship between the visibility through non-irritant smoke and the extinction coefficient is:

$$V = \frac{2}{C_s} \tag{13.23}$$

The visibility through irritant smoke would be significantly less. The work reported by Jin (1997) suggests that the visibility effectively drops to zero for a value of C_s above about 0.55 m^{-1} for a lit

Table 13.2 Mass optical density D_{mass} values obtained experimentally from burning vehicles[a]

Type of vehicle		Mass optical density, D_{mass}: m²/kg
Road	Car (steel)	381
	Car (plastic)	330
	Bus	203
	Truck	76–102
	Subway (steel)	407
	Subway (aluminium)	331
Rail	IC type (steel)	153
	ICE type (steel)	127–229
	Two joined half-vehicles (steel)	127–178

[a] The values given here are recalculated from data presented by Steinert (SP, 1994).

fire-exit sign within irritant smoke. In addition, note that C_s may be simply related to the 'obscura' unit of Rasbash and Pratt (1979), i.e. N obscuras $= 10C_s/2.3$. One obscura corresponds to approximately 8–10 m visibility in non-irritant smoke for an object that is not self-illuminated.

By combining equation (13.20) and equations (13.22) and (13.23) a correlation is obtained between the visibility V and the heat-release rate $Q(\tau)$ in a tunnel at an actual position downstream of the fire:

$$V = 0.87 \frac{uAH_{ec}}{Q(\tau)D_{mass}} \tag{13.24}$$

This equation can be used to determine the visibility in tunnel fires, depending on the fuel load, longitudinal velocity and tunnel area. However, the equation gives only the average value over the cross-section of the tunnel at different locations of x. In addition, it must be remembered that the effective visibility in irritant smoke would be expected to be much less than that calculated using equation (13.24).

There is not much information available in the literature on the mass optical density D_{mass} for vehicle fire loads. Table 13.2 gives values of D_{mass} for different types of vehicle, based on large-scale tests (EUREKA, 1995). Tewarson (1988) presents an extensive list of values of D_{mass} for a variety of fuels.

Example 9
What is the visibility (through non-irritant smoke) at $x = 520$ m for a linearly growing fire ($Q(t) = 200 \times t$ kW) of a HGV fully loaded with piled wood pallets at time $t = 600$ seconds (10 minutes) with a longitudinal ventilation of 2 m/s? The tunnel geometry is $H = 6$ m, $B = 9$ m and $T_a = 20°C$.

Solution. First we calculate $\tau = t - (x/u) = 600 - (520/2) = 340$ s, which means that $Q(\tau) = 200 \times 340 = 68\,000$ kW. According to Table 13.1, D_{mass} for wood is 37 m²/kg and $H_{ec} = 12.4$ MJ/kg. Thus, equation (13.24) yields

$$V = 0.87 \frac{2 \times 6 \times 9 \times 12\,400}{68\,000 \times 37} = 0.46 \text{ m}$$

13.7. Determination of heat-release rates in tunnel fires

In order to be able to determine whether a tunnel fire is fuel controlled or ventilation controlled, it is necessary to know the heat-release rate of the vehicles burning inside the tunnel. In tunnel tests or reconstructions of fire accidents with forced longitudinal ventilation, it is possible to make an estimate of the actual heat-release rate Q (kW) at the measurement point by using equation (13.25) (without correction for production of CO) and oxygen consumption calorimetry (Janssens and Parker, 1992).

13.7.1 Oxygen consumption calorimetry

Oxygen consumption calorimetry uses the equation:

$$Q(\tau) = 14\,330\dot{m}_a \left(\frac{X_{0,O_2}(1 - X_{CO_2}) - X_{O_2}(1 - X_{0,CO_2})}{1 - X_{O_2} - X_{CO_2}} \right) \tag{13.25}$$

where X_{0,O_2} is the volume fraction of oxygen in the incoming air (ambient) ($=0.2095$), and X_{0,CO_2} is the volume fraction of CO_2 measured in the incoming air (≈ 0.00033). X_{O_2} and X_{CO_2} are the volume fractions of oxygen and CO_2 downstream of the fire measured by a gas analyser (dry), and $Q(\tau)$ is the heat-release rate at the measuring station. The difference between $Q(\tau)$ and Q is related to the transport time (x/u). Therefore, there will be some delay in the time history of $Q(\tau)$ and Q, but the absolute values are expected to be the same.

Equation (13.25) has been modified from equation (13.9) by Janssens and Parker (1992). If X_{CO_2} has not been measured, equation (13.25) can be used by assuming $X_{CO_2} = 0$. This will simplify equation (13.25), and usually the error will not be greater that 10% for most fuel-controlled fires. In the derivation of equation (13.25), it is assumed that $\dot{m}_a = \rho_a uA$ and that 13 100 kJ/kg of heat is released per kilogram of oxygen consumed. It is also assumed that the relative humidity (RH) of the incoming air is 50%, the ambient temperature is 15°C, the amount of CO_2 in the incoming air is 330 ppm (0.033%), and the molecular weight of air M_a is 0.02895 kg/mol and that of oxygen M_{O_2} is 0.032 kg/mol. In addition, ρ_a is the ambient air density, u is the average longitudinal velocity upstream of the fire (m/s), and A is the cross-sectional area of the tunnel (m^2) at the same location that u is measured.

13.7.2 CO/CO$_2$ ratio technique

There are other calorimetric methods available to determine the heat-release rate. Grant and Drysdale (1997) estimated the heat-release rate in the largest experimentally measured tunnel fire to date, the HGV test in the EUREKA 499 test series in Norway (French, 1994; EUREKA, 1995). The HGV trailer was loaded with 2 tons of furniture, and the air velocity varied between 2.7 and 2.9 m/s when this peak occurred. The cross-section of the tunnel varied between 28 and 35 m^2. Grant and Drysdale estimated the maximum heat-release rate to be in the range of 121 to 128 MW, where the higher value was obtained using gas concentrations recorded at a measuring station located 30 m from the fire and the lower value using those recorded 100 m from the fire. The method of determining the heat-release rate is based on the CO/CO$_2$ ratio technique developed by Tewarson (1982b):

$$Q(\tau) = \dot{m}_{CO_2} \left(\frac{H_T}{k_{CO_2}} \right) + \dot{m}_{CO} \left(\frac{H_T - H_{CO}k_{CO}}{k_{CO}} \right) \tag{13.26}$$

where H_{CO} is the heat of combustion of CO (kJ/g); k_{CO_2} and k_{CO} are the maximum possible yields of CO_2 and CO (g/g), respectively; H_T/k_{CO_2} and $(H_T - H_{CO}k_{CO})/k_{CO}$ are approximately constant for various types of materials; and $\dot{m}_{CO_2}$ and $\dot{m}_{CO}$ are the generation rates of CO_2 and CO (g/s), respectively. Equation (13.26) can be rewritten as a function of the mass flow rate of air $\dot{m}_a$ (g/s), by assuming that $\dot{m}_a \approx \dot{m}_g$ (the mass flow rate of combustion gases):

$$Q(\tau) = \dot{m}_a \left[\frac{M_{CO_2}}{M_a} X_{CO_2} \left(\frac{H_T}{k_{CO_2}} \right) + \frac{M_{CO}}{M_a} X_{CO} \left(\frac{H_T - H_{CO}k_{CO}}{k_{CO}} \right) \right] \tag{13.27}$$

where $M_{CO_2} = 44$ g, $M_{CO} = 28$ g and $M_a = 28.95$ g. Grant and Drysdale (1997) used mean values (mixed fuel load) of 12.5 and 7.02 kJ/g for H_T/k_{CO_2} and $(H_T - H_{CO}k_{CO})/k_{CO}$, respectively, when they determined the heat-release rate from the HGV test. X_{CO_2} and X_{CO} are the increases above ambient i.e. the 'background' levels should be subtracted from these measured values.

French (1994) has published some experimental data in graphical form from the HGV EUREKA test. According to his graphs, the maximum concentrations of CO_2 and CO can be estimated as 9.4% ($X_{CO_2} = 0.094$) and 0.08% ($X_{CO} = 0.0008$), respectively, around 18–19 minutes into the test. The oxygen measurements 100 m downstream of the fire were not available when the fire reached its peak value (the cable burned off). This means that the lowest oxygen concentration was not measured in the test 100 m downstream of the fire. It is, however, possible to estimate the oxygen concentration based on the CO_2 measurements. The author has, based on X_{O_2} and X_{CO_2} readings from other experiments in model scale tunnels (wooden cribs, heptane, plastics), estimated the lowest X_{O_2} value obtained in the EUREKA test 100 m downstream of the fire to be nearly 0.1 (10%), assuming $X_{CO_2} = 0.094$. The correlation used is based on dry measurements of the gas concentrations.

In the following example the heat-release rates obtained using equations (13.25) (oxygen-consumption technique) and (13.27) (CO/CO_2 ratio technique) are compared, with the aid of the results from the HGV EUREKA test (French, 1994; EUREKA, 1995).

Example 10
The maximum heat-release rate according to Grant and Drysdale (1997), using data measured at 100 m downstream of the fire, was 121 000 kW. According to equation (13.27), the air mass flow rate $\dot{m}_a$ is 67.8 kg/s. Here we used $Q(\tau) = 121$ MW, $M_a = 28.95$ g/mol, $M_{CO_2} = 44$ g/mol, $M_{CO} = 28$ g/mol, $X_{CO_2} = 0.094 - 0.00033 = 0.0937$ and $X_{CO} = 0.0008$, $H_T/k_{CO_2} = 12.5$ kJ/g and $(H_T - H_{CO}k_{CO})/k_{CO} = 7.02$ kJ/g. Now we can use equation (13.25) to estimate the heat-release rate according to the oxygen-consumption calorimetry:

$$Q(\tau) = 14\,330 \times 67.8 \left(\frac{0.2095 \times (1 - 0.094) - 0.1 \times (1 - 0.00033)}{(1 - 0.1 - 0.094)} \right)$$

$$= 108\,300 \text{ kW } (108.3 \text{ MW})$$

If we assume that no CO_2 measurements were performed, i.e. $X_{CO_2} = 0$, we will obtain $Q(\tau) = 118\,200$ kW (118.2 MW), which is 9% higher than the value obtained with the CO_2 measurements. This example shows that similar results are obtained with equations (13.25) and (13.27), despite the inaccuracy in the input data.

13.8. Influence of ventilation on the heat-release rate
The interaction of the ventilation flow and the rate of heat release has been investigated by Carvel *et al.* (2001). They found that the heat-release rate of a fully developed HGV fire, for a tunnel

similar in size to the Channel Tunnel, could increase in size by a factor of four for a ventilation flow rate of 3 m/s, and by a factor of ten for a ventilation flow rate of 10 m/s, in a single-lane tunnel. They also found that the fire-growth rate could increase by a factor of five for a ventilation flow rate of 3 m/s and by a factor of ten for a ventilation flow rate of 10 m/s, in a single lane tunnel. For further details of the study by Carvel *et al.* (2001), see Chapter 11.

The increase in the heat-release rate and fire-growth rate due to increased velocity is the result of more effective heat transfer from the flames to the fuel surface (e.g. tilted flames enhance the flame spread), and more effective transport of oxygen into the fuel bed, leading to enhancement of the mixing of oxygen and fuel. The fire may be locally underventilated under normal conditions (e.g. wooden cribs or densely packed goods), but in a forced ventilation flow the transport of oxygen to the underventilated regions enhances the combustion rate. Beyond a certain air velocity, however, convective cooling can compete with the heat transfer from the flames and the burning rate, and thereby the heat-release rate may be reduced (Egan and Litton, 1986). When, or if, this theoretical turning point occurs has not yet been experimentally documented for HGV fires in tunnels. For HGV fires it is probably not achievable, due to the intensive heat radiated from the flame volume to the fuel surface (DeRis, 2001). Therefore, one could expect that the peak heat-release rate can be estimated by using the heat-release rate per square metre of fuel area under well-ventilated conditions in medium- or small-scale calorimeter tests.

The influence of ventilation on the behaviour of tunnel fires was also addressed in the second Benelux Tunnel fire tests (Directoraat-Generaal, 2002). The cross-section of the second Benelux Tunnel is significantly larger than that of the Channel Tunnel. For a fire load of wooden cribs contained in a housing similar to a HGV trailer, the fire-development rate with ventilation (4–6 m/s) appeared to be about two times faster than the development without ventilation. The peak heat output was about 1.5 times higher. Ingason (2002) has also performed model-scale fire tests in a sloped tunnel (scale 1 : 35, with slopes of 5°, 10° and 30°), where the effects of natural ventilation were investigated for solid materials corresponding to a HGV fire load. Wooden cribs were used as the fire load, and the results indicate that if the cribs are densely packed the increase in the peak heat-release rate can be up to a factor of 1.5, and if not densely packed there was little change in the peak heat-release rates.

13.9. Flame length

The flame length in tunnel fires is very important when considering the fire spread between vehicles. The knowledge about flame length in tunnels is, at present, limited. Presented below is some of the information available in the literature on flame heights in general and flame lengths in tunnels.

It is well known that the height of luminous flames is governed by the fuel supply rate, the entrainment rate, and the fuel geometry and properties. Most liquids and solid fuels burn with luminous diffusion flames, where about 70% of the total energy is released as convective heat and about 30% is released as radiant heat. The characteristic yellow luminosity is the result of emissions from small carbon-based particles formed within the flame (soot particles). The net emissive power of the flame (radiated heat flux) will depend on the concentration of these particles and on the thickness of the flame. Most hydrocarbon fuel fires become optically thick when the diameter of the flames is 3 m or larger. This means that the highest radiant flux from these flames is measured under these conditions. On the other hand, for large hydrocarbon fires, a large portion of the radiant flux can be absorbed by the thick layer of smoke at the outer periphery

Figure 13.8 The cut-off flame height h_{cut} defined by Babrauskas (1980), and the horizontal flame length h_{hor} under a horizontal ceiling. Point A is the location of the flame tip in the absence of a ceiling

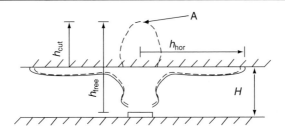

of the flame volume, resulting in lower radiant fluxes to the surroundings. However, the smoke occasionally opens up, exposing the hot flames, and a pulse of radiant heat is emitted to the surroundings (Mudan and Croce, 1988).

There are numerous correlations in the literature for flame heights in the open (see e.g. references (Beyler, 1986; McCaffrey, 1995)). The simplest one is that defined by McCaffrey (1979):

$$h_{free} = 0.2Q^{2/5} \tag{13.28}$$

The flame height is a function of the two-fifths power of the heat-release rate. This equation is valid for axisymmetric fires. There are two factors that must be taken into account when determining the flame lengths in tunnels: the presence of the ceiling and the ventilation.

13.9.1 The effect of the ceiling on flame length

Babrauskas (1980) reviewed the available information on the behaviour of flames under non-combustible ceilings, and showed that the horizontal flame extension can be related to the 'cut-off height' h_{cut}, as shown in Figure 13.8. His analysis relies on a large number of assumptions. He examined how much air must be entrained into the horizontal ceiling flame h_{hor} to burn the excess fuel volatiles, which flow into the ceiling jet. The entrainment rate into the horizontal ceiling jet is based on Alpert's (1971) correlations for unbounded ceiling jets. Babrauskas (1980) calculated the ratio h_{hor}/h_{cut}, i.e. the extension of the 'cut-off' flame above the ceiling, for different scenarios. The most interesting part of his work for tunnel applications is the case with an unbounded horizontal ceiling and that with a corridor. In the corridor case, the flames only extended in one direction. For the unbounded ceiling, the ratio, h_{hor}/h_{cut}, was found to be 1.5 (in all directions), while for the corridor case this ratio was highly dependent on the corridor width. Babrauskas (1980) illustrated this fact using two cases: one corridor of 3 m width, where the ratio h_{hor}/h_{cut} was 1.81; and one corridor of 2 m width, where the ratio was 2.94. Consequently, when there is a fire in a tunnel, it can be expected that the flames will extend along the ceiling to a 'cut-off' length that is increased by a factor of 1.5–3 depending on the tunnel width. One should consider, however, that in these examples the effect of the longitudinal ventilation has not been taken into account.

13.9.2 The effect of longitudinal ventilation on flame length

The presence of a high longitudinal velocity will enhance the turbulent mixing of the oxygen supply with the fuel supply (i.e. the pyrolysed volatile components), and thereby increase the

Figure 13.9 The flame length L_f for fires in tunnels with longitudinal ventilation. B, tunnel width; H, tunnel height

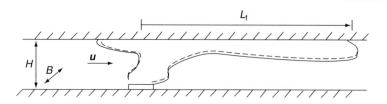

efficiency of combustion. This may result in a shortening of the flame. For moderate longitudinal velocities, however, the flame length may increase as the longitudinal velocity is increased. This flame lengthening can be explained by the fact that, at low air velocities, the pyrolysed volatiles must disperse over a longer distance before entraining sufficient oxygen for complete combustion. As the longitudinal velocity decreases, the flames become less horizontal and interact to a greater extent with the tunnel ceiling, thus further reducing the entrainment of oxygen into the flame. This will also reduce the flame surface area over which entrainment can occur (Rew and Deaves, 1999). If there is no ventilation flow within the tunnel (i.e. only natural convection), the flames will impinge on the tunnel ceiling and spread in both directions, as shown in Figure 13.9. In general we would expect that the entrainment into the flame along the ceiling is a function of the Richardson number.

Rew and Deaves (1999) have presented a model for the flame length for fires in tunnels, which includes the heat-release rate and longitudinal velocity, but not the tunnel width or height. Much of their work is based on the investigation of the Channel Tunnel fire in 1996 and test data from the HGV EUREKA 499 fire test (French, 1994) and the Memorial Tunnel tests (Massachusetts Highway Department, 1995). They defined the horizontal flame length L_f as the distance of the 600°C contour from the centre of the HGV or the pool, or from the rear of the HGV. In the model the flame length from the rear of the HGV is represented by the equation:

$$L_f = 20 \left(\frac{Q}{120} \right) \left(\frac{u}{10} \right)^{-0.4}$$

(13.29)

Note that in the above equation Q is in megawatts. This equation is a conservative fit to limited data obtained from the HGV EUREKA 499 test, as shown in Figure 13.10. Equation (13.29) tells us that the longitudinal ventilation shortens the flame length as the velocity increases. The effects of the velocity can be such that the flame length becomes equal to or less than the flame height given by equation (13.28). This is, of course, not reasonable, as we know that the presence of the tunnel ceiling will prolong the cut-off height by a factor of 1.5 and up to a factor of 3, depending on the tunnel width. Thus, we should expect lower influences of the velocity on the total flame length than equation (13.29) predicts. Furthermore, the work done by Babrauskas (1980) on flame heights under corridor ceilings indicates that the width and the height are important parameters. A possible way of representing both these parameters is to use the hydraulic diameter of the tunnel D_h.

If the data from the HGV EUREKA 499 test and the Memorial Tunnel tests are represented as the ratio of the flame length L_f and the hydraulic diameter of the tunnel D_h, a slightly better linear fit of the data is obtained. The best linear fit is obtained when the longitudinal velocity is

Figure 13.10 Comparison of the flame length data given from the HGV EUREKA 499 test and the Memorial tests 615B, 624B and 502 (Rew and Deaves, 1999)

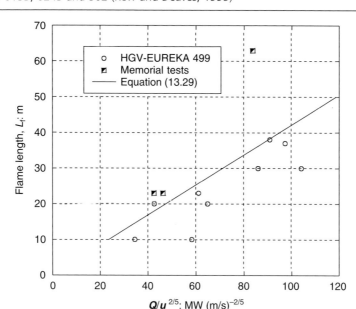

neglected, although the velocity will likely have some effects on the flame length (Figure 13.11). Together, this indicates that the tunnel geometry (width, height, shape) is a parameter that should be included, and the longitudinal velocity is less important than equation (13.29) indicates. The data used here are very sparse, and these results need to be tested by modelling and/or in large-scale tests.

Before leaving this section, it is worth noting the specific result that a flame length of approximately 95 m was estimated in one of the tests in the Runehamar series. Also, in comparing the Runehamar data with equation (13.29), it was found that the factor 20 needed to be changed to 37.9 (Lonnermark and Ingason, 2006) in order to obtain a reasonable fit. In the Runehamar tests a single forced ventilation velocity was used.

13.10. Large fires in tunnels with longitudinal flow
In this section, the situation when the vehicle density is so high that a tunnel fire can potentially spread between vehicles and become ventilation controlled in a tunnel with high longitudinal flow, is discussed more thoroughly.

Experiments in ducts with combustible wall linings show that ventilation-controlled fires occur more readily in narrower passages, having greater ventilation rates, greater fuel loads and larger ignition sources (de Ris, 1970). These factors imply that ventilation control occurs when the turbulent mixing in the combustion zone is high enough to completely engulf the air supply. This, in turn, indicates that the fire, at least for the majority of ordinary road and rail tunnels, has to spread to or involve at least two or more large vehicles with a large fuel load (e.g. train coaches) before it can become ventilation controlled.

Figure 13.11 The non-dimensional flame length as a function of the total heat-release rate for the HGV EUREKA 499 and Memorial Tunnel tests 615B, 624B and 502

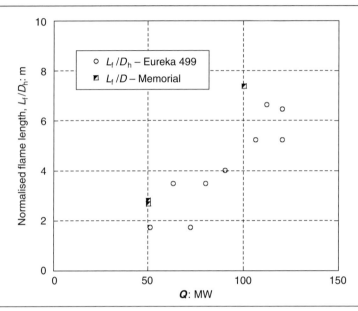

de Ris (1970) has shown schematically how fires become ventilation controlled in ducts with combustible wall linings. The description given by de Ris can be applied to the situation of vehicles fires in a road or rail tunnel. A ventilation-controlled tunnel fire with a relatively high forced longitudinal ventilation rate is shown schematically in Figure 13.12. The burning process can be viewed as stationary.

In order to explain the burning process, we assume five different zones, as shown in Figure 13.12

1. burnt-out cooling zone
2. glowing-ember zone

Figure 13.12 Schematic presentation of the burning process of a ventilation-controlled fire in a tunnel

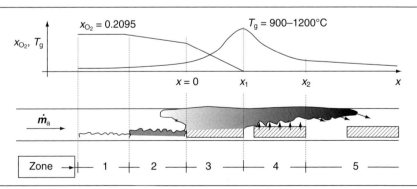

3. combustion zone
4. excess-fuel zone
5. preheating zone.

Provided that there are enough vehicles in the vicinity of the initial fire, these different zones move forwards in a dynamic manner. The burn-out cooling zone involves vehicles that have been completely consumed in the fire and the fire gases have cooled down. The glowing-ember zone contains vehicles at a very late stage of the decay phase (a pile of glowing embers).

The combustion zone, which starts at $x = 0$ in Figure 13.12, contains violently burning vehicles (fully developed fire) where enough fuel is vaporising to support gas-phase combustion. Flaming combustion is observed throughout this zone. The flames cause large heat-transfer rates from the gas to the fuel, and consequently high fuel-vaporisation rates. The gas-phase temperature just beyond $x = 0$ increases rapidly. Simultaneously, the oxygen is rapidly depleted as the temperature reaches a maximum at $x = x_1$, i.e. just beyond the combustion zone.

The excess-fuel zone starts at $x = x_1$, where all the oxygen has been depleted. Fuel vaporises from the vehicles throughout this zone, although no combustion takes place here due to lack of oxygen, up to a point along the tunnel where the gas stream has cooled to the fuel vaporisation (pyrolysis) temperature ($T_{vap} \geqslant 300°C$ for the majority of solid materials). Beyond this point, $x = x_2$, no vaporisation of the vehicle materials occurs, but the gas flows into a 'preheating zone', and loses its heat to the tunnel walls and preheats the vehicle material within this zone.

de Ris (1970) showed that the lengths of the combustion and excess-fuel zones are proportional to the forced ventilation rate when the fire becomes ventilation controlled, and Comitis et al. (1996) showed that the fire will propagate at a constant speed (when ventilation controlled), provided that there is enough combustible material available. This shows the importance of the forced long-itudinal ventilation on fire propagation within tunnels.

Delichatsios (1975) concluded that active burning in a fibreglass-reinforced plastic duct will take place up to a maximum length corresponding to $L_c/D_h = 10$, where D_h is the hydraulic diameter of the tunnel (m) and L_c is the combustion length (m). This means that the entrained oxygen will be engulfed and consumed within this section (combustion zone). This ratio can be used as an indi-cator of the order of magnitude of the length of the combustion zone for ventilation-controlled fires (with forced ventilation) in tunnels with a high vehicle density (e.g. HGVs in a long queue, or a fully developed fire in a train). Assuming that this number is reasonable for tunnels, the combustion zone, for a ventilation-controlled fire with forced ventilation, would not exceed 50–100 m for most common road and rail tunnels (assuming a high vehicle density) where D_h is in the range 5–10 m. It is, however, necessary to investigate experimentally the validity of this ratio for road and rail tunnels.

13.11. Fire spread in tunnels

Here we discuss the situation where there is a risk of fire spread between vehicles. Using the temperature equations given in previous sections, we can estimate the conditions required for a material to ignite at a certain distance from a fire. The possibility of ignition of an object is presently judged by evaluating whether or not the exposed surface would attain a critical ignition temperature. The critical temperature T_{cr} can be estimated as: 600°C for exposure to radiant heat and 500°C for exposure to convective heat in the case of spontaneous ignition; and 300–410°C for exposure to radiant heat and 450°C for exposure to convective heat in the case of piloted ignition. Note that these are approximate values, mostly deduced from experiments on small vertical specimens (Kanury, 1995).

Newman and Tewarson (1983) argue that, at ignition, $T_{cr} \approx T_{avg}$ (the average gas temperature) for duct flow, i.e. when T_{avg} has obtained a critical value, the material at that location will ignite. The critical temperature was estimated from the critical heat flux, $\dot{q}''_{cr} \approx \sigma T^4_{cr}$. This is an over-simplification, as it does not take into account the time dependence of the ignition process. According to Newman and Tewarson (1983), for many materials the critical level of radiated heat is in the range 12–22 kW/m² and the critical temperature $T_{cr} \approx T_{avg}$ is in the range 680–790 K (407–517°C). For example, for red oak the critical radiated heat level is 16 kW/m² ($T_{cr} = 457°C$), for polyethelyene it is 15 kW/m² ($T_{cr} = 447°C$), for polyurethane foam it is 16–22 kW/m² ($T_{cr} = 447–517°C$) and for polystyrene it is 12–13 kW/m² ($T_{cr} = 407–417°C$). The heat flux to the vehicle at a given position from the fire can be estimated from the equation:

$$\dot{q}'' = h_c(T_{avg} - T_a) + \sigma(T^4_{avg} - T^4_a) \tag{13.30}$$

Note that T_{avg} and T_a must be expressed in degrees Kelvin for this equation to be valid. This equation can be used to roughly estimate the heat flux level at different distances from the fire. The average gas temperature T_{avg} is obtained using equations (13.14) and (13.17).

Rew and Deaves (1999) postulated various mechanisms for fire spread based on work concerning the Channel Tunnel fire in 1996.

- Flame impingement: due to a low ventilation rate the flames are deflected by the presence of the ceiling, mainly in the direction of the ventilation flow. The flames visually 'crawl' along the ceiling above the vehicles.
- Surface spread: flame spread across the surface of the fire load.
- Remote ignition: the conditions at vehicles that are not very close to the initial fire reach the point of spontaneous ignition due to the high temperatures produced by the fire.
- Fuel transfer: liquid spread from leaking fuel tanks or debris downwind of the fire.
- Explosion: the explosion of fuel tanks may spread burning fuel to adjacent vehicles.

To this list might also be added the possibility of fire spread via burning brands.

Rew and Deaves (1999) concluded that, although all the mechanisms listed above may have contributed to fire spread in the Channel Tunnel incident, the primary mode of fire spread between wagons was due to flame impingement. They estimated the flame lengths, as presented in Section 13.9, as a function of the heat-release rate and the ventilation flow rate. They assumed that the gas temperature at the flame tip was 600°C and compared this with the criterion given by Heselden (1978) for the ignition of vehicles in tunnels (580°C). This value appears to be very close to the values given earlier for the spontaneous ignition of materials. If the vehicles are exposed for some time to temperatures in this range, they will ignite, without a piloted ignition (i.e. by spontaneous ignition). If the flames are in the vicinity of the pyrolysing material, the material will probably ignite at lower temperatures (piloted ignition).

Beard et al. (1995) developed a more precise method of estimating the fire spread by remote ignition between vehicles in tunnels. It is based on a non-linear model, the first version of which was FIRE-SPRINT A1, which predicts the conditions for fire spread from one object to another within a tunnel with longitudinal ventilation. It does this by identifying unstable states within the system, and associating this instability with the onset of fire spread. The model is based on average temperatures within a control volume, where the feedback of radiant heat from the hot gases in the control volume is essential to the development of the thermal instability. The instability is correlated with a jump in the temperature of the gases in the control volume. This jump is associated with the point at which spread from an initial fire to a target object would be expected. Beard et al. (1995) explain that it should not be assumed that this is necessarily the only possible mechanism for fire spread from one object to another in a tunnel. It is, however, assumed that this thermal instability corresponds to one mechanism of fire spread. The model is conceptually similar to the model FLASHOVER A1 (Beard, 2010), which predicts conditions for flashover in a compartment fire.

The tunnel model has been developed further by Beard, the second version (Beard, 1997) assuming greater flame extension than in the first version. The third version, FIRE-SPRINT A3 (Beard, 2006) assumes greater flame extension than in the second. The A-series models predict the critical rate of heat release for major fire spread in the absence of flame impingement (i.e. effectively assuming spontaneous ignition). Flame impingement has also been considered, and has led to the development of the model FIRE-SPRINT B1 (Beard, 2003). The FIRE-SPRINT models assume a tunnel similar in shape to the Channel Tunnel and a forced longitudinal ventilation. The Channel Tunnel has a cross-section with a perimeter corresponding to a segment of a circle. In the results published by Beard and co-workers (Beard, 1997, 2003, 2006; Beard et al., 1995) a radius equivalent to that of the Channel Tunnel has been assumed, although this is not essential to the model. For more details, see Chapter 16.

Example 11

What is the heat flux $\dot{q}''$ at $x = 100$ m for a linearly growing fire ($Q(\tau) = 200 \times t$) at time $t = 600$ s (10 minutes) with a longitudinal ventilation of 2 m/s? The tunnel geometry is $H = 6$ m, $B = 9$ m and $T_a = 20°C$.

Solution. First, we calculate $\tau = t - (x/u) = 600 - (100/2) = 550$ s, which means that $Q(\tau) = 200 \times 550 = 110\,000$ kW, as the linear fire growth constant is 200 kW/s. Thus, the average temperature can be calculated using equation (13.15) at $x = 0$:

$$T_{avg, x=0}(\tau) = 20 + \frac{2}{3}\frac{110\,000}{130 \times 1} = 566°C$$

(note that the ceiling temperature can be as high as 1000°C). The average temperature at 100 m from the fire at time $t = 600$ s can be calculated using equation (13.14):

$$T_{avg}(100, 600) = 20 + [566 - 20]\,e^{-((0.025 \times 30 \times 100/130 \times 1))} = 326°C$$

where $h = 0.025$ kW/m^2 K, $P = 30$ m, $x = 100$ m, $\dot{m}_a = 1.2 \times 2 \times 6 \times 9 = 130$ kg/s, $T_a = 20°C$ and $c_p = 1$ kJ/kg K. Thus, the heat flux can be calculated from equation (13.30):

$$\dot{q}'' = 0.01 \times (326 - 20) + 5.67 \times 10^{-11}((326 + 273)^4 - (20 + 273)^4) = 9.9 \text{ kW/m}^2$$

This is much lower than the critical radiated heat required to obtain spontaneous ignition of most materials.

It is of interest to compare this simple approach, very approximately, with the results obtained by Beard *et al.* (1995). Using FIRE-SPRINT A1, Beard *et al.* found that the critical heat-release rate for fire spread between HGVs in the Channel Tunnel was 55.2 MW, assuming a distance of 14.2 m between the HGVs, which is approximately the distance from the rear of one HGV in a carrier to the rear of a second HGV in a second carrier behind it in a train shuttle. The ventilation velocity was assumed to be 2 m/s (case 1). When the ventilation velocity was increased to 2.5 m/s, the critical heat-release rate was found to be 72.6 MW (case 2).

Using the same heat-release rates and longitudinal ventilation, and assuming a cross-section $B = 6.7 \times H = 6.7$ m and $A = 45$ m^2, the radiation fluxes at $x = 14.2$ m from the fire obtained using equations (13.14), (13.15) and (13.30) are $\dot{q}'' = 19.2$ kW/m^2 for case 1 and $\dot{q}'' = 22.3$ kW/m^2 for case 2. (The critical heat-release rates for fire spread obtained using FIRE-SPRINT A3 are lower than those found using FIRE-SPRINT A1. Also, the simulations reported by Beard and co-workers (Beard, 1997, 2003, 2006; Beard *et al.*, 1995) assume a tunnel size similar to that of the Channel Tunnel, which has a cross-section of about 38 m^2.) These calculated heat flux values are within the range of critical radiated heat values for different materials given by Newman and Tewarson (1983). Therefore, we can conclude that the results are reasonable for both models but, as there is still a great lack of full-scale experimental data for fire spread in tunnels, this type of calculation should be regarded as approximate at best. In the Runehamar test series, the critical heat-release rates for fire spread to a second object were estimated in some cases (Lonnermark and Ingason, 2006), and these have been compared with the values predicted using FIRE-SPRINT (Beard, 2007).

REFERENCES

Alpert R (1971) *Fire Induced Turbulent Ceiling-Jet*. Report 19722-2. Factory Mutual Research Corp., Norwood, MA.

Atkinson GT and Wu Y (1996) Smoke control in sloping tunnels. *Fire Safety Journal* **27**: 335–341.

Babrauskas V (1980) Flame lengths under ceilings. *Fire and Materials* **4**(3): 119–126.

Babrauskas V (1981) A closed-form approximation for post-flashover compartment fire temperatures. *Fire Safety Journal* **4**(1): 63–73.

Beard AN (1995) Predicting fire spread in tunnels. *Fire Engineers Journal* **55**(9): 13–15.

Beard AN (1997) A model for predicting fire spread in tunnels. *Journal of Fire Sciences* **15**: 277–307.

Beard AN (2003) Major fire spread in a tunnel: a non-linear model with flame impingement. In: Vardy AE (ed.), *Proceedings of the 5th International Conference on Safety in Road and Rail Tunnels, Marseille, France*, 6–10 October 2003, pp. 511–519.

Beard AN (2006) A theoretical model of major fire spread in a tunnel. *Fire Technology* **42**: 303–328.

Beard AN (2007) Major fire spread in tunnels: comparison between theory and experiment. *First International Tunnel Safety Forum for Road and Rail, Nice, France*, 23–25 April 2007, pp. 153–162.

Beard AN (2010) Flashover and boundary properties. *Fire Safety Journal* **45**: 116–121.

Beard AN, Drysdale DD and Bishop SR (1995) Non-linear model of major fire spread in a tunnel. *Fire Safety Journal* **24**(4): 333–357.

Beyler CL (1985) Major species production in solid fuels in a two layer compartment fire environment. In: Grant C and Pagni P (eds), *Fire Safety Science: Proceedings of the 1st International Symposium, NIST, USA*, 7–11 October 1985. International Association for Fire Safety Science, London, pp. 431–440.

Beyler C (1986) Fire plumes and ceiling jets. *Fire Safety Journal* **11**: 53–75.

Beyler C (1995) Flammability limits of premixed and diffusion flames. In: *SFPE Handbook of Fire Protection Engineering*, 2nd edn. National Fire Protection Association, Quincy, MA, pp. 2-147–160.

Bullen ML and Thomas PH (1979) Compartment fires with non-cellulosic fuels. *17th International Symposium on Combustion, The Combustion Institute, Pittsburgh, PA, USA*, pp. 1139–1148.

Carvel R, Beard A and Jowitt P (2001) How much do tunnels enhance the heat release rate of fires? *Proceedings of the 4th International Conference on Safety in Road and Rail Tunnels, Madrid, Spain*, 2–6 April 2001, pp. 457–466.

Carvel RO, Beard AN, Jowitt PW and Drysdale DD (2001) Variation of heat release rate with forced longitudinal ventilation for vehicle fires in tunnels. *Fire Safety Journal* **36**(6): 569–596.

Comitis SC, Glasser D and Young BD (1994) An experimental and modeling study of fires in ventilated ducts, Part I: Liquid fuels. *Combustion and Flame* **96**: 428–442.

Comitis SC, Glasser D and Young BD (1996) An experimental and modeling study of fires in ventilated ducts, Part II: PMMA and stratification. *Combustion and Flame* **104**: 138–156.

de Ris J (1970) Duct fires. *Combustion and Science Technology* **2**: 239–258.

de Ris JL (2004) Personal communication. Factory Mutual Global Research, USA.

Delichatsios MA (1975) Fire protection of fibreglass-reinforced plastic stacks in ducts. Report RC75-T-51, File, Serial No. 22493. Factory Mutual Systems, Norwood, MA.

Directoraat-Generaal (2002) *Project 'Safety Proef' Rapportage Brandproeven*. Ministerie van Verkeer en Waterstaat, Directoraat-Generaal Rijkswaterstaat, Den Haag.

Drysdale DD (1992) *An Introduction to Fire Dynamics*. Wiley, New York.

Egan MR and Litton CD (1986) *Wood Crib Fires in a Ventilated Tunnel*. Report of Investigations RI 9045. US Bureau of Mines.

EUREKA (1995) *Fires in Transport Tunnels: Report on Full-Scale Test*. EUREKA-Project EU499:FIRETUN. Studiensgesellschaft Stahlanwendung, Dusseldorf.

French S (1994) EUREKA 499 – HGV fire test (Nov. 1992). *Proceedings of the International Conference on Fires in Tunnels, Borås, Sweden*, 10–11 October 1994, pp. 63–85.

Grant GB (1988) *Thermally Stratified Forced Ventilation in Railway Tunnels*. PhD Thesis, Department of Civil Engineering, University of Dundee, UK.

Grant GB and Drysdale DD (1997) Estimating heat release rates from large-scale tunnel fires, fire safety science. In: Hasemi Y (ed.), *Fire Safety Science – Proceedings of the 5th International Symposium, Melbourne, Australia*, 3–7 March 1997. International Association for Fire Safety Science, London, pp. 1213 1224.

Guelzim A, Souil JM, Vantelon JP, Son DK, Gabay D and Dallest D (1994) Modelling of a reverse layer of fire-induced smoke in a tunnel. In: Kashiwagi T (ed.), *Fire Safety Science – Proceedings of the 4th International Symposium, Ottawa, Canada*, 13–17 July 1994. International Association for Fire Safety Science, London, pp. 277–288.

Heselden AJM (1978) Studies of fire and smoke behaviour relevant to tunnels. Building Research Establishment/Fire Research Station Report No. CP66.78. BRE, Watford.

Huggett C (1980) Estimation of rate of heat release by means of oxygen comsuption measurements. *Fire and Materials* **4**: 61–65.

Hwang CC and Wargo JD (1986) Experimental study of thermally generated reverse stratified layers in a fire tunnel. *Combustion and Flame* **66**: 171–180.

Ingason H (1995a) *Effects of Ventilation on Heat Release Rate of Pool Fires in a Model Tunnel*. SP Report 1995:55. SP Technical Research Institute of Sweden, Borås.

Ingason H (1995b) Heat release rate measurements in tunnel fires. In: Vardy AE (ed.), *Safety in Road and Rail Tunnels. International Conference, 2nd Proceedings*, 3–6 April 1995, Granada, Spain, pp. 261–268.

Ingason H (2002) Influence of Ventilation Rate on Heat Release Rate in Tunnel Fires. Draft Report. Swedish Fire Research Board (Brandforsk), Stockholm.

Ingason H (2007) Model scale railcar fire tests. *Fire Safety Journal* **42**: 271–282.

Ingason H and Persson B (1999) Prediction of optical density using CFD. In: Curtat M (ed.), *Fire Safety Science – Proceedings of the 6th International Symposium, Poitiers, France*, 5–9 July 1999. International Association for Fire Safety Science, London, pp. 817–828.

Ingason H, Nireus K and Werling P (1997) *Fire Tests in Blasted Rock Tunnel*. Report FOA-R-97-00581-990-SE. FOA, Linköping.

Janssens M and Parker B (1992) In: Babrauskas V and Grayson SJ (eds), *Oxygen Consumption Calorimetry, Heat Release in Fires*. Elsevier Applied Science, Oxford, pp. 31–59.

Jin T (1997) Studies on human behavior and tenability in fire smoke. In: Hasemi Y (ed.), *Fire Safety Science – Proceedings of the 5th International Symposium, Melbourne, Australia*, 3–7 March 1997. International Association for Fire Safety Science, London.

Jin T and Yamada T (1985) Irritating effects of fire smoke on visibility. *Fire Science and Technology* **5**(1): 79–89.

Kanury AM (1995) Flaming ignition of solid fuels. In: *SFPE Handbook of Fire Protection Engineering*, 2nd edn. National Fire Protection Association, Quincy, MA, p. 2-201.

Karlsson B and Quintiere JG (2000) *Enclosure Fire Dynamics*. CRC Press, Boca Raton, FL.

Lee CK, Chaiken RF and Singer JM (1979) Interaction between duct fires and ventilation flow: an experimental study. *Combustion Science and Technology* **20**: 59–72.

Lonnermark A and Ingason H (2006) Fire spread and flame length in large-scale tunnel fires. *Fire Technology* **42**: 283–302.

Massachusetts Highway Department (1995) *Memorial Tunnel Fire Ventilation Test Program – Test Report*. Massachusetts Highway Department and Federal Highway Administration, Boston, MA/Washington, DC.

McCaffrey B (1979) *Purely Buoyant Diffusion Flames:Some Experimental Results*. NBSIR 79-1910. National Bureau of Standards, Washington, DC.

McCaffrey B (1995) Flame height. *SFPE Handbook of Fire Protection Engineering*, 2nd edn. National Fire Protection Association, Quincy, MA, pp. 2-1–2-8.

Megret O and Vauquelin O (2000) Model to evaluate tunnel fire characteristics. *Fire Safety Journal* **34**(4): 393–401.

Mudan K and Croce P (1988) Fire hazard calculations for large open hydrocarbon fires. In: *SFPE Handbook of Fire Protection Engineering*, 1st edn. National Fire Protection Association, Quincy, MA, pp. 2–56.

Newman J (1984) Experimental evaluation of fire-induced stratification. *Combustion and Flame* **57**: 33–39.

Newman J and Tewarson A (1983) Flame propagation in ducts. *Combustion and Flame* **51**: 347–355.

Oka Y and Atkinson GT (1995) Control of smoke flow in tunnel fires. *Fire Safety Journal* **25**(4): 305–322.

Östman B (1992) Smoke and soot. In: Babrauskas V and Grayson SJ (eds), *Heat Release in Fires*. Elsevier Applied Science, Oxford, pp. 233–250.

Parker WJ (1984) Calculation of the heat release rate by oxygen consumption for various applications. *Journal of Fire Sciences* **2**: 380–395.

Persson B (2000a) Internal personal communication. SP Swedish National Testing and Research Institute, Borås, 7 November 2000.

Persson B (2000b) Internal personal communication. SP Swedish National Testing and Research Institute, Borås, 22 September 1999.

Quintiere JG and Rangwala AS (2003) A theory for flame extinction based on flame temperature. *Proceedings of Fire and Materials 2003, San Francisco, CA, USA*, 28–29 January 2003, p. 113.

Rasbash DJ and Pratt BT (1979) Estimation of the smoke produced in fires. *Fire Safety Journal* 2: 23–37.

Rew C and Deaves D (1999) Fire spread and flame length in ventilated tunnels, a model used in Channel Tunnel assessments. *Proceedings of Tunnel Fires and Escape from Tunnels, Lyon, France*, 5–7 May 1999, pp. 385–406.

Saito N, Yamada T, Sekizawa A, Yanai E, Watanabe Y and Miyazaki A (1995) Experimental study on fire behavior in a wind tunnel with a reduced scale model. *2nd International Conference on Safety in Road and Rail Tunnels, Granada, Spain*, 3–6 April 1995, pp. 303–310.

SP (1994) *Proceedings of the International Conference on Fires in Tunnels, Borås, Sweden*, 10–11 October 1994. Swedish National Testing and Research Institute (SP), Borås.

Tewarson A (1982a) *Analysis of Full-scale Timber Fire Tests in a Simulated Mine Gallery*. Factory Mutual Research, Fifth Annual and Final Report (1977–1982).

Tewarson A (1982b) Experimental evaluation of flammability parameters of polymeric materials. In: Lewin M (ed.), *Flame-Retardant Polymeric Materials*. Plenum Press, New York, pp. 130–142.

Tewarson A (1984) Fully developed enclosure fires of wood cribs. 20th *International Symposium on Combustion, The Combustion Institute, Pittsburgh, PA, USA*, pp. 1555–1566.

Tewarson A (1988) Generation of heat and chemical compounds in fires. *SFPE Handbook of Fire Protection Engineering*, 1st edn. National Fire Protection Association, Quincy, MA, pp. 1-179–1-199.

Tewarson A (1995) Generation of heat and chemical compounds in fires. In: *SFPE Handbook of Fire Protection Engineering*, 2nd edn. National Fire Protection Association, Quincy, MA, pp. 3-78–3-81.

Thomas PH (1958) *The Movement of Buoyant Fluid against a Stream and the Venting of Underground Fires*. Fire Research Note No. 351. Fire Research Station, Watford.

Thomas PH (1968) The Movement of Smoke in Horizontal Passages against an Air Flow. Fire Research Note No. 723. Fire Research Station, Watford.

Thomas PH, Bullen M, Quintiere JG and McCaffrey BJ (1980) Flashover and instabilities in fire behavior. *Combustion and Flame* 38: 159–171.

Vantelon JP, Guelzim A, Quach D, Son DK, Gabay D and Dallest D (1991) Investigation of fire-induced smoke movement in tunnels and stations: an application to the Paris Metro. In: Cox G and Langford B (eds), *Fire Safety Science – Proceedings of the 3rd International Symposium, Edinburgh, UK*, 8–12 July 1991. International Association for Fire Safety Science, London, pp. 907–918.

Wu Y and Bakar MZA (2000) Control of smoke flow in tunnel fires using longitudinal ventilation systems: a study of the critical velocity. *Fire Safety Journal* 35(4): 363–390.

Wu Y, Stoddard JP, James P and Atkinson GT (1997) Effect of slope on control of smoke flow in tunnel fires. *Fire Safety Science – Proceedings of the 5th International Symposium, Melbourne, Australia*, 3–7 March 1997. International Association for Fire Safety Science, London, pp. 1225–1236.

Handbook of Tunnel Fire Safety, 2nd edition
ISBN: 978-0-7277-4153-0

ICE Publishing: All rights reserved
doi: 10.1680/htfs.41530.309

Chapter 14
Heat release rates in tunnel fires: a summary

Haukur Ingason and Anders Lönnermark
SP Technical Research Institute of Sweden, Borås, Sweden

14.1. Introduction

The heat release rate (HRR) in a tunnel fire depends on many factors, including the infrastructure of the tunnel, vehicles and their contents, and the ventilation conditions. Furthermore, the separation between vehicles or 'fuel packages' and the nature of the content of each vehicle or fuel package is very important in relation to fire spread. The geometrical and spatial arrangement of the fuels within a fuel package would also be expected to significantly affect fire spread.

Experience from large tunnel fires shows that the HRR is crucial to evacuation success. The HRR is also a key parameter in the design of tunnel safety systems and in considering the structural strength of a tunnel. The design parameters usually involve tabulated peak HRR values in megawatts (PIARC, 1999; NFPA, 2004). In order to obtain an overview of the relevance of these tabulated design data, a summary of all available HRRs in tunnel fires is needed. It needs to be borne in mind that, as a general rule, results from similar experimental tests may vary considerably, and even ostensibly 'identical' tests may produce quite different experimental results (see Chapter 29). This is because not all factors can be exactly controlled. In using results, therefore, caution and circumspection is essential.

Compilations of results have been presented before. Ingason (2001) presented an overview of the HRRs of different vehicles in 2001, but obviously this does not include new data available from large-scale fire tests (Kunikane *et al.*, 2002; Lemaire *et al.*, 2002; Lönnermark and Ingason, 2005) performed after 2001. In 2005, Lönnermark and Ingason (2004, 2005) presented a condensed summary of peak HRRs and corresponding ceiling temperatures from large-scale tunnel fire tests. In 2006, Ingason compiled and described most of the large-scale test data reported in the literature, including HRRs and gas temperatures; and, in the 2008 edition of *The SFPE Handbook of Fire Protection Engineering*, Babrauskas reported HRR curves from various transport vehicles and components. The HRR data presented in this chapter are based on these previous overviews, carried out by the authors and others, with the addition of new data and, where relevant, updating of old data.

HRRs during tunnel tests can be determined using different measuring techniques (see Chapter 13). The most commonly used method is oxygen-consumption calorimetry, but HRRs can also be determined by measuring the weight loss of the fuel, the convective flow or by using carbon dioxide generation calorimetry. The accuracy of these methods depends on the measuring technique, the number and types of probes used for the estimation, and the calculation method used to determine the HRR based on measuring data. Ingason *et al.* (1994) estimated the maximum error in their measurements in the EUREKA 499 test series (SP, 1994; Studiensge-sellschaft Stahlanwendung, 1995) to be of the order of 25% (relative errors conservatively added), whereas in the Runehamar tests Ingason and Lönnermark (2005) estimated the error to be 14.9% (combined expanded relative standard uncertainty with a 95% confidence interval (Axelsson *et al.*, 2001)). This shows that there is a relatively high uncertainty in HRR values measured in tunnels.

14.2. Overview of HRR data

14.2.1 Road vehicles

14.2.1.1 Passenger cars

Numerous measurements of the HRRs of passenger cars can be found in the literature. Table 14.1 gives a summary of HRR measurements for passenger cars and other road vehicles. For each test, the table gives an estimate of the total calorific content, the measured peak HRR, and the time to the peak HRR. All the large vehicles were burned in a tunnel, whereas passenger cars were either burned under a calorimeter hood or in a tunnel.

Figures 14.1 and 14.2 show example graphs of measured HRRs from single passenger cars. Most of the data were extracted from graphs given in the references cited (see Table 14.1). For comparison, the fast fire growth curve (t^2) (Karlsson and Quintier, 2000) is also shown. A short discussion of these tests is given below.

Mangs and Keski-Rahkonen (1994) presented HRRs from three full-scale laboratory tests using typical passenger cars (steel body) manufactured in the late 1970s (Ford Taunus 1.6, Datsun 160 J Sedan, Datsun 180 B Sedan). The experiments were performed indoors under a fire hood (using oxygen calorimetry). The cars were ignited either inside the passenger cabin (0.09 m^2 heptane tray under the left front seat in Test 1) or beneath the engine with an open 0.09 m^2 heptane tray ($\sim$160 kW). Steinert (1994) presented the HRR of a plastic passenger car from a test in the EUREKA 499 test series (SP, 1994; Studiensgesellschaft Stahlanwendung, 1995). The car was a Renault Espace J11-II manufactured in 1988, and the ignition was in a transistor in the console in order to simulate a fire in the cable system. Steinert (2000) also published HRRs of different types of passenger cars in a car park, all with different types of car body (plastic and steel). A total of ten tests were performed, where the aim was to measure the HRR and quantify the risk of fire spread in a car park. The first three tests were carried out using single vehicles, and the other six tests consisted of combinations of two and three passenger cars which were placed beside the ignited vehicle (see Table 14.1). In each case the cars were ignited by dripping flammable liquid onto the front seat of the car, with the front window open.

Shipp and Spearpoint (1995) presented the measured HRRs for two different types of private car (see Figure 14.1): a 1982 Austin Maestro and a 1986 Citroën BX. The Citroën BX was ignited with a small petrol pool fire (5 kW) in the engine compartment, whereas the Austin Maestro was ignited with a small wooden crib in the seat (10 kW). The test was intended to approximately represent a Channel Tunnel shuttle wagon, a calorimeter hood being at each end of the car. Lemaire *et al.* (2002) presented HRR measurements for two passenger cars obtained using two different ventilation rates in the second Benelux Tunnel fire test series (see Figure 14.2). The cars were 1990 Opel Kadetts, with 25–30 litres of petrol in the fuel tank. The HRR was measured both without ventilation and with a ventilation velocity of about 6 m/s. The peak HRR with no ventilation was 4.7 MW, 11.5 minutes into the test. With ventilation, the HRR showed two maxima, the first was about 3 MW, 13 minutes into the test, and the second was about

4.6 MW, 37 minutes into the test. An interesting observation is that the high ventilation rate made it difficult for the fire to spread within the cabin in the opposite direction to the ventilation flow. Lemaire *et al.* (2002) also performed a test using a Citroën Jumper van (see Figure 14.1) during a test of a deluge sprinkler system in the tunnel ceiling. The sprinkler system was activated after about 13.6 minutes into the test. Ingason *et al.* (1997) presented HRR measurements for a Fiat 127 (see Figure 14.2), which was ignited in the engine compartment using an electrical device. The fire was extinguished by fire-fighters 13 minutes into the test.

Joyeux (1997) presented ten HRR measurements from passenger-vehicle fires in a simulated car park. The measurements were made beneath a 10 MW calorimeter, with one car or with two cars. Cars used in the tests included cars manufactured in the 1980s and 1990s by Mazda, Renault, BMW, Citröen BX and Peugeot. In the first seven tests, the first car was ignited with a small petrol tray under the left front seat. In the other tests, the first car was ignited by a petrol tray placed under the car at gearbox level. The HRR results for some of the tests are only available in Table 14.1.

In the experimental tests, the HRRs for a single passenger car (small or large) vary from 1.5 MW to about 9 MW, but the majority of the tests show HRR values less than 5 MW. For two cars, the maximum HRR varies between 3.5 and 10 MW. The World Road Association (PIARC) proposes a maximum HRR for one small passenger car to be 2.5 MW and for one large passenger car to be 5 MW (PIARC, 1999). Based on the data presented here, these values appear reasonable, in very approximate terms. There is a wide range of time taken to reach the peak HRR, this varying between 10 and 55 minutes.

The data presented here show that there is a trend for the peak HRR to increase approximately linearly with the total calorific content of the passenger cars involved in the fire. An analysis of all

Figure 14.1 Experimentally determined HRRs for single passenger cars. (E) Data extracted manually from printed graphs in the literature. For more information, see Table 14.1

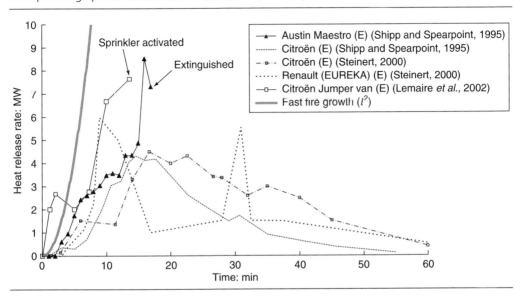

Figure 14.2 Experimentally determined HRRs for single passenger cars. (E) Data extracted manually from printed graphs in the literature. (M) Measured data obtained as numerical values from a measuring file. For more information, see Table 14.1

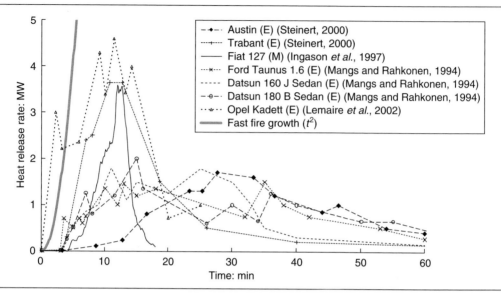

data available shows that the average increase in HRR is about 0.85 MW/GJ. Lönnermark (2005) found this value to be 0.87 MW/GJ. The same value (0.87 MW/GJ) was also obtained for HGVs. It must be realised, however, that this is only a very general guide. For example, in the HGV test in the EUREKA series, the peak HRR reached about 120 MW for a calorific content of about 87 GJ, at a maximum ventilation velocity of 6 m/s. The HRR is usually better represented as a parameter related to the exposed fuel surface.

Shipp *et al.* (2009) presented a series of experiments with passenger cars in a car park environment. The tests were carried out by the Building Research Establishment (BRE) in the UK. Both single cars and multiple cars (side by side or stacked) were burned. The objective was to examine the time to full fire development, and the HRR of a fire starting in the passenger compartment or in the engine compartment of one of the vehicles. The risk of fire spread from car to car was also examined. At the time of writing, the type of cars used has not been published. The results are only available in Table 14.1 and are denoted by BRE test numbers.

14.2.1.2 Buses

Table 14.2 presents the results of three different tests done using buses. Ingason *et al.* (1994) and Steinert (1994) presented a measured HRR for a 25–35 year old, 12 m long Volvo school bus with 40 seats (EUREKA bus, Test 7). Steinert used more coarse measurement points than Ingason *et al.* As can be seen in Figure 14.3, the maximum HRR was measured to be 29 MW by Ingason *et al.* and 34 MW by Steinert. The total calorific content was estimated to be 41 GJ by Ingason *et al.* and 44 GJ by Steinert, which is within an acceptable level of accuracy. The estimated time to the peak HRR was 8 minutes (Ingason *et al.*, 1994) and 14 minutes (Steinert, 1994). The reason for this discrepancy is related to the different methods used to calculate the HRR. Steinert's results for this test are based on convective flow, whereas Ingason *et al.* used oxygen-consumption calorimetry.

The body of the bus was made of fibreglass. Therefore, the roof and walls of the bus (i.e. the fibreglass shell) were totally burned away, down to the bottom edges of the windows. This explains why the fire was able to develop fully without any restriction of oxygen.

In 2008, Hammarström et al. presented the results a large-scale experiment done using a modern coach (Volvo) with 49 seats. The test was carried out from in SP's fire hall under a large hood calorimeter (10 MW). The measured HRR is shown in Figure 14.3, which includes the ultrafast fire growth curve (t^2) (Karlsson and Quintier, 2000) for comparison. The initial fire development was relatively slow. One explanation for the slow fire development is that the fire was started with a 100 kW gas burner in the luggage compartment at the back of the bus, below the passenger compartment. The fire spread to the passenger compartment through the windows on the outside of the bus, and fire growth increased significantly when three windows had broken, approximately 15–16 minutes after ignition. Three different peaks can be observed in the HRR curve given in Figure 14.3. The first peak occurs after about 11 minutes, when the fire broke out on the side of the luggage compartment. The fire then spread under the passenger compartment, and into the passenger compartment through the windows, over the period 15–17 minutes after ignition. As the fire grew in intensity, the situation became intolerable, and the fire was extinguished manually approximately 18.5 minutes after ignition (the last peak of the curve). Due to leakage of the hood used to collect smoke and heat, the final peak was probably higher than the measured peak shown in Figure 14.3. It has been estimated that the HRR was approximately 14–15 MW when the fire was extinguished. At this stage, approximately two-thirds of the passenger compartment was involved in the fire. The authors of the report estimate that the maximum HRR could have been as much as 25 MW had the bus been allowed to continue to burn.

A bus test was carried out in the Shimizu Tunnel using a sprinkler system (Kunikane et al., 2002). The HRR was not measured during the experiment but Kunikane et al. estimated the convective peak HRR based on temperature measurements and the mass flow rate. The peak convective HRR was found to be 16.5 MW when the sprinkler system activated. Kunikane et al. estimated that, if the sprinkler system had not activated, the peak convective HRR would have been approximately 20 MW. If we assume that 67% of the total HRR is convective, we obtain a peak HRR of $20/0.67 = 30$ MW. Two HRR curves are plotted in Figure 14.3, one of data extracted from the information given by Kunikane et al. (2002) for the convective HRR, and one showing the total HRR based on the assumption that 67% of the total HRR is convective.

A bus fire that occurred in the Ekeberg Tunnel in Oslo, Norway, in 1996 (Skarra, 1997), appears to have developed in a similar way to that in the EUREKA 499 bus test (Ingason et al., 1994), despite the large differences in the age and type of buses. In the Ekberg incident, the bus had only been in use for 3 months. In the computational fluid dynamics (CFD) simulation, the bus fire was estimated to have grown to 36 MW in under 6 minutes, remained at a steady state of 36 MW for 4 minutes, and then decayed to 1 MW over the following 12.5 minutes. The total calorific content was estimated to have been 28 GJ.

Sprinkler tests with buses in a garage have been reported (Stephens, 1992). However, in those tests no HRR measurements were carried out, although some temperature measurements were made, and these can give an indication of the fire growth rate. An analysis of these temperature measurements shows that the reported data fit quite well with the HRR data given for the EUREKA bus.

Table 14.1 Large-scale experimental data for road vehicles (passenger cars)

Type of vehicle, model year, test number, longitudinal ventilation (u in m/s)	Tunnel (T) or calorimeter hood (C)	Tunnel cross-section: m²	Calorific content[a]: GJ	Peak HRR: MW	Time to peak HRR: minutes	Reference
Single passenger cars						
Ford Taunus 1.6, late 1970s, Test 1, NV	C		4	1.5	12	Mangs and Keski-Rahkonen (1994)
Datsun 160 J Sedan, late 1970s, Test 2, NV	C		4	1.8	10	
Datsun 180 B Sedan, late 1970s, Test 3, NV	C		4	2	14	
Fiat 127, late 1970s, u = 0.1 m/s	T	8	NA	3.6	12	Ingason et al. (1997)
Renault Espace J11-II, 1988, Test 20, u = 0.5 m/s	T	30	7	6	8	Steinert (1994)
Citroën BX, 1986, NV[b]	C		5	4.3	15	Shipp and Spearpoint (1995)
Austin Maestro, 1982[b]	C		4	8.5	16	
Opel Kadett, 1990, Test 6, u = 1.5 m/s	T	50	NA	4.9	11	Lemaire et al. (2002)
Opel Kadett, 1990, Test 7, u = 6 m/s	T	50	NA	4.8	38	
Renault 5, 1980s, Test 3, NV	C		2.1	3.5	10	Joyeux (1997)
Renault 18, 1980s, Test 4, NV	C		3.1	2.1	29	
Small car[c], 1995, Test 8, NV	C		4.1	4.1	26	
Large car[c], 1995, test 7, NV	C		6.7	8.3	25	
Trabant, Test 1, NV	C		3.1	3.7	11	Steinert (2000)
Austin, Test 2, NV	C		3.2	1.7	27	
Citroën, Test 3, NV	C		8	4.6	17	
Citroën Jumper van, Test 11, u = 1.6 m/s, sprinkler activated after 13.6 minutes	T	50	NA	7.6	NA	Lemaire et al. (2002)
BRE Test No. 7, NV	C		NA	4.8	45	Shipp et al. (2009)
BRE Test No. 8, NV	C		NA	3.8	54	
Two passenger cars						
Citroën BX + Peugeot 305, 1980s, Test 6, NV	C		8.5	1.7	NA	Joyeux (1997)
Small car[c] + large car[c], Test 9, NV	C		7.9	7.5	13	
Large car[c] + small car[c], Test 10, NV	C		8.4	8.3	NA	
BMW + Renault 5, 1980s, Test 5, NV	C		NA	10	NA	

Polo + Trabant, Test 6, NV	C	5.4	5.6	29	Steinert (2000)
Peugeot + Trabant, Test 5, NV	C	5.6	6.2	40	
Citroën + Trabant, Test 7, NV	C	7.7	7.1	20	
Jetta + Ascona, Test 8, NV	C	10	8.4	55	
BRE Test No. 11 (stacked), NV	C	NA	8.5	12	Shipp et al. (2009)
Three passenger cars					
Golf + Trabant + Fiesta, Test 4, NV	C	NA	8.9	33	Steinert (2000)
BRE Test No. 1, NV	C	NA	16	21	Shipp et al. (2009)
BRE Test No. 2, NV	C	NA	7	55	
BRE Test No. 3, NV	C	NA	11	10	

NA, not available; NV, natural ventilation.

[a] Either based on integration of the HRR curve or an estimation of fuel mass and heat of combustion. The error is estimated to be less than 25%.

[b] Cars burned in an approximate mock-up of a Channel Tunnel shuttle wagon with a calorimeter hood/duct at each end.

[c] Small cars include: Peugeot 106, Renault Twingo-Clio, Citroën Saxo, Ford Fiesta, Opel Corsa, Fiat Punto, VW Polo. Large cars include: Peugeot 406, Renault Laguna, Citroën Xantia, Ford Mondeo, Opel Vectra, Fiat Tempra, VW Passat.

Table 14.2 Large-scale experimental data for buses

Type of vehicle, model year, test number, longitudinal ventilation (u in m/s)	Tunnel (T) or calorimeter (C)	Tunnel cross-section: m²	Calorific content[a]: GJ	Peak HRR: MW	Time to peak HRR: minutes	Reference
A 25–35 year old 12 m long Volvo school bus with 40 seats, EUREKA 499, Test 7,[b] u = 0.3 m/s	T	30	41	29	8	Ingason et al. (1994)
A 25–35 year old 12 m long Volvo school bus with 40 seats, EUREKA 499, Test 7,[b] u = 0.3 m/s	T	30	44	34	14	Steinert (1994)
SP laboratory test with a modern tourist bus (Volvo), NV	C		NA	25[c]	NA	Hammarström et al. (2008)
Test in the Shimizu Tunnel, bus, u = 3–4 m/s	T	115	NA	30[d]	7	Kunikane et al. (2002)

NA, not available; NV, natural ventilation.

[a] Either based on integration of the HRR curve, or an estimation of the fuel mass and heat of combustion. The accuracy of these values may vary, depending on the method and available information. The error is estimated to be less than 25%.

[b] The test number sequence in the EUREKA 499 project is given in Studiensgesellschaft Stahlanwendung (1995) and Haack (1994).

[c] Value based on the estimation made by Hammarström et al. (2008).

[d] Value estimated from the convective HRR of 20 MW derived by Kunikane et al. (2002), because a sprinkler system was activated when the convective HRR was 16.5 MW. We assume that 67% of the HRR is convective, and thereby it is estimated that HRR = 20/0.67 = 30 MW.

Figure 14.3 Heat-release rates for buses. (E) Data extracted manually from printed graphs in the literature. (M) Measured data obtained as numerical values from a measuring file. For more information, see Table 14.2

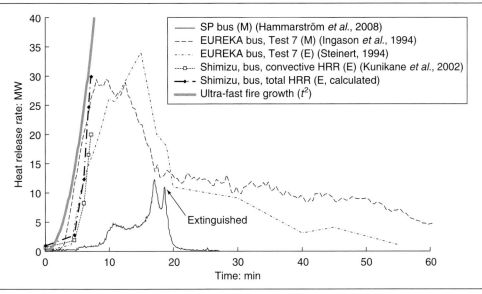

14.2.1.3 Heavy goods vehicles

Some large-scale tests done using HGVs have been reported in the literature (Table 14.3, Figures 14.4 and 14.5). The first one was performed in 1992 in the EUREKA 499 test programme in Repparfjord, Norway (Studiesgesellschaft Stahlanwendung, 1995), the second was done in the Mont Blanc Tunnel in 2000 (Brousse *et al.*, 2001), the third was done in the second Benelux tunnel (Lemaire *et al.*, 2002) in The Netherlands in 2001, and the most recent was done in the Runehamar Tunnel (Ingason and Lönnermark, 2005).

In the EUREKA 499 tunnel test programme, one test (Test 21) was performed with a real HGV loaded with mixed furniture (total approximately 87 GJ) and with varying longitudinal velocity (5–6 and 2–3 m/s). The second test (Test 15) in the EUREKA 499 series was conducted using a simulated HGV-trailer load (mock-up). The test numbers used in the EUREKA test series are given in Studiesgesellschaft Stahlanwendung (1995) and Haack (1994). The fire load, which consisted of densely packed wooden cribs with rubber tyres and plastic materials on top (total approximately 64 GJ), was mounted on a weighing platform. In this test, a natural ventilation rate of 0.7 m/s was obtained. In the Mont Blanc Tunnel, a test with a HGV (truck and a trailer) similar to that which generated the fire in the tunnel in 1999, but with a much smaller amount of transported goods (total approximately 76 GJ; but 35 GJ was consumed in the test, as the fire was extinguished), was used (Brousse *et al.*, 2001, 2002). The fire load in the trailer consisted of 400 kg margarine. In the second Benelux tunnel test series (Lemaire *et al.*, 2002), HGV trailer mock-ups were used. Standardised wooden pallets were arranged in two different configurations (approximately 10 and 20 GJ, respectively) with different longitudinal velocities: natural ventilation (~0.5 m/s), 4–5 m/s and 5 m/s. In the Runehamar test series (Ingason and Lönnermark, 2005), four large-scale tests, each involving a mock-up of a HGV trailer in a road tunnel, were performed. Initial longitudinal ventilation rates within the tunnel were in the range 2.8–3.2 m/s.

Table 14.3 Large-scale experimental data for HGVs

Type of vehicle, model year, test number, longitudinal ventilation (u in m/s)	Tunnel (T) or weight loss (W)	Tunnel cross-section: m²	Calorific content[a]: GJ	Peak HRR: MW	Time to peak HRR: minutes	Reference
A simulated trailer load with a total of 11 010 kg wood (82%[b]) and plastic pallets (18%), Runehamar test series, Test 1, $u = 3$ m/s	T	32	240	202	18	(Ingason and Lönnermark, 2005)
A simulated trailer load with a total of 6930 kg wood pallets (82%[b]) and polyurethane mattresses (18%), Runehamar test series, Test 2, $u = 3$ m/s	T	32	129	157	14	(Ingason and Lönnermark, 2005)
Leyland DAF 310ATi – a HGV trailer with 2 tonnes of furniture, EUREKA 499, Test 21, $u = 3$ to 6 m/s	T	30	87	128	18	Grant and Drysdale (1997)
Simulated trailer with 8550 kg furniture, fixtures and rubber tyres, Runehamar test series, Test 3, $u = 3$ m/s	T	32	152	119	10	(Ingason and Lönnermark, 2005)
Simulated trailer mock-up with 2850 kg corrugated paper cartons filled with plastic cups (19%[b]), Runehamar test series, Test 4, $u = 3$ m/s	T	32	67	67	14	(Ingason and Lönnermark, 2005)
Simulated trailer load with 72 wood pallets, second Benelux tests, Test 14, $u = 1$ to 2 m/s	T	50	19	26	12	Lemaire et al. (2002)
Simulated trailer load with 36 wooden pallets, second Benelux tests, Tests 8, 9 and 10, $u = 1.5$, 5.3 and 5 m/s, respectively	T	50	10	13, 19 and 16	16, 8 and 8	Lemaire et al. (2002)
Simulated truck load, EUREKA 499, Test 15, $u = 0.7$ m/s	T	30	63	17	15	Ingason (1994)
Mont Blanc HGV mock-up test	T	50	35	23	47.5	(Brousse et al., 2001, 2002)
A 3.49 ton pickup truck loaded with 890 kg wooden pallets; ignition in cargo, windows closed, NV	W		26[b]	24[c]	6.6	Chuang et al. (2005–06)

A 3.49 ton pickup truck loaded with 890 kg wooden pallets; ignition on seat, one window open, NV	W	26[c]	21[d]	14.5	Chuang et al. (2005)
A 3.49 ton pickup truck loaded with 452 kg plastic barrels; ignition of the seat, NV	W	25[a]	47[e]	43.8	Chuang et al. (2005)

NA, not available; NV, natural ventilation.

[a] Either based on integration of the HRR curve, or estimation of the fuel mass and heat of combustion. The accuracy of these values may vary, depending on the method and available information. The error is estimated to be less than 25%.

[b] Mass ratio of the total weight.

[c] Estimated from the information given in the reference.

[d] Given in the reference, based on load cell measurements.

Figure 14.4 Heat-release rates for the HGV trailer mock-up tests presented in Table 14.3. (The EUREKA HGV Test 21 was for a real HGV, including the cab.) (E) Data extracted manually from printed graphs in the literature. (M) Measured data obtained as numerical values from a measuring file

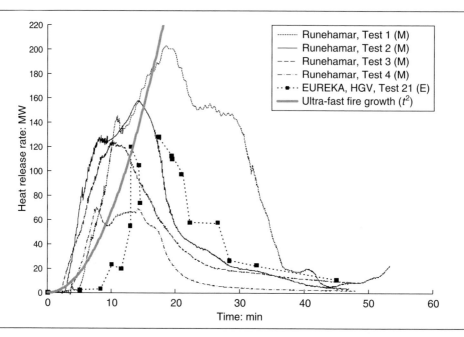

Peak HRRs in the range 66–202 MW were estimated. The peak HRRs were obtained between 7.1 and 18.4 minutes from ignition in the various tests (Figure 14.4). Each HGV trailer mock-up consisted of a steel rack system loaded with a mixed load of wooden pallets and polyethylene pallets (Test T1), wooden pallets and polyurethane mattresses (Test T2), furniture and fixtures with ten truck rubber tyres (Test T3), and paper cartons and polystyrene cups (Test T4). Each load was covered with a polyester tarpaulin in each test, and ignited at the upstream, front, end of the trailer.

The peak HRRs from all the tests so far conducted on HGVs or HGV trailer mock-ups are in the range 13–202 MW, depending on the fire load, ventilation, etc. The time to reach the peak HRR is in the range 10–20 minutes. The fire duration is less than 1 hour for all the HGV trailer tests presented in Figure 14.4. For comparison, the ultra-fast fire growth curve (t^2) is also given in the figure.

It should be noted that the delay time between ignition and a significant rise in HRR varies considerably and will be affected by many factors. For example, in the second Benelux tunnel tests, replacing a canvas cover with an aluminium cover in a HGV trailer mock-up significantly reduced the delay time.

The effect of the ventilation flow rate on the HRR of HGVs has been investigated by Carvel and co-workers (see Chapter 11). As a general rule, ventilation has a considerable effect on both the peak HRR and on the rate of fire growth.

Figure 14.5 Heat-release rates for the HGV trailer mock-up tests presented in Table 14.3. (E) Data extracted manually from printed graphs in the literature. (M) Measured data obtained as numerical values from a measuring file

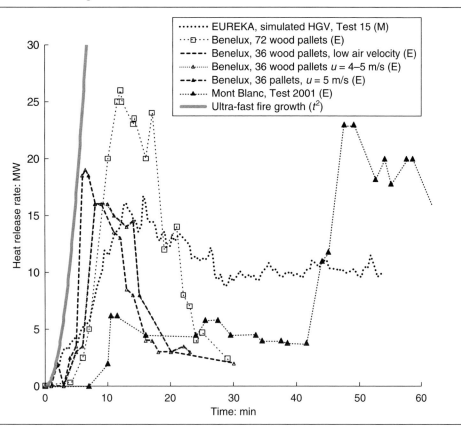

14.2.2   Railway rolling stock

Very few measurements of the HRRs of rail and metro vehicles (rolling stock) have been reported in the literature. The majority of the tests reported are from the EUREKA 499 test series (Studiensgesellschaft Stahlanwendung, 1995). A summary of these tests is given in Table 14.4.

The test results presented in Table 14.4 are based on tests done on single coaches. The peak HRR is found to be in the range 7–43 MW, and the time to reach the peak HRR varies from 5 to 80 minutes. If the fire were to spread between the train coaches, the total HRR and the time to the peak HRR would be much higher than the values given here, although one cannot simply add the HRR for each coach to obtain an estimate of the total HRR. This is because the first coach would not necessarily reach the peak HRR at the same time as the later ones. The EUREKA 499 tests show that there are many parameters that will affect the development of a fire in a train coach. These include the body type (steel, aluminium, etc.), the quality of the glazed windows, the geometry of the openings, the amount and type of combustible interior and its initial moisture content, the construction of wagon joints, the air velocity within the tunnel, and the geometry of the tunnel cross-section. These are all parameters that need to be

Table 14.4 Large-scale experimental data for rolling stock

Type of vehicle, test number, longitudinal ventilation (*u* in m/s)	Calorific content: GJ	Peak HRR: MW	Time to peak HRR: minutes	Reference
Rail				
Joined railway car; two half-cars, one aluminium and one steel, EUREKA 499, Test 11, *u* = 6 to 8 and 3–4 m/s	55	43	53	Steinert (1994)
German Intercity Express railway car, EUREKA 499, Test 12, *u* = 0.5 m/s	63	19	80	Ingason *et al.* (1994)
German Intercity passenger railway car, EUREKA 499, Test 13, *u* = 0.5 m/s	77	13	25	Ingason *et al.* (1994)
British Rail 415, passenger railway car[a]	NA	16	NA	Barber *et al.* (1994)
British Rail Sprinter, passenger railway car, fire-retardant upholstered seating[a]	NA	7	NA	Barber *et al.* (1994)
Metro				
German subway car, EUREKA 499, *u* = 0.5 m/s	41	35	5	Ingason *et al.* (1994)

[a] The test report is confidential, and no information is available on the test set-up, test procedure, measurement techniques, ventilation, etc.

considered during the design of a rail or metro tunnel. A very important factor in the development of a fire is the quality and mounting of the windows. As long as the windows do not break or fall out (and there are no other large openings), the fire will develop slowly. On the other hand, if the windows break, the fire can spread and increase in intensity very quickly. Figure 14.6 gives time-resolved HRR curves for the tests presented in Table 14.4 (with the exception of the British Rail tests). For comparison, the ultra-fast fire growth curve (t^2) is also given for comparison.

14.2.3 HRRs per exposed fuel surface area

Ingason (2006) collected the HRR data for all the large-scale tests reported in the literature and normalised the maximum HRR to the exposed fuel surface area. The fuel surface area was defined as the freely exposed area where release of gasified fuel can occur simultaneously. The reason for normalising to the exposed fuel surface area was that this makes it convenient to compare the maximum HRR between different types of fuel and for different fire conditions. The results may be used to help to estimate the peak HRR in different types of vehicle and with other solid and liquid fuels. Based on this work, the HRR data were divided into three different groups according to fuel type: liquid pool fires, ordinary solid materials (e.g. wooden pallets and wooden cribs), and road and rail/metro vehicles (Table 14.5).

It was concluded, based on the experimental tests considered so far, that the maximum HRR per square metre of fuel surface area in a fuel-controlled fire using different vehicles is approximately 0.3–0.4 MW/m² (Ingason, 2006); although, when HGV trailer mock-ups are included this becomes about 0.3–0.5 MW/m². These values are in line with the HRR per square metre of fuel surface area for the solid materials. The HRR per square metre of fuel surface area for

Figure 14.6 Heat-release rates for the rail vehicle tests presented in Table 14.4. (E) Data extracted manually from printed graphs in the literature. (M) Measured data obtained as numerical values from a measuring file

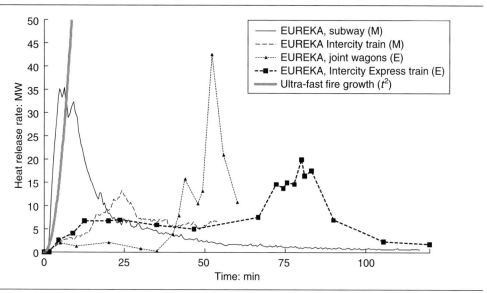

each individual material exhibits a greater variation, but it appears that the variation for mixed materials is not so wide. It is very important to take this observation into account when establishing design fires for tunnels.

It is essential to realise, however, that this is an initial finding, and is based on tests in which the ventilation velocities ranged from about 0.5 to about 6 m/s. In a real-world situation, the ventilation velocity may be higher than 6 m/s (e.g. there may be a natural wind). As a general rule, the total HRR for a single vehicle or item/fuel package in a tunnel fire depends on many factors. Furthermore, the total HRR depends upon the potentiality for spread of the fire from one item to another, i.e. the proximity of items (e.g. vehicles) is of crucial importance. Therefore, it is very important that more large-scale tunnel fire tests using real vehicles are undertaken in order to test these initial observations. Most of the vehicle fire data are available for vehicles that are out of date, and therefore a new large-scale tunnel test series with modern road and rail/metro vehicles is a pressing scientific need.

Tests have been performed to study the influence of a layer of railway Macadam on the HRR of a burning liquid spill (Lönnermark *et al.*, 2008) The test shows that the presence of the Macadam significantly decreases the burning rate of the two fuels tested (heptane and diesel), and that its influence increases with increasing distance from the fuel surface to the upper layer of the Macadam.

14.3. Conclusions

From experimental tests, it appears that the HRRs for single passenger cars (small and large) vary between 1.5 MW and about 9 MW, but the majority of the tests show HRR values less than 5 MW. When two cars are involved, the peak HRR varies between 3.5 MW and 10 MW. There

Table 14.5 Summary of the HRR per per fuel surface area for liquid (pool fires), solid and vehicle fires in tunnels (Ingason, 2006)

Type of fuel	Test series	Exposed fuel area: m²	Maximum HRR per square metre exposed fuel area: MW/m²
Liquid			
Petroleum[a]	Ofenegg (Kommission für Sicherheitsmassnahmen in Strassentunneln, 1965) Zwenberg (ILF, 1976) No. 3 Shimizu (Kunikane et al., 2002)	6.6, 47.5, 95	0.35–2.6
Kerosene[a]	Glasgow (Heselden and Hinkley, 1970)	1.44	1.4
n-Heptane[a]	Eureka (Studiensgesellschaft Stahlanwendung, 1995)	1, 3	3.5
n-60% heptane, 40% toluene	Second Benelux (Lemaire et al., 2002)	3.6, 7.2	1.1–1.6
Low-sulphur No. 2 fuel oil	Memorial (Massachusetts Highway Department, 1995)	4.5, 9, 22.2, 44.4	1.7–2.5
Solid fuel			
Wooden cribs	Eureka (Tests 8, 9 and 10) (Studiensgesellschaft Stahlanwendung, 1995)	140	0.07–0.09
Wooden pallets	Second Benelux (Tests 8, 9, 10 and 14) (Lemaire et al., 2002)	120 (36 pallets) 240 (72 pallets)	0.11–0.16
82% wooden pallets, 18% polyethylene pallets	Runehamar (Test 1) (Ingason and Lönnermark, 2005)	1200	0.17
82% wooden pallets, 18% polyurethane foam mattresses	Runehamar (Test 2) (Ingason and Lönnermark, 2005)	630	0.25
81% wooden pallets and cartons, 19% plastic cups	Runehamar (Test 4) (Ingason and Lönnermark, 2005)	160	0.44
HGV mock-up, furniture	Runehamar (Test 3) (Ingason and Lönnermark, 2005)	240	0.5
HGV, furniture	Eureka (Test 21) (Studiensgesellschaft Stahlanwendung, 1995)	300	0.4
Vehicles			
Passenger cars	Assuming a 5 MW fire in a car (Ingason, 2006)	12–18	0.3–0.4
Passenger car, plastic	EUREKA, Test 20 (Studiensgesellschaft Stahlanwendung, 1995)	17	0.35
Bus	EUREKA, Test 7 (Studiensgesellschaft Stahlanwendung, 1995)	80	0.36
Train	EUREKA, Test 11 (Studiensgesellschaft Stahlanwendung, 1995)	145	0.30
Subway coach	EUREKA, Test 14 (Studiensgesellschaft Stahlanwendung, 1995)	130	0.27

[a] The heat of combustion of petroleum is assumed to be 43.7 MJ/kg, that of kerosene 43.5 MJ/kg and that of n-heptane 44.6 MJ/kg.

is a wide variety in the time to the peak HRR (10–55 minutes). The highest peak HRRs are obtained for the HGV trailers. In the tests conducted so far, the peak HRR values for HGV trailers vary from 13 MW to more than 200 MW, depending on the fire load, ventilation, etc., and the time to the peak HRR is in the range 10–20 minutes. Note that the fire duration was less than 1 hour for all the HGV trailer tests presented. For railway rolling stock the HRRs vary from 7 MW to 43 MW, and the time to reach the peak HRR varies from 5 to 80 minutes.

The HRR per square metre of fuel surface area in fuel-controlled fires of different vehicles, including HGV trailer mock-ups, has been found to be very approximately between 0.3 and $0.5\,MW/m^2$. This value range is in line with the HRR per square metre of fuel surface area for solid materials. The HRR per square metre of fuel surface area of the individual materials shows greater variation, both lower and much higher, but it appears that the variation in the corresponding values for mixed materials is not so broad. These conclusions are based on experimental tests conducted to date, and far more large-scale tests are needed, including replicated tests, to determine the possible ranges of HRRs in tunnel fires.

As vehicles and materials keep changing, the determination of HRRs is an ongoing process. In essence, it must be realised that the total HRR in a fire depends on many factors, and it is essential to employ any results, experimental or theoretical, with great caution and circumspection.

REFERENCES

Axelsson J, Andersson P, Lönnermark A, van Hees P and Wetterlund I (2001) *Uncertainties in Measuring Heat and Smoke Release Rates in the Room/Corner Test and the SBI*. SP Report 2001:04. SP Swedish National Testing and Research Institute, Borås.

Babrauskas V (2008) Heat release rates. In: DiNenno PJ (eds), *The SFPE Handbook of Fire Protection Engineering*, 4th edn. National Fire Protection Association, Quincy, MA, pp. 3-1–3-59.

Barber C, Gardiner A and Law M (1994) Structural fire design of the Øresund Tunnel. *Proceedings of the International Conference on Fires in Tunnels*, 10–11 October 1994. SP Swedish National Testing and Research Institute, Borås.

Brousse B, Perard M, Voeltzel A, Le Botlan Y and Ruffin E (2001) Ventilation and fire tests in the Mont Blanc Tunnel to better understand the catastrophic fire of March 24th, 1999. *Third International Conference on Tunnel Fires and Escape from Tunnels, Washington DC, USA*, 9–11 October 2001.

Brousse B, Voeltzel A, Botlan Y Le and Ruffin E (2002) Mont Blanc Tunnel ventilation and fire tests. *Tunnel Management International* **5**(1): 13–22.

Chuang Y-J, Tang C-H, Chen P-H and Lin C Y (2005) Experimental investigation of burning scenario of loaded 3.49-ton pickup trucks. *Journal of Applied Fire Science* **14**(1): 27–46.

Grant GB and Drysdale D (1997) Estimating heat release rates from large-scale tunnel fires. *Fire Safety Science – Proceedings of the Fifth International Symposium. Melbourne, Australia*, 3–7 March 1997. International Association for Fire Safety Science, London.

Haack A (1994) Introduction to the EUREKA-EU 499 FIRETUN Project. *Proceedings of the International Conference on Fires in Tunnels*. SP Report 1994:54. SP Swedish National Testing and Research Institute, Borås.

Hammarström R, Axelsson J, Försth M, Johansson P and Sundström B (2008) *Bus Fire Safety*. SP Report 2008:41. SP Technical Research Institute of Sweden, Borås.

Heselden A and Hinkley PL (1970) *Smoke Travel in Shopping Malls. Experiments in Cooperation with Glasgow Fire Brigade. Parts 1 and 2*. Fire Research Station, Borehamwood.

ILF (1976) *Brandversuche in einem Tunnel*. Ingenieurgemeinschaft Lässer-Feizlmayr; Bundesministerium f. Bauten u. Technik, Strassenforschung. p. Heft 50, Teil 1 und 2.

Ingason H (1994) Heat release rate measurements in tunnel fires. *Proceedings of the International Conference on Fires in Tunnels*, 10–11 October 1994. Sweden: SP Swedish National Testing and Research Institute, Borås.

Ingason H (2001) An overview of vehicle fires in tunnels. *Fourth International Conference on Safety in Road and Rail Tunnels, Madrid, Spain*, 2–6 April 2001.

Ingason H (2006) Fire testing in road and railway tunnels. In: Apted V (ed.), *Flammability Testing of Materials used in Construction, Transport and Mining*. Woodhead Publishing, Cambridge, pp. 231–274.

Ingason H and Lönnermark A (2004) Recent achievements regarding measuring of time-heat and time-temperature development in tunnels. *1st International Symposium on Safe & Reliable Tunnels, Prague, Czech Republic*, 4–6 February 2004.

Ingason H and Lönnermark A (2005) Heat release rates from heavy goods vehicle trailers in tunnels. *Fire Safety Journal* **40**: 646–668.

Ingason H, Gustavsson S and Dahlberg M (1994) *Heat Release Rate Measurements in Tunnel Fires*. SP Swedish National Testing and Research Institute, Borås.

Ingason H, Nireus K and Werling P (1997) *Fire Tests in a Blasted Rock Tunnel*. Försvarets forskningsanstalt (FOA) (Swedish Defense Research Establishment), Stockholm.

Joyeux D (1997) *Development of Design Rules for Steel Structures Subjected to Natural Fires in Closed Car Parks*. Centre Technique Industriel de la Construction Métallique, Saint-Rémy-lès-Chevreuse.

Karlsson B and Quintier JG (2000) *Enclosure Fire Dynamics*. CRC Press, Boca Raton, FL.

Kommission für Sicherheitsmassnahmen in Strassentunneln (1965) *Schlussbericht der Versuche im Ofenegg Tunnel von 17.5–31.5 1965*.

Kunikane Y, Kawabata N, Ishikawa T, Takekuni K and Shimoda A (2002) Thermal fumes and smoke induced by bus fire accident in large cross sectional tunnel. *The Fifth JSME-KSME Fluids Engineering Conference, Nagoya, Japan*, 19 November 2002.

Kunikane Y, Kawabata N, Takekuni K and Shimoda A (2002) Heat release rate induced by gasoline pool fire in a large-cross-section tunnel. *4th International Conference on Tunnel Fires*, Basel, Switzerland, 2–4 December 2002: Tunnel Management International, Tenbury Wells.

Lemaire A, van de Leur PHE and Kenyon YM (2002) *Safety Proef: TNO Metingen Beneluxtunnel – Meetrapport*. TNO, The Netherlands.

Lönnermark A (2005) *On the Characteristics of Fires in Tunnels*. Doctoral thesis, Department of Fire Safety Engineering, Lund University, Sweden.

Lönnermark A and Ingason H (2005) Gas temperatures in heavy goods vehicle fires in tunnels. *Fire Safety Journal* **40**: 506–527.

Lönnermark A, Kristensson P, Helltegen M and Bobert M (2008) Fire suppression and structure protection for cargo train tunnels: Macadam and HotFoam. *3rd International Symposium on Safety and Security in Tunnels (ISTSS 2008)*, Stockholm, Sweden, 12–14 March 2008. SP Technical Research Institute of Sweden, Borås.

Mangs J and Keski-Rahkonen O (1994) Characterization of the fire behavior of a burning passenger car. Part II: Parametrization of measured rate of heat release curves. *Fire Safety Journal* **23**: 37–49.

Massachusetts Highway Department (1995) *Memorial Tunnel Fire Ventilation Test Program – Test Report*. Massachusetts Highway Department and Federal Highway Administration, Boston, MA/Washington, DC.

NFPA (2004) *Standard for Road Tunnels, Bridges, and other Limited Access Highways.* National Fire Protection Association, Quincy, MA.

PIARC (1999) *Fire and Smoke Control in Road Tunnels.* World Road Association (PIARC), La Defense.

Shipp M *et al.* (2009) Fire spread in car parks. *Fire Safety Engineering* Jun.: 14–18.

Shipp M and Spearpoint M (1995) Measurements of the severity of fires involving private motor vehicles. *Fire and Materials* **19**: 143–151.

Skarra N (1997) *Bussbrannen i Ekebergtunnelen 21.8.1996 (The Bus Fire in the Ekeberg Tunnel on August 21 1996).* Statens vegvesen, Oslo.

SP (1994) *Proceedings of the International Conference on Fires in Tunnels.* SP Report 1994:54. SP Swedish National Testing and Research Institute, Borås.

Steinert C (1994) Smoke and heat production in tunnel fires. *The International Conference on Fires in Tunnels, SP Swedish National Testing and Research Institute, Borås, Sweden*, 10–11 October 1994.

Steinert C (2000) Experimentelle Untersuchhungen zum Abbrand-und Feuerubersprungs-verhalten von Personenkraftwagen. *Zeitschrift, Forschung, Technik und Management im Brandschutz* **4**: 163–172.

Stephens JN (1992) Using sprinklers in bus garages. *Fire Prevention* **253**.

Studiensgesellschaft Stahlanwendung (1995) *Fires in Transport Tunnels: Report on Full-Scale Tests.* Studiensgesellschaft Stahlanwendung e.V., Düsseldorf.

Handbook of Tunnel Fire Safety, 2nd edition
ISBN: 978-0-7277-4153-0

ICE Publishing: All rights reserved
doi: 10.1680/htfs.41530.329

Chapter 15
CFD modelling of tunnel fires

Norman Rhodes
Hatch Mott MacDonald, New York, NY, USA

Nomenclature

C_p	Specific heat at constant pressure
g	acceleration due to gravity
h	enthalpy
k	thermal conductivity
p	pressure
S_{rj}	radiation transfer
t	time
u	velocity in the x direction
$\bar{u}$	mean velocity in the x direction
u'	fluctuating velocity in the x direction
v	velocity in the y direction
w	velocity in the z direction
x_i	coordinate in the i direction
τ_{ij}	shear stress tensor
μ	dynamic viscosity
ρ	fluid density

15.1. Introduction

A fire in a tunnel gives rise to complex three-dimensional flows driven by buoyancy forces created by the energy release of the fire. The behaviour is modified by heat transfer and turbulence, and these factors are in turn influenced by the geometrical nature of the tunnel and any ventilation system that may be in operation. Computational fluid dynamics (CFD) techniques have been developed and applied by specialists for a number of years to analyse fire and smoke behaviour (Rhodes, 1989). They are now frequently used to study and understand smoke behaviour in tunnels (Rhodes, 1994). It is important, therefore, to understand how CFD is applied, the approximations used, and the factors that influence the accuracy of the simulation.

CFD simulations of tunnel fires require a solution of the Navier–Stokes equations with appropriate boundary conditions. Solution methods are embodied in a number of general-purpose and specific CFD codes. The advantage of the CFD approach is that the complex physical interactions that occur in a fire can be modelled simultaneously, and hence their relative influence on the total behaviour understood. However, the fundamental knowledge of all the underlying physics is not complete, and there are inherent assumptions in the mathematical process that give rise to possible inaccuracy. With care, however, these inaccuracies can be reduced to a level where CFD techniques may be regarded as satisfactory for design purposes.

This chapter describes the general basis of CFD codes, outlines the models used in smoke-movement analysis, and attempts to describe some of the uncertainties that can arise. This is not a simple task, as the models themselves are very complex and may respond differently to different physical situations.

It is said that there are two kinds of fluid dynamicists: those who engage in the numerical analysis of fluids, and those who determine fluid behaviour from experiments. It is generally held that:

- nobody believes the results of a numerical analysis of fluid flow except the numerical analyst
- everyone believes the results of the experiment except the experimentalist.

There is, sadly, some truth in this. Experiments are real, while a numerical analysis is a simulation; experiments are difficult to reproduce and ensure correct measurement, while simulations can be repeated in a more controlled way; and much of engineering science is supported by experiment and observation, while simulation with computers is relatively new and requires a different background knowledge.

With these thoughts in mind, the area of validation (understood here as meaning 'comparison between theory and experiment'; see Section 15.5) is discussed in the last part of the chapter (see also Chapter 29). A validated model implies that the numerical parameters have been shown to produce a convergent answer; for example, the solution is 'grid independent', i.e. the grid on which the equations are solved is sufficiently fine to capture the important flow mechanisms, and the results obtained from the model agree with experimental data for the same case.

Clearly, when CFD is applied to a new design, validation is impossible, but the lessons learned from studies of fire experiments should guide the analyst to achieve a degree of reliability over the simulations carried out and confidence in the conclusions drawn.

Validation is a challenging area for tunnel-fire modelling, as the experimental data are far from absolute, given the complexity of the processes. The fire size, for example, is subject to interpretation, and a review of the attempts by modellers to simulate a particular experiment will reveal that different assumptions have been made. However, it will be claimed that a CFD model, correctly applied, will reproduce the qualitative behaviour of fires in a tunnel. The degree to which qualitative agreement is achieved is a matter of opinion, and varies with the modeller and the situation. If validation is sought, then good experimental data are required. If a design is under development, an understanding of the timescales of smoke movement and other life-threatening factors becomes crucial, and a CFD model is a good way of establishing this understanding and, importantly, the influence of design parameters.

15.2. Mathematical overview

The fundamental equations that are used to describe fluid motions are based on the laws of conservation of mass, momentum and energy. The following outline provides the main points, and the reader may find a full derivation of the equations in many textbooks (see, e.g., Versteeg and Malalasekera, 1995). (Note: See also Yeoh and Yuen (2009), which considers large eddy simulation (LES) to some degree.)

The mass-conservation equation is derived by considering the mass balance for a small fluid element having dimensions δx, δy and δz, and considering that:

Rate of increase of mass in a fluid element = Net rate of flow of mass into the element

The rate of increase of mass in the element is:

$$\frac{\partial \rho}{\partial t} \delta x \, \delta y \, \delta z$$

where ρ is the fluid density and t is time. The mass flow rates across the faces of the element are given by the product of density, area and the normal velocity component. For example, in the x direction, these are:

$$\left[\rho u - \frac{\partial(\rho u)}{\partial x} \frac{\delta x}{2} \right] \delta y \, \delta z - \left[\partial u - \frac{\partial(\rho u)}{\partial x} \frac{\delta x}{2} \right] \delta y \, \delta z$$

and similarly for the y and z directions. Equating these terms with the rate of increase of the mass term, and rearranging, yields the mass conservation equation:

$$\frac{\partial \rho}{\partial t} + \frac{\partial(\rho u)}{\partial x} + \frac{\partial(\rho v)}{\partial y} + \frac{\partial(\rho w)}{\partial z} = 0$$

The momentum conservation equations are based on Newton's second law, which states that:

Rate of increase of momentum of a fluid particle = Sum of the forces on the fluid particle

The forces acting on a fluid element include pressure, viscosity and buoyancy forces, in the case of a fire. In Cartesian tensor notation, this can be written as:

$$\frac{\partial}{\partial t}(\rho u_i) + \frac{\partial}{\partial x_j}(\rho u_j u_i) = -\frac{\partial p}{\partial x_j} + \frac{\partial \tau_{ij}}{\partial x_j} + \rho g_i$$

where g_i is the body force term per unit volume in the ith direction, τ_{ij} is the shear stress tensor, and p is pressure. For a Newtonian fluid, the viscous stresses are:

$$\tau_{ij} = \mu \left(\frac{\partial u_i}{\partial x_j} + \frac{\partial u_j}{\partial x_i} \right) - \frac{2}{3} \mu \frac{\partial u_k}{\partial x_k} \delta_{ij}$$

where μ is the dynamic viscosity.

The conservation of energy is based on the first law of thermodynamics, which can be expressed as:

Rate of increase of energy of the fluid element

= Rate of heat addition to the fluid element + Rate of work done on the fluid element

The increase of energy in the fluid element can comprise internal and kinetic energies, and transfer of energy can be due to convection, conduction, radiation, diffusion, and the work done by viscous stresses and body forces. The equation can be written with temperature, enthalpy or internal energy as the main variable. Using static enthalpy, the following equation is obtained:

$$\frac{\partial}{\partial t}(\rho h) + \frac{\partial}{\partial x_j}(\rho u_j h) = \frac{\partial p}{\partial t} + \frac{\partial}{\partial x_j}\left(\frac{k}{C_p}\frac{\partial h}{\partial t} - S_{rj}\right)$$

where h is enthalpy, k is the thermal conductivity, C_p is the specific heat, and S_{rj} is the radiation transfer.

Complete information regarding the three-dimensional flow requires simultaneous solution of the above equations for initial and boundary conditions. However, for most engineering problems, the solution of these equations would require impractical computing resources because of the transient nature of the equations and the fine scales of the flow that would have to be represented. However, as flows encountered in most engineering applications are steady in comparison with the timescales of turbulence, some additional development of the equations is undertaken to provide practical solutions. This is done by recognising that flows can be characterised as having mean values of properties, on which are superimposed fluctuating components. Thus, for example,

$$u = \bar{u} + u' \qquad v = \bar{v} + v'$$

and similarly for pressures, densities, temperatures and other flow properties. The equations of motion for the mean flow are obtained by substituting the above equations into the general equations. For example, the momentum equation becomes:

$$\frac{\partial}{\partial t}(\bar{\rho}\bar{u}_i) + \frac{\partial}{\partial x_j}(\bar{\rho}\bar{u}_i\bar{u}_j) = -\frac{\partial \bar{p}}{\partial x_i} + \frac{\partial}{\partial x_j}\left(\bar{\tau}_{ij} - \bar{\rho}\overline{u_i'u_j'}\right) + \bar{g}_i$$

The above procedure of decomposing the flow into mean and turbulent velocity fluctuations therefore introduces a new term into the mean momentum equation, involving the turbulent inertia tensor $\langle u_i'u_j'\rangle$. This introduces nine additional unknowns, which represent the transport of momentum, heat or mass due to turbulent fluctuations, and can be defined only through a knowledge of the turbulent structure. Hence, in order to solve a turbulent-flow problem, it is necessary to solve for the above unknowns. This is the so-called 'closure' problem.

This problem of closure is a feature of any fluid-flow analysis, not just CFD, and numerous models have been developed to provide a closed set of equations. Prandtl's mixing length model is perhaps the simplest in concept. Here, the turbulent transport is assumed proportional to a length scale and the velocity gradient. This approach is good when the turbulence in the flow can be represented by a single length scale. However, CFD has enabled the development and application of more elaborate turbulence models where this assumption can be relaxed, albeit involving the solution of further transport equations (see, e.g., Rodi, 1993). In the area of smoke modelling, the turbulence model that is most often used is the so-called k–ε model. Here, additional equations are solved for the kinetic energy of the velocity fluctuations and their rate of dissipation.

While, in principle, this can lead to a better realisation of the flow and a more accurate solution of the problem, there are pitfalls of which the user of general-purpose software must be aware in order to make good modelling decisions. For example, the k–ε model assumes isotropic turbulence, i.e. the turbulent fluctuations are of equal magnitude in each direction. In a stratified flow, such as occurs in a fire, this is not quite true, as fluctuations are damped in the vertical direction. Although there are terms in the k–ε equations to account for density stratification, more elaborate methods have been developed to account for this anisotropy (see also Rodi, 1993).

The equations are generally solved using a finite-volume method. This involves the subdivision of the flow domain into grid cells, and the derivation of algebraic finite-domain equations by integration of the differential equations over the volume of each grid cell. The convection, diffusion and source terms in the integrated equations are represented by various approximations, similar in form to finite-difference equations. The resulting algebraic equations are then solved iteratively. The numerical schemes employed in this process can affect the accuracy of the solution, and there are usually a number of alternative options available in a general-purpose CFD programme. As with any other modelling decision, the user has to decide whether there is any advantage in one scheme over another. Generally, simpler schemes will be more convergent but have a lower order of accuracy. The relative importance of this can only be studied by application to a particular problem and comparison with experience or test data. An equally important and related factor is the size of the grid into which the flow is divided. The examination of the sensitivity of the solution to grid refinement is a prerequisite of any analysis.

A potential user of these techniques has two options: either to write a new programme to apply to the specific problem; or to apply a general-purpose CFD programme, of which there are several. The advantage of writing a programme is that every feature which is implemented can be carefully tested and is fully understood. However, such a task may take a long time, and the knowledge required would be prodigious. Applying a general-purpose CFD programme is much quicker, but has the disadvantage that the user does not initially know the limitations, and so has to proceed with caution. The precise details of the internal procedures may be unclear to the user of a general-purpose CFD code. Therefore, apart from an appreciation of the general approximations inherent in the various solution schemes or the desirability of particular schemes for certain types of flow, the user can proceed in relative ignorance of the fine numerical detail. Herein lies a potential for error in the use of CFD, which can only be avoided by applying models to specific experimental cases, verifying that the predictions are in agreement, and thus gaining experience in the correct application of the particular CFD code.

15.3. Physical phenomena in tunnel-fire situations

It has already been mentioned that the flows arising in a fire in a tunnel are three-dimensional and are influenced by buoyancy effects. In addition, the flow velocities and length scales are sufficiently large that the flow is generally turbulent. At the fire, a combustion process takes place, and so there are chemical reactions going on, producing soot particles and combustion products at high temperatures. Some of the heat will be transferred by radiation and some by convection to the tunnel walls. It is possible to formulate models to predict the growth rate of a fire, although a complete knowledge of the properties of the reacting material is required for this to be meaningful. In a tunnel context, this is unlikely to be available, except in simple experiments, such as pool or crib fires. Hence a 'design fire' approach is probably more meaningful in the design process.

CFD models can provide a framework for including all these phenomena in a calculation. However, it must be emphasised that some of the physical processes are not completely understood, and that they are represented by *approximate models*. The more models that are used in a simulation, the more care is required to ensure that they are appropriately applied. Wherever possible, some verification of the simulations should be made.

In summary, the solution of a fire and smoke-movement problem requires the consideration of the above general equations. There is potential for numerical errors due to the solution strategy and decisions over grid density, and also due to inadequate knowledge of the complex and interacting physical mechanisms.

It is the author's opinion that the most important qualification for an analyst to work with CFD tools is a good knowledge of fluid mechanics, and to apply this to tunnels the analyst should have a good understanding of tunnels and their infrastructure. A knowledge of fluid mechanics provides the background for an understanding of the underlying equations, and the ability to critically assess the results of a simulation. A knowledge of the practical aspects of tunnels is necessary to ensure that the right things are modelled. PIARC (1999), for example, gives considerable background on tunnels with regard to smoke management.

15.4.   Application of CFD techniques to tunnel fires

General-purpose CFD codes are now widely used as a basis for simulating fires in tunnels. They facilitate the rapid development of simulations and embody many of the physical models that are needed for such simulations, for example, turbulence models, buoyancy source terms, combustion and radiation models. Their application involves a number of steps at which decisions are made that can affect the quality of the simulation. This section is a guide to the more important considerations.

15.4.1   Geometry and boundary conditions

The geometrical and boundary conditions are first established for the problem. It is necessary to consider how much of the system detail to include and where to apply the various boundary assumptions. That is, what length of tunnel is it appropriate to model in order to obtain the information sought?

With regard to the geometry of the tunnel, it is possible to model a regular cross-section without difficulty. If the cross-section is not regular (e.g. an unlined tunnel driven through rock), then an assessment of the cross-sectional variation is required, together with a determination of what adjustments should be made, if any, to allow for such variations. The length of the tunnel to be modelled will be decided by the overall length of the tunnel and the region of influence of the fire. In a long tunnel, a shorter section might be modelled, but this would require great care in defining the boundary conditions within the tunnel. In a short tunnel, or one where a fire is deemed to occur near a portal, then the stratified smoke layer may well reach the portal, and the consequent effects will need to be accounted for in the model. If there is doubt about the physical process in a particular region, then it is better to rely on the numerical solution in this region and extend the geometrical boundary to a point where the boundary values are more certain or can be shown to have less influence on the solution.

The boundary conditions for the simulation include all the literal boundaries (e.g. walls and portals), and also symmetry planes and any sources and sinks that might be included in the equations.

Whether or not a symmetry plane exists in the problem depends on the scenario. Many experiments utilise a pool fire in the middle of the tunnel as the source of heat. Given a regular geometry, therefore, it might be reasonable to assume symmetry on either side, and therefore model one half of the tunnel, with a corresponding economy in the computation required.

The boundary conditions will specify the assumptions made at portals and ventilation shafts. The pressure, velocity or mass flow may be prescribed. If a boundary is at a portal, then two modelling approaches are possible. The calculation can be extended outside of the tunnel and pressure boundaries set at some distance, so that the behaviour of air and smoke flowing in and out of the portal are calculated by the model. Alternatively, an empirical loss factor might be included at the portal to allow for boundary effects in a more approximate way.

Two other physical considerations relate to the wall friction and wall heat transfer. Conventionally, a wall function would be applied that supplies the shear stress and near-wall turbulence, and the heat-transfer coefficient. The wall temperature may be assumed constant during the transient, or further modelling into the tunnel wall can be carried out to predict the temperature rise at the surface.

The initial velocities of the air in the tunnel, as might be caused by wind, ventilation systems or traffic movements, are also important. These values would be obtained using another calculation method or design assumptions. The wall temperature may cause a flow due to buoyancy if the tunnel has a gradient and the atmospheric air temperature is different. Initial movements of air may well affect the initial smoke behaviour until buoyancy forces arising from the heat source of the fire begin to dominate. If this were found to be an important factor, then parametric studies should be undertaken to assess behaviour for different combinations of wall and atmospheric air temperature.

15.4.2 Modelling the fire

The fire itself can be represented as a heat source in the model; its position, growth rate and maximum size require consideration. The technical approach to modelling the fire might influence, or be influenced by, such considerations.

In general, there are a number of approaches to describing the fire and smoke source in a CFD model. If the source of the fire is known, for example a pool fire where a particular form of fuel is used, then it is possible to utilise a combustion model to represent these processes in some detail. This is an approach that might be used when simulating a fire experiment to compare with theoretical prediction. In the simulation of the combustion, additional partial differential equations would be used to represent the fuel, oxidant and combustion products. However, it should be noted that this approach may introduce additional uncertainties with respect to the ability of the combustion model to correctly predict the process. A common failure appears to be the inability to predict the combustion efficiency, resulting in temperatures that are too high. Several combustion models might be available in the CFD code, and they are unlikely to give identical results. It would be inappropriate, therefore, to use such an approach without some previous study of the capabilities of such models.

When simulating a traffic-related fire, say for a ventilation design, the source of the combustion is much more complicated, as how the vehicle burns will not be known, and nor will the overall calorific value. One approach is to assess the range of combustibility of typical traffic types in different fire-ignition scenarios, and estimate a reasonable fire-development time and fire size. When incorporating this 'design fire scenario' into the CFD model, the vehicle would probably be represented as a blocked region, and the location of the fire source, say the windows of a car, would become an inlet boundary, where hot combustion gases emerge at the flame temperature. The air to support combustion would enter the car from some other source, and ultimately from the outside environment, so locations might be chosen for a sink term for air. The sizes of the source and sink terms are related to the fire size and flame temperature.

As the combustion products are not modelled individually in this approach, a source of 'smoke' can be defined to track their dilution within the simulation. For example, the combustion products may be assumed to have a smoke yield that defines the mass of soot released per mass of fuel burnt. Relating the prescribed heat-release rate to a nominal calorific value for the fuel provides an absolute value for the source term. The 'visibility', which is often the most crucial factor in determining the tenability in the tunnel, can be derived by employing empirical formulae relating the soot concentration to the visibility of illuminated and non-illuminated signs. A similar approach can be taken to derive the concentration of combustion products such as carbon monoxide, carbon dioxide and hydrogen chloride, and so on, depending on the material burning. If there is a knowledge of the nature of the fuel source, then the yields of these various species can be estimated, and their dilution at a particular point inferred. This approach allows an estimate of the local toxicity, from gas concentrations, and obscuration, from an assumed soot yield.

It is self-evident that any or all of these boundary conditions may vary, and it is very important to consider the sensitivity of the simulation results to different boundary values. For example, in a ventilation design exercise, the fire size and growth rate might be varied to ensure that the ventilation system will control smoke in a desired way and to determine the best operation of the system for evacuation and fire-fighting.

15.4.3 Radiation

Of the heat developed in a fire, about 30% can be transferred to neighbouring walls by radiation. Heat can also be redistributed within the smoke-filled region. Radiation models enable such effects to be taken into account. Biollay and Chasse (1994) compared the effect of including radiation in a simulation of the Ofenegg Tunnel fire tests, and showed the importance of including this effect, but commented on the need to prescribe absorption coefficients, and illustrated the sensitivity of the results to this choice. Increasing the complexity of the model produces a higher fidelity in terms of the physical processes, but requires further empirical information. A simpler approach is to enhance wall heat transfer, or simply to reduce the fire output, at least to account for radiative losses at the fire source. Given the uncertainty of potential fire sizes, this approach is reasonable for preliminary design work. As with other assumptions, however, it needs to be justified, and if there is uncertainty about the effect of particular assumptions, then comparative calculations should be made to determine sensitivity.

15.4.4 Necessary precautions

When solutions are obtained, they should be reviewed in the light of questions such as the following.

- What is the influence of grid size, and time step in the case of a transient solution? Have studies been undertaken to confirm the level of accuracy of these parameters? (See Section 15.1 and Chapter 29.)
- What are the critical parameters or boundary conditions, and what influence do they have on the solution?
- Is the solution adequately converged, i.e. have the numerical errors been reduced to insignificant values by sufficient iteration of the equations?
- Is the physics of the problem adequately represented in the model?

To answer these questions is very difficult for a completely new design, and reliance has to be placed on previous experience and validation. This subject is considered in the following section, using a particular example to highlight the main points.

15.5. Validation and verification

In an engineering design context, it is important to know that the results obtained from a model lead to acceptable design decisions. The location and severity of an incident are unknowns, and so the designer might use a model to examine various scenarios. With care, these scenarios will cover all likely fire situations, and the model will provide guidance on the design and required performance of, say, ventilation systems, and some ideas about the management of the emergency response.

The above checklist does not say whether the model predicts the correct answers or whether the results reflect real flow behaviour. To build confidence in the ability to predict flows requires the application of the model to fire situations for which data are available for comparison.

There are two kinds of validation. The first is a very scientific kind, where detailed measurements of a relatively simple flow are made. The boundary conditions are specified in complete detail, and the modeller attempts to reproduce the flow given these boundary conditions. Some of the early work on turbulence-model development is of this kind. For example, for the prediction of an isothermal jet, given the inlet velocity profile and turbulence measurements, the modeller can test the ability of the code and turbulence model to predict the subsequent flow development. It is emphasised that such careful validation forms a severe test for CFD codes from a numerical point of view, and whether or not the physical models represent the flow correctly. They are also useful exercises for would-be modellers to attempt before going on to the greater complexities involved in fire modelling.

The second kind of validation is one where models are applied to full-scale tests. Such tests do not provide the necessary information to specify the initial and boundary conditions with complete accuracy and detail. Thus cases like this should be thought of as 'verifications' rather than stringent 'validations' of models. All fire experiments fall into this category, be they at large or small scale, and modellers have to interpret the experimental results in some way in order to establish the boundary conditions. In the assessment of a model it is advisable to apply the model to several different experimental cases in order to make a better judgement of its predictive capability. A common experience in 'validation' or 'verification' exercises is that different modellers make different assumptions with regard to the representation of boundary conditions, and, even with the same CFD code, arrive at different answers. Therefore, there is also an issue of 'calibrating' the user of the model as well as the model itself. (See also Beard (2000) and Rein *et al.* (2009)).

15.6. Case study: the Memorial Tunnel experiments

A number of fire experiments are available in the literature which form useful application studies. These include experiments at the Ofenegg and Zwenberg tunnels and the EUREKA programme of tests. SP (1994) and Haack (1994) provide further details of these and other experiments.

The Memorial Tunnel experiments (Massachusetts Highway Department/Federal Highway Administration, 1996) are chosen here as a basis for the discussion of various modelling issues. The Memorial Tunnel, located near Charleston, West Virginia, USA, is a two-lane, 853 m long road tunnel through a mountain. It was built in 1953, and abandoned 30 years later. The Memorial Tunnel fire tests were conducted from September 1993 to March 1995, and included tests with fire sizes of 10, 20, 50 and 100 MW, in conjunction with a series of alternative ventilation configurations, including full and partial transverse ventilation, partial transverse ventilation with single-point extraction and with oversized exhaust ports, point supply and point exhaust operation, natural ventilation and longitudinal ventilation with jet fans.

The data from these tests form a useful database for verification of CFD models. An example is chosen from this dataset to illustrate some of the above-mentioned points regarding the application of CFD to such a problem.

The test series dealing with longitudinal ventilation obtained from the use of jet fans is chosen as a specific example. One of the objectives of this series was to determine how many fans have to operate in order to control smoke, and hence to identify the minimum longitudinal air velocity required to control the backlayering of smoke and heated air. (Backlayering is the movement of smoke and hot gases produced by a tunnel fire against the stream of ventilating air along the ceiling or roof of the tunnel.)

15.6.1 Physical situation

The tunnel is 853 m long, ascending from the south to the north portal with a slope of 3.2%. The width of the tunnel is 8.5 m and it is 7.9 m high. A walkway is located on one side of the tunnel, but this is not modelled. At each portal there is a horizontal ceiling, 19.2 m long at the south side and 18.5 m long at the north side.

Twenty four jet fans, located in groups of three, were spaced approximately uniformly within the tunnel. The longitudinal central positions, measured from the south portal, were set at 80, 160, 320, 400, 480, 560, 640 and 720 m. Each jet fan had an inside diameter of 1.35 m and a length of 6.7 m. Moreover, each fan was equipped with a 56 kW motor, rated to deliver 43.0 m^3/s at an exit velocity of 34.2 m/s, and designed to withstand air temperatures of 300°C. As well as various fire sizes, the test plan examined the effect of fan response time, the interval between the onset of the fire and the activation of the ventilation system being set at a 0, 2 or 5 minute response.

The fire source was located 238 m from the south portal, and the height of diesel fuel pan was about 1 m from the floor. The fire size considered in the paper, corresponding to Test 615b, was nominally 100 MW. The ambient air temperature is taken to be 8°C and the tunnel wall temperature to be 12°C. No wind is specified.

Specific instrumentation was installed at various tunnel sections in order to measure air velocity, air temperature and smoke concentration.

15.6.2 Modelling approach

What follows should be regarded as one of several possible approaches to modelling that might be considered for this case, and indicates the preferences of the author. For example, to model the whole length of the tunnel rather than a short section, and to model each of the jet fans and their operation rather than their overall effect. Full details are given in Rhodes (1995).

15.6.2.1 Computational grid

Figure 15.1 shows a preliminary grid. The circular sections of the grid are designed to enable a specific representation of the jet fans at their relative positions. A multi-block grid, a capability of the CFD code used, enables highly skewed cells to be avoided.

Figure 15.1 A preliminary computational grid

Figure 15.1 A preliminary computational grid

The approach to modelling the fans is to represent them as a solid blockage and specify the fan velocity across the outlet area. At the inlet, a similarly sized sink of air is prescribed to accord with mass continuity. In practice, the boundary conditions for the fan are a little more sophisticated, taking into account the density, temperature and smoke concentration that might be drawn into the fan, and ensuring that these values find their way to the outlet stream. In addition, turbulence parameters appropriate for a jet are prescribed. No account is taken of swirl effects in the present model.

There are a number of possible alternatives to this approach. To be more representative of the jet fan behaviour, a momentum-source model could be devised that resembles the fan characteristic and automatically accounts for secondary velocity in the tunnel. This would presuppose a knowledge of such characteristics.

15.6.2.2 Modelling the fire
In an engineering design for a smoke-removal system, the possible fire size and its variation with time, the quantity of smoke and the location of the fire are variables that have to be determined from a risk assessment of the system, taking into account the possible use of the system. In view of this uncertainty, a very common approach is to describe the fire as a source of heat and mass arising from the combustion process. In the case of the Memorial Tunnel experiments, a diesel fuel is used, and so the actual combustion process could be modelled. For this, as discussed above, additional differential equations would be included to calculate the mass fractions of fuel and oxidant, and a combustion model would be used to calculate the local rate of combustion. To be precise, such a model would need to be extremely refined in the burning region, and only a very simple fuel, with a well-known combustion chemistry, would be feasible. However, in the

present example, in which the complete tunnel is modelled, the simpler approach is taken, and the experimentally determined heat input is used and expressed as a source in the mass continuity and enthalpy equations.

15.6.2.3 Other boundary conditions
In the momentum equations, wall friction is imposed in appropriate cells and losses are inserted at the portals. In addition, at the portals, atmospheric external pressure is assumed.

Wind effects and the wall temperature can affect the initial flow in the tunnel. In Test 615b, a small upward flow was measured prior to the test, the magnitude of velocity being about 0.5 m/s. The test was undertaken in February, and the ambient temperature was 8°C. No information is available regarding wind effects, so this initial velocity has been modelled by applying a difference between the ambient temperature and the wall temperature, sufficient to cause the flow due to buoyancy effects.

15.6.3   Steady-state model tests
In the Memorial Tunnel tests, the variation in the tunnel airflow with different numbers of jet fans in operation was measured without any fire. Simulating these measurement conditions is a useful test of the model, ensuring that the overall implementation of frictional effects and the sources and sinks that model the jet fans are correctly represented. A comparison of the measurements and the simulation results is given in Figure 15.2, and it can be seen that the overall operation of the tunnel longitudinal ventilation system is reasonably well modelled.

15.6.4   Transient simulation of Memorial Test 615b
Test 615b was a nominally 100 MW fire. The measured variation in the heat input is shown in Figure 15.3, where it can be seen that the heat release rises over the first 200 seconds of the test

Figure 15.2 Measured and predicted steady-state volumetric flows for different numbers of jet fans in operation

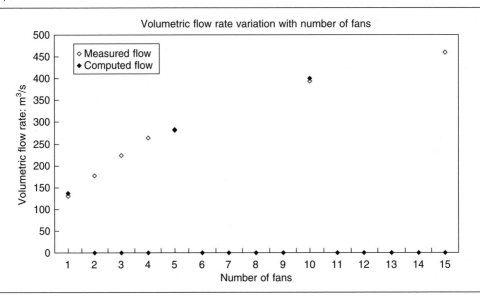

Figure 15.3 Measured heat release rate versus time

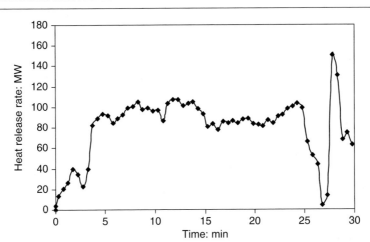

and is then maintained at a level of about 100 MW, dropping slightly at 1000 seconds to about 80 MW. The other physical parameters varied through the test were the number of jet fans in operation at different times during the test (Table 15.1).

A selection of the results of the simulation are shown in Figures 15.4 to 15.8 for predicted mean flow and upper-layer temperatures at several locations along the tunnel. Figure 15.4 shows the mean flow variation, and it can be seen that, with no ventilation, during the first 2 minutes the flow begins to increase in magnitude from about 50 m^3/s up to 100 m^3/s. The flow is overpredicted by the model, reaching almost 200 m^3/s. When the jet fans are activated, the flow reverses and, apart from a slight overshoot in magnitude at about 4 minutes, the flow is reasonably well predicted. Stopping one of the six jet fans at 14 minutes is apparent in the results, and the rate of flow reversal when the fans are stopped is also well predicted, although the flow for this no-ventilation condition is overpredicted.

Figure 15.5 shows the upper-layer temperature at a distance of 29 m from the fire. The rapid rise in temperature is accurately predicted, although some overshoot is observed in the prediction. The temperature falls equally rapidly after the start of the jet fans. The measured value indicates that the smoke backlayer is pushed back to this location after about 10 minutes and remains under

Table 15.1 The number of jet fans in operation at different times during the test

Elapsed time: min:s	Ventilation
0 to 2:00	No ventilation
2:00 to 14:00	6 jet fans
14:00 to 22:00	5 jet fans
22:00 to 25:53	6 jet fans
25:53 to 26:03	No ventilation
26:03 to 34:19	3 jet fans reversed

Figure 15.4 Test 615b: tunnel flow rate versus time

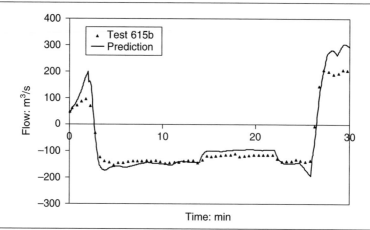

control until one of the fans switches off at 14 minutes, when the temperature increases. This behaviour is observed in the prediction, although the temperature is not brought as low as the measured value, indicating that backlayering is not quite under control.

Figure 15.6 shows the temperature variation at 62 m from the fire on the uphill side. The overall behaviour is similar to that at 29 m. The peak temperature is somewhat overpredicted in the initial rise, although general behaviour is well predicted. The backlayer is brought under control by the six jet fans, but the temperature rises again prior to the reduction to five fans.

Figures 15.7 and 15.8 show the temperature conditions at 107 m and 189 m upstream of the fire. They indicate a broadly similar pattern, where the peak temperatures are higher than measured and the control of backlayering is not as good as indicated by the measurement.

Figure 15.5 Test 615b: temperature 29 m upstream of the fire

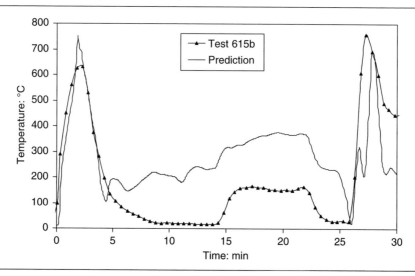

Figure 15.6 Test 615b: temperature 62 m upstream of the fire

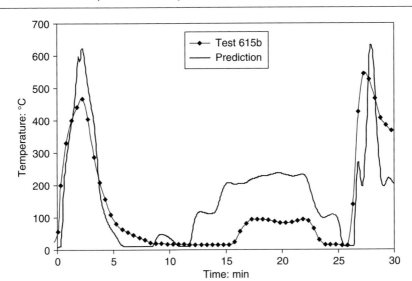

The higher predicted temperatures in the tunnel must be caused by too much heat being assumed to be in the air and smoke, either because too much heat is put in, for example by assuming too high a combustion efficiency, or the heat losses due to radiative or convective heat transfer have been assumed to be too low. This is a situation where parametric studies would enable some judgements to be made about possible causes of the differences, bearing in mind that the experimental values may have their own inaccuracies.

Figure 15.7 Test 615b: temperature 107 m upstream of the fire

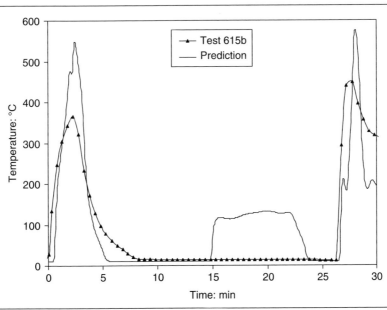

Figure 15.8 Test 615b: temperature 189 m upstream of the fire

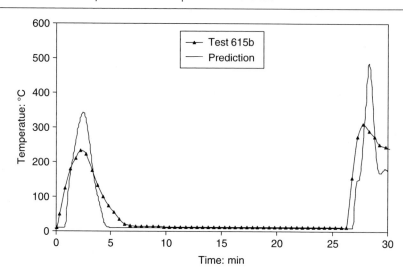

15.7. Concluding remarks

CFD provides a powerful tool for the analysis of complex fire situations in tunnels. The details of tunnel geometry and ventilation systems can be represented, taking advantage of the three-dimensional prediction capability of modern CFD codes. The physical behaviour of the fluid is represented by means of mathematical models, and it is possible to extend the description of the fluid to include effects such as turbulence, buoyancy, combustion and heat transfer by radiation and convection, all in a single simultaneous calculation.

Fluid flow processes are complex and, in order to judge the quality and accuracy of a simulation, an understanding of fluid mechanics and experience of the particular class of flows simulated is necessary. When applied to the design of a new ventilation system, CFD, used appropriately, can help to provide an understanding of the general fire and smoke behaviour, from which it may be possible to develop a more optimal configuration of the system. During this process, consideration of the fire scenarios that might occur and the fire science that these imply is necessary, as is the development of a knowledge of the sensitivity of the variables. It is important to understand how CFD is applied, the approximations used, and the factors that influence the accuracy of the simulation.

REFERENCES

Beard AN (2000) On *a priori*, blind and open comparisons between theory and experiment. *Fire Safety Journal* **35**: 63–66.

Biollay H and Chasse P (1994) Validating and optimising 2D and 3D computer simulations for the Ofenegg Tunnel fire tests. *8th International Symposium on Aerodynamics and Ventilation of Vehicle Tunnels, Liverpool, UK*, July 1994.

Haack A (1994) Introduction to the EUREKA-EU 499 FIRETUN Project. In: Ivarson E (ed.), *Proceedings of the International Conference on Fires in Tunnels*. SP Swedish National Testing and Research Institute, Borås.

Massachusetts Highway Department/Federal Highway Administration (1996) *Memorial Tunnel Fire Ventilation Test Program*. Parsons Brinckerhoff, New York.

PIARC (1999) *Fire and Smoke Control in Road Tunnels*. Report 05.05.B. World Road Association (PIARC), La Defense.

Rein G, Torero JL, Jahn W, Stern-Gottfried J, Ryder NL, Desanghere S, Lázaro M, Mowrer F, Coles A, Joyeux D, Alvear D, Capote JA, Jowsey A, Abecassis-Empis C and Reszka P (2009) Round-robin study of *a priori* modelling predictions of the Dalmarnock Fire Test One. *Fire Safety Journal* **44**: 590–602.

Rhodes N (1989) *Prediction of Smoke Movement: An Overview of Field Models*. ASHRAE paper CH-89-12-4. American Society of Heating, Refrigerating and Air-Conditioning Engineers, Atlanta, GA.

Rhodes N (1994) Review of tunnel fire smoke simulations. *8th International Symposium on Aerodynamics and Ventilation of Vehicle Tunnels, Liverpool, UK*, July 1994.

Rhodes N (1995) CFD Predictions of Memorial Tunnel Fire Test No. 614. *Proceedings of the 2nd International Conference on Safety in Road and Rail Tunnels, Granada, Spain*, 3–6 April 1995.

Rodi W (1993) *Turbulence Models and their Application in Hydraulics – A State of the Art Review*. A.A. Balkema, Rotterdam.

SP (1994) *Proceedings of the International Conference on Fires in Tunnels*. Report No 1994:54. SP Swedish National Testing and Research Institute, Borås.

Versteeg HK and Malalasekera W (1995) *An Introduction to Computational Fluid Dynamics: The Finite Volume Method*. Longman, Edinburgh.

Yeoh GH and Yuen KK (2009) *Computational Fluid Dynamics in Fire Engineering: Theory, Modelling and Practice*. Butterworth-Heinemann, London.

Handbook of Tunnel Fire Safety, 2nd edition
ISBN: 978-0-7277-4153-0

ICE Publishing: All rights reserved
doi: 10.1680/htfs.41530.347

publishing

Chapter 16
Control volume modelling of tunnel fires

David Charters
BRE Global, Watford, UK

16.1. Introduction

Control volume modelling is an approach for predicting various aspects of fires in tunnels and is often known as 'zone' modelling. The approach works by dividing the tunnel fire 'system' into a series of control volumes (or zones). Each control volume represents a part of the system that is homogeneous in nature, i.e. it is assumed to have the same properties throughout (e.g. temperature, velocity, density, species concentration). Conservation equations are applied to each control volume to predict how 'source terms', such as the fire or plume, and processes between control volumes, such as radiative and convective heat transfer, affect the properties of the control volumes. Figure 16.1 shows how the tunnel-fire domain can be divided up into different control volumes (Charters *et al.*, 1994).

16.2. Limitations

Control volume modelling has the capacity to predict various aspects of fires in tunnels, and has been used in many fields of engineering. Indeed, control volume modelling has been developed and applied to fire in compartments since the 1970s (Quintiere, 2003). However, any radical departure by the fire system from the conceptual basis of the control volume model can seriously affect the accuracy and validity of the approach. All models make assumptions, both conceptual and numerical, and it is essential for results to be interpreted in the light of these assumptions. Likewise, all models have limitations, and it is extremely important that users are aware of these. See also Chapter 29.

16.3. Application of control volume modelling to tunnel fires

Control volume modelling of tunnel fires is built on three types of model: conservation equations, source terms, and mass and heat transfer submodels. These models are described briefly below, and extensive review of control volume equations as applied to a compartment fire can be found in the literature (Quintiere, 1989).

The kinds of assumptions that are typical of control volume modelling include

- all properties in the control volume are homogeneous
- the gas is treated as an ideal gas (usually as air)
- combustion is treated as a source term of heat and mass
- mass transport times within a control volume are instant
- heat transfer to tunnel contents, such as a vehicle, is neglected

Figure 16.1 A typical tunnel-fire control volume model

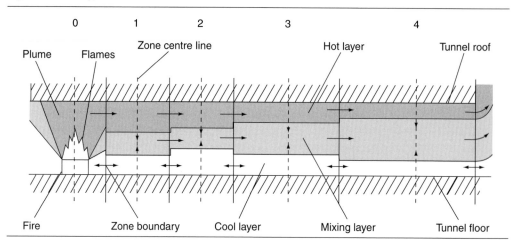

- the cross-section of the tunnel is constant and the tunnel is horizontal (although, there are empirical models that modify critical ventilation velocity for slope, and there is experimental evidence that slopes make little difference up to 5°)
- the pressure in the tunnel is assumed to be constant
- frictional effects at boundaries are not treated explicitly.

When applying control volume equations to tunnel fires, consideration should be given to the unique nature of some fire phenomena in tunnels. For example

- an assumption that hot-layer properties are homogeneous along the length of the tunnel will only be tenable for very short tunnels
- ambient and forced ventilation flows in tunnels may affect air entrainment in plumes
- the relative velocities of hot and cold layers may mean that shear mixing effects at the interface may not be negligible.

16.3.1 Conservation equations

The main basis of control volume modelling is the conservation of fundamental properties. The concept of conservation is applied to mass, energy and momentum. However, momentum is not normally applied explicitly, as the information needed to calculate velocities and pressures is based on assumptions at the boundaries.

The conservation of mass for a control volume states that the rate of change of mass in the volume plus the sum of the net mass flow rates out of the volume is zero:

$$A\frac{d(\rho z)}{dt} + \sum m_o = 0 \tag{16.1}$$

where A is the area of the control volume, ρ is the density of the gas in the control volume, z is the height of the control volume, t is time, and m_o is a net mass flow rate out of the control volume. Figure 16.2 shows how this differential equation for the conservation of mass can be applied to a control volume (Rylands *et al.*, 1998).

Figure 16.2 The mass flow between control volumes

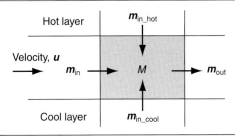

The mass contained within the control volume at time t is given by the equation (Rylands $et\ al.$, 1998):

$$M(t) = M(t - \Delta t) + \Delta t(m_{in} - m_{out} + m_{in\_hot} + m_{in\_cool}) \tag{16.2}$$

where M is the mass in the control volume, Δt is a small interval of time, m_{in}, m_{out}, m_{in_hot} and m_{in_cool}, are the mass flow rates into and out of the control volume due to convection and shear mixing effects.

Similar equations can be used for the conservation of species (i.e. mass concentrations such as that of carbon monoxide) and energy (Quintiere, 2003). The mass concentration of a species Y can be conserved by modifying equation (16.1):

$$\rho z A \frac{dY}{dt} \sum m_o(Y - Y_{cv}) = \omega \tag{16.3}$$

where Y is the mass concentration of a species in the flow, Y_{cv} is the mass concentration in the control volume, and ω is the mass production of that species due to the fire.

The mass concentration of a species depends strongly on plume entrainment (mass and heat transfer submodels, see below) and the mass production rate of that species by the fire. The mass production rate of a species will vary significantly, depending on the type of fuel and the air/fuel ratio near the fire. For example, the mass production rate of carbon dioxide depends on the rate of heat release, the amount of carbon in the fuel, and the air/fuel ratio. Limited knowledge of the detailed chemical reactions in 'real fires' combustion means that a simplified one-step chemical model and studies of stoichiometry can be used to estimate species production. However, uncertainties around the application of stoichiometry mean that most control volume models use a mass conversion factor of the mass flow rate of burnt fuel. For carbon monoxide these mass conversion factors can vary tremendously, depending on the type of fuel and whether the fire is fuel or ventilation controlled. Therefore, results of large-scale fire tests are used to derive mass conversion factors for carbon monoxide.

The conservation of energy for a control volume combines equation (16.1) and the equation of state, $p = \rho RT$, such that:

$$\rho C_p z A \frac{dT}{dt} - zA \frac{dp}{dt} + C_p \sum m(T - T_{cv}) = m_f \chi \Delta H - Q_{net\_loss} \tag{16.4}$$

where C_p is the specific heat capacity of the gas in the control volume, p is the pressure in the control volume, T is the temperature of gas in the flow, T_{cv} is the temperature of gas in the control volume, m_f is the rate at which fuel is volatilised, χ is the combustion efficiency, ΔH is the heat of combustion (taken as positive), and Q_{net_loss} is the net rate at which heat is lost to the boundary.

This model assumes that there is sufficient oxygen to react with the fuel, with a combustion efficiency factor that needs to be adjusted for incomplete combustion. There is also difficulty in dealing with combustion in vitiated layers. Often in control volume models, the rate of heat release is a user input from which the mass flow rate of fuel is derived.

If the rate of temperature change in the control volume is low, the first term can be neglected. This leads to a simpler quasi-steady analysis for growing fires. In well-ventilated conditions, most transport tunnels are well ventilated in terms of fire dynamics, and the second term can also be neglected. This leaves an enthalpy term for flows into and out of the control volume and two source terms. For flows out to the control volume, the lumped parameter assumption means that $T = T_{cv}$.

16.3.2 Source terms

The main source terms in fire modelling are the rate of heat release and the mass flow rate of fuel. Where the fire source is known and well controlled, such as a gas burner, precise values can be used. In most fire safety situations, the rate of heat release and the mass flow rate of fuel is the result of a spreading fire over a variety of materials and surfaces. In these circumstances, an empirical model to give the rate of heat release at time t can be used (Law, 1995):

$$q_f = \alpha(t - t_o)^2 \tag{16.5}$$

where q_f is the rate of heat release of the fire, α is the fire growth coefficient, and t_o is the time between ignition and fire growth (incubation period). Alternative relationships have also been developed for large tunnel fires (Ingason, 2005, 2009).

Similar source term models exist for the mass flow rate of fuel. For example:

$$m_f = \frac{q_f}{\Delta H \chi} \tag{16.6}$$

Given the conversion rate of fuel to carbon monoxide (a function of the materials and how well ventilated the fire is), the mass concentration of carbon monoxide can be estimated.

These empirical models provide approximations, and so consideration should be given to ensuring that values for the coefficient of fire growth, combustion efficiency and mass conversion rates are appropriate.

16.3.3 Mass and heat transfer submodels

Mass and heat transfer models are an essential feature of control volume models. These models may include

- entrainment in plumes
- flows through openings
- mixing between layers
- convective heat transfer to surfaces
- radiative heat transfer
- conductive heat transfer
- other effects.

Figure 16.3 Plumes in natural and forced ventilation. (a) Natural ventilation (b) longitudinal ventilation

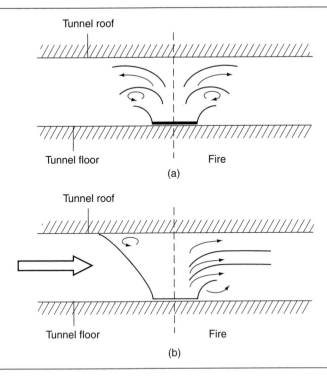

Entrainment in plumes has been shown to have a critical effect on the tenability and development of fire hazards. There is a range of models available to predict flame heights and mass entrainment in plumes (Law, 1995; Heskestad, 2003a). It is essential that the model used is appropriate to the physical situation of the real fire. For example, entrainment in an axisymmetric plume may be appropriate for a fire in the floor in the middle of a tunnel section, but may not be appropriate for a spill plume from a vehicle window or a plume attached to one of the tunnel walls. The effect on plume properties of any ambient ventilation, and for situations where flames reach the hot layer, should also be taken into account. A schematic illustration of a tunnel fire plume in natural and longitudinal ventilation is given in Figure 16.3. The diagram of longitudinal ventilation shows a small region of backlayering to the left side of the fire. If the longitudinal ventilation is greater than the critical velocity (see Chapter 10), the backlayering regions would no longer exist.

Flows through openings are crucial to the modelling of fires in compartments, such as the doors and windows of vehicles, flows into and out of tunnel ventilation system openings, and entrance/exit portals. Detailed models for flows through openings and tunnel ventilation systems can be found in the literature (Haerter, 1991; Emmons, 2003).

Mixing between layers can occur in one of three ways

- a cold flow injected into a hot layer
- shear mixing associated with lateral layer flows (including backlayering)
- mixing due to tunnel wall flows.

Cold flows injected into a hot layer can be resolved by computational fluid dynamics or physical modelling. Shear mixing of layers has been studied to a certain extent, and some correlations are available for counterflow between layers and backlayering against forced ventilation. Mixing due to wall flows has been studied in compartment fires. None of these three phenomena is as critical as the primary buoyant mixing in the plume (Emmons, 1991; Vantelon et al., 1991).

Convective heat transfer to surfaces is one of the main processes of heat loss between the hot-layer control volume and the tunnel lining. Convective heat transfer to ceilings has been studied extensively. Convective flows will vary along the tunnel walls and ceiling, depending on their position with respect to the fire. Where convective heat transfer is dependent on local boundary-layer temperatures (rather than the hot-layer control volume temperature), an adiabatic wall temperature approach can be used. Convective heat transfer for ceilings and walls in tunnel fires has not been developed, and so most control volume models use natural convection correlations (Alpert, 2003; Atreya, 2003).

Radiative heat transfer theory is generally sufficient for control volume models of tunnel fires. Grey-body radiation from uniform-temperature hot gas layers can be predicted, although emissivity values require careful consideration. For radiation from flames, empirical data are used, because the complex temperature distributions for radiation from flames and the role of soot are not well understood (Heskestad, 2003b; Tien et al., 2003).

Conductive heat transfer through the tunnel linings should be balanced with the radiative and convective heat transfer from the hot gas layer control volumes. This entails a numerical (or graphical) solution to a set of partial differential equations. Most control volume models consider conductive heat transfer in one dimension only, and that the tunnel linings are thermally thick, or conductive heat transfer to an infinite ambient heat sink (Carslaw and Jaeger, 1959; Rockett and Milke, 2003).

A schematic illustration of the kinds of heat-transfer process that may be relevant to fires in tunnels is given in Figure 16.4.

From the above it can be seen that other effects can easily be incorporated in control volume tunnel fire models, as long as there is an adequate description of the phenomenon and there are appropriate data.

16.4. Application of control volume models in tunnel fire safety
16.4.1 Tunnel fire growth
Fire growth is normally modelled using a general parabolic fire-growth model, such as equation (16.5), and/or specific empirical data from fire tests (Law, 1995; PIARC, 1999). In control volume models, thermal radiation and piloted ignition are normally considered as the dominant mechanisms for the fire growth between items/vehicles. This is not always the case, however (see below).

Observations of tunnel fire experiments have indicated that, when flames impinge on the tunnel ceiling, their length is extended down the tunnel due to the lower levels of oxygen available for combustion in the hot layer. The length of these horizontally extended flames has been estimated at between 5 and 10 times what would have been their vertical length. This can have a significant implication for fire spread in tunnel fires.

Figure 16.4 Typical tunnel fire heat-transfer processes

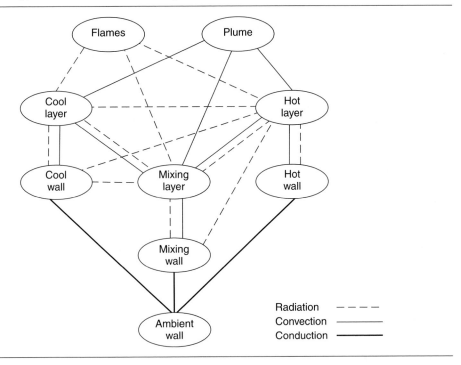

The development of a non-linear control volume model for fire spread in tunnels was initiated in the 1990s, and this model has been applied to several situations (Beard, 1995, 1997, 2001, 2003, 2004, 2006; Beard *et al.*, 1995). Two sets of models have been created under the generic title FIRE-SPRINT (an acronym of Fire Spread in Tunnels), and they continue to be developed (see Chapter 10). The models consider a burning object in a tunnel (Figure 16.5), and a longitudinal flow driving smoke and flames to one side of the fire and over a target object some distance

Figure 16.5 Vertical sections through tunnel and control volume (FIRE-SPRINT A3) (From Beard (2001), with permission of Independent Technical Conferences Ltd and the University of Dundee)

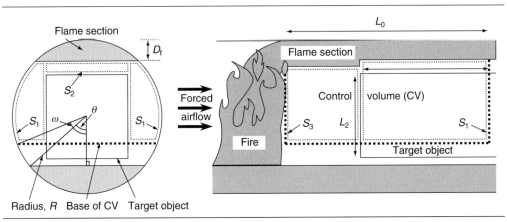

away. The A series of models assumes that there is no flame impingement from an initial fire onto a target object and that fire spread is by spontaneous ignition. The B series assumes that there is flame impingement. There are three models in the A series and one model in the B series. In the A series, FIRE-SPRINT A1 (i.e. model A, version 1) assumes (Beard, 1995; Beard et al., 1995) that there is no flame downstream of the initial fire, and FIRE-SPRINT A2 assumes (Beard, 1997) that flame extends downstream over the target object but without impinging on it. FIRE-SPRINT A3 assumes (Beard, 2001, 2006) a deeper flame section between the initial fire and the target, but without impingement. FIRE-SPRINT B1 is similar to FIRE-SPRINT A3, but assumes that there is persistent flame impingement on the target (Beard, 2003, 2004).

The models use the conservation of energy and mass to develop, depending on the model, one or more differential equations for one or more control volumes between the fire and the target object. For the A series, the temperature T in the control volume is assumed to be given by:

$$\frac{dT}{dt} = \frac{(G - L)}{(C_v \rho V)} \qquad (16.7)$$

where G is the rate of gain of energy of the control volume, L is the rate of loss of energy from the control volume, V is the volume of the control volume, C_v is the specific heat at constant volume of the gases in the control volume, and ρ is the density of hot gases in the control volume.

In the B series models, there is a second control volume and a second temperature. The models identify the onset of instability with major fire spread in a tunnel; in this case, spread from an initial fire to the 'target object'. The models enable the identification of the thermo-physical and geometrical conditions that lead to such sudden fire spread.

16.4.1.1 Caveat: uncertainty of input
As with all models, the results obtained depend on the conceptual and numerical assumptions. In any real-world application there will be uncertainty in the numerical input, and different values should be used for input parameters to get some idea of how sensitive the results are to variations in input. In FIRE-SPRINT A3, for example, sensitivity has been found to the assumed length of the control volume (Beard, 2006) (see also Beard (2003, 2007), and Chapter 29).

16.4.1.2 Caveat: comparison with experiment
In comparing theoretical predictions from any model with experimental results, it is very important to be explicit about any assumptions made and to make it clear how the comparison has been conducted (Beard, 2000).

Each model has been used to predict the critical heat-release rate (HRR) for fire spread from one object to another in a tunnel with forced longitudinal ventilation. Figure 16.6 shows an example of the predicted critical HRR for such fire spread as it varies with the separation distance between the target and the initial fire obtained using FIRE-SPRINT A3. The target length is 7.75 m and the control volume is assumed, here, to extend to the end of the target. In this case the tunnel is assumed to be similar in size to the Channel Tunnel, with a forced longitudinal ventilation velocity of 2 m/s. Analysis of fire spread, via spontaneous ignition, to a heavy goods vehicle (HGV) in the Channel Tunnel has shown that the predicted critical HRR is approximately 30–40 MW, and increasing ventilation velocity (above 2 m/s) reduces the critical HRR. This assumes a separation of about 6.45 m. FIRE-SPRINT A3 assumes that there is no flame impingement and that fire spread is by spontaneous ignition. (The term 'remote ignition' may be regarded as including

Figure 16.6 Predicted critical HRR for fire growth versus separation. Data obtained using FIRE-SPRINT A3, and assuming a tunnel similar in size to the Channel Tunnel and a ventilation velocity of 2 m/s (From Beard (2001), with permission of Independent Technical Conferences Ltd and the University of Dundee)

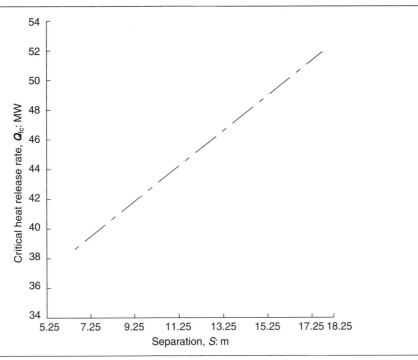

mechanisms other than spontaneous ignition, e.g. spread by burning brands.) The model FIRE-SPRINT B1 assumes that there is flame impingement and, using this model, the general indication is that including the assumption of flame impingement greatly reduces the calculated critical HRR for fire spread by the order of 60–70%. Which of the models (i.e. FIRE-SPRINT A1, A2, A3 or B1) is appropriate in a given case depends on the degree of fire development and whether or not flame impingement is likely to take place.

16.4.2 Smoke movement

Recent fires in tunnels have illustrated how, for fires in tunnels, the products of combustion are confined in all but one or two directions, and this may result in rapid smoke movement and threat to life. Figure 16.1 shows the typical control volumes associated with a three-layer tunnel fire model. A review of tunnel fire control volume models indicated a number that are being used to predict smoke movement in tunnel fires (Charters et al., 1994; Bettis et al., 1995; Altinakar et al., 1997; PIARC, 1999; Riess and Bettelini, 1999; Dériat and Réty, 2004; Michaux and Vauquelin, 2004; Suzuki et al., 2008).

The main differences between the various models are

- two layer (hot and cool), three layer (hot, mixing/shear and cool) or multi-layer
- fundamental conservation equations applied to control volumes, or integration of differential equations

Figure 16.7 Typical temperature versus distance results for varying times (From Charters *et al.*, 1994, with kind permission from Springer Science + Business Media)

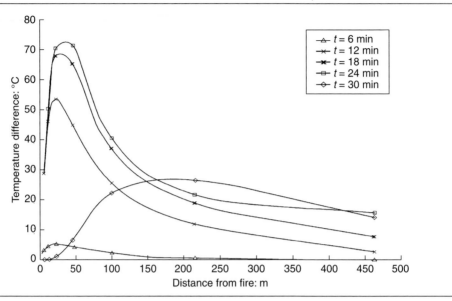

- transient or steady state
- bespoke tunnel fire model, or modified compartment fire model
- ventilation (natural, longitudinal, transverse or smoke extract).

There are also likely to be significant differences between the source term and the heat and mass transfer submodels used in each of the models, and these should be reviewed when selecting a model. For example, some models incorporate modified plume submodels to address flame impingement on the hot layer or tunnel roof.

As with all modelling, predictions from control volume models should be compared with measurements taken from a wide range of tunnel-fire experiments. A number of tunnel-fire experiments are referenced in the literature (Ingason, 2008). Figures 16.7 and 16.8 show typical results from a transient, three-layer, tunnel fire control volume model (Charters *et al.*, 1994). Figure 16.7 shows the transient temperatures at head height at different distances from the fire. Figure 16.8 compares a typical vertical temperature profile with experimental data (Charters *et al.*, 1994, 2001).

Control volume models are being increasingly used to inform design decisions regarding tunnel fire safety, and the following two sections indicate two typical applications.

16.4.3 Tunnel fire tenability

By analysing the response of people escaping a tunnel fire in relation to the hazards that the fire creates, it is possible to gain a much better understanding of the way in which a tunnel and its fire precautions will perform (Charters, 1992). Control volume models are particularly well suited to providing information on fire hazards so that tenability can be assessed.

Figure 16.8 Typical control model results compared with experimental data (From Charters *et al.*, 1994, with kind permission from Springer Science + Business Media)

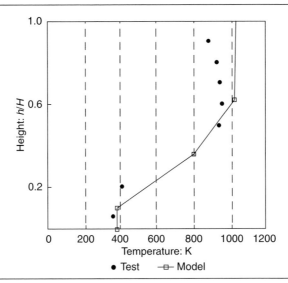

The main hazards from tunnel fires are (Purser, 2003)

- heat (hyperthermia)
- toxic gases and hypoxia
- thermal radiation.

The level of visibility in a tunnel fire, although not directly hazardous, may reduce the ability of occupants to find their way, and thus increase their exposure to other fire hazards.

Tenability to heat and many toxic gases is strongly dose related. This led to the idea of a person experiencing a 'concentration integral' for toxic gases and the concept of a 'fatality fraction'. With this concept, death corresponds to a fatality fraction of 1 (Beard, 1981). Unconsciousness was assumed to occur at a value of the fatality fraction between 0 and 1. The related concept of fractional effective dose (FED) was developed, and FED techniques are used (Purser, 2003). The FED is defined as the fraction of the dose that would be expected to result in a specific effect (e.g. incapacitation) in 50% of the population exposed to that dose. For example, the heat FED for incapacitation per second can be calculated as:

$$F_{\text{heat}} = \frac{1}{\left(60\,e^{(5.2 - 0.027T)}\right)} \tag{16.8}$$

where T is the temperature of the smoke (°C). This relationship indicates that a temperature of 100°C could be tolerated for 12 minutes and a temperature of 60°C could be tolerated for 35 minutes.

For toxicity (of carbon monoxide) the FED per second, for incapacitation, is given by (Purser, 2003):

$$F_{\text{toxicity}} = \frac{C}{90} \tag{16.9}$$

where C is the percentage concentration of carbon monoxide. This relationship indicates that 0.5% carbon monoxide could be tolerated for 3 minutes. A maximum concentration criterion of 1% carbon monoxide is also applied, and consideration should be given to selecting a carbon monoxide conversion factor that will be conservative.

For thermal radiation, a criterion of $2.5\,kW/m^2$ as a threshold is often taken. Radiation at this level causes intense pain, followed by burns within about 30 seconds (Charters *et al.*, 2001). However, below this value, thermal radiation can be tolerated over extended periods of time (typically minutes), and above this level tolerability is measured in seconds.

The visibility criterion may be either 5 or 10 m, depending on the nature of the tunnel and the ease of way-finding within it. For example, for a single-track rail tunnel with a clearly marked walkway and cross-passage doors, a 5 m visibility criterion may be appropriate. Similarly, for a three-lane road tunnel with clearly marked cross-passage doors, a 10 m visibility criterion may be appropriate.

Therefore, the hazard versus time output for each control volume from the smoke movement model can be integrated with a semi-infinite stream of people moving away from the fire. As each individual moves down the tunnel, he moves from one control volume to the next. Their dose can then be calculated based on the level of hazard in each control volume as he passed through it.

To aid the visualisation of the distance/time domain, tenability maps have been developed (Charters, 1992). Figure 16.9 shows a tenability map for a 4 MW fire that breaks out of a rail carriage, with an ambient air flow of 0.5 m/s (Charters *et al.*, 2001). The vertical axis is the distance from the fire in one direction and the horizontal axis is the time from ignition of the fire. The parallel dashed lines show the locus of people escaping, at a walking speed of 1 m/s, starting 20 m from the fire, at various times after ignition. The thick dashed curve shows the locus of

Figure 16.9 Typical tenability map

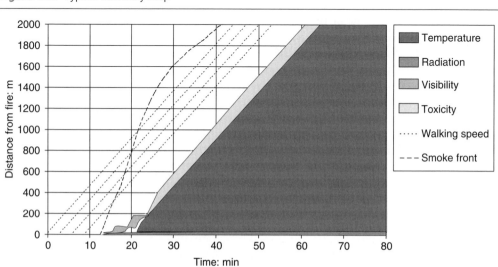

Figure 16.10 A tunnel fire scenario with a safe outcome

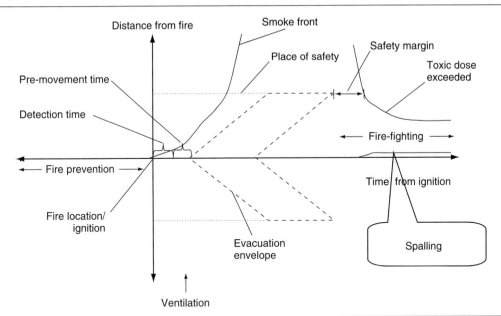

the smoke front from the fire. The various hashed and shaded areas show where the various visibility, temperature and toxicity criteria are exceeded. The area to the left of these areas shows the space/time envelope in which evacuation is tenable. This work showed that considering a range of fire scenarios is important in reducing fire risk, and the selection of a single large fire scenario may not lead to the minimisation of risk.

Tenability maps can be used in two main ways

- for the fire safety design of a tunnel system
- to quantify consequences for a fire risk assessment.

Figures 16.10 and 16.11 show how tenability maps can be used in practice (Salisbury *et al.*, 2001). In these figures, the 'pre-movement' time is the time between the discovery of a fire and the start of egress travel.

Figure 16.10 illustrates how a tunnel fire with short detection and pre-movement times and reduced fire hazards (due to longitudinal ventilation being 'critical', i.e. sufficient to prevent significant backlayering) can have a safe outcome. The occupants are predicted to move within the evacuation envelope and reach a place of safety (i.e. a portal, cross-passage or vertical escape door). A factor of safety is present, where the time taken for toxicity to lead to incapacitation is longer than the time taken for the last person to reach safety. Critical ventilation allows fire-fighting upwind of the fire, and reduces the temperature downwind of the fire by dilution, thus reducing the extent of spalling.

Figure 16.11 illustrates how longer detection and pre-movement times with greater fire hazards can lead to casualties. The people evacuating during the latter half of the evacuation envelope are predicted to become incapacitated by smoke toxicity before reaching a place of safety.

Figure 16.11 A tunnel fire scenario with casualties

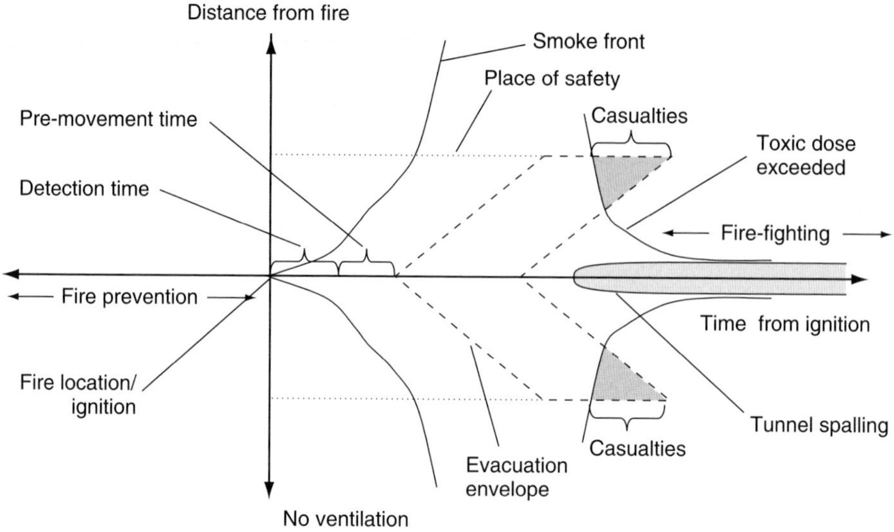

16.4.4 Quantitative fire risk assessment

Historical data indicate that tunnels are a relatively safe form of transport system, but that they can also experience high-consequence/low-frequency fire events. These are naturally of concern to society, and one of the more rational ways of addressing these concerns is through quantitative fire risk assessment (Charters, 1992; Scott *et al.*, 2001). Control volume modelling is particularly well suited to the prediction of fire hazards and consequences of the wide range of fire scenarios that can occur.

Quantitative fire risk assessment involves taking an 'initiating event', such as the ignition of a vehicle, and assessing all the other events that might happen to lead to safe 'outcomes' (e.g. suppression using first-aid fire-fighting equipment, through to an uncontrolled HGV fire).

Figure 16.11 shows how control volume models can predict the consequences of the many fire scenarios required for quantitative fire risk assessment. These, 'what if' scenarios may involve a variety of variables such as, inter alia

- fire size and growth rate
- ventilation regime
- number of people
- escape distance
- tunnel cross-sectional area
- head height above tunnel floor
- walking speed.

The output of a quantified fire risk assessment of societal risks (events that can cause multiple fatalities) is normally shown on a *F–N* curve. A typical *F–N* curve is shown in Figure 16.12. The vertical axis indicates the cumulative frequency of events that have consequences of more

Figure 16.12 A typical F–N curve

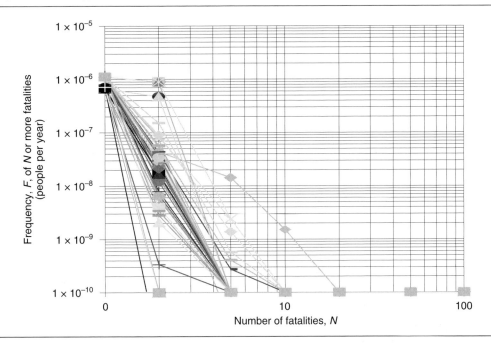

than or equal to the values on the horizontal axis. *F–N* curves can be used to assess the accept-ability of risks against risk acceptance criteria, or by comparison with the level of risk from code-compliant tunnel design solutions. These techniques are increasingly being used by designers, operators and regulators of tunnels (Vardy, 1992, 1995, 1998, 2001; PIARC, 1999; Fraser-Mitchell *et al.*, 2007).

16.5. Summary
Control volume models have been shown to be very valuable in understanding fire development and smoke movement in a tunnel. They have the potential to make a significant contribution to the fire risk decision-making process.

REFERENCES
Alpert RL (2003) Ceiling jet flows. *The SFPE Handbook of Fire Protection Engineering*, 3rd edn. National Fire Protection Association, Quincy, MA, Section 4, Chapter 2.

Altinakar MS, Weatherill A and Nasch P-H (1997) The use of a zone model in predicting fire and smoke propagation in tunnels. In: *9th International Symposium on Aerodynamics and Ventilation of Vehicle Tunnels, Aosta, Italy*. BHR Group, Cranfield.

Atreya A (2003) Convection heat transfer. *The SFPE Handbook of Fire Protection Engineering*, 3rd edn. National Fire Protection Association, Quincy, MA, Section 1, Chapter 3.

Beard AN (1981) A stochastic model for the number of deaths in a fire. *Fire Safety Journal* **4:** 169–184.

Beard AN (1995) Prediction of major fire spread in a tunnel. In: Vardy AE (ed.), *Second International Conference on Safety in Road and Rail Tunnels, Granada, Spain, 3–6 April 1995*. ITC Ltd/University of Dundee, Dundee.

Beard AN (1997) A model for predicting fire spread in tunnels. *Journal of Fire Sciences* **15**: 277–307.

Beard AN (2000) On *a priori*, blind and open comparisons between theory and experiment. *Fire Safety Journal* **35**: 63–66.

Beard AN (2001) Major fire spread in tunnels: a non-linear model. In: Vardy AE (ed.), *Proceedings of the 4th International Conference on Safety in Road and Rail Tunnels, Madrid, Spain*, 2–6 April 2001. ITC Ltd/University of Dundee, Dundee.

Beard AN (2003) Major fire spread in a tunnel: a non-linear model with flame impingement. In: Vardy AE (ed.), *5th International Conference on Safety in Road and Rail Tunnels, Marseille, France*, 6–10 October 2003. ITC Ltd/University of Dundee, Dundee.

Beard AN (2004) Major fire spread in a tunnel assuming flame impingement: effect of separation and ventilation velocity. *5th International Conference on Tunnel Fires, London, UK*, 25–27 October 2004.

Beard AN (2006) A theoretical model of major fire spread in a tunnel. *Fire Technology* **42**: 303–328.

Beard AN (2007) Major fire spread in tunnels: comparison between theory and experiment. *1st International Tunnel Safety Forum for Road and Rail, Nice, France*, 23–25 April 2007.

Beard AN, Drysdale DD and Bishop SR (1995) A non-linear model of major fire spread in a tunnel. *Fire Safety Journal* **24**: 333–357.

Bettis RJ, Diash N, Jagger S and Linden PF (1995) Control of smoke movement close to a tunnel fire. *Proceedings of the 2nd International Conference on Safety in Road and Rail Tunnels, Granada, Spain*, 2–6 April 1995.

Carslaw HS and Jaeger JC (1959) *Conduction of Heat in Solids*, 2nd edn. Oxford University Press, Oxford.

Charters DA (1992) Fire risk assessment of rail tunnels. In: Vardy AE (ed.), *Proceedings of the 1st International Conference on Safety in Road and Rail Tunnels, Basel, Switzerland*, 23–25 November 1992. ITC Ltd.

Charters DA, Gray WA and McIntosh AC (1994) A computer model to assess fire hazards in tunnels (FASIT). *Fire Technology* **30**(1): 134–154.

Charters DA, Salisbury M, Scott P and Formaniak A (2001) To blow or not to blow? In: Vardy AE (ed.), *Proceedings of the 4th International Conference on Safety in Road and Rail Tunnels, Madrid, Spain*, 2–6 April 2001. ITC Ltd/University of Dundee, Dundee.

Dériat E and Réty J-M (2004) Stratified flows generated by railway tunnel fires: initial conditions for a two-layer model. *Proceedings of the 5th International Conference on Tunnel Fires, London, UK*, 25–27 October 2004.

Emmons H (1991) The ceiling jet in fires. In: Cox G and Langford B (eds), *Fire Safety Science: Proceedings of the Third International Symposium*, 1st edn. Routledge, Oxford.

Emmons H (2003) Vent flows. *The SFPE Handbook of Fire Protection Engineering*, 3rd edn. National Fire Protection Association, Quincy, MA, Section 2, Chapter 3.

Fraser-Mitchell J, Trijssenaar-Buhre I and Waymel F (2007) UPTUN: a risk assessment methodology for fires in tunnels. *Proceedings of Interflam 2007, London, UK*, 3–5 September 2007. Interscience Communications, New York.

Haerter A (ed.) (1991) *Aerodynamics and Ventilation of Vehicle Tunnels*. Elsevier Applied Science, Oxford.

Heskestad G (2003a) Fire plumes. *The SFPE Handbook of Fire Protection Engineering*, 3rd edn. National Fire Protection Association, Quincy, MA, Section 2, Chapter 1.

Heskestad G (2003b) Fire plumes, flame height and air entrainment. *The SFPE Handbook of Fire Protection Engineering*, 3rd edn. National Fire Protection Association, Quincy, MA,

Section 2, Chapter 1.

Ingason H (2005) Fire development in large tunnel fires. In: Gottuk DT and Lattimer BY (eds), *Fire Safety Science – Proceedings of the Eighth International Symposium, Beijing, China*, 18–23 September 2005. Interscience Communications, New York.

Ingason H (2008) State of the art of tunnel fire research. *Proceedings of the Ninth International Symposium for Fire Safety Science, Karlsruhe, Germany*, 21–26 September 2008. Interscience Communications, New York.

Ingason H (2009) Design fire curves for tunnels. *Fire Safety Journal* **44**: 259–265.

Law M (1995) *Relationships for Smoke Control Calculations*. Technical Memorandum (TM) 19. Chartered Institution of Building Service Engineers, London.

Michaux G and Vauquelin O (2004) A model to evaluate the smoke confinement velocity in the case of a tunnel fire. *Proceedings of the 5th International Conference on Tunnel Fires, London, UK*, 25–27 October 2004.

PIARC (1999) *Fire and Smoke Control in Road Tunnels*. World Road Association (PIARC), La Defense.

Purser DA (2003) Toxicity assessment of combustion products. *The SFPE Handbook of Fire Protection Engineering*, 3rd edn. National Fire Protection Association, Quincy, MA, Section 2, Chapter 6.

Quintiere JG (1989) Fundamentals of enclosure fire zone models. *Journal of Fire Protection Engineering* **1**(3): 99–119.

Quintiere JG (2003) Compartment fire modelling. *The SFPE Handbook of Fire Protection Engineering*, 3rd edn. National Fire Protection Association, Quincy, MA, Section 3, Chapter 5.

Riess I and Bettelini M (1999) The prediction of smoke propagation due to road tunnel fires. In: *Proceedings of the International Conference on Tunnel Fires and Escape from Tunnels, Lyon, France*, 5–7 May 1999. ITC Ltd.

Rockett JA and Milke JA (2003) Conduction of heat in solids. *The SFPE Handbook of Fire Protection Engineering*, 3rd edn. National Fire Protection Association, Quincy, MA, Section 1, Chapter 2.

Rylands S, David P, McIntosh AC and Charters DA (1998) Predicting fire and smoke movement in tunnels using zone modelling. In: Vardy AE (ed.), *Proceedings of the 3rd International Conference on Safety in Road and Rail Tunnels, Nice, France*, 9–11 March 1998. ITC Ltd/University of Dundee, Dundee.

Salisbury M, Charters DA and Marrion C (2001) Application and limitations of quantified fire risk assessment techniques for the design of tunnels. *Proceedings of the Third International Conference on Tunnel Fires and Escape from Tunnels, Washington, DC*, 9–11 October 2001. Tunnel Management International, Tenbury Wells.

Scott P, Faller G and Charters DA (2001) Cost–benefit analysis of tunnel safety systems – Under which circumstances are complex systems justified? In: Vardy AE (ed.), *Proceedings of the 4th International Conference on Safety in Road and Rail Tunnels, Madrid, Spain*, 2–6 April 2001. ITC Ltd/University of Dundee, Dundee.

Suzuki K, Tanaka T and Harada K (2008) Tunnel fire simulation model with multi-layer zone concept. *Proceedings of the Ninth International Symposium for Fire Safety Science, Karlsruhe, Germany*, 21–26 September 2008. Interscience Communications, New York.

Tien CL, Lee KY and Stretton AJ (2003) Radiation heat transfer. *The SFPE Handbook of Fire Protection Engineering*, 3rd edn. National Fire Protection Association, Quincy, MA, Section 1, Chapter 4.

Vantelon JP, Guelzim A, Quach D, Son DK, Gabay D and Dallest D (1991) Investigation of fire induced smoke movement in tunnels and stations: an application to the Paris Metro. In:

Cox G and Langford B (eds), *Fire Safety Science: Proceedings of the Third International Symposium*, Edinburgh, 8–12 July, 1991 1st edn. Elsevier Applied Science, London.

Vardy AE (ed.) (1992) Risk assessment and risk management. *Proceedings of the 1st International Conference on Safety in Road and Rail Tunnels, Basel, Switzerland*, 23–25 November 1992. ITC Ltd, Parts 5 and 6.

Vardy AE (ed.) (1995) Risk assessment and risk management. In: *Proceedings of the 2nd International Conference on Safety in Road and Rail Tunnels, Granada, Spain*, 2–6 April 1995, ITC Ltd, Part 3.

Vardy AE (ed.) (1998) Hazards and associated risks. In: *Proceedings of the 3rd International Conference on Safety in Road and Rail Tunnels, Nice, France*, 9–11 March 1998. ITC Ltd, Part 1.

Vardy AE (ed.) (2001) Hazards and risks. In: *Proceedings of the 4th International Conference on Safety in Road and Rail Tunnels, Madrid, Spain*, 2–6 April 2001. ITC Ltd/University of Dundee, Dundee, Part 5.

Handbook of Tunnel Fire Safety, 2nd edition
ISBN: 978-0-7277-4153-0

ICE Publishing: All rights reserved
doi: 10.1680/htfs.41530.365

publishing

Chapter 17

One-dimensional and multi-scale modelling of tunnel ventilation and fires

Francesco Colella
Dipartimento di Energetica, Politecnico di Torino, Italy; BRE Centre for Fire Safety Engineering, University of Edinburgh, UK
Vittorio Verda
Dipartimento di Energetica, Politecnico di Torino, Italy
Romano Borchiellini
Dipartimento di Energetica, Politecnico di Torino, Italy
Guillermo Rein
School of Engineering, University of Edinburgh, UK

17.1. Introduction

In underground structures, the use of ventilation in a fire incident is crucial in order to limit smoke propagation, and to maintain safe conditions during evacuation and fire-fighting.

It is important to note that a tunnel and its corresponding ventilation plant constitute a single complex system. The thermo-fluid-dynamic behaviour of this system is affected by several internal and external factors, such as the barometric pressure at the portals, the tunnel slope, the set-point of the ventilation system, and the traffic conditions. Moreover, in the case of a fire it is necessary to consider the fire dynamics, the production of smoke, and the heat exchange with the tunnel linings. In order to take all these factors into account, it is necessary to consider the whole system in the analysis and not simply a part of it (Chasse and Apvrille, 1999). This means that the size of the computational domain can be up to several kilometres long, and the resulting computing time very long. Furthermore, comprehensive studies of tunnel ventilation flows and fires very often require the analysis of different scenarios in order to determine the emergency procedures and the optimal ventilation strategy. In most of these cases, the simulation should cover the entire event development, from ignition to completion of evacuation. Depending on the accuracy required and the computational resources available, different numerical tools can be adopted.

The overall behaviour of the ventilation system can be approximated using one-dimensional (1D) fluid dynamics models, under the assumptions that all the fluid-dynamic quantities are effectively uniform in each tunnel cross-section, and significant gradients are present only in the longitudinal direction. One-dimensional models have low computational requirements, and are particularly attractive for parametric studies where a large number of simulations have to be conducted. The assumption of uniform fluid-dynamic quantities in each tunnel cross-section makes 1D modelling unsuitable for simulating the fluid behaviour in regions characterised by high temperature or velocity gradients. Therefore, their accuracy relies mainly on the calibration constants and ill-defined semi-empirical parameters. The main features and the applicability of such models are discussed in Section 17.2.

These assumptions can be greatly relaxed, in principle, and the resulting accuracy can be greatly increased by using computational fluid dynamics (CFD). CFD modelling of fire phenomena within tunnels suffers from limitations due to the large size of the computational domain, and the high aspect ratio between longitudinal and transversal length scales, which leads to very large meshes. The number of grid points escalates with increasing tunnel length, resulting in very large computing times that often become impractical for engineering purposes, even for short tunnel (less than 500 m long). An assessment of the mesh requirements for tunnel flows by Colella *et al.* (2009, 2011) grid-independent solutions are achieved for mesh densities greater than 4000 and 2500 cells/m for ventilation- and fire-induced flows, respectively.

High computational costs arise when the CFD model has to take into account boundary conditions in the tunnel exterior, or flow characteristics in locations far away from the region of interest. This is the case with tunnel portals, ventilation stations or jet-fan series located long distances away from the fire. In these cases, even if only a limited region of the tunnel has to be investigated with CFD, an accurate solution of the flow fields requires that the numerical model includes all the active ventilation devices and the whole tunnel layout. Nevertheless, CFD remains the most powerful method to predict the flow behaviour due to ventilation devices, large obstructions or fire, but it requires much larger computational resources than simpler 1D models. In the last two decades, the application of CFD as a predictive tool in fire safety engineering has become widespread. The results are still limited due to the difficulties of modelling turbulence, combustion, buoyancy and radiation (Wang *et al.*, 2005), but it may be argued that great achievements have been made. There are, however, still important issues associated with using CFD models in a reliable and acceptable way (see Chapters 15 and 29).

As CFD suffers from limitations when dealing with large domains, it is practical to use it only to model a small tunnel portion, and to account for the effects of the remaining portion of the tunnel by means of 1D models. This class of hybrid models, known as 'multi-scale models', represents a way of avoiding both the large computational cost of complex models and the inaccuracies of simpler models. For tunnels, these models are usually based on the coupling between multi-dimensional (e.g. CFD) and 1D (e.g. network model) solvers, the latter providing the boundary conditions for the former, and vice versa. The main features of multi-scale models for tunnel applications are described in Section 17.3.

17.2. One-dimensional models

The main advantage of using 1D models for the analysis of a complex network system, constituted by a tunnel and its ventilation ducts, is that it allows a complete and compact description of the system. This characteristic has two major consequences: (a) it is possible to define with adequate precision the boundary conditions, such as the ambient conditions at the portals; and (b) it is suitable for applications requiring the computation of a large number of scenarios, such as during the assessment of safety strategies for complex tunnels.

17.2.1 Limitations of one-dimensional models

The intrinsic limitations of 1D models are due to the fact that the flow quantities are assumed to be homogeneous in each cross-section, and thus they are identified with a unique value for each of the variables pressure, velocity, temperature, smoke concentration, etc. This assumption makes 1D models unsuitable for simulating the fluid behaviour in regions characterised by high temperature or velocity gradients. Regions characterised by high temperature gradients are typically encountered close to the flames or in the regions where well-defined smoke stratification is found. In the case of small fires, smoke stratification is maintained for long tunnel sections, and determines a particular propagation of the smoke front. In fact, the smoke front proceeds with greater velocity than if it were to occupy the entire cross-section. Thus, if 1D models are applied to fire events (particularly small fires where the smoke is stratified), it is necessary to introduce corrections to account for non-homogeneity caused by the stratification (Riess et al., 2000), otherwise the calculated propagation velocity of the smoke front would be significantly under-predicted. Once the smoke away from the flames has occupied the entire, or nearly entire, tunnel cross-section, the conditions are close to homogenous, and the prediction of the propagation velocity is reasonably accurate.

Regions with high velocity gradients are also typically encountered close to ventilation devices (e.g. jet fans), where the fan thrust produces highly three-dimensional (3D) flows, and the assumption of flow homogeneity fails. Usually, 1D models describe the behaviour of such regions on the basis of empirical correlations that must be calibrated for the specific tunnel layout. Indeed, the jet-fan thrust curve provided by the manufacturer only applies to the isolated jet fans, and it does not describe the behaviour once installed in a particular tunnel gallery. The accuracy of 1D results is based on calibration constants (e.g. friction coefficients), which need to be input to the model and are case specific. Usually, these are determined via small-scale experiments. However, problems with using 1D modelling would be expected to arise if: (a) a real obstruction in a tunnel were to be different in shape from the one used in the small-scale test; and (b) the intersection between two galleries is different from the one tested at the small scale or available in the literature. Caution is required. In the 'near field' then, CFD may be employed (see Section 17.3.4.1).

17.2.2 Historical developments

The first reported codes for the digital calculation of fluid networks were formulated in the late 1950s. They were mainly developed to design mine ventilation systems and, in the late 1960s, they became a fundamental part of any ventilation planning (Greuer, 1977). In spite of the fact that an increasing number of attempts were made during the early years to adapt such network calculation codes to the simulation of fire scenarios, none of them had progressed sufficiently. A first significant attempt at including the effect of fire in a network system calculation was made in the late 1970s, when Greuer and co-workers produced a tool that could be used for the steady-state calculation of network systems, providing temperature velocity and pollutant distributions (Greuer, 1977). The solution was computed using a Hardy-Cross-like method (Tullis, 1989). The numerical method adopted was based on the solution of the longitudinal momentum equation on closed airway loops, the definition of which was not straightforward, as erroneous loop definitions could lead to slow convergence.

In the 1990s and 2000s, several national institutions proposed contributions to the subject. Models such as MFIRE (US Bureau of Mines, 1995), WHITESMOKE (Cantene, 2010), ROADTUN (Dai and Vardy, 1994; West et al., 1994), RABIT and SPRINT (Riess and Bettelini, 1999; Riess et al., 2000), Express'AIR and SES (NTIS, 1980; Parsons Brinckerhoff Quade & Douglas, 1980; Schabacker et al., 2002) are now commonly used to perform complete studies of tunnel ventilation systems and fires.

MFIRE, developed by the US Bureau of Mines, performs steady-state fluid-dynamic simulations of underground network systems. The model has been tested by Cheng *et al.* (2001) using experimental data from a small-scale network consisting of 27 branches of about 0.1 m diameter and 1–2 m length, and then applied to simulate a hypothetical fire outbreak in the Taipei Mass Rapid Transit System. The simulations were designed to investigate the direction and rate of air flow, temperature distribution and emergency ventilation responses. The same theoretical approach has been used by Ferro and co-workers (Ferro *et al.*, 1991; Borchiellini *et al.*, 1994) and Jacques (1991). The former presented a 1D computer model for tunnel ventilation. The model was designed to deal with a complex tunnel network, including phenomena such as the piston effect from moving vehicles and the distribution of pollutant concentration. The model can be used to perform steady-state calculations. A similar approach was used by Borchiellini *et al.* (1994), who presented numerical simulations of an urban tunnel 2.5 km long.

The Subway Environmental Simulation code (SES), developed by Parsons Brinckerhoff Quade & Douglas Inc. (1980), is a 1D simulation tool capable of predicting ventilation scenarios in tunnel networks. The tool includes a simplified sub-model to predict the occurrence of backlayering as a function of the fire size and ventilation conditions (Parsons Brinckerhoff Quade & Douglas, 1980; Kennedy, 1997). The sub-model, based on a Froude scaling analysis, has been calibrated using small-scale experiments, and therefore its validity for large-scale data remains uncertain. The experiments were conducted in a 10 m long tunnel with a $0.09\,\mathrm{m}^2$ cross-section, and the fire source was represented by the tunnel wood lining (Lee *et al.*, 1979). No information on the heat release rate (HRR) of the fire was made available.

A more recent application of a 1D model for the analysis of fire scenarios in tunnels is represented by the code SPRINT. This must not be confused with the model FIRE-SPRINT, which is quite different (see Chapter 16). The code SPRINT performs time-dependent analysis of fire scenarios in tunnels, and handles gravity-driven smoke propagation due to thermal stratification. The latter effect is accounted for by superposition of the mean flow velocity, and the front velocity is estimated on the basis of a semi-empirical correlation. The model, compared with experimental data measured during the 1995 Memorial Tunnel Fire Ventilation Test Program (Riess *et al.*, 2000) and in the Mont Blanc Tunnel, has been applied to simulate real tunnel-fire scenarios.

In a recent application, for cold flows (no fire) a 1D model has been used in an optimisation procedure to determine the aerodynamic coefficients in a 1.8 km long highway tunnel (Jang and Chen, 2002). The optimisation, performed on the basis of detailed experimental measurements, was able to provide the pressure-rise coefficients of the jet fans, the wall friction coefficient and the averaged drag coefficients of small and large vehicles.

17.2.3 Typical mathematical formulation for one-dimensional models

The vast majority of 1D models for tunnel applications identified in the literature are based on a generalised Bernoulli formulation (Fox and McDonald, 1995). Most of the models are designed to account for buoyancy effects, transient fluid-dynamic and thermal phenomena, piston effects and transport of pollutant species. They are usually designed to handle the complex layout typical of modern tunnel ventilation systems (especially transversely ventilated tunnels) on the basis of a topological representation of the tunnel network.

17.2.3.1 Topological representation

The topological structure of complex flow distribution systems, such as pipelines, tunnels and mines, may be described using a matrix representation and graph theory (see, e.g. Chandrashekar and Wong (1982); and for application to tunnel systems, Ferro et al. (1991) and Borchiellini et al. (1994)). This representation relies on two concepts: node and branch. A node is a section where state properties such as temperatures, pressures, mass or molar fractions, are defined. A single value of these properties is defined at a node. A branch is an element bounded by two nodes and characterised by geometrical properties, such as length and cross-section, together with flow and thermal properties, such as roughness, wall temperature and thermal resistance. Branches are associated with mass flow rates, velocities and heat fluxes. A conventional flow direction is also selected for each branch, so that inlet and outlet nodes are defined. Negative flows would then refer to flows directed from the outlet towards the inlet.

The network is described through the interconnections between nodes and branches. Multiple branches can join at the same node, which plays the role of 'flow splitter' or junction. In graph theory, the incidence matrix $\mathbf{A}$ is used to express the interconnections. This matrix is characterised by a number of rows equal to the total number of nodes, and a number of columns equal to the number of branches (in some analyses, the incidence matrix is defined as the transpose of that presented here). The general element A_{ij} is 1 if the ith node is the inlet node of the jth branch, it is -1 if the ith node is the outlet node of the jth branch, and it is 0 in the other cases. A typical layout of a tunnel network system is presented in Figure 17.1, where $G_{\text{ext},i}$ is the mass flow rate exchanged in a node with the external environment. The corresponding incidence matrix $\mathbf{A}$ is:

$$
\mathbf{A} =
\begin{array}{c}
\begin{array}{ccccccc}
 & j & j+1 & j+2 & j+3 & j+4 & j+5
\end{array} \\
\begin{array}{c}
i \\ i+1 \\ i+2 \\ i+3 \\ i+4 \\ i+5 \\ i+6
\end{array}
\begin{pmatrix}
+1 & 0 & 0 & 0 & 0 & 0 \\
-1 & +1 & -1 & +1 & 0 & 0 \\
0 & -1 & 0 & 0 & 0 & 0 \\
0 & 0 & +1 & 0 & 0 & 0 \\
0 & 0 & 0 & -1 & +1 & 0 \\
0 & 0 & 0 & 0 & -1 & +1 \\
0 & 0 & 0 & 0 & 0 & -1
\end{pmatrix}
\end{array}
\tag{17.1}
$$

17.2.3.2 Fluid dynamics model

Modelling a flow system requires that continuity and momentum equations are written with spatial dependence on one single coordinate, which, in the case of tunnels, is the longitudinal coordinate x. Starting from the Navier–Stokes equations, as described in their classical form in Chapter 15, after eliminating the y and z spatial dependences, and neglecting the viscous stress term, which loses its significance in a 1D formulation, the equations become:

$$
\frac{\partial \rho}{\partial t} + \frac{\partial (\rho u)}{\partial x} = S_{\text{Mass}}
\tag{17.2}
$$

$$
\frac{\partial (\rho u)}{\partial t} + \frac{\partial (\rho u u)}{\partial x} = -\frac{\partial p}{\partial x} + \sum S_{\text{M}x}
\tag{17.3}
$$

where ρ is the fluid density, u is the longitudinal velocity, S_{Mass} is the mass source term per unit volume, p is the pressure, and $S_{\text{M}x}$ is the momentum source term per unit volume acting along the longitudinal direction; x and t represent the spatial and temporal coordinates, respectively.

Figure 17.1 Example of the network representation of a tunnel, showing branches between nodes (Reproduced from Colella *et al.* (2009), with permission from Elsevier)

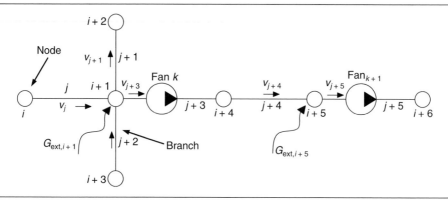

After rearrangement, equation (17.3) can be expressed as:

$$\frac{\partial(\rho u)}{\partial t} + uS_M + \rho u\frac{\partial u}{\partial x} = -\frac{\partial p}{\partial x} + \sum S_{Mx} \tag{17.4}$$

The momentum source term contains all the terms related to the chimney effect, wall friction, and losses due to flow separation at the portals and after obstacles. The pressure rise due to fan operation and the piston effect can be also accounted for. The buoyancy term can be made explicit and included in the spatial derivative on the left-hand side of equation (17.4). Furthermore, neglecting the distributed mass source term, equation (17.4) can be rearranged to give:

$$\frac{\partial u}{\partial t} + \frac{\partial(u^2/2)}{\partial x} + \frac{1}{\rho}\frac{\partial p}{\partial x} + g\frac{\partial z}{\partial x} = \frac{1}{\rho}\sum S_{Mx} \tag{17.5}$$

where z represents the vertical elevation and g the acceleration due to gravity. Equation (17.5) can be solved only after discretising the computational domain in branches interconnected by nodes. Such discretisation generates control volumes, allowing the integration of the momentum and continuity equations. In particular, equation (17.5) integrated along a branch j (which is also a streamline) that extends from node $i-1$ to node i, can be rearranged as:

$$\int_{i-1}^{i}\frac{\partial u}{\partial t}\,dx + \left(\frac{u_i^2 - u_{i-1}^2}{2}\right) + \int_{i-1}^{i}\frac{dp}{\rho} + g(z_i - z_{i-1}) = \int_{i-1}^{i}\frac{1}{\rho}\sum S_{Mx}\,dx \tag{17.6}$$

Once the integration over the branch j has been performed, and after making explicit all the sources of momentum (due to fans, friction and piston effect), the following equation is obtained:

$$\frac{du_j}{dt}L_j + \left(\frac{u_i^2 - u_{i-1}^2}{2}\right) + \frac{p_i - p_{i-1}}{\rho_j} + g(z_i - z_{i-1}) + \frac{\Delta P_{\text{fan},j}}{\rho_j} + \frac{\Delta P_{\text{piston},j}}{\rho_j}$$

$$-\frac{1}{2}\left(f_j\frac{L_j}{D_{h,j}} + \beta_j\right)u_j^2 = 0 \tag{17.7}$$

where ρ_j and u_j are the average density and velocity in the branch j, f_j and β_j are the branch major and minor friction coefficients, L_j is the branch length, $D_{h,j}$ is the branch hydraulic diameter, and $\Delta P_{\text{fan},j}$ and $\Delta P_{\text{piston},j}$ represent the pressure gains inside the branch due to the fans and the piston effect, respectively. The subscript j refers to the generic branch where the equation is applied. The subscripts i and $i-1$ refer to the inlet and outlet nodes of branch j. The same notation has been maintained throughout this chapter.

The introduction of the total pressure P allows equation (17.7) to be written in a more compact way:

$$\rho_j \frac{du_j}{dt} L_j + (P_i - P_{i-1}) + \Delta P_{\text{fan},j} + \Delta P_{\text{piston},j} - \frac{1}{2}\rho_j \left(f_j \frac{L_j}{D_{\text{h},j}} + \beta_j \right) u_j^2 = 0 \qquad (17.8)$$

The final formulations of the continuity and momentum equations for steady-state calculations (setting to zero the transient derivative) are:

$$\sum_j A_j \rho_j u_j = G_{\text{ext},j} \qquad (17.9)$$

$$P_i - P_{i-1} - \frac{1}{2}\left(f_j \frac{L_j}{D_{\text{h},j}} + \beta_j \right) \rho_j u_j^2 + \Delta P_{\text{fan},j} + \Delta P_{\text{piston},j} = 0 \qquad (17.10)$$

where $G_{\text{ext},j}$ is the mass flow rate exchanged in a node with the external environment, and A_j is the branch cross-section. The interested reader can refer to Colella (2010) for a complete derivation of equations (17.9) and (17.10). (See also Jacques (1991) and Borchiellini *et al.* (1994).)

The pressure rise due to fans is commonly represented as a generic second-order polynomial, known as a 'fan characteristic curve':

$$\Delta P_{\text{fan},j} = a_j + b_j u_j + c_j u_j^2 \qquad (17.11)$$

where a_j, b_j and c_j are the characteristic curve coefficients specific for each tunnel layout and obtained from empirical correlations. Alternatively, in some works, the pressure rise due to fans is represented as:

$$\Delta P_{\text{fan},j} = n_j \rho_j \frac{A_{\text{F},j}}{A_j} K_j u_{\text{f},j} (u_{\text{f},j} - u_j) \qquad (17.12)$$

$$\Delta P_{\text{fan},j} = \eta_j \frac{\dot{W}_{\text{e},j}}{A_j u_j} \qquad (17.13)$$

where, for branch j, n_j is the number of operating fans, $A_{\text{F},j}$ is the fan discharging area, $u_{\text{f},j}$ is the fan discharging velocity, K_j is the pressure-rise coefficient, η_j is the fan efficiency, and $\dot{W}_{\text{e},j}$ is the fan electric power.

The piston-effect term can be evaluated following the expression proposed by PIARC (1995), which includes the characteristics of the vehicles and the air velocity in the tunnel:

$$\Delta P_{\text{piston},j} = \varepsilon_j \frac{A_{\text{v},j}}{A_j} \frac{\rho_j}{2} \left[N_{1,j} (u_{1,j} - u_j)^2 - N_{2,j} (u_{2,j} + u_j)^2 \right] \qquad (17.14)$$

where ε_j is the aerodynamic factor of the vehicles in the generic branch j (multiple terms should be considered for each kind of vehicle), $A_{\text{v},j}$ is the vehicle cross-section, $N_{1,j}$ and $N_{2,j}$ are the number of vehicles moving in the same direction and in the opposite direction to branch j with corresponding velocities $u_{1,j}$ and $u_{2,j}$.

Typical calibration constants to be input in the 1D model such as friction coefficients, jet fan pressure-rise coefficients or aerodynamic factors of the vehicles can be found in the literature (Ferro *et al.*, 1991; Jang and Chen, 2002). In addition, a large set of semi-empirical correlations for modelling the hydraulic resistance in complex intersections between or among galleries, shafts or large obstructions is given by Idelchik (2001).

The final formulations of the continuity and momentum equations for time-dependent calculations are:

$$\sum_j \frac{A_j L_j}{2\Delta t}(\rho_j^t - \rho_j^{t-1}) + \sum_j A_j \rho_j^t u_j^t = G_{\text{ext},\,j}^t \tag{17.15}$$

$$\rho_j \frac{u_j^t - u_j^{t-1}}{\Delta t} L_j + P_i^t - P_{i-1}^t + \Delta P_{\text{fan},\,j}^t + \Delta P_{\text{piston},\,j}^t - \frac{1}{2}\left[\left(f_j \frac{L_j}{D_{\text{h},\,j}} + \beta_j\right)\rho_j u_j^2\right]^t = 0 \tag{17.16}$$

The summations contained in equation (17.15) are extended to all the branches *j* connected to the node *i*. Note that the superscripts *t* and *t−1* refer to the current time step and the preceding time step, respectively, while Δt is the integration time step. The interested reader should refer to Colella (2010) for a complete derivation of equations (17.15) and (17.16).

In both steady-state and transient formulations, the continuity and momentum equations can be solved by using an iterative solution algorithm known as the SIMPLE algorithm (Versteeg and Malalasekera, 2007). The method, based on a 'guess and correct' procedure, has been rearranged for a 1D network model and is described elsewhere (Colella *et al.*, 2009; Colella, 2010).

17.2.3.3 Thermal model
In this section, only the general features of the thermal problem are described; the interested reader can find details about the method in Calì and Borchiellini (2003).

The problem is further complicated by the definition of the temperatures in the nodes. Whereas pressures in nodes are uniquely defined, temperatures are not. In the case of a flow junction, two flows at different temperatures can converge in the same node, and the total mass flow rate exits at the average temperature. The temperature resulting from perfect mixing of the mass flow rates entering a node can be defined as the node temperature.

The thermal analysis requires the solution of the energy equation. The general formulation, as presented in Chapter 15, must be simplified, eliminating the spatial dependences with the exception of the *x* coordinate, and neglecting the time derivative of the pressure, which is negligible for low Mach number flows (Versteeg and Malalasekera, 2007). The resulting equation must be solved for each branch and each node.

In the case of branches, the energy-conservation equation can be written in differential form as:

$$\lambda \frac{\partial^2 T}{\partial x^2} - U\frac{\Omega}{S}(T - T_{\text{e}}) + q_{\text{v}} = \rho c\frac{\partial T}{\partial t} + \rho c u\frac{\partial T}{\partial x} \tag{17.17}$$

where λ is the fluid thermal conductivity, *U* is the global heat-transfer coefficient between the fluid and the wall, Ω is the perimeter of the tunnel cross-section, *S* is the cross-section, T_{e} is the deep soil temperature, *c* is the specific heat, and *u* is the fluid velocity along *x*. The term q_{v} accounts for heat generation in the control volume (i.e. due to fire or vehicles). This term is discussed in the next section. In general, the first term, representing the heat conduction along the longitudinal coordinate, can be neglected if compared with the other terms in equation (17.17).

The energy equation in the nodes becomes an enthalpy balance. All the mass flows exiting a node are considered to have the same specific enthalpy, which is the result of an ideal mixing of all the flows entering the node. This mixing enthalpy is the enthalpy of the node.

Boundary conditions are set in all the inlet nodes. In these nodes the external temperature is imposed. In order to resolve the energy equation, it is possible to use numerical methods such as the finite-difference method, the finite-volume method or the finite-element method (Ferziger and Peric, 2002).

The estimation of the heat generation due to fire is a complex process, and 1D models cannot provide any accurate result. Therefore, a time function of the HRR is required as input to the simulations. The HRR function can be derived from available experimental data or from design prescriptions. Alternatively, the approach presented by Carvel (2008) or by Ingason (2009) can be adopted. On the basis of a wide analysis of tunnel fire experiments, Carvel (2008) proposed a two-step linear approximation of the fire growth phase: during the first step the fire would grow slowly up to 1–2 MW, while during the second step the growth rate would be significantly higher (up to 15 MW/min). Ingason (2009) has described a number of different design curves using different mathematical expressions.

17.2.4 A case study: the Fréjus Tunnel

As a case study, the 1D model is applied here to the Fréjus Tunnel (Borchiellini *et al.*, 2006). The aim of the analysis is to define ventilation strategies to be used in the case of fire, and to illustrate the approach.

The Fréjus Tunnel is a two-way link between Italy and France, with a total length of 12 870 m. A diagram of the typical cross-section and ventilation system of the Fréjus Tunnel is shown in Figure 17.2. The diameter of the semicircular sections is around 6 m. Ventilation is fully transversal. Ordinary ventilation is operated by introducing fresh air along the tunnel through six independent L-shaped fresh air ducts, which have two fans at each end (see Figure 17.2b, bottom right). Each duct serves a tunnel section of 2 km (numbered T_1 to T_6) and is connected to an independent ventilation station (numbered AF1 to AF6). Fresh air openings are installed every 5 m. Emergency ventilation is operated using the fresh air ducts and the three U-shaped extraction ducts (see Figure 17.2b, top right), which connect the ventilation stations (AV1 to AV2, AV3 to AV4, and AV5 to AV6). The extraction dampers are installed every 130 m. Note that in

Figure 17.2 The Fréjus Tunnel: (a) cross-section; (b) ventilation system layout

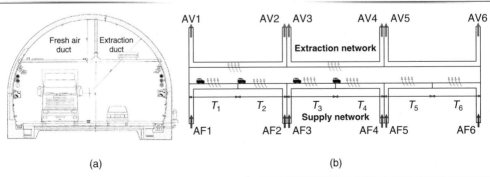

(a) (b)

Figure 17.2b the arrows do not necessarily show the actual positions of the inlets/outlets. The French portal is located on the left-hand side ($x = 0$ m) with the Italian on the right-hand side ($x = 12\,870$ m).

The goal of the ventilation system is to create a zero-velocity section as close as possible to the fire position, with a positive velocity upstream of the fire and a negative velocity downstream of the fire. Here, 'upstream' and 'downstream' refer to a predefined positive direction from France to Italy. In these conditions the smoke does not tend to propagate along the tunnel, but is extracted close to the section where it is produced.

The most effective strategy to be applied in the case of fire depends on the location of the fire along the tunnel and the pressure difference between the two portals. The pressure difference, assumed to be positive when inducing air flow from the French towards the Italian portal, can reach values of several hundreds of pascals, which is typical for long tunnels through the Alps. Furthermore, given a certain positive pressure difference between the portals, the ventilation scenario to be adopted depends on the location of the fire.

If a fire occurs in the first tunnel section (between the French portal and the central section), ten extraction dampers over the fire (five upstream and five downstream) are opened. In addition, if the pressure difference is large, fresh air is supplied downstream of the fire to enhance smoke confinement (the 'opposite supply' strategy). In addition, some fresh air is supplied all along the tunnel, mainly in order to prevent smoke passing into the fresh air ducts, which may be used as an escape route.

If a fire occurs in the last section of the tunnel (between the central section and the Italian portal), air extraction in the first portion of the tunnel is used to oppose the effect of the pressure difference between the portals (the 'opposite extraction' strategy). Similar strategies are also defined for a negative pressure difference.

In order to implement these conceptual strategies, it is necessary to provide the operators with adequate guidelines, in the form of tables or algorithms relating the pressure difference and the fire position, to the extraction dampers to be opened, and the set point of each group of fans. This information is obtained through undertaking multiple numerical simulations using the 1D approach. However, the model must be calibrated and tested (compared with experimental data) before being used as a simulation tool.

17.2.4.1 Example

An example of 1D model testing for a specific ventilation scenario is presented here. The global network, built for the purpose, is composed of 650 branches and 450 nodes. The volumetric flow rate of the fans at full charge ranges between 222–257 and 117–124 m^3/s for supply and extraction units, respectively. The tunnel friction coefficient has been established, on the basis of experimental measurements, to be around 0.017.

The first comparison between the numerical predictions and the experimental data is for a steady-state scenario with only the supply fans operating at 30% of full charge. Using the classification given by Beard (2000), this was a *blind* comparison; i.e. some of the experimental data being used for comparison were used as input data (see Chapter 29). The pressure difference across the tunnel was negligible. This is confirmed by the computed and measured velocity profiles, which are

Figure 17.3 The longitudinal velocity distribution along the tunnel computed using the 1D model (line) compared with experimental data (squares) recorded in the Fréjus Tunnel. Positive velocity indicates flow from the French portal to the Italian portal

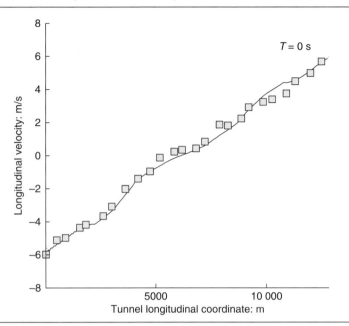

almost symmetric with respect to the tunnel central section (Figure 17.3). An accurate agreement between prediction and experimental data has been achieved.

The second comparison is for an 8 MW fire located in tunnel section T_3. Given the high pressure difference between the portals ($\sim$1000 Pa) and the location of the fire, an opposite supply ventilation strategy has been adopted. The strategy consists of localised extraction over the fire zone and an enhanced fresh air supply in the region downstream of the fire. The fans are started at $t = 0$. The computed horizontal velocity profiles established along the tunnel are compared with the measured data in Figure 17.4. Overall, a reasonably good agreement is obtained. The 1D model seems to predict well the overall flow behaviour of the tunnel ventilation system.

17.2.5   Concluding remarks on one-dimensional models
The results obtained using 1D models can be used to assess whether the overall ventilation conditions are acceptable for proposed fire-safety strategies. In the case of transversally ventilated tunnels, the assessment can be made by determining the presence of a cross-section characterised by a zero longitudinal flow velocity as close as possible to the fire. In the case of longitudinally ventilated tunnels, 1D models can be used to assess whether or not the tunnel ventilation system is able to guarantee a super-critical ventilation velocity in the fire zone, and therefore prevent backlayering.

One-dimensional models can be used to determine the temperature and pollutant distributions along the tunnel and the possible ventilation strategies to be adopted in order to achieve acceptable conditions. As already stated, the effects of traffic on the propagation of pollutants

Figure 17.4 The longitudinal velocity distribution along the tunnel computed using the 1D model (lines) compared with experimental data (squares) recorded in the Fréjus Tunnel. The data are for four different times

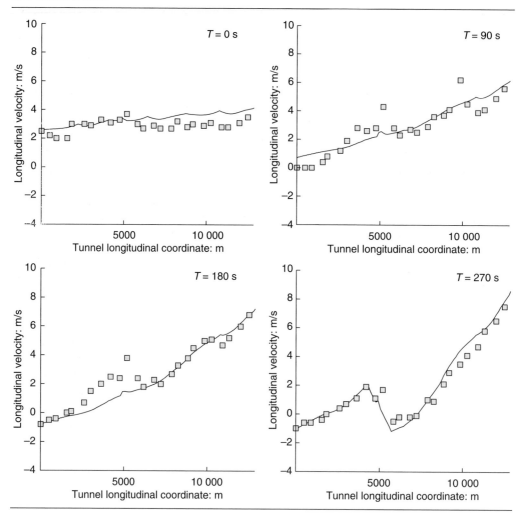

are difficult to model with any accuracy, so 1D models provide an approximation that may be regarded as acceptable in some cases.

However, 1D models are not suitable for simulating the fluid behaviour in regions characterised by high temperature or velocity gradients, as typically encountered in the vicinity of the fire plume, ventilation devices or interconnections of galleries. In order to deal with such complex flow conditions, 1D models mainly rely on empirical correlations or calibration constants defined on the basis of experimental measurements or detailed calculations.

17.3. Multi-scale models

CFD calculations are now part of many designs, assessments and investigations. However, a caveat is necessary here. CFD models have been shown to predict the critical ventilation velocity and backlayering distance to an acceptable level of accuracy (deviation from experimental results less than 30%) (Wu and Bakar, 2000; Vauquelin and Wu, 2006; Van Maele and Merci, 2008). However, the predicted local field data (i.e. velocity and temperature fields) in the vicinity of the fire source (i.e. the near fire field) can be affected by higher errors (Karki and Patankar, 2000; Wu and Bakar, 2000; Vega et al., 2008).

CFD modelling usually requires very large computational domains that extend to the tunnel regions where the operating ventilation devices (jet fans or axial fans) are located. Among the CFD studies that can be found in the literature (Colella, 2010), the two most important are reviewed here. Karki and Patankar (2000) used a CFD model to simulate the ventilation flows within a tunnel fire. The model results showed good agreement with experimental values measured in the Memorial Tunnel Fire Ventilation Test Program (Massachusetts Highway Department, 1995). The Memorial Tunnel data have also been used by Vega et al. (2008) to test a CFD model that includes ventilation devices (e.g. fans).

CFD is the most accurate technique available for resolving the flow pattern in high velocity and temperature gradient regions localised in areas within the vicinity of the fire and areas close to ventilation devices, dampers or junctions of galleries. However, there is evidence that these high-gradient regions, the *near-field* regions, do not extend far from the locations where they are generated (Colella et al., 2009, 2011). After a certain distance, the temperature and velocity gradients become low, and the flow can be represented, with reasonable accuracy, using a 1D assumption. The low-gradient regions are called the *far-field* regions.

On the basis of these observations, and for the sake of an efficient allocation of computational resources, CFD may be applied for the modelling of near-field regions, while the far-field regions are simulated using a 1D model. This form of hybrid model allows a significant reduction in the computational time required, as the more time-consuming tool (CFD) is applied to only a limited portion of the domain. However, there are some caveats to the use of multi-scale models, and these are discussed in the following text.

From now on, it is assumed that CFD can be used to produce sufficiently accurate results in the near field, including near the fire, so as to produce acceptably accurate results for the multi-scale model. For a more detailed explanation on the use, accuracy and limitations of CFD models, see Chapters 15 and 29 of this handbook.

In the multi-scale approach, the CFD and the 1D models exchange flow information at the 1D–CFD interfaces. However, it is important to realise that deciding on the locations of the interfaces requires care (see Section 17.3.5). There are two general coupling options. The simplest is one-way coupling (or superposition). For example, in the case of inclined tunnels, it is possible to evaluate the global chimney effect using a 1D model of the entire tunnel (Merci, 2008). Then, a CFD analysis of specific tunnel portions can be run using the 1D results as boundary conditions. This approach does not represent true multi-scale modelling, as there is no coupling of the CFD results and the 1D flow in any way. This would be equivalent to assuming that the flow behaviour in the high-gradient regions does not affect the bulk tunnel flow.

A two-way coupling of 1D and CFD models, 'proper' multi-scale modelling, consists of a physical decomposition of the problem into two parts: a section of the tunnel is simulated using a CFD model and the remaining sections are simulated using a 1D model. The advantage of multi-scale modelling lies in including in the flow-field calculations the effect of the fire on the entire ventilation system, and vice versa. If the solver is able to receive the two sets of equations, the problem can be solved at once. In most cases there is a different solver for each model, and therefore iterative calculations are necessary in order that the two solvers continuously exchange information at the boundary interfaces.

Previous work on multi-scale modelling reported in the literature has focused on fluid flow systems. Examples include the simulation of blood flow in the circulatory system (Formaggia et al., 2001), the computation of gas flows in the exhaust ducts of internal combustion engines (Motenegro, 2002), and the characterisation of the flow pattern around high-speed trains moving through tunnels (Mossi, 1999). Multi-scale methods were cited as possible techniques for simulating tunnel ventilation flows and fires by Rey et al. (2009). Other applications of multi-scale modelling to tunnel ventilation systems and tunnel fires have also been reported (Colella et al., 2009, 2010, 2011).

17.3.1 Coupling strategies

The solution to the multi-scale problem requires coupling of the 1D and CFD models. The iterative solver procedure requires a continuous exchange of information between the models at the interfaces during the computations (Figure 17.5). The coupling algorithm, operating at the interfaces between the 1D nodes and the CFD grid (nodes i and $i+1$) has to perform the following operations at each iteration step.

1. Provide the CFD model with the pressure and temperature boundary conditions at the interfaces i and $i+1$ calculated by the 1D model.
2. Run the CFD model of the near field.
3. Integrate the flow velocities at the interfaces i and $i+1$ to calculate the global mass flow rate and fix them as boundaries for the 1D model.
4. Average the temperature at the interfaces i and $i+1$ to provide the boundary conditions for the 1D model.
5. Run the 1D model of the far field.
6. Return back to step (1) until convergence is reached
7. Proceed to the following time step (for transient calculations).

This method of coupling is called *direct coupling*. It allows a significant reduction in the computational time compared with the full CFD calculation of the same scenario. However, the timescale

Figure 17.5 Coupling procedure between 1D and CFD models

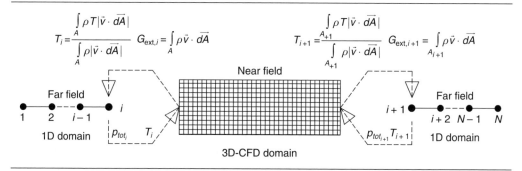

for *direct coupling* calculations is controlled by the computational time required to solve the CFD portion of the model (i.e. the near field), which can be from a few minutes to many hours, depending mostly on the size of the CFD domain and fire size.

Most comprehensive ventilation studies require only bulk flow velocities and temperatures within the tunnel domain. In this situation, 1D models could provide acceptable global results, but the accuracy of the predicted behaviour of the near field is very limited. In these cases, CFD can be used to study the behaviour of the near field under a wide range of bulk flow conditions at the inlet and outlet boundaries of the domain. Usually, the total pressure and mass flow rate can be used as averaging quantities at the boundaries. The results when expressed in terms of bulk flow values, can be easily introduced into the 1D model, improving its prediction capabilities. This approach is known as *indirect coupling*. The method can be used to calculate the characteristic curves for a section of tunnel containing, for example, a jet fan, without the need for calibration constants or experimental results. The prediction capabilities of the 1D model can also be improved by including an expression for the pressure losses induced by a fire, calculated using CFD tools.

17.3.2   The effect of the interface location

One of the most important issues when using a multi-scale model for tunnels is related to the location of the interfaces between 1D and CFD models. These boundaries must be located in regions of the domain where the temperature or velocity gradients are negligible and the flow behaves largely as 1D. These boundaries dictate the length of the CFD domain, and the required length is case specific, in general depending on the tunnel geometry, installation details of ventilation devices, the presence of obstacles, etc. Only broad guidelines are given here.

A multi-scale model has been used by Colella *et al.* (2009) to study the discharge velocity cone generated by a jet fan installed in a modern 1.5 km long tunnel. The analysis of the effect of the interface location showed that the multi-scale solution was within 1% of the full CFD solution when the length of the CFD domain containing the jet fan was more than 20 times the tunnel hydraulic diameter.

In another study (Colella *et al.*, 2011), a multi-scale model was used to simulate several fire scenarios, with fire sizes ranging between 10 and 100 MW in a longitudinally ventilated, 1.2 km long tunnel. The analysis of the effect of the interface location showed that the multi-scale solution was within 1% of the full CFD solution when the length of the CFD domain containing the fire was more than 13 times the tunnel hydraulic diameter.

When simulating tunnel-fire scenarios for which the ventilation flow is below the critical velocity, it must be ensured that the backlayering does not extend upstream of the inlet boundary. If this condition does not apply, the hypothesis of a 1D flow pattern at the interfaces would not be valid, and the results obtained would be inaccurate.

17.3.3   Case study: ventilation flows in a modern tunnel

In this section the application of the multi-scale methodology (both direct and indirect coupling approaches) is illustrated using the calculation of cold ventilation flows in a modern tunnel – the Dartford Tunnel, UK. The tunnel is 1.5 km long and carries unidirectional traffic in two lanes. The central portion of the tunnel is flat, while the initial and the final segments present a descending and ascending layout. The cross-section is about 40 m² with slight variations along the length of the tunnel. For more details, see Colella *et al.* (2010).

The tunnel is equipped with a semi-transverse ventilation system. A shaft with axial extraction fans is located around 150 m from each tunnel portal. The fresh air is supplied into an invert under the roadway by means of two axial fans, and is introduced into the main tunnel void through supply grills located along the side of the roadway between the two extraction shafts. The tunnel is equipped with 14 pairs of jet fans installed in two groups of seven pairs in the inclined sections of the tunnel. The fan pairs are spaced 50 m apart. According to the manufacturer, each jet fan provides a volumetric flow rate of 8.9 m³/s, with a discharge flow velocity of 34 m/s.

17.3.3.1 Calculation of the discharge cone of the jet fans with direct coupling

The assessment of the performance of the tunnel ventilation system requires a detailed study of the flow field around the jet fans, in particular an analysis of the jet fan discharge cone. These details are fundamental to understand how a jet-fan flow interacts with its surroundings and how the installation detail (e.g. niches and pairing) affects the resulting thrust and the overall performance of the system.

The analysis was performed using a multi-scale model with direct coupling and cold flow. A 300 m long CFD model of the near field was directly coupled to a 1D representation of the rest of the tunnel, extending all the way to the portals and including the vertical shafts. A sketch of the coupling is given in Figure 17.6. The numerical predictions were compared with experimental flow measurements made in the tunnel. The longitudinal air velocities were measured at six different tunnel locations, at 20 m intervals, starting 20 m downstream of the jet-fan discharge surface. At each location, the velocity was measured at nine points across the section, as show in Figure 17.7.

The comparison between the predicted and measured velocities is presented in Figure 17.8. As for the comparisons described above, this comparison was conducted on a blind basis (Beard, 2000). In Figure 17.8 the predicted values are represented by the lines, the continuous lines representing the velocity profiles calculated in the middle of the tunnel cross-section (Profile 1 in Figure 17.7) and the dashed lines representing the velocity profiles calculated along the vertical lines corresponding to Profile 2 in Figure 17.7. The measured velocity values in Figure 17.8 are numbered 1–9, following the same pattern as in Figure 17.7.

Figure 17.6 Multi-scale coupling procedure for the representation of the jet-fan discharge cone (Reproduced from Colella *et al.* (2009), with permission from Elsevier)

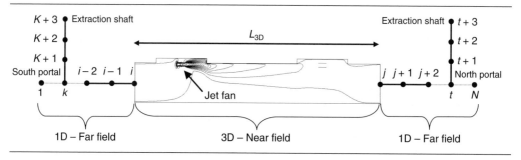

Figure 17.7 Layout and general dimensions of the tunnel cross-section including the points 1–9 where the air velocities were measured (dimensions are given in mm). Each profile corresponds to a vertical line in the tunnel cross-section, e.g. Profile 1 corresponds to points on the vertical line through positions 2, 5 and 8 (Reproduced from Colella *et al.* (2009), with permission from Elsevier)

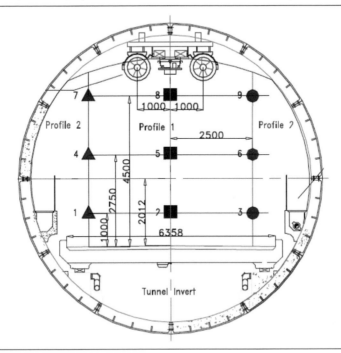

There is good agreement between the experimental measurements and the numerical predictions for all sections. This shows that multi-scale modelling with direct coupling may be used to obtain accurate results.

17.3.3.2 Assessment of ventilation-system performance with indirect coupling

The assessment of the ventilation-system performance required a comprehensive study of the multiple emergency ventilation strategies. In particular, the study aimed to assess the consequences of changing the number of operating fans during an emergency. For this task, the use of indirect coupling was sufficiently accurate, as only bulk flow quantities are required. This approach was preferred to the traditional 1D modelling approach using calibration constants for the fan characteristic curves as that method provided unreliable results. In particular, the traditional 1D formulation overpredicted the ability of the fans to produce thrust, as it did not account for the losses introduced by the peculiar installation details (i.e. niches, pairing, tunnel section layout, etc.).

The CFD grid of the jet-fan near field was built following the available tunnel geometry information, and the jet-fan installations were represented in detail. Several runs of the CFD model were conducted for different pressure drops across the domain boundaries. The flow through the domain was calculated assuming perfect periodic behaviour of a jet-fan series. This assumption proved to be accurate (see Colella *et al.*, 2009, 2011). The results can be presented in terms of

Figure 17.8 Comparison of the predicted (lines) and experimentally measured (symbols) horizontal velocities. The profiles and numbers refer to the locations in the tunnel section shown in Figure 17.7 (Reproduced from Colella *et al.* (2009), with permission from Elsevier)

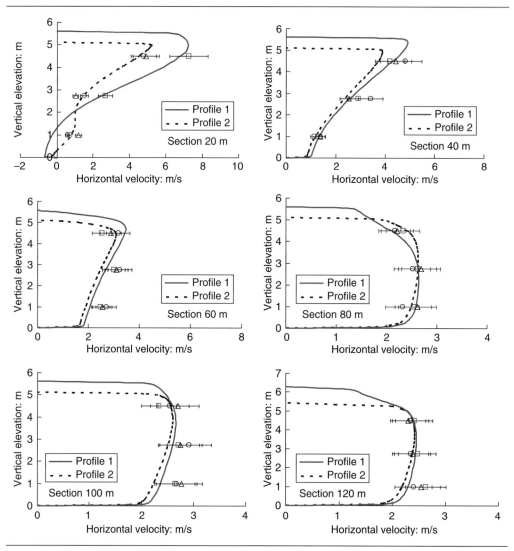

the bulk flow velocity versus the pressure difference across the domain (Figure 17.9). Comparison with CFD simulations without niches shows that the niches have a significant effect on the resulting thrust of the jet fans.

The results of this CFD study of the near field were then coupled to the 1D model built for the rest of the tunnel. Specifically, the computed curves were used as the characteristics of any branch of the 1D model containing a jet-fan pair, instead of the classical momentum formulation as described by equation (17.10), which is not able to adequately describe complex 3D behaviour.

Figure 17.9 Jat fan characteristic curve of pressure rise versus flow velocity in a 50 m long tunnel portion. Results were calculated using CFD of the near field containing a jet-fan pair, with and without modelling the niches (Reproduced from Colella *et al.* (2009), with permission from Elsevier)

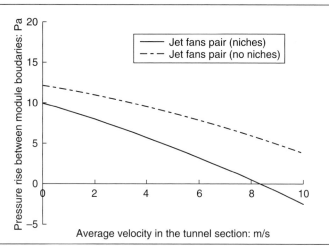

The results were compared with the measured average flow velocities in the central section of the tunnel (Figure 17.10). It can be seen from Figure 17.10 that the results obtained with the multi-scale model with indirect coupling compare well with the experimental results.

17.3.4 Case study: tunnel fire coupled with a ventilation system

A multi-scale modelling approach has been used to simulate the temperature and velocity fields in a 1200 m long tunnel containing a 30 MW fire (Colella *et al.*, 2011). The modelled tunnel is 6.5 m high, with a standard horseshoe-shaped cross-section and an area of about 53 m^2. The tunnel is longitudinally ventilated and equipped with 10 jet fan pairs spaced 50 m apart near both tunnel

Figure 17.10 Comparison of the average flow velocities, for various numbers of operational jet-fan pairs, measured experimentally and predicted using the multi-scale model with indirect coupling (Reproduced from Colella *et al.* (2009), with permission from Elsevier)

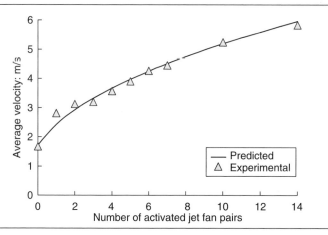

Table 17.1 Settings of the ventilation system for the four fire scenarios considered

Scenario	Fire size: MW	Jet fan pairs				
		1 and 2	3	4	5	6–10
1	30	On	On	Off	Off	Off
2	30	On	On	On	On	Off
3	30	On	On	On	On	On
4	30	Off	Off	Off	Off	On

portals. The jet fans are rated at the volumetric flow rate of 8.9 m³/s with a discharge flow velocity of 34 m/s. No vertical shafts are installed. This tunnel is similar, but not identical, to the one considered in the previous sections.

The fire is located in the middle of the tunnel. The objective of the emergency ventilation strategy is the same as that used in most longitudinally ventilated tunnels, i.e. to push any smoke away from the incident region, in the direction of traffic flow, and to prevent upwind spread (i.e. no backlayering). The vehicles downstream of the fire are assumed to leave the tunnel safely. Four different ventilation scenarios were considered, as shown in Table 17.1.

17.3.4.1 Calculation of the fire near-field
In this section 'near field' is taken to mean 'near the fire'. The calculations were performed using direct coupling. A 400 m long CFD model of the near field was used, while the rest of the tunnel domain (including the jet fans) was modelled using a 1D representation. A sketch of the coupling is shown in Figure 17.11.

The bulk flow rates predicted using the multi-scale approach are compared with predicted values obtained using the full CFD model in Table 17.2. The results show excellent agreement. The deviations range between 0.01% and 2%. It is worth noting that the simple representation conducted with a traditional 1D model systematically overpredicts the capability of the ventilation system (by up to 30% for scenario 1). This tendency is expected to worsen for scenarios involving higher HRRs.

Figure 17.11 Multi-scale coupling procedure for the representation of the fire near-field region

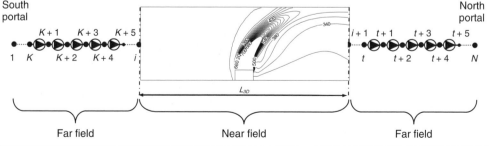

Table 17.2 Accuracy of the multiscale simulation for the four fire scenarios considered. Comparison of full CFD, multiscale (direct and indirect coupling) and 1D model results. The last column refers to a traditional 1D modelling approach

Scenario	Full-scale CFD	Multi-scale, direct coupling		Multi-scale, indirect coupling		1D model	
	Mass flow rate: kg/s	Mass flow rate: kg/s	Deviation: %	Mass flow rate: kg/s	Deviation: %	Mass flow rate: kg/s	Deviation: %
1	216	220	2.09	223	3.4	277	28
2	301	301	0.01	306	1.4	356	18
3	435	434	0.15	446	2.4	490	13
4	299	296	1.08	300	0.3	351	17

The results are also presented in Figure 17.12 as temperature and horizontal velocity fields in the tunnel longitudinal plane. For the sake of simplicity, only the results for the first scenario are presented. The rest of the scenarios can be found in Colella *et al.* (2011).

The multi-scale field predictions are in excellent agreement with the full-CFD predictions. In particular, there are no important differences in the temperature fields. Some local deviations

Figure 17.12 Comparison of results for the fire near field obtained using multi-scale and full CFD simulations for a fire of 30 MW and ventilation scenario 1 (see Table 17.1). The longitudinal coordinates start at the upstream boundary of the corresponding CFD domain

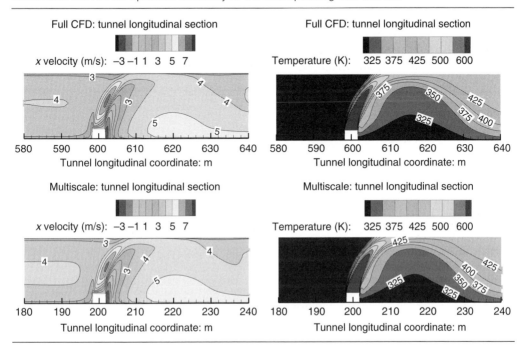

Figure 17.13 Characteristic curves of the total pressure drop in a 400 m long tunnel section versus the mass flow rate for different fires (calculated using CFD)

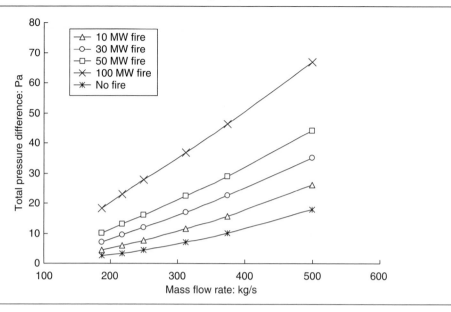

are observed in the horizontal velocity fields, due to flow perturbations from the operating jet fans.

The assessment of the ventilation-system performance in the case of fire requires a comprehensive study of the multiple emergency ventilation strategies. In particular, the study aimed to assess the consequences of changing the number of operating fans during a fire emergency. For this task, the use of indirect coupling was sufficiently accurate, as only bulk flow quantities are required. This approach was preferred to the traditional 1D model, which does not account for the pressure losses in the fire-plume region. In this region the flow behaves in a complex manner, and using a simple 1D model would lead to an underestimation of the fire-induced pressure gradients (see Table 17.2).

Different CFD simulations were carried out for various total pressure differences across the near-field domain including the fire. The characteristic curves were constructed, representing the total pressure difference across the near-field boundaries as a function of the computed global mass flow rate. The computed 30 MW fire curve is presented in Figure 17.13 together with other curves for other fire sizes (10, 50 and 100 MW). The fire-induced pressure gradients are higher for larger fire sizes.

17.3.4.2 Assessment of the ventilation performance in fire scenarios with indirect coupling

The results are compared with full-scale CFD data for the same scenarios in Table 17.2. The predictions made using the multi-scale model with indirect coupling show deviations of only 0.3% to 3.4% from the CFD values. The predictions based on a simple 1D representation of the fire region show deviations of 9.5% to as much as 28%. A more comprehensive description of the multi-scale results obtained for the tunnel-fire scenarios is given in Colella *et al.* (2011).

17.3.5 Concluding remarks on multi-scale models

The use of multi-scale techniques to simulate tunnel ventilation flows and fires is a relatively new computational approach. The multi-scale model, built by coupling 1D and CFD representations of the fluid-dynamic phenomena, allows a significant reduction in the computational time, as the more time-consuming tool (CFD) is used to simulate only a small part of the domain (the near field) characterised by high velocity and temperature gradients. The regions within the domain where the flow behaves as a fully developed channel flow are modelled using a 1D model, which is computationally effective for this type of flow and provides sufficiently accurate results. The numerical coupling of the models takes place at the 1D–CFD interface, where the two models provide the required boundary conditions to each other.

The reduction in the computational complexity when using multi-scale models is significant. In both the presented examples for modern 1.2–1.5 km long tunnels, the computational time was reduced by 40 times. With direct coupling, less than 2 hr of computing time was required, while with indirect coupling only a few seconds were needed to obtain the results once the characteristic curves were computed. These times are to be compared with the 48–72 hr required by the full CFD representation.

The adoption of the multi-scale approach requires care regarding the location of the 1D–CFD interfaces. Previous studies have shown that the position of the interfaces can have a significant impact on the accuracy of the solution. Better predictions are achieved when the interfaces are located in regions where the flow behaves in a 1D fashion. The attainment of this flow pattern depends on the specific tunnel layout, the jet-fan characteristics, the jet-fan installation details and the presence of any other obstacles.

All simulations of tunnel ventilation flows contain uncertainty; for example, due to a lack of knowledge of the real flow conditions at the portals (atmospheric pressure and wind), effective wall roughness, fire load, fire geometry, throttling effects of vehicles, and various other perturbing factors. Given this, the authors consider that the error introduced by using the multi-scale approach is within the uncertainty range of CFD calculations, and may be regarded as acceptable for engineering purposes. On the other hand, the significantly lower computational time required when using multi-scale models is an asset, and offers the advantage that many ventilation scenarios may be considered. As with all models, how multi-scale models are used and how the results obtained are interpreted is of crucial importance (see Chapters 15 and 29). Furthermore, some caveats on the use of such models have been mentioned, or implied, in the sections above. It is asserted here that the multi-scale modelling approach represents the most practical way to perform acceptably accurate simulations of coupled ventilation flows and fire in tunnels longer than a few kilometres, when the limitation of the computational time required becomes seriously restrictive.

REFERENCES

Beard AN (2000) On *a priori*, blind and open comparisons between theory and experiment. *Fire Safety Journal* **35**: 63–66.

Borchiellini R, Ferro V and Giaretto V (1994) Transient thermal analysis of main road tunnel. In: *Aerodynamics and Ventilation of Vehicle Tunnels*. Mechanical Engineering Publications, London, pp. 17–31.

Borchiellini R, Ferro V, Giaretto V and Verda V (2006) Use of large extraction openings for the emergency ventilation in the Fréjus Tunnel. *Tunnel Protection and Security against Fire and other Hazards, Turin, Italy*, 15–17 May 2006, pp. 179–186.

Calì M and Borchiellini R (2003) District heating networks calculation and optimization. *Encyclopedia of Life Support Systems (EOLSS)*. Available at: http://www.eolss.net (accessed 22 June 2011)

Cantene (2010) WHITESMOKE User's Guide, www.CANTENE.IT

Carvel R (2008) Design fires for tunnel water mist suppression systems. *Proceedings of 3rd International Symposium on Tunnel Safety and Security, Stockholm, Sweden*, 12–14 March 2008.

Chandrashekar M and Wong FC (1982) Thermodynamic system analysis – a graph-theoretic approach. *Energy* 7(6):539–566.

Chasse P and Apvrille J-M (1999) A new 1D computer model for fires in complex underground networks. *Proceedings of the 1st International Conference on Tunnel Fires and Escape from Tunnels, Lyon, France*, 3–7 May 1999. Independent Technical Conferences Ltd, Bedford, pp. 201–222.

Cheng LH, Ueng TH and Liu CW (2001) Simulation of ventilation and fire in the underground facilities. *Fire Safety Journal* **36**: 597–619.

Colella F (2010) *Multiscale Analysis of Tunnel Ventilation Flows and Fires*. PhD Thesis, Dipartimento di Energetica, Politecnico di Torino. Available at: http://hdl.handle.net/1842/3528 (accessed 22 June 2011).

Colella F, Rein G, Borchiellini R, Carvel R, Torero JL and Verda V (2009) Calculation and design of tunnel ventilation systems using a two-scale modelling approach. *Building and Environment* **44**: 2357–2367.

Colella F, Rein G, Carvel R, Reszka P and Torero JL (2010) Analysis of the ventilation systems in the Dartford tunnels using a multi-scale modelling approach. *Tunnelling and Underground Space Technology* **25**: 423–432.

Colella F, Rein G, Borchiellini R and Torero JL (2011) A novel multiscale methodology for simulating tunnel ventilation flows during fires. *Fire Technology* **47**: 221–253.

Dai G and Vardy AE (1994) Tunnel temperature control by ventilation. *Proceedings of the 8th International Symposium on the Aerodynamics and Ventilation of Vehicle Tunnels, Liverpool, UK*, 6–8 July 1994, pp. 175–198.

Ferro V, Borchiellini R and Giaretto V (1991) Description and application of a tunnel simulation model. In: *Aerodynamics and Ventilation of Vehicle Tunnels*. Elsevier Applied Science, London, pp. 487–512.

Ferziger JH and Peric M (2002) *Computational Methods for Fluid-dynamics*, 3rd edn. Springer-Verlag, Berlin.

Formaggia L, Gerbeau JF, Nobile F and Quarteroni A (2001) On the coupling of 1D and 3D Navier–Stokes equations for flow problems in compliant vessels. *Computer Methods in Applied Mechanics and Engineering* **191**: 561–582.

Fox RW and McDonald AT (1995) *Introduction to Fluid Mechanics*, 4th edn. Wiley, Singapore.

Greuer RE (1977) *Study of Mine Fires and Mine Ventilation. Part I: Computer Simulations of Mine Ventilation Systems under the Influence of Mine Fires*. Department of the Interior, Bureau of Mines, Washington, DC.

Idelchik IE (2001) *Handbook of Hydraulic Resistance*, 3rd edn. Begell House, Redding, CT.

Ingason H (2009) Design fire curves for tunnels. *Fire Safety Journal* **44**: 259–265.

Jacques E (1991) Numerical simulation of complex road tunnels. *Proceedings of the 7th International Conference on the Aerodynamics and Ventilation of Vehicle Tunnels*, pp. 467–486.

Jang H and Chen F (2002) On the determination of the aerodynamic coefficients of highway tunnels. *Journal of Wind Engineering and Industrial Aerodynamics* **89**(8): 869–896.

Karki KC and Patankar SV (2000) CFD model for jet fan ventilation systems. *Proceedings of the 10th International Symposium on Aerodynamics and Ventilation of Vehicle Tunnels, Boston, MA, USA*, 1–3 November 2000.

Kennedy WD (1997) Critical velocity – past, present and future (revised 6 June 1997). *American Public Transportation Association's Rail Trail Conference, Washington, DC.*

Lee CK, Chaiken RF and Singer JM (1979) Interaction between duct fires and ventilation flow: an experimental study. *Combustion Science & Technology* **20**: 59–72.

Massachusetts Highway Department (1995) *Memorial Tunnel Fire Ventilation Test Program. Comprehensive Test Report.* Massachusetts Highway Department and Federal Highway Administration, Boston, MA/Washington, DC.

Merci B (2008) One-dimensional analysis of the global chimney effect in the case of fire in an inclined tunnel. *Fire Safety Journal* **43**: 376–389.

Mossi M (1999) *Simulation of Benchmark and Industrial Unsteady Compressible Turbulent Fluid Flows.* PhD thesis, Mechanical Engineering Department, École Polytechnique Fédérale de Lausanne, Switzerland.

Motenegro G (2002) *Simulazione 1D-Multi D di Flussi Instazionari e Reagenti in Sistemi di Scarico ed Aspirazione in Motori a Combustione Interna.* PhD thesis, Dipartimento di Energetica, Politecnico di Milano, Italy.

NTIS (1980) *User's Guide for the TUNVEN and DUCT Programs.* Publication PB80141575. National Technical Information Service, Springfield, VA.

Parsons Brinckerhoff Quade & Douglas (1980) *Subway Environmental Design Handbook. Volume II: Subway Environment Simulation (SES) Computer Program Version 3 Part 1: User's Manual.* Draft for US Department of Transportation. Parsons Brinckerhoff Quade & Douglas Inc. Honolulu, HI.

PIARC (1995) *Tunnel Routiers: Emission, Ventilation, Environnement.* World Road Association (PIARC), La Defense.

Rey B, Mossi M, Molteni P, Vos J and Deville M (2009) Coupling of CFD software for the computation of unsteady flows in tunnel networks. *Proceedings of the 13th International Symposium on Aerodynamics and Ventilation of Tunnels, New Brunswick, NJ, USA*, 13–15 May 2009.

Riess I and Bettelini M (1999) The prediction of smoke propagation due to tunnel fires. *Proceedings of the 1st International Conference on Tunnel Fires and Escape from Tunnels, Lyon, France*, 3–7 May 1999. Independent Technical Conferences Ltd, Bedford.

Riess I, Bettelini M and Brandt R (2000) Sprint – a design tool for fire ventilation. *Proceedings of the 10th International Symposium on Aerodynamics and Ventilation of Vehicle Tunnels, Boston, MA, USA*, 1–3 November 2000.

Schabacker J, Bettelini M and Rudin Ch (2002) CFD study of temperature and smoke distribution in a railway tunnel. *Tunnel Management International* **5**(3).

Tullis JP (1989) *Hydraulics of Pipelines.* Wiley, New York.

US Bureau of Mines (1995) *Fire Users Manual Version 2.20.* Department of the Interior, Bureau of Mines, Washington, DC.

Van Maele K and Merci B (2008) Application of RANS and LES field simulations to predict the critical ventilation velocity in longitudinally ventilated horizontal tunnels. *Fire Safety Journal* **43**: 598–609.

Vauquelin O and Wu Y (2006) Influence of tunnel width on longitudinal smoke control. *Fire Safety Journal* **41**: 420–426.

Vega MG, Arguelles Diaz KM, Oro JMF, Ballesteros Tajadura R and Morros CS (2008) Numerical 3D simulation of a longitudinal ventilation system: Memorial Tunnel case. *Tunnelling and Underground Space Technology* **23**: 539–551.

Versteeg HK and Malalasekera W (2007) *An Introduction to Computational Fluid Dynamics, the Finite Volume Method*, 2nd edn. Pearson Prentice Hall, Glasgow.

Wang L, Haworth DC, Turns SR and Modest MF (2005) Interactions among soot, thermal radiation, and NO_x emissions in oxygen-enriched turbulent non-premixed flames: a computational fluid dynamics modelling study. *Combustion and Flame* **141**: 170–179.

West A, Vardy AE, Middleton B and Lowndes JFL (1994) Improving the efficacy and control of the Tyne Tunnel ventilation system. *Proceedings of the 1st International Conference on Tunnel Control and Communication, Basel, Switzerland*, 28–30 November 1994, pp. 473–489.

Wu Y and Bakar MZA (2000) Control of smoke flow in tunnel fires using longitudinal ventilation systems – a study of the critical velocity. *Fire Safety Journal* **35**: 363–390.

Handbook of Tunnel Fire Safety, 2nd edition
ISBN: 978-0-7277-4153-0

ICE Publishing: All rights reserved
doi: 10.1680/htfs.41530.391

Chapter 18
Non-deterministic modelling and tunnel fires

Alan Beard
Civil Engineering Section, School of the Built Environment, Heriot-Watt University, Edinburgh, UK

18.1. Introduction

Non-deterministic modelling includes the three key categories: (1) probabilistic models, (2) statistical models and (3) points schemes. In addition, there is also the category that might be called 'checklists', which may be purely or partially qualitative. While decision-making in relation to tunnel fire safety is often made on a non-deterministic basis (although there may be elements of deterministic modelling), details of such models (their conceptual and numerical assumptions) are generally not available in the public domain, and this is not desirable or acceptable. Details of all models, whether deterministic or non-deterministic, used for safety decision-making should be made available in the public domain and subjected to peer review as well as comment by those outside the immediate process. With this in mind, this chapter will give a very brief account of some kinds of non-deterministic models. However, relatively little detail exists in the published literature on non-deterministic modelling in relation to tunnel fire safety; and little has been subjected to rigorous independent review and comment. For more, see Beard and Cope (2008) and Chapters 22, 25, 29 and 30.

18.2. Probabilistic models

A probability is a pure number between 0 and 1, inclusive, which gives an indication of how likely an event is to occur: 0 meaning the event is impossible and 1 meaning the event is certain. The basics of probability theory may be found in many textbooks, and will not be gone into here. It is vital to ensure that the correct logic is used (i.e. correct Boolean relationships) and that any significant dependencies are taken account of. It is easy to make mistakes. Probabilistic models have been constructed in relation to tunnels in the form of logic trees, including their submodels; see below. Outwith the direct context of logic trees, there have been relatively few probabilistic models created in relation to tunnels. Those that have been produced have tended to be in the area of construction (e.g. see Oreste, 2005). An application to tunnel signage may be seen in Du and Pan (2009).

18.3. Statistical models

Statistical models do not call upon the concept of probability explicitly, although statistical information may be used as part of a probabilistic model. Statistical models may deal with, for example, numbers of fatalities as an annual variation. 'Frequency models' would be a part of this class, where a frequency may correspond to, for example, the number of failures per year.

Such frequencies may be related to probability, and, in general, 'frequency models' may be cast into probabilistic form. Care must be taken to use the correct logic in any models and to clearly distinguish quantities that have dimensions (e.g. the number of fatalities/year) from quantities that are dimensionless (e.g. probabilities).

18.4. Logic trees

Logic trees may be cast in probabilistic or 'frequency model' terms. Events may be combined leading to a 'top event', as, for example, the logic tree constructed for a hospital fire (Beard, 1983a) and for a fire barrier (Beard et al., 1991). Such trees have their origin in the 'fault trees' that were developed in the nuclear and aerospace industries. Logic trees may also be constructed leading away from a 'top event' to various possible consequences, and these are generally called 'event trees'. In such a case, the 'top event' essentially becomes, logically, more like a 'central event'. In risk assessment, there will often be more than one 'top/central event', and such events may be identified with the 'crucial events' in Chapter 5.

In tunnel safety risk assessment, a logic tree (equivalent to a 'fault tree') may be constructed, leading to a crucial event, and an event tree may be constructed leading away from it, for example the 'train accident risk model' used as part of the risk assessment for the Channel Tunnel Rail Link in the UK (Small, 2004). The TUNprim model (Kruiskamp and Hoeksma, 2001; SAFE-T, 2006) calculates the 'expected value' (average number of deaths per year) and the 'societal risk' (F–N curve). The work of SAFE-T (2006) also describes other models such as the TUSI model (which calculates the number of traffic incidents, accidents and fires) and the QRAM model described in Chapter 22. There also exists the spreadsheet-based model developed under the aegis of the Highways Agency in the UK (Fraser-Mitchell, 2006) and the model TRAFFIC (Powell and Grubits, 1999), as well as other models.

Sub-models for the constituent probabilities need to be constructed, and this may consist of putting in a value based on historical statistics. However, relying purely on historical statistics is generally inappropriate, as such statistics may be for systems quite different to the one under consideration; also, systems change with time. For example, before the Mont Blanc Tunnel fire of 1999, the tunnel had been regarded as relatively safe because in 34 years of operation only 17 fires had been reported, 12 of which had been extinguished by the drivers and five by the operators. However, the system had changed and, in particular, vehicle construction and traffic patterns had altered, especially with regard to heavy goods vehicles. Bearing this in mind, the disastrous fire of 1999 becomes much less surprising.

Sub-models for the constituent probabilities of event trees or fault trees may be of many different types; for example, stochastic models. Tunnel fires are, crucially, time-varying incidents, and models such as stochastic models may help to represent this. Such submodels may, in principle, be very complex. (See also Chapters 22 and 29.)

18.4.1 Artificial neural networks

Artificial neural networks have been presented in the literature as being suitable for systems where there are a large number of factors. They have been devised in many areas, and have had limited application to fires in buildings (e.g. see Okayama, 1991; Wilson et al., 2000). They have been applied to a very limited extent in relation to tunnel construction (e.g. see Suwansawat and Einstein, 2006).

18.4.2   Checklists

Checklists are a very simple form of heuristic technique, and may be regarded as an elementary form of assessment; they may also form part of a more complex system of assessment. They are basically a list of 'things to look for' compiled over time, usually by pooling experience. Checklists would generally be 'systematic': see Chapter 5. See, for example, checklists developed by the National Aeronautics and Space Administration (NASA) in the USA (Hammer, 1972). A development of the checklist is the 'check-table': an example of a check-table, from a NASA source, can be seen in Hammer (1972).

Checklists have been used for many years, and may be regarded as playing a part in leading to the concept of a points scheme.

18.5.   Points schemes

Points schemes were originally developed by the insurance industry as a way of deciding on premium charges, especially for industrial risks. They have been in use since at least the 1970s. Essentially, points are assigned to various factors corresponding to hazard or protection. The points are then combined through a relatively simple mathematical formula or prescription, to produce a number that represents the risk, in some way. Usually, the points assigned depend to some extent on 'objective' features (e.g. the number of doors) and to some extent on 'subjective' assessments (e.g. the capacity of occupants). The conditions are often associated with various codes of practice or standards or guidance notes. The association of the points scale with given conditions may be pre-set by an expert or group of experts.

The points assigned to various factors may be combined in different ways, i.e. via addition, subtraction, multiplication or division. The way in which points are combined is decided by the devisers of the scheme.

Once an estimate has been made of the 'risk' for a given system, and a point or points associated with it, it may be decided what is to be regarded as 'acceptable'. If the risk is seen as too high, then the scheme may be used again to find the effects on the estimated risk of making various changes to the system, for example, putting in extra fire-fighting measures. An option would then be accepted that brought the risk within the acceptable range. ('Acceptable' raises the question of acceptable to whom?) For examples of points schemes for non-tunnel applications, see Stollard (1984) and Chi et al. (2008/2009).

In a scheme for hospital fire safety (Stollard, 1984), the final point representing 'risk' is a number between zero and 500, which is the maximum, corresponding to every feature within a guidance document being applied. The acceptable range is taken to be 350 or above.

A points scheme that has been devised for application to tunnels can be seen in Fusari et al. (2003). Also, a points scheme approach has been applied to existing tunnels in Europe under the rubric EuroTAP (European Tunnel Assessment Programme), organised under the aegis of the German Automobile Association (ADAC). This gives a final 'point' for each tunnel, corresponding to categories ranging from 'very good' to 'very poor'.

Advantages of points schemes are as follows.

■   They are relatively easy to use.

- They provide a way of distilling the 'expert judgement' of a single person or group of people into a quantitative model (i.e. during the development of the scheme).
- A large number of different factors may, in principle, be taken into account, ranging from human behaviour to hardware.

Disadvantages of points schemes are as follows.

- It is unclear what the 'points' associated with a given condition actually mean in the real world. That is, what events do they correspond to? This may be compared with, say, a probabilistic model, where numbers correspond to probabilities of events, such as 'ignition', 'fatality', etc.
- The mathematical logic by which points are combined is not firmly based in mathematical theory but is, in essence, up to the model developer; for example, whether to subtract or divide. Compare this with a probabilistic logic tree, where probabilities are combined through logic gates having a logic grounded in probabilistic theory.
- The 'transformation question'. While a model based directly on events such as ignition, death, injury. etc., may, in principle at least, be compared with what happens in the real world, the same cannot be said of points schemes (Beard, 1983b). How do the points associated with a given system compare with the actual fire experience for that system? Do systems/tunnels with similar numbers of points actually associate with similar fire experiences in terms of ignition, fire spread, fatality, injury, etc.? If they do not, then this shows an inconsistency. Also, if different systems, having similar points, are found to associate with similar fire experiences, is the relation between the points and the occurrence of real-world events uniform? For example, does a small change at some interval in the range of points produce a large change in the real events experienced? This would be a form of sensitivity. However, sensitivity of one form or another exists in many types of models, if not all.
- Different users may come up with different results: this is true of most models, if not all.

It is also important to be aware that different experts may come up with very different answers when asked to estimate the same thing: see the section on Bayesian techniques below. This would influence both the construction and use of points schemes.

18.5.1 Bayesian methods

It is very often the case in risk assessment that there are sparse data, i.e. very few data directly relevant to the case under consideration. Even if there is a lot of historical information going back over many years, such information may not be directly applicable to the case of concern, because the system changes. (See, for example, the case of the Mont Blanc Tunnel fire, above.) However, sparse information (or relevant but not directly applicable information) may be valuably used via the Bayesian approach, in which expert judgement also plays a part. One or more experts makes a 'prior estimate' in relation to a parameter or variable of concern, and this is adjusted through a process including estimation of the 'likelihood' of each piece of 'evidence' regarded as relevant, for example, results from an experimental test. As more and more relevant data become available over time, then this can be 'fed into' the process, and estimates of the parameter or variable updated. After successive updates, the empirical data will, in principle, tend to dominate, and the importance of experts' estimates will decline. The alternative, in a situation where there is relatively little directly applicable 'hard' information, is to use experts' estimates outside of the framework of the Bayesian method, or to use historical statistics (or other data), which may well not be directly applicable to the current case and may produce unrealistic estimates, especially bearing in mind future possible changes in the system. The Bayesian approach provides a systematic way of combining data and the expertise of experts such that, as new information becomes available, it can be incorporated. In principle, if a very large amount of data were to be incorporated over time, then the influence of the original experts' estimates would decline to zero. See also the chapter 'The influence of tunnel ventilation on fire behaviour'. The Bayesian approach may be used to estimate parameters or variables for deterministic models as well as play a part in non-deterministic models.

18.5.2   Caveats in relation to the application of Bayesian methods

As with any modelling technique it is necessary to be very careful in application and to be aware of caveats. In the case of Bayesian methods, this means, *inter alia*, (1) possible institutional, cultural or national bias in the production of prior estimates (any estimate by an expert will be based on their own knowledge and experience); (2) bias in the estimation of likelihoods. For example, in the study of Carvel *et al.* (2001) it had initially been hoped that the members of an expert panel would be able to acceptably provide likelihood values, even though they had also provided prior estimates. However, after a short pilot study it became clear that some people considered any evidence as support for their own prior estimates. It was therefore decided that a person who was independent of all the prior estimates would be the most appropriate person to estimate these values. This independent person was familiar with all the evidence.

18.6.   Caveats in general

In non-deterministic risk analysis, Ferkl and Dix (2011) have described what they refer to as the 'seven most deadly sins'. To these might be added an eighth sin, that of not being explicit, comprehensive and clear about what has been done in an analysis, especially with regard to assumptions made and logic assumed.

18.7.   Concluding comment

All models, whether deterministic or non-deterministic, have conceptual assumptions; in addition, quantitative models have numerical assumptions. Given the uncertainty that exists in modelling systems associated with tunnel fire safety, it is necessary to see all results from all models in light of the assumptions made. A questioning frame of mind is essential in both creating and using results from models. Given this, they may play a valuable part in tunnel fire safety decision-making.

REFERENCES

SAFE-T (2006) Work package 5: Harmonised Risk Assessment. Framework Programme 5, funded by the European Union, 2006. Available at: http://www.mep.tno.nl/SAFET/ Workpackages.

Beard AN (1983a) A logic tree approach to the St. Crispin Hospital Fire. *Fire Technology* **19**(2): 90–102.

Beard AN (1983b) Letter to the Editor on points schemes. *Fire Technology* **19**(1): 69–70.

Beard AN and Cope D (2008) Assessment of the Safety of Tunnels. Available via the website of the European Parliament under the rubric of Science & Technology Options Assessment (STOA).

Beard AN, Burke G and Finucane MT (1991) Assessing the adequacy and reliability of fire barriers in nuclear power plants. *Nuclear Engineering and Design* **125**: 367–376.

Carvel RO, Beard AN, Jowitt PW and Drysdale DD (2001) Variation of heat release rate with forced longitudinal ventilation for vehicle fires in tunnels. *Fire Safety Journal* **36**: 569–596.

Chi JH, Chen CT and Chen JW (2008/2009) Research on fire risk assessment for factory-type buildings. *Journal of Applied Fire Science* **18**: 61–77.

Du Z and Pan X (2009) Application research of visual cognition probabilistic model on urban tunnel's sign. In: *Proceedings of the International Conference on Measuring Technology and Mechatronics Automation, Zhangjiajie, China*, 11–12 April 2009, vol. 3, pp. 490–493.

Ferkl L and Dix A (2011) Risk analysis – from the Garden of Eden to its seven most deadly sins. In: *Proceedings of the 14th International Symposium on Aerodynamics and Ventilation of Tunnels, Dundee*, 11–13 May 2011. BHR Group, Cranfield.

Fraser-Mitchell J (2006) *Risk Assessment for Road Tunnels*. UK Road Tunnel Operators Forum, London.

Fusari S, Giua S and Luciani F (2003) Priority criteria for investments in tunnel safety. In: *Proceedings of the 5th International Conference on Safety in Road and Rail Tunnels, Marseille*, 6–10 October 2003. Tunnel Management International, Tenbury Wells, pp. 365–374.

Hammer W (1972) *Handbook of System and Product Safety*. Prentice-Hall, NJ.

Kruiskamp D and Hoeksma J (2001) TUNprim: a spreadsheet model for the calculation of the risks in road tunnels. In: *Proceedings of the 4th International Conference on Safety in Road and Rail Tunnels, Madrid*, 2–6 April 2001. Independent Technical Conferences, Tenbury Wells, pp. 129–137.

Okayama Y (1991) A primitive study of a fire detection method controlled by artificial neural net. *Fire Safety Journal* **17**: 535–553.

Oreste P (2005) A probabilistic design approach for tunnel supports. *Computers and Geotechnics* **32**: 520–534.

Powell S and Grubits S (1999) Tunnel design with TRAFFIC – tunnel risk assessment for fire incidents and catastrophes. In: *Proceedings of the International Conference on Tunnel Fires and Escape from Tunnels, Lyon*, 5–7 May 1999. Independent Technical Conferences, Tenbury Wells.

Small L (2004) Rail safety. *Arup Journal* **1**: 10–11.

Stollard P (1984) The development of a points scheme to assess fire safety in hospitals. *Fire Safety Journal* **7**: 145–153.

Suwansawat S and Einstein H (2006) Artificial neural networks for predicting the maximum surface settlement caused by EPB shield tunnelling. *Tunnelling and Underground Space Technology* **21**: 133–150.

Wilson PD, Dawson CW and Beard AN (2000) Flashover prediction – an evaluation of the application of artificial neural networks. In: *Proceedings of ESREL 2000, Foresight and Precaution, Edinburgh*, 15–17 May 2000. Balkema, Rotterdam, pp. 711–715.

Part IV

Fire safety management and human factors

Handbook of Tunnel Fire Safety, 2nd edition
ISBN: 978-0-7277-4153-0

ICE Publishing: All rights reserved
doi: 10.1680/htfs.41530.399

Chapter 19
Human behaviour during tunnel fires

Jim Shields
University of Ulster, UK

19.1. Introduction

In 1995, the author presented a paper entitled 'Emergency egress capabilities of people with mixed abilities' at the 2nd International Conference on Safety in Road and Rail Tunnels (Shields and Boyce, 1995). The paper addressed many issues including the prevalence of disability, the mobility of people with physical and sensory impairments, evacuee egress capability profiles, the use of public and private transport by disabled persons and emergency planning for tunnels. It was pointed out that despite recent work there was little usable data on the emergency egress capability of disabled persons. The paper argued that tunnels by their very nature, coupled with different transport modes, added new dimensions to the already complex process of emergency evacuation from fire. It was also suggested that it was not unreasonable to assume that people involved in fire and other emergencies in tunnels would not be familiar with the tunnel, provisions of means of escape or the locations of places of refuge. Further, the evacuation routes were likely to be obstructed. It was postulated that the conceptual models of human behaviour in fire in buildings might not be valid for fires in tunnels and other subterranean spaces. The time required, for example, for people with disabilities and people injured in vehicle crashes to disembark from vehicles could be up to 20–30 times longer than the disembarkation times for people without disabilities or injuries. In summary, the paper appealed for better understanding of human behaviour in fire in tunnels, in order that it be given due consideration in future tunnel designs. Unfortunately, in 1995 the presentation seemed to fall largely on deaf ears. Some delegates in discussion suggested that commuters and other travellers could be relied upon to assist people with disabilities to evacuate from tunnels when necessary. Then, fires in tunnels was not a major issue. Fires in tunnels, because of recent tragic events, are now a major concern. Unfortunately, until recently there has been little interest in or advancement of our understanding of human behaviour in fires in tunnels.

19.2. Some recent tunnel fires

In recent times there have been a number of very serious tunnel fires, as discussed in Chapter 1. A few tunnel fire incidents will be briefly reviewed here with respect to human behaviour in fire. It is clear from available reports on tunnel fires that in some cases insufficient consideration may have been given in the design of the tunnel to the life safety potential of tunnel users in a fire emergency. It is also apparent that some of the behaviours of people threatened by fire in a tunnel were foreseeable. With the aid of hindsight, it is possible to reflect on particular events and draw conclusions, which should ensure, as far as possible, that such events do not happen again.

However, a fundamental behavioural concept (Leslie, 1996) is often overlooked in our engineering model of the built environment:

$$R = f(E)$$

i.e. R, the responses, actions and behaviours of individuals in time and space are a function of E, the environment in which they find themselves. This basic concept is often overlooked as we engineer our built environments. It must be understood that this notion of 'occupancy' serves as a potential bridge between engineering and environmental psychology, i.e. do we shape our environments or do they shape us? Built environments are not without their innate complexities, which influence the behaviours of the end users, especially in fire emergencies.

Developing the notion of complex environments, there are transit tunnels, road tunnels, rail tunnels and other tunnels (e.g. canal or waterway tunnels). Further, tunnels have distinguishing features, i.e. they may be single-bore unidirectional, single-bore bi-directional, double/triple-bore unidirectional, etc. Thus, for fire emergencies in tunnels, there may be a serious complexity issue that must be understood and addressed at the design stage, i.e. the characterisation of the setting; the developing fire scenario; traffic flows; human behaviour; fire evacuation; and rescue dynamics. These complex relations can be addressed by the designers and managers of systems, if they understand and apply the basic behavioural concept outlined previously. Some specific past fires will now be described, with particular reference to human behaviour aspects.

19.2.1 Caldecott Tunnel fire – 7 April 1982

The Caldecott Tunnel is a triple bore unidirectional tunnel. On 7 April 1982, there was a pile-up inside the tunnel involving a passenger car, a bus and a gasoline tanker truck (Lake, private communication; Carvel and Marlair, 2005). After the accident, vehicles continued to enter the tunnel. The accident was witnessed by a mother and son. The mother used the emergency phone to alert the control room of a nascent fire – but the emergency phone connection failed. The mother returned to her vehicle, entered it, stayed there and later perished. Her vehicle was less than 15 m from an unmarked cross passageway. The son back tracked to warn others and was enveloped by smoke; he survived. In all, there were seven fatalities. Five people perished in their vehicles, which they chose not to abandon.

19.2.2 BAKU underground railway/metro fire – 28 October 1995

On 28 October 1995, a fire occurred in the underground railway/metro of Baku, the capital of Azerbaijan, which claimed the lives of 289 people with another 265 being severely injured (Hedefalk et al., 1998; Carvel and Marlair, 2005). As the train pulled out from Uldus Station passengers, in car number 5, smelled smoke. They did not perceive themselves to be in any immediate danger. Later, passengers in car number 4 observed white smoke, which soon turned black and became irritating. A malfunction of some sort caused the train to stop only 200 m from Uldus Station. The driver reported the incident and demanded the power be cut off. When the train stopped the tunnel filled with smoke. The passengers started to evacuate. As the evacuation proceeded, it was made more difficult because the directional mode of the ventilation system was changed – to the detriment of the evacuees.

19.2.3 Channel Tunnel fire – 18 November 1996

On 18 November 1996, a goods train travelling from France to England stopped in Running Tunnel South with a fire on board (Allison, 1997; Carvel and Marlair, 2005). There were 33 people on board the train – 31 passengers and two crew in the amenity coach. The passengers and crew survived, but smoke inhalation and the presence of smoke made their safe evacuation difficult. This tunnel fire is well documented, as it was the subject of an inquiry into the incident. It is clear from the report of the inquiry into the incident that the safety systems did not all work as they should. What is significantly different, however, is that the flow of traffic in the Channel Tunnel could be controlled and used to best advantage. A rescue train could, because of the tunnel design, reach a point adjacent to the fire-stricken train, and rescue the fire-threatened passengers.

19.2.4   Mont Blanc Tunnel fire – 24 March 1999

The Mont Blanc Tunnel is a single-bore, bi-directional tunnel linking France and Italy. At the time of the incident on 24 March 1999, the traffic flow was described as of average intensity, i.e. 4–5 vehicles per minute (Lake, private communication; Carvel and Marlair, 2005). At 10:53 hours, a heavy goods vehicle (HGV) loaded with margarine and flour pulled into a lay-by some 6300 m from the French portal. As the driver exited the vehicle, it burst into flames. The driver evacuated himself towards Italy – the airflow in the tunnel was forcing the smoke towards France. Thirty-eight people perished in the fire. Twenty-seven of the victims were found in their vehicles, two were found in vehicles other than their own, and nine victims were found outside of vehicles. Despite the road traffic signals going to red, vehicles continued to enter the tunnel. Some cars were able to execute 'U' turns and exit – but the HGVs could not.

The tunnel had the usual safety features, including CCTV and fire alarm facilities. However, the CCTV-activated video recording did not operate on the French side, and the fire detection system on the Italian side had been disabled the day before the incident because of false alarms. Two fire-fighting teams entered the tunnel at 10:57 hours, but were so hampered by smoke that they were ordered by the control centre to a shelter located at lay-by 17, where they remained for more than 7 hours before being rescued. At 11:10 hours, fire-fighters entered the tunnel without reporting to the control centre – of these six fire-fighters, only four had breathing apparatus, i.e. appropriate for building fires but not tunnel fires. They had to find shelter in an unpressurised room at lay-by 12. They were there for 5 hours. The leader of this group died shortly after being taken from the tunnel. A second fire-fighting team entered the tunnel to rescue the first fire-fighting team. They had to give up and take shelter at lay-by 5. A motorcycle patrolman together with another car driver took refuge at lay-by 20. Both perished in the fire.

19.2.5   Tauern Tunnel fire – 29 May 1999

The Tauern autobahn is one of the main traffic arteries through the Alps, linking northern and southern Europe. The Tauern Tunnel is a single-bore bi-directional tunnel, i.e. vehicular traffic through the tunnel is in contra-flow. Typical features were included in the design of this tunnel: ventilation, fire detection, facilities for the fire service, emergency call points, first-aid fire-fighting equipment and CCTV continuous monitoring.

Immediately prior to the incident there was a considerable volume of traffic flow, which was influenced by manually operated road traffic signals at construction works (Lake, private communication; Eber, 2001; Carvel and Marlair, 2005). At 04:50 hours the tunnel fire alarm activated. The duty officer could not detect anything visually via CCTV. The duty officer contacted the motoring police, and a general alarm was raised. At 04:41 hours the traffic lights were on red, and traffic had backed up. A lorry rammed into the back of two stationery cars, forcing them underneath a lorry in front of them that was carrying paint. A severe fire developed. Twelve people died in this incident, eight of whom were involved in the original road traffic accident.

As the fire developed, two people remained in their vehicles: they perished. One driver returned to his vehicle to retrieve documents. He then joined the two people who remained in their vehicles, and he also perished. Another person perished some 800 m away from the fire origin. Within seconds of the fire starting, the CCTV cameras were immersed in dense black smoke and rendered useless. Drivers ignored the road traffic signals and continued to pass the red signal and enter the tunnel; some tried to turn their vehicles around but could not. Drivers refused to leave their cars, some took photographs, others looked around. Sixteen lorries and 24 cars burned inside the tunnel, forcing the fire brigade back. People travelled over 800 m to escape – three people took shelter in a telephone niche and survived. The tunnel roof had to be propped up because of fear of collapse.

19.2.6   Kitzsteinhorn Funicular Tunnel fire – 11 November 2000

The Kitzsteinhorn Funicular (Kaprun) was the first Alpine underground railway in the world, with a 2.2 km-long inclined tunnel. On 11 November 2000, at approximately 09:00 hours, a fire occurred in the ascending train (Carvel and Marlair, 2005): while still out in the open air, smoke was seen coming from the rear of the train. Those who saw the smoke did not perceive that they were in any imminent danger. As the train entered the tunnel, flame was seen and reported – but still the passengers did not perceive that they were in any real imminent danger. Shortly afterwards, the train was stopped, the doors were opened by an attendant and nearly all the passengers disembarked from the train. As the fire and smoke spread upwards, the passengers evacuated upwards away from the fire, only to be overcome by the smoke. The inclined tunnel behaved as a flue. A few passengers evacuated to safety down the tunnel, going past the fire. In the upper terminal, five employees were totally unaware of the fire moving towards them. Similarly, the driver and passengers in a downward moving train were totally unaware of the fire. They died in the fire, as did three of the five employees in the upper terminal. There were 155 fatalities in total.

Only 12 people in the tunnel survived. They survived by evacuating downwards and out of the inclined tunnel. So, another feature to be considered in the characterisation of tunnels and associated fire hazards is the angle of inclination.

It is also clear from this incident that many of the concepts and theories with respect to human behaviour in fires in buildings are not valid for tunnel fires. That is, as in this case, moving directly away from a fire is not necessarily the best thing to do.

19.2.7   St Gotthard Tunnel fire – 24 October 2001

The St Gotthard Tunnel is a single-bore bi-directional tunnel. On 24 October 2001, a collision occurred between two HGVs, which were travelling in opposite directions (Lake, private communication; Carvel and Marlair, 2005). A serious fire resulted. The tunnel had the usual safety provisions, but traffic was flowing in opposite directions. Both drivers of the vehicles involved in the initial incident survived, 11 others were victims of the fire. All of the victims were some distance from the origin of the fire, for example, up to 2000 m; all died from smoke inhalation, and some never left their vehicles. It has been suggested that all of the victims had sufficient time to evacuate the tunnel safely had they chosen to do so.

19.2.8   Ted Williams Tunnel fire – 19 May 2002

The Ted Williams Tunnel is a twin-bore unidirectional tunnel in central Boston, USA. On 19 May 2002, a passenger bus had a fire in its engine compartment (Lake, private communication; Carvel and Marlair, 2005). The bus stopped, and the passengers disembarked and boarded the bus in front, which proceeded to exit from the tunnel. As the bus drove towards the tunnel exit, it was partially enveloped in smoke. This was because the tunnel ventilation system forced the smoke to flow in an easterly direction. East-bound traffic behind the bus on fire could not proceed. People left and locked their vehicles, blocking the tunnel for fire-fighters, and back-tracked – walking westwards to safety. Some people stayed with their vehicles.

19.2.9   Selected quotes from reports and summary of major tunnel fires

To complete this section, a summary of major fire incidents is presented in Table 19.1 together with some quotes from reports on recent tunnel fires.

Channel Tunnel fire, 11 September 2008 – a lorry driver's eyewitness account:

> There was this lorry in flames and a series of explosions. There were about 20 of them. Everything was exploding around us – tyres, fuel tanks. Then there was this smoke that stopped us from seeing and breathing properly. The door of our coach was locked and impossible to open. We saved ourselves by breaking a window with a hammer. We left the train through this window. It's really the moment we panicked the most. We felt we were really stuck. We were incredibly lucky.
>
> (*Daily Telegraph*, 2008)

Santa Clarita Tunnel fire, 2007:

> Firefighters have made it about 50 feet inside the interstate tunnel where two people died in a massive fire that erupted from an apparent big rig collision. Teams have been moving in 10 feet at a time – what's making it so tough, is the fire burned so intense that a number of vehicles are burned down to the core.
>
> (CNN, 2007)

Viamala Tunnel fire, 2006:

> After all, in the Viamala tunnel in 2006 nine people died in smoke and fire because they simply could not get away from the scene of the accident in time. The Viamala tunnel, by the way, is only 742 metres long and this shows that even the short tunnels have a risk potential that should not be underestimated.
>
> (EuroTAP, 2011)

> The collision in the 750 m long Viamala tunnel between a car and a bus carrying a local ice hockey team was the worst incident of its kind since 2001 when eleven people died after two trucks crashed in the north-south Gotthard tunnel.
>
> (Swissinfo, 2006)

Frejus Tunnel fire, 2005:

> There are now two bodies, we don't know if we will find any more – rescue workers were struggling to put out the fire and one kilometre of the tunnel remained unreachable because of thick black smoke. As many as 20 people suffering from smoke intoxication had been rescued. The incident comes six years after 39 people were killed in a fire in the Mont Blanc tunnel about 100 km north of Frejus.
>
> (*China Daily*, 2005)

From the early 1970s onwards, major fires in tunnels have occurred with some regularity. It may be that the number of tunnels in use has significantly increased, the nature of goods, etc.,

Table 19.1 Summary of major tunnel fires

Tunnel	Country	Cause	Injuries/ fatalities	Vehicles destroyed	Duration of fire	Year
Channel	France/UK	Freight shuttle	None	Rolling stock	18 hours	2008
Burnley	Australia	collision	3 dead	2 HGVs, 3/4 cars	1 hour	2007
Santa Clarita	USA	collisions	2 dead 10 injured	2 trucks, other vehicles	4 hours	2007
Viamala	Switzerland	collisions	9 dead 5 injured	2 cars, 1 bus	4 hours	2006
Frejus	France/Italy	collision	2 dead 21 injured	3 trucks, other vehicles	6 hours	2005
Ted Williams	USA	Bus engine fire	None	Bus	2 hours	2002
St. Gotthard	Switzerland	collision	11 dead	2 HGVs, cars	48 hours	2001
Howard Street	USA	derailment	None	Rolling stock	5 days	2001
Hong Kong Cross Harbour	China	car fire	None	Car	2 hours	2000
Kaprun	Austria	Unknown	155 dead	Train	Not recorded	2000
Tauern	Austria	Collision	12 dead	2 HGVs, 4 cars	15 hours	1999
Mont Blanc	France/Italy	HGV overheating	39 dead	23 HGVs, 11 other vehicles	2.2 days	1999
Channel	France/UK	HGV overheating	Several injured	10 HGVs, carriers	7 hours	1996
Isola delle Femmine	Italy	2 collisions	5 dead 20 injured	1 tanker, 1 bus, 19 cars	Not recorded	1996
Baku	Azerbaijan	Electrical failure	300+ dead	Tube train	Not recorded	1995
Pfänder	Austria	Collision	3 dead (by crash)	1 HGV, 1 van, 1 car	1 hour	1995
Huguenot	South Africa	Gearbox	1 dead	1 bus	1 hour	1994
Serra a Ripoli	Italy	Vehicle out of control	4 dead	4 HGVs, 11 cars	2.5 hours	1993
Gumefens	Switzerland	Collision	2 dead	2 HGVs, 1 van	2 hours	1987
Kings Cross	London, UK	Escalator fire	31 dead	N/A	6 hours	1987
L'arme	France	Collision	3 dead	Car and trailer	Not known	1986
Summit	UK	Derailment	None	Train	3 days	1984
Pecorile	Italy	Collision	8 dead	Not known	Not known	1983
Caldecott	USA	Collision	7 dead	1 tank, 4 cars, 2 HGVs, 1 bus	3 hours	1982
Kajiwara	Japan	Gearbox	1 dead	2 HGVs	1.5 hours	1980
Nihonzaka	Japan	Collision	7 dead	127 HGVs, 46 cars	7 days	1979
Velsen	Netherlands	Collision	5 dead	2 HGVs, 4 cars	1.5 hours	1978
Holland	USA	HGV shedding load	66 injured	10 HGV, 13 cars	4 hours	1949

transported has changed, the volume of traffic has greatly increased, or perhaps authorities having jurisdiction have just been too complacent. For the majority of the public using tunnels, there is no evidence of an increased perception of risk with respect to fire. Hence, no change in behaviour.

The graphic eyewitness accounts of those involved in tunnel fires convey with chilling realism their experiences of fire in unfamiliar environments, and reinforce the need for individuals to be aware of the risks and possible appropriate actions.

19.2.10 Comment

This brief review of the few tunnel incidents presented here is sufficient to raise a number of pertinent issues that should influence tunnel design and management, fire emergency incident management, and provide insights into end-user behaviours in a fire emergency.

It is clear that tunnel fires can be very severe, so severe that even emergency rescue services can become entrapped. The geometry of tunnels aids the rapid spread of fire and smoke, such that occupants many hundreds of metres away from the fire source can quickly become victims. Tunnels are not compartmented as buildings are with respect to fire. The provision of refuges or places of relative safety must reflect the potential severity of real natural fires that are likely to occur. People should not die in refuges provided for their safety.

For tunnels that have relatively steep gradients, consideration may have to be given to evacuation reversal behaviours, i.e. evacuate downwards, past the fire source, through the smoke rather than evacuate upwards to be overcome by the smoke flowing up the tunnel. There is much to ponder here.

Travellers, for example, vehicle drivers with destination affiliation, will drive into tunnels even under emergency conditions – they bring an entirely different dynamic to the notion of evacuation from fire.

Many people died in their vehicles, giving rise to the notion of object/place affiliation.

In human behaviour in fire within buildings the notion of familiarity has taken root. It has its origins in research on movement towards the familiar in the context of person and place affiliation in a fire entrapment setting (Sime, 1985). However, the notion is now interpreted in terms of buildings, i.e. the occupant response may be influenced by familiarity with the access, egress and fire safety systems of a building. The concept of familiarity has not yet been applied to tunnels, and it is difficult to envisage how this might be done beyond users knowing when they entered the tunnel and that they are progressing towards the exit. It is not a 'walk about, have a look around' type of environment. Thus, the level and degree of user familiarity, even for frequent users, must be less than that experienced by the occupants of buildings. Even the language is different, in that we do not refer to people in tunnels as occupants, which is what they are.

In summary, from a brief review of these few tunnel fire incidents, it is clear that

- people often do not perceive themselves to be in immediate danger
- safety systems can and do fail
- fire and smoke can spread very rapidly in tunnels

- people may exhibit property/vehicle/destination affiliation
- in some circumstances, as the situation deteriorates, clustering behaviours may occur, i.e. groups of people occupying vehicles
- egress routes, pedestrian and vehicular, will become blocked
- people do not evacuate immediately
- traffic will continue to enter tunnels experiencing emergencies unless they are prevented physically or otherwise
- systems can be used in such a way as to impede evacuation and increase the hazard to evacuees
- managed systems offer the greatest fire safety potential
- tunnels are complex environments, and user familiarity cannot be assumed
- vehicle driver gender differences may influence behaviours
- rescue services may not be able to reach entrapped people because of the severity of the fire and road traffic chaos (e.g. drivers attempting 'U' turns)
- refuge areas may not have the fire endurance necessary in order to ensure the survivability of their occupants – people have died in so-called refuge areas
- serious consideration must be given at the tunnel design stage to the use of credible, realistic, fire scenarios
- evacuation strategies employed must take into account the seat of the fire, tunnel design, the tunnel gradient and fire emergency information systems.

19.3. Towards understanding human behaviour in tunnel fires

Tunnels and other subterranean environments are often perceived by users as complex, confining, visually monotonous and boring. The environments created have no natural lighting, reduced visibility and often poor visual accessibility. Persons confined by such environments may feel claustrophobic and fear entrapment. In other words, tunnel and subterranean environments for many people are alien environments, that is, places where they would rather not be, i.e. not like buildings that they use daily and generally inhabit without undue psychological stress. If tunnels and similar environments are so different from buildings that the public in general access, it follows that our understanding of human behaviour in fire in buildings may not be valid for, or applicable to, subterranean environments. However, we can usefully draw on some of the concepts used in human behaviour in fire in buildings to inform our understanding of human behaviour in fire in tunnels. For the purposes of this chapter, the notion of 'occupancy' is introduced, to focus attention on human behaviour in fire in tunnels. Occupancy has been defined as 'the constraints on, conditions and possibilities of knowledge and action afforded by the social, organisational and physical locations occupied by people over time' (Sime, 1999).

This notion of occupancy introduces a people–environment paradigm where goal orientation is afforded by setting (Figure 19.1). For our purposes, the setting is a tunnel, any tunnel – transit, rail, road, waterway.

Simply put, if we are to understand and predict human behaviour in fire in tunnels, we must understand the nature of the occupancy, as depicted in Figure 19.1. The setting – the tunnel itself – will induce specific occupant behaviours. In everyday usage, for example, a road tunnel in addition to being a mere conduit is also an information system. Introducing a fire emergency into a tunnel environment produces setting and fire dynamics, which induce specific occupant behaviours. These can include destination, person, place and possession affiliations, which must be understood if future disasters are to be avoided.

Figure 19.1 Fire (F), setting (S) and occupant (O) relationship

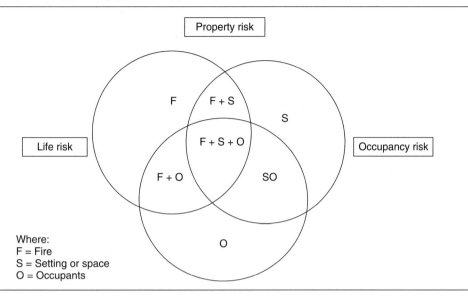

Where:
F = Fire
S = Setting or space
O = Occupants

19.3.1 Setting (tunnels)

Tunnels will vary with respect to the nature of the traffic passing through them, traffic flows (i.e. unidirectional or contra-flow), in addition to tunnel geometry. Tunnel management and fire safety provision will also vary with respect to the associated risks and safety case protocols. Thus, in attempting to understand human behaviour in fire in tunnels, it is essential that the 'environment' created by the tunnel and as impacted by the fire is fully understood. Different types of tunnels (e.g. road, transit, rail and waterway tunnels) will, by the very nature of the environments they create, induce different sets of behavioural responses in a fire emergency (Leslie, 1996), i.e.

$$R_1 = f(S_1)$$

Thus, a particular reflex response (R_1) may be related to some stimulus provided by the setting (S_1), i.e. human behaviour in fire is treated in context. In considering the characterisation of tunnels in the context of occupancy, a number of issues will be discussed by way of illustration.

In the 1986 Channel Tunnel fire (Allison, 1997) referred to earlier, the impact of management in terms of controlling traffic, the use of safety systems, the deployment of rescue personnel, the use of service tunnels, etc., significantly contributed to the safe evacuation of all of the people at risk. The evacuation of the people was not without its difficulties, but management control was exercised to maximum benefit. (Although the result might have been very different, had the fire origin been much closer to the amenity coach.) However, in the 2008 Channel Tunnel fire (see quotation in Section 19.2.9), lorry drivers saved themselves by breaking a window with a hammer and climbing out. Each emergency is unique, calling for appropriate response actions from all concerned, especially those in danger of entrapment.

It is not unusual for people to enter a building when the fire alarm is sounding. This can be associated with a low perception of imminent threat or danger. Usually, such people are swept out of the building by the flow of evacuating people. It is not unusual for car drivers to ignore motorway warning signs to slow down, just as it was not unusual in the UK for some train drivers to sometimes proceed against red signals. In behavioural terms, the car and train drivers had acquired learned irrelevance to the warning signs, i.e. they had been exposed to them so frequently before that they learned to ignore them – they had acquired learned irrelevance (McClintock *et al.*, 2001).

As far as road tunnels with high-density traffic flows are concerned, more radical means may have to be introduced to prevent vehicles entering tunnels that are experiencing serious ongoing emergencies, including fire. This may mean pursuing other forms of stimuli to induce the required behaviours or it may mean the introduction of physical barriers to prevent vehicular access to tunnels in emergencies.

Consider tunnels with steep gradients: in a fire emergency, what evacuation strategies are incorporated into the initial designs? In the Kitzsteinhorn Funicular Tunnel fire (Schupfer, 2001), passengers evacuated upwards, only to be engulfed in smoke. How can the setting be used as an information system to ensure that people, if necessary, evacuate downwards to safety? The challenge for the setting and management is to provide the information needed by the occupants, i.e. when they most need it in rapidly changing circumstances.

In road tunnels, most information is provided for vehicle drivers, for example, exit route information. However, emergency exit information in the event of fire is generally not provided for vehicle drivers – it is provided for pedestrians. Emergency exit signage is minuscule in comparison with exit signage for drivers. It seems to the author that in a fire emergency in tunnels, emergency exit information, directional or otherwise should be immediately available. Thus, static emergency exit signs to which tunnel users have already acquired learned irrelevance through constant exposure to the same kind of emergency exit signs in buildings must be replaced by emergency information systems that attract attention and convey the notion of safety to evacuees, and are thus compelling to use. A different concept for emergency exit signage is required. Emergency exits cannot be permitted to remain anonymous in an emergency: they have to have sufficient magnetism in order to attract and capture the potential users.

If refuges and places of relative safety are to be provided, they have to have sufficient fire endurance, at least to ensure the survivability of the occupants for the design fires and fire scenarios used in the design of the tunnel.

The setting in a global sense is a tunnel, with its individual characteristics. However, within any tunnel there will be a collection of micro-settings that will influence behaviours: a refuge space, a lay-by, a coach, people-carrier, etc. These micro-settings, within a global setting, can be identified, and their potential influences on behaviours considered.

19.3.2 Fires in tunnels

In fire safety engineering, the selection of fire scenarios requiring analysis is critical. The number of possible fire scenarios (including types of vehicles) in tunnels could be very large, and it is probably not feasible or practicable to quantify them all. Thus, the potentially large set of possible fire scenarios must be reduced to a finite set of design fire scenarios that are amenable to analysis.

The characterisation of a tunnel fire scenario involves a description of fire initiation, growth and suppression/extinction together with the propensity for rapid fire and smoke spread. The foregoing includes interaction amongst fire dynamics, tunnel geometry, tunnel inclination and proposed fire protection provisions for the tunnel. In addition, the consequences of each fire scenario need to be considered, particularly with respect to the life safety potential of the tunnel users. It follows that for each fire scenario there should be a corresponding life safety/ evacuation scenario for analysis.

Having identified the design fire scenarios, it is necessary to describe the assigned characteristics of the fire scenario upon which quantification can be based, i.e. the design fire or fires. The design fires chosen need to be appropriate for the fire safety design objectives. Correspondingly, for each design fire there has to be a design life safety/evacuation strategy.

Design fires are usually depicted by way of time-dependent variables, for example, the rate of heat release and the rate of smoke production.

Information is available on heat outputs, from burning cars for example. However, experience suggests that other design fires need to be considered. We do know that tunnels by their very nature provide uninterrupted flue-like conduits along which fire and smoke can rapidly spread. Much work on vehicle fire data generation and tunnel fire modelling has been done with respect to tunnel fires (Ingason, 2003). However, tunnel designers now have to couple their increasing knowledge of tunnel fire dynamics and modelling in terms of appropriate design fire scenario selection and evaluation with corresponding life safety/evacuation strategies where information on human behaviour in fire in tunnels is somewhat sparse. It is a difficult design judgement to make – especially remembering that in the Viamala Tunnel fire in 2006, nine people died because they simply could not get away in time (see the quotation in Section 19.2.9). There needs to be time for people to get out or be rescued, whatever their physical and mental characteristics.

There is, of course, much more to the characterisation of tunnels than can be discussed here but it is noted that people have died in tunnel fires in places of so-called 'relative safety', i.e. refuges.

19.3.3   Human behaviour in fire

In order to provide for the fire safety of occupants in buildings, it is necessary to have some understanding of the factors and conditions that may influence the response and behaviours of people exposed to life-threatening fires. The prediction of human responses and underlying behaviours is one of the most complex areas of fire safety engineering. At present, our understanding of human behaviour in fire in buildings is limited compared with other areas of fire safety engineering, such as fire dynamics and structural fire engineering. Consequently, it is difficult to predict with certainty the behaviours of people in fire in buildings. If our understanding of human behaviour in fire in buildings is limited, our understanding of human behaviour in fire in tunnels is very limited indeed. However, it is important to recognise and acknowledge the extent to which our understanding of human behaviour in fire in tunnels is lacking, as it gives focus and thrust to future research endeavours.

There has been, though, some significant recent research on human behaviour in tunnel fires. Frantzich and Nilsson (2004) undertook evacuation experiments to investigate walking speeds and behaviours in smoke-filled tunnels using three different types of guidance systems. This study found that persons following a wall without emergency exits incorporated in a smoke-filled tunnel may not notice exits on the opposite side in a real fire, with disastrous consequences. It was also found that a large proportion of the participants in the study did not use the emergency exits, even though they saw the emergency signs inside the tunnel. The integration of human behaviour into the improvement of French road tunnels was articulated by Tesson and Lavedrine (2004), who sought to derive lessons from annual training exercises required of tunnel operators in France. From the training scenario described in their paper, the essential observations are: people at the fire location may come near to or pass the burning vehicle in spite of smoke; nobody helps to extinguish the fire, nobody assists the passengers of the burning vehicle; people will walk on the carriageways and do not attempt to get safe; and people are not aware of the danger they are exposed to. This paper also introduced research work in progress on egress that is discussed in Chapter 20. Human behaviour in tunnel accidents was tentatively explained by way of possible moderators and modulations in terms of risk perception, cognitive appraisal and stress, coping strategies and disengagement reduction (Eder *et al.*, 2009). It was suggested that because drivers rely on visual impressions their perception is narrowed.

Despite the above recent research, it is evident that in terms of beginning to really understand human behaviour in tunnel fires we have still hardly scratched the surface. Consequently, it is still possible to usefully draw on knowledge of human behaviour in fire in buildings to inform and shape thought about human behaviour in tunnels. In fire safety engineering, time is the metric used to assess the life safety protection of the occupants for a given fire scenario. Figure 19.2 conceptualises and illustrates the parallel time lines for fire development and occupant life safety (i.e. evacuation). Fire cues received by the occupant of any space may be visible flame/smoke or fire alarm activation. The cues received by the occupants initiate a process of recognition and interpretation, which in turn will result in a decision-making process. The recognition and response time periods are often referred to as pre-evacuation activity times (PEATS), i.e. activities undertaken before evacuation is commenced.

In Figure 19.2, the concept of available safe egress time (ASET) is introduced. ASET is a function of the fire growth and development, i.e. it is the time period between the ignition of a fire and the onset of untenable conditions in the near- and far-field tunnel areas. The required safe egress time (RSET) is a function of the occupants, i.e. how much is needed to ensure that the occupants can safely evacuate the fire-threatened spaces. Tenability analysis for ASET will include the time for a smoke layer to descend and threaten occupants, smoke layer gas temperatures and exposure to harmful fire gases. RSET involves estimating the time for people to recognise and respond to a fire threatening a space, i.e. PEAT and the time required for people to evacuate to safety. Often in fire-threatened buildings the PEAT times, i.e. finishing a task, collecting belongings, etc., are much longer than the actual travel time. An additional complexity is that, in tunnel fires, in addition to providing for mixed-ability occupants in the normal sense, consideration may need also to be given to those accidentally injured. Time is of the essence in any fire emergency, i.e. the time for the fire to develop and threaten the occupants and the time for the occupants to reach a place of safety. Insufficient attention to the human factors associated with fire and evacuation scenarios can result in inaccurate estimation of evacuation times and possible failure in achieving the desired condition

$$RSET \ll ASET$$

i.e. safe evacuation. However, the time differential between completion of an evacuation and the onset of untenable conditions in the near and further fire threatened fields should provide an

Figure 19.2 Tunnel hazard development

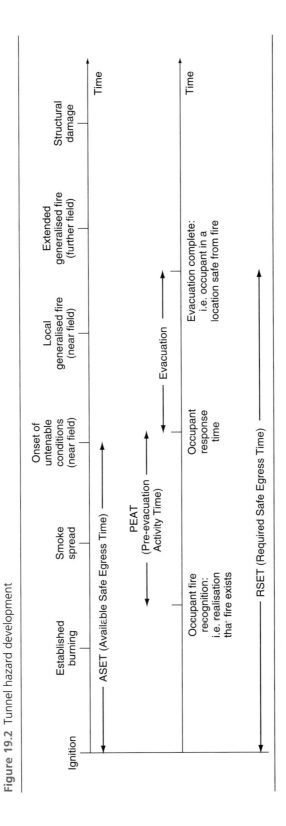

acceptable margin of safety. For the safety of occupants, other strategies, for example, defend in place, may be employed. Although this is unlikely in tunnel design, with the exception of places of refuge, the same principle applies to the survivability of the occupants.

19.3.4   Occupant characteristics

Human input into the life safety evaluation process is seen in the form of so-called 'occupant characteristics'. Occupant characteristics, which must be taken into account when predicting human behaviour in fire in buildings, include the number and distribution of occupants, gender, age, physical and sensory capabilities, familiarity with the building, social and cultural factors, role and responsibility commitment to activities, involvement with others, previous experience, knowledge of fire, and training. All of the foregoing, and more, may be applicable to the prediction of human behaviour in tunnel fires.

19.3.4.1   Number and distribution of occupants

The number of users of tunnels and their distribution at any point in time is obtainable from survey data. The number and distribution of the occupants, hence occupant density, is a function of the tunnel and mode of transport. The number of occupants and their distribution density within the tunnel or tunnel segment will affect travel speeds in an evacuation, as will obstacles such as abandoned motor vehicles.

19.3.4.2   Gender

In fires in buildings, some behaviours related to gender have been identified, for example, females are more likely to alert or warn others and evacuate whereas males may be more likely to fight the fire. These behaviours, distilled from work in the 1970s, relate mainly to residential environments. With regard to tunnel fires, there does not seem to have been any serious research conducted to determine the behaviours, if any, that may be associated with gender. Object/possession/attachment affiliation could also be gender associated.

19.3.4.3   Age

The global trend in advanced countries towards ageing populations suggests that this factor must receive due attention. Further, there is evidence of the increasing mobility of elderly people via motor vehicles and other forms of transport. Kose (1998), in assessing the performance of individuals related to age, used three categories: sensory skills, decision-making and action (mobility, speed of travel, etc.). It is sufficient here to note that elderly people may have relatively poor resistance to the debilitating effects of the smoke and heat associated with tunnel fires. In the absence of firm conclusions with respect to age, 'common sense' must inform tunnel design decisions, i.e. tunnel designers know that the tunnels they are designing will be used one way or another by elderly people.

19.3.4.4   Physical and sensory capabilities

It has been established that in advanced countries in the world a significant proportion of the population has physical and/or sensory impairment (Boyce and Shields, 1999a, b). It is also well known that a significant proportion of the impaired population is out and about in the community every day using buildings (Boyce and Shields, 1999b). It follows that will also be the case for tunnels. Further, large numbers of impaired people, people with various illnesses and medical conditions, travel in groups through tunnels to places of pilgrimage. Tunnel designers must be aware of these facts. Some studies have addressed this issue with respect to the occupants of buildings. No research has been undertaken with respect to tunnel fires. Additionally, it has been suggested by Eder *et al.* (2009) that drivers suffer sensory deprivation, and perceptually rely on visual stimuli. This very important issue has not been well researched with respect to tunnel fires.

19.3.4.5 Familiarity

The notion of 'familiarity' has become enshrined in the fire safety literature, and unfortunately is used with different meanings. The affiliative model proposed by Sime (1985) predicts that in a situation of potential entrapment, people would move towards familiar people and places, and it was suggested that future research should examine the degree to which people's direction of movement is influenced by affiliative ties in different fire scenarios. Place affiliation, it was argued, was not sufficiently addressed in escape route design. Many years later, not much has changed.

More commonly today, familiarity of the occupants with a building and its fire safety systems is linked with occupant response, i.e. it is assumed that all occupants have knowledge of the building and its fire safety systems.

Further dilution of the affiliative model occurs when familiarity is linked with exit choice, i.e. occupants will, in an emergency, exit a building by the way they entered it. Sime's (1985) notion of entrapment is expediently forgotten. Clearly, with respect to buildings, clarity is needed on the true meaning of familiarity and its application in fire safety engineering. Not only, as suggested by Sime (1999), does the fire influence familiarity-orientated behaviour, the setting also potentially exerts a significant influence.

With regard to tunnel fires, it is safe to assume that few, if any, occupants/users are familiar in any meaningful sense with the tunnel environment. It might be argued that people caught in a fire in a tunnel would migrate towards one of the portals, perhaps the one they entered by – which in the hierarchy of definitions of familiarity must be the bottom rung. The safe assumption in tunnel design is simply that the occupants/users are unfamiliar with the environment and the fire safety systems. In addition, the provision of static exit signs to indicate emergency exit routes in tunnels does not mean that they will be noticed or that the exits will be used to leave the tunnel (McClintock *et al.*, 2001). Research in this area with respect to tunnel fires is needed to determine the factors that influence the direction of movement in a fire emergency.

19.3.4.6 Social affiliation

The behaviour of occupants of buildings can be significantly influenced by whether they are alone or in groups. People often travel in groups: large/small, family or other groups. In an emergency the tendency for groups to re-establish may considerably extend the pre-evacuation activity times, and hence expose some groups to greater risk. There are exceptions to the rule. In the Caldecott Tunnel fire (Lake, private communication; Carvel and Marlair, 2005), a mother stayed with a vehicle and perished but her son evacuated the area, leaving his mother, and survived.

19.3.4.7 Object/activity attachment

Object and/or activity attachment can significantly influence the behaviour of people in different settings. In a fire emergency in a tunnel, a car owner may be very reluctant to abandon a shiny new car, or a commercial driver to leave a HGV – their livelihood.

Activity attachment can be work or livelihood related, creating strong bonds, which would have to be severed in some way in a fire emergency if the occupants of the spaces under severe fire were to survive, for example, users of road tunnels whose livelihood depends in getting from A to B within a certain time.

19.3.4.8 Culture

Culture can and does influence people's behaviours, and will do in a fire emergency, particularly where there is a well-established hierarchical decision-making structure. For most practical purposes, cultural influences may be subsumed into social affiliation and role and responsibility, but in tunnels there is no hierarchical management-type structure. The people involved in the fire incident will be of mixed abilities and culturally diverse (Shields *et al.*, 2009).

19.3.4.9 Role and responsibility

The roles and responsibilities of occupants of buildings will influence behaviours in a fire emergency, and consequently the behaviours of others. Occupants with 'ownership' of projects, processes, etc., will behave differently from occupants bereft of ownership tendencies.

19.3.4.10 Commitment

People generally have some good reason to be in a particular place at a particular time. Commitment to an activity can be weak or very strong, for example, eating a meal in a restaurant, catching a flight or travelling through a tunnel. McClintock *et al.* (2001) have shown that staff intervention was sufficient to overcome the commitment tendencies of shoppers. On the other hand, car drivers, intent on a destination, may have a level of commitment that, short of the use of physical barriers, could be difficult to overcome.

19.3.4.11 Alertness

As discussed previously, commitment to activities and object/activity attachments, together with several affiliations, can affect the awareness of the occupants of buildings and tunnels to a developing fire situation. Other factors affecting alertness include wakefulness, use of medication and alcohol.

19.3.4.12 The concept of 'panic'

One model of human behaviour in fires, developed from accounts analysis of survivors, has three headings, representing phases of response: recognition, coping and escape (Canter *et al.*, 1980). It was clearly revealed that many of the survivors' decisions as to what to do and when to do it were based on ambiguous cues and incomplete knowledge as to what might be happening elsewhere in the buildings. Thus, from the perspective of people caught up in a fire, their behaviours were rational. However, behaviours associated with disasters such as crushing/crowding/piling up at a fire exit were often attributed to panic. The term 'panic', which is usually associated with 'panic attacks', is a clinical anxiety disorder. It is a period of intense fear or discomfort in which several of the following systems develop rapidly and peak within 10 minutes:

- palpitations, pounding heart or accelerated heart rate
- sweating
- trembling or shaking
- sensations of shortness of breath or smothering
- feeling of choking
- chest pain or discomfort
- nausea or abdominal distress
- feeling dizzy, unsteady, light-headed or faint

- derealisation (feeling of unreality) or depersonalisation (being detached from oneself)
- fear of losing control or going crazy
- fear of dying
- paraesthesias (numbness or tingling sensations)
- chills or hot flushes.

For a diagnosis, at least one attack should be followed by 1 month (or more) of one (or more) of the following

- persistent concern about having additional attacks
- worry about the implications of the attack or its consequences (e.g. losing control, having a heart attack, 'going crazy')
- a significant change in behaviour related to the attacks.

A detailed review and critique of the concept of panic with respect to fire (Sime, 1980) drew on earlier work (Quarantelle, 1957). It was found that the attribution of 'panic' to people, when they deviate from rule-based so-called rational fire code design solutions, had very little support once human behaviour in fires in buildings was understood from the perspective of the people at the time. The behaviour was much more understandable when interpreted in terms of delays in warning the occupants and the common ambiguity of a fire situation in its early stages. People exposed to fire can and do experience significant stress at the time and afterwards. They often do not receive the benefits of counselling, which many undoubtedly deserve. The notion of panic as applied to the behaviour of occupants of buildings exposed to fire is retrospective, and imposed, as it were, by outside distant non-involved observers. The question is, 'Is it valid for the behaviours of people exposed to fire in tunnels?' Fahy *et al.* (2009) have suggested that panic, as an irrational behaviour, is rarely seen in fires. However, the evidence drawn on to form this view is wholly from accounts of fires in buildings. It would be of interest to know behaviours exhibited by people involved in tunnel fires and whether or not there was any evidence of panic.

Further, it would be very useful, as we continue the study of human behaviour in fires in tunnels, to afford survivors of fire incidents an opportunity to characterise panic from their 'own' perspective and to explore how their notion of panic experience, if any, may differ from the clinical definition.

19.4. Responding to a developing emergency

For an individual to respond to a developing emergency, for example, a fire in a tunnel, that individual must receive some indication (i.e. a cue) that there is something 'happening', recognise that the cue is providing information, interpret the cue (i.e. 'There is a fire, I am in danger') and respond appropriately. Of course, people may not receive the cue for different reasons, they may not recognise it for what it is and they may misinterpret the cue.

Every fire produces cues, i.e. smoke, heat, light, sound, etc. The setting will provide cues in response to automatic or manual detection of an incipient fire, people will alert others and, in tunnels, traffic congestion, etc., may also serve as a cue.

For a particular setting, the most appropriate and effective cue should be employed. For fires in buildings, a number of strategies have been suggested to improve the likelihood of building occupants correctly interpreting fire alarm cues (Proulx and Sime, 1991), including:

- voice alarm systems
- providing occupants with as much information as possible
- live voice instruction.

The fire safety engineering design and analysis for tunnels should result in the choice and use of fire detection and alarm systems that are the most effective given the setting and fire scenario with respect to occupant cue reception, recognition, interpretation and response. Given the foregoing, it must be understood that it is very unlikely that occupants will leave a fire-threatened space immediately – this is well understood for buildings, and there is nothing to suggest that there will not be delays by occupants before evacuating from tunnels.

19.5.   Recent developments

Developments in the field of human behaviour research for fire in buildings have been presented as passing through different phases (Canter *et al.*, 1980; Pauls, 1998). Phase 1 was the commencement of work in the 1950s (Pauls, 1998), followed by phase 2 in the 1970s, which saw the drying up of research funding. In the third phase in the 1980s, the *SFPE Handbook of Fire Protection Engineering* (SFPE, 1988) was published, which included contributions on human behaviour in fire. The 1990s saw the commencement of the fourth phase, i.e. a resurgence in the field coinciding with the global transition of national fire regulations from prescriptive towards performance-based fire safety regulation. It is a sobering thought that after many decades of work there is a paucity of useful and robust human behaviour occupancy-specific data for use in fire safety engineering. It is the view of the author that with respect to understanding and quantifying human behaviour in fires in tunnels, we are in phase 1. Although, now, some work has been done, there is still no 'definitive work' on human behaviour in fires in tunnels, and this needs to change.

In a literature review of human behaviour in tunnel fires, Steyvers *et al.* (1999) use the concept of occupancy. They discuss traffic behaviour, the tunnel setting in terms of the environment created, and fires in tunnels. The notion of person, place and property affiliation is introduced, as is destination affiliation, which can account for the fact that instructions to drivers are often disobeyed. Measures to enhance the environmental cues necessary to driving performance in tunnels are given, but not much is offered with respect to understanding human behaviour in fire in tunnels.

The aim of the ACTEURS project was set out by Noizet and Richard (2003): to improve understanding of how tunnel users behave in both normal and emergency situations. From interviews with 30 tunnel professionals, a perspective of the tunnel user as seen by tunnel professionals was obtained. The tunnel user was seen as a fundamental player in tunnel safety, but

- the user is not involved in the preparation of the safety system
- the user has to come to terms with approaches and systems devised and applied by others on the basis of assumptions drawn from experience and intuition
- the user is insufficiently informed about the tunnel, its facilities, etc.
- the user is not clearly informed of what is expected from them in different situations.

In order to obtain some insights into evacuation behaviours, a 'tunnel fire' behavioural test was conducted by TNO in the Netherlands (Boer, 2002; Carvel and Marlair, 2005). Participants were invited for a test on driving behaviour in a tunnel. Part way through the test a burning vehicle inside the tunnel was simulated. The events were captured on CCTV for analysis. What emerged from the behavioural test were

- motorists tended to remain with their vehicles, even with smoke dispersing through the tunnel
- 5 minutes into the test the evacuation was at a standstill, with motorists returning to their vehicles
- a public announcement 'explosion danger' had a noticeable effect, i.e. people evacuated the tunnel
- motorists in their cars enveloped in smoke remained in their vehicles irrespective of the public announcements
- passivity is a major problem.

In dealing with rail tunnels, several personnel problems were identified

- staff are not well trained
- staff are reluctant to raise the alarm
- staff are afraid that they might cause passengers to panic, despite research showing that 'panic' is not a useful concept (see earlier).

In recent years, work has also been carried out as part of the UPTUN project.

The outcomes of the above work and other research are presented in detail in Chapter 20.

19.6. Concluding remarks

- Human behaviour in fire research in buildings has, since 1956, progressed through distinct phases.
- The basic behavioural fire response model of
  - recognition
  - testing
  - definition
  - evaluation
  - commitment
  - reassessment
  is valid for human behaviour in tunnel fires.
- The affiliation model developed from human behaviour in fires in buildings is applicable to tunnel fires.
- The affiliation model may need further development to include
  - place affiliation
  - person affiliation
  - object/possession/vehicle affiliation
  - destination affiliation.
- In tunnel design, the basic behavioural concept $R = f(E)$ must be applied.
- A theory of occupancy must be developed for tunnels.
- Occupant/users life safety must be a design issue.
- Generic sets of predictable behaviours for occupancy specific tunnels are needed.
- There is a need to develop a set of occupancy-specific, credible, fire scenarios to test tunnel designs and management.

- Credible fire scenarios must be coupled with credible rescue and evacuation scenarios.
- Consideration must be given to the large number of users with physical and sensory disabilities.
- Consideration must be given to the elderly and the young.
- Fire and smoke can spread very rapidly in tunnels.
- In fire emergencies, people do not evacuate immediately.
- Pre-evacuation activity times (PEATS), i.e. the time elapsed before evacuation actually commences, will be considerably longer than the actual evacuation time.
- Occupants will not immediately perceive themselves to be in real danger.
- Users will not be familiar with the tunnel and its safety systems.
- Users will have acquired learned irrelevance to tunnel signage – new signage concepts need to be developed.
- Traffic will continue to enter road tunnels in an emergency.
- Refuges provided for life safety purposes must be designed and constructed so as to remain as places of safety during a tunnel fire.
- Users may not be able to evacuate a tunnel by the way they entered – they have to find a means of escape.
- Occupancy-specific user profiles are needed for each type of tunnel.
- Group behaviours may be more important in tunnel fires than in building fires.
- The setting, taken together with the users, i.e. individuals, family groups, organised group travel, etc., will greatly influence behaviours.
- In certain circumstances, clustering behaviour will occur.
- There is much similarity between human behaviour in fire in tunnels and human behaviour in fire in buildings, i.e. a body of knowledge exists to draw upon.
- The concept of panic with respect to fires in tunnels needs to be revisited.

REFERENCES

Allison R (1997) *Inquiry into the Fire on Heavy Goods Vehicle Shuttle 7539 on 18 November 1996*. HMSO, London.

Boer LC (2002) *Behaviour by Motorists on Evacuation of a Tunnel*. TNO Report TM-020C034. TNO, Amsterdam.

Boyce KE and Shields TJ (1999a) Towards the characterisation of building occupancies for fire safety engineering: prevalence, type and mobility of disabled people. *Fire Technology* **35**(1): 35–50.

Boyce KE and Shields TJ (1999b) Towards the characterisation of building occupancies for fire safety engineering; capabilities of disabled people moving horizontally and up an incline. *Fire Technology* **35**(1): 51–67.

Canter D, Breaux J and Sime J (1980) Domestic multiple-occupancy and hospital fires. In: Canter D (ed.), *Fires and Human Behaviour*. Wiley, Chichester.

Carvel R and Marlair G (2005) A history of fire incidents in tunnels. In: Beard A and Carvel R (eds), *The Handbook of Tunnel Fire Safety*, 1st edn. Thomas Telford, London, pp. 3–41.

China Daily (2005) Available at: http://www.Chinadaily.com.cn (accessed 5 June 2005).

CNN (2007) CNN news. Available at: http://www.cnn.com (accessed 13 October 2007).

Daily Telegraph (2008) Available at: http://www.telegraph.co.uk (accessed 12 September 2008).

Eber G (2001) The Tauern Tunnel incident, what happened and what has to be learned. *Proceedings of the 4th International Conference on Safety in Road and Rail Tunnels, Madrid, 2–6 April 2001*. ITC, Tenbury Wells, pp. 17–27.

Eder S, Brutting J, Muhlberger A and Pauli P (2009) Human behaviour in tunnel accidents. *Proceedings of the 4th International Symposium on Human Behaviour In Fire, Cambridge*, 13–15 July 2009. Interscience Communications, London.

EuroTAP (2011) EuroTAP – The Future of Tunnel Testing. Available at: http://www.Eurotestmobility.com/eurotap.php (accessed 22 June 2011).

Fahy RF, Proulx G and Aiman L (2009) Panic and human behaviour in fire. *Proceedings of the 4th International Symposium on Human Behaviour in Fire, Cambridge*, 13–15 July 2009. Interscience Communications, London.

Frantzich F and Nilsson D (2004) Evacuation in a smoke filled tunnel. *Proceedings of the 3rd International Symposium on Human Behaviour in Fire, Belfast*, 1–3 September 2004. Interscience Communications, London.

Hedefalk J, Wahlstrom B and Rohlen P (1998) Lessons from the BAKU subway fire. In: Vardy A (ed.), *Proceedings of the 3rd International Conference on Safety on Road and Rail Tunnels, Nice*, 9-11 March 1998. ITC, Tenbury Wells, pp. 15–27.

Ingason H (2003) *Proceedings of the International Symposium on Catastrophic Tunnel Fires, Borås*, 20–21 November 2003. Swedish National Testing and Research Institute, Stockholm.

Kose S (1998) Emergence of aged populace: who is at higher risk in fires? In: Shields TJ (ed.), *Proceedings of the 1st International Symposium on Human Behaviour in Fire*. University of Ulster, Belfast.

Lake JD (private communication) Mitigating the impact of human factors in tunnel emergencies. Personal communication.

Leslie JC (1996), *Principles of Behavioural Analysis*, Prentice-Hall, Englewood Cliffs, NJ.

McClintock T, Shields TJ, Reinhardt-Rutland H and Leslie JC (2001) A behavioural solution to the learned irrelevance of emergency exit signage. In: Shields TJ (ed.), *Proceedings of the 2nd International Symposium On Human Behaviour In Fire*. University of Ulster, Belfast.

Noizet AS and Richard F (2003) ACTEURS: Improving understanding of road tunnel users with a view to enhancing safety. In: *Fifth International Conference on Safety In Road And Rail Tunnels, Marseille*, 6–10 October 2003.

Pauls J (1998) Arsonal perspective on research consulting and codes/standards development in fire-related human behaviour, 1969–1997, with emphasis on space and time factors. In: Shields TJ (ed.), *Proceedings of the 1st Human Behaviour in Fire Symposium*. University of Ulster, Belfast, pp. 71–82.

Proulx G and Sime JD (1991) To prevent panic in an underground emergency: why not tell people the truth?. *Proceedings of the 3rd International Symposium on Fire Safety Science, Edinburgh*, 1991, pp. 843–852, Elsevier Applied Science.

Quarantelle EL (1957) The behaviour of panic participants. *Sociology and Social Research* **41**: 187–194.

Schupfer H (2001) Fire disaster in the tunnel of the Kitzsteinhorn Funicular in Kaprun on 11 Nov. 2000. In: *4th International Conference on Safety in Road and Rail Tunnels, Madrid*, 2–6th April 2001. Unpublished paper.

SFPE (1988) *The SFPE Handbook of Fire Protection Engineering*, 1st edn. National Fire Protection Association, Quincy, MA.

Shields TJ and Boyce KE (1995) Emergency egress capabilities of people with mixed abilities. In: Vardy A (ed.), *Proceedings of the 2nd International Conference, Safety in Road and Rail Tunnels, Granada*, 3–6 April 1995. ITC, Tenbury Wells, pp. 347–354.

Shields TJ and Boyce K (2000) A study of evacuation from large retail stores. *Fire Safety Journal* **35**: 25–49.

Shields TJ, Boyce KE and McConnell N (2009) The behaviour and evacuation experiences of WTC 9/11 Evacuees with self-designated mobility impairments. *Fire Safety Journal* **44**: 881–893.

Sime DJ (1980) The concept of panic. In: Canter D (ed.), *Fires and Human Behaviour*. Wiley, Chichester.

Sime DJ (1985) Movement towards the familiar – person and place affiliation in a fire entrapment setting. *Environment and Behaviour* **17**(6): 697–724.

Sime JD (1999) Understanding human behaviour in fires – an emerging theory of occupancy. Inaugural Guest Lecture. University of Ulster, Belfast.

Steyvers FJJM, de Waard R and Brookhuis KA (1999) Aspects of human behaviour in tunnel fires – a literature review. *Proceedings of the International Tunnel Fire and Safety Conference, Rotterdam*, 2–3 December 1999. Brisk Events, Leusden, Paper No. 2.

Swissinfo (2006) Available at: http://www.Swissinfo.ch (accessed 18 September 2006).

Tesson M and Lavedrine S (2004) Integration of human behaviour in the improvement of safety in French road tunnels. *Proceedings of the 3rd International Symposium on Human Behaviour in Fire, Belfast*, 1–3 September 2004. Interscience Communications, London.

Handbook of Tunnel Fire Safety, 2nd edition
ISBN: 978-0-7277-4153-0

ICE Publishing: All rights reserved
doi: 10.1680/htfs.41530.421

Chapter 20
Egress behaviour during road tunnel fires

Alain Noizet
SONOVISION Ligeron®, Saint Aubin, France

20.1. Introduction

Looking back at the major road tunnel fires of recent years (e.g. those in the Mont Blanc Tunnel in 1999, the Tauern Tunnel in 1999 and the St Gotthard Tunnel in 2001), what is clear is, in each case, the vital importance of people immediately adopting egress and safety behaviours appropriate to the structure involved. With a few rare exceptions, the people who lose their lives during such incidents are those who find themselves trapped in a cloud of deadly toxic fumes or near a source of heat greater than the human body can withstand. Some of the time, these behaviours are the result of decisions that were, in retrospect, inappropriate (e.g. waiting next to a damaged vehicle) or an inability to find safety facilities that were nonetheless present in the tunnel due to poor design of the tunnel. It should be remembered, however, that people have actually died in so-called 'refuges': for example, during the Mont Blanc Tunnel fire of 1999. Also, people have died because they simply could not move away quickly enough: for example, in the Viamala Tunnel fire of 2006.

The lessons learned from such dramatic events demonstrate that road tunnel safety design is meaningless unless it takes into account all human behaviours that can be observed on such occasions; otherwise, the best efforts of the designers and tunnel operators to reinforce the safety of their structures are doomed to failure. An emergency exit is of no use unless it is perceived and understood to be a means of escape by those who need it in an emergency situation. A safety guideline is pointless if drivers don't have the opportunity to get to know it, or are not given it at the relevant moment, or, for that matter, if they don't understand it. In these areas, the 'ergonomic' approach – adapting the road tunnel system and its safety procedures or installations to its end-users – can be an effective way forward. This chapter follows that approach by focusing on a particular category of road tunnel users – the driver; however, it needs to be borne in mind, therefore, that others may be present, such as passengers.

This chapter aims to highlight several aspects of the egress behaviour of drivers confronted with a road tunnel fire. It discusses the various natural or triggered behaviours adopted by drivers in the first minutes of a fire, between the moment it breaks out and that moment they reach safety, and searches for insights that will help road tunnel designers and operators to gain a clearer understanding of these behaviours. The aim is to give them the means to inhibit non-adaptive behaviours and to favour those appropriate to the situation.

20.2. Scientific literature about tunnel egress behaviours

The scientific literature on tunnel egress behaviours is relatively scarce and usually quite recent (e.g. Steyvers *et al.*, 1999; Tesson and Lavedrine, 2004; Eder *et al.*, 2009; Tesson, 2010). There is, however, much more material on egress behaviours in emergency environments in general (Sime, 1985; Proulx, 1994; Ozel, 2001; Bryan, 2002) and on how to design environments that take into account these behaviours (Proulx, 1998, 1999; SFPE, 2002). However, although we know much more than we did about human behaviour in building fires, there is, even there, no room for complacency: there is still a considerable amount that we do not know.

It must be noted that those studies treat generic standard cognitive behaviours considered valid for any human being regardless of their age and/or abilities. Regarding the elderly and disabled, some studies have mentioned their speed of movement, with some even including mathematical models for evacuation (Schadschneider *et al.*, 2009). In fact, most of the studies concerning emergency evacuation from a confined environment indicate that the elderly or people encountering additional difficulties are usually helped by other evacuating users present on the site (Sime, 1985; Proulx, 1994; Chertkoff and Kushigian, 1999). However, these aspects are not discussed in the following sections, and this needs to be borne in mind.

Among the studies on egress behaviours in emergency environments in general, three recent European research projects have addressed road tunnels. Through their work, and the additional data collection initiatives conducted to develop a body of knowledge specific to tunnels, they have become fundamental references on the subject. Findings from these projects need to be tested further over the coming years.

- The work of TNO Human Factors, 1999–2002. The Human Factors division of the Dutch research organisation TNO has produced several studies on driver egress behaviour in tunnels (Boer, 2002). Its aim was to determine whether emergency exit mechanisms were sufficiently adapted to drivers' natural egress behaviours. Four fieldwork studies were conducted on road users in Rotterdam's Second Benelux Tunnel, involving a survey on how drivers thought they would act in a tunnel disaster, a series of individual simulations in which 69 participants demonstrated what they would do if there were an emergency in the tunnel, and several series of evacuation exercises in the tunnel under real conditions, including one under conditions of very dense smoke. This thorough and ambitious study generated an unequalled quantity of information about egress behaviours during tunnel fires.
- The ACTEURS project, 2003–2006. The French project ACTEURS (Améliorer le Couplage Tunnel-Exploitant-Usagers pour Renforcer la Sécurité), funded by France's Ministry of Equipment (as it then was), brought together several road tunnel operators (APRR, AREA, ATMB, GEIE-TMB, ESCOTA and SFTRF), a firm specialising in the human factors aspects of risk management (DEDALE), and representatives of French government bodies (the Centre d'Etude des Tunnels (CETU) and the Directorate for Road Traffic and Safety (DSCR), both part of the Ministry of Equipment). The aim of the project was to study and understand the constraints represented by drivers (Workpackage 1 – Noizet and Paries, 2004) so that they could be better integrated into safety improvement initiatives by the regulators and the tunnel operators (Workpackage 2 – Noizet and Mourey, 2005a). Report 3 from Workpackage 1 looked at driver behaviours in tunnel emergencies (Noizet and Mourey, 2005b). It consolidated the results of a review of the scientific literature, a series of interviews, a questionnaire survey of 620 drivers, a review of accident reports and a collection of witness statements.
- Workpackage 3 of the UPTUN project, 2002–2006. The European research project UPTUN (Cost-effective, Sustainable and Innovative Upgrading Methods for Fire Safety in Existing Tunnels) brought together 41 partners from 19 EU member states under the 5th

Framework Programme of the European Commission. The aims of the UPTUN project were twofold: to develop innovative methods and technologies for use in tunnels and to contribute to the implementation of a tunnel safety evaluation process. Workpackage 3 specifically addressed the question of human behaviours in tunnel accidents and incidents in two deliverables: Deliverable 3.2 (Papaioannou and Georgiou, 2008) reviewed the known responses of drivers, operators and rescue teams to tunnel disasters; Deliverable 3.3 (Martens, 2008) presented a set of means and methods for the optimal management of such responses in a tunnel emergency.

The data presented in the following pages draws heavily on the results of these fundamental studies of egress behaviours in tunnels. The author would like to take this opportunity to thank all of their contributors and to salute the quality and accessibility of the deliverables produced by these projects.

20.3. Understanding the determinants of human behaviour

Understanding egress behaviours in road tunnels requires some understanding of the factors that shape human behaviour in any context. This section explores the fundamentals of human behaviour, focusing on those that play a crucial role in road tunnel evacuation situations. The determinants of human behaviour are examined from three perspectives: the cognitive processes underlying behaviour, behavioural adaptation to emergency situations and the social determinants of individual behaviours.

20.3.1 Human behaviour from the cognitive perspective

Figure 20.1 summarises the literature on human performance in a situational context (Amalberti, 1996; Baars, 1997; Edworthy, 1998). It represents, in a deliberately simplified form, the cognitive processes that determine an individual's behaviour when faced with a given situation.

Figure 20.1 Cognitive processes that determine behaviour in a given situation

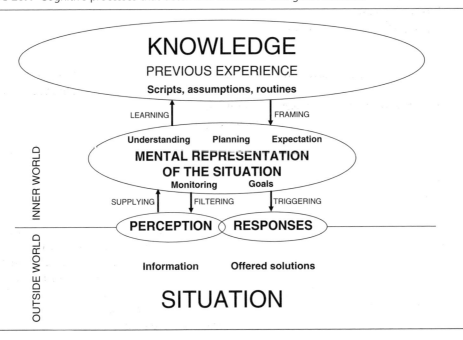

To act on a situation, we, as humans, first need to understand it. The *mental representation* that we construct of the situation in which we find ourselves is central to this understanding. It draws both on the *information* coming from the situation and on all the *knowledge* available and accessible at that moment. It should be noted that 'knowledge', as the term is used here, is that as understood by the person at the time, and some of this 'knowledge' may be inaccurate or false. This knowledge is of different types.

- Generic, conceptual knowledge (declarative knowledge) for describing and understanding the outside world.
- *Non-tested assumptions*, which compensate for lack of tested knowledge on a given subject. By 'non-tested assumptions' is meant hypotheses about the world and its operation that are not supported by formal empirical observation but considered to be facts by the cognitive system. By 'tested' is meant a hypothesis that is supported by formal empirical observation.
- Procedural knowledge (or know-how): the body of action and control procedures stored in memory and mobilised to act on the outside world. Within this category, *scripts* function like blocks of knowledge that include a framework for interpreting situations, along with the associated responses. Based on our *previous experience*, they act as internalised solutions, preformatting our behavioural responses. *Routines* are scripts that, as a result of repeated experience, are triggered automatically whenever the situation arises.

Behavioural responses result from our understanding of the situation and from the process of selecting actions that will influence the situation effectively. They are also heavily dependent on the possibilities offered by the environment in which we find ourselves – as the saying goes: if there's no solution, there's no problem.

Once they have been defined, the action goals will shape the behaviours. A system of expectations about likely results will then structure the mechanisms for monitoring and assessing the results obtained.

Perceptual processes are also driven by the mental representation of the situation. Filtering comes into play to extract from the environment the information needed to maintain an understanding of the situation and to control the actions taken.

An essential element to be taken into consideration is that people's mental *resources* are *limited in capacity* and distributed across a range of mental activities (perception, understanding, decision-making, planning, action, monitoring, etc.), generated by the need to act in a given environment. To address this problem of limited resources, the human cognitive system has, in the course of the evolution of the species, developed a number of strategies.

- *The principle of economy*: whenever the situation permits, humans seek to limit the cognitive processing cost by favouring the mental operations that require the least effort. Rasmussen (1986), in his SRK (skill-based/rule-based/knowledge-based) model, distinguishes between three modes of cognitive functioning, depending on the situation encountered:
  - *Skill-based*: this mode, adapted to day-to-day situations (e.g. driving), automatically triggers the appropriate action routines. This is the least resource-intensive mode of functioning, and is therefore the one favoured by the cognitive system.
  - *Rule-based*: this mode, adapted to familiar or 'normally chaotic' situations (e.g. responding to ordinary road traffic events), starts by analysing the situation, then selects

available scripts and applies them to the situation – akin to a functioning mode based on 'if X then Y ...' reasoning. Checking the validity of the scripts as they are applied makes this mode of functioning more costly than the previous one.

– *Knowledge-based*: this mode, adapted to new or surprising situations (e.g. tunnel fires) for which people do not have ready-made solutions, constructs a response to the situation by building on available knowledge and information. It is the most resource-intensive mode of functioning.

The cognitive system also makes use of thought simplification mechanisms that can reduce the cost of mental processes. For example, when faced with a problem or a choice to be made, the individual will give priority to the first solution that emerges, so long as it seems to make sense in the light of the known elements of the situation and the options available (Zsambock and Klein, 1997).

■ *The principle of relevance*: only knowledge and information deemed useful to meet the needs of the situation are considered and integrated into the mental representation.

■ *The principle of sufficiency*: achieving the best performance, as judged by an external observer, is rarely sought by human beings apart from those in competitive situations; instead, people tend to look for a solution that provides global, acceptable and sufficient performance, allowing them to achieve their objectives considering the perceived situation requirements. The resulting behaviour could be perceived, by an external observer (e.g. a tunnel operator) as non-optimal and less effective than it could be (Amalberti, 1996).

Another key consideration is that, in crisis situations, people do not just manage the objective risks (*external risks*) that result from the laws of physics and which are imposed upon them from the outside (e.g. the toxicity of fumes) – they also have to deal with *internal risks* (*cognitive risks*) that arise from the situation

■ the risk of losing control over the situation due to a lack of skills or situation awareness; the risk of losing the ability to manage the situation due to a lack of internal resources or a lack of internal resources management

■ the risk of 'losing face' in front of others or being involved in emotionally charged exchanges (e.g. challenging someone else's decision).

These two types of risk are cumulative, and are managed simultaneously. Several phenomena are involved in this process.

■ Most people have, in the course of their lives, built up a wealth of experience in managing internal risks, and have developed solutions to help control them. By contrast, their knowledge of external risks is tenuous, and these are assigned a low probability of occurrence.

■ Due to people's lack of experience of external risks, these are often poorly understood. When it comes to fire, for example, Canter (1990) demonstrated that people have false representations of how fire develops: when the fire is limited, people do not feel in danger, as they fail to realise how quickly it can spread.

Amalberti (1996), through his analysis of these phenomena, showed that people faced with a delicate situation have a natural tendency to prioritise internal risk management. As a result, people will generally accept additional external risk, which they see as improbable and uncertain (e.g. the development of a tunnel disaster) in order to cope with the internal risk they perceive in a situation (of not reaching their original destination, of choosing the wrong egress solution, of not taking other people (e.g. family members) with them, etc.).

20.3.2 Behaviour in emergency situations

In emergency situations, anyone faced with an evident risk (burning, smoke intoxication, death, etc.) and subject to intense time pressure will feel stress and anxiety regardless of culture, age, gender or past experience (Ozel, 2001). This is not an abnormal or negative reaction but a necessary condition for stimulating an appropriate response to an emergency situation.

This emotional state generates psychological mechanisms designed to maximise the benefits to those involved (attaining their goals and preserving their physical integrity) and to reduce the cognitive tension induced by the danger and the change in their behavioural repertoire.

The main mechanisms for optimising behavioural response in emergency situations are as follows.

- Processing information more quickly, notably by reducing the quantity of information assessed in determining behaviour (Miller, 1960). Bryan (1991), for example, showed that in fire situations only 7–8% of people actually notice the existence of emergency exit signs.
- Focusing exclusively on the clear, unambiguous elements of the situation (Martens, 2008).
- Preferring known or familiar solutions (Sime, 1985), such as getting out via the tunnel entrance.
- Rejecting options with negative connotations (Bickman *et al.*, 1977), for example, ignoring an emergency exit, mistaken for a doorway to a technical area usually off-limits to the public.
- Automatic validation of the decisions taken, to avoid losing time by reassessing one's own decisions. This mechanism is generally followed by a straightforward execution of the action (example: escaping with determination and without hesitation).

The key prerequisite, however, for optimising behaviours in an emergency situation remains the fact that the emergency situation is actually perceived as such. In fact, due to the principle of economy mentioned earlier, people are extremely reticent to abandon their objectives – and the script they are currently following – in order to adjust their mode of operation. The trade-off between the cost of changing one's mode of operation and the benefits of doing so must be obvious to the person before they will change (Edworthy, 1998). For this, the dangers of the situation must be perceived as such, clearly and unambiguously. This is why, in tunnel fires, so many people stay inside their vehicles until they are completely engulfed in smoke (Martens, 2008).

Finally, in emergency situations, the effort of negotiating this trade-off – attempting to maximise the benefits while minimising the cognitive tension induced by the situation – accounts for the phenomenon of reluctance to abandon property, observed by many authors (Martens, 2008), which takes the form of hesitation in abandoning one's vehicle and/or its contents. Beyond even the sometimes quite powerful sentimental attachment that every human being feels for their possessions, what is happening here is a process of coming to terms with potential loss. In the case of tunnel fires, the abandonment dilemma is often resolved by a compromise solution: retrieving a travel bag or laptop computer in exchange for abandoning the vehicle (Noizet and Ricard, 2004).

20.3.3 Social determinants of individual behaviours

In tunnels and elsewhere, in emergencies and under normal conditions, individual behaviours can only be explained in the context of the collective, social behaviours inherent to situations where several individuals are present in the same environment.

The main social determinants to be taken into account when analysing egress behaviours in road tunnel fires are as follows (Noizet and Ricard, 2004; Martens, 2008).

- The *phenomenon of affiliation* refers to the strong affective commitments that exist between the members of the same family or social group (e.g. groups of bikers, tourist coach passengers). Because of these commitments, the members of such groups tend to act in the same way, in the group interest, avoiding anything that might split up the group – even if this means waiting in the tunnel while the fire spreads.
- Individuals tend to modify their own behaviour in order to conform to what they perceive as the majority behaviour of the group around them. This *phenomenon of conformism*, familiar to psychosociologists, is a behavioural constant that few of us are able to shrug off. The phenomenon is even stronger when the concerned individuals are confronted with a situation they cannot understand. For an individual, the sight of group movement (e.g. people leaving their cars) offers a ready-made solution – a solution, moreover, that has clearly been validated as relevant by the majority. By acting in conformity with the group, the individual in return validates the collective action.

The TNO Human Factors team (Martens, 2008) put the spotlight on this phenomenon when they observed a tunnel evacuation situation in response to a (simulated) heavy goods vehicle (HGV) fire. In this situation, designed to test the effectiveness of a public address system for tunnels, the participants were not given any prior evacuation instructions. The authors were thus able to verify that as soon as one person reacts, and the other participants perceive that reaction, the behaviours become collective. Three groups formed spontaneously by means of this process: the first group (9% of participants) evacuated spontaneously and immediately, before hearing any announcement; the second group (9% of participants) also evacuated spontaneously, but only several minutes after noticing the vehicle in difficulty; the third group (82% of participants) evacuated seconds after the announcement.

- As soon as such a collective is formed, voluntarily or otherwise (e.g. a group of drivers suddenly confronted with a tunnel fire), it tends to organise around the behaviour of a few individuals who naturally take *leadership* of the situation. The notion of leadership refers to an individual's ability to influence the behaviours of group members non-coercively and to get them to converge on a common goal (e.g. deciding in which direction to evacuate). Leadership can be status based (e.g. a policeman), signified by a uniform, or it can emerge in someone as a result of the specific circumstances of the situation (e.g. because they have the required skills or experience, or natural authority). In evacuation situations, such as tunnel emergency egress, this will often be someone who knows the tunnel, its means of evacuation and/or the procedure to follow to reach safety. This individual will have a character trait that helps them to clearly express their knowledge to the other occupants, and to persuade the group to follow their instructions. This person will take charge of the group from the outset to lead the evacuation. The ACTEURS project (Noizet and Mourey, 2005b) points to several witness statements from tunnel emergency survivors confirming this phenomenon. Sometimes a person may emerge as a 'leader' who actually makes decisions that are poor with regard to life safety, for example, moving in a wrong direction. Following a 'leader' in this case would be undesirable.

20.3.4 What do these determinants tell us about tunnel evacuation issues?

The key elements for understanding and controlling egress behaviours in tunnel fires can be summarised as follows.

- In the context of an emergency, awareness of danger is decisive in getting people to reassess their risk management strategy and adjust their behaviour.
- In accordance with principles of economy and sufficiency, we should expect drivers to adopt the simplest and most familiar solutions (e.g. sheltering inside their vehicles, trying to leave by the tunnel entrance).
- The quality of the decisions taken depends on the quality of knowledge, the quality of previous experience, and the quality of information given locally.
- Egress behaviours are, with very few exceptions, collective behaviours structured around a situational leader.
- The final outcome ultimately depends on the solutions offered by the environment, and on their being perceived and understood as solutions by the tunnel occupants.

20.4. Accounting for the specifics of road tunnel fires

Emergencies and egress behaviours in tunnels are very different from emergencies and egress behaviours observed in other environments. Some authors maintain that behavioural models based on studies conducted in other contexts (e.g. building evacuation) have no validity for addressing the issue of egress behaviours in road tunnels – but this is to ignore the fundamentals outlined above, and the modalities common to all fire emergencies. The fact remains, however, that the control of road tunnel egress behaviours must be based on the correct identification of the specific context in which such behaviours are expressed. Indeed, this applies to all occupancies, including tunnels. That is, we need to be more occupancy specific when trying to understand human behaviour in fire.

20.4.1 The specific features of road tunnels

The ACTEURS project (Noizet and Mourey, 2005b) identified several characteristics of tunnel fires that need to be taken into consideration when discussing egress behaviours.

- A road tunnel is a confined environment with an entrance, an exit, and a limited number of shelters (safety areas) or emergency exits (egress routes). What are the solutions offered by the tunnel for the protection of drivers (including drivers with disabilities/mobility impairments) and other people in the case of fire?
- Each tunnel has its own safety system. This system is the result of decisions made about the tunnel layout, structural engineering (one-way/two-way, curved, on a gradient, etc.), equipment (automatic fire detection systems, sprinklers, public address system, etc.) and operation (e.g. whether emergency facilities are sited nearby). Each road tunnel is therefore unique, and is characterised by a specific set of evacuation procedures. In many ways, the only thing that one road tunnel has in common with another is the drivers and other people who pass through them, and it is down to them to adapt to the specifics of each tunnel. This suggests a degree of familiarity with tunnel fire safety specifics, which most drivers and tunnel users will not possess.
- Tunnel fires may appear to present a localised threat, but their consequences (toxic fumes or gas, heat and radiation) can spread considerable distances along the tunnel and at

various speeds. Indeed, for a given road user, the distance to the locus of the accident or incident can be considerable. A fire several hundreds of metres away (or even, for certain tunnels, several kilometres away) does not become an objective reality for drivers until they see smoke appear, or until they can feel the heat radiating from the fire. Additionally, the accident or incident that started the fire, or the fire itself, may block the lanes, bringing the traffic to a halt. Unless the tunnel has an alarm system, the resulting traffic jam will seem no different from the usual peak-time congestion (a familiar situation for urban drivers). As a result, the first cues of a tunnel fire may be very unclear, depending on where fire breaks out in the tunnel.

■ We know from previous major tunnel fires and exercises that there is only a limited amount of time available to escape from the toxic fumes. This places severe time constraints on the evacuation procedures. Moreover, it may take several minutes for the response or rescue teams to intervene in the tunnel. More often than not, drivers are left to face the situation by themselves for several minutes. An instruction sometimes given in crisis situations, telling people to 'wait for assistance' is inappropriate in the tunnel environment.

20.4.2 The specific features of road tunnel users

A specific feature of road tunnels is that they are technically complex environments, presenting a non-negligible degree of risk – a fact that most tunnel users are largely unaware of.

In the ACTEURS project, a survey was conducted at the exit of each of the main tunnels in the French Alps, involving 620 drivers (Noizet and Ricard, 2004). The survey clearly revealed how little the drivers knew about the tunnels, their equipment and the solutions available in the event of fire. Among the most telling findings, the survey showed that

■ respondents could cite, at most, three or four tunnel safety facilities
■ 12.5% of respondents were unaware of the existence of emergency exits in the tunnels they had just driven through
■ about one-third of respondents did not know where the emergency exits led to, and thought that they would not be safe there
■ a number of responses also suggested major representation errors about the tunnel (e.g. concerning emergency exits leading to outside the Mont Blanc Tunnel).

Other ACTEURS project initiatives (observations of tunnel journeys, qualitative interviews, collecting witness statements) additionally showed that drivers

■ pay little attention to leaflets distributed at the tunnel entrance, or to any radio messages that may be broadcast inside the tunnel (of 620 respondents, almost 45% did not tune into the broadcast, although they knew they could)
■ pay little attention to the tunnel and its facilities as they drive through, and tend to mistake telephone stations and other alcoves for emergency exits
■ predominantly transfer behaviours acquired on the rest of the road network to the tunnel environment, in order to compensate for their lack of knowledge and to some extent to define their responses in the tunnel.

The road tunnel user is, above all, a driver for whom passing through the tunnel is a necessary means to an end – reaching their destination. With a few rare exceptions, they are not experts in tunnel systems. Their usual 'tunnel usage script' is incomplete, and is limited to passing through the structure. Unlike the tunnel operators or rescue teams, the motorist has a script that is unlikely to include the possibility of being blocked and faced with a fire. This is a basic datum of road tunnel evacuations, and it in large part determines driver behaviours. Before assuming the status of 'pedestrian evacuating a tunnel', drivers have a mental journey to make – accepting that they are no longer in a normal tunnel usage situation, relinquishing their previous objectives, and finding a relevant way to respond, alone, to the new situation. In this context, the information received locally, and the egress solutions offered by the tunnel, take on critical importance.

20.5. A generic model for egress behaviour in tunnel fires

Egress behaviours in fire situations have been described in several models (Bickman *et al.*, 1977; Canter, 1990; Proulx, 1994). Most of these distinguish between a number of stages (from three to five, depending on the model) that describe the mindsets and behaviours of the occupants from the moment the fire breaks out to the moment they reach safety. These models represent, in more or less similar fashion, the mental journey made by drivers to get to safety, and the factors that influence egress behaviours: (1) interpreting the situation and recognising the need to evacuate; (2) identifying the solutions offered by the tunnel environment and selecting one; (3) deciding to evacuate the vehicle; and (4) getting to safety.

Without rejecting the validity and usefulness of these approaches, the TNO Human Factors team (Martens, 2008) recommended (as part of the UPTUN project) an alternative, simplified model based on behaviours that can actually be observed, for example by using video recordings. The pragmatic model of egress behaviour that they proposed differentiates three stages (Figure 20.2): (1) waiting in the car; (2) hesitating (the time between opening the car door and starting to head for the emergency exits), and (3) walking towards the emergency exits. In this case, the starting point of this model corresponds to an initial time t_0 related to each individual history relating to the event: it refers to the moment when the individual indentifies the occurrence of an incident (through one or more cues such as stopped vehicles in front, smoke, fire or alarm) that triggers an evacuation behaviour.

The model developed under the ACTEURS project (Noizet and Mourey, 2005b) distinguishes between four stages that differ according to the level of awareness of danger, the type of decisions made, the actions taken and the perceived degree of time pressure:

1. stage 0: inception of the event
2. stage 1: perception and recognition of the alerting cues (e.g. fire, smoke, heat, noise, alarm)

Figure 20.2 The UPTUN three-stage model for a motorist who leaves a tunnel on foot after being stuck behind a burning HGV (Martens, 2008; http://www.uptun.net)

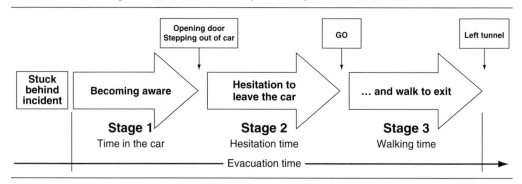

3. stage 2: decision-making
4. stage 3: movement.

The advantage of this model is that it integrated, at the time of its design, all of the available knowledge on tunnel egress behaviours (experience feedback, additional data collection methods). Each evacuation stage is therefore associated with a set of factors capable of optimising or hindering drivers' egress behaviours.

The models from the UPTUN and ACTEURS projects have been tested by comparison with lessons learned from previous events and observational data from tunnel evacuation exercises, and this indicates that it is probably reasonable to use the models as a basis for the simplified generic model of road tunnel fire evacuation proposed below. Picking up on the principle recommended by TNO Human Factors, this model distinguishes four stages based on observable behaviours:

1. stage 0: arrival at the scene of the fire
2. stage 1: analysis of the situation from the vehicle
3. stage 2: search for the appropriate response in or near the vehicle
4. stage 3: movement towards the tunnel evacuation routes.

20.5.1 Stage 0: arrival at the scene of the fire

The motorists are in a normal driving situation in the tunnel. Fire breaks out, and the first consequences of the event generally take the form of congestion, possibly with fumes emanating from the stricken vehicle. See Table 20.1.

Table 20.1 Stage 0

Driver main activity	Driving, fulfilling travel objectives	
Mode of operation	Skill based: the active script is the one for driving in tunnels under normal conditions	
Anxiety (level/object)	Low/related to the tunnel, and perhaps concern about being late	
Factors known to facilitate evacuation	Drivers:	■ knowledge of the tunnel and its safety facilities[a] ■ knowledge of tunnel evacuation procedures[a] ■ previous experience of fire in a tunnel or other environment[a] ■ other drivers' reactions (group effect)
	Tunnel:	■ evacuation solution clearly visible and distinct from the rest of the tunnel environment[a] ■ prior information on the tunnel and what to do in the event of fire[a]
Factors known to hinder evacuation	Drivers:	■ reduced vigilance (fatigue, medications, alcohol, drugs)[a] ■ reduced mobility[a]
	Tunnel:	■ absence of evacuation solutions[a] ■ evacuation solutions poorly signed, hard to distinguish from the rest of the environment[a]

[a] Valid for all stages.

Insights derived from previous events or studies on tunnel egress behaviours:

- If one of the lanes remains open, many drivers will continue straight towards the fire.
- If one of the drivers starts a reverse or U-turn manoeuvre, most of the others will follow suit.
- Drivers often stop too close behind the vehicle in front, during an incident, increasing the chance of fire spread.
- Drivers often stay in their vehicles too long.

20.5.2 Stage 1: analysis of the situation from the vehicle

The drivers are in their cars, having come to a halt perhaps near to, perhaps some distance away from, the stricken vehicle. (Drivers often stop too close behind the vehicle in front.) The fire is starting to spread. The direct signals (fire, smoke, heat, smell) and indirect signals (alarms, behaviour of other drivers and passage of emergency vehicles) are building up. Anxiety

Table 20.2 Stage 1

Driver main activity	■ Searching for additional information and analysis of the situation ■ Trade-off between internal and external risk management ■ Trade-off between costs and benefits of changing behavioural register ■ Abandonment of active script and construction of a new behavioural response
Mode of operation	Knowledge based: most drivers do not have a script for the situation they are faced with. High cognitive cost
Anxiety (level/object)	Very high/related to the initial ambiguity of the situation and the tension induced by possible danger and by the need to change behavioural register
Factors known to facilitate evacuation	Drivers: ■ other drivers' reactions (group effect) Tunnel: ■ functional tunnel alarm system ■ consistency between different types of signals ■ public address system: giving clear, explicit messages on the nature of the event, the associated dangers, and what to do ■ broadcasting alarm messages to car radios
Factors known to hinder evacuation	Drivers: ■ perceptual handicap ■ 'friendly fire syndrome' (impression of harmlessness) ■ counter-productive assumptions about safety (e.g. 'I'm safe inside the car') ■ the 'Wait for assistance' script sometimes associated with emergencies
	Tunnel: ■ absence of a tunnel alarm system ■ no message dissemination system ('dumb tunnel') ■ lack of clear, accurate, information on the nature of the danger and the degree of urgency (e.g. 'Please wait') ■ inadequate level of alarm systems

levels are mounting. These signals may not necessarily be perceived by the drivers, and may even be rejected by them (denial behaviour). When they are perceived, the drivers try to interpret them, to understand precisely what they mean. They seek to disambiguate (i.e. remove the ambiguity of) the signals before embarking on any action. When they accept the signals as an indication that they are in a different kind of situation, they engage in a process of evaluating the reality of the danger at hand. This evaluation will lead them to assess the relevance of changing behavioural registers (i.e. changing the possible range of behaviours in mind) and abandoning their initial objectives. See Table 20.2.

Insights from experience feedback or studies on tunnel egress behaviours.

- Having an emergency exit nearby makes for faster evacuation (Boer, 2002).
- A short, simple message, delivered by a non-recorded human voice, makes for faster evacuation (Proulx and Sime, 1991).
- Drivers with previous experience of tunnel evacuations start walking towards the exits within seconds of stopping in the tunnel (confidential study).

Table 20.3 Stage 2

Driver main activity		- Searching for more information in the tunnel and from other drivers - Assessing the situation and the reality of the danger - Adjusting the trade-off between the costs and benefits of evacuating - Resolving the abandonment dilemma - Devising an action plan
Mode of operation		Knowledge based: most drivers do not have a script for the situation they are faced with. High cognitive cost
Anxiety (level/object)		Very high/related to the proximity of danger, the lack of an established action plan, the dilemma of abandoning the vehicle, and possibly fears about loved ones or friends
Factors known to facilitate evacuation	Drivers:	- other drivers' reactions (group effect)
	Tunnel:	- public address system: giving clear, explicit messages on what to do - broadcasting messages to car radios - presence of operating or emergency personnel
Factors known to hinder evacuation	Drivers:	- strong affective attachment to the vehicle and its load - 'friendly fire syndrome' (impression of harmlessness) - counter-productive assumptions about safety (e.g. 'I'm safe inside the car')
	Tunnel:	- effectiveness of smoke/fire control mechanisms (clear environment = sense of security) - no message dissemination system ('dumb tunnel') - lack of clear, precise, evacuation instructions - inadequate level of alarm systems

20.5.3 Stage 2: search for the appropriate response in or near the vehicle

The emergency situation is accepted as such. The decision to evacuate towards the emergency exits has not yet been made. In or near their vehicles, the drivers are looking for additional information to determine the appropriate course of action for the situation in which they find themselves. They explore their immediate environment, observe the behaviour of other drivers, perhaps entering into exchanges with them to agree on what is to be done. In parallel, they continue to assess the reality of the danger they face, before starting to move towards the evacuation routes. Some of them engage in immediate situation management activities: calling the emergency services, fighting the fire, getting their families together, informing other drivers, organising the evacuation … It is also at this stage that drivers resolve their dilemma about abandoning their vehicles and the possessions that they contain. See Table 20.3.

Insights from experience feedback or studies on tunnel egress behaviours.

- Drivers are generally reluctant to act on the causes of the fire (Papaioannou and Georgiou, 2008).
- Most drivers remain in their cars even when they are engulfed in smoke (Boer, 2002).
- An official instruction such as 'Danger of explosion' or 'Please leave the tunnel' systematically leads to evacuation within seconds (Boer, 2002).

20.5.4 Stage 3: movement towards the tunnel evacuation routes

The decision to head for the evacuation routes has been made. An evacuation solution (place, path) has been selected. In most cases, this will be the solution adopted by the other evacuating

Table 20.4 Stage 3

Driver main activity	Moving collectively towards the chosen evacuation route	
Mode of operation	Skill based: applying a previously established plan of action	
Anxiety (level/object)	High/mitigated in part by having a solution and being engaged in action, but the proximity to danger remains, along with uncertainty about where the evacuation is heading	
Factors known to facilitate evacuation	Drivers: Tunnel:	■ other drivers' reactions (group effect) ■ smoke/fire control ■ evacuation route signalling system (lighted signs) ■ audio guidance system to indicate evacuation routes (when smoke is out of control) ■ presence of operating or emergency personnel
Factors known to hinder evacuation	Drivers: Tunnel:	■ reduced mobility ■ size of group (notably with tourist coaches) ■ smoke out of control ■ absence of evacuation route signalling system (evacuation routes hard to see, or hard to distinguish from the rest of the environment) ■ absence of audio guidance system (when smoke is out of control)

drivers. This movement is part of a collective behaviour (group effect), in which people look out for each other and assist those in need. The movement towards the evacuation routes is massive and determined.

When the conditions are seriously degraded by smoke, the movement strategy consists in feeling one's way along the sidewall of the tunnel or ground markings (white line). In the absence of an audio guidance system, the chances of reaching an emergency exit are slim. See Table 20.4.

Insights from experience feedback or studies on tunnel egress behaviours.

- Collective evacuation movements tend to structure around a driver who takes leadership of the group.
- People tend to evacuate along the traffic lanes.
- People keep the emergency exit doors open, in order to monitor the situation.
- Some people go back into the tunnel after reaching an emergency exit in order to retrieve personal items.

20.6. Conclusion: taking action to optimise egress behaviours in road tunnel fires

This chapter has concentrated on drivers. However, it needs to be remembered, of course, that other people (i.e. passengers) pass through tunnels as well. That is, there will, in general, be people of different mobilities and ages present in a tunnel fire. When a road tunnel fire breaks out, everything (or almost everything) depends on what happens in the first few minutes. In most cases, during these first few minutes, drivers have to fend for themselves. Tunnel safety is often based on the assumption that drivers will immediately understand the danger signals coming from the event, or from the safety systems, and that they will adopt the appropriate behaviours without delay: stopping immediately, leaving the car, going to the nearest shelter or emergency exit and, in general, making good use of all of the safety facilities and equipment in the tunnel. The lessons learned and the actual behaviours described in this chapter suggest that these hypotheses about drivers' responses to tunnel fires are often unrealistic.

The human processes and mechanisms described above do, however, offer tunnel operators and designers a number of concrete lines of enquiry that can help them to optimise driver egress behaviour in the event of a tunnel fire.

- *Enrich drivers' prior knowledge.* Every communication campaign aimed at the general public about tunnels, tunnel facilities and what to do in the event of fire helps to generate scripts in the minds of the tunnel users on how to use road tunnels and how to evacuate them in a fire; these scripts make it easier for people to recognise the alarm signals and identify evacuation solutions. A driving simulator study carried out under the UPTUN project (Martens, 2008), for example, demonstrated that giving drivers prior information on what to do in the event of fire, in the form of a leaflet (in this case the draft European Union leaflet) significantly improved their (stated) egress behaviours in fire situations. All kinds of media can be used to provide information on tunnels: TV programmes, radio messages, leaflet distribution, communication campaigns at motorway service areas or at tunnel entrances, etc. Workpackage 2 of the ACTEURS project (Ricard and Noizet, 2006), for instance, led in 2006 to the creation of a national tunnel safety training programme, integrated into basic driver training. The general theory section of the French driving examination now includes questions about road tunnels.

- *Favour the emergence of leaders capable of guiding tunnel evacuations.* The most appropriate strategy here is to target professional drivers (e.g. HGV and taxi drivers), giving them, as part of their professional training, the knowledge required to help organise an evacuation in the event of a tunnel fire. Workpackage 2 of the ACTEURS project has tested out just such an approach (Ricard and Noizet, 2006).
- *Enrich on-site information on the emergency situation and instructions.* Broadcasting warnings and announcements in the tunnel remains one of the most effective ways of giving drivers the information they need to take stock of the impending emergency, decide to evacuate and identify the available egress solutions. A few simple rules can reinforce the effectiveness of these systems:
  - use different types of alarm, both visual and auditory
  - be clear about the nature and gravity of the reason for evacuation
  - be directive in the instructions given to drivers
  - contextualise the messages as far as possible (specifying the tunnel and segment affected)
  - make sure that messages are audible and comprehensible under all tunnel operating conditions (e.g. forced ventilation).
- *Design for evacuation.* Drivers faced with a tunnel fire need simple and, in particular, visible evacuation facilities. The evacuation routes must stand out from the rest of the tunnel environment. Systems must be put in place to capture a driver's attention in the event of a fire (dynamic signage) and to compensate for poor visibility in the tunnel (marker lights, sound beacons). Sound beacons, for example, are particularly effective in helping drivers reach emergency exits in smoke-filled tunnels (Boer, 2004). Deliverable D33 of the UPTUN project proposed and evaluated other innovative systems designed to fulfil these objectives.

A few years ago, very little was known about human behaviour in tunnel fires, but a good start has now been made. However, there is much still to be done, such as determining the characteristics of mobility-impaired drivers and passengers (e.g. finding out how long it actually takes a severely mobility-impaired person to get out of their vehicle). A single 'pre-egress/delay' time does not apply to all.

In conclusion, we should remember one of the basic tenets of all ergonomic approaches: any solution designed for a particular user must be tested on that user. The only way to ensure that the solutions designed to facilitate driver egress actually work is to put them through a genuine experimentation and testing process.

(Author's note: for a very brief summary on egress from tunnels, see Ingason and Wickstrom (2011).)

REFERENCES

Amalberti R (1996) *La conduite des systèmes à risques.* PUF, Le Travail Humain, Paris.
Baars BJ (1997) *In the Theatre of Consciousness: The Workspace of the Mind.* Oxford University Press, Oxford.
Bickman L, Edelman P and McDaniel M (1977) *A Model of Human Behaviour in a Fire Emergency.* NBS-GCR-78-120. National Bureau of Standards, Center for Fire Research, Washington, DC.

Boer LC (2002) *Behaviour by Motorists on Evacuation of a Tunnel*. Report, Centre for Tunnel Safety, TNO Human Factors, Amsterdam.

Boer LC (2004) Directional sound evacuation from smoke-filled tunnels. In: *Safe and Reliable Tunnels. Innovative European Achievement. 1st International Symposium, Prague*, 4–6 February 2004. HOJ Consulting, Brunnen.

Bryan JL (1991) Human behaviour in fire. In: *Fire Protection Handbook*, 17th edn. National Fire Protection Association, Quincy, MA.

Bryan J (2002) Behavioural response to fire and smoke. In: *SFPE Handbook of Fire Protection Engineering*, 3rd edn. National Fire Protection Association, Quincy, MA.

Canter D (1990) *Fire and Human Behaviour*, 2nd edn. David Fulton, London.

Chertkoff JM and Kushigian RH (1999) *Don't Panic: The Psychology of Emergency Egress and Ingress*. Praeger, Westport, CT.

Eder S, Brutting J, Muhlberger A and Pauli P (2009) Human behaviour in tunnel accidents. *Proceedings of the 4th International Symposium on Human Behaviour in Fire, Cambridge*, 13–15 July 2009. Interscience Communications, London.

Edworthy J (1998) Warnings and hazards: an integrative approach to warnings research. *International Journal of Cognitive Ergonomics* **2**: 3–18.

Hutchins E (1994) *Cognition in the Wild*. MIT Press, Cambridge, MA.

Ingason H and Wickstrom U (2011) Tunnel fire safety research in Europe. *Industrial Fire Journal* **82**: 30–33.

Martens MH (2008) *Human Factors Aspects in Tunnels: Tunnel User Behaviour and Tunnel Operators*. UPTUN, Workpackage 3: Human Response, Deliverable 3.3. EC, Brussels.

Miller JG (1960) Information input overload and psychopathology. *American Journal of Psychiatry* **116**: 695–704.

Noizet A and Mourey F (2005a) *Les comportements des usagers en situation de traversée normale des tunnels*. Projet ACTEURS Lot 1, Rapport de recherche n° 2. Ministère de l'Ecologie, du Développement durable, des Transports et du Logement, Paris.

Noizet A and Mourey F (2005b) Crisis situation in tunnels – what kind of behaviours can we expect from drivers? Some results from the French ACTEURS project. *Proceedings of the 3rd International Conference on Traffic and Safety in Road Tunnels*. Hamburg, 18–20 May 2005.

Noizet A and Paries J (2004) *Point sur ce que savent les professionnels des tunnels des connaissances et comportements des usagers des tunnels*. Projet ACTEURS Lot 1, Rapport de recherche n° 1. Ministère de l'Ecologie, du Développement durable, des Transports et du Logement, Paris.

Noizet A and Ricard F (2004) The ACTEURS project: results from the survey of drivers using the tunnels operated by the French alpine motorway companies. *Proceedings of the 5th International Conference in Tunnel Fires, London*, 25–27 October 2004. SP, Borås.

Ozel F (2001) Time pressure and stress as a factor during emergency egress. *Safety Science* **38**: 95–107.

Papaioannou P and Georgiou G (2008) *Human Behaviour in Tunnel Accidents and Incidents: End-Users, Operators and Response Teams*. UPTUN, Workpackage 3: Human Response, Deliverable 3.2. EC, Brussels.

Proulx G (1994) Human response to fires. *Fire Research News* **71**: 1–3.

Proulx G (1998) The Impact of voice communication messages during a residential highrise fire. In: *Human Behaviour in Fire – Proceedings of the 1st International Symposium, Belfast*, 31 August–2 September 1998. University of Ulster, Belfast, UK.

Proulx G (1999) Occupant response to fire alarm signals. In: *National Fire Alarm Code Handbook*. National Fire Protection Association, Quincy, MA.

Proulx G and Sime J (1991) To prevent 'panic' in an underground emergency: why not tell people the truth? In: Cox G and Langford B (eds), *Fire Safety Science: Proceedings of the 3rd International Symposium*. Elsevier Applied Science, London.

Purser DA and Bensilum M (2001) Quantification of behaviour for engineering design standards and escape time calculations. *Safety Science* **38**: 109–125.

Rasmussen J (1986) *Information Processing and Human–Machine Interaction: An Approach to Cognitive Engineering*. Elsevier, Amsterdam.

Ricard F and Noizet A (2006) The ACTEURS project on the behaviour of tunnel users. A study on change management within an industry. *Proceedings of the 13th World Congress and Exhibition on Intelligent Transport Systems (ITS) and Services, London, 8–12 October 2006*. Ibeo Automotive Systems, Hamburg.

Schadschneider A, Klingsch W, Klüpfel A, Kretz T, Rogsch C and Seyfried A (2009) Evacuation dynamics: empirical results, modeling and applications. In: Meyers RA (ed.), *Encyclopedia of Complexity and System Science*. Springer-Verlag, Berlin.

SFPE (2002) *SFPE Engineering Guide to Human Behaviour in Fire*. National Fire Protection Association, Quincy, MA.

Sime J (1985) Movements toward the familiar: person and place affiliation in a fire entrapment setting. *Environment and Behaviour* **17**(6): 697–724.

Steyvers F, de Waard D and Brookhuis K (1999) Aspects of human behaviour in tunnel fires – a literature review. *International Tunnel Fire and Safety Conference, Rotterdam, 2–3 December 1999*. Brisk Events, Leusden.

Tesson M (2010) Recent research results on human factors and organizational aspects for road tunnels. *Proceedings of the 4th International Symposium on Tunnel Safety and Security, Frankfurt, 17–19 March 2010*. SP, Borås, pp. 191–202.

Tesson M and Lavedrine S (2004) Integration of human behaviour in the improvement of safety in French road tunnels. *Proceedings of the 3rd International Symposium on Human Behaviour in Fire, Belfast, 1–3 October 2004*. Interscience Communications, London.

Zsambock CE and Klein G (1997) *Naturalistic Decision Making*. Lawrence Erlbaum, Mahwah, NJ.

Handbook of Tunnel Fire Safety, 2nd edition
ISBN: 978-0-7277-4153-0

ICE Publishing: All rights reserved
doi: 10.1680/htfs.41530.439

Chapter 21
Recommended behaviour for road tunnel users

Michel Egger
Conference of European Directors of Roads, Paris, France

21.1. Introduction

On 24 March 1999, a truck loaded with margarine and flour caught fire in the Mont Blanc Tunnel between Chamonix (France) and Aosta (Italy). The blaze spread rapidly to other vehicles, with the result that 39 people died due to the development of intense heat and smoke. On 29 May 1999, a collision took place in the Tauern Tunnel in Austria, between a lorry that collided with four light vehicles and another lorry loaded with different spray-cans, standing in front of a traffic light inside the tunnel. The collision caused a fire, which spread rapidly. Twelve people died: eight due to the collision and four due to the smoke.

These two tragic accidents brought the risks in tunnels to the fore, and led political leaders to get involved, although concern for safety in road tunnels did not start with these dramatic fires: designers, contractors and operators had accumulated experience over many years, and a number of countries already had regulations.

On 24 October 2001 in the Gotthard Tunnel, Switzerland, a heavy goods vehicle (HGV) crossed over to the oncoming lane. An approaching HGV tried to avoid a head-on collision by turning sharply to the left. A lateral collision ensued that caused a huge fire involving five HGVs. Eleven people died, intoxicated by the fast-developing smoke, as they stayed inside the tunnel instead of using the safety exits. This last tunnel fire shows how important it is for road users to behave correctly when there is a fire in a tunnel: all the victims had ample time to save themselves from the deadly smoke by using the nearby safety exits.

For both new and renovated road tunnels, structural and technical safety installations have to comply with national and international recommendations, regulations and standards. These safety installations can only be fully effective if they are well operated and combined with an efficient emergency service and correct behaviour on the part of road users. In this connection, traffic control and monitoring by the police or other relevant authorities have a preventive effect. However, even permanent and intensive efforts on the part of road construction authorities and traffic police cannot fully eliminate the occurrence of accidents and fires in tunnels. A chain of unfortunate circumstances could also give rise to disasters in other tunnels, such as the fires in the Mont Blanc and Tauern Tunnels.

21.2. Safety and risks in road traffic

To ensure safety in road traffic, the necessary structural, technical and organisational measures need to be taken so that incidents can be prevented as far as possible and so that their impact can be kept to a minimum. All safety measures have to correspond to the latest technology and have to apply to all concerned, i.e. to road users, traffic control and emergency services, infrastructure and vehicles. Taking account of the limited funds that are available, the measures to be implemented first are those that most efficiently reduce the risks.

There is no such thing as absolute safety in traffic, for it is in the nature of traffic that incidents will occur, some of which have grave consequences for people, the environment and property. Dealing with these residual risks is not just a technical matter, it is also a political and social issue.

The main causes of road accidents are the incorrect behaviour of road users, inadequate installations on the road network, vehicles with technical defects and other faults (e.g. defective electrical systems and brakes, overheated engines) and problems with cargoes (e.g. unstable loads, chemical reactions). According to a report published by the OECD (1999), the incorrect behaviour of road users is the main cause of 95% of all accidents.

Incidents primarily endanger road users. The number of cases in which people who are not actually on the road are exposed to danger (e.g. due to the release of toxic gases when vehicles carrying dangerous goods are involved) is very low. As far as the environment is concerned, it is primarily the road surface water, collected by road drainage systems and groundwater in the close vicinity of roads, that is affected.

In the case of road accidents in which no fire is involved, speed and the number of vehicles involved primarily determine the extent of harm to road users. In accidents in which fire or the transport of dangerous goods is involved, it is mainly the quantity of explosive, inflammable, toxic or water-polluting substances that is the determining factor. With respect to protection of the environment, the most important factors are distance from the groundwater or surface water, the type of drainage, and the conditions allowing access for emergency services.

Due to the fact that tunnels are enclosed spaces, fires that occur in them result in poor visibility and the spread of smoke and toxic gases along the tunnel, the rapid development of high temperatures and a reduction in the level of oxygen in the air. The extent of harm to road users in the event of a fire in a tunnel is therefore far greater than is the case on open roads.

In view of this, it is essential to provide adequate facilities for road users to escape or be rescued by emergency crews.

This means that there should be enough escape routes, and that the ventilation system needs to be fast and efficient, particularly in tunnels with bi-directional traffic. These prerequisites also apply in the event of an accident that does not involve fire, but which results in the release of toxic gases.

Fires in tunnels not only endanger the lives of road users, they can also cause damage to structural components, installations and vehicles, with the result that the tunnel concerned may have to be closed for a considerable length of time (Figure 21.1). The capacity of a fire expressed in terms of the heat release rate can differ greatly depending on the type of vehicle and load (see Chapter 14).

Figure 21.1 Damage to the tunnel and vehicles following a tunnel fire © Ti-Press/Polizia Cantonale

21.3. Objectives for safety in road tunnels

The following objectives have been set for attaining the optimal level of safety in road tunnels.

- Primary objective: *prevention* – to prevent critical events (i.e. *crucial events*, see Chapter 5) that endanger human life, the environment and tunnel installations.
- Secondary objective: *reduction of the consequences* of events such as accidents and fires – to create the ideal prerequisites for:
  - people involved in the incident to rescue themselves
  - the immediate intervention of road users to prevent greater consequences (a fire can generally be easily extinguished immediately after it breaks out, but 10 minutes later it will have developed into a full blaze)
  - ensuring efficient action by emergency services
  - protecting the environment
  - limiting material damage.

In the event of an incident, the first 5–10 minutes are decisive – less time if there is an explosion – when it comes to people saving themselves and limiting damage. The prevention of critical events (i.e. *crucial events*, see Chapter 5) is therefore the number one priority, which means that the most important measures to be taken have to be of a preventive nature (Figure 21.2).

The level of safety in tunnels is influenced to varying degrees by a variety of factors that can be collectively summarised in four main groups, as shown in Figure 21.3 (see also Chapters 19 and 20).

Figure 21.2 Evacuation into a safety tunnel

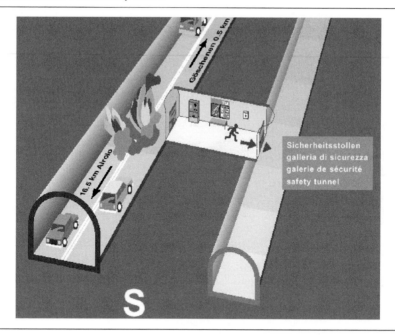

In the section that follows, a closer look is given to road users as a factor influencing safety in road tunnels.

21.4. Road users as a factor influencing safety in road tunnels

In-depth analyses of incidents on our roads show that an accident is the consequence of one or more faults in a complex system involving drivers, vehicles, the road and its surroundings.

Figure 21.3 Four factors influencing safety in road tunnels

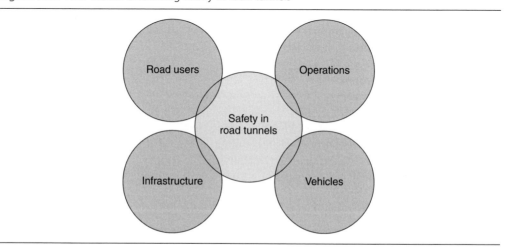

However, the principal factor in bringing about a road accident is human error, so efforts to increase the level of road safety have to be primarily aimed at preventing human errors. The second step is to ensure that errors that may still be made by drivers do not give rise to grave consequences.

There is no doubt that it is easier to rectify technical faults than it is to influence human behaviour. Nonetheless, there are various ways in which it is possible to directly or indirectly influence the way people act, and these include education, driving instruction and the provision of information, as well as regulations, police enforcement and penalties for traffic violations.

Basically, the driving rules that apply in tunnels are the same as those for open roads, i.e. maintaining a safe distance, observing speed limits and maximum loads, thoroughly securing all loads and warning other road users in the event of a breakdown or congestion. And, even more than on open roads, it is recommended that drivers listen to their radio while in tunnels so that they are able to receive traffic reports as well as possible specific instructions. However, there are a number of additional traffic regulations that apply especially to tunnels

- overtaking is forbidden if there is only one lane in each direction
- no turning or reversing is allowed, if not specifically instructed by tunnel officials
- headlights must be used, even in lit tunnels
- no stopping is allowed in a tunnel, except in an emergency, in which case the engine must be switched off immediately.

The following rules for correct behaviour should apply when driving in a tunnel or in the event of a vehicle breakdown, traffic congestion, an accident or a fire in a tunnel.

- Correct behaviour whilst driving through a tunnel:
  - Switch on headlights.
  - Take off sunglasses.
  - Observe road signs and signals.
  - Keep an appropriate distance from the vehicle in front.
  - Switch on the radio and tune to the indicated frequency (Figure 21.4).
- Correct behaviour in the event of traffic congestion:
  - Switch on warning lights.
  - Keep your distance even if moving slowly or stopped.
  - Switch off your engine if traffic is completely stopped.
  - Listen to possible messages on the radio.
  - Follow instructions given by tunnel officials or variable message signs.
- Correct behaviour in the event of a breakdown or accident (own vehicle):
  - Switch on warning lights.
  - Switch off the engine.
  - Leave your vehicle.
  - If necessary and possible, give first-aid help to injured people.
  - Call for help from an emergency point (Figure 21.5).
- Correct behaviour in the event of a fire (own vehicle):
  - If possible, drive your vehicle out of the tunnel.
  - If not possible:
    - ☐ Pull over to the side and switch off the engine.

Figure 21.4 It is recommended that drivers in tunnels listen to the radio

- □ Leave the vehicle immediately.
- □ Put out fire using the vehicle's extinguisher or one available in the tunnel.
- □ If extinction of the fire is not possible, move away without delay to an emergency exit.
- □ Call for help from an emergency point.

Figure 21.5 Emergency station with fire extinguisher, emergency phone and hydrant

Figure 21.6 Emergency exit

- Correct behaviour when stopped behind a fire (other vehicle):
    - Do not park too close behind the vehicle in front (say 50 m or more).
    - Pull over to the side and switch off the engine.
    - Switch on warning lights.
    - Leave the vehicle immediately.
    - If necessary and possible, give first-aid help to injured people.
    - Put out the fire using the vehicle's extinguisher or one available in the tunnel.
    - If extinguishing the fire is not possible, move away without delay to an emergency exit (Figures 21.6 and 21.7).

Figure 21.7 Emergency exit signage

21.5. Proposed measures for road users

21.5.1 Information campaigns

Information campaigns regarding safety in tunnels should be regularly organised and implemented in collaboration with the principal partners involved.

These information campaigns should cover the correct behaviour of road users when approaching and driving through tunnels, especially in the case of vehicle breakdown, congestion, accidents and fires.

Information on the safety equipment available and proper road user behaviour in tunnels should be displayed in rest areas before tunnels and at tunnel entries when the traffic is stopped (e.g. at tolls).

21.5.2 Driving tests

Driving tests for all categories of vehicles should include specific questions concerning correct behaviour for road users in the event of a vehicle breakdown, congestion, an accident or a fire in a tunnel.

Special instructions concerning careful and responsible driving should become an integral part of compulsory driving lessons and should include teaching correct behaviour in special situations, for example, in congestion in tunnels, in the event of a vehicle breakdown, an accident or a fire in a tunnel.

It is not possible, of course, to train drivers how to behave correctly in practice in the event of congestion, a vehicle breakdown, an accident or fire in a tunnel. However, it is both possible and advisable to include questions in the theoretical section of the driving test that deal with these particular situations.

21.5.3 Drive out burning vehicle

If a vehicle catches fire, it is strongly recommended that the driver drives his vehicle out of the tunnel whenever possible (self-help principle).

This recommendation should be made to road users both in information campaigns and as part of their driving instructions.

21.5.4 Roadside checks

Roadside checks of HGVs should be intensified and harmonised at the international level. The necessary funding should be made available to the authorities concerned.

It is advisable to intensify checking measures through various methods, for example, X-raying transported loads, setting up online connections to IT systems and using detection devices. Coding vehicles and their loads via GPS would also make it possible to track them at any time.

Wide-ranging regulations already exist with respect to the transport of dangerous goods, and apply to senders, carriers, vehicle owners, drivers and recipients. However, they can only be effective if they are also duly enforced.

21.5.5 Tests for professional drivers

Truck, coach and bus drivers should be tested periodically with respect to their knowledge of safety-relevant aspects of vehicles and equipment.

All truck, coach and bus drivers are required to possess adequate knowledge of safety-relevant aspects of vehicles and equipment so that they are able to use them when necessary. Specific aspects of correct behaviour in road tunnels should be incorporated into driving instructions in the future. In particular, all drivers should be trained in the correct use of a fire extinguisher.

Efforts at the international level to introduce a periodical test (at least *every 5 years*) for truck, coach and bus drivers are to be supported.

21.5.6   Test for dangerous goods drivers
Instruction of drivers of vehicles carrying dangerous goods should include specific aspects of behaviour in tunnels.

All drivers of vehicles transporting dangerous goods already have to undergo special instruction leading to a test, and successful candidates are awarded a certificate. They are required to attend a follow-up course and take another test every 5 years in order to renew this certificate. The initial training of new drivers as well as the follow-up courses for experienced drivers should include instruction on safety in tunnels.

21.5.7   Regulations for dangerous goods transport
Regulations governing the transport of dangerous goods through tunnels should be rationalised at the international level. (See also Chapter 22.)

The OECD and the World Road Association (PIARC) have finalised a proposal (OECD, 2001) to

- create five dangerous goods cargo groupings accepted at the international level and which should be used to regulate the authorisation of the transport of dangerous goods in road tunnels
- carry out a quantitative risk analysis and run a decision support model, taking into account both the itinerary including the tunnel and any alternative itineraries, before the decision is made to authorise or not all or part of the transport of the dangerous goods through each tunnel.

The following are recommended.

- To include the five dangerous goods cargo groupings proposed by the OECD and PIARC in the appropriate UN and/or UNECE legal instruments so that their use becomes compulsory for tunnel regulations regarding dangerous goods.
- To create a new sign to be placed at tunnel entrances, indicating which groupings of dangerous goods are allowed/prohibited, with reference to the five dangerous goods cargo groupings to be created.
- To perform a quantitative risk analysis as proposed by the OECD and PIARC before deciding on tunnel regulations regarding dangerous goods.
- To study the possibility of classification as dangerous goods of certain liquids or easily liquefied substances with energy values comparable with that of hydrocarbons.
- To consider operating measures for reducing the risks involved in the transport of dangerous goods in tunnels (declaration before entering, escort, etc.), on a case-by-case

basis. Regulations may require the formation of convoys and accompanying vehicles for the transport of certain types of particularly dangerous goods; however, these measures are also dependent on sufficient space being available in front or in advance of the tunnel as well as available operational means.

- To study the possibility of introducing automatic detection of dangerous goods transport (e.g. by electronic devices carried on vehicles).

21.5.8 Overtaking

In certain cases, it should be possible to prohibit trucks from overtaking in tunnels with more than one lane in each direction.

In most tunnels in which there is only one lane in each direction, overtaking is already prohibited for all vehicles; in tunnels with more than one lane in each direction, a ban on overtaking for trucks could in some cases lead to an improvement in road safety, for example, in tunnels with a gradient of over 3%. This measure should be accomplished by placing corresponding traffic signs and possibly variable message signs at appropriate locations. However, incorporating a complete overtaking ban for trucks in tunnels into the highway code is unlikely to achieve the desired effect, so a general ban is therefore not recommended.

21.5.9 Distance between vehicles

For safety reasons, road users should maintain an adequate distance from the vehicle in front of them under normal conditions and also in the event of a breakdown, congestion, an accident or a fire in a tunnel.

The applicability and effectiveness of introducing a minimum compulsory distance between vehicles should be examined on a case-by-case basis.

Traffic regulations require road users to maintain sufficient distance from the vehicle in front of them so that they are able to stop in good time if the vehicle in front should suddenly brake. This distance (say 50 m) should always be maintained in a tunnel, even when traffic is stopped.

Introducing a compulsory distance between vehicles of 100 m in all tunnels is neither necessary nor advisable. However, it should be possible to prescribe a compulsory distance between lorries in certain cases. Trials have been undertaken in France on measurements of distances between vehicles.

21.5.10 Speed limit

In order to maintain a uniform traffic flow, it is recommended that outside built-up areas the maximum speed of trucks in tunnels should not be systematically reduced to 60 km/h.

According to existing traffic regulations, there is a fixed speed limit for trucks on most kinds of roads. If this limit were to be systematically reduced to, for example, 60 km/h in tunnels, this would mean that all vehicles in a tunnel with only one lane in each direction would have to adjust their speed to that of the slower trucks, and this in turn would increase the risk of congestion.

In tunnels with more than one lane in each direction, a speed limit of 60 km/h for trucks would not enhance the degree of road safety, but only disturb the homogeneity of the traffic flow.

21.5.11 Access to the profession

The rules on access to the profession of road transport operator and their implementation should be reinforced and harmonised at the level of professional qualifications, financial standing and good repute. The checking of compliance with these rules both on the roadside and at transport enterprises should also be intensified.

21.5.12 Driver information systems in the event of a fire

Internationally harmonised systems (sirens, flashing lights, etc.) should be developed and implemented to inform drivers in the event of a fire that they have to leave their vehicles without delay and proceed immediately to the nearest emergency exit in the tunnel.

Drivers are not always aware of the extent of the danger posed by a fire in a tunnel, above all when they are at some distance from the fire. In addition to messages transmitted by radio, systems have to be developed that inform drivers that their lives are in danger and that they should leave their vehicle. Such systems should also clearly indicate the emergency exits and the direction of the closest emergency exit.

21.6. Conclusions and outlook

The potential risks that are prevalent in road tunnels need to be taken seriously, but they should not be allowed to give rise to panic. Stretches of road through tunnels are among the safest, as can be seen from the fact that generally fewer incidents occur in tunnels than on open stretches of road. The main reasons for this are not difficult to find: stretches through tunnels are virtually unaffected by weather conditions and lighting conditions remain constant.

On the other hand, if an incident occurs in a tunnel, the impact is often much greater than on open stretches. This fact clearly justifies the comprehensive work carried out by all the parties mentioned at the beginning of this chapter.

Safety in road tunnels is not simply a question of efficient operation and sound infrastructure. It also depends to a great extent on the behaviour of road users and on the condition of vehicles on the road. It is therefore essential that road users be constantly made aware of correct behaviour in road tunnels, partly through education and information campaigns, but also as part of their driving instruction. In the event of an incident, detection and the ability of road users to rescue themselves are of the utmost importance.

In addition to the above-cited measures, the behaviour of people in tunnels (claustrophobia, etc.) should be the subject of in-depth studies. The behaviour of road users can change considerably when they are driving through a long tunnel (e.g. due to boredom, claustrophobia, etc.), and this has a negative impact on safety.

REFERENCES

OECD (1999) *Programme de Recherche en Matière de Transport Routier et Intermodal. 'Stratégies de Sécurité Routière en Rase Campagne'.* Organization for Economic Co-operation and Development, Paris.

OECD (2001) *Safety in Tunnels: Transport of Dangerous Goods through Road Tunnels.* Organization for Economic Co-operation and Development, Paris.

Handbook of Tunnel Fire Safety, 2nd edition
ISBN: 978-0-7277-4153-0

ICE Publishing: All rights reserved
doi: 10.1680/htfs.41530.451

Institution of Civil Engineers

publishing

Chapter 22
Transport of hazardous goods

Benjamin Truchot and Philippe Cassini
Institut National de l'Environnement Industriel et des Risques (INERIS), Verneuil-en-Halatte, France
Hermann Knoflacher
Technical University of Vienna, Austria

22.1. Introduction

The development of transport throughout Europe promotes an increase in the transport of 'dangerous goods', i.e. flammable liquids/gases, chemicals, etc. To prevent crossing towns with such goods, tunnels are being increasingly used. However, this significant growth in dangerous goods transport increases the probability of accidents involving such goods. At the same time, major fires in tunnels, such as took place at Mont Blanc or Tauern, have brought about an increasing public awareness of risk in tunnels.

Because of the serious consequences that may result from dangerous goods transit in terms of injuries or death – not only in tunnels – road transport regulation needs to be clear and as simple as possible. To reach such an objective, the new version of the European Agreement concerning the International Carriage of Dangerous Goods by Road (ADR) (see below) is based on a grouping system to separate dangerous goods according to possible consequences. This regulation needs to be uniform all over Europe to simplify decisions when creating routes for hazardous goods.

The foregoing not only applies to road tunnels but also to rail tunnels. The use of rail tunnels is emerging as a solution to the inconvenience of modern-day transport. This is illustrated by two major European rail tunnel projects: the Gotthard Base Tunnel and the Lyon–Turin rail liaison. These two Alpine tunnels are projected to be longer than 50 km, and at considerable depth.

22.2. Road tunnel classification
22.2.1 The situation concerning the road transport of hazardous goods in the European Union

The road transport of dangerous goods is now regulated by Directive 2008/68/EC of 24 September 2008 (EU, 2008). This Directive merges Directives 94/55/EC (EU, 1994) and 96/49/EC (EU, 1996) concerning, respectively, dangerous goods transport by road and by rail. It describes the application of the ADR (Economic Commission for Europe Inland Transport Committee, 2008).

Following the working party on Land Transport of the Council of the European Union, specificities about road tunnels were defined in the ADR version published in January 2009. In particular, Appendix A was modified. Section 1, Chapter 1.9, gives the restrictions for dangerous goods

451

transport in tunnels, and section 3, chapter 3.2, provides the restriction code for each type of goods. Signalisation for these tunnels is prescribed in chapter 8.6 in Part B of the ADR.

This Directive must be combined with, for tunnels in the Trans-European Road Network (TERN), Directive 2004/54/EC of 29 April 2004 on minimum safety requirements (Council of the European Union – Working Party on Land Transport, 2002; Economic Commission for Europe Inland Transport Committee, 2008). The 2004 Directive defines specific measures relating to dangerous goods transport in tunnels longer than 500 m. Section 3.7 of Annex 1 of the 2004 Directive deals with the transport of dangerous goods, and states:

> The following measures shall be applied concerning access to tunnels for vehicles transporting dangerous goods, as defined in the relevant European legislation regarding the transport of dangerous goods by road:
>
> - Perform a risk analysis in accordance with Article 13 before the regulations and requirements regarding the transportation of dangerous goods through a tunnel are defined or modified.
> - Place appropriate signs to enforce the regulation before the last possible exit before the tunnel and at tunnel entrances, as well as in advance so as to allow drivers to choose alternative routes.
> - Consider specific operating measures designed to reduce the risks related to some or all of the vehicles transporting dangerous goods in tunnels, such as declaration before entering or passage in convoys escorted by accompanying vehicles, on a case by case basis further to the aforementioned risk analysis.

22.2.1.1 The situation in the professional engineering world for roads

In the 1990s, the joint Organisation for Economic Co-operation and Development (OECD) and the World Road Association (formerly known as the Permanent International Association of Road Congresses, and still using the abbreviation PIARC) Scientific Expert Group ERS2 developed objectives within the OECD Road Transport Research Programme (OECD, 1992, 2001; OECD/PIARC, 1996). In order to apply these objectives, dedicated models have been developed by consultants and tested by the expert group at the end of the 1990s. This chapter will present the governing methodology and the quantitative risk assessment (QRA) road model (QRAM), developed during this research work. It also presents the experience acquired by some EU member states during the first few years of the use of the QRAM and the recent development in the QRA model as the French tentative development of a geographical information system (GIS)-interfaced QRAM.

22.2.2 Harmonised groupings of dangerous goods

Following the work conducted in the 1990s and during the years following 2000 by different working groups of the OECD and PIARC, the ADR was developed for road tunnels based on groups of dangerous goods. It must, however, be kept in mind that the groupings prescribed in the ADR are quite different from those initially proposed by the working groups.

The objective of this chapter is to present the harmonised grouping classification. It is based on the ADR and, more particularly, on Part A, section 1.9.5 (Economic Commission for Europe Inland Transport Committee, 2008).

22.2.2.1 Objectives of harmonised regulations

The topic of harmonised groupings was submitted to interested bodies in charge of international regulations for the transport of dangerous goods in 2005. In addition to the OECD and PIARC committee members, it was circulated amongst the UN Sub-Committee of Experts on the

Transport of Dangerous Goods, the UN Economic Commission for Europe Working Party No. 15 on the Road Transport of Dangerous Goods and the European Commission Technical Committee on the Transport of Dangerous Goods. Following those proposals, this grouping description was introduced in the regulations applying to dangerous goods.

A harmonised grouping system enables one to have a homogeneous regulation for all countries. This is of great importance, considering the possible consequences of incidents involving dangerous goods in tunnels. The best way to create respect for a regulation is to make it simple and homogeneous. Such a simplification makes hauliers' work simpler: when establishing a vehicle's journey, they have the same regulation for authorised and banned goods in tunnels all over Europe. Consequently, this should help to limit the possibility of mistakes in the journey that may cause a vehicle carrying dangerous goods to go into a tunnel from which it has been banned. Harmonising regulations should help to meet the following objectives

- facilitate the organisation of international transport, and thus eliminate technical barriers to trade and rationalise national transport operations
- improve safety, because harmonised regulations would be easier to comply with and easier to enforce.

22.2.2.2 General principles of the groupings
Appendix A, Chapter 1, Section 1.9, of the ADR defines the groups of materials that are used in the restriction of dangerous goods in tunnels. Having harmonised regulations does not mean that the same regulations should apply to all tunnels, nor even that two similar tunnels in two different places should have the same regulation. The only essential aspect is that the regulations should be expressed in the same way everywhere in Europe, which means that they should refer to the same lists of dangerous goods that are authorised or banned. A group refers not only to the nature of the transported goods but also whether they are transported in bulk or packaged form and the possible presence of different dangerous goods in the same vehicle ('transport unit' in regulatory terms).

The basis of the system is that the definition of the groupings of dangerous goods is the same for all tunnels in all countries all over Europe. The general idea of the system is to split dangerous goods into five groupings. The number of groupings is reasonably low, to make the system practicable.

The system is based on groupings A, B, C, D and E, ranked in order of increasing restrictions. Grouping A is the largest one: it contains all goods that are authorised for road transport, including the most dangerous ones. Grouping E is the most restrictive: it contains only goods that do not require a special marking on the vehicle – i.e. the least dangerous ones. Then, all goods in grouping E are included in grouping D, all goods in grouping D are in grouping C, and so on. Transport units with goods in grouping E cannot easily be banned because there is no way for authorities to differentiate them from vehicles that do not carry dangerous goods.

22.2.2.3 Grouping system
This section describes the methodology used for the groupings separation.

MAIN RISKS IN A TUNNEL
Banning from a tunnel those dangerous goods that are authorised for transport in the open can only be justified where the risk of serious accidents (e.g. involving numerous victims or

unacceptable damage to the tunnel) is greater in the tunnel than in the open. Consequently, dangerous goods that do not pose a serious threat to people or to the tunnel structure should not be considered for such a decision. It is assumed, for example, that liquids that are dangerous by contact only cannot cause serious harm to people or serious damage to the tunnel structure. Considering that three types of effects can hurt people and the tunnel structure, i.e. overpressure, toxicity and thermal effects, the grouping system is based on the three major hazards in tunnels that may lead to numerous victims and possibly serious damage to the structure. These are

1. explosions
2. releases of toxic gas or volatile toxic liquid
3. fires.

The main consequences of these hazards, and the efficiency of possible mitigating measures, are assumed to be roughly as follows.

1. Large explosions. Two levels of large explosions have been distinguished:
   (a) 'Very large' explosions: typically, the explosion of a full cargo of liquefied petroleum gas (LPG) in bulk that is heated by a fire, for example a boiling liquid expanding vapour explosion (BLEVE) followed by a fireball (this is often referred to as a 'hot BLEVE'). Other explosions can also have similar consequences. Such an explosion induces both pressure and thermal effects.
   (b) 'Large' explosions: typically, the explosion of a full cargo of a relatively non-flammable compressed gas that is heated by a fire (i.e. a BLEVE with no fireball, referred to as a 'cold BLEVE'). Such an explosion produces only pressure effects.
   A 'very large' explosion would be expected to kill all the people present in the whole tunnel or in an appreciable length of tunnel and to cause serious damage to the tunnel equipment and possibly its structure. The consequences of a 'large' explosion would be expected to be more limited, especially with regard to damage to the tunnel structure. There is generally no possibility of mitigating the consequences, particularly in the first case. It is, however, important to keep in mind that a BLEVE is not an instantaneous phenomenon, and the potential short delay (of the order of minutes) may allow people to evacuate.
2. Large toxic gas releases. A large release of toxic gas can be caused by a leakage from a tank containing a toxic gas (compressed, liquefied, dissolved) or a volatile toxic liquid. It would be expected to kill all the people near the release and in the zone where ventilation (either natural or mechanical) moves the gas. A part of the tunnel may be protected, but it is not possible to protect the whole tunnel, especially in the first few minutes after the accident. Additionally, ventilation systems are commonly designed for fire, and may not be able to manage a toxic cloud, which can also affect people outside the tunnel (Truchot *et al.*, 2010).
3. Large fires. Depending on the tunnel geometry, traffic and equipment, a large fire will have consequences varying in importance, ranging from a few victims and limited damage to several tens of victims and serious damage to the tunnel.

It is assumed that the order of these hazards, namely explosion, toxic release (gas or volatile toxic liquid) and fire, corresponds to decreasing consequences of an accident and increasing efficiency of the possible mitigating measures. The ADR grouping system is based on this hierarchy. These main characteristics are the skeleton around which the QRAM, developed by the Institut National de l'Environnement industriel et des Risques (INERIS) (overall model: open and

Table 22.1 Groupings of dangerous goods

Grouping	Description	Category
A	All goods, including all dangerous goods authorised on open roads	Least restrictive
B	All goods in grouping A except those that may lead to a very large explosion ('hot BLEVE' or equivalent)	
C	All goods in grouping B except those that may lead to a large explosion ('cold BLEVE' or equivalent) or a large toxic release (toxic gas or volatile toxic liquid)	
D	All goods in grouping C except those that may lead to a large fire	
E	No dangerous goods (except those that require no special marking on the vehicle and some specific goods: 2919, 3291, 3331, 3359, 3373)	Most restrictive

tunnel section) and WS-Atkins (tunnel section), was constructed. The model was developed on the basis of the evaluation of these risks by considering pan-European scenarios.

DESCRIPTION OF THE SYSTEM

From the above assumptions, a system with five groupings was proposed in the ADR (see Table 22.1). In this system, for example, a tunnel authorised to accept grouping A goods would be allowed to admit vehicles carrying the most dangerous goods, whereas tunnels authorised to accept only grouping E goods would not be allowed to admit vehicles carrying any dangerous goods.

Table 22.2 defines the substances and types of load (packages, bulk, tank) to be included in each tunnel group. It is based on the 'restructured ADR', which came into force on 1 July 2009. When a transport unit carries dangerous goods of more than one class, the most restrictive grouping shall apply to the whole load.

CONSISTENCY WITH THE QUANTITATIVE RISK ASSESSMENT MODEL (QRAM) AND THE DECISION SUPPORT MODEL (DSM)

The DSM has been developed by 'Consultancy within Engineering, Environment and Economics' (COWI A/S, Denmark) in order to follow up these important objectives and ensure consistency with the QRAM. It is important to ensure consistency of the grouping system with the QRAM and the DSM.

- The QRAM must incorporate accident scenarios representative of each of the groupings: if the groupings allowed in a tunnel change, the scenarios taken into account must be different so that the risk indicators produced by the QRAM differ and make it possible to discriminate between groupings.
- The DSM must process the results from the QRAM (and other data) so as to provide decisions expressed as the optimal groupings of goods to be allowed in a tunnel. The DSM might be used, for example, by a regulatory body. As mentioned previously, there are some differences between the initially proposed groupings and that prescribed in the ADR. Consequently, some adjustments of the DSM model are required. The DSM is also described briefly by Kroon and Kampmann (2003).

Table 22.2 Groupings of dangerous goods carriages permitted to be transported through each category of road tunnel (Economic Commission for Europe Inland Transport Committee, 2008).

Class	Groupings[a,b]				
	A	B	C	D	E
1	Unrestricted	Restricted for ▪ explosives belonging to compatibility groups A and L ▪ divisions 1.1, 1.2 and 1.5 above 1000 kg maximum permissible net mass of explosives, except compatibility groups A and L	Restricted for: ▪ loading in group B ▪ divisions 1.1, 1.2 and 1.5, except group A and L, division 1.3 compatibility groups H and J ▪ division 1.3 compatibility groups C and G above 5000 kg maximum permissible net mass of explosives	Restricted for: ▪ loading in group C ▪ division 1.3 (compatibility groups C and G)	Restricted for: ▪ goods of this class
2	Unrestricted	Restricted for: ▪ flammable gases (classification codes F, TF and TFC) in tanks	Restricted for: ▪ loadings in group B ▪ classification codes 2A, 2O, 3A and 3O in tank ▪ toxic gases (classification codes T, TC, TO and TOC) in a tank	Restricted for: ▪ loadings in grouping C ▪ flammable and toxic gases (classification codes F, FC, T, TC, TF, TO, TFC and TOC)	Restricted for: ▪ goods of this class
3	Unrestricted	Restricted for: ▪ classification code D (UN 1204, 2059, 3064, 3343, 3357 and 3379)	Restricted for: ▪ loadings in grouping B ▪ PG I for classification code FC, FT1, FT2 and FTC in tanks	Restricted for: ▪ loadings in grouping C ▪ all dangerous goods of class 3 in bulk or in a tank	Restricted for: ▪ goods of this class
4.1	Unrestricted	Restricted for: ▪ classification code D, DT self-reactive substances Type B (UN 3221, 3222, 3231 and 3232)	Restricted for: ▪ loadings in grouping B	Restricted for: ▪ loadings in grouping C ▪ self-reactive substances of types C to F and UN 2956, 3241, 3242 and 3251	Restricted for: ▪ goods of this class

4.2	Unrestricted	Restricted for: - packing group (PG) I in a tank	Restricted for: - loadings in grouping B	Restricted for: - loadings in grouping C - PG II in bulk or in a tank	Restricted for: - goods of this class
4.3	Unrestricted	Restricted for: - PG I in a tank	Restricted for: - loading in grouping B	Restricted for: - loading in grouping C - PG II in bulk or in a tank	Restricted for: - goods of this class
5.1	Unrestricted	Restricted for: - PG I in a tank	Restricted for: - loading in grouping B	Restricted for: - loading in grouping C	Restricted for: - goods of this class
5.2	Unrestricted	Restricted for: - type B (UN 3101, 3102, 3111 and 3112)	Restricted for: - loading in grouping B	Restricted for: - loading in group C - organic peroxides of types C to F	Restricted for: - goods of this class
6.1	Unrestricted	Unrestricted	Restricted for: - PG I in a tank	Restricted for: - loadings in grouping C - PG I for classification codes TF1 and TFC - toxic by inhalation entries (UN 3381 and 3390) - PG II in bulk or in a tank - PG III for classification codes TF2 in bulk or in a tank	Restricted for: - goods of this class
6.2	Unrestricted	Unrestricted	Unrestricted	Unrestricted	Restricted for: - goods of this class
7	Unrestricted	Unrestricted	Restricted for: - UN 2977 and 2978	Restricted for: - loadings in grouping C	Restricted for: - goods of this class
8	Unrestricted	Unrestricted	Restricted for: - PG I of classification code CT1, CFT and COT in a tank	Restricted for: - loadings in grouping C - PG I for classification codes CT1, CFT and COT, - PG I of classification codes CF1, CFT and CW1 in bulk or in a tank - PG II of classification code CF1 and CFT in bulk or in a tank	Restricted for: - goods of this class

Table 2.2 Continued

Class	Groupings[a,b]				
	A	B	C	D	E
9	Unrestricted	Unrestricted	Unrestricted	Restricted for: ■ classification codes M2 and M3	Restricted for: ■ goods of this class

[a] Notes:

(1) The table reads across, from the least restrictive grouping A to the most restrictive grouping E.

(2) UN 2919, 3291, 3331, 3359 and 3373 are authorised in tunnels of class E. For dangerous goods 2919 and 3331, restrictions to the passage through tunnels in category E may, however, be part of a special arrangement approved by competent authorities on the basis of section 1.7.4.2 of the ADR.

[b] Key:

Class   The classification of goods by the type of risk involved has been drawn up to meet technical conditions while at the same time minimising interference with existing regulations. The definitions of the class in the ADR indicate which goods are dangerous and in which class, according to their specific characteristics, they should be included.

Class 1   The explosives class is divided into six divisions (1.1 to 1.6), depending on the type of hazard they present, and these goods are assigned to one of 13 compatibility groups (A to H, J, K, L, N, S) that identify the type of explosives substances and articles that are deemed to be compatible.

Class 2   Substances and articles are assigned to one of the following groups according to their hazardous properties:

A: asphyxiant
O: oxidising
F: flammable
T: toxic
TF: toxic, flammable
TC: toxic, corrosive
TO: toxic, oxidising
TFC: toxic, flammable, corrosive
TOC: toxic, oxidising, corrosive.

Class 3   Flammable liquid and liquid desensitised explosives:
Classification code D: 'liquid desensitised explosives'
Flammable liquids shall be assigned to one of the following packing groups according to the degree of danger they present for carriage.

Class 4.1   Flammable solids, self-reactive substances and solid desensitised explosives:
D: solid desensitised explosives without subsidiary risk
DT: solid desensitised explosives, toxic
Self-reactives.

Class 4.2   Substances liable to spontaneous combustion:

Substances liable to spontaneous combustion (pyrophoric) are assigned to packing group I

Self-heating substances and articles in which spontaneous combustion or a rise in temperature is observed are assigned to packing group II or III on the basis of test procedures of the *Manual of Tests and Criteria*.

Class 4.3   Substances which, in contact with water, emit flammable gases:

Packing group I: substances presenting high danger

Packing group II: substances presenting medium danger

Packing group III: substances presenting low danger.

Class 5.1   Oxidising substances:

Packing group I: substances presenting high danger

Packing group II: substances presenting medium danger

Packing group III: substances presenting low danger.

Class 5.2   Organic peroxides.

Class 6.1   Toxic substances:

Packing group I: highly toxic substances

Packing group II: toxic substances

Packing group III: slightly toxic substances

TW: toxic substances that, in contact with water, emit flammable gases:

     TW1: liquid

TF: toxic substances, flammable:

     TF1: liquid

     TF2: liquid, used as pesticides

     TFC: toxic substances, flammable, corrosive.

Class 6.2   Infectious substances.

Class 7   Radioactive material.

Class 8   Corrosive substances:

CF: corrosive substances, flammable:

     CF1: liquid

CW: corrosive substances which, in contact with water, emit flammable gases:

     CW1: liquid

CT: corrosive substances, toxic:

     CT1: liquid

     CFT: corrosive substances, flammable, liquid, toxic.

Class 9   Miscellaneous dangerous substances and articles:

M2: substances and apparatus which, in the event of fire, may form dioxins

M3: substances evolving flammable vapour

M10: elevated temperature substances, solid.

Table 22.3 Goods representative of each grouping in the QRAM

Grouping of goods	Representative goods for the QRAM	Restrictions
A	LPG in bulk and in cylinders, carbon dioxide in bulk, ammonia/chlorine[a] in bulk, acrolein in bulk and cylinders, motor spirit in bulk, HGV without dangerous goods	Least restrictive
B	Carbon dioxide in bulk, ammonia/chlorine[a] in bulk, acrolein in bulk and cylinders, motor spirit in bulk, LPG in cylinders, HGV without dangerous goods	
C	Motor spirit in bulk, LPG in cylinders, acrolein in cylinders, HGV without dangerous goods	
D	LPG in cylinders, acrolein in cylinders, HGV without dangerous goods	
E	HGV without dangerous goods	Most restrictive

[a] Chlorine is considered in countries where its transport is allowed in appreciable quantities on roads.
HGV, heavy goods vehicle.

Table 22.3 lists the goods chosen for the development of the OECD/PIARC QRAM. Because of the involvement of some individuals in both regulation and QRA construction, this model was built in accordance with the 2009 version of the ADR despite differences between the proposed and prescribed groupings. The model is thus representative of the five groupings described above. The OECD/PIARC DSM considers the various groupings that may be allowed in the tunnel.

22.2.2.4 Conclusions concerning the grouping system for dangerous goods

A harmonised regulation for dangerous goods transport in tunnels makes its application simpler and, consequently, should help to reduce the risk of mistakes that may lead to the transit of dangerous goods through tunnels from which they are banned. The new regulation provides a list of five groupings of dangerous goods that can be allowed or banned in a tunnel. The decision of 'authorised' or 'banned' with respect to a grouping in a given tunnel is left to the authority in charge of the tunnel, and depends not only on the tunnel characteristics but also on the risks associated with an alternative route.

The groupings are based on the different types of hazards than can affect people in tunnels that lead to three different risks. They are consistent with the QRAM and the DSM developed by the OECD and PIARC, so that these models can be used to help make decisions on the groupings to be allowed in a given tunnel.

It is an essential element of the system that it be transparent and widely recognised. To this end, each road tunnel needs to be categorised on the basis of the grouping of goods permitted through the tunnel. In addition to regulatory notices, toll bylaws, etc., it is recommended that each tunnel should be signed, to indicate which grouping is permitted to be transported through the tunnel. New signalisation has been introduced to indicate to drivers the restrictions on dangerous goods transit.

Categorising road tunnels according to this grouping system should help to ensure a greater level of compliance. Not only will tunnel operators and hauliers be able to utilise a simple and straight-forward regime that is accessible and easily understood but enforcement bodies will also be able to carry out random checks on the approaches to tunnels without having to familiarise themselves with complex international agreements or tunnel bylaws.

22.3. Roads tunnel hazard quantification

22.3.1 The quantitative risk assessment model

Authorising or banning dangerous goods from a tunnel must be based on a comparative study between two routes, with and without the tunnel. It is not legitimate to ban dangerous goods from a tunnel if the hazard is greater using the alternative way. A risk analysis may be carried out to assist with making decisions on tunnel requirements regarding dangerous goods. The following is based on the QRAM manual (Cassini *et al.*, 2003a, b), and summarises the main characteristics of this model. It was largely developed between 1996 and 2000. Recent develop-ments primarily concern compatibility with new computer operating system versions and aim to make the software more user-friendly.

The transport of dangerous goods through tunnels implies special risks to road users, physical structures, the environment and people residing near tunnels or detour roads. Transport autho-rities have to decide whether or not dangerous goods transport is permitted on certain routes. If permitted, the safest and most practical manner for these dangerous goods has to be decided. A QRA model can assist decision-makers by providing risk estimates for different types of dangerous goods, tunnels and transport scenarios.

22.3.1.1 Problem description

Risk is characterised by two aspects

1. probability
2. consequences.

Quantification of risk is difficult due to the fact that probabilities for traffic accidents are low and, consequently, the probability of an accident involving dangerous goods is even lower. However, the consequences of such an accident can be dramatic. Numerous factors and variables influence the probability and consequences of accidents involving dangerous goods both inside and outside tunnels. Even with expert knowledge, it is difficult to assess risk for all circumstances, environ-ments, weather conditions, etc. Computer simulations are a tool that can help to develop a rational approach to the problem.

QRA models have been used for many years to estimate the risk of dangerous goods transport for different transport conditions on the open road. Some OECD member countries (Norway and, to a certain extent, France) also developed QRA models for road tunnels. There was a need, however, for a comprehensive model to deal with both tunnels and the open road. Due to the complexity of such a model, this was best carried out through international cooperation. The result is a unique tool that can be used in all countries: the QRAM.

22.3.1.2 Purpose

The purpose of the QRAM is to quantify the risks due to the transport of dangerous goods on given routes of the road system. A comparison of one route including a tunnel with an alternative route in the open can be made. The QRAM was developed based on the following components

- indicators
- accident scenarios
- evaluation of accident probability
- determination of physiological consequences, and structural and environmental damage
- evaluation of consequences (for open-road and tunnel sections)
- uncertainty/sensitivity analysis
- testing by comparison with empirical data.

The QRAM methodology is as follows

- choose a small number of representative goods
- select a small number of representative accident scenarios involving these goods
- determine the physical effects of these scenarios (for open-road and tunnel sections)
- determine the physiological effects of these scenarios on road users and the local population (fatalities and injuries)
- take into account the probability of escape and sheltering
- determine the associated probability of occurrence.

22.3.1.3 Indicators

The consequences of an accident are fatalities, injuries, destruction of buildings and structures, and damage to the environment. As in every modelling process, simplifying indicators are necessary to describe the effects of the behaviour of the system. The following indicators are considered as the output of the QRAM

- societal risk
- individual risk
- estimation of structural damage
- estimation of environmental damage.

22.3.1.4 Societal risk

A common way to describe societal risk is to calculate F–N (frequency–number of victims) curves. F–N curves illustrate the relationship between accident frequency and accident severity. On the abscissa, the number of victims (fatalities in the example in Figure 22.1, but the model may also be used to provide information about injuries), N, is shown on a logarithmic scale. On the ordinate, the corresponding yearly frequencies $F(N)$ for the occurrence of accidents with N or more victims are plotted (also on a logarithmic scale). For each given situation (population, traffic, dangerous goods traffic, route, weather, etc.) one F–N curve represents the 'societal risk'. As an illustration, F–N curves computed for different scenarios in the Austrian Tauern Tunnel are shown in Figure 22.1.

22.3.1.5 Individual risk

Here, the individual risk indicator has been taken to refer to the risk of fatalities or injuries to the local population due to an incident. (See also Chapter 30, where it is shown that the term 'individual risk' may be used in different ways.) Individual risk is expressed as a frequency per year. It could also be expressed in terms of return time – that is, the average number of years between two accidents with the considered consequence (fatality, injury). The QRAM calculates

Figure 22.1 Example *F–N* curves for different scenarios: test case ammonia and chlorine releases. Frequency (*F*) indicates the yearly frequency of accidents with *N* or more fatalities. (Reproduced with permission from the QRAM reference manual, Table 4.6-1 (Cassini *et al.*, 2003a))

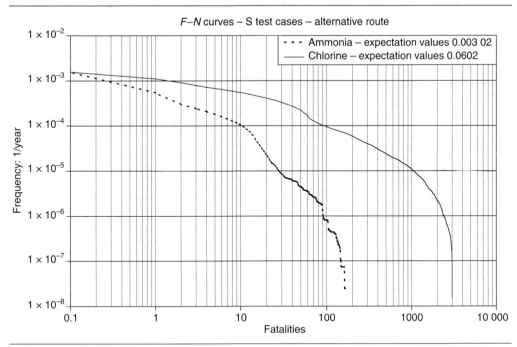

the spatial allocation of risk. Two-dimensional maps containing the individual risk for the surroundings of the analysed route can be drawn, as shown in Figure 22.2. In the example shown, the alternative route for dangerous goods transport avoiding the Kaisermühlen Tunnel (on the A22 in Vienna, parallel to the River Danube) involves crossing the River Danube by bridge, driving through a populated area and crossing the river again by a second bridge. The individual risk can be calculated for the residents or the workday population.

22.3.1.6 Structural and environmental damage
In addition, the QRAM calculates rough estimates of structural and environmental damage.

22.3.1.7 Accident scenarios
A complete assessment of the risks involved in transporting dangerous goods would require the consideration of all kinds of dangerous materials, or, more precisely, all the groupings in accordance with the ADR, all possible meteorological conditions, all possible accidents, sizes of breaches, vehicles fully or partially loaded, and many other variables. Since all circumstances are impossible to consider, simplifications have to be made.

To deal with this problem, the following aspects of the model development have to be taken into consideration

- choice of accident scenarios to be studied
- evaluation of accident probability

Figure 22.2 Two-dimensional map of individual risk: Kaisermühlen Tunnel (Vienna) case study (Reproduced with permission from the QRAM reference manual, Cassini *et al.*, 2003a)

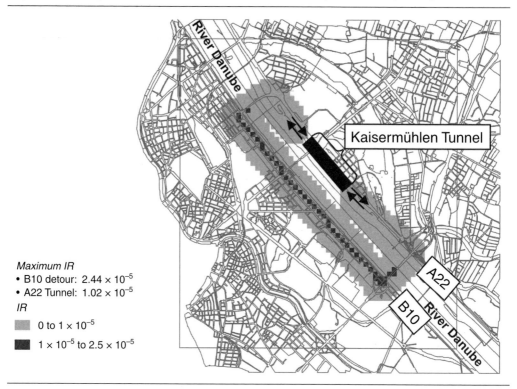

Maximum IR
- B10 detour: 2.44×10^{-5}
- A22 Tunnel: 1.02×10^{-5}

IR

0 to 1×10^{-5}

1×10^{-5} to 2.5×10^{-5}

- determination of physiological consequences, and structural and environmental damage
- evaluation of consequences (open-road and tunnel sections)
- uncertainty/sensitivity analysis.

22.3.1.8 Choice of accident scenarios to be studied

The following aims at selecting relevant scenarios for QRAM purposes that are defined with the help of a previous selection of a few dangerous goods. One criterion that can help one to select or ignore a scenario is the following: the probability–consequence product should not be negligible in either the open road or the tunnel, or both. The scenarios listed in Table 22.4 have been chosen to represent the variety of hazardous goods transported.

As shown in Table 22.4, only a limited number of scenarios are taken into account. Two scenarios relate to fires of medium and important intensity involving HGVs without dangerous goods. These scenarios represent a serious risk in tunnels. The other scenarios involve dangerous goods. The scenarios have been selected to represent various groupings of dangerous goods (see Section 22.2.2) and were chosen to examine different severe effects: overpressure, thermal effects and toxicity. Each scenario is based on a different event tree.

The following shows examples for two scenarios. The first is a HGV fire without dangerous goods while the second is the BLEVE of an LPG cylinder.

Table 22.4 Main characteristics of the 13 selected scenarios (QRAM 3.60)

Scenario No.	Description	Tank capacity	Size of breach: mm	Mass flow rate: kg/s
1	HGV fire 20 MW	–	–	–
2	HGV fire 100 MW	–	–	–
3	BLEVE of LPG in cylinder	50 kg	–	–
4	Motor spirit pool fire	28 tonnes	100	20.6
5	VCE of motor spirit	28 tonnes	100	20.6
6	Chlorine release	20 tonnes	50	45
7	BLEVE of LPG in bulk	18 tonnes	–	–
8	VCE of LPG in bulk	18 tonnes	50	36
9	Torch fire of LPG in bulk	18 tonnes	50	36
10	Ammonia release	20 tonnes	50	36
11	Acrolein in bulk release	25 tonnes	100	24.8
12	Acrolein in cylinder release	100 litres	4	0.02
13	BLEVE of liquefied CO_2	20 tonnes	–	–

Reproduced with permission from the QRAM reference manual, Table 4.9-1 (Cassini *et al.*, 2003a).
VCE, vapour cloud explosion.

HEAVY GOODS VEHICLE WITHOUT DANGEROUS GOODS

A scenario that must be considered, especially in a tunnel, is fire. Then, the questions to be answered are the following:

- What kind of fire can develop, considering the flammability and the fire load of the HGV and its goods?
- What are the associated probabilities of occurrences?

Concerning these probabilities, an event tree can be drawn as illustrated in Figure 22.3. As shown in the event tree, the associated probability of the HGV fire is:

$$P_{fire} = P1 + P2 = (P1.1 \times P1.2) + (P2.1 \times P2.2)$$

(This assumes the 'OR' gate to be mutually exclusive.)

Figure 22.3 Event tree for a HGV fire. *Ignition in an accident may start on the HGV itself, or on a nearby damaged vehicle (Reproduced with permission from the QRAM reference manual, Section 4.1 (Cassini *et al.*, 2003a))

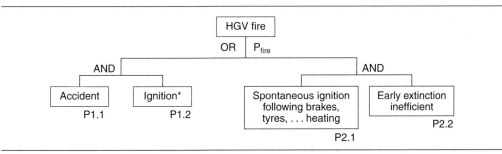

Considering the development of fire involving vehicles loaded with 'non-dangerous' goods, one may have two situations.

1. A partially unloaded HGV (or a HGV containing relatively non-flammable materials) that would lead to a fire of approximately 20 MW (scenario 1).
2. A HGV loaded with materials that are more flammable than in scenario 1, which could lead to a 100 MW fire (scenario 2).

(Note: a HGV fire in a tunnel could lead to a heat release rate much greater than 100 MW – see Chapter 14.)

LPG IN A CYLINDER (50 KG)

A related but more relevant scenario involving these types of dangerous goods and conditioning and leading to important consequences is a BLEVE. The calculation for its probability of occurrence can be made according to the event tree illustrated in Figure 22.4.

The formulae establishing probabilities of occurrence of this phenomenon for a cylinder are:

$$P_{bleve} = P11 \times P12$$

$$P11 = P8 + P9 + P10$$

Figure 22.4 Event tree for a BLEVE of a cylinder (Reproduced with permission from the QRAM reference manual, section 4.2.1 (Cassini *et al.*, 2003a))

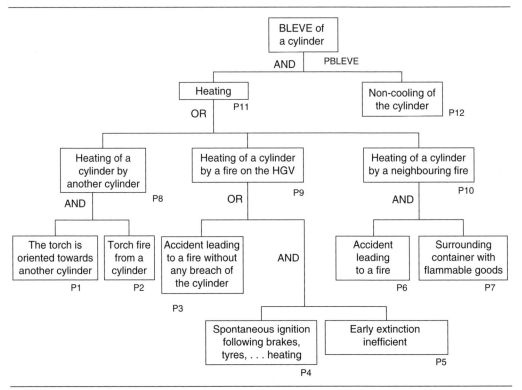

P8 = P1 × P2

P9 = P3 + (P4 × P5)

P10 = P6 × P7

Thus,

P11 = (P1 × P2) + (P3 + P4 × P5) + (P6 × P7)

and

P_{bleve} = P12 × [(P1 × P2) + (P3 + (P4 × P5)) + (P6 × P7)]

(once again, this assumes the 'OR' gates to be mutually exclusive).

It is assumed that a torch from a cylinder cannot generate a BLEVE from the same cylinder. The reasons for this are

- the internal pressure drop due to torch firing will partially compensate for the internal pressure rise due to heating
- the total duration of the torch fire will probably not be sufficient to create the internal temperature and pressure conditions required to generate a BLEVE.

Additional questions related to this scenario (scenario 3 of the QRA: see Table 22.4) are

- How many cylinders will create a BLEVE?
- Will the BLEVEs be simultaneous or will a sequence of successive BLEVEs occur?

First, it should be pointed out that a BLEVE of at least one cylinder is likely in the case of a fire surrounding the dangerous goods.

Additional physical considerations lead to the assumption that simultaneous or long sequences of successive BLEVEs are very unlikely: a BLEVE of one cylinder will probably push the neighbouring cylinders in the vicinity away – in some instances, a significant distance from the fire.

Regarding the effects of a BLEVE, two parameters are of major importance

- the mass involved in the phenomenon
- the duration of the fireball, which has a direct influence on the time of exposure of the people in the vicinity.

Assumptions made concerning these two parameters are

- the mass of LPG to consider relative to scenario 3 is the whole mass of one single cylinder (simultaneous BLEVEs are very unlikely)
- the total duration of the fireball cannot exceed the duration of two or three fireballs that would occur successively (a long sequence of successive BLEVEs is very unlikely).

For scenario 3, two calculations have been performed

- for the BLEVE of one cylinder containing 50 kg of propane
- for the successive BLEVEs of three cylinders.

Table 22.5 Conservative approach for successive BLEVEs. ε is a small distance compared with the radius R of the fireball

Scenario	Single BLEVE	Three successive BLEVEs
Mass involved: kg	50	3×50
Total duration of the phenomenon: s	2.4	7.2
Radius R of the fireball: m	11	11
Lethality 100%	R	R
Lethality 50%	$R + \varepsilon$	$R + \varepsilon$
Lethality 1%	$R + 2\varepsilon$	$R + 4.3$ m

Reproduced with permission from the QRAM reference manual, Section 4.2.1 (Cassini *et al.*, 2003a).

The results are summarised in Table 22.5. In the case of three successive BLEVEs, a conservative approach was adopted. The results of the two calculations given in the table illustrate that the effects of one BLEVE and three successive BLEVEs of cylinders are of the same order of magnitude. That is why, in a simple approach, one may consider, as scenario 3, the BLEVE of a single cylinder.

Finally, the risk assessment implies the evaluation of a number of fatalities and injuries that a BLEVE may produce. Figure 22.5 shows the fatalities/injuries taken into account in the model, depending on the distance of people from the centre of the BLEVE.

Figure 22.5 shows that it is difficult to build *F–N* curves for injuries only. Thus, it was decided to build *F–N* curves for fatalities only, and for fatalities plus injuries, and then to deduce expected values for injuries only by calculating the difference between the expected values for fatalities plus injuries and the expected values for fatalities only.

Figure 22.5 Example of the probability of fatalities and injuries plotted against distance for the BLEVE of a single cylinder (Reproduced with permission from the QRAM reference manual, Figure 6.2.1 (Cassini *et al.*, 2003a))

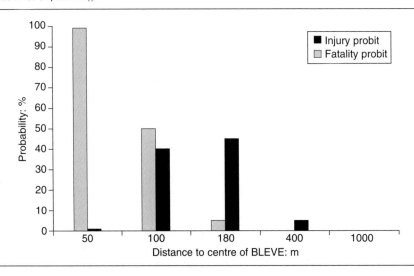

22.3.2   Risk reduction measures for road tunnels

A list of risk reduction measures classified according to the main purposes has been developed by the Scientific Expert Group (established by the OECD and PIARC); these are discussed below.

22.3.2.1 Objectives and contents

When vehicles carrying dangerous goods are allowed into a tunnel, a number of measures can be employed to reduce the probability and mitigate the consequences of an accident. The circumstances that led to this project were as follows

■ no systematic description of the measures was readily available
■ many measures are costly during construction or operation, or both, while their effectiveness is usually not well known
■ it is very difficult to decide whether, and in which case, each measure should be implemented.

For these reasons, a significant part of this project was devoted to risk reduction measures, with the following objectives

■ to review all possible risk reduction measures, make a detailed description, and analyse their advantages and disadvantages
■ to objectively analyse the effectiveness of the measures and thus give the basis for an assessment of their cost-effectiveness, taking advantage of the QRAM.

No systematic cost-effectiveness analysis has ever been reported for measures to reduce the risks of dangerous goods transport in road tunnels. The main reason is that the effects of measures on risks are very difficult to assess. To develop a methodology for such an assessment has been a major issue for the project. It is not worth developing tools to estimate the cost of measures since reasonably accurate cost estimations can be performed by specialised consultants for tunnel construction or refurbishment.

22.3.2.2 Review of the risk reduction measures

The review of risk reduction measures included a literature survey, a questionnaire sent to several tunnel operators and a tentative ranking of the measures. In addition, complementary information provided by PIARC was available on a number of measures related to fire and smoke control.

22.3.2.3 Identification of the risk reduction measures

Measures restricting the transport of dangerous goods (such as prohibition, limitation of quantities transported, or restriction of transit times) are not considered here. These measures require the consideration of alternative routes as well as the tunnel. The decisions on these measures require the use of the QRA and decision support models.

Box 22.1 lists the 27 measures that have been examined. Because of the complexity in dealing with such measures, the measures that were taken into account are only those given by the OECD as requirement for the QRA model. As many measures have several functions, the classification is somewhat arbitrary, and is based on their main function. It takes into account the fact that most fatalities generally occur before the emergency services arrive. Some measures appear a second time in italics, to show a second important purpose. The measures to ensure

Box 22.1 List of risk reduction measures classified according to their main purpose

Measures to reduce the probability of an accident
- Tunnel cross-section
- Speed limit
- Road surface (friction)
- Alignment
- Prohibition to overtake
- Escort
- Lighting (normal)
- Distance between vehicles
- Truck checks

Measures to ensure communication and/or information
(*a*) Alarm, information, communication of operator and rescue services
- Closed-circuit television
- Automatic fire detection
- Automatic vehicle identification
- Automatic incident detection
- Radio communication (services)
- *Emergency telephone*

(*b*) Communication with users
- Emergency telephones
- Alarm signs/signals
- Loudspeakers
- *Radio communication (users)*

Other measures to mitigate the consequences of an accident
(*a*) Evacuation or protection of users
- Emergency exits
- *Lighting (emergency)*
- Failure management
- Smoke control
- Fire-resistant equipment

(*b*) Reduction of accident importance
- Fire-fighting equipment
- Drainage
- Action plan
- Rescue teams
- *Road surface (non-porous)*
- *Escort*

(*c*) Reduction of the consequences on the tunnel
- Fire-resistant structure
- Explosion-resistant structure

communication and/or information mainly aim at reducing the consequences of an accident, but can also have an effect on its probability by informing the users when a first incident has occurred. Since the situation varies considerably amongst tunnels, it is necessary to develop the measures for each situation.

Since the publication of the QRAM, there have been new developments in tunnel safety. One of the most important concerns water mist use in tunnels. However, there is not yet an international consensus on this topic, which is, today one of the most important mitigation measures in tunnels.

22.3.3 EU member states' experiences of the Qualitative Risk Assessment Model

22.3.3.1 The Austrian experience

A tunnel safety board was set up by the Austrian Federal Ministry of Transport, Innovation and Technology after the fatal accident in the Gleinalm Tunnel (Austria, 7 August 2001). QRAM results were presented and discussed by this expert committee. In May 2002, the committee agreed on threshold values for the QRA. The committee also suggested that QRA be made obligatory for all tunnels and that the threshold values be incorporated into the Austrian tunnel regulation framework.

On behalf of the Austrian Federal Ministry of Transport, Innovation and Technology, 13 Austrian tunnels have been analysed using the QRAM (Figure 22.6 and Table 22.6). The results have been discussed with experts and practitioners from the fields of tunnel ventilation and tunnel traffic management. In recent years, new ventilation systems and emergency ventilation regimes have been installed in several Austrian tunnels. Currently, the capability of the QRAM to include these innovations is considered to be limited. As a result, the expert group suggested that the representation of the ventilation system be improved in the QRAM. Additionally, several controlled fire tests have been carried out in existing Austrian tunnels. It is intended that the data gathered from these experiments will be used in the ongoing development of the model.

Figure 22.6 Austrian tunnels analysed using the QRAM (Cumulated frequency means frequency of N or more fatalities)

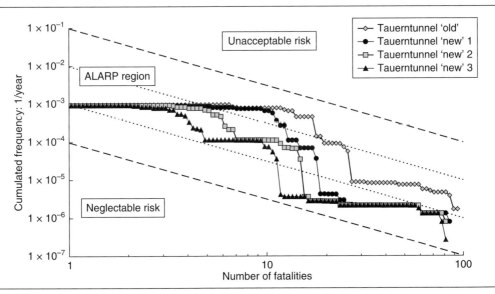

Table 22.6 Austrian tunnels analysed using the QRAM

Name	Length: km	Ventilation	Average daily traffic	HGV share: %
Amberg Tunnel	2.978	Semi-transverse	2 221	8.58
Bosruck Tunnel	5.500	Transverse	7 365	38.50
City Tunnel, Bregenz	1.311	Longitudinal plus extraction	12 911	3.69
Gleinalm Tunnel	8.320	Transverse	14 068	10.70
Gräbern Tunnel	2.144	Semi-transverse	16 505	18.93
Herzogberg Tunnel	2.007	Semi-transverse	16 118	19.80
Kaisermühlen Tunnels	2.020	Longitudinal	84 644	10.55
Karawanken Tunnel	7.864	Longitudinal and transverse	5 106	10.41
Lainberg Tunnel	2.278	Longitudinal	9 435	20.20
Plabutsch Tunnel	9.919	Transverse	20 681	15.40
Schönberg Tunnel	2.988	Longitudinal	8 448	12.94
Tanzenberg Tunnels	2.384/2.476	Semi-transverse/longitudinal	21 479	11.19
Tauern Tunnel	4.601	Transverse	13 200	21.00

From 2001 to 2003, the QRA of a road tunnel chain on a proposed new highway was carried out on behalf of the Viennese Municipal Department 'Urban Development and Planning'. One of the difficulties in this study was the calculation of the risk for a tunnel with a changing cross-sectional aspect. The QRAM available at the time of the study was unable to do this.

However, the QRAM was found to be very useful in testing different regulations for the transport of dangerous goods through Austrian road tunnels. This was found to be of great value when comparing the requirements of the EU Commission with those of the Austrian national standards. The differences could be solved by using the QRAM method since it is accepted by both partners.

Several improvements in tunnel equipment have been made in recent years, the database has been extended and more parameters are now available. This new knowledge should be used to improve the existing QRAM. It will be important to collect and coordinate the different experiences with the model at national and international levels after a few years. There are also differences in the threshold values between road and rail tunnels which might be an issue for future research and discussion.

22.3.3.2 The French experience
USE OF THE OECD/PIARC QUALITATIVE RISK ASSESSMENT MODEL WITHIN THE FRENCH REGULATORY FRAMEWORK

Following the Mont Blanc Tunnel fire accident (39 fatalities), the French regulations covering road tunnel safety issues were changed, and currently rely on two new Circulars. One of them, Circular 2000-82 of November 2000, sets up a new procedure concerning the definition of the restrictions for the transit of dangerous goods through road tunnels. This Circular was abrogated, but the heart of the document, i.e. the technical instruction in the appendix was kept, and is still applicable. The main requirement is for an evaluation of the risk levels obtained for a given set of alternative routes existing around a tunnel. The evaluation implies the choice of the least risky route on the basis of a QRA. In the case of an obviously low-risk tunnel, the risk level assessment process can be based on a qualitative engineering judgement. Where a full QRA procedure is needed, the recommendation was made to use QRAM 3.2 or other comparable model that has been through a similar amount of testing. QRA is to be considered for all road tunnels with a length of more than 300 m that are covered by the regulation.

After 10 years, this new legal procedure has led to the evaluation of the quantified risk level of more than 20 tunnel routes and their alternative routes. This procedure has been shown to give a good level of applicability in the overall decision process of the risk management associated with dangerous goods transport on roads.

These studies have been performed by four different consultants who had attended a preliminary 1-day basic training course (Cassini *et al.*, 2003b). The experience gained during this training period and during the performance of the risk studies allowed the consultants to familiarise themselves with the capabilities and limits of the model, and also provided an indication of how the results have to be interpreted or presented.

The French experience was also used to suggest a slightly simplified legal procedure by the end of 2003, and suggestions have been accepted by the Comité d'Evaluation de la Sécurité des Tunnels Routiers (CESTR, Committee for Road Tunnel Safety Assessment). The main modification is a simplified evaluation process for the tunnel risk level and the implementation of a rough comparison with predefined risk level criteria. This first-step comparison routine allows one to decide whether or not it is necessary to perform a complete comparative and quantitative risk study.

Thus, the 10 years' experience acquired in France regarding the practical application of the OECD/PIARC QRAM has confirmed its efficiency as a tool for specifying the restrictions concerning the transit of dangerous goods through road tunnels. At present, it is intended that the QRAM 3.60 software be distributed by PIARC at the international level and the training of users be organised by PIARC, with the help of the model developers and skilled bodies.

A NEW RESEARCH DEVELOPMENT FRAMEWORK ON THE BASIS OF THE ORIGINAL OECD/PIARC QRAM

In parallel with the application of the QRAM within the French regulatory framework, INERIS (the leading developer of the original model) launched a new research project in order to facilitate the generation of data necessary for a given risk study. A new version of the model is in development. These new developments have been funded by the French Ministry of Ecology and Sustainable Growth, the French Ministry of Equipment, Transport and Buildings, and the French Ministry of Tourism and the Sea.

On the basis of the original OECD/PIARC model, INERIS has followed up the development process in order to enhance the user interface performance of the original model and to extend its capacities to the study of longer routes. In order to reach these goals, a two-step development process was decided on

1. to enhance the user interface performance of the original model
2. to extend the capacity of the model to study long routes by collecting all the needed road characteristics and population data with the help of a GIS.

The main objective of the basic user interface development was to separate the data collection from the calculation phase. By doing this, it is now possible to reset the data collection process,

Figure 22.7 Structure of the QRAM with an enhanced user interface (Adapted with permission from the QRAM reference manual (Cassini *et al.*, 2003a))

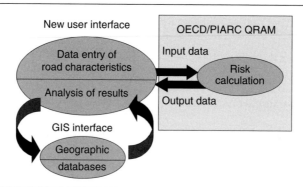

modify the data set and launch the calculation when all the elements of the data set have been collected. Figure 22.7 shows the new structure of the user interface. This new interface allows the user to manage the various options tested for a given route more easily. For example, modifying 'section characteristics' or 'traffic vehicle hypotheses' is now easier. Moreover, the optimised data collection procedure allows the developers to work on data collection without changing the calculation process itself, which had been tested during the OECD/PIARC project. This development was also necessary for the GIS user interface implementation and development.

As regards the GIS user interface, the main objective was to reduce the time spent by users for the data collection process, especially in the case of long routes. This was considered to be a strategic goal, because during a risk study it is essential to spend as much time as possible on the other phases, namely traffic study, the risk analysis phase and the evaluation of options required by the decision-making process regarding potential dangerous goods transit restrictions. Thus, the time saved on data collection can lead to better risk analysis and better assessments of the possible restrictions. Figure 22.8 shows an example of a test case located to the north of the French Alps.

The two tested routes (for GIS development testing purposes only) were from Bourgoin-Jallieu to Chambery, (1) via the Dullin and the Epine Tunnels and (2) via Grenoble. The test focused on the long route via Grenoble (122 km), illustrating the efficiency and quality of the new GIS interface. The quality of the data collection using GIS has been tested and compared with the manual process (the original OECD/PIARC model), leading to the conclusion that the GIS interface provides

- a tremendous time saving (typically a 1:100 ratio) for road and local population data gathering for long routes
- the capability to evaluate the risk level for a long route and of detecting the more risky location.

The new possibilities offered by the GIS interface might also help users to compare long road and rail routes (rail alternatives implying generally longer routes).

An example of *F–N* curves produced by the original OECD/PIARC model coupled to the GIS user interface is given in Figure 22.9.

Figure 22.8 An example of the GIS-interfaced QRAM with Bd CARTO – IGN: (a) the selection of a long route around Lyon with Chronovia; (b) development of one- and two-dimensional selection (Cassini *et al.* (2003b))

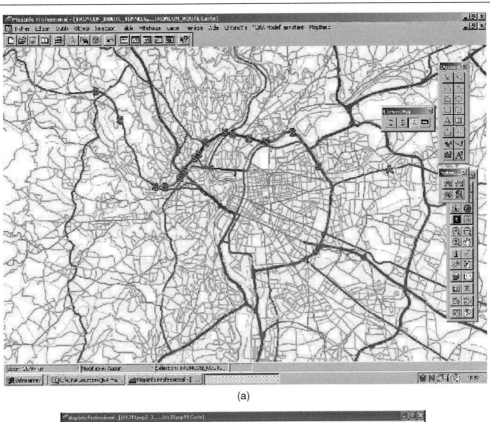

(a)

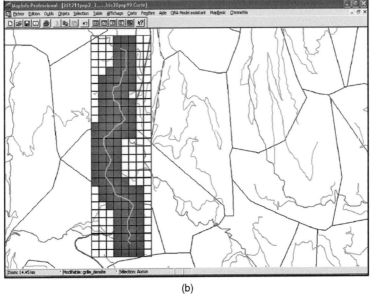

(b)

Figure 22.9 An example of GIS-interfaced QRAM *F–N* curves (Cumulated frequency is the frequency of *N* or more fatalities; EV, expected value for (fatalities + injuries)/year)

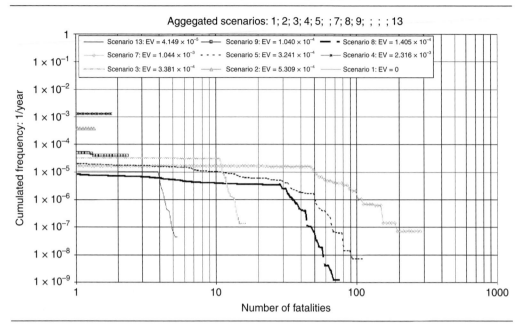

Theoretically, it is possible to gain from the French experience for future practice in all European countries, but the *sine qua non* is the existence of properly harmonised GIS databases. Such databases do not exist at present, but this may change as a result of ongoing projects in this area.

To conclude, the OECD/PIARC QRAM can now operate with a GIS interface (limited at present to French databases), and adequate risk assessment pertaining to transport problems on long road routes is now achievable. In particular, risk studies, such as QRAs, related to the transport of dangerous goods at a regional scale, with or without the involvement of tunnels, are now feasible.

22.4. Rail transport and road/rail inter-modality
22.4.1 Rail transport of hazardous goods in the EU

The general regulation of the rail transport of dangerous goods is based on EU Directive 96/35/EC of 3 June 1996 (EC, 1996a) on the appointment and vocational qualification of safety advisers for the transport of dangerous goods by road, rail and inland waterways and by Directive 96/49/EC of 23 July 1996 (EC, 1996b) and its amendments, on the approximation of the laws of the EU member states with regard to the transport of dangerous goods by rail. The latter directive leads to the application of the Regulations Concerning the International Transport of Dangerous Goods by Rail (RID) within and across the member states (Article 1). Thus, the RID constitutes the annex of this Directive, and the main recommendations concern the following parts

- Part 1 – General requirements
- Part 2 – Special requirements relating to the various classes
- Part 3 – Appendices (miscellaneous requirements about goods, recipients, tests, etc.).

In these two Directives, there are no general recommendations about the transit of dangerous goods through tunnels.

Such regulations/recommendations are addressed in the following documents.

- Commission Decision 2002/732/EC of 30 May 2002 (EC, 2002) on the Technical Specification for Interoperability (TSI) relating to the infrastructure subsystem of the trans-European high-speed rail system referred to in Article 6(1) of Council Directive 96/48/EC (EC, 1996c). In fact, this TSI is not specific to dangerous goods but refers to fire events. The specifications mainly concern the limitation of the initiation, propagation and the effects of fire and smoke in the event of fire.
- UNECE (draft recommendations 30 July 2003), *Report of the Ad Hoc Multidisciplinary Group of Experts on Safety in Tunnels (Rail) on its Fourth Session* (UN Economic Commission for Europe, Inland Transport Committee, 2003).
- UIC leaflet 779-9, *Safety in Railway Tunnels* (UIC, 2003), which has been taken into account for the elaboration of the UNECE document.
- CEN prEN 45545 (CEN, 2010), with regard to the fire resistance of railway passenger wagons (no specific indications for dangerous goods).

The UNECE and UIC publications appear to be the most specific documents on rail tunnel safety, but they are not enforced in law. In law, TSIs concerning the infrastructure subsystem give the minimum safety requirements for rail tunnels within the scope of the interoperability purpose and only for fire events. It is also important to note that the causes of dangerous goods accidents may depend on railway incidents, and thus Directives 96/48/EC (EC, 1996c) 2001/16/EC (EC, 2001a) giving TSIs as well as the Project Directive for the safety of European railways setting up the common safety objectives (CSO), methods (CSM) and indicators (CSI) might need to be considered.

All these documents concerning safety in rail tunnels or for the transport of dangerous goods by rail constitute a complex set of references, which are hard to use efficiently and are not sufficiently harmonised with the road mode of transport.

22.4.2 The situation in the professional engineering world for rail transport

Compared with the road mode, the question of dangerous goods transport by rail follows quite a different regulatory scheme, mainly based on prescription.

The main prescriptive regulations relative to the transport of dangerous goods through rail tunnels are included in the following documents.

- UNECE – R C1-10 (for new tunnels only):
  - A tunnel to be designated as high risk, if operating conditions permit, to segregate trains carrying dangerous goods in bulk from passenger trains.
  - The train driver to have information about dangerous goods carried when going through a tunnel, and the operator must be able to pass information on the dangerous goods to fire services.
  - Risk analyses of safety measures involving dangerous goods to be based on cost–benefit considerations of various options; for example, freight trains carrying dangerous goods are diverted on routes without tunnels or diverted on routes without a densely populated area.

- UNECE – S C4-15 (for new tunnels only):
  - The 'operating company should normally provide such information [list of wagons, goods being carried and wagon order] to the network operator before the departure of the train carrying dangerous goods'.

Within the UNECE prescription (draft document), there is no specific recommendation or standard indicated for the carriage of dangerous goods through existing rail tunnels. A global recommendation to raise the safety level is made, if possible, with non-structural measures, in order to limit the cost. The use of QRA approaches and performance objectives appears to be a better way to raise the safety level for existing tunnels.

In the framework of the present regulations, Directive 96/49/EC suggests that for some safety aspects not addressed in the RID, EU member states can propose specific safety measures (Article 1.2). These cover the following areas

- the organisation of the rail transport system
- the order of the wagons in a train for intra-state transport
- the operation of safety rules for the nodes of the rail transport system such as marshalling yards or parking places
- the training of the operators' staff and the management of information related to dangerous goods
- the specific rules concerning the transport of some dangerous goods in passenger trains.

Article 6.11 should also lead EU member states to make use of the QRA approaches for the rail mode in order to define specific safety rules for the regular local transport of dangerous goods from a given industrial site and on a given well-defined route. Thus, the application of this Article also implicitly entails the identification of the less risky route.

Reaching an appropriate balance for the different modes in use (i.e. road/rail) for the transport of goods is also a major objective of the EU, as expressed in a White Paper on transport (EC, 2001b). This new view leads one to consider the railway mode as well as the road mode, in order to choose the less dangerous route for the transport of goods. In some cases, an alternative route can result in a combination of various modes. For this objective, it is important to be able to assess and compare the risks related to hazardous goods transport, even though the perception of 'common' accidental risks is often not related to dangerous goods transport concerns. Nevertheless, reasonable assessment of the risks involving dangerous goods is required in the overall decision-making process for upgrading or building railway infrastructure, taking into account the interactions with the road network. Thus, in the same way as for the risk assessment process pertaining to road transport mode, using the QRA approach could lead to better management of global risks pertaining to dangerous goods as a complement to the RID prescriptive approach.

22.4.3 A new qualitative risk assessment model for rail

Even though the present regulations do not recommend QRA models for the rail mode, some objectives are, to lesser or greater degrees, related to QRA application, and should involve this type of model in the near future. Considering such a possibility, INERIS has developed a model for the rail transport of dangerous goods. This new model has been developed taking into account the same global philosophy on which the road OECD/PIARC model was based. In this way, road and rail models may produce compatible assessments and thus allow the

Figure 22.10 The multimodal platform for dangerous goods risk assessment (adapted from Knoflacher and Pfaffenbichler, 2001). In 'road/rail' transport, the entire truck is transported by rail, whereas in 'piggy-back' transport only the inter-modal transport unit is transported by rail, not the truck

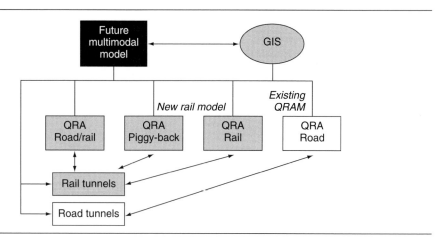

study of multimodal transport systems. This new development work has been funded by the French Ministry of Ecology and Sustainable Growth and by the French Ministry of Equipment, Transport and Buildings. Such models could contribute in the future to refining the distribution in transport modes, and could be used to promote new recommendations for the mitigation of the risks from dangerous goods at the local or European level.

The long-term principle of a global assessment of risks on multimodal routes is shown schematically in Figure 22.10.

During the development of the INERIS QRA rail model, two points served as guiding ideas: compatibility with the QRA road model and specificity of rail transport. Risk assessment, level of details, and global approach have to be similar to allow comparison between modes, and therefore the general philosophy of the road model has been retained. This philosophy is based on a choice of a limited number of hazardous goods, a choice of a limited number of scenarios, assessment of probability and assessment of physical and physiological effects. However, capacities engaged, accident development scheme, and traffic characteristics are different for the two modes. These differences have been assessed with the help of a systemic approach to the railway mode. In particular, frequencies of accident scenarios take into account the probability of an incident involving conditional probabilities, to model a major accident involving hazardous goods. The frequencies are also connected to the local properties of the railway lines' environment and to the type of operating system.

The overall risk assessment relies on the establishment of a typical (simplified) scenario tree considering a major rail accident involving dangerous goods (Figure 22.11). The accident scenarios have been set up considering the French rail traffic of dangerous goods and also the scenarios taken into account in the road model. Using a methodology compatible with that of the road case, the proposed scenarios to be assessed in the rail model have been accepted by the steering committee of the project. Table 22.7 compares the road and rail scenarios.

Figure 22.11 Simplified scenario tree governing the new QRA rail model (broadly indicative only)

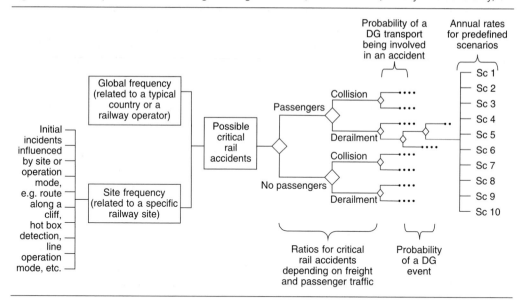

These new developments allow the possibility of making comparative risk assessments among modes for ten typical dangerous goods scenarios in each mode such as torch fire, BLEVE, VCE, pool fire and toxic releases which are the main potential hazards with LPG, motor spirit, chlorine and ammonia transport.

Further research and development is still to be supported at the European level in order to promote harmonised QRA rail models taking into account common safety objectives, commonly accepted methods, indicators and also technical specifications for inter-operability.

Table 22.7 Available scenario for quantitative risk comparison between transport modes

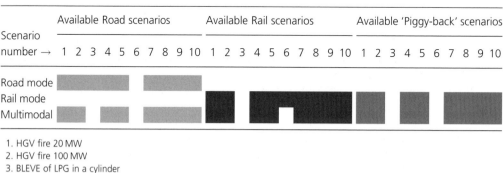

1. HGV fire 20 MW
2. HGV fire 100 MW
3. BLEVE of LPG in a cylinder
4. Motor spirit pool fire
5. VCE of motor spirit
6. Chlorine release
7. BLEVE of LPG in bulk
8. VCE of LPG in bulk
9. Torch fire of LPG in bulk
10. Ammonia release

22.5. Conclusions

The volume of 'hazardous goods' transported through tunnels is growing rapidly. However, there is a lack of knowledge about how to organise this transport in a responsible way and how to find the right balance between economic demands and safety needs. The scientific and research work done so far is the first step on the way to a more rational and responsible treatment of this problem.

Tunnels, as part of the transport system, contribute to the range of options for the transport of people and goods. However, particular problems arise with regard to the transport of 'dangerous goods' through tunnels. Unfortunately, there is no simple or uniform solution, and a risk assessment is necessary for each individual tunnel, based on rational scientific methodology, which is now available from OECD/PIARC research work. This tool has proved efficient during its years of use in Austria and France, but there is still a need to extend this experience to other states.

There are still too many assumptions, which may or may not be justifiable, and not enough sound engineering data. Administrations, tunnel operators and experts from fire brigades and research institutes could help to further the use of and improve the road model. A compatible model for comparisons with rail has been developed, but still needs to be tested at the French and European levels.

REFERENCES

Cassini P, Hall R and Pons P (2003a) *Transport of Dangerous Goods through Road Tunnels. Quantitative Risk Assessment Model (Version 3.60). Reference Manual.* OECD/PIARC/EU, Paris (CDROM).

Cassini P, Hall R and Pons P (2003b) *Transport of Dangerous Goods through Road Tunnels. Quantitative Risk Assessment Model (Version 3.60). User Guide.* OECD/PIARC/EU, Paris (CDROM).

CEN (2010) *CEN prEN 45545 – Railway applications – Fire protection of railway vehicles. Harmonised Standard Complying with an EC Directive: DI 96/48/CE 01/07/1996 and DI 93/38/CE 01/06/1993.* European Committee for Standardization, Brussels.

Council of the European Union – Working Party on Land Transport (2002) Proposal for a Directive of the European Parliament and of the Council on minimum safety requirements for tunnels in the Trans-European Road Network. Inter-institutional file 2002/0309 (COD).

Defert R, Cassini P and Kordek M-A (2001) L'utilisation du modéle EQR Routier. *Formation INERIS*, 3 and 26 October 2001.

EC (1996a) Council Directive 96/35/EC of 3 June 1996 on the appointment and vocational qualification of safety advisers for the transport of dangerous goods by road, rail and inland waterway. *Official Journal of the European Communities* **L 145**: 0010–0015.

EC (1996b) Council Directive 96/49/EC of 23 July 1996 on the approximation of the laws of the Member States with regard to the transport of dangerous goods by rail. *Official Journal of the European Communities* **L 235**: 0025–0030.

EC (1996c) Council Directive 96/48/EC of 23 July 1996 on the interoperability of the trans-European high-speed rail system. *Official Journal of the European Communities* **L 235**: 0006–0024.

EC (2001a) Directive 2001/16/EC of the European Parliament and of the Council of 19 March 2001 on the interoperability of the trans-European conventional rail system. *Official Journal of the European Communities* **L 110**: 0001–0027.

EC (2001b) White Paper submitted by the Commission on 12 September 2001: European Transport Policy for 2010: Time to Decide (COM(2001) 370 final. (Not published in the *Official Journal of the European Communities.*)

EC (2002) Commission Decision of 30 May 2002 concerning the technical specification for interoperability relating to the infrastructure subsystem of the trans-European high-speed rail system referred to in Article 6(1) of Council Directive 96/48/EC (notified under document number C(2002) 1948).

Economic Commission for Europe Inland Transport Committee (2003) Information document No. 15. *Proceedings of the 74th session of Working Party on the Transport of Dangerous Goods,* May 2003. Available at: http://www.unece.org/trans/main/itc/ (accessed 4 July 2011).

Economic Commission for Europe Inland Transport Committee (2008) ADR applicable as from 1 January 2009. European Agreement concerning the International Carriage of Dangerous Goods by Road. United Nations, New York and Geneva 2008. Available at http://www.unece.org/trans/danger/publi/adr/ (accessed 4 July 2011).

EU (1994) Council Directive 94/55/EC of 21 November 1994 on the approximation of the laws of the Member States with regard to the transport of dangerous goods by road. *Official Journal of the European Communities* **L 319**.

EU (1996) Council Directive 96/49/EC of 23 July 1996 on the approximation of the laws of the Member States with regard to the transport of dangerous goods by rail. *Official Journal of the European Communities* **L 235**.

EU (2004) Directive 2004/54/EC of the European Parliament and of the Council of 29 April 2004 on minimum safety requirements for tunnels in the Trans-European Road Network. *Official Journal* **L 167**.

EU (2008) Directive 2008/68/EC of the European Parliament and of the Council of 24 September 2008 on the inland transport of dangerous goods. *Official Journal of the European Union* **L 260**.

Knoflacher H and Pfaffenbichler P (2001) A quantitative risk assessment model for road transport of dangerous goods. *Proceedings of the 80th Annual Meeting of the Transportation Research Board, Washington, DC,* 7–11 January 2001.

Kroon IB and Kampmann J (2003) Decision support for tunnel design and operation. *Proceedings of the 5th International Conference on Safety in Road and Rail Tunnels, Marseille,* 6–10 October 2003, pp. 111–120.

OECD (1992) Strategies for transporting dangerous goods by road: safety and environmental protection. OECD Road Transport Research Programme, Directorate for Science, Technology and Industry, Environment Health and Safety Division, Environment Directorate. *Proceedings of an OECD Meeting, Karlstad,* 2–4 June 1992.

OECD (2001) *Safety in Tunnels – Transport of Dangerous Goods through Road Tunnels.* Organisation for Economic Cooperation and Development, Paris.

OECD/PIARC (1996) Transport of dangerous goods through road tunnels – risk assessment and decision-making process: methodologies, models, tools. *Meeting, Oslo,* 11–13 March 1996.

Ruffin E, Bouissou C and Defert R (2003) Elaboration d'un modéle d'évaluation quantitative des risques pour le transport multimodal des marchandises dangereuses. BCRD, Ministère de l'Ecologie et du Développement Durable, AP2000, Convention No. 2000-0102, Rapport Final Synthétique, INERIS/Direction des Risques Accidentels, August 2003.

Truchot B, Oucherfi M, Quezel-Ambrunaz F, Duplantier S, Fournier L and Waymel F (2010) Are the tunnel ventilation systems adapted for the different risk situations? *International Symposium on Tunnel Safety and Security (ISTSS), Frankfurt,* 17–19 March 2010.

UIC (2003) *Safety in Railway Tunnels*, 1st edn. Leaflet 779-9. Union Internationale des Chemins de fer, Paris.

UN Economic Commission for Europe, Inland Transport Committee (2003) *Report of the Ad Hoc Multidisciplinary Group of Experts on Safety in Tunnels (Rail) on its Fourth Session.* TRANS/AC.9/8. UNECE, Geneva.

Handbook of Tunnel Fire Safety, 2nd edition
ISBN: 978-0-7277-4153-0

ICE Publishing: All rights reserved
doi: 10.1680/htfs.41530.485

Institution of Civil Engineers

publishing

Chapter 23
A systemic approach to tunnel fire safety management

Jaime Santos-Reyes
SARACS Research Group, SEPI-ESIME, Instituto Politecnico Nacional, Mexico City, Mexico
Alan Beard
Civil Engineering Section, School of the Built Environment, Heriot-Watt University, Edinburgh, UK

23.1. Introduction

Highly publicised accidents in road and rail tunnels have highlighted the need for tunnel designers and operators to address safety proactively. Fire safety still tends to be addressed in isolation, though any accident is an emergent property of a system; that is, it results from the interaction of parts of a system. There is a need for a systemic approach to understand the nature of fire risk in tunnels. This chapter presents a tunnel fire safety management system (TFSMS) model that aims to help maintain fire risk within an acceptable range in road or rail tunnels. It is hoped that this approach will lead to more effective management of fire safety in tunnels. (See also Chapters 5, 29 and 30.)

Fire may be regarded as the biggest threat to human life in road and rail tunnels. Vehicle or train fires in tunnels may have far more devastating effects than fires in the open air involving similar vehicles, not only with regard to life safety but also in terms of property damage. This is due to the effects of high heat release rates and dense smoke. Moreover, if the tunnel is not properly ventilated, these conditions make emergency response efforts difficult, if not impossible. Therefore, the greatest challenge for road and rail tunnel safety management is to try to prevent such accidents happening and to diminish the impact of such incidents.

Several major fire accidents in tunnels, such as the fire in the Channel Tunnel (UK–France, 1996) (Channel Tunnel Safety Authority (CTSA), 1997), the Mont Blanc Tunnel fire (France, 1999) (Bell, 1999), the Kaprun Tunnel fire (Austria, 2000) (Philip *et al.*, 2000), the fire in the Gleinalm Tunnel (Austria, 2001) (Leiding, 2001) and the St Gotthard Tunnel (Switzerland, 2001) (Boyes, 2001) have highlighted the need to address fire safety proactively. Overall, in the past it seems that both tunnel designers and operators have tended to address safety by focusing on direct technical aspects and looking for immediate causes of incidents or accidents after they have taken place. While this is important, of course, there is a need to look beyond immediate causes. Hitherto, safety management in tunnels seems to have been unable to deal with the systemic nature of fire risk in an effective way. Traditionally, fire risk has been treated in a fragmentary way; however, such an approach will ultimately fail to understand the nature of fire safety. That is, the causes of fires in tunnels may be found in the complexity of the relationships implicit in the system involving, *inter alia*, the tunnel designers, tunnel owners, tunnel operators, regulators, contractors and tunnel users. In order to gain a full understanding and comprehensive

485

awareness of fire risk in a given situation, it is necessary to try to consider, in a coherent way, all aspects that may come together to produce fire and its effects (Beard, 1985); that is, there is a need for a systemic approach. 'Systemic' may be regarded as trying to see things as a whole, and this is different to 'systematic' (Beard, 1999). To be systematic is to be 'methodical' or 'tidy'. While it is certainly desirable to be systematic, to be systemic is much more than this. To be systemic is to try to see failure as a result of the working of a system. This is very difficult, and there is no 'automatic' way of doing it; it is not a matter of running through a checklist. Nevertheless, the intention of this work has been to adopt a systemic approach. This may include, for example, the design or redesign of a road/rail tunnel through 'inherently safer design' principles, or by designing a safety management system at the very beginning of the tunnel design process. It may also include, say, the consideration of physical characteristics (e.g. number of tubes, length, cross-sectional area) of the tunnel in the design process; this is because they may play an important role in an emergency situation. A cross-sectional area may determine, for example, whether emergency walkways can be provided at the side of the carriageway to facilitate means of escape in case of a tunnel fire. The application here has been to fire safety, but the approach is general. Both the safety management system and the operation of a tunnel in general need to be considered at the start of the design stage: not as a 'bolt-on' afterwards.

The TFSMS described has a fundamentally *preventive potentiality* in that if all the systems and connections which are described are in place and working effectively in carrying out their functions, then the probability of an accident would be expected to be less than otherwise. The model is intended to provide a *sufficient* set of features or properties (covering 'structure' and 'process') to achieve the aim of maintaining risk within an acceptable range. An analysis of the Channel Tunnel fire of 1996 is currently in progress, using this approach, and a preliminary paper has been published (Santos-Reyes and Beard, 2011).

23.2.   A tunnel fire safety management system model (TFSMS)

The TFSMS model described here builds on the viable system model developed and proposed by Beer (Beer, 1979, 1982, 1985) and the failure paradigm method proposed by Fortune and Peters (Fortune, 1993; Fortune and Peters, 1995). It is derived from the fire safety management system model described by Santos-Reyes and Beard (Beard and Santos-Reyes, 1999; Santos-Reyes and Beard, 2001). In earlier work, it has been considered in relation to offshore fire safety (Santos-Reyes and Beard, 2001) and, to some extent, to safety management on the railways (Santos-Reyes and Beard, 2003). The 'viability' of a fire safety management system is defined as the probability that it will be able to maintain the risk within an acceptable range for a given time period. While it would be desirable to calculate the viability, the TFSMS is capable of being used without doing this. It may be used to assist in the design of the safety management system of a new tunnel system, or as a 'template' to compare with an existing safety management system with a view to improvement. The model may also be employed as a tool for examining past accidents or 'near misses'. This may be achieved by using it as a basis for comparison to see how features in a 'real-world' system were deficient and led to an accident or 'near miss'. In this way, it has been applied to the Paddington railway disaster (UK, 1999) to try to gain 'learning points' for safety management (Santos-Reyes and Beard, 2006). While the model described in this chapter has been applied to fire safety, it is capable of being applied to safety as a whole. The TFSMS consists of the following fundamental characteristics

■   a structural organisation that consists of a 'basic unit' in which it is necessary to achieve five functions associated with systems 1 to 5 (described below)

- a 'recursive' (i.e. layered) structure
- relative autonomy
- four principles of organisation
- the concept of the maximum risk acceptable (MRA) and the acceptable range of fire risk
- the concept of the 'viability' of a fire safety management system.

In addition, it may be augmented by various submodels. A brief description of each of the above characteristics of the model is given below, together with possible communication and control submodels and a fire safety performance planning framework. Also, various 'paradigms' have been described by Fortune and Peters (1995), including 'human factors paradigms', for example, the 'group-think' paradigm. Such paradigms may be valuably employed as part of the overall approach.

23.3. Structural organisation of the TFSMS
23.3.1 Key systems of the TFSMS model
The TFSMS model has a 'basic unit' in which it is necessary to achieve five functions associated with systems 1 to 5. System 1 consists of various operations within an organisation that deal directly with the organisation's 'core' activities (Figure 23.1). An organisation's operations, for example, rail/road tunnel operations and emergency services, could form part of system 1, as shown in Figure 23.1. The entirety of systems 2–5, represented as a box in Figure 23.1, is referred to as the 'management unit' for system 1. It should be noted that the zigzag lines connecting the operations of system 1 indicate physical interdependency among the operations. Furthermore, each operation, as identified with a subsystem of system 1 (e.g. A1, as in Figure 23.1) performs five functions associated. Lines of communication and information flow are indicated, including a 'hot line' directly from a lower recursion to system 5 at a higher recursion. The general principle is that systems 2–5 facilitate the function of system 1, as well as ensuring the continuous adaptation of the whole organisation. The five key systems and functions are now described.

System 1, tunnel fire safety policy implementation, implements tunnel fire safety policies in the tunnel organisation's operations. The various operations or subsystems of system 1 are also responsible for implementing the organisation's fire safety policy. The number of operations that form part of system 1 will depend on the specific organisation being modelled. In this particular case, for example, for a road/rail tunnel operator, three subsystems called A1, A2 and A3 could form part of system 1, as shown in Figure 23.2. The 'traffic management' in relation to road tunnels would be one of the main activities of the management unit of system 1. The management unit should manage the traffic volume and traffic speed as well as the nature of vehicles (e.g. those carrying 'hazardous materials') on a day-to-day basis. In the case of a fire incident in the tunnel, the 'emergency procedures' should be implemented by the management unit.

System 2, tunnel fire safety coordination, coordinates the activities of the operations of system 1. System 2, along with system 1, implements the fire safety plans received from system 3. It informs system 3 about routine information on the performance of the operations of system 1. To achieve the plans of system 3 and the needs of system 1, system 2 gathers and manages the safety information of the operations of system 1. An example of the action of system 2 could be coordination of the teams and organisations involved in a tunnel fire, for example the tunnel operator, emergency response team, fire service, ambulance service, local authority and police.

Figure 23.1 A Tunnel Fire Safety Management System model (Subsystems A1, A2 and A3 are for illustration purposes only)

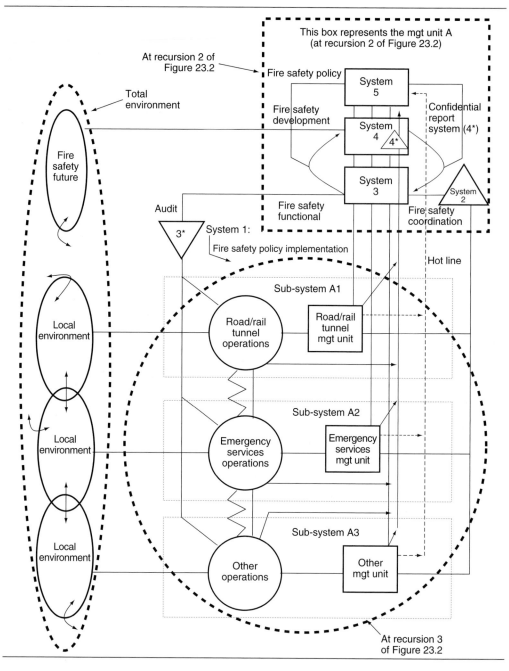

System 3, tunnel fire safety functional, is responsible for maintaining fire risk within an acceptable range in system 1 in a direct way, and ensures that system 1 implements the fire safety policy of the organisation. It achieves its function on a day-to-day basis according to the fire safety plans received from system 4. System 3 requests from systems 1, 2, and 3* information about the fire

Figure 23.2 Recursive (layered) structure for the TFSMS (The right-hand side is to be viewed as an approximate 'cylinder'. The ellipses correspond to circles and the parallelograms correspond to squares, viewed obliquely)

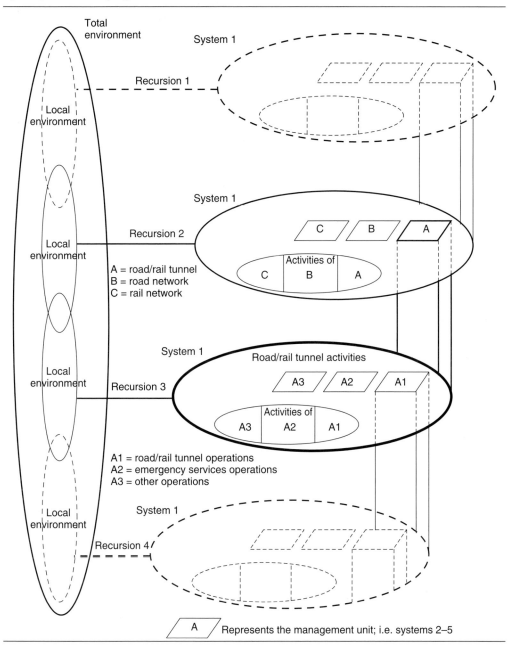

safety performance of system 1 to formulate its fire safety plans and to communicate future needs to system 4. It is also responsible for allocating the necessary resources to system 1 to accomplish the fire safety plans of the organisation. *System 3*, tunnel fire safety audit,* is part of system 3, and its function is to conduct audits sporadically into the operations of system 1. System 3* intervenes in the operations of system 1 according to the fire safety plans received from system 3. System 3 needs to ensure that the reports received from system 1 reflect not only the current status of the operations of system 1 but are also aligned with the overall objectives of the organisation. The audit activities should be sporadic (i.e. unannounced) and they should be implemented under common agreement between system 3* and system 1. A corrective action during a tunnel fire, either through system 2 or through the command channel (i.e. the channel that connects systems 3 and 1, see Figure 23.1), could be an example of the action of system 3. Furthermore, system 3 should allocate resources for the training of train personnel and operators, etc., in relation to an emergency response. On the other hand, continual assessment of the adequacy and the functioning of the engineering services and fixed installations of the tunnel (e.g. mechanical ventilation systems, CCTV, fire-fighting equipment, sprinklers, manual and automatic equipment) is an example of the action of system 3*.

System 4, tunnel fire safety development, is concerned with fire safety research and development for the continual adaptation of the organisation. By considering strengths, weaknesses, threats and opportunities, system 4 can suggest changes to the safety policies of the organisation. This function may be regarded as a part of effective fire safety planning. First, system 4 deals with the safety policy received from system 5. Second, it senses all relevant threats and opportunities from the wider physical and socio-economic environment of the organisation, including the tunnel 'fire safety future environment' (this is explained in more detail later). Third, system 4 deals with all the relevant needs of the performance of system 1 and its potential future. Finally, it deals with the confidential or special information communicated by system 4*. *System 4*, *tunnel fire safety confidential reporting*, is part of system 4, and is concerned with reports or causes of concern, from anyone involved in the entire 'system', about *any* matter, of any kind – some of which may require the direct and immediate intervention of system 5. Such causes of concern may include, for example, finding a broken rail – not just 'personal mistakes'. It is essential that system 4* deals with reports in a genuinely confidential way, and be seen to do so by all employees. System 4* should be an extremely valuable source of information, but only if the person reporting the cause of concern can be assured that there will be no repercussions at a personal level. People will be unwilling to report on 'near misses' or possible 'mistakes' if there is a chance of repercussions. This means that the mechanism for receiving the confidential reports needs to be independent of the 'mainstream management' and does not pass on names of those reporting. Likewise, system 4* will only be effective if reported causes of concern are genuinely considered and, if necessary, acted upon. (Related to the issue of reporting 'causes of concern' is the concept of a 'blame-free' ethos, whereby a person is encouraged to admit errors without fear of blame. This would require a genuine commitment on behalf of management not to 'blame' a person who openly comes forward in this way. If this were the case, in practice, then a 'blame-free' policy would be desirable. A capacity in the system for confidential reporting is essential, even if the management espouses a 'blame-free' policy.)

To fulfil the fire safety policy of the organisation, system 4 should respond proactively to the threats and opportunities regarding the physical tunnel system. For any change in the physical characteristics of the tunnel or personnel, system 4 should reassess the consequences of such changes and incorporate them into the new procedures. Another example of the action of system 4 is concerned with the development of new methods of training in relation to emergency response.

Finally, *system 5, tunnel fire safety policy*, is responsible for deliberating fire safety policies and for making normative decisions. According to alternative fire safety plans received from system 4, system 5 considers and chooses feasible alternatives that aim to maintain fire risk within an acceptable range throughout the life cycle of the road/rail tunnel operations. It also monitors the interaction of system 3 and system 4, as represented by the lines that connect the loop between systems 3 and 4 in Figure 23.1. An example of the policies of system 5 is to address the problem of how to anticipate fire accidents in tunnels. These policies should also address principles of 'inherently safer design' as well as 'protection' in existing or new tunnels. Furthermore, system 5 should promote a good 'safety culture' throughout the organisation and among tunnel users. Appendix 23A gives examples of systems 1–5.

23.3.2 Recursive structure within the TFSMS

The TFSMS model is described here at two levels of recursion only. However, the two levels should be seen in the context of Figure 23.2. In principle, there would be levels above recursion 2 and below recursion 3. The top right-hand-side dotted-line square box in Figure 23.1 is the TFSMS management unit at recursion 2 (i.e. level 2) of Figure 23.2. The dotted-line circle represents the tunnel operations that form part of system 1 (i.e. equivalent to the operations of trains/vehicles, ventilation systems, physical installation of the tunnel, etc.). This is depicted as being at recursion 3 of Figure 23.2. However, in principle, the 'basic unit' may be replicated for every operation of system 1, as implied in Figure 23.2. Likewise, the 'systems' at recursion 2 may be seen as part of a 'basic unit', the management unit of which is at recursion 1. The TFSMS is intended to be able to maintain fire risk within an acceptable range at each level of recursion, but this safety achievement at each level is conditional upon the cohesiveness of the whole organisation.

23.3.3 Relative autonomy

Employing a concept derived from Beer (1979), who refers to *autonomy*, the TFSMS favours *relative autonomy* and the incorporation of local safety problem-solving capacity. Relative autonomy means that those involved in bringing about each operation of system 1 of the TFSMS are responsible for that activity with minimal intervention of systems 2, 3, 4 and 5. The organisational structure of the TFSMS allows decisions to be made at the local level. Decision-making is distributed throughout the whole organisation. This means that decision-making involves a set of decision-makers in each operation of system 1 and at each level of recursion. These decision-makers should be relatively autonomous in their own right and act relatively independently based on their own understanding of tunnel fire safety and their specific tasks. However, it must be recognised that they have interdependence with other decision-makers of other operations of system 1. Therefore, each operation of system 1 should be endowed with relative autonomy so that the organisational safety policy can be achieved more effectively. Relative autonomy must not be confused with isolation: it must be within an adequate system of control and communication. Given this, relative autonomy implies allowing people to make safety-related decisions of importance being accepted as within their competence, without fear of reprisal.

An example is seen in the Piper Alpha oil rig disaster of 1988 (Cullen, 1989). If the offshore installation manager of the Claymore Platform had had the power to shut down oil flow without fear of reprisal, this would have stopped the back-flow into Piper Alpha: back-flow helped to fuel the fire. In a tunnel context, a tunnel manager should have the power to restrict or control traffic flow through a tunnel, or even shut the tunnel, if risk is considered to be too high, without fear of reprisal. This links to institutional and socio-economic pressures (see Chapter 30).

These aspects of organisational structure, which have a role in making organisations more rather than less effective, are poorly understood in the tunnel fire safety literature.

23.3.4   The TFSMS and its environment

The TFSMS relies on five functional imperatives, and the extent to which the TFSMS structural organisation accommodates contextual constraints determines its ability to adapt. The organisational structure of the TFSMS is shown as interacting in a defined way with its environment through the operations of system 1, and through system 4, as illustrated in Figure 23.1. The environment, both socio-economic and physical, is understood as being those circumstances to which the TFSMS response is necessary. The TFSMS also needs to respond to necessary internal matters, for example inadequate training. System 4 deals with the total socio-economic and physical environment of the TFSMS, into which a road or rail tunnel designer and operator are embedded. The dotted-line elliptic symbol in Figure 23.1 represents the total environment of the TFSMS. System 4 also deals with the fire safety 'future environment', which is also embedded into the total environment of the TFSMS. The fire safety 'future environment' is concerned with threats and opportunities for the future development of fire safety. On the other hand, the operations of system 1 interact with local environments with which the operations of the organisation must deal. These local environments are embedded into the total environment of the TFSMS, as illustrated in Figure 23.1. For example, the road and rail tunnel designers and operators are embedded within a wider socio-economic and physical structure that will constrain the way they can develop. There are various important socio-economic and physical characteristics that need to be taken into account. These characteristics can be segregated thus: first, physical characteristics, such as the geography of the area (e.g. Alpine road tunnels will be affected by the topography of the Alps), weather conditions and public utilities; second, economic characteristics such as the level of employment, train operators, tunnel users, cost/profits, and other types of industrial and commercial considerations; finally, socio-political characteristics, such as regulators and social organisations. The demands and needs inherent in these characteristics will suggest and condition patterns of structural organisation of the TFSMS. Organisations need to pay more attention not only to these characteristics but also to the complexity, stability or uncertainty of changing technologies.

Apart from confronting demands for its 'products', an organisation faces an environment upon which it is dependent for finance, workforce and materials; that is, for its resources. The total environment of the organisation has a certain pattern of resource availability to which the organisation has to relate. The supply of resources to the organisation changes over time, forcing it to make organisational adaptations. These adaptations may involve merging departments, changing the location of decision-making, introducing new procedures and so on. These changes may have significant impacts on the fire safety performance of the whole organisation. Similarly, local or institutionalised environments are also characterised by the socio-economic and physical characteristics of the total environment of the organisation; that is, physical, economic and socio-political.

Whenever a line appears in Figure 23.1, which represents the TFSMS model, it indicates a channel of communication, except for the lines that connect the balancing loop that connects systems 4 and 3. There is a particular concern in the TFSMS about the nature of these channels and the information that flows in the communication channels. These channels of communication should obey four organisational principles in order to increase the viability (see Appendix 23B). These organisational principles are understood as responding appropriately to the weaknesses and strengths, threats and opportunities as presented in the wider and local environments of the TFSMS; the channels of communication, and the necessary transducers translating information when it crosses boundaries of the system, must be designed according to the requirements of Ashby's law of requisite variety (Beer, 1979, 1982, 1985), and these principles must be put into effect without time lags. In Figure 23.1, the bi-directional arrows in the total environment of the TFSMS indicate the interactions among the local environments, as well as the interaction of these local environments with the total environment.

Figure 23.1 shows a dashed line directly from the management units of system 1 (road/rail tunnel management unit) to system 5, representing a direct communication or 'hot line' for use in exceptional circumstances, for example during an emergency. Also shown in Figure 23.1 are lines with an arrow from system 1 to system 4* and system 5, representing a fire safety confidential reporting system. Both channels, the 'hot line' and the confidential reporting system, represent 'initially' one-way communication channels, but they may become two-way communication channels between systems 1 and 5 and 1 and 4*, respectively.

23.4.   Communication and control in the TFSMS
23.4.1   Proactive commitment to safety

The safety approach of an organisation can be reactive or proactive and less or more committed to safety. Additionally, organisations may focus purely on 'technical' aspects of accidents. For example, active fire protection systems have been developed with the purpose of controlling and mitigating fire incidents. However, the degree of risk in an organisation or system is an emergent property resulting, *inter alia*, from the interrelated activities of the people who design it, manage it, operate it and use it. Humans, individually, in teams and in organisations decide the technical aspects. Direct technical aspects are, of course, vitally important; however, in order to gain a deeper understanding of risk it is necessary to go beyond purely technical considerations. People who are involved in the road/rail tunnel life cycle, such as tunnel designers, tunnel constructors, tunnel operators, maintenance contractors, legislators, regulators, fire brigade and tunnel users, make decisions that effectively contribute to creating a system that has risk implicit within it. These factors, which are the potential, but not the obvious or explicit, causes of incidents or accidents, are known here, following Reason (1997), as 'latent factors' – factors that do not have an immediate effect in bringing about an accident. It is claimed by Turner (1976) that common (i.e. commonplace) causal failures form part of an 'incubation process' in a sequence of disaster development. Moreover, according to Reason, latent factors accumulate unnoticed until a precipitative event or 'trigger' leads to the onset of an incident, accident or disaster. For example, a faulty fire extinguisher or inadequate training of staff in relation to a 'permit to work system' could exist 'in the background' for a long time without a harmful incident occurring. However, eventually such factors may contribute to an accident or disaster. Reason's use of the concept of 'latent factor' tends to concentrate on organisational and directly 'human' aspects. However, the concept may also be applied to directly technical aspects, such as a piece of faulty wiring that may be undiscovered for a long time, as well as to organisational/human aspects. Ultimately, all directly technical aspects, including, say, the existence of faulty wiring, result from human action and decision-making, and so, in that sense, all factors are ultimately 'human factors'.

Traditionally, tunnel operators may not be explicitly aware of latent factors, but they may look for immediate causes of incidents or accidents after they have taken place. There is a tendency to divide 'causes' into separate objects and events. This division is, of course, useful and necessary to cope with fire risk, but it is only part of understanding risk. The immediate causes of incidents or accidents as readily observed or understood are known here as immediate factors. The 'incubation' (Turner, 1976) period of a 'latent failure' before the immediate failure appears is known as the 'latent period'. All parts that constitute an organisation can be seen as inter-dependent and inseparable parts of the organisation as a whole. Moreover, these constituents are all interconnected, interrelated and interdependent, in that they cannot be understood as isolated entities but only as integrated parts of the organisation as a whole. Loss is therefore seen as a systemic failure, not a result of a single cause. Clearly, addressing latent failures is as important as focusing on immediate failures in accidents. An assiduous application of the TFSMS model should assist in bringing latent factors to the surface.

23.4.2   Internally and externally committed systems

In addition to latent and immediate factors, the distinction between directly technical and directly human factors should be explicitly realised so that effective safety objectives, plans and measures of performance can be set. Failure comes about as a result of a conjunction of causal factors, not because of a single 'cause'; all aspects, both 'hard' and 'soft' play a part. Accommodating this requires knowledge about the degrees to which the technical and human factors are 'committed' to safety and linked together into a coherent whole. Moreover, the way in which they are implemented into the operations of a particular organisation should be understood clearly. These two aspects lead to the concepts of internally committed systems (ICSs) and externally committed systems (ECSs). The distinction between these two aspects may be a source of insight into the ways safety can be approached, as well as the ways in which these two aspects differ from each other. Table 23.1 illustrates some characteristics of ICSs and ECSs.

ECSs refer to the safety performance of systems that are committed to a particular purpose, func-tion or objective based on external reasons or motivation. This definition addresses both directly technical aspects and humans. For example, tunnel installations are designed to accomplish a well-defined objective, while the evacuation procedures in the case of a fire are formulated by safety-related professionals to be followed by tunnel users. Here, the performance of the road/rail tunnel satisfies the purpose of the tunnel designer, and the tunnel users 'satisfy' the safety manager's 'purpose'. Of course the safety manager's 'purpose' (i.e. to avoid death/injury) would also correspond with the motivation of the users. However, tunnel users are generally not expected to be involved in the design of evacuation procedures. It is arguable that users of

Table 23.1 Internally and externally committed systems

External commitment	Internal commitment
Tasks in the organisation are defined by others	Employees participate in defining tasks
The behaviour required to perform tasks with an acceptable degree of risk is defined by others	Employees participate in defining the behaviour required to perform tasks with an acceptable degree of risk
Safety performance goals are defined by the management of the organisation or by others	The management and employees of the organisation jointly define safety performance goals
Others define the importance of the safety performance goals	Employees participate in defining the importance of the goals

tunnels should, as far as possible, be involved in designing evacuation procedures. Traditionally, after an accident, an inquiry or assessment process is conducted to determine the immediate causes (a better term is 'causal factors') so that understanding may be adjusted. Goals for improvement may be defined to address the newly found failure, so that the same failure would be less likely to occur again. Moreover, very often it is assumed by organisations that the absence of incidents or accidents or other negative outcomes is an indication of good safety management. For example, many organisations traditionally use lost time injuries (LTIs) and deaths as the basis to measure the effectiveness of their safety management. However, the absence of negative incidents does not necessarily indicate good safety management. Also, it may lead to complacency.

Addressing immediate causal factors is essential, but insufficient: a proactive approach is necessary. However, often tunnel designers and operators still tend to comply with externally imposed safety objectives (i.e. existing regulations, standards or procedures) in a narrow way. It is generally accepted that safety is better assessed and managed by addressing in advance the hazards of the organisation's operations. This is usually done through (1) identification of hazards, (2) assessment of the significance of hazards and (3) hazard management by prevention, control and mitigation. Having an open mind and creative thinking are important parts of 'hazard identification'. It is also essential to learn from former accidents worldwide. In particular, 'near misses' are a vital form of information, which should be used far more than they are at present. However, organisations are often still only committed to complying with existing regulations, standards or procedures, in a narrow sense, and this is basically reactive mentality. Of course, a regulation may demand or imply a proactive approach; however, only complying with externally imposed regulations or existing standards does not necessarily mean that the operations of an organisation will have an 'acceptable risk'. Moreover, organisations still tend to focus on immediate factors, and have very little understanding and appreciation of latent factors and internally committed systems.

The idea of an ECS may produce important insights into the ways fire risk can be addressed, but it is fundamentally incomplete. Directly human factors, including organisational factors, may have quite dramatic risk consequences. There may be a substantial gap between safety objectives as defined by regulators, standards or procedures, or the management of the organisation and what may be achieved in a real-world situation. It is necessary to introduce the idea of ICSs. An ICS is a system that is committed to a particular purpose or objective based on its own conscious awareness, reasons or motivation. In other words, an ICS refers to the critical awareness of self-reflective human beings regarding their purposes and the implications of their actions for all those who might be affected by the consequences. (A classic example of the need for ICSs is seen in the Piper Alpha offshore disaster of 1988 (Cullen, 1989), where the 'permit to work' system had become degraded to being merely a formality rather than being genuinely seen as a vital part of safety management.) This means that all those involved in the life cycle of the operations of the organisation should be committed to addressing safety proactively and anticipating incidents or accidents, motivated by their own objectives or purposes. This freedom to achieve safety objectives is, however, limited by the safety policy, plans, standards and procedures of the organisation. Individuals, teams, groups and departments that perform the operations of an organisation should not only be assigned tasks but should have both authority and responsibility by their understanding of safety and their specific tasks. They should be endowed with authority in their daily tasks by their knowledge of how to perform their tasks properly and in an acceptable way with regard to risk. This knowledge involves their knowledge of risk itself and the skills required for performing a specific activity. In other words, individuals, teams, groups and departments that constitute a road/rail tunnel organisation should have more involvement with fire safety in their daily tasks. Top and line management should encourage the development of ICSs. The more the management of an organisation wants internal commitment from its employees, teams and departments, the more it must try to involve employees in defining safety objectives, specifying what these are and how to achieve them, and setting fire safety targets. Also, systems must be created to enable tunnel users to become far more aware of the fire-related aspects of a tunnel they are using. There is no safety vision, strategy or policy that can be achieved without able and committed employees. However, it is unrealistic to expect the management of an organisation to allow total autonomy to employees. Indeed, total autonomy would not be desirable. Given this, 'internal commitment' and 'relative autonomy' should form essential parts of a total fire safety management system.

23.4.3 Procedures: 'alive' and 'dead'

A procedure that is 'alive' is one which is genuinely seen as a means to an end, not simply an end in itself, a formality. An example is seen in the Piper Alpha disaster mentioned above. The 'permit to work' system is intended to allow an operation to take place at an acceptable time and to ensure that others are aware of it. On Piper Alpha, the 'permit to work' system had become 'dead' and seen as a formality, and not as a vital part of controlling risk. How to make a procedure 'alive' rather than 'dead' may be regarded as one of the key questions in safety management. A keen awareness of 'relative autonomy' and 'internal commitment', and genuinely attempting to put these into effect, may help.

As a part of considerations of this kind, the concept of 'familiar role' may be mentioned. Research suggests that people tend to adopt a 'familiar role' (see Chapters 19–21). As part of this, they may follow a 'script': people may 'get into a rut' and fail to 'think outside of the box'. This is thought to have been the case during the King's Cross Underground station fire, London, in 1987 (Fennel, 1988). To try to address this now, apparently, London Underground varies the jobs of employees. For example, a person may spend some time on a platform and some time on a task elsewhere. The reasoning behind this is that people's activities may be considered as 'multi-tonic' rather than 'mono-tonic'.

23.4.4 Safety culture

The 'safety culture' is a way of referring to the degree to which safety awareness does or does not permeate the system. It may be 'good' or 'bad'. As top management has the ultimate decision-making power, the degree to which the safety culture of top management is 'good' or 'bad' will tend to dominate the system. (The attitude of top management in the case of the Piper Alpha oil rig disaster (Cullen, 1989) is a classic example.) Safety culture is not something that top management should expect 'others' to worry about. If top management has a 'sloppy' attitude towards safety culture, then this would be expected to permeate the organisation. Some indicators of the degree to which top management *may* be genuinely committed to a good safety culture or not may be seen, *inter alia*, in the attitude to

1. system 4* (i.e. the confidential reporting system)
2. relative autonomy
3. internal commitment.

Also, if the organisation espouses a 'blame-free' ethos, then one may consider whether or not this is genuinely put into effect.

23.4.5 Paradigms for communication and control

Specific 'paradigms' have been put forward that are intended to act as 'templates', giving essential features for effective communication and control (see Appendix 23C). Communication failure is thought to have been a direct causal factor in a salient historical example: the sinking of the *Mary Rose*, the English flagship, in 1545, during a battle. The ship sank while carrying out a sharp turn with the gun-ports open. Water came in through the gun-ports. Research suggests that an order to close the gun-ports seems to have been ineffective because a large percentage of the crew (probably more than half) were foreign and could not speak English. Also, during the Channel Tunnel fire of 1996 (CTSA, 1997), two security guards saw a fire on the train before it entered the tunnel. Although they reported it to their supervisor, who passed on the information, the system as a whole was insufficient to stop the train before it had travelled a significant distance into the tunnel. This may, perhaps, not have been a failure of communication *per se* but a failure to act adequately upon receipt of information. Further, there was significant communication failure during the King's Cross Underground station fire of 1987, which had not been adequately addressed at the time of the London bombings in 2005 (*Metro*, 2011). In the King's Cross fire (Fennel, 1988), the radio system used by London Underground staff was incompatible with that used by the emergency services.

23.5. Fire safety performance

Figure 23.3 shows a framework that is intended to help organisations plan their fire safety performances. It consists of four levels of fire risk, four kinds of fire safety plans and four kinds of fire safety indices.

23.5.1 Risk levels

Four different levels of fire risk can be specified to plan and measure the fire safety performance of an organisation, as shown in Figure 23.3. The first, current, fire risk level or 'current achievement level' (CAL) is a continuously fluctuating value. The second, 'without (significant) extra investment level' (WEIL), is a relatively static value. On the other hand, the 'minor extra investment level' (MINEIL)

Figure 23.3 Fire safety planning

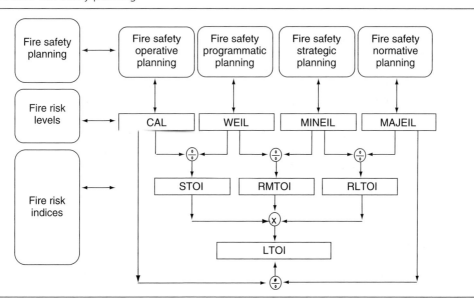

and the 'major extra investment level' (MAJEIL) are static values until the operations of the organisation, or the organisation itself, have structurally changed. It should be stressed that when defining these fire risk levels, it is necessary to consider both ICSs and ECSs for immediate factors and latent factors. A description of these fire risk levels is given below.

23.5.2 Current achievement level

The CAL is the fire risk level an organisation is managing at the current time with existing resources and under existing constraints. The CAL represents the day-to-day fire risk level in the operations of the organisation. It expresses a continuously fluctuating risk level, since it corresponds to risk arising from the continuous activities of the operations of the organisation. The continuous variation of the fire risk level requires organisations to have up-to-date fire safety information. If fire safety information is out of date, then the decisions of the organisation may be irrelevant: organisations need to manage fire safety continuously.

23.5.3 Without (significant) extra investment level

The WEIL can be defined as the fire risk level an organisation could achieve, in the very short term, with existing resources and under existing constraints, if resources and technology were better organised. This is a fire risk level that an organisation is capable of achieving in each operation, given the limitations imposed on any operation or activity by other operations or activities of the organisation.

23.5.4 Minor extra investment level

The MINEIL can be defined as the fire risk level an organisation ought to be able to achieve if a minor investment were made to eliminate some shortcomings in capacity or technology. This is a medium-term and relatively static level.

23.5.5 Major extra investment level

The MAJEIL can be defined as the fire risk level an organisation would be capable of achieving if a major investment in new equipment, new technology or people were made to eliminate current constraints. The MAJEIL is a long-term and static level until structural changes in the operations of the organisation, or the organisation itself, have come about.

The MAJEIL and the MINEIL fire risk levels are better than the WEIL, and the WEIL in turn is better than the current fire risk level. The fire safety performance of an organisation cannot rise to its desirable level without some kind of investment. It cannot even rise to the WEIL unless resources and technology are well organised.

23.6. Fire safety plans

Fire safety can be planned according to the four risk levels as described above. Planning is understood here as a continuous process of decision-taking, whereby resource allocations are made, so that the future fire safety performance of an organisation may be better. Fire safety should be planned to address the concepts of both ICSs and ECSs in relation to both directly technical and latent ('background') factors. Four kinds of fire safety plans are identified below.

23.6.1 Operative planning

Planning fire safety from the CAL is referred to as *operative* fire safety planning. In this kind of planning there is no fire safety performance improvement or change but it accepts the existing status of fire safety as it is in the organisation's operations.

23.6.2 Programmatic planning

Planning fire safety from the WEIL is called *programmatic* planning. This kind of planning sets new short-term objectives for improving the fire safety performance and tries to achieve them. However, these objectives can be achieved without significant extra investment; that is, they can be achieved with essentially existing resources and under existing constraints.

23.6.3 Strategic planning

Planning fire safety from the MINEIL is referred to as *strategic* fire safety planning. It sets new medium-term objectives for fire safety performance improvement, but they can only be achieved with some minor investment to eliminate current constraints.

23.6.4 Normative planning

Planning fire safety from the MAJEIL is termed *normative* planning. This planning process involves setting long-term fire safety objectives. To accomplish these fire safety objectives, organisations will need to commit major investment to develop new technologies, new equipment or processes. These plans may offer major benefits for the organisation and tunnel users.

23.7. Fire risk indices

It can be seen from Figure 23.3 that the four levels of fire risk are combined as three ratios, to form short-, medium- and long-term indices. Moreover, Figure 23.3 shows how these three indices are combined to create an overall 'fire risk objective' index of the whole organisation. It should also be emphasised that these fire risk indices involve deriving indices for both ICSs and ECSs relating to both directly technical and latent factors. A brief description of these indices is given as follows.

23.7.1 Short-term objective index

The ratio of the WEIL to the CAL is called the *short-term objective index* (STOI). The STOI is in a continual state of change, since the CAL is understood as a fluctuating value, whereas the WEIL, on the other hand, is a relatively steady value.

23.7.2 Relative medium-term objective index

The ratio of the MINEIL to the WEIL is referred to as the *relative medium-term objective index* (RMTOI). The RMTOI requires minor investment. This is a relatively static index since the minor fire safety performance level is a medium-term fire safety goal.

23.7.3 Relative long-term objective index

The ratio of the MAJEIL to the MINEIL is called the *relative long-term objective index* (RLTOI). This index involves major investment. The MINEIL and the MAJEIL fire safety levels are static measures. The resulting index will also be static.

23.7.4 Long-term objective index

The overall *long-term objective index* (LTOI) of an organisation is defined as the ratio of the MAJEIL to the CAL; equivalent to the product of the STOI, RMTOI, and RLTOI indices. A working example is given in Figure 23.4 and below:

1. Fire risk indices:

$$STOI = WEIL/CAL = 10^{-4}/10^{-3} = 0.1$$

Figure 23.4 Example

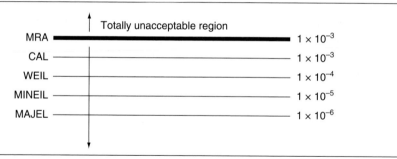

$$RMTOI = MINEIL/WEIL = 10^{-5}/10^{-4} = 0.1$$

$$RLTOI = MAJEIL/MINEIL = 10^{-6}/10^{-5} = 0.1$$

2. Long-term objective index:

$$LTOI = STOI \times RMTOI \times RLTOI = 0.1 \times 0.1 \times 0.1 = 0.001 = 0.1\%$$

Evidently a *medium-term objective index* could be defined and found in a similar way. However it is done, for effective tunnel fire safety management it is important to think in terms of short-, medium- and long-term objectives, and to carry this thinking through on a continuous basis. The case is never 'closed' because the real-world system is always changing.

23.8. The maximum risk acceptable (MRA), the acceptable range of fire risk and the viability

The critical aspect of the TFSMS model is its ability to maintain fire risk below, and preferably well below, a MRA in the face of events. The MRA is understood as the level of fire risk above which the risk is totally unacceptable, as illustrated in Figure 23.5. An acceptable fire risk, on the other hand, is one which is below the MRA and, as a general rule, well below the MRA. The viability of the TFSMS is then the probability that it will be able to keep fire risk within an acceptable range for a given time period. This probability depends, *inter alia*, upon the MRA level set and the 'environmental drivers', for example economic and political. The viability must always be seen in relation to the MRA level set. It would be possible to achieve a relatively high viability if the MRA were set at a high level, corresponding to a relatively

Figure 23.5 The MRA and the acceptable range of fire risk

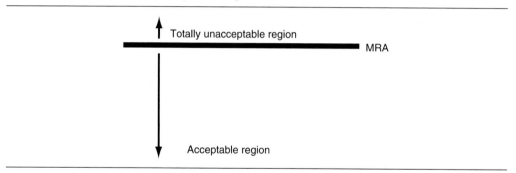

high risk. The lower the level of MRA that is set, then the more difficult it is to achieve a high viability. It is possible to use the TFSMS model without calculating the viability. However, calculations of viability would be desirable in the longer term and should be valuable when attempting to compare performances of the safety management system for a given tunnel, over time; and for comparing performances of safety management systems for different tunnels. In the latter case, a common MRA would need to be set.

23.9. Conclusion

A model for a TFSMS has been presented. The TFSMS aims to maintain fire risk within an acceptable range in the operations of a road/rail tunnel in a coherent way. If the features of the model (i.e. the systems, their associated functions and interconnections) are in place and working effectively, then the probability of an accident should be less than otherwise. In this way, the TFSMS has a *fundamentally* preventive potentiality. The model is intended to provide a *sufficient* set of features (including structure and process) to achieve the aim of maintaining risk within an acceptable range. The model should be read in conjunction with Chapters 5, 29 and 30. The idea of the viability of a safety management system has been introduced, this being the probability that the safety management system will be able to maintain fire risk within an acceptable range for a given period of time. The model may be employed without calculating the viability, however. It is capable of being applied proactively in the case of a new tunnel or an existing tunnel system as well as reactively. In the latter case, a past failure, whether disastrous or not, may be examined using the TFSMS model as a 'template' for comparison. In this way, lessons may be learned from past accidents. It may also be employed as a 'template' to examine an existing safety management system, outwith any specific preceding accident. In the case of new tunnels, the safety management system should be considered at the very beginning of the design stage, and not as a 'bolt on' at the end. It is hoped that this approach will lead to more effective management of fire safety in tunnels. As already indicated, an analysis of the Channel Tunnel fire of 1996 using this approach is in progress, and a preliminary paper has been published (Santos-Reyes and Beard, 2011).

Appendix 23A: Examples of systems 1–5
System 1: safety policy implementation

System 1 implements safety policies; it contains a set of subsystems or operations that deal directly with the 'core' activities of an organisation. For example, system 1 may contain the following subsystems or operations.

- A transport system (system 1 may contain the four modes of transport), e.g.:
  System 1: 'rail operations', 'road operations', 'sea operations' and 'air operations'.
- A railway system (system 1 may contain two subsystems or operations), e.g.:
  System 1: 'rail infrastructure operations' and 'train operations'. Each of the operations can be decomposed into other subsystems or operations, depending on our level of interest. For example, 'rail infrastructure operations' can be decomposed into 'track operations', 'signalling operations', 'tunnel operations', etc.
- The Channel Tunnel system (system 1 may contain three subsystems or operations), e.g.:
  System 1: 'running Tunnel South operations', 'running Tunnel North operations' and 'Service Tunnel operations'.
- An oil and gas field system (assuming that an oil and gas field consists of four offshore platforms, then system 1 will consists of four subsystems), e.g.:
  System 1: 'offshore platform-A operations', 'offshore platform-B operations', 'offshore platform-C operations' and 'offshore platform-D operations'.

The safety management units (SMUs) of system 1 are responsible for maintaining risk substantially below the MRA in the subsystems or operations (see above), and for informing system 3 whether the safety plans are being accomplished or not. For example

- the SMUs are responsible for the continuous monitoring of the performance of their physical installations, for example, the performance of oil and gas–water separation systems, the gas compression system, emergency shutdown systems, ventilation systems, communication systems, fire and gas detection systems, active fire protection systems and passive fire protection
- the SMUs should be responsible for the performance of the employees who perform the activities in the installations they manage, i.e. maintenance personnel, production process personnel, safety officers, train drivers, etc.

System 2: safety coordination

System 2 coordinates the activities needed to be implemented in the subsystems or operations of system 1. For example

- in the case of a fire incident in a tunnel, the systems in place (fire detection systems, fire fighting systems, traffic management systems, etc.) should provide the necessary information to those involved in the fire response management so that they act in a coordinated way to prevent the undesirable events
- good communication should be guaranteed amongst the key players in case of a fire incident in a tunnel, for example, the train driver, control centre, fire-fighting crew and rescue vehicles inside the tunnel.

System 3: safety functional

System 3 is responsible for sustaining an acceptable level of risk in system 1 in a direct way. System 3 conducts risk assessments and communicates safety plans to system 1. For example

- procedures and plans to operate, for example, the rail/road/metro tunnel to prevent a fire incident, for example, procedures for preventive maintenance of the systems in place for dealing with a fire incident
- procedures and plans to adequately respond to a fire incident, for example, well-trained personnel
- procedures and plans to deal with the aftermath of a fire incident
- allocation of the necessary resources for the implementation of the above safety plans and procedures.

System 3*: safety audit

System 3* conducts audits sporadically into the operations of system 1. For example

- audits of the state of rail/road/metro tunnels, oil and gas–water separation systems and gas compression systems of an offshore platform, telecommunication facilities, emergency shutdown systems, fire and gas detection systems, active fire protection systems, passive fire protection systems, ventilation and smoke exhaust systems, automatic controls for ventilation and smoke exhaust systems, telephone systems and mobile telephones, emergency lighting, traffic management systems, public address systems, etc.
- audits of the performance of incident response personnel, traffic controllers, maintenance personnel, production and construction personnel, safety officers, inspectors, and suppliers.

System 4: safety development

System 4 deals with strengths and weaknesses, opportunities and threats to the whole system. For example

- identifying current and possible new regulations that may have a significant effect on the system
- understanding the trends in new technology regarding the design of road/rail/metro tunnels, offshore platforms (inherently safer design, etc.)
- 'scanning' the future and identifying threats and opportunities regarding extreme weather conditions, for example, the severe weather conditions that severely disrupted Eurostar services in the run-up to Christmas in 2009 (*The Times*, 2009), where heavy snow caused the trains to lose power and the services were cancelled for 3 days
- responding proactively to the identified threats and opportunities regarding natural disasters such as floods and volcanic eruptions, for example, the global disruption caused by the volcanic ash cloud over northern Europe in 2010 (Carter, 2010).

System 4*: safety confidential report

System 4* deals with confidential incident reports or causes of concern, as well as any particular or critical fire safety information that may require direct and immediate intervention of the corporate management. For example

- it must be perceived by all members of the workforce as being completely independent of management, thus giving the necessary assurance of confidentiality.

System 5: safety policy

System 5 deliberates safety policies and makes normative decisions. For example

- these policies should promote safety culture throughout the organisation and also address principles of fire protection in the existing or new installations
- the fire safety policies deliberated and decided by system 5 for implementation should reflect the needs of the employees of the whole system (rail/road/metro tunnels, offshore platforms, etc.) about both directly related and not directly related safety issues.

Appendix 23B: The four organisational principles
The first principle of organisation

This is derived from Ashby's law (Beer, 1979): 'Managerial, operational and environmental varieties, diffusing through an institutional system, tend to equate; they should be designed to do so with minimum damage to people and to cost' (i.e. for a viable system).

An example could be an evacuation system designed to save lives in the case of a fire or explosion; then, the evacuation capacity must be at least as great as the number of possible evacuees. (On a ship, the number of *physical* life-boat spaces may be greater than the number of evacuees, as some life-boat spaces may become ineffective because of, for example, wind-driven smoke.) In terms of tunnels, the capacities of egress routes, refuges, etc., must be at least as great as the number of people who may need to escape. In fact, 'greater than the number of people who may need to escape' would be sensible, to allow for the possibility that some spaces/routes may become ineffective. Another example is provided by the sinking of the *Titanic* in 1912. It seems that at the time the regulation governing the number of life-boats was based on tonnage rather than the number of passengers (Channel 4, 2008).

Figure 23.6 The basic elements of a TFSMS, illustrating the four key information channels

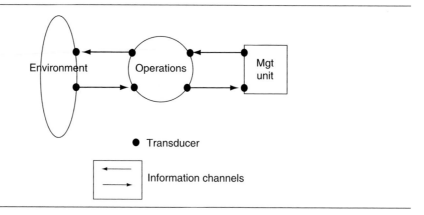

The second principle of organisation

This is derived from Shannon (Beer, 1982): 'The four directional channels carrying information between the management unit, the operation, and the environment must each have higher capacity to transmit a given amount of information relevant to variety selection in a given time than the originating subsystem has to generate it in that time.' (As shown in Figure 23.6.)

As an example, the channels carrying procedures of evacuation must have sufficient specificity so as to reduce ambiguities or eliminate unclear instructions.

The third principle of organisation

'Wherever the information carried on a channel capable of distinguishing a given variety crosses a boundary, it undergoes transduction; and the variety of the transducer must be at least equivalent to the variety of the channel.'

For example, in the case of means of escape for tunnel users, a transducer might be a fire safety instruction leaflet. This would 'transduce' between the person making up the evacuation rules and the people the rules are aimed at; the notice must be comprehensive and clear.

The fourth principle of organisation

'The operation of the first three principles must be cyclically maintained through time, and without hiatus or lags.' (That is to say, they must be adhered to continuously.)

Appendix 23C: Control and communication paradigms
Control paradigm

The regulatory system of the TFSMS is based on the paradigm shown in Figure 23.7. The management or controller and the system or organisation under control are inseparable in the TFSMS model. The sources of control are spread throughout the whole structure of the TFSMS rather than localised within a separate system.

As shown in Figure 23.7, the feedback control model of the TFSMS consists of various elements, such as the 'operations', 'comparator', 'reactive adjuster', 'proactive adjuster', 'input changer A' and 'input changer B'. The management plans or sets fire safety objectives. These fire safety

Figure 23.7 Feedback control

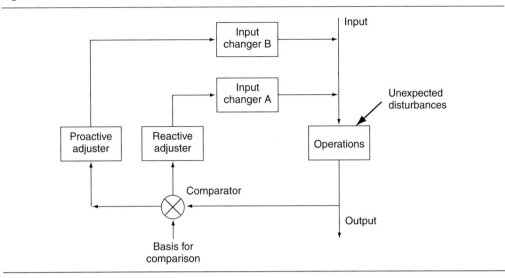

objectives are represented in the comparator. The function of this comparator is to compare the 'actual output' with the planned fire safety objectives. Thus, this control model can detect any deviation from the planned fire safety objectives through the comparator. The 'reactive adjuster' involves adjusting the input to the 'operations' through the 'input changer A'.

The 'proactive adjuster', on the other hand, is intended to manage fire safety proactively. In other words, the proactive adjuster involves anticipating any deviation from the fire safety objectives of the organisation proactively. In order to do so, the proactive adjuster involves modifying the input through the 'input changer B'. This process can be accomplished through modelling fire risk for the whole system. If the TFSMS is able to do so, then it can be said that the TFSMS is an adaptive system.

A simple example of how to control a system through this feedback control model could be a heavy goods vehicle (HGV) carrying hazardous materials. The truck and its route represent the 'operation', with the dispatcher as the management. The dispatcher schedules the journey of the HGV, and this schedule is represented by the 'comparator', as shown in Figure 23.7. The dispatcher monitors the schedule of the HGV and detects any deviation of the actual truck schedule from the planned schedule. If a deviation is detected, the dispatcher may elaborate a new plan, the 'reactive adjuster', in order to achieve the planned schedule of the HGV. Also, as another example, if a HGV is scheduled to drive through a region where a snowstorm is expected or already in progress, the dispatcher could 'anticipate' by assessing the impact of such an event on the planned route, and evaluate alternative routes. This process of changing plans is the function of the 'proactive adjuster'.

Communication paradigm
The TFSMS communication paradigm is governed by the four organisational principles listed in Appendix 23B, as suggested by Beer (1982). However, a complementary communication paradigm could be the communication paradigm suggested by Fortune and Peters (1995).

Figure 23.8 A bi-directional communication paradigm (from J. Fortune and G. Peters, *Learning From Failure*, © 1995, Wiley-Blackwell. Reproduced with permission)

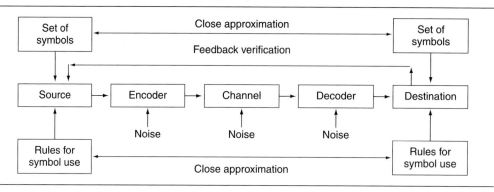

Figure 23.8 shows a dynamic two-way process of communication in which the sender's message can be used to modify subsequent messages. An example of the application of the communication paradigm can be found in Santos-Reyes and Beard (2006).

REFERENCES

Beard AN (1985) Towards a systemic approach to fire safety. *Proceedings of the 1st International Symposium on Fire Safety Science, Washington, DC*, 7–11 October 1985.

Beard AN (1999) Some ideas on a systemic approach. *Civil Engineering and Environmental Systems* **16**(1): 197–209.

Beard AN and Santos-Reyes J (1999) Creating a fire safety management system for offshore facilities. *Facilities* **17**(9/10): 352–361.

Beer S (1979) *The Heart of Enterprise*. Wiley, Chichester.

Beer S (1982) *Brain of the Firm*. Wiley, Chichester.

Beer S (1985) *Diagnosing the System for Organisations*. Wiley, Chichester.

Bell S (1999) 1,000°C blaze in alps tunnel kills 35. *The Times*, 27 March 1999.

Boyes R (2001) Nine killed in Alps tunnel as lorries crash. *The Times*, 25 October 2001.

Carter H (2010) Ferries and Eurostar lay on extra capacity to help Britons beat shutdown. *The Guardian*, 18 April 2010.

Channel 4 (2008) *The Unsinkable Titanic*. Channel 4, 3 November 2008 (television programme, UK).

Channel Tunnel Safety Authority (CTSA) (1997) *Inquiry into the Fire on Heavy Goods Vehicle Shuttle 7539 on 18 November 1996*. HMSO, London, May.

Cullen WD (1989) *The Public Inquiry into the Piper Alpha Disaster*. HMSO, London.

Fennel D (1988) *Investigation into the King's Cross Underground Fire*. HMSO, London.

Fortune J (1993) *Systems Paradigms: Studying Systems Failures*. Open University Worldwide, Milton Keynes.

Fortune J and Peters G (1995) *Learning from Failure*. Wiley, Chichester.

Leiding M (2001) 26 hurt in third tunnel coach crash. *The Daily Telegraph*, August 14, 2001.

Metro (2011) 7/7 problems were pointed out in 1987. *Metro* (UK newspaper), 9 February 2011.

Philip S, Leiding M, Coman J, Booth J and Foggo D (2000) 170 Perish in Alpine tunnel inferno. *Daily Telegraph*, 12 November.

Reason J (1997) *Managing the Risks of Organisational Accidents*. Ashgate, Aldershot.

Santos-Reyes J and Beard AN (2001) A systemic approach to fire safety management. *Fire Safety Journal* **36**: 359–390.

Santos-Reyes J and Beard AN (2003) A systemic approach to safety management on the railways. *Civil Engineering and Environmental Systems* **20**: 1–21.

Santos-Reyes J and Beard AN (2006) A systemic analysis of the Paddington Railway Accident. *Journal of Rail and Rapid Transit* **220**: 121–149.

Santos-Reyes JR and Beard AN (2011) A preliminary analysis of the 1996 Channel Tunnel fire. *Proceedings of the 3rd International Tunnel Safety Forum, Nice*, 4–6 April 2011. Tunnel Management International, Tenbury Wells.

The Times (2009) Thousands face tunnel chaos as bad weather halts Eurostar. *The Times*, 21 December 2009.

Turner BA (1976) The organisational and inter-organisational development of disasters. *Administrative Science Quarterly* **21**: 378–397.

Handbook of Tunnel Fire Safety, 2nd edition
ISBN: 978-0-7277-4153-0

ICE Publishing: All rights reserved
doi: 10.1680/htfs.41530.509

Chapter 24
Road tunnel operation during a fire emergency

John Gillard
Formerly of Mersey Tunnels, UK

Since the first edition the world has moved on. Equipment, communications and computer hardware and software have improved and will continue to do so. The capability to support and to a degree replace human activity has changed. However, the processes remain fundamentally the same and the intelligence to create and manage them is still required, whether it be human or artificial. The extent to which new technology can replace human beings in any tunnel operation, both normal and emergency, will be specific to each tunnel.

This chapter deals with tunnel safety from the tunnel operator's point of view. The general introduction discusses the organisation, arrangements and standards. The concept of stakeholders in safety is then followed by a review of the factors that influence safety, and then the nature of incidents.

After considering the role of the emergency services and the need for close liaison, planning and training, the chapter concludes with an idealised consideration of how incident response can be managed and actioned.

24.1. General introduction

To begin, it is worth recalling that it is often said that every tunnel is unique and must be dealt with individually. Whilst this may be true in some respects, it is equally true that standardisation and harmonisation are beneficial and should be pursued as far as possible.

Best practice results from learning from the widest possible experience, and combining this learning with a thorough understanding of local factors and their influence, particularly any restrictions and limitations that they impose.

There are no agreed mandatory international standards, although there exists the 2004 EU Directive 'Minimum Safety Requirements for Tunnels in the Trans-European Road Network'. Mandatory minimum standards apply in some countries, but this is not always the case. In some countries that have national standards, these may apply only to newly constructed tunnels owned by the national government, which may be only those on the primary or 'trunk' highway network. Also, such standards may apply to construction and equipment, to maintenance, to operation, or to all of these areas. There does not appear to be an ideal or even a preferred model.

However, where mandatory minimum standards apply, it may not be sufficient to merely meet their requirements. Safety culture has a core ethic of 'as safe as is reasonably practicable', which can also be defined as 'best practice possible within the constraints of local conditions'.

The legal framework within which a tunnel is owned and operated is also vitally important. The owner and the operator are often the same body. Many are national or local governments.

Tunnels are usually necessary to overcome obstacles, mountains or water, and obstacles often also form convenient boundaries between national or regional governments. It is essential in such situations that at the outset a single joint body is established with the total cooperation and agreement of the parties involved, and that this joint body must have complete and clearly defined constitutional terms of reference, resourcing, roles and responsibilities for the whole-life design, construction, maintenance and operation of the tunnel to proper standards and consistency throughout its entire length.

Where the owner is not the operator, the same principles must apply. The owner must create or appoint the operator. The operator's functions and responsibilities must be clearly defined in all physical and legal respects, and the relationship between owner and operator must be equally clearly specified and legalised.

It cannot be too strongly emphasised that any tunnel must be under the control of a single operator with uniform standards, procedures and processes throughout.

It is also vitally important that the standards, procedures and processes for any particular tunnel, whilst conforming to any applicable mandatory standards, should fully take into account the geographical and traffic environment and the availability or limitations, as the case may be, of local support resources.

24.2. The stakeholders in tunnel safety

There are many stakeholders in the whole-life safety of a tunnel, generally collected into five groups as follows

1. regulators and enforcers
2. tunnel designers
3. the tunnel operator
4. the emergency services
5. the tunnel users.

The interaction amongst the stakeholders is illustrated graphically in Figure 24.1.

24.3. The factors that influence tunnel operational safety

Many factors have an influence on the overall safety of a tunnel, but they can be collated into four main groups. All of the factors and the groups have a very interactive relationship. The groups can be illustrated as in Figure 24.2.

Some of the more important individual factors are listed below. The lists are not in priority order or exclusive:

1. Tunnel design and equipment standards:
   – tunnel highway geometry and length
   – tunnel cross-section
   – lane widths and carriageway markings
   – lighting quality

Figure 24.1 Tunnel safety stakeholders

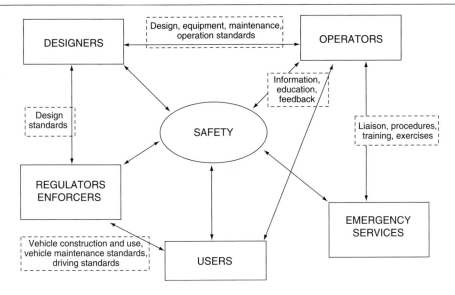

- air quality (ventilation)
- single or bi-directional traffic flows
- second or separate service tube
- emergency escape routes and distances
- communications equipment
- surveillance and traffic-monitoring equipment
- portal marshalling area facilities

Figure 24.2 Groups of factors influencing tunnel safety

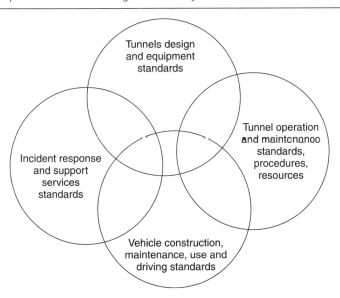

    – incident detection and alarm equipment
    – tunnel management and control centre.
2. Tunnel and maintenance operation standards:
    – proper planned preventative maintenance of the whole
    – proper emergency/contingency plans for critical failures
    – traffic management and control
    – traffic surveillance and incident detection
    – rapid vehicle breakdown response and recovery
    – control arrangements for hazardous loads
    – reliable, accurate, concise communications protocols and procedures
    – well-trained staff and good human–systems interfaces
    – good cooperation and mutual understanding with local emergency services
    – jointly agreed procedures, training and exercises for immediate response and subsequent support.
3. Vehicle construction and use, and driving standards:
    – nationally enforced proper standards
    – surveillance for defective vehicles on approach
    – surveillance for oversize vehicles on approach
    – surveillance for hazardous loads on approach
    – surveillance for improperly loaded vehicles on approach
    – surveillance for poor driving standards
    – facilities to stop and check anything questionable
    – power to apprehend and detain
    – driver education, particularly relating to tunnel conditions.
4. Incident response and support service standards:
    – quality and quantity of on-site tunnel operators incident response resources
    – quality and quantity of external emergency services resources
    – proximity and speed of response to all portals and onward to the scene.
    – speed of deployment to all locations
    – safe method of deployment
    – staff training (by or to the same standards as the emergency services)
    – working relationships and jointly agreed procedures
    – orientation, wayfind, training and exercises
    – communications equipment
    – emergency traffic management plans to ensure access.

This chapter is concerned with incident response, which is the last of the preceding groups. However, the importance of the first three in generally contributing to the overall safety of operation and incident prevention, and in more specifically contributing to the ability to detect and respond effectively to any incidents that do occur is very important.

Safety has two dimensions, which are reflected in the risk assessment process. These are

1. the chance of a harmful event occurring
2. the seriousness of the consequences of the event.

Statistics show that, historically, the initial cause of accidents is related to driver error or vehicle defects in approximately 95% of cases. Thus, the chance of a harmful event, or incident, being

Figure 24.3 The development of an uncontrolled incident

caused by a properly designed tunnel or its equipment (assuming that the tunnel and equipment are adequately maintained and operated) is relatively small. However, the seriousness of the consequences of an incident is related both to the circumstances of the initial incident and to how quickly it grows and develops. This is particularly true in the enclosed environment in tunnels. The speed and quality of detection and response are vitally important to impose control in the early stages, whilst this is safely possible.

Figure 24.4 The development of a controlled incident

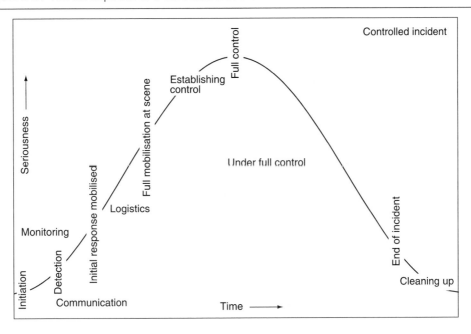

Although the initial consequences of an incident in a tunnel and on the open road are often generally much the same (in a tunnel with a controlled environment they may even be less), the subsequent developments in an enclosed environment can be much worse than the initial incident itself. The consequences of restricted access and escape plus containment of toxic gases and/or heat and smoke in the case of fire have been repeatedly demonstrated by previous events. Figures 24.3 and 24.4 help to illustrate the key difference between the seriousness–time relationship that is typical of tunnel incidents, controlled and uncontrolled.

24.4. The nature of incidents

Major incidents that result in almost instantaneous death, serious injury, explosion or immediate major fire are relatively rare in developed countries, and it is tragic when they do occur.

Minor incidents, although reducing in frequency as vehicle design, construction and maintenance, and driver education all gradually improve, are still relatively common. And it has been illustrated that, without rapid intervention, a minor incident can easily become the start of a chain reaction, leading to disaster. A fairly common obstruction such as a stationary vehicle is an obstacle waiting to be hit by an inattentive or fatigued driver. A minor accident can often be the cause of others. A minor fire can develop within minutes. A minor fire that is not brought under control can grow. On the open road, such situations rarely lead to disaster, but in the enclosed environment of a tunnel they easily can.

Thus, the characteristics in most tunnel incidents are that they normally have relatively small beginnings, and develop with time. Time becomes a crucial factor. Minimising the likelihood of incidents happening in the first place is one issue. Preventing them from developing, to control the seriousness of the consequences, is the second.

Time in this scenario has multiple elements that may be concurrent or consecutive according to circumstances – time to

- initially detect and react
- mobilise an initial response
- diagnose and determine further needs
- mobilise further support action as required
- protect human life
- establish overall control of the incident
- completely deal with the incident
- re-establish normal operation.

Each incident is unique. However, many incidents share similarities or commonality to some degree. Attempts to analyse and group incidents are often based on physical characteristics such as explosion, fire, flood, liquid chemical spillage, or hazardous gas or vapour escape.

From the tunnel operator's point of view, it is equally important to analyse incidents with regard to consequence. In this respect, consequence has three elements

1. the immediate consequence at the initial scene of the primary incident
2. the secondary consequences, which can subsequently develop at the scene (and which may spread if not brought under control)

3. the tertiary consequences, which can subsequently develop away from the scene, and seriously hamper access to deal with the primary and secondary situations.

A typical example illustrates the point: (1) a relatively minor collision or a vehicle with a fault; (2) the resulting fire that develops until either brought under control or burnt out; and (3) the resultant traffic congestion, which both hampers the response to (1) and (2) and magnifies the economic loss.

A further complication in the case of vehicle tunnels is that 'the scene' may itself be mobile.

There is a direct cause-and-effect relationship through the consequence sequence (1)–(2)–(3). There is also often the reactive relationship (3)–(2)–(1).

What is also very important is that whilst (1) is often either instantaneous or only relatively briefly transient, (2) and (3) are usually very time-dependent. Although there may be very little that can be done to control the seriousness of (1) (other than prior prevention and protection measures), much can be done to control the seriousness of (2) and (3), and the effectiveness will depend largely on resources and time.

Most important of all is the fact that the relative seriousness of (2) and (3) following (1) is often very much greater.

The foregoing applies generally to all incidents and harmful events. If, however, we exclusively consider tunnels, it becomes clear that elements (2) and (3) can gain even more prominent relative importance.

Although incidents may be very variable in their causes and characteristics, the objectives in dealing with them are always the same, and these are very clear. In priority order, they are

- to protect and save human life
- to establish initial control of the incident and prevent escalation
- to minimise environmental damage
- to minimise physical infrastructure damage
- to restore operational service as quickly as possible, if necessary on a temporary or reduced basis and to minimise use/economic/community loss.

24.5. Liaison between the tunnel operator and emergency services

Where time is of the essence and speed of response and intervention are critical, it follows that planning and resources plus training and exercises must be made available.

The tunnel operator must liaise very closely with the neighbouring emergency services and, in partnership, they must carefully assess their needs and put in place arrangements which will adequately provide for those needs and ensure the highest possible level of cooperation and coordination. This liaison must be structured so that it can deal with all issues, and, to be most effective, have direct interfaces at two levels, i.e. a strategic group and a tactical group.

The strategic group includes appropriate senior managers, and deals with policy, roles and responsibilities, and resources, and briefs a tactical working group. The agreement of relative roles and responsibilities and interfaces is particularly important. Where the interface lies, between the tunnel operator and the emergency services, will depend upon local circumstances and the tunnel operator's resources. At one extreme, the operator may have no emergency response resources and be totally dependent on external support. But it has already been agreed that 'time is of the essence', and, with due regard to risk and economies, it is highly desirable that the operator has sufficient trained and equipped resources to make an initial intervention to save and protect life and to prevent or at least minimise escalation.

A further role of the strategic group is to ensure that the emergency services have a clear understanding of the tunnel infrastructure, its capabilities and its limitations, and agree with the operator all essential safety controls, which are to be incorporated into procedures by the tactical group. This is particularly relevant to the potential performance of the ventilation system and its operation in any fire emergency situation. This, of course, may vary according to traffic conditions and fire location. There must be clear understanding of the options available in any given set of circumstances and a clear decision-making process, to avoid confusion creating additional danger.

The tactical group includes appropriate senior operation staff to deal with

- emergency procedures
- communications
- studies and assessment (including operational, risk assessments)
- inspections
- emergency equipment
- familiarisation, orientation and wayfinding
- training
- exercises.

The emergency procedures must cover the initial response role and responsibilities of the tunnel operator's staff, the agreed response of the emergency services, including approach routes, rendez-vous and handover arrangements, and ongoing roles and responsibilities after handover. The communications procedures must be precise and clear, to eliminate misunderstanding. Simple clear systems must be defined and agreed to describe locations. The capability and limitations of communications equipment must be clearly understood.

Studies, assessments and risk assessments must be used to ensure a clear shared understanding of the tunnel and its environment and the factors that may affect incident response.

Inspection needs must be identified, planned and carried out to ensure that facilities are always available and functional whenever required.

Emergency equipment provision and performance capability and limitations needs to be clearly understood between the parties.

Familiarisation, orientation and wayfinding are very important. Training visits for emergency service staff need to be regular. Emergency signage throughout, and durable route/location diagrams at points of entry for complex underground structures, must be considered.

Training is an issue for both the emergency services and the tunnel operator's staff. Emergency services personnel will benefit from training in the underground confines of the tunnel environment in simulated incidents. The tunnel operator's staff must be trained in the roles identified for them, including traffic management, breakdown recovery, communications, medical first aid, crash rescue and basic initial fire-fighting. Such training should ideally be to the same standards, employing the same methods, albeit at a more preliminary and basic level, to those used by the emergency services. It may even be provided by the emergency services trainers. This will ensure use of the same techniques and procedures and a smooth almost seamless handover on arrival of the external support.

Exercises are also necessary to test the procedures, arrangements and physical resources, and should be held regularly and as often as economically justifiable. Minor exercises can be arranged relatively inexpensively with minimal disruption to tunnel operation to test specific aspects. Major exercises require a great deal of planning and resources, can be very expensive, and will normally require total tunnel closure for an extended period to provide for incident scenario preparation, the exercise itself and the subsequent clean-up. Where adequate resources are available, a major exercise every 2 years should be the aim.

All exercises should be carefully recorded and debriefed, and the results fed back into the strategic and tactical processes as an aid to continuous improvement.

24.6. Incident response

How should a tunnel operator respond to an incident or emergency?

The following sets out a sequence of events in an idealised situation. It would be better formatted diagrammatically as an 'event tree' or as a process flow chart; however, the page size and format do not facilitate that here.

Each tunnel operator can construct such a diagram according to tunnel characteristics and layout, local circumstances, equipment and internal resources plus external/support resources.

Incidents and emergencies take many forms. There is no universal definition of 'tunnel incident' or 'tunnel emergency'. Even if there were, in the early stages of development of most incidents or emergencies there is rarely sufficient information to be able to make confident, precise quality judgements and decisions. Also, most incidents and emergencies have relatively small or minor beginnings, and develop with time if not dealt with and brought under control, and the manner in which they develop will depend upon the nature of the incident itself.

The safest approach is to regard anything whatsoever that adversely effects the normal, safe operation of the tunnel as an incident, to regard it as capable of growing into a major emergency or even potentially a disaster, and by responding to every incident with a clearly defined and well-rehearsed rapid-response procedure. It cannot be too strongly emphasised that the time and speed of response is vitally important. What is achieved in the first 5, 10 or 15 minutes will often determine the outcome. This is graphically illustrated in Figures 24.3 and 24.4.

The sequence is based upon the assumption that the tunnel has a traffic control centre with surveillance and adequate communications facilities plus (either combined or separate) an engineering function control facility. The sequence also attempts to be comprehensive.

It is also important to note that all communications, decisions and actions must be recorded and timed.

24.6.1 The passive stage

An incident is detected by any combination of

- an automatic alarm system (fire, stationary vehicle, etc.)
- CCTV surveillance (human observation)
- a verbal report (emergency telephone)
- a verbal report (exiting vehicle occupant).

24.6.2 The active stage

The duty traffic controller requests mobile tunnel patrol vehicles to attend from both directions if possible, and passes to them all information available. If the tunnel does not have its own mobile patrol vehicles, a substitute arrangement must be in place and be activated.

The controller continues to collect information by all means available and to pass it to the mobile. The first mobile arrives at the scene, assesses the nature and seriousness of the incident, passing a report back to the controller as it progresses.

As soon as the appropriate information has been passed, this first mobile will start to deal with the incident in the most appropriate manner and be assisted by others as they arrive, all in accordance with the incident response procedure.

The principle objectives of this initial action are twofold:

1. To collect the best quality information possible, to support decision-making, as follows:
   - Is fire already involved or a potential risk?
   - Are casualties involved? (Number and seriousness)
   - Is vehicle recovery required? (Number, type, weight)
   - Is load shift, load shedding, liquid spillage or gas escape involved?
   - What are the details, hazard protection and clean-up requirements?
   - Is a bus involved? (Number and condition of passengers. Is a replacement vehicle required?)
   - Is terrorist action involved?
   - What is the estimated total time required to deal with the incident?
2. To mobilise trained personnel and equipment to the scene to provide immediate first response as required in respect of:
   - medical first aid
   - fire-fighting
   - crash rescue
   - breakdown recovery
   - removal of obstructions or debris from the tunnel
   - traffic management and safety in the immediate vicinity
   - evacuation.

24.7. Decisions and actions

At the earliest possible time during the above described passive and active information-gathering process, decisions and actions must be taken. Equally, if any change occurs as the situation develops, previous decisions and corresponding actions must be revised appropriately.

Table 24.1 The five sets of decisions and actions

Decision	Action
1. Level of the traffic management plan required	
(a) Transient local control For very minor incidents that can be cleared up immediately (e.g. debris on the roadway, a broken-down vehicle that can be towed out by the attending patrol)	■ Officers at the scene stop all traffic clear of the scene using patrol vehicles with emergency hazard lights for protection ■ Advise the tunnel controller (TC) of actions being taken ■ Reopen using convoy escort to control traffic speed
(b) Stop all traffic and hold in position For minor incidents that can be dealt with and cleared within an estimated 10-minute duration (a straightforward incident, but where attending patrol officers and vehicles require further assistance that is immediately available and can be deployed without delay)	■ The TC to stop all traffic from entering the tunnel or portal approach, by illuminating 'stop' signs. Stop all traffic inside the tunnel, by illuminating 'stop' and 'stop engine' signs ■ The TC calls further assistance to the scene as required. 'Talk in' by radio. Liaise with officers in the tunnel by radio, using CCTV, to clear stationary traffic past the incident where safe and advantageous to do so ■ The TC advises the engineering controller of stationary traffic in tunnels plus any other relevant details ■ The TC advises emergency services operation control centres of 'tunnel closed' and to use alternative routes ■ The TC advises public media (radio broadcast 'traffic and travel'-type services) of traffic conditions and the estimated duration of the incident
(c) Stop all traffic and initiate the traffic diversion plan For incidents that require more time to clear (breakdown of a large vehicle, vehicle crash, damaged vehicles, shed loads, leakages, spillages, injuries, fatalities, fire, terrorist action or threat)	■ All as above, as appropriate and relevant, plus the following ■ The TC advises the local police control room of the need for the agreed wide-area traffic plan, to divert traffic away from tunnel approaches and to minimise consequent congestion ■ Additional actions as required, specific to the nature of the incident (e.g. advise the local police control room of the need to clear a route for ambulances if numerous casualties)
(d) Incident concluded	■ Incident dealt with ■ The TC and mobile patrols restore normal traffic arrangements in accordance with the procedure when safe to do so, using convoy escort to control traffic speed ■ The TC advises public media and emergency services that normal conditions are restored

Table 24.1 Continued

Decision	Action

2. Fire

(a) Fire is known or suspected to be present, or a significant risk

- The TC advises the local fire service in accordance with the agreed procedure, confirming approach routes and rendezvous points
- The TC advises tunnel response units en route to scene
- The TC activates the level (c) traffic management plan
- The TC activates emergency ventilation in accordance with the procedure (which may depend on the fire location and traffic conditions)
- The TC advises the tunnel engineering controller of fire alarm details/location, and to:
  - confirm emergency ventilation performance
  - evacuate any maintenance staff in the tunnel
  - put maintenance staff on support standby
  - await further information regarding ventilation action[a]
- The first response unit to arrive at the scene reports full details of the incident and fire back to the TC, and commences 'first response'. Report details must include information on any 'hazardous goods' involved
- The TC passes fire information to the fire service control centre, which relays this to vehicles and the 'incident commander' en route
- The tunnel duty senior officer attends the rendezvous point, hands over incident command to the fire service senior officer, and establishes on-site communications link using the agreed procedure and protocol
- Tunnel response units at the scene 'First response' actions are:
  - to save/preserve/protect human life (first aid, crash rescue, evacuate instructions, put smoke hoods on and rescue injured/immobile people, etc., as required)
  - to extinguish or at least control the fire, to allow escape and rescue
  - to provide best possible information back to the TC

Table 24.1 Continued

Decision	Action
	– to remain at the scene only as long as it is safe to do so – to hand over to the public fire service ■ The fire service set up their incident response plan and take full control of the incident in accordance with agreed procedures and move from rendezvous to fight fire ■ The tunnel duty officer takes a support role, in liaison with the fire service incident commander. Ensure good communications links. Call in stand-by personnel as required. Arrange additional support service and equipment as required
(b) Major incident or disaster is declared	■ Advise the appropriate local 'civil emergency planning agency' ■ Set up GOLD/SILVER/BRONZE[b] incident management arrangements ■ Set up a public and media information service ■ As soon as the incident is declared safe by the incident commander, the TC liaises with engineering control, the breakdown service and tunnel maintenance personnel, to remove all arisings and clean up. Note: depending on the nature and seriousness of incident, dynamic risk assessments may be required prior to commencing all or part of this operation ■ On completion of the clean-up, carry out a safety inspection ■ If/when safe to do so, restore normal traffic operations
3. Human injury or fatality (a) Human injury or fatality is known, suspected or a significant risk	■ The TC activate Level iii Traffic Management Plan ■ The TC advises all Emergency Services (Ambulance, Fire and Rescue, Police) and activate plan as for fire ■ On arrival, Ambulance Service and Paramedics report to agreed Rendezvous Point and Officer in Charge for time being (either Tunnels Duty Senior Officer of Fire Service Incident Commander) and receive instructions in accordance with agreed Procedure ■ TC will liase and establish communication and support

Table 24.1 Continued

Decision	Action
(b) Multiple serious injuries or fatalities are involved	■ Ambulance/Paramedic service request additional support and facilities as required ■ The TC liaises and support to provide location and services for mobile Triage, Treatment, Mortuary facilities as may be required
(c) Major incident or disaster is declared	■ The TC advise appropriate 'Civil Emergency Planning Agency' ■ Set up Gold/Silver/Bronze incident management arrangements ■ The TC set up public and media information service

4. Hazardous spillage

(a) Spillage of hazardous liquids or solids is known or suspected	■ The TC advises the local fire service in accordance with the agreed procedure
	■ The fire service responds as for fire
	■ The TC activates the level (c) traffic management plan
	■ The TC advises the tunnel engineering controller
	■ The tunnel engineering controller:
	  – evacuates any maintenance staff in the tunnel
	  – isolates pumps
	  – puts relevant maintenance staff and equipment on standby
	■ On arrival of the fire service, carry out a dynamic risk assessment and agree the safest method of dealing with the spillage
	■ Advise and liaise with the engineering controller
	■ On conclusion, clean all contaminated areas, drainage and sumps and safely dispose of arisings in accordance with any applicable controls or regulations
	■ Carry out a final safety inspection and tests
	■ Reopen as per the traffic management plan

(b) Escape of hazardous gas or vapour	■ All actions as for hazardous spillage ■ Activate the emergency ventilation procedure to protect life during complete evacuation ■ The TC, engineering controller and fire service incident controller agree the safest further deployment of tunnel ventilation to minimise human and environmental risk in containing/dispersing the escape, if still ongoing

Table 24.1 Continued

Decision	Action
5. Suspicious package, bomb or terrorist threat[c] (a) Information warning or threat received	■ The TC activates the level (c) traffic management plan ■ The TC advises the local police in accordance with the agreed procedure ■ The TC advises other emergency services to be on alert ■ The police act in accordance with their own terrorist alert plan (which appropriate security-cleared tunnel operator staff should be aware of) ■ The police advise the TC of the action plan for response and dealing with the threat, according to the level and nature of the threat and risk ■ The TC activates support and liaison arrangements in accordance with the agreed procedure

[a] Tunnel ventilation settings must only be further changed in accordance with agreed plans and with the full knowledge and agreement of the incident personnel via the incident commander.

[b] In the UK, local authorities are responsible for major incident and disaster emergency planning. Emergency plans are categorised in terms of seriousness and the level of resource and decision-making response required. For major incidents and disasters, command and control is described by three levels:

Level 1

'Bronze': operational control of the response – established at or as close as safely possible to the scene.

Level 2

'Silver': tactical decision-making, communication and support to the operational response – established at a suitable site with necessary facilities, nearby the scene.

Level 3

'Gold': strategic control centre, to which senior officers with the authority to procure and commit any necessary resources to deal with a major incident would report – established at the nearest emergency services headquarters.

[c] Arrangements and procedures for dealing with incidents of this type are normally classified as 'restricted' or 'secret', as special and/or military intelligence and expertise relating to national security will be involved. Nevertheless, appropriate arrangements and procedures with the relevant agencies should be agreed and put in place.

The decisions and actions are to a very large degree sequential, and can be defined within a procedure and, as stated earlier, illustrated by flow charts or event trees to suit the circumstances and resources and facilities of the tunnel operator and the tunnel controller, and the external emergency services and the procedures that have been agreed with them.

Also, there are 'sets' of decisions and actions, and each may be present or absent according to the nature of the incident. A typical draft that can be adapted and tailored for most tunnels is given in Table 24.1. It is grouped into five 'sets'

1. traffic management (which is the core, and common to all sets)

2. fire
3. human injury or fatality
4. hazardous spillage
5. terrorist activity.

For clarity, each set is listed separately in Table 24.1, but it is important to emphasise that each sequential set is initiated as soon as the need is known, and the whole process runs concurrently and is fully integrated.

Handbook of Tunnel Fire Safety, 2nd edition
ISBN: 978-0-7277-4153-0

ICE Publishing: All rights reserved
doi: 10.1680/htfs.41530.525

publishing

Chapter 25
Tunnel fire safety and the law

Arnold Dix
University of Western Sydney, Australia

25.1. Introduction

When fires occur in tunnels causing injury, death and damage to property, legal enquiries will be made of the adequacy of design, maintenance, regulation and operations of the tunnel.

This introduction to law provides an overview of some general legal principles and international trends in law as they apply to professionals engaged in tunnel design and operation. There are trends, themes and norms in law that span all countries in all jurisdictions. This chapter is a brief introduction to those trends, themes and norms. The chapter deals with both the responsibilities of the individual working within a project and the legal position of their employer.

The section on corporate crime highlights the significant variations around the world in the treatment of corporations for potentially criminal acts. This chapter is not a substitute for specific legal advice. The administration of justice is different in different countries while also changing with time. This spatial and temporal variability in legal principles demands timely and competent legal advice. This chapter can only serve as a means to help identify the circumstances in which such advice should be sought.

25.2. Legal investigations follow incidents

In most countries there is a requirement that an incident occurs prior to the legal system's investigative processes and procedures being activated. As would be expected, the greater the losses, whether it is by way of infrastructure or human injury and death, the more resources come to bear on the legal investigative process.

Because tunnel infrastructure is so expensive, it is inevitable that tunnel incidents can cause severe consequences – and tunnel professionals' decisions may be subjected to intense legal scrutiny.

Whether it is the fires in the Mont Blanc, Kaprun, Tauern, Burnley and Channel Tunnels, the Heathrow tunnel collapse, etc., or more minor incidents, it is inevitable that aspects of the original tunnel design will be scrutinised and aspects of the operational environment and decision-making process examined in a subsequent legal investigation.

25.2.1 Legal powers

In each legal jurisdiction in the world the government grants the legal system the power to

- reallocate wealth
- deprive a person of their liberty or their right to work
- compel a person to do something.

In each country, such powers are not limited to human individuals but include companies. The legal system exercises these powers on the basis of its findings as to facts and law.

For tunnelling professionals, it is essential that the favourable facts about a person's conduct are accepted by the legal inquiry as appropriate. This chapter is intended to help tunnelling professionals prepare themselves for the presentation of material that explains the basis of their designs, decisions and actions.

25.2.2   Standard to be applied to professionals

In all jurisdictions the standard to be applied to the tunnelling professional is based upon an analysis of the process by which a decision was made – not the outcome of the decision. In other words, in determining whether a tunnelling professional has reached the required legal standard, they are not judged with the benefit of hindsight.

There is, however, one important exception to this rule. There are trends in some countries to make some types of offences *absolute liability offences*. That is, the legal system deems the professional involved to be responsible whatever the circumstances. Such laws are increasingly popular because they avoid the difficulties of legal argument, as the facts of the incident prove the charge.

It is generally recognised that a professional in the tunnelling field has to determine which of, often, many hundreds or even thousands of factors is most important and must be managed, and how best to manage them. In such circumstances, it is the process by which the decisions are made that is often more important than the decisions themselves. It is inevitable that following a tunnel incident, many of these prior decisions will be subjected to critical review and analysis in the context of the subsequent incident.

When events transpire that cause damage, injury or death, it is expected that the reasons for the event will be sought. Often, the actions of individuals and organisations are identified as likely causal factors leading to the unwanted event. It does not logically follow that because damage, injury or death has occurred, any individual or organization has done anything wrong. Some things simply occur because an unlikely event occurred in circumstances where it was perfectly reasonable to take those risks. The fact that any person or corporation failed to adequately manage the risk that ultimately caused the adverse events to occur does not mean anyone is necessarily legally responsible for the adverse event.

Being able to demonstrate this fact is central to the requirement of sound and responsible decision-making prior to an incident.

In practical terms, these requirements may arise in circumstances such as where the board of a tunnel operations company unreasonably defers critical maintenance for safety equipment or when a tunnel manager refuses to appropriately consider a report from the tunnel's safety officer.

Recording the reasoning for a decision at the time it is made is usually a prudent measure.

25.2.3   Criminal responsibility

Tunnelling professionals are not exempt from the operation of what can broadly be described as *criminal* law in all countries. Should their conduct be such that they either knowingly, recklessly (i.e. they don't turn their mind to consider the question of whether someone will be injured or killed), or intentionally engage in conduct that endangers or actually injures and kills other people, there is a very real prospect that the legal system within that jurisdiction will interfere with their liberty, impose a financial burden, remove their right to practice in their chosen profession, or any combination of these consequences.

These concepts of criminal punishment are in the process of expansion in a number of jurisdictions around the world at present. This expansion is most clearly manifest in the area of corporate manslaughter or 'corporate killing'. Currently, this concept is enjoying recognition in a number of the common law countries.

The basis for the development of corporate criminal punishment is that organisations should, like individuals, take responsibility for the consequences of their criminal acts. The general theme of these developments in the law is that senior managers and/or members of boards or other controlling bodies of corporate organisations that engage in activities which result in death or serious injury can be held responsible for the crimes.

Furthermore, the individuals within those organisations in positions that would influence their conduct, such as 'members of boards and senior managers', can themselves be held personally responsible for the conduct of the corporation and be severely punished. These new criminal laws arise out of the development of an area of law known as corporate governance, which is intended to regulate the behaviour of large organisations.

This trend towards corporate responsibility for organisational misbehaviour has implications for the tunnelling sector. For example, an engineering design team that has inappropriately lowered the cost of a tunnelling project by unprofessionally reducing the level of safety within a tunnel may, subsequent to an event, find itself being pursued for corporate crime as well as subject to a more traditional legal investigation.

25.2.4   Engineering – the professional's legal position

As in all areas of human endeavour, management of an individual's or company's exposure to the exercise of legal powers in relation to a tunnelling project requires a number of steps to be taken, which include

- identification of actions that may be subsequently scrutinised through a legal process
- prioritisation of the importance of each of those actions
- understanding the way in which the legal system within a particular jurisdiction deals with people implicated in damage to property, injury and death
- familiarisation with the contractual basis upon which services are being provided
- familiarisation with the regulatory environment (independent of the contract) within the jurisdiction
- determining what requirements will be *implied* by the legal system within the country performing the tasks (independently of what may be written)
- determining the likely range of consequences that may arise in the event of legal scrutiny of the process.

Having ascertained the basic legal landscape within the jurisdiction, it is important to consider how best to conduct oneself professionally so as to best be able to explain the basis of one's decision-making process.

Inevitably, consideration must be given to the following range of documents.

- Agreements. These should be drafted in a way that reflects the norms of the environment in which any dispute will be resolved and articulates what will be required of professionals and also any specific instructions to professionals about limitations that are placed upon their conduct or restrictions on what matters they may or may not consider.
- Notes. In all jurisdictions, notes that are taken at the time a decision-making process is being conducted or of conversations, events and directions can often be used to assist an independent investigation to understand the basis for a particular course of action.
- E-mails. These have emerged as a highly critical form of documentation. Because e-mails are often written with little regard to their content, they can introduce a high level of vulnerability into a decision-making process, particularly if the decision is controversial either intellectually, politically or economically. Accordingly, there should be protocols developed to ensure that e-mails do not inadvertently provide a mechanism to bring professional conduct into disrepute.
- Online forums. Discussion threads in special-interest or social networking groups such as LinkedIn are often used to discuss contemporary issues. Such discussions can be used as evidence in court proceedings.
- Diaries. These are a traditional form of record that documents the personal recollections (often completed on an hourly or perhaps daily basis). Although one-sided (and usually self-serving), they are nonetheless excellent reminders of the process by which a decision was made. Diaries also have the advantage of allowing annotation of thoughts that might otherwise have not been expressed or communicated.
- Correspondence. This is an excellent way of documenting a process by which a decision is made. Because it is from one person to at least one other person and usually signed, it combines a record of the actual processes that have been conducted with a degree of consideration due to its formal nature.
- Expert reports. These are particularly useful, although they can introduce a degree of uncertainty into what should be a comparatively simple decision. There is often a temptation to 'hide' reports that are unfavourable: this should be avoided. Where expert reports are hidden, it is implicit that there is an expert who maintains a memory of having written the report and may have an ongoing concern that the report has never been acted upon. It is far better to explain why an expert report is not being followed than to ignore an expert report. Do not attempt to destroy or hide the expert report (or the expert).

25.2.5 Corporate criminal responsibility

In law, individuals and corporations are both treated as separate legal identities or 'persons'. There is an international trend towards pursuing corporations where their conduct has caused death and/or injury. An excellent analysis of these legal developments was published in late 2003 (Forlin and Appleby, 2003).

The international communities of most countries are demanding that organisations be held accountable for their conduct and not escape criminal prosecution due to legal technicalities.

Most common are criminal cases against company directors, senior managers and officers, while companies as a whole are also increasingly being pursued.

Law reform on corporate liability has commenced in most countries. In the European Union, it began in 1988, when the Committee of Ministers, Council of Europe, made recommendations on liability of enterprises having legal responsibility for offences committed in the exercise of their activities. Countries are embracing these changes in a variety of ways.

25.2.5.1 France
In France, the concept of corporate liability was introduced by the French Penal Code of 1994 (Act No. 2000-647). The French law does not require the demonstration of 'fault' for the corporation to be held liable. All that is required is that the act must

- be a criminal offence and
- have been committed by an individual of the corporation in the name of the corporation.

25.2.5.2 Italy
In Italy, corporations have only recently been exposed to criminal prosecution. Under the Decreto Legislativo 8 June 2001 N231 (Gobert and Mugnai, 2002), a company can be prosecuted whether an individual can be identified or not. Under the Decreto Legislativo the corporation can be prosecuted for deaths caused by its directors, subsidiaries and managers. The Decreto Legislativo also anticipates death through operation of the corporation's policy and even organisational incompetence. That is, a corporation may be prosecuted because of the potential for death within its operations, even if death has not actually occurred. Importantly, Italian organisations can defend criminal prosecutions on the basis that the organisation met the requirements of due diligence in its corporate governance.

25.2.5.3 The Netherlands
The Netherlands has recognised corporate criminal responsibility for over 50 years. The way in which the legal system embraces corporate crime considers not who committed the act but instead whether the corporation engaged in the particular activity, and then subsequently whether such activity would normally be condoned. As such, this approach focuses on the corporate personality when determining the guilt of the corporation.

25.2.5.4 Germany
German law does not recognise corporate crime. Criminal charges are normally brought against individuals within corporations. This approach is almost opposite to the Dutch criminal justice system. Accordingly, in Germany the conduct of the individual regularly receives more detailed investigation when there are criminal inquiries.

25.2.5.5 Norway
Criminal liability for corporations was introduced as law in 1991 through amendment of the Norwegian General Civil Penal Code of 1902 (Act No. 10). Under Norwegian law, the corporation may be liable even where no individual can be identified as having committed a criminal offence. In determining what penalty is applied to a corporation, the Norwegian courts are required to consider whether the criminal conduct provided a benefit to the corporation and whether refinement of the corporation's internal governance systems would be an appropriate remedial action that would reduce the likelihood of the criminal behaviour occurring again.

25.2.5.6 The USA

Corporate criminal liability has been established and pursued in the USA for over 100 years. Under US law, companies can be held liable for the actions of their subsidiary companies – even if the criminal acts took place before the companies were taken over and, in certain circumstances, even overseas (*United States* v. *von Schaik* (1904, CC NY), 134 F 592; Union Carbide Corporation Gas Plant Disaster at Bhopal, India December 1984, 809 F 2d 195). US courts impute knowledge to corporations even where no single employee knew or behaved in such a manner as to attract criminal liability to themselves. Importantly for the tunnelling sector, independent contractors have been held criminally responsible despite contractual arrangements being put in place in an attempt to shield the principal corporation from responsibility (e.g. *United States* v. *Parfait Powder Code*, 163 F 2d 1008 (7th Cir, 1947)). At a state level within the USA, the laws with respect to corporate criminal liability vary, and are an adjunct to the US federal laws on the subject.

25.2.5.7 Australia

Under Australia's common law and, since 1995, pursuant to the Commonwealth Criminal Code, corporations can be held criminally liable. Specific laws for corporate manslaughter have as yet not been introduced at State levels although the Australian Capital Territory has enacted such provisions.

The Australian Criminal Code Act 1995 (Section 12) attributes the acts of agents, employees and officers to the corporation. Furthermore, approval of those acts can be proven by the criminal prosecution demonstrating that

> a ... the body corporate board of directors, intentionally, knowingly or recklessly carried out the relevant conduct, or expressly, tacitly or impliedly authorised or permitted the commission of the offence; or
> b ... that a high managerial agent of the body corporate intentionally, knowingly or recklessly engaged in the relevant conduct, or expressly, tacitly or impliedly authorised or permitted the commission of the offence; or
> c ... that a corporate culture existed within the body corporate that directed, encouraged, tolerated or lead to non compliance with the relevant provision; or
> d ... that the body corporate failed to create and maintain a corporate culture that requires compliance with the relevant provisions.

However, because Australia is a federation, crimes such as manslaughter are regulated by the Australian states, and, despite current attempts to impose corporate liability for manslaughter, it has as yet not been widely introduced.

25.2.5.8 The UK

In the UK, the legal system has long recognised that corporations may be guilty of criminal offences. However, it is a prerequisite to sustaining such a criminal prosecution that an individual has also committed a criminal offence (*Walker* v. *Bletchley Fletoms Limited*, 1 (1937) 1 all ER 170). There has been considerable pressure placed upon the government to reform the law so as to allow corporate criminal prosecution in circumstances where an individual has not been found guilty of the criminal offence.

25.2.6 Evidence

In all countries, a legal investigation requires the consideration of evidence. How that evidence is considered varies from jurisdiction to jurisdiction. Understanding the nature of evidence is fundamental to developing an appropriate strategy for documenting robust decision-making in a project.

Most importantly, there are two types of evidence: material that can be used and material that cannot be used. From a legal perspective, the first critical issue is whether or not a piece of evidence can be brought before the person or people making the legal investigation or not.

If the evidence is able to be brought to their attention, it is often known as admissible evidence. If it is not able to be brought to their attention, it is usually considered inadmissible evidence. It is essential that all strategies designed to document key decisions with respect to tunnelling be in the most appropriate form of evidence. Although the rules of admissibility of evidence vary substantially from country to country, it is fundamental that for documentary evidence

- the author be identified
- the origin of the document be independently provable
- the date of the document be provable.

In the case of both documentary evidence and oral evidence, it is essential that care be taken to identify the person whose evidence is used.

Using an appropriate person to either create a document or give oral testimony can fundamentally affect how problems are perceived in a legal investigation. It is the process of evidence collection by which decisions are made, and therefore it is crucial that the evidence which goes before a legal decision-maker be appropriate in terms of content, expression and presentation.

If information is not properly brought before a legal inquiry, there can be no expectation that the legal inquiry will be able to deal with the problem that must be resolved appropriately. Most tunnel legal investigations are conducted by lawyers assisted by engineers.

25.2.7 Opinion evidence

It is inevitable that there will be conflicting opinions with respect to issues in dispute. For this reason, obtaining advice from people with experience of a particular issue is essential. The level of knowledge on a particular subject is critical in all legal investigations in determining the weight to be placed upon that opinion. The more experienced and independent the person providing the opinion, the greater weight is likely to be placed upon it for the purposes of resolving a dispute.

In many circumstances, the opinions relied upon by a legal inquiry are from people identified to be external to a project and, therefore, implicitly, not under the influence of unseen pressures.

It is well understood that in major infrastructure construction projects there are often factors that, due to time, cost and even personalities, drive the decision-making process within the project. When documenting the reasons for a decision within a project, it must be borne in mind that such factors are of little relevance (if any) to appropriate decision-making, and particularly the process by which tunnelling professionals are required, by law, to meet a minimum standard independently of the pressures under which they are working.

25.3. Legal investigations scrutinise past decisions

It is inevitable that accidents will occur. Accordingly, this chapter seeks to best prepare the engineer or tunnel decision-maker for taking reasonable steps to ensure that the basis for the decisions which are made with respect to the design and operation of the tunnel are defensible.

From a legal perspective, it is extremely desirable to defend a decision using information created *before* an incident took place. Examples of material that may reasonably be relied upon subsequently to the actual decision as justification for making decisions include

- international standards
- contractual requirements
- professional guidelines
- computations
- memoranda
- reports
- directives
- known best practice

from the *time* the decision was made.

It is always a difficult task to determine the state of world knowledge on a technical question in retrospect. Articulation of the information relied on at the time of making the decision is therefore often critically important to demonstrating the appropriateness of a decision.

In law, the criteria used to determine the appropriateness of a decision is almost universally the state of knowledge at *the time the decision was made – **not** in hindsight, not with the wisdom gained following an event*.

Understanding this fundamental legal truth provides an informed basis for making, and subsequently demonstrating, that a decision made was appropriate.

25.3.1 First principle of tunnelling decisions

Those responsible for the design and operation of a tunnel must bring to the task due professional skill and care.

For so long as there have been engineers there have been engineering failures rewarded by punishment. In Babylonian times the engineer responsible for a collapsed housing complex was put to death. Except in the most extreme circumstances it is unlikely that an engineer would be put to death today; however, the theme from those ancient Babylonian times remains.

In each country, the professional standard that is required to be discharged by the tunnelling professional may vary; however, at the very least the following apply:

- A person who accepts a task to design or operate aspects of a tunnel must bring to that task appropriate expertise.
- In determining what level of expertise is appropriate, regard may be had to:
  - the qualifications of that person

- the experience of that person
- the minimum standards of a person able to discharge the professional requirements
- the contractual requirements
- the requirements imposed by operation of the laws within that country
- the standards imposed by treaties, conventions or international agreements to which that country is a signatory
- national conventions
- national practices
- international practices
- publications of internationally recognised bodies, for example,
  □ the World Road Association (PIARC)
  □ the United Nations Economic Committee for Europe (UN, 2002)
  □ the National Fire Protection Association (NFPA, 2001, 2003).

Each project must have its technical issues determined on their merits. It is normally not sufficient for a tunnelling professional to copy another project.

Increased international communications on tunnelling issues coupled with new learning and regulatory intervention mean that it is inevitable that there are conflicts between differing standards and practices in tunnels. It is therefore essential to explain why a particular regulatory platform has been chosen for a project. In some instances, the explanation may be that it is to comply with national standards. However, in many instances other approaches may be adopted such as using international standards, amalgamations of standards or even novel approaches.

25.3.2 Economic considerations

Recent drives for economic efficiency throughout the world have placed increasing pressure on tunnelling projects to increase their economic efficiency. This has resulted in phenomena such as designing for the lowest bid.

Where a tendering process is used to select the design, tenderers sometimes adopt a lowest-bid policy in order to get the work. Subsequently, they attempt, by way of variations and other elaborate techniques, to improve the design of the tunnel projects. Designing for the lowest bid is most likely to impact on fire life safety issues. Fire and other life safety issues are likely to result in legal action against designers, constructors and operators of tunnels.

Tunnels are always extremely costly, usually more than expected. Professional standards must be maintained while appropriately addressing the cost issues of projects. An allegation of a failure to maintain professional standards is one of the simplest allegations to make against a tunnelling professional after an accident. The best way to defend such a decision is to have documented the decision at the time it was made.

25.3.3 Risk analysis

The emerging area of risk analysis and its various sub-disciplines poses fertile grounds for potential future legal action. Risk analysis, in no matter what form, is a way of documenting a decision-making process. By documenting the decision-making process, it is thought that the decisions themselves will be better through a combination of transparency and forced consideration of various alternatives.

Unfortunately, many of the events for which design decisions are made are high-consequence low-probability events for which conventional risk analysis is intellectually vulnerable. Whether it is qualitative or quantitative analysis that is undertaken, there remains the ability to manipulate the process to achieve an outcome by distorting the risk analysis process. There are many examples of risk analysis that have been conducted to justify the departure from engineering norms to more cost-effective (cheaper) solutions.

In many circumstances, this is appropriate; however, the abuse of this decision-making technique to justify what would otherwise be unjustifiable engineering decisions is likely to be the most fertile ground for future legal proceedings in tunnel projects.

Nonetheless, recent tunnelling initiatives have all adopted the risk-based approach to design and operation. For example, EU Directive 2004/54/EC on trans-European road tunnels, the United Nations Economic Committee for Europe's analysis of rail tunnels (UN, 2002) and the *Joint Code of Practice for Risk Management for Tunnel Works in the UK* (British Tunnelling Society and the Association of British Insurers, 2003) all invoke risk-based analysis for aspects of the decision-making process in tunnelling projects.

There is no agreed risk management methodology – such a global methodology does not exist. In the author's opinion, there will never be an agreed global methodology because risk assessments reflect the unique social and ethical values of a community, including the value that the community places on a human life.

25.3.4 Quantitative risk assessment

Because lawyers in all countries deal with damages and injuries after an event has occurred, it is a comparatively easy task to argue that a quantitative or qualitative risk assessment was done incorrectly – because the event has actually occurred. Such an argument is highly persuasive even if it is intellectually incorrect.

Because of this method of criticising the decision-making process, it is essential that any risk-based analysis of various options be clearly articulated. In articulating the reasoning, care must be taken to explain the decision-making approach in terms that will not be intellectually offensive to the legal system of a country.

For example, in the Netherlands it is accepted that computations can be made using the value of a human life (which is calculated from time to time for citizens of the Netherlands); however, in other countries such a process would be absolutely unacceptable. Specific legal advice should be taken as to the appropriateness of this type of decision-making tool for a particular jurisdiction.

Often, the practical solution to this dilemma is to use the tool (as is appropriate in good engineering) but to be careful with the language that is used to describe the outcome. For example, if the nominal value of a human life for the purposes of a quantitative risk-based decision-making analysis is US $1 million, and the analysis determines that option A would reduce the theoretical loss of life risk by one life per year, and option B would reduce the theoretical risk by one life every 10 years, care should be taken in expressing the result. The analysis results could be expressed in terms such as 'The quantitative risk assessment suggests that investment in option A would result in a tenfold benefit in saving human lives than option B.' It should not be reported as: 'It is considered not worth spending the money on option B because it will only save the life of one person every 10 years.' (In many jurisdictions, such a proposition would be responded to by 'But engineer, what if that life is that of your child?')

This fundamental misunderstanding of the role of risk analysis in decision-making – and its role of differentiating the benefits between comparative investment strategies – is often mistaken by lawyers as evidence that engineers designed a project to kill a predetermined number of people by justifying a foreseeable number of deaths in a project.

Great care must be taken with the management of documentation and process to ensure that the decision-making process is not misconstrued and its outcomes wrongly interpreted (see also Chapters 18, 22, 29 and 30).

25.3.5 Economics
Tunnelling projects are often being built in countries to stimulate their economies through infrastructure investment. In circumstances where the client for a tunnel is attempting to drive an economy through infrastructure investment, tunnelling professionals must be very careful to ensure that they are not seen to conspire with the government or other controlling organisation in delivering a tunnelling outcome that is of a lower standard than what would generally be considered an acceptable minimum standard.

There are circumstances in which such tunnels can be properly built and legal liability avoided by tunnelling professionals. However, they do require particular attention to detail and clear articulation of what is considered acceptable and unacceptable by the client.

Such requests for tunnels usually arise in economies that are experiencing an economic recovery following political or social reforms. Although there exists a desire to improve infrastructure, there is often inadequate funding, a lack of specific tunnelling expertise, and comparative political instability to run and maintain an advanced tunnel design.

Because the tunnelling community is a mobile international community, there is a very real risk that a consultant will inadvertently be drawn to providing advice on a project that he understands would not meet the standards within his country of origin, but does not articulate this to the client – subsequently, if an event occurred where there is significant damage to the infrastructure, injury or death, the consultant will be open to legal criticism.

Because tunnels remain operational for comparatively long periods of time (upwards of 100 years) the legacy of an earlier design may become the subject of international scrutiny long after completion. This is particularly so given the rapid expansion and massive infrastructure investment in the Asia/Pacific region, South America and Eastern Europe.

25.3.6 Issues to beware of
There are a number of recurrent issues in tunnelling that stand out from others as potential future subjects for independent legal investigations. These are (in no particular order of importance)

- the fire size used for the design of tunnels
- the assumptions with respect to human behaviour in the event of a fire (or other scenario requiring a change in human behaviour in tunnels), for example evacuation or going to a point of comparative safety

- the use of fire suppression systems
- the implications of the fire suppression systems for other safety systems
- the actual tenability of environments following fire or other life-threatening events underground
- performance engineering as a means of cost cutting
- cross-passages in terms of both spacing and design
- signage
- computed smoke volumes
- smoke extraction and both emergency and non-emergency states
- effects on human beings of changes in tenability of environment
- health impacts of vehicle emissions both within and external to tunnels for internal-combustion-powered vehicles
- maintenance regimes for existing equipment
- refurbishment of existing tunnels
- adequacy of emergency exercises
- traffic predictions
- combining practices and standards from different countries and operational environments to create new standards and operational practices
- inappropriate use of models.

25.4.   Conclusions

Tunnelling professionals provide their advice on how best to ensure that the underground environment meets appropriate levels of safety and performance for their clients. However, as with all areas of human endeavour, it is inevitable that incidents occur that may cause damage to the tunnels, injury to people and, in some circumstances, death. No matter how well built and operated an underground tunnelling environment there remains a residual risk of damage, injury and death.

Throughout the world, the legal system seeks to ensure that a minimum standard of expertise is provided for these tunnels by the tunnelling professions and that where that standard is not met and harm occurs, society is able to punish those who have wronged and, where possible, reinstate people to the position they had before the harm they suffered at the expense of those who have wronged.

It is also the objective of a legal system to try to promote future practices and conduct that would not result in the similar circumstances causing damage, injury or death in the future.

It is fundamental to the proper administration of justice that professionals involved in tunnelling anticipate independent legal analysis of their conduct, and document the process and procedures they have followed in order to come to their decisions.

In most circumstances, the ability to clearly explain the basis upon which decisions are made in tunnelling projects is fundamental to protecting both the individual and their organisation from adverse findings by a legal investigation.

It is important to distinguish between adverse findings against an individual or an organisation and recommendations for change.

It is not only inevitable but highly desirable that legal investigations make recommendations for changes in practice, conduct and procedures following damage, injury and death.

Being able to communicate the basis for decisions prior to such events occurring is the best way to protect an individual or an organisation from adverse findings from such a legal investigation.

It is not illegal for a tunnelling professional to incorrectly predict the future. It is, however, contrary to the laws of all countries to give the impression that considered expert advice has been provided and given very serious consideration when it has not.

REFERENCES

British Tunnelling Society and the Association of British Insurers (2003) *Joint Code of Practice for Risk Management for Tunnel Works in the UK*, 1st edn. British Tunnelling Society, London.

Committee of Ministers, Council of Europe (1988) *Recommendation on Liability of Enterprises having Legal Personality for Offences Committed in the Exercise of their Activities.* Recommendation No. R(88) 18. EC, Brussels.

Forlin G and Appleby M (eds) (2003) *Corporate Liability: Work Related Deaths and Criminal Prosecutions.* Tottel, London.

Gobert J and Mugnai E (2002) Coping with corporate criminality; some lessons from Italy. *Criminal Law Review* 619.

NFPA (2001) *NFPA 502: Standard for Road Tunnels, Bridges, and Other Limited Access Highways.* National Fire Protection Association, Quincy, MA.

NFPA (2003) *NFPA 130: Standard for Fixed Guideway Transit and Passenger Rail Systems.* National Fire Protection Association, Quincy, MA.

UN (2002) *Safety in Railway Tunnels.* UIC Codex 779-9. UN Economic Committee for Europe, Geneva.

Part V

Emergency procedures

Handbook of Tunnel Fire Safety, 2nd edition
ISBN: 978-0-7277-4153-0

ICE Publishing: All rights reserved
doi: 10.1680/htfs.41530.541

Chapter 26
Emergency procedures in road tunnels

David Burns
Formerly of Merseyside Fire Brigade, UK
Michael Nielsen
Formerly of Tyne & Wear Fire Brigade, UK; Provided updated chapter

26.1. Introduction

This chapter examines the current practice concerning operational issues that arise when dealing with emergencies in road tunnels. The key factors that affect safety in road tunnels and that have a bearing on emergency procedures are discussed, and the role currently played by emergency intervention should an incident occur is considered.

Emergency response can be considered in terms of traffic management, security and policing, fire-fighting and emergency medical assistance. These actions can be effected by dedicated teams, or by the local and/or national emergency services. In this chapter, the term 'emergency response team' is used to cover all types of emergency response.

Tunnels have many common features, and all pass through or under obstructions, such as rivers, mountains or even built up areas. In general, there is little natural ventilation, little or no natural light and, more often than not, access to and from the tunnel is via the portals (i.e. there are no other means of escape). In tunnels sited on major international routes, tunnel users may be tired from travelling, may not be alert, and may be foreign nationals who cannot understand safety advice or announcements. Some tunnels have 24-hour multi-mode monitoring and surveillance as well as providing dedicated emergency response teams, whereas others have nothing. Some tunnels have different operators from different countries responsible for different sections of the tunnel. In tunnels, the functioning of communications devices, such as two-way radios and cellular telephones, can be compromised. Smoke and hot gases may stratify quickly, and other products harmful to health may be released in an accident and cannot easily disperse.

26.2. Managing safety in tunnels

It can be argued that tunnel safety is dependent on three factors.

- Tunnel design: the basic layout of the tunnel, including all its engineering services and fixed installations; the provision of means of safe egress and access for fire-fighters and rescuers.
- Tunnel management: operational management, traffic management and engineering management; integrated systems that ensure a rapid response to any emergency.
- Emergency response: early intervention to reduce the impact of an emergency and reduce fire development.

Safety in a tunnel is achieved through a mix of these three factors, and it should be stressed that both tunnel design and tunnel management will significantly affect emergency response, both in terms of the nature of that response, and in terms of its role in ensuring the safety of those using the tunnel. Each factor must be examined when considering the nature of the emergency response.

26.3. Tunnel design

One of the most basic of design features is the length of the tunnel. Generally speaking, the longer the tunnel the greater the problems associated with access under fire conditions. If a long tunnel is smoke-logged, there can be enormous difficulties associated with getting fire-fighters to the scene of the fire, given both the duration of operation of breathing apparatus and the capacity of the fire-fighters themselves. However, tunnels do not need to be particularly long to provide the potential for disaster involving considerable loss of life. For example, the fire in the Tauern Tunnel in 1999 occurred within 1 km of the tunnel portal, the St Gotthard tunnel is over 16 km long, but the fire in 2001 occurred 1.5 km from one of the portals, and in the 1999 Mont Blanc Tunnel fire all fatalities occurred within 850 m of the initial fire.

Another important feature is the number of bores, or tubes, the tunnel is provided with. In single-tube road tunnels, traffic normally flows in two directions through the single tube (i.e. bidirectional flow). Two-tube tunnels, on the other hand, can operate with traffic flowing in one direction through each tube (i.e. unidirectional flow). However, even in two-tube tunnels, road layouts can be changed so that bidirectional flow takes place in order, for example, to accommodate peak (or 'tidal') flows of traffic or maintenance work. The implications of this are mentioned further below.

One problem associated with bidirectional flow is that, should an accident occur, be it a fire or a collision of some sort, it is likely that traffic travelling in both directions will be affected, and may be brought to a standstill, with traffic queues in both directions. This makes intervention by emergency response teams more difficult because of the traffic build-up. This difficulty is magnified greatly if heat and smoke are present.

There is also the risk of a head-on collision when bidirectional traffic flow is in operation. Experience has shown that the likelihood of serious road traffic accidents is reduced in two-tube tunnels when operating in unidirectional flow mode. It is also the case, therefore, that the likelihood of fires occurring as a result of road traffic accidents is also reduced.

The provision of service tunnels and/or cross-passages in two-tube tunnels can facilitate a safe means of escape for tunnel users, as well as providing a safe route to the scene of operations for emergency response teams via the unaffected tube, provided they are adequately protected from the ingress of smoke and heat. However, service tunnels and cross-passages do not guarantee safety. Emergency refuges were provided in the St Gotthard Tunnel; despite this, 11 people perished in a fire in 2001.

As stated above, carrying out maintenance in a tube can result in that tube being closed, effectively turning the tunnel into a single-tube tunnel, with bidirectional flow. It is suggested that tunnel operators give consideration to the time of day when such maintenance is carried out, and perhaps avoid periods of high traffic volume. Clearly, there may be cost implications arising from this action. Some tunnels operate tidal flow systems to accommodate traffic flows at peak periods. For example, in two-tube tunnels where each tube has more than one traffic lane, one tube can operate in unidirectional flow mode while the other tube operates in bi directional flow mode.

Clearly, when this mode of operation is put into practice, some of the problems associated with single-tube tunnels will be apparent.

The dimensions of the tunnel, such as length and cross-sectional area, should also be taken into account. They will, among other factors, play a role in determining the rate of heat build-up and the rate of smoke stratification. Clearly, the tunnel dimensions do not play a role in isolation, but have to be considered together with the ventilation characteristics of the tunnel. Secondly, the tunnel cross-sectional area may determine whether emergency walkways can be provided at the side of the carriageway to facilitate means of escape. It is also the case that the amount of head-room will determine whether jet fans can be fitted. These characteristics will, in turn, affect the ease with which an emergency response can be made.

The point has to be made that, once constructed, it is virtually impossible to make major changes to the design of a tunnel. Some design changes can be made, as in the case of the Mont Blanc and Fréjus Tunnels; however, the massive financial costs of such projects may well render such changes impractical, even if they are technically feasible.

26.3.1   Engineering services and fixed installations

Although engineering services and fixed installations can be regarded as a subsection of tunnel design, they can often be altered or upgraded in a way that the basic layout of the tunnel cannot. However, such works are still very expensive. In any case, they will have a bearing on the effectiveness of emergency response.

Mechanical ventilation systems are a major factor in ensuring safety, both of tunnels users and emergency response teams. Ventilation systems include those required to provide a supply of fresh air and to exhaust foul air during the day-to-day operation, often using ducts running above or below the road decks. It is worth noting that changes to the design of the Mont Blanc and Fréjus Tunnels, which run between France and Italy, have enabled the fresh air ducts to be used both as a means of escape and as a means of access for emergency response teams.

Additional systems for use purely in the event of an emergency (e.g. jet fans) may be provided to force heat, smoke or noxious gases in one direction or another. However, basic tunnel design will affect the provision and use of these jet fans. As mentioned above, there must be sufficient headroom to allow the fans to be fitted. If there is bidirectional traffic flow in operation, either in a single-tube tunnel or in one tube of a two-tube tunnel, the use of jet fans becomes difficult. This is because the occurrence of fire often results in stationary traffic building up in both directions. The operation of the fans in these circumstances would result in heat and smoke being directed onto a line of standing vehicles and their occupants.

It should not be forgotten that all tunnels have natural ventilation characteristics, which are determined by the tunnel dimensions and inclination, topographical features and prevailing winds. The effect of natural ventilation has to be factored into the emergency response contingency planning process.

Examples of other fixed installations and engineering services include: warning systems for drivers, such as radio data systems (RDS) transmissions, CCTV, fire-fighting equipment, fire-suppression systems, and manual and automatic fire alarm systems. The provision of carbon monoxide monitoring equipment, commonly found in road tunnels, can also provide an effective fire alarm system.

26.4.   Tunnel management
26.4.1   Traffic management

Traffic management can be regarded as the second most important factor in determining safety in tunnels. It involves managing

■   traffic volume
■   traffic speed
■   the nature of the vehicles using the tunnel, including those carrying hazardous loads.

It is much easier to effect change in respect of traffic management than tunnel design. Restricting the speed of traffic can reduce both the likelihood of an accident occurring and the severity of that accident. Furthermore, the risk of a major conflagration can be reduced if restrictions are placed on the nature of the traffic using the tunnel. All of this will have a bearing on how effective an emergency response will be.

With regard to traffic volume, it is interesting that some tunnel operators have noted that the higher the traffic volume, the larger the number of accidents. Should a fire occur as a result of an accident during periods of high traffic volume, the number of people exposed to the hazard will also be high. One compensating factor is that heavy traffic volumes tend to reduce traffic speed and the nature of accidents tends to be less serious. Serious traffic accidents, on the other hand, tend to occur when traffic speeds are higher. As high speeds tend to be achieved when traffic volumes are light, the potential for a large number of vehicles to become involved in any fire or release of harmful products arising from an accident is lower.

Traffic may be restricted to cars and light commercial vehicles. It is thought that the heat output of a fully developed car fire is unlikely to pose a serious threat in terms of fire spread and development, especially when compared with that of buses and commercial vehicles. The heat output from a car fire is relatively low, and the products of combustion are more easily dealt with by ventilation systems. In terms of the emergency response, such fires can be approached and dealt with much more easily than can those involving large commercial vehicles.

The heat output from fires involving commercial vehicles and buses is known to be much greater than that for cars, although to what extent remains unclear. For example, during a fire in 1997 in the St Gotthard Tunnel involving a car transporter carrying eight new passenger cars, the peak fire size was estimated to be 22 MW (Henke and Gagliardi, 2000). However, Norwegian engineers estimated that the heat output from an articulated, single-deck bus fire in a tunnel in Oslo in 1996 had reached some 36 MW after only 6–10 minutes and averaged 13.4 MW (Skarra, 1997).

Clearly, the size of a fully developed fire in a commercial vehicle will depend on the nature of its load. It is worth noting that the lorry involved in the Mont Blanc Tunnel fire in 1999 was carrying flour and margarine, and one of the initial two lorries involved in the St Gotthard Tunnel fire in 2001 was carrying tyres – products that are not normally regarded as being particularly hazardous. Indeed, the heat output of a fully developed large goods vehicle fire has been estimated as 100 MW, and as 300 MW for a large hydrocarbon road tanker (Broekhuizen, 1999) (see also Chapter 14). Further research may be required into what may be considered a hazardous load, together with research into the fire loading of commercial and passenger transport vehicles arising from the nature of their construction. With regard to emergency response, the heat output from such fires is such that fire-fighting teams are unlikely to approach the incident, let alone make any realistic attempt to put the fire out. In such cases, fires tend to burn themselves out after all the available 'fuel' has been consumed.

The involvement of hazardous substances (e.g. toxic or corrosive substances) can increase the severity of an incident further still. However, the point should be made that tunnels may also contain and ameliorate the effects of such substances, and prevent them escaping into the wider environment. It should also be borne in mind that, even if restrictions are placed on the nature and speed of traffic using a tunnel, effective enforcement will be required to ensure compliance. However, the level of enforcement required will depend on local conditions and may be more or less effective.

26.5. Emergency response

Irrespective of good tunnel design and traffic-management measures, there is still potential for incidents to occur. It is important that those agencies charged with protecting the public have the plans, procedures, equipment and training to deal with reasonably foreseeable emergencies.

In the past, reliance has been placed on emergency response as a key means of ensuring the safety of tunnel users in the event of an emergency. However, it can be argued that emergency response is the last step in ensuring safety in tunnels, coming after prevention and self-rescue. Emergency response and assisted rescue only take place once a failure of engineering or management systems has occurred.

It has been stated that there are two basic types of emergency response teams: the normal emergency services and special tunnel emergency teams (Luong, 2000). In this chapter, the latter are referred to as 'rapid response teams'.

26.5.1 Contingency planning

It is likely that, if an emergency were to occur within a tunnel, the response will involve several emergency services, and contingency planning and operational protocols need to take account of this. Pre-planning and preparation is vital if assisted rescue and fire-fighting is to be effected quickly. Contingency plans are nearly always drawn up by the agencies required to deal with the various types of event that occur. However, unless these plans fit together seamlessly, they may be of little value.

It is common practice for each emergency service to draw up its contingency plans separately, according to its defined role. The provision of copies of one agency's plans to the others in the hope that each will comprehend what the other will do when an emergency occurs is unlikely to ensure a coherent and coordinated approach to incident management. It is suggested that contingency plans should be jointly developed by all the key agencies involved. Indeed, to be effective, a multi-agency task group should be established to jointly develop the plans. Each agency must not only comprehend the role and purpose of the other agencies, but the activities of one agency must complement the activities of the others if maximisation of effort is to occur. Coordinating these planning activities is much easier said than done, especially when dealing with relatively large organisations, such as police forces and fire brigades, where staff changes occur frequently.

What constitutes the key agencies will vary from one situation to another, but it is suggested that in the main they would consist of the tunnel authority or site operator, the fire service, the police and the emergency medical services. It should also be noted that major incidents involving large-scale evacuation will also involve the local or national disaster-management authorities, and these should be included in the planning process. Documentation and audit trails need to be maintained in order to demonstrate that everyone has been provided with the appropriate information, skills and equipment to carry out their role.

Plans need to take account of

- the tunnel design and facilities
- the nature of the traffic using the tunnel
- the accident/incident-detection process
- the communication of relevant information to the emergency services and/or dedicated tunnel rapid response teams
- the role of each agency involved
- clear lines of command and control
- rendezvous points and triage points
- command and control points
- rescue and evacuation procedures
- fire-suppression tactics appropriate to tunnels, and the employment of safe systems of work, including dealing with hazardous goods
- the provision of the appropriate equipment by both the tunnel operator and the emergency services
- the provision of information to the public in the event of an emergency – arrangements should be made to inform the travelling public of what action they should take during an emergency before any such emergency occurs.

Traditionally, contingency planning has been structured around the above-mentioned points, taking into account the resources that are available from each agency, as opposed to being structured around planning for the events that may take place. One drawback of this approach is that the plans can appear more conceptual than real, and understanding can be more difficult to grasp. This problem can be overcome by undertaking scenario planning. Using this method, various scenarios are drawn up, using the 'what if ...' principle in order to determine the nature and magnitude of the emergencies that could occur within the tunnel, and the consequences and the resources required to deal with each of those scenarios.

In simple terms, the process involves completing a simple risk assessment by calculating the likelihood and consequence of each scenario that can be foreseen, given the particular circumstances of the tunnel. Clearly, tunnel design, engineering services, fixed installations and traffic management will be important factors when assessing risk and determining the possible emergency scenarios. The scenarios are then placed in order of priority, the highest priority being the scenario posing the greatest risk (i.e. the greatest product of consequence and likelihood). A contingency plan is then drawn up to deal with each scenario.

The process may well identify scenarios that cannot be dealt with adequately by the existing resources. If this is the case, a feedback loop must be provided to ensure that the matter is addressed by one or more of the following means

- the provision of additional emergency resources
- changes to tunnel design
- changes to engineering services and fixed installations
- changes to traffic management

or, in certain circumstances

- acceptance by society via the political process that a particular scenario cannot be dealt with and the risk is accepted.

A further major benefit of this approach, especially if exercises are carried out to test scenario plans, is that it leads to the build up of operational experience of dealing with the same or a similar emergency before it occurs for real. When a real emergency does take place, all those who have been involved in the planning process will more readily recognise and adopt the tactics that will be effective in dealing with the incident, and thus recognition-primed decision-making (see Section 26.6), will be improved.

It should be noted that safe systems of work when dealing with emergencies, together with the prevailing safety culture and philosophy, may differ between countries and even between different emergency services within a given country. However, tunnel designers and operators should be aware that the emergency services will increasingly refuse to expose their employees to undue risk in order to compensate for inadequacies in tunnel design, facilities or management. Safe tunnel operation is a matter for the tunnel operator, not the emergency services. As mentioned above, ultimately, it is for society, via the political process, to determine what level of risk, both to travellers and to members of emergency response teams, is acceptable and what is not.

26.5.2 Reconnaissance visits

For contingency plans to be truly effective, it is essential that all relevant agencies are familiar with the tunnel layout, its design features (e.g. ventilation systems and drainage systems), and any fixed installations fitted specifically to cope with emergency situations. Indeed, personnel from those agencies need to visit the site on a regular basis. If this is not practicable, then other measures must be put in place to ameliorate this problem. This may include the provision of information packs containing contingency plans, or standard operational procedure cards readily available for emergency service personnel and situated on police and fire department vehicles or located at predetermined rendezvous or on-site control points. It may also involve having tunnel operator's personnel readily available to meet and greet incoming emergency services' response teams. In particular, the fire brigade needs to have an in-depth knowledge of the risk, as it is they who will be expected to enter the tunnel complex under the arduous conditions of heat and poor visibility due to smoke.

The frequency of reconnaissance visits by the emergency services will depend on the following factors

- their particular role during an emergency
- the size, layout and design of the tunnel, together with the traffic-management systems employed
- the turnover of personnel within the emergency services who are likely to respond to incidents within the tunnel (if the tunnel is sited in a metropolitan area where emergency

service employees are frequently rotated between bases, the frequency of visits should be greater than if personnel are static)

■ the degree of complexity of the contingency plans
■ whether or not maintenance works in the tunnel mean that normal procedures have to be changed.

26.5.3   Testing of contingency plans

Once the contingency plans have been determined, it is important that they are tested to ensure they are fit for purpose. One way of doing this is to hold regular exercises to test the plan against the appropriate range of scenarios. This should indicate whether the plan ties together the activities of the various agencies involved, and indicate whether or not the plan would work in practice. Where possible, the exercises should also test the effectiveness of the engineering and fixed installations, in particular, the ventilation systems, for any given scenario. It goes without saying that large fires that would result in extensive damage or disruption to tunnel operations cannot be set in tunnels in order to test the effectiveness of engineering systems. However, many facilities, for example, communication systems, can and should be tested during exercises.

The frequency of the exercises will depend on several factors, including the nature and the complexity of the tunnel, the number of envisaged scenarios that are to be tested, and the turnover of staff in the agencies involved. As stated above, if scenario planning is adopted, there may be a need for a greater number of exercises to be carried out in order to test the plan for each scenario. Any changes to plans should also be tested by carrying out further exercises. In addition to testing contingency plans, exercises provide an excellent means of providing hands-on experience for emergency service personnel and the tunnel operator's staff in emergency procedures and in dealing with incidents.

As is the case with contingency plans, documentation and audit trails must be provided to show evidence that all agencies can demonstrate they are competent to carry out their role. One recent tunnel fire involving major loss of life was characterised by an absence of exercises and training in tunnel incidents by the emergency services.

Notwithstanding the arguments made above in favour of carrying out practical exercises, the difficulties of arranging the attendance of all necessary personnel from the various agencies involved can mean that large-scale, on-site exercises are impractical or impossible. It may be that the only realistic way of organising a multi-agency exercise is to use tabletop scenarios. Of course, there is nothing to prevent individual agencies practicing their particular role using on-site exercises.

26.5.4   Feedback

It is essential that the planning process contains a feedback loop to enable the plan to be amended in light of experience from real-world incidents or from exercises. It is suggested that the task group formed to determine plans should be given the role of planning review. This review group should meet at regular intervals, and should hold positions in their respective organisations that allow them to authorise changes to the plan. There also needs to be a feedback loop that provides information to those responsible for enforcing any legislative requirements concerning the design and/or operation of the tunnel. The precise arrangements for this will vary from location to location, and according to the legislative requirements that apply.

26.5.5 Equipment provision

The nature of the environment within tunnels will normally require the provision of general safety equipment, such as first-aid fire-fighting equipment for use by members of the public. It is also likely that specialist equipment may also need to be provided for use by the emergency services (e.g. extended-duration breathing apparatus or communications equipment). Interoperability of equipment, in particular, communications equipment, will need to be considered in some detail, as the lack of such interoperability has led to command and control problems in numerous major emergency incidents.

It is very important that all agencies know who is responsible for providing the equipment and that all equipment provided is compatible with that used by others. This information should be fully documented. For example, it is pointless providing first-aid fire-fighting equipment with extinguishing media that are highly effective if this adversely affects or is affected by the extinguishing media that will be used by the fire brigade upon their arrival.

It is suggested that each tunnel operator and emergency service, particularly the fire brigade, establish a task group to determine equipment provision for first-aid fire-fighting and medical equipment to be used by members of the public, equipment provided by the tunnel operators for use by their personnel, together with specialist equipment provided by the emergency services. If carried out properly, the contingency and scenario planning process will identify the situations where additional equipment will need to be provided by either the tunnel operator or the appropriate emergency service.

The issue of the testing of specialist equipment will also require addressing. Regular testing must take place, and all agencies must know who is responsible for testing and maintenance. Too often, specialist kit is provided at great expense, only to be left to deteriorate over a period of months or even years. Even if equipment is maintained, unless everyone is aware of it and knows they can rely on it in the event of an emergency, it may simply be ignored. The testing and maintenance of equipment must be recorded and an appropriate audit trail established.

26.5.6 Location of emergency response teams

If the situation demands assisted rescue, rapid response teams sited at the tunnel portals are the most effective way of getting a first line of response to the scene in the minimum time. These teams may be provided on site, either by the tunnel operator, normally at the tunnel portals, or, perhaps, by the emergency services. It is appreciated that in remote, rural locations, provision of on-site rapid response teams may not be feasible. In other circumstances, it may be the case that local fire stations are sited close enough to the portals to provide an effective rapid response.

It should be noted that in the St Gotthard Tunnel incident, emergency teams were deployed within one minute of the crash occurring (Ananova News Agency, 2001), and the incident occurred only 1.5 km from the tunnel portal (BBC News Online, 2001). This indicates that, unless tunnel design allows self-rescue to be effected, if fire growth is rapid, the work of the emergency services will be restricted to casualty recovery, rather than rescue.

26.5.7 Rapid response teams

Rapid response teams sited at the tunnel portals bring enormous benefits to effective emergency response, the speed of response being one major benefit. The teams can be equipped with vehicles that carry fire-fighting, rescue and medical first-aid equipment, and be equipped with breathing apparatus. Being dedicated to dealing with tunnel emergencies, they will have an in-depth knowledge of the tunnel complex, its facilities and engineering systems, and the traffic management arrangements that apply. Their use means that trained professionals are on scene within a matter of minutes and, upon arrival, they have the means to deal with incidents such as small fires. In addition, their rapid attendance means they are in a position to control or extinguish many small fires before they develop further.

Rapid response teams are also able to act as 'guides' to tunnel users, giving information and directions to motorists, and are able to do this before the use of any other communications media, such as matrix signs or RDS radio transmissions. Not only do they act as guides for tunnel users, they also act as guides for emergency service personnel, who will generally be less familiar with the tunnel. Furthermore, they should be able to establish communications links between the emergency services, tunnel traffic control and the engineering control facilities before, or shortly after, the arrival of the emergency services.

26.5.8 Incident management

In order to ensure assistance is provided as quickly as possible, rendezvous points and forward control points must be established as described in the contingency plan. If the accident becomes a major incident, a predetermined off-site control point will also need to be established. A major incident can be described as an emergency requiring the deployment of considerable resources of multiple agencies, such as police, fire, ambulance, hospitals and local authorities. This control point is referred to in the UK as a 'gold' control. It may be some considerable distance from the scene of operations, perhaps sited at the headquarters of one of the emergency services.

The purpose of an off-site, or gold, control is to perform the role of a strategic command point that coordinates the response of all the agencies involved. To be effective, personnel attending the strategic control must be top-tier managers able to take policy decisions. Unfortunately, some emergency services merely send liaison officers to the strategic control, and deploy their senior commander at the scene of the incident. This is not an effective approach to incident management, as it means strategic decision-making is delayed by the requirement for liaison officers to transmit information to their respective on-site commanders for decisions to be made. Communications delays often mean that the junior liaison officers end up making strategic decisions, for which they may lack the experience or expertise, because of the unavailability of more senior commanders.

Whether or not an off-site control point is established, there will also be a requirement for an on-site control point, where the command posts of the emergency services attending will normally be sited. The site should be identified in advance and its location should be chosen taking the following factors into account

- it should not be adversely affected by any products of combustion issuing from the tunnel
- it should not be affected by the run-off of hazardous substances or water from fire-fighting that may issue from the tunnel following an accident
- it should provide good communication links with the tunnel control room – the tunnel engineering/traffic control room should be adequately protected to allow it to function in the event of a major fire or other emergency
- it should be able to accommodate all appropriate emergency services

■ it should be able to accommodate the communications systems of all the services attending and the tunnel operator without adverse affects.

The on-site control should act as a tactical control and liaison point for each of the services attending. Although each service will normally have its own command, control and incident-management structure, everyone should act in accordance with the predetermined contingency plan.

26.6. Integration of design and management with emergency response

Although the design stage comes well before the emergency response, it is necessary to briefly revisit this issue. This is because it must be stressed that the design and construction of certain safety features does not necessarily result in a 'safe' tunnel. Design is only one factor in the safety equation – the design of the tunnel. However, design is nothing without a comprehensive understanding of those features, and this is a two-way process.

Firstly, although general knowledge and understanding of how members of the public react in fire or other emergencies is considerable, it remains incomplete and the subject of much debate amongst design engineers. Therefore, it can be argued that, in the absence of such a complete understanding, 'safety' cannot be fully designed-in to a tunnel (see also Chapters 19–21).

The design features must be understood by all agencies involved in tunnel safety, the operators and the various emergency services over the entire lifespan of the tunnel. If this understanding does not exist, then very expensive design features may not even be used in the event of an emergency. An example of this would be the provision of a complex ventilation system for use by the fire brigade in the case of fire. It may be that the system is capable of effectively ventilating the tunnel, and preventing smoke stratification and heat build up. However, if fire brigade personnel are unaware of how the system operates, or do not possess (or have failed to maintain) the competence to operate the system or to make tactical fire-fighting decisions that utilise the system in a coherent way, it is unlikely to be used effectively. Design engineers must be aware that, for the emergency services, the tunnel will be only one of a multitude of hazardous sites they have to deal with.

Notwithstanding the comments made above with respect to contingency planning, training and reconnaissance visits, both tunnel design engineers and tunnel operators must understand that employee turnover will inevitably lead to a dilution of specialist knowledge of the tunnel site, and this must be constantly replenished by training and familiarisation of staff. It is often not possible for the emergency services to commit large amounts of time and other resources to only one of a number of risk sites they are required to deal with.

26.6.1 The incident commander

Moreover, both design engineers and the tunnel operator must understand the capabilities of the fire and rescue agencies under emergency conditions. They must appreciate that, in an emergency, information may be sparse, incomplete and/or inaccurate. The incident commander will be processing a great number of different pieces of information at the same time, and will be required to make many decisions, almost simultaneously. The situation is one of chaos, with everyone needing to communicate with the incident commander and receive instructions, and all at the same time. The number of people that the incident commander/manager has to deal with is known as the 'span of control'. In the early stages of an incident, when relatively few incident managers have been deployed, the span of control is often greater than the incident commander can cope with.

It is also the case that, in the early stages of an incident, when the emergency services have just arrived, it is likely that the incident commander/manager present will be relatively junior. He is likely to have access only to rudimentary command and control facilities, because the more sophisticated, dedicated command and control support facilities have not yet arrived. Initially, there are unlikely to be the fire-fighting and rescue resources present to deal with the incident. These resources have to be mobilised, and take time to arrive. Clearly, in remote locations the time span between the mobilisation and arrival of sufficient resources to deal with the incident may be considerable. Indeed, the nature or the magnitude of the incident may be such that neither the emergency services nor anyone else possesses the means to deal adequately with it. For example, the fire brigade may not have the means to reach a fire inside a tunnel because reaching the scene in conditions of extreme heat and smoke may be beyond the endurance of human beings, even if they have the appropriate equipment, such as long-duration breathing apparatus.

It is thought that a span of control of between three and five lines of communication is the maximum one person can deal with effectively, no matter how capable or well trained he or she may be. The incident commander/manager will initially be fully involved in gathering and processing information coming from diverse sources, prioritising information and actions to be taken, formulating a plan of action, and deploying resources accordingly. However, an incident commander/manager is unlikely to have the capacity to comprehend complex engineering systems and give instruction as to their operation unless those systems are simple to operate or operate automatically.

It could also be the case that the first incident commander/manager on scene would not possess an in-depth knowledge of the installation, the engineering systems, or the smoke dynamics in the tunnel. If manually operated engineering systems are provided, which require careful considera-tion and a degree of rational analytical decision-making before operation, although technically capable of ameliorating the situation the systems are likely to be too complex to use in the early stages of an emergency.

It is worth remembering that, during the near chaotic situation that occurs at the start of all emergencies, the incident commander/manager is unlikely to possess the information required or be able to process that information quickly enough to make rational decisions. To use the terminology of command and control, the incident commander/manager will use 'recognition-primed' decision-making to make satisfactory decisions to deal with the incident, not analytical decision-making processes to make optimal decisions. Put another way, in a time-critical situation, the incident commander/manager will carry out such actions as they think are likely to ameliorate the situation, based on previous experience. They will not have the time or the capacity to assess fully all the resources available to them, generate a list of all the various options open to them, analyse all those options and select the 'optimum' solution.

At the later stages of an incident, when command and control facilities are in place, both on-site utilising tactical command points or, perhaps, off-site strategic controls, the situation is likely to be less time critical, and incident commanders/managers are able to engage in rational–analytical decision-making, generate a list of options and select what they see as the 'optimum'. At this point, they will be able to comprehend and make use of manual engineering systems. In any case, the tunnel operators must provide the emergency services incident commanders/managers with expert knowledge about and advice on the tunnel facilities and characteristics. Clearly, tunnel operators must also possess the competence to operate effectively the systems provided by the design engineers.

26.6.2 The need for realistic expectations

It is vital that design engineers appreciate just what an unfolding emergency is like to deal with. It is often the case that the capabilities of the emergency services in the early stages of an incident are vastly overestimated. Resources take time to be mobilised, and command and control systems need time to become established. It should also be noted that it is not unknown for the emergency services themselves to overestimate their own capabilities.

On the other hand, the emergency services also have to accept that, very often, designs cannot be changed to suit their needs and their particular operational practices. Design changes may not be practicable in new-build tunnels and may well be impossible in existing ones. Both the emergency services and the tunnel operators have to manage the risk as it exists, not how they would wish it to be.

It must also be remembered that tunnels are vital arteries through which the economic lifeblood of a region flows. They do not exist to keep the emergency services in employment. Emergency services managers have to appreciate that risks posed by tunnels have to be balanced against the social and economic advantages that tunnels can bring. It should also be borne in mind that, although tunnels can produce a high-risk environment, the alternatives are unlikely to be risk free. Serious accidents do frequently occur on surface roads, resulting in loss of life. Reference has already been made to the fact that tunnels can contain some types of incident and thus make them easier to control, ensuring they pose less risk to the environment than if they were to occur on the surface.

26.7. Conclusions

The key factors in tunnel safety are the basic design of the tunnel, traffic management and emergency response. In the case of road tunnels, evidence is mounting that two-tube tunnels using unidirectional flow are inherently safer than single-tube tunnels with bidirectional flow, in that accidents are less likely to occur and the results of those accidents affect a smaller number of tunnel users. Moreover, two-tube tunnels can facilitate rapid and effective emergency intervention in a way that single-tube tunnels cannot.

However, there are many single-tube tunnels in existence and, in most cases, it is impossible or impractical to change their basic design. Attention must be focused on engineering systems and traffic management if the risk posed to users is to be reduced. Emergency response has a role to play, but that role is a limited one. Prevention and self-rescue are the only effective methods of ensuring adequate levels of safety. Rapid intervention to reduce the spread of fire is crucial to ensuring tunnel safety, and this may be provided by fire-suppression systems or first-aid fire-fighting. However, safe egress and early evacuation are vital in reducing the risk to life, and must always be considered before allowing any personnel into the tunnel during a fire incident. Unfortunately, events have shown time and time again that unplanned and unforeseen events occur that can nullify the safety features of any structure, tunnels included, and situations will arise when an emergency response is required and expected by the public. For this response to be as effective as it can be, and to ensure that those called to deal with the incident carry out their tasks in a safe manner, considerable effort must be put into contingency planning, testing the plans, providing equipment, and formulating incident management processes that ensure 'safe' systems of work. Too often the capability of the emergency services to resolve situations is over-estimated, both by tunnel designers and operators, and by the emergency services themselves.

Design features, engineering systems, traffic management and emergency response must be integrated into one coherent approach in an attempt to ensure the safety of the tunnel users and emergency service personnel in a way that is understood by all. If this does not occur, recent history has shown that tunnels will represent a high-risk and dangerous environment. Nevertheless, despite the risks posed, the benefits that tunnels bring must not be forgotten, and we may wish to ponder what the 'body count' would be if they did not exist.

REFERENCES

Ananova News Agency (2001) *Fire Rages on in Alpine Tunnel.* Story filed 25 October, 2001. Available at: http://www.ananova.com/news (accessed 22 June 2011).

BBC News Online (2001) *Forensic Experts Search Tunnel Wreckage.* Story filed 26 October 2001. Available at: http://www.bbc.co.uk/news (accessed 22 June 2011).

Broekhuizen JM (1999) Handling tunnel fires by the Rotterdam Fire Brigade. *International Tunnel Fire & Safety Conference, Rotterdam, The Netherlands,* 2–3 December 1999, pp. 1–11.

Henke A and Gagliardi M (2000) Lessons learned from the St Gotthard road tunnel fire. *Tunnel Management International* **2**(9): 14–19.

Luong Y (2000) *OECD/PIARC ERS2 Project – Task 3 Summary Report.* Internal Report No. CL3002-R1. WS Atkins Consultants Limited, Epsom, p. 30. This report is publicly available via OECD/PIARC.

Skarra N (1997) *The Bus Fire in the Ekberg Tunnel on August 21, 1996.* Norwegian Public Roads Administration, Oslo County Roads Office, Oslo.

Handbook of Tunnel Fire Safety, 2nd edition
ISBN: 978-0-7277-4153-0

ICE Publishing: All rights reserved
doi: 10.1680/htfs.41530.555

Chapter 27
Emergency procedures in rail tunnels

John Olesen
Slagelse Fire Brigade, Denmark
Graham Gash
Kent Fire Brigade, Kent Fire and Rescue Service, UK; Provided updated chapter

27.1.   Introduction

Even though the number of accidents in rail tunnels is few, one cannot be certain that an incident will not occur tomorrow. A review of past experience reveals that many accidents could easily develop into life-threatening events, and that appropriate plans and procedures should be in place to ensure that the consequences of any accident are reduced to an acceptable level.

In this chapter, the procedures that are generally planned for use in accidents in rail tunnels are described, and the similarities and differences between procedures are discussed. A general description is given of the structure of the emergency plan and its division into levels, as some items that can be included in the planning phase are proposed.

There is no common standard for tunnels or emergency procedures. However, there are some prevailing factors that should form the a foundation of all procedures, regardless of tunnel design or other facilities.

27.2.   Standard operational procedures

At present, there is no common international practice for the preparation or management of contingency plans for accidents in rail tunnels. This lack of common practice is evidenced in many more or less obvious factors, and in general it is difficult to make an immediate comparison between the emergency services and procedures deployed in different countries. The geographic and demographic characteristics, etc., of the country coupled with its unique historic traditions all play a role in the building of emergency services, plans and procedures.

27.2.1   Differences

There are great differences in the rail tunnels that have already been built and those which will be built in the future. The basic design can be a one-, two- or three-tube tunnel (including a service tunnel), a combined rail and road tunnel, or a metro system. Tunnels can be built with no, one or several stations, and can carry one- or two-way traffic. Tunnels vary in length from under 1 km to over 50 km, and naturally they vary widely in age. There is a wide range of distances between emergency escape routes for passengers and between access routes for emergency services. Some tunnels are full of technical installations and equipment, while others are, in principle, just a tunnel, without any relevant emergency installations.

Furthermore, there are wide differences in the operation of tunnels. The volume of traffic can vary from a few trains a day to one every minute. There is also a wide variety of operational factors and restrictions, with some tunnels carrying only passenger trains, and some carrying in addition vehicle (cars) and goods trains, which may, among other things, be transporting 'dangerous goods'. There may, of course, be an extremely serious fire in a tunnel arising from goods that are not designated as 'dangerous' or 'hazardous'.

In many countries, there are no special additional requirements for emergency services due to the presence of a tunnel in a particular district, and emergency procedures are thus designed based on what is available rather than what is desirable. If there is no legislation within an area in relation to emergency services, procedures will often depend on the geographic position and the capacity of the emergency services. Some tunnels are in the immediate vicinity of the responsible emergency services, say within a 10 minute drive, while some tunnels are in uninhabited areas, with, say, a 30 minute or more driving distance for the nearest emergency service.

Incident procedures for tunnels that cross council, county or national borders are also dependent on how legislation and standard procedures function in and between the individual authorities, including the national boundaries. In some countries the normal emergency services are used for rail tunnel incidents, whereas in other countries special services are located beside or near tunnels. Some countries exclude, often because of the distance, the regular emergency services as an essential factor in emergency procedures, while in other countries emergency response teams are always established beside tunnels. In still other countries, the amount of emergency response facilities varies from tunnel to tunnel and is determined according to demand, and some countries set limits for when there should be any special preparation at all. For example, in some countries all tunnels under 1 km are exempt and no special incident response is established.

The greatest differences between rail and road tunnels are typically that rail tunnels have a considerably longer approach way and a greater collective fire load, special vehicles are required for transport, and there are potentially many people in the same area. However, trained personnel are normally on hand to help in an emergency situation in a rail tunnel, and normally it will be known what vehicles (trains) are in the tunnel.

There is a large difference in the likelihood of an incident occurring in a road tunnel and a rail tunnel. For road tunnels it is only a matter of when an incident will occur, and this frequency has a great impact on deciding what safety conditions and procedures should form the basis of the complete safety concept for a road. However, as incidents in rail tunnels happen so rarely, it can be difficult to work out a suitable emergency procedure, because it is often difficult to argue for the imposition of special safety precautions and need for special emergency procedures, particularly with regard to cost–benefit analysis.

It can be said that it takes a fire to prevent a fire. It is for this reason that every opportunity should be taken during the design phase to encourage and, where possible, enforce the tunnel contractor, owner or operator, to provide services for the emergency response crews to use during an emergency. A cut in project finance should not result in a reduction in safety systems.

27.2.2 Common features
Many different methods and combinations of methods are used to secure an acceptable level of safety in rail tunnels. There is complete agreement that the highest priority is to avoid accidents, both in tunnels and as at other sites, and that the second most important consideration is to ensure

that the consequences of any incident are kept low. Thus, while the probability of an incident occurring is often low, in safety planning the prevention of incidents takes priority over reducing the possible consequences of an incident, even though it is highly likely that the consequences of an incident will be severe.

However, there is broad agreement that safety planning should include an evaluation both of the probability that an accident will occur and of the possible consequences of an accident. There is also broad agreement that a risk should be divided into different consequence groups, including life and health, environment and the value of property, so that it is possible to make a scale of comparison. For this reason, it is in increasingly the case that risk is quantified using objective measurable criteria. In some countries, in addition to risk analysis as a preventive measure, individual accidents are analysed, soon after their occurrence, with a view to gaining a better understanding of safety procedures and the materials and personnel necessary for different types of response.

A number of terms are used to describe the phases, activities, factors, etc., of safety work (e.g. design, precautions, intervention, response, operation, recovery, procedures, facilities, installations, equipment). All these terms can have different values and meanings when used in different countries, and this can lead to confusion, especially when making comparisons between the situations in different countries. At the same time, however, this multitude of words and expressions illustrates the complexity of formulating a safety concept and the emergency procedures that extend from it. Any combination of terms and expressions can be used at different levels of the safety planning phase, but the principles and work methods that are current within a particular country should be used in the first instance.

27.2.3   General principles

While there are a number of different operational methods, some fundamental principles are applied in the majority of tunnels. The characteristics of some of the most commonly used designs and principles for ventilation strategy, evacuation and intervention by the emergency services are described in general terms and illustrated below. It should be emphasised that the examples are very general, and do not give specific detail.

27.2.3.1 One tube, two tracks

In tunnels containing two tracks the distance between the emergency exits and the access entrances is typically more than 1000 m, and there are often longer 'tunnels' from the outside down to the main tunnel or tunnels, where the only exit and access route is through the portals (Figure 27.1).

Emergency procedures are, in the main, based on self-rescue, and many of these tunnels have poor conditions for rescue services, which are often stationed a long way from the tunnel. The emergency services are usually faced with difficult means of access, and use normal rescue vehicles up to the nearest access entrance. Ventilation, if any, is often started in a direction based on messages received from the train driver, or may be determined by the natural ventilation of the tunnel.

27.2.3.2 Two tubes

Tunnel systems with two tubes, some with crossovers, are typical for newer rail tunnels, especially for tunnels with heavy traffic volumes. There are often many built-in safety provisions and the

Figure 27.1 One tube, two tracks

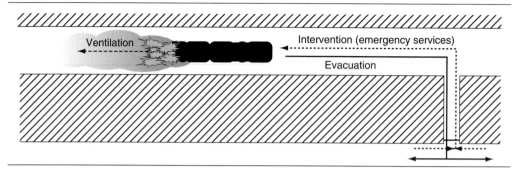

opposite tunnel tube is normally regarded as a safe haven, which provides good possibilities for effective emergency procedures (Figure 27.2). Clear terminology is essential for command and control purposes, and the terms 'incident tunnel' and 'non-incident tunnel' are recommended.

These tunnels give better possibilities for both self-rescue and assisted rescue, which is typically carried out through cross-passage tunnels spaced 100–500 m apart (but the spacing can be greater). In the case of fire, the normal procedure is to evacuate people in the direction opposite to the ventilation direction and over into the opposite tube, from where incident rescue teams, who travel on special rail vehicles, are launched. The provision of emergency transportation should form part of the emergency-plan procedures, and should comprise approved 'man rider' facilities. Emergency service personnel should not be transported in the same compartment as equipment, or on open flat-bed wagons, and the planning should take account of this.

27.2.3.3 Two tubes with a central service tunnel
Tunnels containing two tubes and a service tunnel are not particularly widespread, and a service tunnel would not be built merely to create a basis for effective emergency procedures. The design is, however, a good example of how to establish solutions that take account of operation and maintenance, and create a good framework for working out emergency procedures (Figure 27.3).

Emergency response and evacuation in these, often long, tunnels is typically carried out via the service tunnel, and, if further response is necessary, the opposite tube is also used. The provision of emergency transportation should be arranged as part of the emergency plan procedures.

Figure 27.2 Two tubes

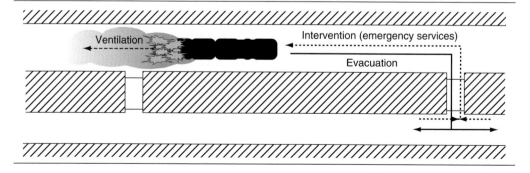

Figure 27.3 Two tubes with a central service tunnel

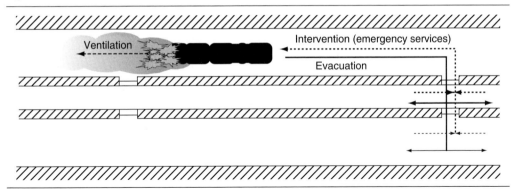

The ventilation strategy in tunnels of this type, especially in longer tunnels, varies to a larger degree than in smaller tunnels, and often special ventilation systems, which cannot be generalised, are used. Consideration of when ventilation should be started, and the speed and direction of the ventilation, typically leads to greater discussion for these tunnels. The use of the ventilation system is a specialist matter, and operators must be aware that it can work 'for' or 'against' emergency crews if not used correctly. If set incorrectly, the consequences for those making an intervention could be disastrous. A fire service liaison officer should be present in the rail control or ventilation control centre to ensure that the needs of the emergency services are fully accommodated.

27.2.3.4 Metro systems

The number of underground railway/metro systems is increasing, and most large cities now have a metro system of one type or another. These systems are often complex, and a characteristic of many of them is that they are extended or rebuilt in relation to other infrastructure, including traffic needs. Therefore, a metro system can consist of a tunnel system that in age can stretch over several generations, and therefore can have different design and safety provisions in different parts of the system. This gives rise to great challenges in relation to safety concepts, both with regard to the tunnels and to the underground stations (Figure 27.4).

The ventilation strategy is highly complex in these systems, and can often affect the entire metro system, including the stations. Therefore, this further consideration must be included in

Figure 27.4 Underground railway/metro system

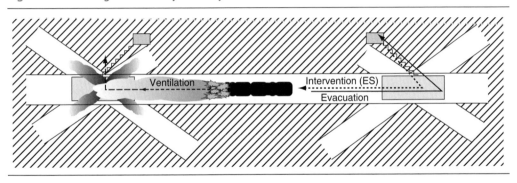

Figure 27.5 Standard procedures

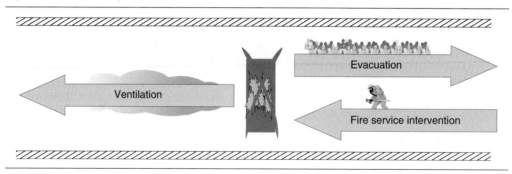

emergency planning. The emergency procedure will normally include direct assisted response, while at the same time focusing on all the affected areas, including the stations and the complete tunnel system. It is becoming more widely accepted that a 'division' should be secured between tunnels and stations, so that smoke in the tunnel system does not immediately affect the stations, and vice versa.

Entrance to and evacuation from metro systems in an emergency is more often planned to take place via stations or special emergency shafts. The tunnels are often very narrow, with minimal or no emergency walkways in the tubes, so evacuation and fire-fighting are often especially difficult.

27.2.3.5 Common principles
The above descriptions of design and emergency procedures are very general and will vary from tunnel to tunnel. As far as it is possible to generalise about standard procedures, the most used procedures are built on the following bases (Figure 27.5)

- to remove smoke from the accident zone as quickly as possible
- to evacuate away from the fire to a safe haven (in the opposite direction to the direction of the ventilation)
- to undertake an assisted-rescue response from a starting point of a safe haven, beginning the operation working in the same direction as the ventilation.

In the above, no consideration is given to where the fire is situated on the train (front/rear, inside/outside), the effect of natural or mechanical ventilation (speed/direction, starting time), or the short- and long-term operational handling viewed as a whole (tactical, technical). In a particular incident, for example, a fire-fighting response may be made in the direction opposite to the ventilation direction, and/or the direction of mechanical ventilation may be reversed in order to undertake a search and rescue operation once the smoke-free areas have been searched.

Note: As stated above, the use of the ventilation system is a specialist matter, and operators must be aware that it can work 'for' or 'against' crews if not used correctly. Incorrect use of the ventilation system could have disastrous consequences for those making an intervention.

27.2.4 Emergency procedures – where, when and why?
Obviously, not all tunnels require the same special emergency procedures, and these will vary widely in scope and content. However, safety and emergency procedures, including risk

assessment, should be considered in every case. While, the emergency procedure for an accident in a 10 m tunnel in an urban area, with one train per week, could be based on 'help yourself', the procedures for a large tunnel with heavy traffic should be well thought out and include contingency plans and procedures. It is important that the planning of emergency and safety procedures is based on both general principles and a specific evaluation for each tunnel, and the prioritising of the procedures may be done by dividing possible scenarios into categories.

The following section deals with those tunnels where it is clear that there should be emergency procedures. The aim is not to prescribe a model for the preparation of emergency procedures, but to offer some suggestions of considerations that can be included in the planning phase.

27.3. Contingency planning

The safety of rail tunnels depends on both preventive and remedial arrangements, and includes maintaining the operational conditions for tunnel use. These arrangements and conditions have a decisive influence on the emergency procedures that form the basis of a rescue response, as well as the aims that these procedures are based on.

Many aspects of emergency procedures are of a technical nature, while some are resource based. Some parts of an emergency procedure can be put into operation right at the start of an incident, while some are initiated in connection with self-rescue and the arrival of external rescue units. Fundamentally, the focus when planning safety procedures for rail tunnels should be on two main aims: (1) to avoid accidents, and (2) to control the consequences of an accident.

There is broad agreement that the prevention of accidents should be the priority, because a 100% achievement of this goal would render the other target superfluous. Unfortunately, it is impossible to guarantee that an accident will not happen and, while the number of accidents in rail tunnels is relatively limited, experience has shown that the consequences can be severe, and thus effective contingency planning is essential.

A lack of emergency procedures, or a lack of knowledge about them, including uncertainty regarding formal agreements, procedures, command relationships and communication channels, has in many cases led to confusion, and delayed response to an incident, as well as leading to a number of other problems. The response to a large and complex accident involves many different emergency response teams, as well as other authorities and companies, which may have no tradition of working together. To maximise the effect of the emergency response, and thereby reduce the consequences of an accident, it is vital to have well thought out plans in place and effective coordination of all the organisations involved. This can be achieved by involving in the contingency planning all agencies that would be involved in an emergency response. The plans should be regularly tested to ensure their effectiveness and the understanding of them by all the agencies involved.

27.3.1 Planning in general

When formulating safety and emergency plans and procedures, it is important to start from the objective of achieving an overall safety policy that includes an acceptable rescue concept corresponding to an acceptable goal. The concept should be established on the basis of any existing laws and regulations, discussions with the tunnel owner, who is often directly responsible for the fixed 'built-in' safety measures, and discussions with the emergency services, who have the responsibility for the task on the day when an accident happens.

All possible solutions should be discussed carefully, and should, in addition to design and structural measures, include principles for self-rescue and assisted external rescue, including the quality and quantity of, and the economic responsibility for, the expected response. The agreed solutions and wider planning should cover equipment, possible special rail/road vehicles and training, and this will then form the basis of a further and more detailed preparation of plans and procedures.

Once a decision has been reached on an overall concept, based on the agreed aims, more detailed planning can be done. Planning should focus primarily on possible responses, and to a lesser degree on causes. It is important that planning is not thwarted by disagreements about the design, including built-in safety precautions or the concept as a whole. Starting with the main agreements described in the main contingency plan, the involved parties can undertake their own planning for procedures, using the 'language' and methods normally used within the individual organisations (Figure 27.6).

All possible scenarios should be included in the planning work, and here the 'what if . . .' approach is a good method. When undertaking scenario planning, it could be sensible and positive to focus on procedures alone. Who does what, when, for what reason, and with what competence?

Some examples of more specific questions that should be considered for inclusion in the detailed procedure planning are as follows.

- How is the first information about the incident received?
- What actions should be taken, and by whom, when there is an indication of an accident which has not been verified?
- How is the information about the extent of the incident and the correctness of this information to be confirmed?
- Who decides that the reported incident should be treated as real and decides the category of the accident?
- Who decides what actions should be undertaken at the accident location?
- Who makes the decision about ventilation, including speed and direction, and on what basis?
- Who decides whether the train should continue its journey, and under what conditions?
- If the train wagons can be uncoupled, who makes the decision and who carries out the action?
- If it is considered necessary to switch off the power, who takes the decision and how is it done?
- How can it be confirmed that the power is really disconnected?
- If overhead catenary equipment is used, who will earth the system so that it can be deemed to be safe to work on or near to it?
- Who has an overview of maintenance and the temporary potential danger in the tunnels?
- When should the alarm be raised, and by whom?
- Who should be called and where should they meet? Who decides this, and using what criteria?
- Which information should be given, to whom, at what time and how?
- Which technical equipment should be put into operation, and who should be informed of this, and how?
- When should evacuation start, on whose orders, and on what basis?

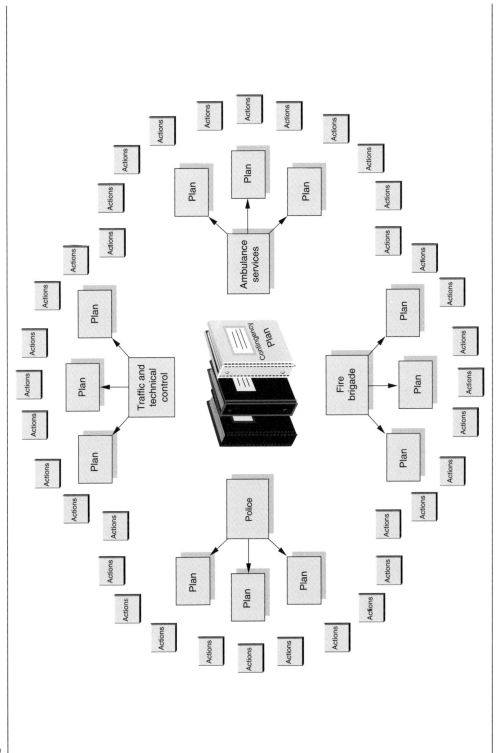

Figure 27.6 Plan of actions

- Who controls self-rescue and who decides the direction of rescue and which emergency escape routes to use?
- Who gives permission for intervention by the emergency services?
- When and how should the personnel be assisted and relieved?
- How will the passengers and the injured be brought out of the tunnels?
- What should be done with the passengers and the injured when they come out of the tunnels?
- Who will control the area surrounding the tunnels, and how?
- Who will establish an effective logistics system, and how?
- Are there any procedures for handling the media?
- Are there procedures for informing the public and the next of kin, including psychological crisis help?
- How and when should special resources people be called in?
- Who decides when the power should be reconnected in the different areas?
- Who decides when to re-establish normal operation in all or part of the tunnel system?
- Have any plans been made for alternative operation?
- Who is in command of what, and when?
- How is command to be maintained or changed?

It is important to be aware that plans in themselves do not reduce the consequences of an accident. Plans are written on paper and merely describe the conditions that should form the basis of the favourable completion of the response. During the preparation of plans, the focus should continually be on what it is that one wants to achieve and how the users of the plans will interpret them.

In an emergency, the people involved are under stress, and a mistake can have serious consequences. A quick response to an accident and rapid collection of information is essential for the favourable completion of the response, and one must be ahead of the accident at all times. Simplified action lists can help ensure that the planned procedures are put into action at the right time and place and in the right way. The testing of such plans is essential.

27.3.2 Simplify

When an accident occurs, there is no time to start looking at the contingency plan and begin reading from page one, about statement of policy, economic responsibility, etc. Plans should be laid out in simple terms, and describe the concrete actions that should be carried out by the various agencies. In the initial phase of the accident in particular, when lengthy discussions between the involved parties is not possible, it is essential that all parties work in accordance with the main plan.

The simplified action plans can take the form of short action lists containing definite information about starting the ventilation system, alerting and attendance of emergency units, communication and information that should be gathered immediately, etc. It is essential that there is coordination of the action lists. For example, if, in the planning phase, the fire service decided that before entering the tunnel it should have information about ventilation direction, power lines status and carbon monoxide levels, this should be an item on the control centre's action list, so that the information is ready when the fire service requests it. Good pre-planning and safety exercises involving all parties will ensure that this occurs.

Imagine the following example. The operator receives an alarm about a fire in a goods train. He starts the technical and traffic-relief systems, which have been agreed and described in an action list, and the fire brigade service responds to the alarm in accordance with their action plan and action lists. The operator begins, because of what he considers important, to compile information on the technical status, such as the ventilation, traction power, lights, traffic situation and gas measurements. However, when the fire officer arrives, he has one question: 'What is on the train?' It could be 'dangerous goods'.

It is very clear that for the fire brigade such information is essential, but it may not be so for the operator, who has a tendency to think along more technical lines. On the other hand, the fire officer might also be so focused on the 'dangerous goods', that he forgets, for example, about the power. While people can think for themselves, it makes sense to put all the basic factors on a simple action list, and personnel can then use their energy and resources on those unexpected things that it is not possible to prepare for, or which take an unexpected turning – of which there will be many in an emergency situation. Such concerns can be addressed or reduced by formulating a comprehensive and agreed action plan. The operator, who is there to assist the emergency services, must not be overloaded with other tasks that will distract him or her from their primary duty.

Action lists may be obligatory (e.g. 'disconnect power') or decision-making lists (e.g. 'consider disconnection of power'). Deciding which actions should be obligatory and which not can be a great dilemma. For example, should an operator start the ventilation system simply because it is stated on a general or specific obligatory action list that he should do so, or should he make a considered decision based on the information received from train personnel about where the fire is and their evaluation of the situation? Opinions are many, and problematic situations are not discussed further in this chapter. However, everything else being equal, the benefits of putting as many actions as possible on obligatory action lists outweigh the disadvantages. This is particularly so in the initial phase of an accident, when quick actions are necessary, and there is little time for discussion and a more formal management and command structure has not been established. Once the emergency response units are on site, and more senior officers and other relevant staff have arrived and established the situation, more situation-oriented decisions can be taken.

No matter what methods are used, it is important to keep it simple and keep it short, in order to keep it safe.

27.3.3 Testing

Plans, procedures and action lists cannot be considered finished until they have been tested to ensure they are fit for purpose. Testing can be done by means of different types of exercise, and the exercises themselves should be part of the tests. It might be suitable that plans and procedures are initially tested in small procedural exercises, wherein the use of 'what if ...' approach is continued. Each unit should test its own plan, but the compatibility of the plans of all units should also be tested if they are to have any value. Multi-agency exercises, whether table-top or actual exercises, form the final stage of testing of procedures, and these must be conducted regularly.

One should be aware that emergency plans are often made by 'experts' within the different organisations, and the plans will be clear to them, as they have been through the complete planning phase a number of times. Therefore, when undertaking small-scale exercises to test plans and procedures, it is sensible to involve different representatives of organisations, in particular those who will actually use the plans, who can provide useful feedback on whether the plans are fit for purpose.

It is not advisable to begin the testing of plans and procedures by undertaking major exercises, without having first undertaken smaller procedural exercises. A major exercise that is undertaken before 'pre-testing' of components of the overall plan will lack so much and have so many faults that it can lead to frustration among those who will be required to use the plan in an actual accident, as well as raise the danger of giving a negative image to the public and the media. The faults and shortcomings of a plan should, as far as is possible, be uncovered by smaller, procedural exercises, and the discoveries made therein used as the basis for necessary adjustments in the plans.

The testing of plans should not only be done around a desk, but practical tests and exercises should also be undertaken to ensure that the technical systems and installations work properly and can be operated according to purpose. It is also reasonable to undertake full-scale exercises to test the complete contingency plan and the procedures arising from it. It is often the case that the smaller tests and partial exercises have gone very well, but when all components are put into practice at the same time a lot of important faults and shortcomings are uncovered. It may not be that the individual procedures themselves are at fault. For example, if the technical operator's handling of the ventilation system is tested one day, and the next day the operator participates in a table-top exercise, the idea that he and the technical system are geared up for handling an accident may be reinforced. However, during a full-scale exercise, where a multitude of alarms are activated and the control screen lights up like a Christmas tree, the operator might have to undertake incident-related work and at the same time maintain contact with his staff throughout the system, and this may uncover the need to make adjustments to plans, procedures, technical operations, etc.

27.3.4 Acceptance

For emergency procedures to be effective, they should both be of good quality and accepted by all the agencies that will be required to put them into action. The acceptance of procedures does not have a direct influence on the procedures themselves, but it is nevertheless important that there is acceptance of the decisions taken in relation to the different scenarios and targets, and the plans and procedures that will be used in the event of an accident. There should be a common understanding with regard to the procedures, even if they are not regarded as rational in relation to an individual organisation's priorities.

For example, the plan may state that, when there is no longer a danger to life, the restoration of traffic will take a higher priority than the long-term, non-life-saving response and operation. If this order of priorities has been decided upon, it should be accepted by all, including those whose further response will be made more difficult, and it is important that everyone understands, and works to achieve, the common objectives.

It also needs to be accepted that there is a limit to when human intervention can be regarded as a helping factor. The fire service is often thought of as the last port of call, and is expected to solve extreme situations when everything else has failed. However, there is a limit to when intervention by the fire service in a tunnel incident is possible. Plans should therefore not over-estimate capacity or ability, and should not prescribe unrealistic procedures that cannot be implemented in practice. It is better to accept these limits in the planning phase and devise alternative solutions and procedures, than to include unworkable procedures that will fail at the time of a real incident.

27.3.5 Training

Plans and procedures are of little value if they are not repeatedly practised in training. This subject could form a chapter in itself, but in this section just a few general points will be raised.

It is a fact that training is one of the most important factors in relation to emergency procedures, and that intensive training and the quality of training has a decisive influence on the quality of the emergency response. Only with relevant training can it be expected that response personnel (train personnel, operators, etc.) will function in accordance with the planned procedures, and that the collective resources are brought into use with the greatest possible effect (optimal utilisation of facilities, equipment, etc.).

There are a number of educational and training methods which can be implemented in various locations and with the participation of different personnel groups and different organisations. Training should be tailored to the particular needs of an individual organisation, depending on its role in the emergency response, including its tasks and responsibilities.

It must be assumed that the personnel within a particular organisation are already trained in the plans and procedures that are part of their day to day work, and are trained in these to an acceptable level. This training that is provided in connection with the emergency plan should concentrate primarily on those conditions that are different or which will place supplementary demands on job-specific training. The level of training should reflect the challenges that personnel may have to face, and include stipulated goals and success criteria. For example, a stipulated goal may be that the response should be carried out with the same degree of success as would be expected to be attained in a similar accident outside a tunnel. Although the personnel will have been trained to be competent to achieve a successful response in non-tunnel conditions, the different challenges posed by tunnel conditions mean that personnel should receive further and supplementary training to reach an equivalent degree of success in this situation.

So, what type of training should be provided? First, everyone in the involved agencies must be familiar with the emergency procedures. Experience from accidents has shown that, while the planned procedures have been adequate in the main, some personnel who should have acted in accordance with these procedures either did not know of their existence or their content. Thus, personnel must be trained again and again.

Besides procedure training, reconnaissance visits should be an essential part of training. Familiarisation with the tunnel complex and the surrounding area is of vital importance for the success of the response. The first visit should not be in smoke and extreme heat on the day the accident happens. However it can be difficult to get admission to a tunnel, which of course is built to operate, and there may not be many periods when the tunnel is not in operation. Furthermore, if such periods do exist, they are often at night, and it is difficult to fit in a visit with the individual organisation's planning. To compensate for this, one should make use of training materials (e.g. videos) that can give a reasonable insight into possible incidents. However, as a minimum, the first response teams should have on-site visits.

Figure 27.7 (a) The full-scale training tunnel at Korsør, Denmark; (b) an exercise in the Great Belt Tunnel

(a)   (b)

Irrespective of the personnel's knowledge of the tunnel complex and the surrounding area, maps should be available, with relevant information about each component, as these can help relevant key people in the event of an accident. In addition, individual organisations should be familiar with each other's procedures and materials, and these aspects should be included in training. As a minimum, this familiarity is obligatory for those who are involved in key roles and who are involved as decision-makers during an incident.

Table-top exercises are a relatively cheap form of training that can provide education in many areas (procedures, tactics, command and control, dilemmas, etc.). They are also a good means of meeting the partners with whom one will have to work during a real accident, and of ensuring the ongoing and dynamic development and adjustment of procedures.

Practical exercises should also be included in the training, and should focus on the particular challenges that will be faced in implementing the safety procedures, and problems that commonly arise during real accidents. For example, some of the practical challenges that will be faced by fire-fighters in a tunnel incident are the long and difficult approach ways, the large heat loads, the physical strain on the fire-fighters and the great logistical demands. Training undertaken in suitable premises (Figure 27.7) should ensure that, in addition to procedure planning, fire-service staff are prepared to face and tackle these challenges in an effective way, and, of great importance, are aware of their own limitations.

Despite the inconvenience and expense, the benefits of a full-scale exercise cannot be over-estimated. Such exercises allow the emergency procedures, including the organisations involved and the technical system taken as a whole, to be properly evaluated. Mistakes and shortcomings that are not evident in small-scale exercises are often uncovered in full-scale exercises, and the outcome of a full-scale exercise will certainly be very different from that of a table-top exercise, even for the same scenario.

Although not the primary goal, full-scale exercises help to keep the different plans and organisations 'in shape', and in the period leading up to the exercise there is greater activity around a lot of emergency-relevant factors, especially if the media are invited. Emergency systems are double checked, emergency plans are polished and checked by involved personnel, the surrounding area (including meeting points, triage points, entrance points, etc.) is put in order, and so on.

Finally, it should be mentioned that the conduct of the passengers at an accident will have a great influence on the outcome, and the possibility of training the public should not be neglected. Of course, in this case one cannot use traditional training methods, but short, clear, instructions or information displayed on trains and at stations is a method that should be considered. This may be combined with short, clear, instructions or information given at various intervals over the train's public address system.

27.3.6   Review: the system changes

A tunnel is often built to have a lifespan of 100 years or more, and unfortunately the same sometimes applies to procedures. However, procedures should be reviewed regularly, and keeping them up to date must be a dynamic process (see also Chapter 5).

It is often the case for a new tunnel system that brilliant plans and procedures are put in place and tested, training is undertaken, and all is well. As time passes, there are changes in personnel, the operation of the tunnel changes gradually, and the supervision of precautionary activities and the attitude to safety changes bit by bit. There may not have been even a very minor accident in the tunnel, and the initial scenario planning may come to be seen as a doomsday prophecy from the past and a waste of time. The emergency plans become dusty and get a fixed in a place on a shelf. Then the accident strikes, and one wonders why the procedures did not work.

To achieve the goal of a high level of safety and up-to-date procedures, the topic of safety should be subject to continuous review. The relationship between preventive and remedial arrangements should be reviewed regularly, and how the theoretical goals fit the practical implementation evaluated, including the possible consequences of an accident, while still maintaining focus on the possibility of improvement. Moreover, those responsible (including the emergency services) should keep an eye on any changes in the operational use of the tunnel and relate these to the existing emergency procedures and collective rescue concept. It is important that this process of review and revision does not become static or fade out. Accidents, near-miss events and exercise experience in other tunnels should also be included in evaluations.

It is, therefore, sensible to have a formalised group that continuously evaluates the emergency plans and procedures, and ensures that they are up to date, including ensuring that the relevant organisations are aware of and trained in the procedures, so that they can be put into action at the right time and place, and in the correct way with the greatest possible effect. It is important that legal duties are placed on the tunnel operator to maintain any systems provided for emergency-service use for the life of the tunnel, and enforcement action should be taken against the operator/owner if the systems are allowed to fall into a state of disrepair.

27.4.   Considerations

There are a number of key factors (design, management, facilities, equipment, scenarios, objectives, etc.) that have an important influence on emergency procedures, and thus these procedures vary from tunnel to tunnel. To achieve the best possible emergency response it is essential that planning focuses on which factors can positively or negatively affect the response and in which phases of an incident these factors assert themselves. This analysis should form part of the basis of the collective safety concept, and be an important part of the planning.

27.4.1 Objectives

The statement of safety policy should set out the goals and objectives to be addressed. The policy statement should include an explicit commitment to safety on the part of the top-level management of all the organisations involved.

However, there is a number of possible goals, and various opinions with regard to how the consequences of an accident should be controlled, including what priority the assisted response should have. Before goals can be set it is essential that all the parties involved recognise the risk, regardless of the probability of its occurrence. The immediate consequences of an accident can generally be described as

- fatalities and injuries
- damage to the tunnel
- infrastructure and traffic problems.

Goals are sometimes defined such as 'best possible' or as 'consequences of an accident shall be kept to an acceptable level of risk', and there is often great disagreement about the extent of the possible consequences. In the design phase, the normal goals are

- to control the immediate consequences of an accident
- to make self-rescue possible
- to make assisted rescue possible.

These goals comprise a good start for a basic general design. However, to be able to formulate goal-oriented emergency procedures it is necessary to define some more response-oriented and precise goals, including partial goals.

An evaluation should be made of which scenarios to consider and what response is necessary to achieve a given goal, including partial goals for the individual parts of the scenario. To set some acceptable goals it is also essential to analyse the assumptions and conditions that underlie the basis of the goal response. That is, what are the possibilities of controlling the consequences and, most importantly, what can actually be achieved in reality? In this planning, the focus should be on the following questions (Figure 27.8).

- What is the scenario – what is the immediate consequence?
- What will one achieve with the response – when and how will one achieve it?
- What are the objectives?

The scenario, response and goal are all variable parameters, and if the content of one of the boxes in Figure 27.8 is changed it will probably affect the content of one or both of the other boxes.

Emergency procedures and objectives can be set during the design phase, but they can also be set after the tunnel has been built. It should be noted that undertaking this work during the design phase can often save the owner a lot of money, as it can be very expensive if changes have to be made to the tunnel and its safety installations after building has been completed.

If it is not possible to reach the stated objectives, the simplest option is to change the objectives, keep quiet about it, or blame someone else. As an example of an aim that has been shown to fail,

Figure 27.8 Planning focus

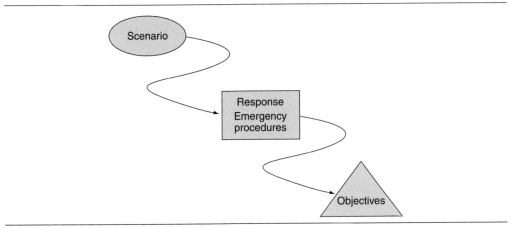

one could mention evacuation time. In several tunnels, it has been shown that it is not possible to achieve the expected evacuation time, after which the only acceptable economic solution has been to change the stated objective.

When planning to achieve the 'best possible' response, given an actual socio-economic situation, it is essential that each element of the emergency procedures and response is evaluated, in order to get the most for the money available in relation to the general objectives, and to create a broad unity of agreement on these.

27.4.2 Phases and key factors
27.4.2.1 Key factors
The formulation of emergency procedures depends on the tunnel design as well as several other key factors, and, depending on how these factors are weighted and work together as a whole, the emergency procedures and the subsequent outcome will achieve a degree of success. The degree of success of emergency procedures can be defined according to the success of the completed response.

The following key factors should be considered when formulating emergency procedures (Figure 27.9)

1. precautions and fixed installations
2. equipment and materials
3. personnel and organisation
4. education and training.

PRECAUTIONS AND FIXED INSTALLATIONS
These are facilities incorporated within the tunnel structure or transport system to assist in mitigating the immediate effect of a tunnel incident

- installations that should ensure quick alarms (fire or smoke, gas detection, CCTV, emergency telephones, etc.)

Figure 27.9 Key factors in formulating emergency procedures

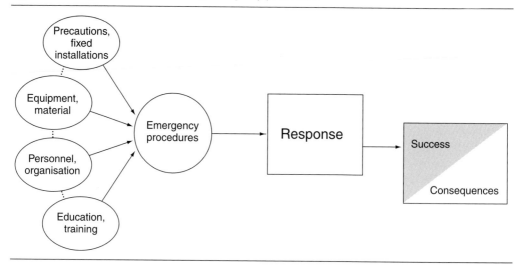

- installations that should help self-rescue (ventilation, emergency exits, fire-resisting escape doors and shelters, walkways, escape signs, tunnel lights, loudspeakers, etc.)
- installations that can be used by incident response teams (fire hydrants, communication systems, power points, drainage systems, etc., and several of the facilities mentioned above).

EQUIPMENT AND MATERIALS
These are materials and equipment that is available outside the tunnel or in the trains that operate in the tunnel and which can be used in case of an accident.

- Materials and equipment that can be used in the initial response (fire extinguishers, megaphones, public address (PA) system, etc.).
- Response vehicles, trains and other materials and equipment that can be used in the later phases of the response (evacuation train, other special rescue trains, infrared cameras, communication systems, etc.). Any equipment provided must be safe, suitable and sufficient, and comply with industry standards.

PERSONNEL AND ORGANISATION
These are the personnel included in the response and have a role in connection with the response in its entirety.

- Personnel on trains and, if there are any, tunnel personnel, traffic and technical operators, etc.
- Response units (fire service, police, ambulance service, etc.).

An important key factor is how the involved people are organised, both within the individual organisations involved and, more importantly, in general across individual organisation boundaries.

Figure 27.10 The phases of an emergency response

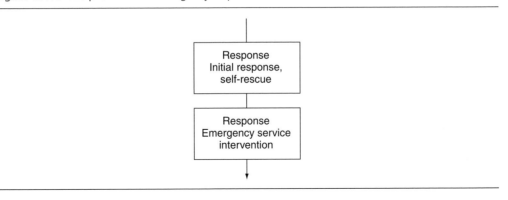

EDUCATION AND TRAINING
These are the education and training that the involved individuals have received

- qualifications of the operators and the train personnel
- qualifications of the emergency service personnel.

To plan for an optimal response it is important to focus on which factors can have a positive or negative impact on the emergency procedures and the response and the time during the response when these factors assert themselves.

27.4.2.2 Phases of the response
If one looks at key factors (1) and (2) (precautions and fixed installations, and equipment and materials), they normally become effective in two different phases of the collective response (Figure 27.10)

- the initial response: self-rescue, inclusive detection and alarms
- the emergency service intervention and response.

THE INITIAL RESPONSE: SELF-RESCUE, DETECTION AND ALARMS
This is the most decisive phase, especially in terms of saving life, and covers fire detection and alarm raising, and the actions that unfold in the immediate aftermath of the accident. If the responsibility for initial evacuation rests with the train operator or tunnel operator, this should be formally recorded so that all parties know where the responsibility lies.

The success of the initial response depends on many things, including key factor (1) (precautions and fixed installations), and the outcomes of this first phase form the basis for the next phase, where the external rescue teams are set in action.

EMERGENCY SERVICE INTERVENTION AND RESPONSE
This phase should be an immediate extension of the rescue response, and should deal primarily with saving the system. It may be too late to rescue people who are directly threatened by fire and smoke.

The degree of success of this phase depends on key factor (2) (equipment and materials), but also depends to a large degree on the success achieved in the initial response. The consequences of this

Figure 27.11 Key factors and phases

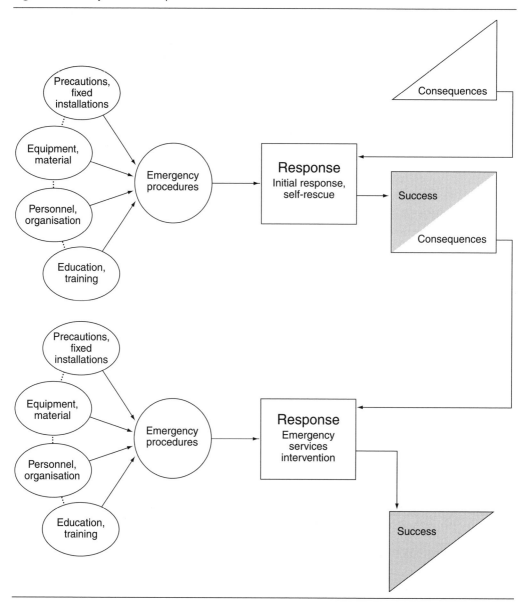

second response phase are an expression of the collective success of the overall response to the incident (Figure 27.11).

27.4.2.3 Partial goals in the response phases
It is clear from Figure 27.11 that each individual organisation's planning should have a connection to each other organisation's partial goals, including the degree of success that is expected, if the common goal is to be achieved.

When planning for the operator's response, including the response of technical and train personnel, the plan should take as its starting point the expected immediate consequences of the incident. Procedures should take into account scenarios and objectives, and include precautions, equipment, personnel resources, the organisation of personnel, and education and training levels. What is the possible scenario and what are the immediate consequences of it? What are the success criteria for receiving alarms and their immediate processing? For example, what are the criteria for informing those responsible for the technical installations (e.g. the ventilation operator) and the success criteria for the wider and collective response?

When the emergency response teams are planning their specific procedures, they should focus on scenario planning in relation to the immediate consequences, with the 'consequences' they expect they will have to handle as the starting point. However, if the initial response, including alarm and self-rescue, achieves such a positive degree of success that the emergency services are not required, the planning is finished as far as they are concerned.

To be more specific, here are some examples. If the train personnel have a partial goal which states that at an evacuation they should make a total evacuation of the threatened area, the fire brigade's plan should correspond with that. On the other hand, if this is not a goal for the train personnel, then it should be focused on in the fire brigade's planning, i.e. one of their first tasks should be evacuation, and equipment, training, etc., should be arranged accordingly. It is very important to ensure that all parties know where the statutory duty and responsibility rests for each activity, in order to prevent any misunderstandings during the incident.

As another example, for some tunnels the distance to the nearest fire brigade is so great that, in practice, the fire brigade is not considered an essential factor in the collective response. In such a case, one should hope that the partial goals for the immediate response, and the conditions they should be carried out under, ensure that the collective consequences are at an acceptable level. However, instead of keeping the fire brigade out because of the distance to the tunnel, it could be considered to place the fire station near the tunnel so that it could be put into action almost immediately. Obviously, if this option is taken, the partial goals will change. Such matters will need to be discussed, agreed and financed during the planning phase of the project, so that suitable measures can be put in place.

With regard to stating goals, the term 'best possible' is too vague. If a plan is to be effective and balanced, goals should be definite, as should the partial goals of the individual organisations and the associated success criteria in the different phases. In this connection, the risk analysis, possible responses and the different degrees of goal fulfilment should be considered, including how the different organisations commit themselves in relation to the collective rescue concept and their individual partial goals.

When a tunnel is new, training intensity is often high. For example, train personnel are often well trained to handle an evacuation situation. Should the emergency procedures be changed after 10 years, when all the personnel may have changed and training not been maintained? The partial goals are still the same, but can the current personnel achieve them, and what does this mean for the second response phase and for response success as a whole?

27.4.3 Time

Time is always important in connection with an accident, and in tunnels the response time is so significant that it should be included as an essential link in the planning work.

Figure 27.12 The timing of actions

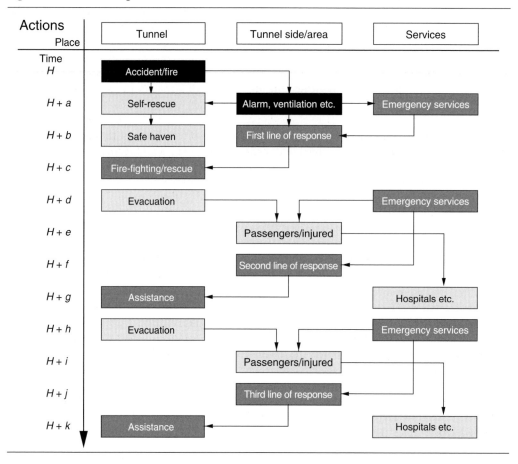

A flow chart showing times and actions can be of benefit when evaluating the need for pre-cautions, equipment, personnel, and the organisation and training of personnel. The chart can also be used when formulating partial goals, as some partial goals will be essential in both the individual phases and in regard to the complete response. An example of a flow chart is given in Figure 27.12, which shows the timing of actions in the tunnel, actions in the surrounding area (surveillance functions, etc.) and the actions of the incident response services. Many examples and considerations can be related to the flow chart in Figure 27.12, but just a few are mentioned here.

Suppose that an accident happens at H hours; alarm and evacuation starts. The fire service receives the alarm, drives to the approach way and makes an intervention. According to the scheme they will be at the accident scene at $H + c$ (say within 30 minutes). Which task should they undertake, under which conditions should they be sent in, what training will they need to reach the partial goal, and should the tunnel be isolated?

If the fire brigade arrives after 30 minutes, what training should the train personnel receive in order to deal with this situation?

If the first fire incident team goes into action, say after 30 minutes, how long should they be in action, and when can they expect assistance? It is strongly recommended that the initial weight of response of each responder team be sufficient to fully undertake an intervention, rather than commit to an intervention and then have to wait a long time for back up resources.

In relation to objectives, the flow chart can be used in reverse. As an example, consider the 'golden hour', which is a generally accepted term for the period when the possibility of successful life-saving treatment is high. After an hour, the chances of a good medical outcome drops off dramatically. Looking at the chart, the possibility of professional medical treatment will be at $H + e$, if injured people are evacuated out of the tunnel almost immediately, or at $H + g$ if they are treated in the tunnel (with assistance) or if professional treatment can be provided by the first line of response.

As other examples, the police force (or ambulance service) should be ready to receive evacuees at $H + e$, the first line of response should be capable of handling the accident from $H + c$ until an expected relief at $H + g$, and the train personnel should be ready to deal with the accident from H until $H + c$.

All the above factors should be included in the planning to ensure the most effective emergency procedures and the optimal effect of the response. The main focus should be on what will be achieved and for what reason.

27.4.4 The bottom line

There are many methods for providing an acceptable degree of safety, and most of the developed procedures can generally be regarded as good. However, more focus should be put on the outcome and the bottom line.

The starting point for determining the bottom line is important, although not an objective in itself. The most important thing is the outcome, and so this must be the focus for the bottom line in the planning work.

Regardless of the tunnel design, precautions, facilities, etc., the procedures should always provide the best possible basic response overall. This is summarised in Figure 27.13, which brings together Figures 27.8 and 27.11. It is equally important that responders demand the best possible active and passive services at the tunnel planning stage. Such safety systems should not be placed at risk when finances come under pressure and there is a need to make savings. Those responders with a statutory duty and legislative powers will need to enforce their requirements, while other partners will need to provide support to ensure that the safety systems are retained.

27.5. Conclusion

On the whole, it can be said that there is no general standard for when and where emergency procedures should be provided, or how extensive they should be. Emergency procedures range from non-existent to well thought out and well-tested procedures. Several even include special units, established as a direct link in the safety policy for one or several tunnels.

Regardless of the safety level, it is essential that decisions regarding the need for and the level of safety arrangements, including emergency services and procedures in relation to a tunnel's collective safety concept, are made in a concrete and reasoned way. A plan should be prepared, laying out the objectives for the expected degree of success of the response, and including all the involved organisations; the focus should be on the objectives.

Figure 27.13 The bottom line

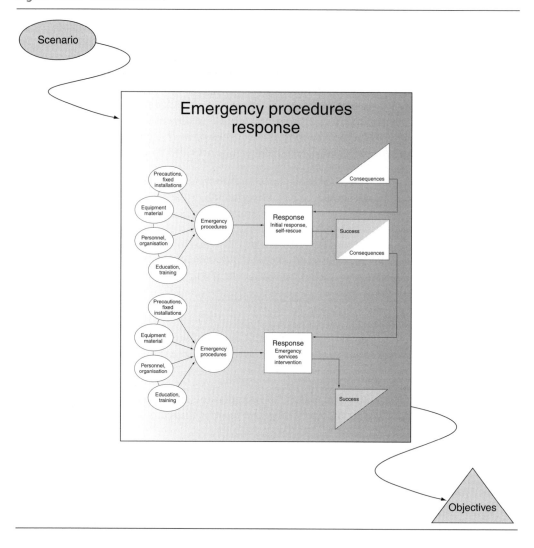

Again regardless of the safety level decided upon, it is important that the emergency procedures provide favourable conditions for the response as a whole and contribute to the success of the response. The emergency procedures must be good, consistent, simple, easy to apply, not confusing, and well tried and tested, and the emergency services and other personnel should be familiar with them to the degree that they can be implemented almost automatically.

27.6. A detailed example: emergency procedures in the Great Belt Tunnel, Denmark

27.6.1 Introduction

The Great Belt Tunnel, which was opened in 1997, unites Denmark's two largest islands, and is a very important link in the European rail network (Figures 27.14 to 27.16). The 18 km fixed link consists of: a rail tunnel between Zealand and the island of Sprogø; the East Bridge, a 6.8 km

Figure 27.14 Map of Denmark showing the location of the Great Belt Tunnel

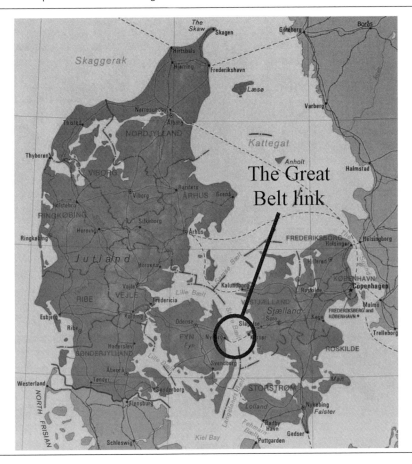

road bridge between Zealand and Sprogø; and the West Bridge, a 6.6 km combined rail and road bridge between Sprogø and Funen. The Great Belt Tunnel is 8 km long and consists of two tubes. Cross-tunnels to the opposite tube, with fireproof doors, are located at 250 m intervals. There is a walkway on both sides of the tracks, strong tunnel lights that can be switched on when required, and emergency exit signs, which are always lit. There are 80 ventilators, which can give a maximum flow of approximately 5 m/s, and fire hydrants and power points every 125 m.

In Denmark, there is no tradition of tunnel building. Therefore, there is no legislation that ensures safety arrangements in a tunnel, and there are no special laws that the authorities should provide extended emergency services for responses in tunnels.

In the preparation of a safety concept for the Great Belt Tunnel, the relevant emergency services were involved from an early stage, and there was constructive cooperation between all the parties involved. This ensured that acceptable emergency procedures were prepared which, compared with those for other tunnels, having been given a relatively high priority. The purpose of this detailed example is to describe the overall emergency procedures that were formulated and the training that was deemed necessary to achieve the objectives.

Figure 27.15 The island of Sprogø

Figure 27.16 The Great Belt Tunnel

27.6.2   Emergency procedures

During the building of the tunnel the focus was on the safety, with the following aims

- to avoid accidents
- to control the immediate consequences of an accident
- to enable self-rescue
- to enable help from emergency services.

At an early stage in the design and construction all the involved parties agreed on a basic rescue concept, which is described in brief below.

The basic rescue concept is that when an accident happens, initially there is a quick alarm, followed by an evacuation carried out by the train personnel. In the case of fire, there is a 'drive-out' policy, but for other accidents, or if drive-out is not possible, the train personnel will make an evacuation to a 'safe haven'. For this they will use the walkways, which are placed on both sides of the rail tracks, and the cross-tunnels, which are located at every 250 m.

The fire service will use special rescue vehicles, which can drive on both rail and road, to undertake the first emergency service intervention in the tunnel. In the second phase of the response, which includes the transport of passengers (out) and the emergency services (in), the nearest trains are called in and emptied on both sides of the tunnel, and then made available for use in the response. If the fire-service leader in the tunnel considers it necessary, further emergency services are sent into the tunnel, possibly just to a limited area, and passengers are brought out of the tunnel on the same trains.

Based on this general concept, the responsible authorities and companies, in cooperation with the relevant interests, decided on the emergency procedures that should be used in the case of an accident.

The collective incident response at the Great Belt Link and the following emergency service response are complex. Several response authorities and companies are involved, each of which has a number of competencies and responsibilities, which are described in part in their individual plans and in part in the collective and coordinated plans.

The individual plans and the collective contingency plan were prepared from the starting point of the basic design and rescue concept, and the individual response authorities' normal procedures and task plan. The collective plans and procedures have an effect on the individual plans, and lay down the expected chronological order of events.

The remainder of this section gives a description of the intended execution of the procedures for the alarm phase, self-rescue and evacuation, and the external rescue response.

27.6.2.1 Alarm phase

If an accident occurs in the tunnel, the train personnel report it to traffic-control staff, with whom the train driver is in continuous contact via the train radio. If the train driver is not able to give the alarm from the train, it can be given by a member of train staff or a passenger from one of the emergency telephones that are situated at all signals and cross-tunnels. In addition, the alarm may be set off automatically via one of the technical alarms, which are part of the traffic and

Figure 27.17 Computer-based accident programme (the example shown not a real example)

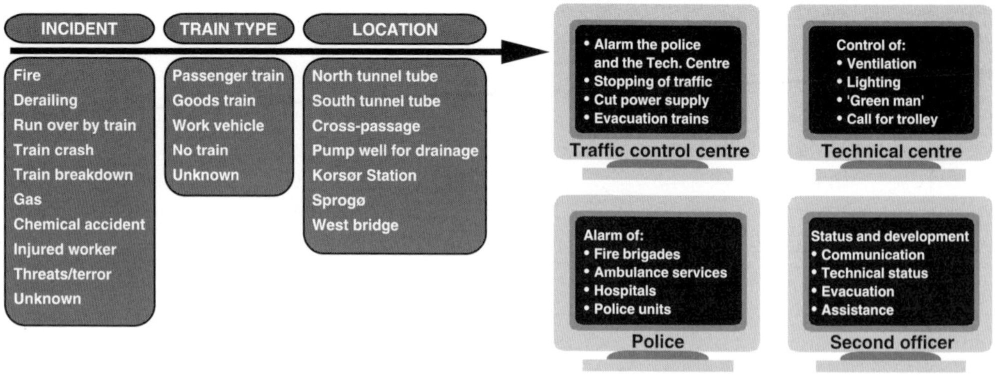

INCIDENT	TRAIN TYPE	LOCATION
Fire	Passenger train	North tunnel tube
Derailing	Goods train	South tunnel tube
Run over by train	Work vehicle	Cross-passage
Train crash	No train	Pump well for drainage
Train breakdown	Unknown	Korsør Station
Gas		Sprogø
Chemical accident		West bridge
Injured worker		
Threats/terror		
Unknown		

Traffic control centre
- Alarm the police and the Tech. Centre
- Stopping of traffic
- Cut power supply
- Evacuation trains

Technical centre
Control of:
- Ventilation
- Lighting
- 'Green man'
- Call for trolley

Police
Alarm of:
- Fire brigades
- Ambulance services
- Hospitals
- Police units

Second officer
Status and development
- Communication
- Technical status
- Evacuation
- Assistance

technical control. A similar alarm is set off if there is an unscheduled train stop for which the control room has no information about the cause.

When the report of an accident is received in the traffic control centre, a special computer-based accident programme is activated. The operator can choose from different defined possibilities. The verbal report is documented as a specific alarm report, which initially includes information about the incident (what has happened), the type of train (which train is involved) and the location (which tunnel tube, pinpointing the accident).

The report is then sent electronically in a network to the technical control, which controls the technical conditions in the tunnels (including ventilation), and to the police, who are responsible for alerting the incident response team. As a result of the report content, there is simultaneous activation of a specific actions list for all those who have been alerted (Figure 27.17). There are several combinations for possible accidents; for example, 'fire–passenger train–northern tunnel tube'. According to the exact combination, the electronic action list contains the tasks that the individual responder should undertake. All the tasks have been well thought out by those responsible in the planning phase. (*Note*: It is essential that operators do not get side-tracked by other tasks that divert them from their primary notification or liaison duties.)

Meanwhile, in the tunnel, the train personnel maintain contact with traffic control, and provide supplementary information on the status in the tunnel, inform other train personnel about the response status, who then pass the information on to the passengers.

The supplementary information received by traffic control is continually input into the accident programme, and is sent on to technical control and the police, who pass it on to the emergency services, who have already been called out and are on their way to the tunnel.

The composition of the response team is decided on the basis of the first alarm. The supplementary information is used to prepare the response units for further decision-making in relation to initiating the necessary response, including the choice of approach way and the location where the intervention should start.

27.6.2.2 Self-rescue and evacuation

In the period immediately after the alarm, the train personnel will start the evacuation of the train. During the initial evacuation every effort is made to bring all the passengers to safety in the non-incident tunnel. If there are any injured, they may be left behind; the train guard can decide whether he can spare personnel to render first aid, if conditions permit. In addition, the train personnel, first in connection with the report of the accident and then continuously, keep traffic control informed about the train's precise position in the tunnel, the type and extent of the accident, the number of injured and the total number of passengers.

Once the evacuation has been completed, the train guard, insofar as is possible and if it is safe to do so, will remain in the non-incident tunnel, where he will receive the fire service when it arrives and give further detailed information on the extent, type and development of the accident.

When the passengers have reached the non-incident tunnel the train crew will, until the police arrive, keep the passengers informed via megaphone about the evacuation train and any other details they have about the response.

Incident cards have been prepared for the train personnel that describe what they should do in the case of an accident.

In the passenger trains that run on this line there is information available to the passengers on the safety concept, including an easy to understand illustrated brochure about the tunnel, focused on the principles of self-rescue.

27.6.2.3 External rescue response (emergency services)

The fire brigade is the first agency to approach the accident location. This is accomplished in special vehicles that can drive on both roads and rail tracks (Figure 27.18). The special tunnel

Figure 27.18 A road–rail vehicle

vehicles are equipped with state-of-the-art equipment to deal with fire extinguishing, rescue tasks, accidents involving 'dangerous goods' and ambulance tasks. The vehicles are constructed of special materials so that they can immediately tackle all possible types of accident. During long incidents they are supplemented with material from outside, depending on the actual situation.

This first response team is ready to drive into the tunnel 15 minutes after the alarm has been received, and the crew is specially composed such that it can meet the demands of the prepared objectives.

The technical control personnel and a second-in-command fire officer, who is responsible for the area outside the tunnel for the duration of the response, are in contact with the incident commander inside the tunnels. This communication is established and functioning before the first fire-service units enter the tunnel. On arrival the fire-service staff become responsible for all technical changes in the tunnel, which are accomplished in cooperation with the incident commander in the tunnel and the technical operator. The fire-service leaders are also in direct contact with traffic control with whom they agree when the special tunnel vehicles should go on the tracks and about disconnecting the power if necessary. During this time the incident commander receives up-to-the-minute information from the control centre, which has contact with train personnel inside the tunnel.

When the first line of response arrives at the accident site, they immediately start work to prevent further damage occurring.

Outside the tunnel a large number of supplementary rescue personnel arrive, including fire-fighters from the surrounding fire stations, the police, ambulances and emergency hospital staff. The nearest passenger trains on both sides of the tunnel are cleared of passengers, and the train staff are sent to aid the emergency response, in particular the evacuation.

An important question for the incident commander inside the tunnel is the matter of when, from a safety standpoint, it is possible to get more units into the tunnel. Unless it is a very small accident, they will be needed.

As a starting point, it is planned that the non-incident tunnel tube, if not threatened by the accident, is used as a safe haven. Therefore, the incident commander will, after this tube has been inspected and evaluated in relation to needs and safety, request that the assisting rescue personnel, emergency hospital units and the police be transported into the tunnel on the evacuation train, which should be ready within 30 minutes after the alarm. When the supplementary personnel arrive in the tunnel, the long-term response begins as appropriate to the specific situation (Figure 27.19).

The general concept is that the evacuation train comes into the tunnel, bringing assisting rescue units and the police, and then takes passengers and the injured out of the tunnel in order of priority.

Outside the tunnel, the ambulance service will have established triage points, etc., and the police coordinate the response in the surrounding area in general, including the transport on the rescue train.

Figure 27.19 Response principles

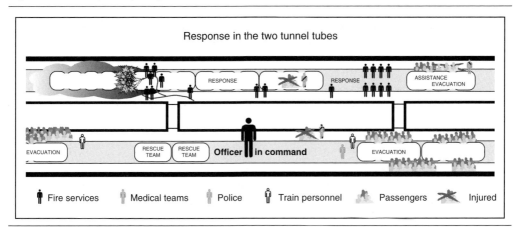

27.6.3 Education and training

Plans and procedures are of little value if they are not 'trained in', again and again. Therefore, there is continuous education and training of all those who may be involved in the planned emergency procedures, and training is regarded as very valuable.

Education and training is undertaken in part by the individual authorities, and in part collectively through regular exercises involving several or all of the involved parties. Training is implemented according to the grade that personnel have within their own organisation, and is generally divided into three levels: manual (e.g. fire-fighters), middle management (e.g. junior officer) and senior management (e.g. commanding and senior officers). The training is also carried out according to the existing level of training of an individual, and is divided into training of new personnel, refresher courses and advanced courses.

The complete education and training programmes, which vary between the different authorities, are very complex. In general, the focus of all the authorities is that the training should ensure that every individual is in a position to fulfil the role he or she has been selected for in relation to the planned emergency procedures. Education and training takes place in three different ways: independent exercises by individual authorities, cooperative exercises, and annual full-scale exercises.

27.6.3.1 Independent exercises by individual authorities

Each individual authority undertakes independent exercises and training. As an example, the general content of the training provided by the fire service is described below.

The first line of response comes from Korsør Fire Brigade. To be included in the response team, fire-fighters have to complete a one-week special tunnel-training course. The course concentrates on material, plans and procedures, tactical guidelines, and training in using special breathing apparatus, including the individual fire-fighter's capacity and limitations.

After qualification, each fire-fighter, in addition to their general and legally required training, is required to participate in 6–8 days of training annually. In these courses they learn routines in different scenarios, and try out new routines and methods. In addition, special task groups and resource people receive specific education and training.

To be included in a response team assisting unit (second line of response), fire-fighters initially participate in a one-day course where they are trained in the special plans and procedures and tactical guidelines. Following that, assisting-unit fire-fighters participate in two exercises annually, to practice in what they have learned and perhaps learn some new routines. Assisting units comprise fire-fighters from eight different fire brigades/council areas.

All unit leaders go through the same training as the fire-fighters. In addition, they receive regular education in staff work, where tactical guidelines are discussed and tried out regularly in tactical table-top exercises. The leaders gain experience by working as instructors and as incident commanders during exercises for the other units.

27.6.3.2 Cooperative exercises

The cooperative exercises are very important, especially for the leading personnel and for relevant resource people. There are regular tactical table-top exercises and staff and communications exercises that involve different levels of staff and organisations. The exercises are in different aims and a variety of training methods is used.

27.6.3.3 Annual full-scale exercises

Every year there is a full-scale exercise in the Great Belt Tunnel, which is closed for about 5 hours. Several inexpedient factors have been uncovered by these exercises, including some that could not have been revealed by undertaking smaller independent exercises outside the tunnel. It has been given a high priority that these possibly inadvisable conditions should be improved (Figure 27.20).

The holding of the exercises at all levels puts a sharp focus on improving the emergency procedures, including synchronising the individual organisations' plans and action lists. The emergency response thus has a general cycle. First, individual organisations carrying out individual training in relation to their plans, which stem from the rescue concept and the common emergency procedures. These procedures and the associated training are then tested in the cooperative exercises and the annual exercise. As a result of the experience gained in these exercises, the current plans and procedures are adjusted, and then training courses are held in the revised and adjusted plans and procedures, including further training in the existing

Figure 27.20 Exercises and improvements

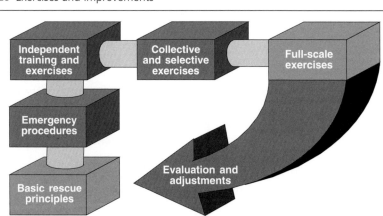

skills. This cycle is, in principle, carried out once a year, and ensures the dynamic development of the emergency procedures, and the education and training, and strengthens the collective response.

27.6.4   Conclusion on the example

It is often the case that an emergency plan is established for a new tunnel, and thereafter no resources are dedicated to update the plans and procedures. However, it is vital that the updating of these plans and procedures is a dynamic process.

For the Great Belt Tunnel, the emergency services have designed some special objectives for an emergency response in the event of an incident. There is both a main objective and some partial objectives, and a number of conditions have been described that must be fulfilled in order to meet these objectives. These conditions are to do with the plans and procedures, materials, personnel, training and other factors.

To ensure that the level of the emergency services response is maintained and the natural development of the safety concept, the emergency response is monitored closely by Korsør Fire Brigade. The brigade regards it as an essential task that the emergency response continues to be revisited with regard to changing political demands, risk analysis, new rescue materials, improved procedures, etc.

The authorities and companies have formed a Coordination Group for the Great Belt Link, which is responsible for the development, coordination and maintenance of the response concept as a whole, including the emergency procedures, and this should ensure an optimal response in the event of an accident occurring in the tunnel.

Handbook of Tunnel Fire Safety, 2nd edition
ISBN: 978-0-7277-4153-0

ICE Publishing: All rights reserved
doi: 10.1680/htfs.41530.589

Chapter 28
Fire and rescue operations in tunnel fires: a discussion of some practical issues

Anders Bergqvist
Stockholm Fire Brigade, Sweden
Håkan Frantzich
Lund University, Sweden
Kjell Hasselrot
BBm Fire Consulting, Stockholm, Sweden
Haukur Ingason
SP Technical Research Institute of Sweden (SP), Borås, Sweden

28.1. Introduction
This chapter is based on work that was performed as the first of three subprojects in the 'Rescue Work in Tunnels and Underground Facilities' group project, financed by the Swedish National Rescue Services Agency. It was originally published in Swedish.

28.2. Reference assumptions
The material in this chapter is based on a fictitious accident, the conditions of which have been defined in order to provide a framework for the work of the project group. The scenario of the accident is extensive, both in terms of the size of the fire and of the number of people exposed to it. The reference assumptions have been based on careful evaluations of what can occur when tackling a fire in a train in a tunnel. In this specific case, the fire is presumed to be in a rail tunnel, but the basic approaches to rescue work depend on the same fundamental principles for all types of rescue work in similar environments. The results from the analysis of this accident must then be modified and applied to the actual conditions applying in each specific accident.

The initial event is a fire starting in one of the coaches of a passenger train carrying about 240 people. The fire, in the first coach behind the locomotive, was first observed by a passenger, who applied the emergency brake. At that point, the train was just before a tunnel, and its stopping distance was such that it came to rest in the tunnel, with the front of the locomotive about 300 m from the tunnel exit (Figure 28.1). The fire started in the interior fittings in the coach, but spread only relatively slowly within the coach, taking about half an hour to become fully developed. This had the effect of breaking all the windows, so that the carriage became completely enveloped in flames.

The direction of the wind, determined by the external conditions, was against the original direction of travel of the train, i.e. northwards (Figure 28.2). In addition to the driver, there were three further rail staff on the train. They were able to use the train's public address system, so that they could speak to the passengers.

Figure 28.1 A section through the tunnel, showing the train and space for evacuation on each side

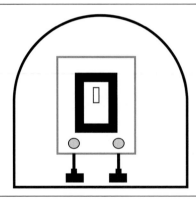

After the emergency braking, all the doors were intact and could be used to leave the train. It was also assumed that none of the passengers was seriously injured by the emergency stop, and therefore they could all evacuate the train by themselves. Evacuation was started after the train crew had instructed the passengers over the public address system to leave the train. This means that the action of the train crew is important in determining the progress of events.

The work of tackling a fire in a tunnel is determined largely by how extensive the fire is and by what is burning, the design of the tunnel itself, and the number of people threatened by the fire. Information on these points is usually not known at the time of starting the rescue work.

28.3. An accident has occurred and rescue work is in progress

An unlikely event has just occurred and the fire and rescue services have been notified that a train is on fire and has stopped in the long local rail tunnel.

On the way to the site, with the first rescue group consisting of five people, various questions arise. The train is in the tunnel and is on fire, but where in the tunnel and how much of the train is on fire? In which direction is the smoke in the tunnel flowing and where are the passengers?

Figure 28.2 The position of the train in the tunnel

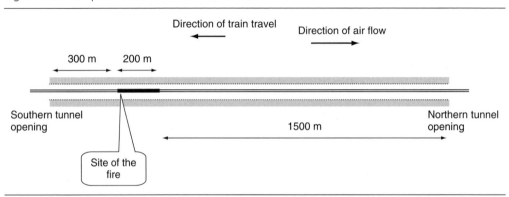

The first step is to choose the correct tunnel opening and decide how to start dealing with the fire in order to provide the best chance of saving everyone on the train.

Further questions come thick and fast.

- Where is everyone on the train? Are they alive? What will need to be done to get them out?
- What is the size of the fire, and what must be done to put it out?
- What are the risks to the fire and rescue personnel entering the tunnel? Can the fire suddenly flare up? Could the tunnel collapse?
- How far can fire-fighters wearing breathing apparatus advance?
- How can the smoke best be cleared out of the tunnel?

At this stage, the incident commander contacts the emergency service headquarters to see if there is any more information from the person who raised the original alarm or from the train company. At the same time he calls for more personnel, ambulances and police. Finally, bearing in mind the current wind direction, he decides to instruct all resources on their way to the site to reach the site on the road that passes the southern tunnel opening.

He decides to make the first approach to the fire from the southern opening, basing this choice largely on the current wind direction, which is from south to north. Investigations have shown that this is a wise decision, which experienced fire officers would choose if they had a good overview of the situation. However, in real life, the alarm is often initially raised by someone who has seen the smoke, which results in the fire services being directed to where the smoke is coming from, which is not necessarily always the best place from which to start a rescue operation. In the case of a tunnel fire, fire-fighters are restricted in the range that they can operate over while wearing breathing apparatus. It can be more effective, in such cases, to attempt to reach the fire from other tunnel openings, in order to avoid the need to work in smoke-filled spaces.

Having reached the tunnel opening, about 10 minutes after the alarm was raised, the incident commander finds that the actual tunnel opening itself cannot be reached by vehicles but must be approached on foot over the last few metres. Two people are visible outside the tunnel, but no smoke is visible.

The two people at the tunnel opening tell the commander that the fire is in the first passenger coach behind the locomotive, that the train is about 300 m into the tunnel, and that they are the only ones who have come out of the tunnel in this direction.

At this stage, the commander decides to send in a group of fire-fighters wearing breathing apparatus in order to find out more about the situation in the tunnel and about the situation at the train. By the time they enter the tunnel, about 25 minutes have passed since the alarm was first raised. The group wearing breathing apparatus is instructed to enter the tunnel, without hoses, and to continue until they meet smoke, to try to obtain a picture of the situation and of what has happened, and then report to the incident commander.

On their way to the site, and after their arrival, the crew has attempted to obtain a better picture of what has happened, and what is happening, in the tunnel. The incident commander needs answers to several questions in order to plan the work properly.

A general problem in all fire-fighting and rescue situations is that the way in which the work is started must be based on the information available at the time of assessing the situation and deciding what to do. Important decisions usually have to be made at an early stage of the work. What makes the work of dealing with fires in tunnels particularly complicated is that it is very difficult to obtain an overview of the site, which means that any information for assessing the situation, or deciding on the direction from which to tackle the fire, is very limited.

The information that the fire and rescue services need in order to deal with such a situation is.

- How long is the tunnel?
- Where in the tunnel is the train?
- Where on the train or in the tunnel is the fire?
- Where are the passengers and crew from the train?
- Are there people in the tunnel and, if so, where?
- Are there any other trains in the tunnel?
- What is the size of the fire and how is it progressing?
- What is the air direction in the tunnel?
- What is the slope of the tunnel?
- What points of access are there to the tunnel?

One of those with whom the incident commander establishes contact is the tunnel operator, in the form of the railway company, in order to obtain information. He learns that there are thought to be about 240 passengers on the train.

The incident commander now has a group of fire-fighters wearing breathing apparatus in the tunnel, and directs the next arriving crew to the other end of the tunnel. From them, about 40–45 minutes after the first alarm, he learns that smoke is coming out of that end of the tunnel. Meanwhile, the group in the tunnel has reported that, before they reached the train, they were met by heavy smoke at the roof of the tunnel, and have been forced to stop due to a lack of safety back-up. They have no fire hose with water with them in order to provide safer working conditions in the smoke-filled environment.

At this point, the fire and rescue service have a number of options to assist evacuation and life-saving using a combination of the following methods

- fire-fighters wearing breathing apparatus can extinguish the fire, and thus remove the danger to those escaping from the fire
- the crew wearing breathing apparatus can help people to escape from the tunnel
- fire-fighters can begin to extinguish the fire and the tunnel can be ventilated from the outside, in order to assist evacuation
- active measures can be taken to assist those leaving the tunnel once they are outside the tunnel.

The incident commander must now decide on the direction of approach and which of the above tactical methods to apply.

28.4. Breathing apparatus operations in complicated environments

When wearing breathing apparatus and in a smoke-filled area, it can be very difficult for fire-fighters to move and carry out their work. Problems are likely to be encountered due to the lack of visibility in the smoke and the limited working time available, as determined by the amount of air carried. Further complications arise due to the fire itself and its direct consequences, in terms of high temperatures and possible thermal radiation from the fire and fire gases. In addition, the tunnel must be searched for people escaping from the fire, with the further burden of the heavy physical load of the necessary equipment, and, last, but absolutely not least, the pulling of fire hoses, which often have to be water-filled.

Many investigations, including those by Danielsson and Leray (1999) and Lennmalm (1998), have shown that such work is very demanding, and is severely limited. The investigations have shown that effective action time is restricted by the physical tiredness of the fire-fighters wearing breathing apparatus, resulting from high body temperatures, the heavy work of pulling the hoses, lack of visibility in the smoke and weight of equipment having to be carried. There is also the limitation imposed by the amount of air in the breathing bottles, which is normally sufficient for about 30 minutes' work. Sometimes, the working duration and range of the work can be reduced by high temperatures, although this is unusual with work in tunnels, and is probably likely to be encountered only in the immediate vicinity of the fire. This has been shown by the temperature calculations carried out in the above projects.

If the rescue personnel find injured people during this work, it is extremely heavy work to get injured or unconscious people out of the tunnel. Tests have shown that, in a smoke-free environment, two firemen can carry a person on a stretcher about 300 m in a tunnel before they themselves are so physically worn out that they cannot, in principle, perform any further useful work (Tunnelexercise TYCO, 1999).

There is little information available concerning the rate of advance of a group wearing breathing apparatus in a tunnel. Tests carried out in industrial premises, which could approximately be compared with other breathing apparatus operations with long access distances, such as tunnels, showed that the average rate of advance of the group was about 6 m/minute.

Tests carried out by the Malmö Fire Service have shown that the maximum working distance for a group wearing breathing apparatus and carrying hoses is 125 m, after which the hose becomes too heavy to pull. If fire-fighters are likely to have to work at greater distances, there needs to be a network of fire hydrants in the tunnel, although this raises the question of whether this is placing too great a level of risk on the safety of the fire-fighters. Instructions for fire-fighters using breathing apparatus in Stockholm allow only a maximum penetration distance of 75 m.

Table 28.1 shows the results of tests carried out by the Stockholm Fire Service on the working distances of fire-fighters with and without breathing apparatus and with and without

Table 28.1 Working distances for fire-fighters

Test	Conditions	Rate of advance: m/minute
1	Smoke, dry hose	4.3
2	No smoke, pulling hose	18
3	No smoke, not pulling hose	80
Industrial premises	–	6

Figure 28.3 The exercise site

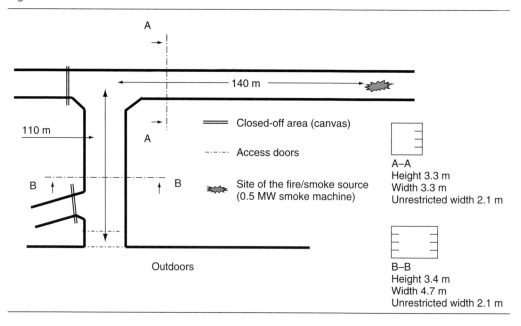

water-filled hoses, in a situation as shown in Figure 28.3. The tests were carried out in a rock tunnel in central Stockholm. The first 135 m of the tunnel floor are asphalted, after which the floor is coarse macadam. The first 60 m of the tunnel slopes gently downwards.

Putting together the results of these tests, and assuming that all the fire-fighters were wearing breathing apparatus containing 2400 litres of air, the maximum one-way distance into the tunnel would be as shown in Table 28.2. This calculation is based on an air consumption of about 62 l/minute, and assuming the fire-fighters would not use their reserve air supply for planned work.

The results of these investigations show that there seems to be only very limited knowledge about such work among Swedish fire services. The differences between one fire service and another with regard to their perceived values of efficacy and limitations indicate that knowledge and experience are very limited. The investigation shows major differences in views on how to perform demanding operations while wearing breathing apparatus, and there are wide differences in the views on the distances over which such personnel can work and how long it takes to carry out such work.

Applying the results shown in Table 28.2 to where those evacuating the train are likely to be, it can be seen that any attempt by fire-fighters wearing breathing apparatus to reach them from the northern end of the tunnel would be very difficult.

Table 28.2 The maximum one-way distance into the tunnel

	Exercise 1	Exercise 2	Exercise 3	Industrial premises
Maximum distance (one way): m	58	243	1080	80

Figure 28.4 The fire service arrives at a fully developed fire in a train outdoors, and starts to tackle it

28.5. Extinguishing extensive fires in tunnels

There is little knowledge or experience of putting out major fires in tunnels. Once a fire has started in a carriage or locomotive, the work of putting it out will be very difficult, due to the thermal radiation, the fire gases and the physical obstacles presented by the rail vehicles, all combining to make it difficult to reach the fire with extinguishant.

The conventional extinguishing systems used by the fire services, which are based on various types of water-based system, depend on it being possible to reach the heart of the fire with the extinguishant in order to put the fire out (Figure 28.4). This is difficult when the fire-fighters, wearing breathing apparatus, cannot approach the burning locomotive or carriage because of the heat radiated from the fire. In addition, the spread of the fire to other carriages in the train will further complicate the work.

28.6. The rescue work continues

From the southern end of the tunnel, the group wearing breathing apparatus has progressively approached the train, and on their way have found no further people escaping from it. Their impression is that the seat of the fire is in a carriage behind the locomotive. Around and behind the locomotive the entire cross-section of the tunnel is filled with smoke, resulting in zero visibility. An attempt to advance further towards the locomotive is prevented by the heat from the fire.

During this period, the incident commander has realised that it would be very complicated, and probably ineffective, to look for survivors evacuating the train in the smoke-filled section of the tunnel between the train and the northern end of the tunnel. Up to this point, the chosen tactical approach has been to save those on the train by helping them to escape from it, and attempting to improve conditions in the tunnel by putting out the fire. When the commander has been able to

Figure 28.5 How the fire services see the situation

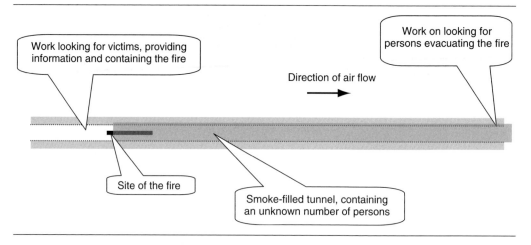

Work looking for victims, providing information and containing the fire

Work on looking for persons evacuating the fire

Direction of air flow

Site of the fire

Smoke-filled tunnel, containing an unknown number of persons

assess the information from the fire-fighters in the tunnel, he decides that a new method of working is needed in order to rescue anyone in the tunnel between the fire and the northern end of the tunnel. This is based on the assumption that there is no one in the tunnel between the fire and the southern end of the tunnel, and that any survivors are between the fire and the northern end. The commander also assesses that any such people cannot be helped by groups wearing breathing apparatus and that these groups cannot effectively tackle the fire (Figure 28.5).

Based on this assessment, the incident commander decides on a new tactical approach, using ventilation from the outside of the tunnel to assist evacuation. This will be done by using fans at the northern end of the tunnel to reverse the direction of air flow in the tunnel, with the aim of removing the toxic smoke in the northern end of the tunnel, where any survivors are likely to be. This will mean, of course, that the southern end of the tunnel will be filled with smoke, but this is acceptable, as this end of the tunnel has already been searched for survivors. At the same time, fire-fighters will be withdrawn from the southern end. When the smoke has been cleared from the northern end of the tunnel, rescue personnel will be sent in to look for and treat anyone in this part of the tunnel. Any such people are likely to be in need of advanced care, although for most of them the immediate environmental threat will be reduced, as shown in Figure 28.6.

Before using ventilation as a method of tackling the situation, a number of questions need to be answered in order to ensure that it will not worsen the situation for those attempting to get away from the train. First, where are they in the tunnel (which is a key question)? Second, how will the fire be affected by the higher ventilation in the tunnel? And, third, how can the ventilation be changed in practice? All this requires more knowledge of the site of the fire, its size and opportunities for fire spread.

While the operation in the tunnel itself to deal with the fire is in progress, a parallel operation is being carried out to deal with those expected to escape from the tunnel. This second organisation includes medical personnel and facilities to look after the injured. This will not be described further here, as it is described thoroughly in other literature, such as the normal standard operation procedures of the fire and emergency services.

Figure 28.6 The situation in the tunnel after application of the ventilation measures

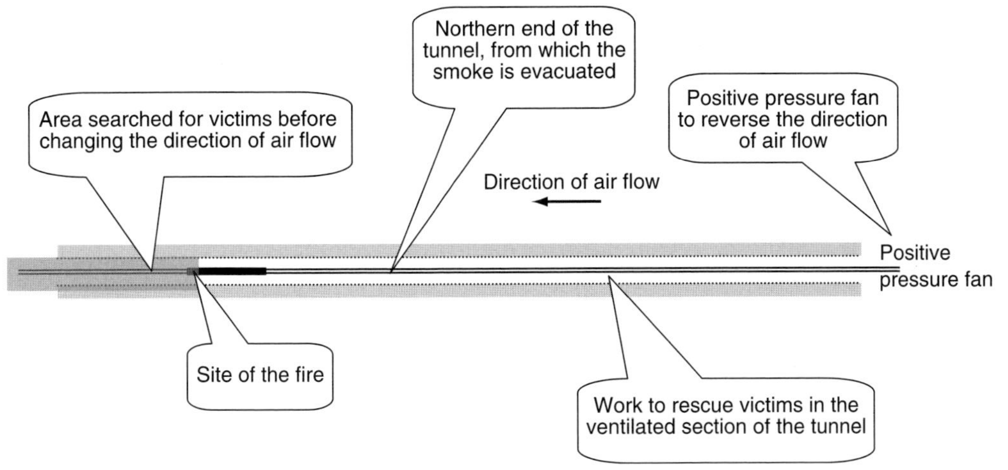

Northern end of the tunnel, from which the smoke is evacuated

Positive pressure fan to reverse the direction of air flow

Area searched for victims before changing the direction of air flow

Direction of air flow

Positive pressure fan

Site of the fire

Work to rescue victims in the ventilated section of the tunnel

28.7. The main problems in dealing with a fire and rescue situation in a tunnel and proposals for dealing with them

When the project work that forms the background to this chapter was carried out, many different proposals and ideas on how to tackle fires in tunnels were put forward. It became apparent that there was a considerable lack of experience and knowledge on how to tackle fires in rail tunnels. The following discussion is structured around the problem areas, in order to be able efficiently to apply various solutions to the particular problems associated with a tunnel fire. Various proposals for solutions to the problems are discussed under the headings.

The discussion is based on the scenarios, the calculated spread of fire gases, evacuation from fires, and the implementation of rescue as used in this project.

28.7.1 It is difficult to obtain an overview of the accident site, resulting in a severe lack of information on what is happening

The incident commander in charge of tackling a fire in a rail tunnel is faced with many problems and difficulties, all of which need to be dealt with. When the alarm is first received, it is necessary to decide quickly what sort of accident is involved and what is happening. On the way to the site, the commander must decide which approach route is best in order to be the most effective. However, even deciding this requires answers to the following, and other, questions.

- Where is the train in the tunnel?
- Where is the fire on the train, and how severe is it?
- Which way is the smoke flowing in the tunnel?
- Where, in the tunnel and/or on the train, are those who need rescuing?

When a decision has been (quickly) made on from which end of the tunnel to start work, further information is needed. At this initial stage of the work, it is probably the obtaining of information, and its interpretation, that is most critical to ensuring that the work gets off to the best possible start.

Rail tunnels are normally relatively long and large, in which respect they differ considerably from the normal fires tackled by fire services, occurring in apartments or houses, and for which it is relatively easy to quickly grasp a situation overview. However, in the case of fires in tunnels, it is very difficult to get any impression at all of what is really happening and why smoke is coming out of the tunnel. This difficulty creates major coordination problems for the operational management of the work at the site. The work is considerably complicated by the relatively large geographical areas and distances involved, as the fire and rescue services may need to attack the fire from several different points along the tunnel. The fire in the Mont Blanc Tunnel showed how important proper coordination of the rescue work is. In that case, coordination between the French and the Italian fire services failed, as did that between the various rescue services and the tunnel operator, and between the tunnel operator bases at each end of the tunnel. If such a large operation as this is to operate properly, it is vital that the work has been properly planned and equally properly carried out. In turn, this requires careful pre-planning of the work and facilities that will be needed in the real event. In general, there is poor planning for accidents and rescue situations in tunnels by tunnel operators and fire and rescue services in the Nordic countries.

The following questions should be considered.

- Where is the fire and what is its size? These are important aspects, but they can be completely unknown in the case of fires in tunnels.
- The effect of the wind on the fire can be decisive, both in terms of its direction and its effect on the fire. This was a most important factor in the Kaprun mountain railway fire in November 2000.
- Are there still people in the tunnel, and, if so, where? This is often an unknown factor, which is decisive in determining the emphasis of the work at the site.
- How will coordination be arranged between the various parties involved in any such extensive rescue operation?
- Can the fire and rescue services find their way to the site, to the tunnel openings and to access roads, and how will they know where they are in the tunnel?
- How will these services be able to reach the tunnel openings in order to treat anyone escaping from the fire and finding their way out of the tunnel?
- How well prepared are the fire and rescue services to use all the necessary equipment? This is a factor that is often forgotten. Local fire services, for example, often find it very difficult to keep up to date with information on everything for which special knowledge is required.

Information must be presented in such a way that it can be assessed and used by the rescue services' decision-makers working under stress at an accident site.

28.7.1.1 Proposal for dealing with these problems

Tunnels must be fitted with various forms of monitoring and surveillance equipment, and the information from it should be presented in such a way that it can provide answers to the questions that the fire and rescue services will have in order to decide how best to deal with the fire. This equipment can be carbon monoxide, carbon dioxide and oxygen detectors, as are used in the Channel Tunnel, air flow meters (wind gauge), CCTV, heat detectors, combinations of temperature and smoke detectors, and indicators for showing the positions of trains. Locomotives and carriages should be fitted with detectors to provide early alarms to the train crew, and to assist the fire and rescue services in assessing the size and extent of the fire.

The emergency response must be properly planned, so that it can be carried out smoothly when an accident occurs. This planning includes details and information on equipment, technical systems and proposals for the actual carrying out of the work.

28.7.2 Dealing with an extensive evacuation or life-saving situation involving large numbers of people

Dealing with such a situation involves overcoming a number of difficulties. For a start, merely getting to where people are evacuating from the train and/or to the tunnel openings can present a whole set of problems. Once there, it will be necessary to deal with the injured and get the uninjured well clear of the site. It may also be necessary to go into the smoke-filled tunnel to rescue survivors or to start fighting the fire.

One of the important factors to be determined, whether in prior planning or at the time of the event, is what means of evacuation there are in the tunnel. Is the tunnel designed and/or of sufficient size for those involved in a fire to be able to get away from the fire, or is it the intention that they should be assisted by the fire services? Is evacuation through emergency exits, or only through the tunnel openings? Which of these strategies is reasonable, and what is the capacity of the local fire services in the event of an evacuation or life-saving situation? These are all important questions that should be considered when planning and making preparations for safety and rescue in the event of a tunnel fire.

In a tunnel where it is not possible to control the direction of air flow, it can be difficult to decide in which direction it is best to attempt to evacuate. How should, or can, the operator of the tunnel or the train deal with or influence an evacuation?

How those caught in a fire will react is a difficult area to foresee. It has been shown on several occasions – including the fire in the Ekberg Tunnel in 1996 (see Chapter 1) and in the experience of the authors in connection with several fires in Stockholm – that drivers are very unwilling to leave their cars in order to evacuate a tunnel through a separate emergency exit.

In the worst case, the evacuation may be directed by people who, in their normal work, are not accustomed to dealing with critical or uncommon accident situations. This can mean that their decisions may complicate an evacuation, depending on their ability and procedures for handling the situation. The fire at King's Cross underground station in London in 1987 showed that safety in a complicated system of this type was entirely dependent on the way in which the station staff acted. The staff had not been trained to deal with such a situation, and had not taken part in any exercises, which also contributed to this catastrophe.

By far, the most important parameter in determining whether those caught in a fire can escape seems to be the distance to a safe environment, i.e. to places where the toxic fire gases and high temperatures cannot reach those escaping from the fire. This distance must not be particularly long if those caught in the fire are to be able to get there without injury, as has been shown in various evacuation calculations (Bergqvist et al., 2001).

In the situation of a fire in a train in a tunnel, much indicates that it is basic human reactions that will drive people to escape from the fire, without thinking about the spread or direction of smoke. It seems as if it is the fire and its flames that drive evacuation more than the spread of smoke, which means that those escaping from the fire will leave in any direction available, regardless of the wind direction. As a result, some will go downwind, and may have a long distance to go before they reach safety outside the tunnel.

If there are emergency exits in the tunnel, it is important that they are so designed that they can, and will, actually be used by those escaping from the fire. Sufficient time needs to be available for escape by people of different ages and mobilities. Considerable thought must be put into the design of emergency exits so that they can be found in the circumstances of a situation in which evacuation is necessary.

A questionnaire survey of Swedish fire services (Bergqvist, 1999) showed that one of the first things that fire-service personnel would do in such a situation is to investigate the situation in the tunnel in order to get a better idea of what is happening, and possibly also to attempt to start putting out the fire. The result of starting in this way from the southern end of the tunnel is that those escaping from the fire would be moving away from one end of the burning train, in the smoke, while the fire-fighters were approaching the train from the other end. If their attempt to put the fire out was unsuccessful, it would mean that their efforts up to that point had not helped evacuation.

The results of evacuation calculations (Bergqvist et al., 2001) show that, in certain cases, people can survive in fire smoke for a relatively long time, although they might be unconscious and unable to escape further. This occurred in the accident and fire in the Norwegian Seljestad Tunnel in 2000 (see Chapter 1), where two women and two children were trapped in a smoke-filled road tunnel for over an hour before fire-fighters wearing breathing apparatus could get them out and revive them.

The results of the evacuation, together with an awareness of what means are available in the tunnel to survive in the smoke, make it important that the right responses are made, as there is a relatively high probability that there will still be people alive in the smoke between the burning train and the northern end of the tunnel. It is important that work on putting the fire out should start as quickly as possible, as there is a real risk that the chances of survival of anyone in the tunnel would be considerably reduced if the fire were to spread to more carriages in the train.

In the case of the reference scenario, a maximum of 240 people are likely to evacuate the train. Many of them will go towards the northern end of the tunnel, thus finding themselves in a smoke-filled environment. Any effort intended to assist evacuation by sending fire-fighters with breathing apparatus into the tunnel would probably be ineffective in terms of increasing the number of survivors.

Swedish fire and rescue services (and probably other such services in other European countries) do not have the capacity to deal with this type of emergency situation. Evacuating people already affected by smoke, and others who are unconscious, would require considerable resources. Clean air would be needed to be put into the tunnel in order to reduce the toxicity of the smoke and then to transport the survivors out of the tunnel.

It is difficult, in other words, for the fire and rescue services personnel to get those involved in the fire out of the tunnel: instead, the best hope rests with people being able to get themselves out. This means that, in long tunnels, consideration must be given to emergency exits, leading either to another tunnel bore or to the open air. This provides a shorter path to a safe environment than does evacuation to the tunnel mouths.

This then raises the question of whether these emergency exits have any value. If we look at the calculated visibility distances in this project, we find that they are extremely short (of the order of only a few metres) in many of the scenarios. It can therefore be difficult even to see an emergency exit, if it is on the opposite side of the track from where people are. In other words, the design of the emergency exits, and the way in which they are indicated, are extremely important.

One way of saving many lives would be to use a method of tackling the fire and ventilating from the outside of the tunnel to push away the smoke in order to facilitate evacuation. However, to do this successfully and safely requires a lot of information about the fire and where those escaping from it are (see below). The decision regarding the correct direction to drive out the smoke and fire gases is very important, and at present there is little chance of being able to decide this on analytical grounds. If important information is lacking, there can be a risk that ventilation will make the situation worse for those in the tunnel. The Kaprun fire, where the chimney effect in the sloping tunnel created a very strong air flow, making conditions difficult for those attempting to escape the train, showed how a powerful flow of air can affect a fire.

The last alternative available to the rescue services is to have an efficient site organisation for dealing with the many injured. This method of working must be used as a complement to the three other methods, or perhaps is the only one practical for saving lives. The major fires in recent years (Mont Blanc, Tauern and Kaprun) have shown that the rescue services themselves have not been able to save many lives in the tunnels, but have been restricted mainly to helping those who have escaped from the tunnel by their own efforts.

The overall conclusion of these various points seems to be that, in the case of large fires, the rescue services should concentrate on providing ventilation to facilitate evacuation of those escaping from the fire.

28.7.2.1 Proposal for dealing with these problems

In order to reduce the time taken for evacuation, the train crew should be trained and rehearsed in dealing with passengers in accident situations. They should also have access to technical equipment to assist the evacuation. The fire and rescue services should develop methods of enabling fire-fighters wearing breathing apparatus to assist those escaping from the smoke. The fire and rescue services need to develop improved methods of assisting evacuation by driving the smoke away from those fleeing from the train, together with improved methods of extinguishing the fire and thus eliminating the underlying threat (ventilation and extinguishing are discussed later in this chapter). As described above, the work of assisting evacuation requires a good knowledge of conditions in the tunnel, and this needs to be available to the fire officer from the start and throughout the rest of the work.

Research is also needed into the design of emergency exits, in order to ensure that they are more easily seen and found, even though smoke may be dense and visibility restricted.

28.7.3 The tunnel is full of fire gases and safety equipment has to be used when attempting to fight the fire

In a tunnel fire, the fire gases will not dissipate in the same way as in a normal house fire. The fire will result in a substantial evolution of smoke, filling the tunnel. Most of the fire gases will be carried away horizontally in the wind direction. Over the first 100–150 m from the fire, the smoke will tend to be layered, gradually filling the whole cross-section of the tunnel with increasing distance from the fire. This smoke will affect those escaping from the fire, while making it more difficult for fire-fighters to enter the tunnel to look for survivors. In the Kaprun fire, six people were overcome by smoke at the top station, 3600 m from the fire, preventing them from escaping from the station. The Zell am See fire-fighters, wearing breathing apparatus, got them out, but only three survived – a drastic demonstration of the toxicity of fire smoke.

Some of the smoke will also spread against the direction of the wind due to the backlayering effect, filling the tunnel with smoke for about 150 m upwind. This will considerably complicate approaching the train from this direction.

If they are to be able to move in the tunnel, the fire-fighters will have to protect themselves against the toxic gases, which means that their overall efforts will be severely restricted by the need for air for themselves and the reduction in their speed of progress in smoke.

The Kaprun fire showed that the progress of the fire is strongly affected by the slope of the tunnel, which means that the ability of the fire brigade to influence the development of the fire declines as the slope of the tunnel increases.

Our investigations have shown that the greatest danger to those fighting the fire is presented by the combination of poor visibility and the effect of the toxic fire gases. Surprisingly, it is likely that, early in the fire scenario, conditions in the tunnel will not be as hot as is common in more normal fires, due partly to the cooling effect of the tunnel walls. This means that explosive development of the fire in the tunnel is unlikely to present the greatest danger. However, such an explosive development could occur in a carriage or locomotive, and if this were to occur there would be a very high level of thermal radiation, which would limit the ability to approach the train. This effect was clearly seen in the fire in the Tauern Tunnel, and prevented the fire-fighters from getting closer to the train than about 100 m, as the heat from the fire and fire gases was too high.

28.7.3.1 Proposals for dealing with these problems

To be able to plan properly how to rescue those involved in the fire, the fire service needs to know the wind direction in the tunnel. It also needs to be able to constantly monitor the spread of the smoke, so that appropriate measures can be taken. This includes an assessment of environmental conditions in the tunnel.

Advanced rescue operations need to be carefully prepared and planned. This includes the means and ability to ventilate the tunnel and control the direction of the fire gases. Fire services need to improve their knowledge of their abilities and limitations in dealing with extensive, long-duration, breathing apparatus operations in tunnels.

28.7.4 Difficulty in assessing the risks to which the fire and rescue personnel are exposed when working in the tunnel

As it is very difficult to forecast the progress of a fire if it is not known how much material is on fire, any risk assessment of a tunnel fire becomes very difficult. How can the risk of the fire overwhelming the train, and developing very high fire gas temperatures, be assessed? It is difficult, too, to assess what will be the effect of flames and heat on the tunnel walls and roof, and how these will be affected by the shock of extinguishing with cold water. The fire officer needs constantly to assess the risks, in order to reduce the risks to those in the tunnel.

In Sweden, the use of breathing equipment for tackling fires and rescuing victims is controlled by the National Board of Occupational Health and Safety's regulations (1995). These regulations describe what the person in charge must do. Section 11, for example, states that: 'The person in charge must ensure that the risks to which a person wearing breathing apparatus is exposed are reasonable, in relation to what can be achieved by the work', while the supplement to the section says that 'It is important that the risks of such work should be assessed before the work is started'. Other countries almost certainly have similar regulations to ensure safe working conditions for fire-fighters. The key to making this risk assessment is knowledge of the fire, of what has occurred, and of the means and limitations available to the rescue service.

The Mont Blanc and Tauern fires showed the problems of tunnel collapse. During the Mont Blanc fire, one of the fire-fighters was killed and several were in direct danger for a long period. This indicates that those in charge of the work had not correctly assessed the risks (or were not able to do so), and had misjudged them. The lives of several of the fire-fighters in the Tauern Tunnel were probably saved as a result of experience from the mistakes made in the Mont Blanc Tunnel fire about 2 months before, which meant that they discontinued their active internal work in time, and thus avoided exposing themselves to extreme risks.

Practical fire trials carried out in Sweden by the Telia telephone company (Bengtson and Lundin, 1995) have also shown the considerable risks due to the collapse of tunnels as a result of fire when large blocks of rock collapse.

The risk of collapse of the tunnel in a fire seems to be highest at a later stage of the fire, by which time the tunnel should have been evacuated, in connection with active extinguishing work in the tunnel.

All types of rescue work on or near rail tracks involve considerable risks with regard to electricity and other trains. These risks can, of course, also be present when working in tunnels.

28.7.4.1 Proposals for dealing with these problems

Fire and rescue services need to develop more comprehensive guidelines for how to tackle various types of situation. Such guidelines must be drawn up in conjunction with the operator of the site or facility concerned. There needs to be detailed accident and rescue planning, based on an objective view of the abilities and limitations of the fire and rescue service.

In order to be able correctly to assess the real-time risk situation while work is in progress, the incident commander must have a good knowledge of the capacity of his service to deal with the particular type of situation. He should know clearly what the service can do, and what it cannot do. With this knowledge, he knows the limitations of the framework within which he can act.

Those likely to be in charge of such situations need to carry out rehearsals of this complicated risk-assessment process in an accident situation. This training and rehearsal needs to be carried out with an eye to the various situations that the people concerned may be called upon to face.

The incident commander must be actively supported in order to enable him to assess the risk situation. This support can be in the form of various checklists or outline plans, which can provide a better basis for tackling the work. There should be an active analysis of the risks to which the personnel are exposed, with the results being used to actively reduce these risks. While the fire is being tackled, the incident commander must be able to overview the site and the work.

It is equally important that the owner of the site or facilities should be aware of what the fire and rescue services can do, so that other parts of the safety system can be provided with the necessary abilities and capacities.

28.7.5 Large, extensive sites, involving long and complicated routes to reach the fire from a safe outdoor environment

Many rescue attempts in recent years – of which the fires in the Mont Blanc, Tauern and Kaprun tunnels have attracted the most attention – have demonstrated these conditions. The fire and rescue personnel risked their lives, with actual loss of life in the Mont Blanc fire.

Just how much can a group wearing breathing apparatus do under such conditions, and how far into a smoke-filled tunnel can it penetrate? These are important questions, to which insufficient attention seems to have been paid when planning rescue work in complicated situations.

Depending on how the work is organised, it may be necessary for the fire and rescue personnel to go a long way into the tunnel. If the tunnel is full of smoke, this is going to impose severe limitations on the work. The 1995 National Board of Occupational Safety and Health regulations on the use of breathing apparatus defines such use, in these circumstances, as penetrating into dense fire smoke, with the aim of tackling the fire, saving life, etc. This is normally done in the form of a group, of two fire-fighters, wearing breathing apparatus and other protective clothing, going into the tunnel and a group leader stationed outside the smoke. All this means that a breathing apparatus group cannot penetrate any great distance into the tunnel before it must start to retreat towards a safe environment. It also means that, if any greater distance is to be covered, there must be several groups with breathing apparatus, so that replacement groups can penetrate further into the tunnel without having to spend time searching the sections already searched by previous groups. All this means that this type of work is very demanding on resources. At the same time, the greater distances mean that the groups with breathing apparatus are further away from safety if something should occur while they are in the tunnel.

Investigations that have been carried out show that incident commanders are very uncertain about how fires of this type should be tackled, and that the use of groups with breathing apparatus will be started in accordance with the procedures for, and using the experience from, apartment and house fires. This means that the Swedish fire and rescue services need more information on how to carry out advanced breathing apparatus group work, and what the potentials and limitations are.

Emergency exits in tunnels can be used as routes to reach the site of the fire for rescue work. It is therefore important that any emergency exits are designed for, and suitable for use as, rescue access points.

28.7.5.1 Proposals for dealing with these problems

Most importantly, the fire and rescue services should develop procedures capable of dealing with such conditions as could arise in a tunnel, and the threats presented by these conditions, and the measures that must to be taken in the event of an accident. This means, for example, that the present standard procedure for safe working using breathing apparatus needs to be revised, as fires in tunnels cannot be compared with fires in, for example, large and extensive storage buildings.

Equipment needs to be developed to help overcome the difficulties of working in tunnels whilst wearing breathing apparatus. This could be equipment that can assist determining location in a smoke-filled environment, vision equipment to see through smoke, and equipment that provides fast, safe transport through the tunnel to look for survivors. The latter equipment should also be able to carry other equipment into the tunnel and carry survivors out of it.

The rescue services need improved equipment and procedures for the supply of water, and means of getting water to the fire. Water supply should be provided by fixed fire hydrants in the tunnel, in combination with lighter and more flexible hose equipment to connect to the hydrants at the site of the fire. There also needs to be simple means of marking the approach routes that have been used, in order to improve the safety of getting out of the tunnel when hoses are not being used. This could be provided, for example, by the breathing apparatus group leaving some form of illuminated 'rope' behind itself as it advances, in order to assist the next group and to assist retreat. Better and more effective guidance and location systems for breathing apparatus groups needs to be developed and used for such work. It should be designed to reduce the load on those using it, which means that there should be items such as ice jackets and lighter equipment.

Many factors indicate that it is the risk of smoke poisoning and the poor visibility in combination with long access routes that presents the main danger to those tackling the fire; because they are, in general, unable to get very close to the fire due to the high temperature radiation from the fire. This distinguishes such fire-fighting from more normal house and apartment fires, for which it is high fire gas temperatures, and the associated risk of flashover, that poses the greatest risk. Awareness of this means that there is a need for providing more air to extend the operating time of breathing apparatus, coupled with some means of improving visibility in smoke. This would also improve the safety margins before the air runs out in the breathing apparatus when carrying out extensive work. It should also make it possible to improve the ability to deal with failure of breathing apparatus equipment, so that working methods and equipment suit the type of work required in dealing with tunnel fires.

28.7.6 Control of the fire gases in order to facilitate evacuation, rescue and fire-fighting

Rescue work in dealing with a fire in a tunnel can involve considering the use of fans to drive away the fire gases. Getting the toxic gases away from those escaping from the train seems to be an effective means of facilitating evacuation.

However, there are considerable problems in deciding how to apply such ventilation in order to facilitate evacuation, rescue work and possible fire-fighting work. The application of forced ventilation could assist the fire, and this needs to be considered and dealt with. Many factors indicate that the fire power increases with increasing air velocity, particularly when solid materials are on fire. There is an upper limit to this power increase, as finally the cooling effect of the air takes over, similar to blowing out a candle, but this is not a consideration that could be realistically applied to a fire of this type. Trials have shown that the fire power can increase by a factor of two or three, depending on the air velocity, the disposition of the burning materials and their type (Ingason,

2001). It is not clear what the actual upper limit of maximum fire power would be, but it is clear that the air velocity does have a significant effect on the progress of the fire.

Attempting to make the right decision regarding the application of ventilation can be very difficult in an accident situation. The wrong decision can have extreme consequences, as was shown by the underground railway fire in Baku, which killed 289 people and injured 265. The fact that these numbers were so high is probably partly due to the incorrect use of ventilation during the fire (see Chapter 1).

The backlayering phenomenon, which can cause problems by considerably complicating attempts to reach the train, will be affected by the progress of the fire and by the air velocity in the tunnel. Today, many tunnels are being built based on the concept of being able to blow away all smoke upstream of the fire, with the intention of providing the fire and rescue personnel with free access to the fire. However, a problem arises in this context: if the longitudinal air velocity does not exceed a certain critical velocity, there is a risk of the smoke 'backing' upstream, to produce back-layering. The problem with this lies in the fact that backlayering results in the fire-fighters being exposed to thermal radiation from the hot fire gases in the top of the tunnel. As the smoke that is backing up is at a higher temperature than the air which the fire-fighters have behind them, they are exposed to thermal radiation, which can be very troublesome, despite having good visibility forwards towards the fire. A 15 MW fire in a tunnel with a 7 m roof height, with an air velocity of 1.5 m/s, can cause smoke to back up over about 70–90 m, while a 35 MW fire in the same tunnel can cause the smoke to back up over 90–110 m. The heat from the hot gases above the fire-fighters increases as they get nearer to the fire, to such an extent that the situation can even-tually become intolerable. Tackling this requires other methods, such as introducing water spray into the airstream or increasing the air velocity by the use of mobile fans.

The main factors that determine how far the fire gases extend upstream are primarily

- the roof height (the higher the roof, the further the smoke will extend upstream)
- the site of the fire (the larger the fire, the greater the backlayering)
- the width of the fire (the wider the fire, the less the backlayering)
- the slope of the tunnel (the steeper the slope in the wind direction, the greater the backlayering).

Investigations have shown that an air velocity of about 2–2.5 m/s can reduce the backlayering phenomenon (Ingason, 2001).

The design of a tunnel will have a considerable effect on the feasibility or efficacy of forced venti-lation. In tunnel systems with many different linked bores, such as in an urban underground railway system, it will be difficult to establish an air flow through the tunnel. This has been found in investigations carried out in the Stockholm underground railway (personal communica-tion, B. Wahlström, Stockholm Fire Protection Team). The ventilation problem seems to be considerably simpler in tunnels that consist of a single bore, with just one entrance and one exit.

The natural wind conditions in a sloping tunnel can vary, depending on whether it is summer or winter. For example, the air temperature in a railway tunnel longer than about 1 km is often rela-tively constant throughout the year, probably in the range 0–10°C, depending on the geographical position and size of the tunnel. During the colder parts of the year, the natural draught tends to flow with the slope of the tunnel. Conversely, if it is very warm outside, the natural draught tends to flow against the slope of the tunnel. An external wind acting on the tunnel mouth can easily overcome these natural flows, so that the direction of air flow in the tunnel is determined by the local wind direction.

After a while, the thermal forces resulting from the fire can overcome the natural ventilation direction, which means that the direction of flow of the smoke can suddenly change during the fire, complicating both evacuation and rescue work.

One of the conclusions from tests conducted in a model tunnel was that thermal fire gas ventilation was not particularly effective in ventilating the fire gases out of tunnels (Ingason and Werling, 1999). This is because the fire gas temperature falls rapidly with increasing distance from the seat of the fire, and after about 100–150 m the temperatures are relatively low, somewhat below 100°C. This means that shafts down to the tunnel do not work very effectively, due to an insufficient chimney effect. Large quantities of the fire gases simply pass the shafts, rendering them ineffective.

Taken together, all this means that we cannot put forward any suggestions for the type of ventilation to be used, or for its direction. In order to decide on the most appropriate direction, the fire service must have access to all the necessary information needed to make the right decision.

28.7.6.1 Proposals for dealing with these problems

More fundamental work is required. We need to know more about how ventilation can affect the ways of tackling tunnel fires. At present, the knowledge of how ventilation should be applied to a fire in a tunnel is insufficient.

The fire and rescue services need to develop working methods, and to obtain the important information needed, to combine searching the tunnel with the use of fans to clear smoke away from those evacuating the train and those engaged in active work to extinguish the fire. As part of this, there needs to be investigation of how combinations of ventilation and the provision of extinguishant can be applied, and how this affects the prospects for escape and for extinguishing the fire.

28.7.7 Communication between personnel in a safe environment and those in the tunnel using breathing apparatus

Communication is vital for two reasons: partly it is needed for coordination of extensive, complicated work, and to quickly obtain all necessary feedback, and partly for the safety of the personnel concerned. The availability of effective communication is also decisive for the safety both of those in the tunnel and for the incident officers outside the tunnel.

The 1995 National Board of Occupational Safety and Health regulations specify that working radio communication is required. The third section states: 'In addition, communication radio, a fire helmet, protective clothing, protective gloves, fire boots and a fire belt must be used'. Furthermore, the supplement to the third section states that: 'From this, it follows that the two people with breathing apparatus and the incident commander must each have their own communication radio'. It is likely that the normal radio sets used for communication with those wearing breathing apparatus will not work in tunnels, as their range is insufficient for this purpose.

In tunnels where fans are installed, and are intended to be used in connection with a fire, their noise level will cause problems, being so high that it will make normal communications significantly more difficult.

If radio communication is to be relied upon, it needs to be planned in advance at each site. The range of normal radio communication is such that it requires additional systems in order to be able to work in tunnels. Systems that utilise leaky cable antennae within the tunnel do seem to work.

Various types of mobile systems are available today, intended to extend radio communication in applications where the normal radio links as used by breathing apparatus crews do not have sufficient range. One example of such equipment is the cross-band repeater, used by the Stockholm Fire Service. However, in practice, it has been found to be very difficult to get such equipment to work at an incident site, as it is not particularly user-friendly.

28.7.7.1 Proposals for dealing with these problems
Tunnels where there might be a need for the use of breathing apparatus in order to tackle incidents should be fitted with a system to ensure reliable radio communication with the users of the breathing apparatus.

There is also a need to develop improved mobile systems for extending communication where there is no permanently installed system.

28.7.8 Getting water to the site of the fire
Many tunnels are long, which makes it difficult to get water from the fire tender to a position where it is needed. There are two problems involved: the practical problems of pulling hoses into a smoke-filled tunnel, and the hydraulic problems of pumping water a long distance through hoses.

Experience indicates that getting water to where it is needed can be very difficult. It has been found that transporting water in fire hoses is subject to hydraulic limitations in obtaining sufficient water flow rate and pressure. There is also the purely physical limitation set by the need for fire-fighters to pull water-filled hoses after them. Hoses have to be filled with water, so that the fire-fighters can quickly tackle any flashovers (e.g. within vehicles).

Effective fire-fighting using the type of equipment currently employed by Swedish fire and rescue services requires a water flow rate of at least 300 l/min, with a pressure of about 60 m WG at the nozzle, which are very demanding requirements. In the case of large fires, it is likely that the amount of water that can be pumped at these levels would be insufficient to deal with the fire.

The 1995 National Board of Occupational Safety and Health regulations for the use of breathing apparatus are very clear that a reliable supply of water must be available in connection with such work. The reason for this seems to be to ensure that fire-fighters wearing the breathing apparatus are in a position to deal with flashovers. However, as mentioned previously, the risk of flashovers does not seem to be a major threat to the safety of crews wearing breathing apparatus in tunnels. (Strictly speaking, flashover can only occur in a compartment. However, flashover may occur within a vehicle, e.g. a train carriage, in a tunnel. Extensive burning of all items within a section of tunnel might be thought of as related to flashover within that section; however, it would not be flashover as such.)

28.7.8.1 Proposals for dealing with these problems

Tunnels through which passengers are carried should have a permanently installed hydrant system in order to simplify the problem of getting water to the site of a fire and to provide the same level of safety as would be provided by a normal fire-fighting water system for use by crews wearing breathing apparatus.

The fire and rescue services should develop procedures for running out hoses in such situations, taking into consideration of the particular problems and dangers likely to be encountered.

28.7.9   Extinguishing the fire in the desired manner

Traditional methods of fighting the fire, using hoses and normal jet nozzles, may be insufficient if it is difficult to approach the fire in order to tackle it properly. Other methods may be needed.

Small fires are generally easy to extinguish, which means that early detection and early tackling must be the objective in dealing with all types of fire. It is also important to extinguish the fire as early as possible in order to reduce the risk of it spreading.

Fire-fighters are faced with major problems when having to tackle large fires while wearing breathing apparatus, as was shown clearly by the fire in the Tauern Tunnel, where fire-fighters could not get closer than about 100 m to the fire. This was probably due to thermal radiation, and meant that it was not possible to reach the heart of the fire or effectively extinguish it. Nor were the fire-fighters able to reach the fire in the Channel Tunnel in 1996, but could only restrict its size and prevent it spreading against the direction of the wind. The rest of the train, from the point where the fire started and downwind thereof, was completely destroyed (personal communication, I. Muir, Kent Fire Brigade).

A point that is not considered in more detail in this chapter, but which needs to be considered before an accident occurs, is the importance of fighting a fire not only to save lives but also to retain the use of important tunnels. If this realisation is only made at the time of an accident, the fire and rescue services will be under considerable pressure to save the tunnel without having the necessary safety systems or being prepared for property salvage. A clear example of this is presented by the fire in a cable tunnel in north-west Stockholm during the spring of 2001 (Stockholm Fire Brigade, 2002), where a lot of work was put into extinguishing the fire in order to enable repair work to be started.

28.7.9.1 Proposals for dealing with these problems

Sprinklers in locomotives and carriages could be one way of improving fire protection in trains. Early detection, in combination with an automatic fire-fighting system, would radically improve the ability to tackle fires on trains. Such a physical protection system needs to be complemented by training the train crew to tackle fires effectively.

Improved investigation of the types of extinguishing systems and extinguishants to be used in these contexts must be carried out, as there is at present a considerable lack of information on what works effectively. One area that needs to be investigated is how combinations of extensive ventilation and extinguishing systems work in conjunction with each other.

28.8.   Proposed model for tackling fires in single-bore tunnels

When the first fire crew arrives at the site it is very unlikely to have an overview of the situation, which means that it is not possible to attempt to assess what has happened and/or how the

Figure 28.7 The situation when the fire and rescue services start their work

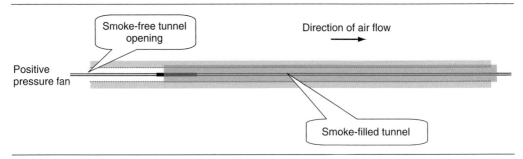

situation is likely to develop. At the same time there is strong pressure to start rescue work in the tunnel. The first approach should therefore be through the smoke-free end of the tunnel, with the prime objectives of investigating what has happened, assisting evacuation and, if possible, attacking the fire. The natural direction of flow of the air through the tunnel can be assisted by a fan, in order to improve the safety of conditions in the tunnel and reduce any backlayering effect. This will also prevent a fast, unplanned, change in air flow direction (Figure 28.7).

If the resources are available, this should be followed up as soon as possible by sending a group wearing breathing apparatus into the smoke-filled end of the tunnel. Its job will be to investigate what has happened, and to rescue any people found in the immediate vicinity of the tunnel opening.

Meanwhile, work should continue from the smoke-free end of the tunnel, attempting, if possible, to tackle the fire, and checking whether anyone is attempting to escape from the fire at this end of the tunnel. There is a possibility that the thermal radiation is so high that it limits the ability to approach the fire. If it is too high, work will have to be concentrated on assisting rescue by other means.

When it is no longer possible to do anything useful from the smoke-free end of the tunnel, the direction of air flow should be reversed, in order to assist rescue of people in the smoke-filled section of the tunnel (Figure 28.8).

When the direction of air flow has been reversed, conditions for survivors in the previously smoke-filled end of the tunnel improve. At the same time, it also becomes easier for the fire and rescue

Figure 28.8 Ventilation to assist rescue

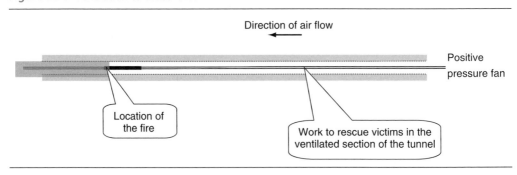

services to enter the tunnel and assist victims. As the originally smoke-free end of the tunnel has already been searched, the risk of making the situation worse in this end of the tunnel is reduced.

28.9. Conclusions

Any safety work associated with tunnels must start by identifying the particular conditions for the specific tunnel concerned. This can then be followed by decisions on what to do and the preparation of a safety concept. This is important, as there can be considerable differences in the most appropriate solutions for different tunnels. The fire and rescue services must be seen as an integral part of the overall safety systems of mass transport systems.

What comes out in the report is that safety solutions for passengers, based on the assumption that the fire and rescue services will be able to participate actively in evacuation, seem to be unrealistic, and are unlikely to succeed in the event of an accident. In general, evacuation concepts should be based on externally unassisted evacuation.

On the basis of the typical scenarios analysed here, it is clear that emergency exits should be less than 300 m apart if they are to provide safe exits. The routes to them must also be so designed that they can be found, even in the conditions of poor visibility that are likely to be encountered as a result of a fire in a tunnel. Bear in mind, too, that these emergency exits should also serve to provide the primary means of access for the fire and rescue services.

There needs to be a defined boundary indicating what types of fire and of what size it is feasible to tackle by manual fire-fighting. This is important so that future safety systems for tunnels can be more appropriately sized and designed.

The fire and rescue services should understand the difficulties of performing rescue operations in tunnels, and not plan to carry them out in the same way as if they were apartment or house fires. It seems likely that the first approach should not be through the end of the tunnel out of which smoke is coming, but via the smoke-free end of the tunnel. Groups wearing breathing apparatus entering the smoke-filled end of the tunnel will be involved in complicated and difficult work, possibly resulting in misapplication of the resources available, so that it is not possible to carry out other actions that could result in a more efficient rescue.

Insufficient knowledge is available on how combinations of breathing apparatus groups, fire-fighting and ventilation, all with the aim of saving lives, should work together, and of how these particular methods should be applied to a tunnel fire. More resources need to be applied to this subject in order to develop realistic models that can be used by the fire and rescue services in connection with tunnel fires. Various attack concepts should be investigated for different types of tunnel.

In general, rail tunnels in Sweden seem to be quite well designed in these respects, in the form of single-bore tunnels with two-way traffic. The model for dealing with fire situations in this type of tunnel is based on the assumption that the fire and rescue services will perform their work in a planned and considered manner. Models for dealing with other types of tunnel should be developed from this starting point. There is a need to look at this model, and others, more deeply, and to carry out tests to see how they actually work in practice.

In order to tackle such incidents in the most effective way, the planning should be based on the likely accident scenarios, followed by the development of a system to identify the type of accident and to determine which approach way is best used. This is important as the lack of full information at the time of arriving at the site is critical. Many factors indicate that it is this shortage of information available to the incident commander, failing to give him or her sufficient information to make a proper analytical decision, that has a significant effect on the rescue work. This means that any measures intended to assist the presentation of correct information to the incident commander will improve the ability of the fire and rescue services to perform an effective rescue operation.

REFERENCES

Bengtson S and Lundin K (1995) Heat release rates in telecommunication tunnels. *Proceedings of the International Conference of Fires and Engineering, Asiaflame 95*, 15–16 March 1995, pp. 127–138.

Bergqvist A (1999) *Räddningsinsatser i tunnlar och undermarksanläggningar. Förstudie om läget i Norden.*

Bergqvist A, Frantzich H, Hasselrot K and Ingason H (2001) *Räddningsinsatser vid tunnelbränder. Probleminventering och miljöbeskrivning vid brand i spårtunnel.*

Danielsson U and Leray H (1999) *Fysisk belastning vid räddningsarbete i tunnlar*, FOA-R-01192-720-SE. FOA Defence Research Establishment, Tumba.

Ingason H (2001) An overview of vehicle fires in tunnels. *Proceedings of the Fourth International Conference on Safety in Road and Rail Tunnels, Madrid, Spain*, 2–6 April 2001.

Ingason H and Werling P (1999) *Experimental Study of Smoke Evacuation in a Model Tunnel.* FOA-R99-01267-311-SE. FOA Defence Research Establishment, Tumba.

Lennmalm B (1998) *Räddningsinsatser i industribränder – Rökdykarstudier.* Brandforsk Project No. 419-925.

Stockholm Fire Brigade (2002) Investigation report of the cable tunnel fire 2002-05-29.

Tunnelexercise TYCO (1999) *Cooperationexercise at Södertörn* (report from the fire and rescue service perspective).

Part VI

Tunnel fire safety decision-making

Handbook of Tunnel Fire Safety, 2nd edition
ISBN: 978-0-7277-4153-0

ICE Publishing: All rights reserved
doi: 10.1680/htfs.41530.615

Institution of Civil Engineers

publishing

Chapter 29
Problems with using models for fire safety

Alan Beard
Civil Engineering Section, School of the Built Environment, Heriot-Watt University, Edinburgh, UK

29.1. Introduction

This chapter relates to other chapters in this Handbook, in particular the chapters on decision-making (see Section VI) and modelling (see Section III). Since the early 1970s much has taken place in the area of computer-based models related to fire risk, and such models are now being used in relation to tunnel fire safety. Overall, the intention in producing such models has been to enable more accurate assessment of the fire risk implicit in a given case with particular kinds of design, contents, users and facility operators (if any). They are also intended to assist in estimating some of the effects of making changes to a system, whether it corresponds to a new design or an existing facility. In the 1970s, the dominant way of thinking about 'fire modelling' was probabilistic, although a lot of deterministic modelling also took place. In the 1980s there was a tendency to replace the probabilistic approach with the deterministic one, although probabilistic modelling did continue. During the 1990s and into the 21st century, a revival of interest in probabilistic modelling has come about, and it may be that, eventually, a fusion of the two will take place, the best being gained from each approach. Overall, this would be a desirable development. An example of this is afforded by the approach to the estimation of fire risk in buildings of reference (Beck, 1994). This model combines both probabilistic and deterministic concepts within a single framework. A similar approach has not yet emerged in tunnel fire modelling in a significant way. If 'fire models', i.e. models pertaining to fire safety, are to be incorporated into tunnel fire safety decision-making in a coherent way, then a framework will need to be such as to enable a decision to be made as to whether the level of risk is within an acceptable range or not, *with an acceptable degree of reliability*. The part played by deterministic models in such a framework, along with probabilistic models, is one of the questions which emerges. The Bayesian approach, which combines 'engineering judgement' with empirical evidence in a systematic way, and which has already been applied in tunnel fire modelling (e.g. see Carvel *et al.*, 2001), may well come to play an important part in such a framework.

The essential problem which comes about is this: how may fire models, using the term in the broadest sense, be acceptably employed as part of fire safety decision-making? The term 'fire model' is often used to imply a deterministic model of fire development. However, it may be argued that the term should include all models that may play a part in fire safety decision making, including probabilistic models and evacuation/human behaviour models. This broad use of the term is the way in which it will be used in this chapter. A primary theme that arises is the need to avoid the danger of seeing a model in isolation, as 'the answer'. Results from a fire model must always be seen within context; that is, the conditionalities need to be realised in an explicit way and other knowledge and experience taken account of in decision-making. Fire models should only ever be used in a supportive role.

29.2. Models and the real world

There are many different definitions of 'model' in existence. Essentially, the purpose of a model is to help us to understand the real world, and it does this by trying to 'represent' the real world in some way. However, a model may be realistic in some ways only, not in all ways. For example, Figure 29.1a shows a statue of a sphinx (a human-headed winged animal, generally a bull or lion) from a palace in ancient Assyria (now northern Iraq). Ignoring the unrealistic head and wings, as a model of a bull or lion from the front it is realistic in that it shows two legs, which is what would be seen from the front of a real animal. Looking at this model directly from the side (Figure 29.1b), it is realistic in that it shows the animal to have four legs. However, viewed from an oblique angle, as in Figure 29.1c, the model displays five legs, which is unrealistic. (Actually, Figure 29.1a shows a bull and Figures 22.1b and 22.1c show a lion: the stance of the sphinx is identical in both cases.) The general point is this: a model may be realistic in one way or even in two ways, but it may not be realistic in a third way. That is, we must not allow ourselves to assume that, because a model seems to be good with respect to predicting one variable (e.g. temperature in the far field) that it will necessarily be good at predicting another variable (e.g. temperature in the near field or gas concentrations). This effect was found in a study by Beard (1990).

29.2.1 Assumptions made in theoretical models

In conducting decision-making on risk, there are very likely to be possible sequences of future events that are not known about or possible sequences that are known about but which are not taken into consideration for one reason or another. Also, assumptions may be made that are not realistic, in relation to what has been considered in an assessment.

Figure 29.1 Sphinx: (a) front view, (b) side view and (c) oblique view

(a)	(b)	(c)

For example, if one had been carrying out an assessment, 30 years ago, of passenger death/injury/ illness (d/i/i) associated with air travel, the two main causes of death/injury/illness might have quite reasonably been taken as either being due to a crash or a fire. The probability, $P(d/i/i)$, might then have been reasonably calculated on such a basis. Over recent times, however, it has come to be realised that there are at least two other significant causes of death, injury or illness in relation to air travel, i.e. (1) deep-vein thrombosis and (2) cross-infection due to recirculated air; the assessment should be altered to include these possibilities. The probability $P(d/i/i)$ would then be different and, possibly, very different to that resulting from the earlier calculation. (Following the attack on the World Trade Towers in New York in September 2001, and subsequent events, it might even be argued that hijacking of an aircraft should also be included as a possibility.)

As well as readily plausible sequences of events, it must be realised that the seemingly 'incredible' may happen. In the tunnel context, for example, before the Channel Tunnel fire of 1996, it is likely that the possibility of a fire developing to involve 13 heavy goods vehicle (HGV) carrier wagons, including ten becoming damaged beyond repair, would have been regarded as 'incredible'. It is the case, though, that this is what happened within 3 years of the Channel Tunnel being opened for rail travel.

There is a general need to question assumptions, and this has been found in relation to escape from aircraft. In 1985, there was a fire on a Boeing 737 at Manchester Airport that resulted in 54 deaths. According to the guidance of the time, all the passengers should have been able to evacuate without harm because tests had shown that evacuation from such an aircraft could take place in less than 90 seconds, and this time was available. However, this assumption was shown to be wrong for the Manchester fire, because it does not take account of what people actually do in the real world. Following the fire, practical trials were carried out that were as realistic as possible, and this has led to recommendations for the redesign of aircraft exit areas.

Because it is effectively certain, in real-world cases, that there will be possible sequences that have not been included in an assessment, it is always prudent to err on the side of caution in assessment and decision-making in an attempt to allow for the unanticipated.

In order to try to overcome this problem, there is a need for creative thinking when carrying out risk assessments. Concrete measures may be taken by specifically attempting to consider, inter alia, (1) 'near misses', (2) 'buried' research (i.e. published research which is not generally realized to be relevant to a case under consideration) and (3) the precautionary principle. It is also essential to learn from those who have direct practical experience of the system concerned. Near misses represent an extremely valuable source of information about how a system might fail – yet these vital accounts of how the real-world system of concern is performing (and almost failing) are usually not employed in a productive way to genuinely inform our risk assessments. We need to be able to incorporate information derived from near misses into our risk assessments. Sometimes, research results, which may be of considerable relevance in a particular case, are available in the open literature, but the results are not noticed or are ignored. An example of this is afforded by the King's Cross Underground railway/metro station fire of 1987, in London, in which 31 people lost their lives (Fennel, 1988).

As well as attempting to take account of near misses and buried research, the precautionary principle may also be used to try to allow for the fact that we do not know all possible sequences that might occur in a tunnel. The precautionary principle was originally devised for environmental protection, but may be applied to tunnel fire risk. The application of the principle to risk in general has been put forward by the Commission of the European Union. These issues have been discussed further by Beard (2004), who gives additional references.

29.3. Kinds of theoretical models

As already implied, the two kinds of mathematical models used most are probabilistic and deterministic. Both probabilistic and deterministic concepts have been applied to the assessment of fire risk in tunnels.

29.3.1 Probabilistic models

In principle, probabilistic models are desirable to the degree that they allow conceptually for the complex nature of the real world: see, for example, the stochastic model of Beard (1984/1985), which predicts the expected number of deaths if a fire were to start on a bed in a hospital ward. The system does not calculate a single possible course but considers a number of 'chains' representing different possible developments, ranging from a small fire to flashover. A probability is computed for each chain. In principle, the complex nature of the real world may be taken into account with this type of model. Unfortunately, the transition probabilities and distributions necessary as input are usually very difficult to decide upon accurately and to relate to the thermophysical, geometrical and occupant characteristics in a particular case. One problem is the general lack of relevant data, and this certainly applies in the tunnel fire case. This problem is related to the challenge indicated in the introduction above, where the possible value of Bayesian methods was mentioned. Relatively little probabilistic modelling has been reported in the open literature with respect to tunnel fires. Apart from the Bayesian work reported in Chapter 11, some examples of probabilistic modelling are given by Geyer et al. (1995), Powell and Grubits, (1999) and Knoflacher (2001). (See also Chapter 18.)

29.3.2 Deterministic models

The main types of deterministic models that have been constructed and applied to tunnel fires are control volume models and models based on computational fluid dynamics (CFD) (see Chapters 15 and 16 for more on these). In the realm of fire safety science and engineering, CFD-based models are often called 'field models'. Control volume models assume uniform conditions within relatively large control volumes. For example, a single temperature or gas concentration may be associated with each control volume at a given time.

However, while probabilistic models allow for a range of possible developments, deterministic models calculate a single possible development for a given input. Deterministic models may be valuable to the degree that they may help to shed light on particular processes; for example, the CFD-based models FLOW-3D (Simcox et al., 1988) (now updated as CFX) and JASMINE (Cox et al., 1989) both showed, qualitatively, the 'trench effect' that is thought to have been responsible for the disastrous development in the King's Cross Underground station fire of 1987. (The trench effect created close adherence of the fire gases to the escalator floor, which led to preheating and produced the very dramatic spread of fire seen in the disaster (Drysdale and MacMillan, 1992).) It may be the case, though, that while a deterministic model may, in some cases, give important qualitative insights into the fire dynamics in a given case, the predictions from a deterministic model may be quite different to the results of experimental observation. This was the case, for example, in simulations conducted as part of an assessment of one CFD-based model and three control volume models for the Home Office in the UK (Beard, 1990).

Predictions from deterministic models are not poor in all cases, of course. As a general rule, there is a scarcity of data from tunnel fire experiments with which to compare predictions from deterministic models. It is generally difficult, therefore, to know whether predicted values of variables (such as temperature or smoke intensity) are 'good' or 'bad' quantitatively. Further, a poor (or good) correspondence between a theoretical prediction and the results from a single experiment must be regarded with great caution: as considered below, results from experimental tests that are intended to be 'identical' are often very different. Comparison between prediction and experiment for deterministic models is considered further in Section 29.5.3. Also, see the other chapters covering different types of deterministic models.

29.3.3　Testing of probabilistic models

Direct testing of the predictions of a probabilistic model is very difficult. By definition, the predictions are probabilities of various events (e.g. death), and it is necessary to be able to collect sufficient data to constuct 'empirical' distributions with which to compare predictions. Data used for comparison need to be at the same 'abstraction level'. The abstraction level corresponds to the definition of the case being considered in the probabilistic model, and everything that conforms to that definition effectively specifies the variability that forms the random nature of the case under consideration. When constructing any probabilistic model, it is necessary to specify the abstraction level explicitly. For example, a particular abstraction level might correspond to 'a hospital', and there would be a particular probability of fatality from fire and a mean number of fatalities per year associated with this. A lower abstraction level might be 'a non-psychiatric hospital', and a lower level again might be 'a ward in a non-psychiatric hospital'. Different abstraction levels would correspond to different probabilities and means. With respect to tunnels, a very high abstraction level would correspond to 'a tunnel'; a lower level might be 'a road tunnel'; and a lower level again would be 'a bi-directional road tunnel'. The key point is that in constructing a probabilistic model it is essential to specify the abstraction level explicitly and clearly and, when comparing with empirical data, for only those data which satisfy the definition of the abstraction level to be used.

29.3.4　Variability in results from using models

Different users may produce quite different results, even when using the same probabilistic models, the same data base and applying it to the same case. In a European study (Hawker, 1995), it was found that risk estimates produced by different users differed by 'several orders of magnitude'.

A similar point may be made about deterministic models. Different users may employ very different input in applying the same model to the same case, producing different results. Also, the knowledge and experience of the user becomes crucial. In a 'round-robin' study (Rein et al., 2007, 2009), seven teams using the same CFD model (the Fire Dynamics Simulator, FDS) independently predicted the results for an experimental test case, without having been given the experimental results before-hand. (This was an a priori study (Beard, 2000).) Overall, the predictions were not at all good: there was generally a wide scatter amongst the predictions and, also, predictions usually compared poorly with experimental results.

Further, there may be considerable differences in results from applying different deterministic models, including CFD models, to the same case. Different users, applying different deterministic models to the same case, may produce very different results (Hostikka and Keski-Rahkonen, 1998). Also, Tuovinen et al. (1996) show significant differences found by the same user in applying two different CFD models to the same case. This gives a flavour of the kinds of problems that exist in using models as part of fire safety decision-making (see also Section 29.8 on 'knowledgeable use'). Testing of the predictions of deterministic models by comparison with experimental results is also considered later in this chapter.

29.3.5 Statistics and false inferences

The British nineteenth-century prime minister Disraeli is quoted as having stated that 'there are three kinds of lies: lies, damned lies and statistics'. Today, few would doubt that: while statistics may represent vital sources of information, they are also capable of giving a false impression. Statistics, in isolation, are simply numbers. However, statistics are always within a context and always undergo interpretation. It is vital to beware of any statement involving context-free statistics. An illustration of how statistics may be misleading, or misinterpreted, is seen in the use of the 'average'. A person may have their head in a hot oven and their feet in ice, with an average body temperature of 37°C: they would not, however, be in a healthy state. Going further, there are different kinds of 'average'; for example, 'arithmetic average' and 'geometric average'. Also, the median (i.e. the point at which 'half are above and half below') is often referred to as an 'average'. The most commonly used average is the arithmetic average; for example, the arithmetic average of the data set 5, 6 and 9 is $(5 + 6 + 9)/3$, i.e. 6.67. These different 'averages' will give different numbers for a set of data: the arithmetic average for salary in a neighbourhood may be £30 000, but the median may be £20 000. Also, the average may represent a case that is, in direct terms, physically meaningless, for example, to say that the average number of children per family in a country is 1.9 is physically meaningless in the sense that it is not possible to have 0.9 of a child. (Although it does imply 19 children per 10 families, which does have physical meaning.) Concentrating on averages may tend to mask the fact that real-world cases may cover a very wide spectrum. Standard incandescent electric light bulbs have an average life-time of about 1000 hours. However, in one particular case, a light bulb was reported to have been in use since 1938, corresponding to nearly 600 000 hours of use (*The Times*, 2003).

In the tunnel fire case, a list of fire incidents known to the authors has been collected in Chapter 1. In reality, this must be regarded as a list of 'significant' fire incidents, as there are certain to have been other (smaller) fires outwith the records gleaned for that chapter. According to the data in the chapter, the (arithmetic) average number of fatalities, given a fire in a tunnel, whether it is fatal or not, is very approximately eight; similar values are found whether one includes underground railway/metro systems and their stations or not. However, concentrating solely on that average masks the fact that many fires would have resulted in zero fatalities, and many would have resulted in large numbers of fatalities (including three fires since 1994 that each resulted in more than 150 deaths). While the average is certainly important as one measure, it is essential not to be mesmerized by it. It is important, also, not to be tempted to assume a causal connection when one may not exist. The philosopher David Hume has emphasised that conjunction (or association) must not be assumed to correspond to causation (Hume, 1740), although a causal connection may exist. For example, in some areas of Denmark there is a correlation between the number of storks' nests on a house and the number of children in the family that lives in the house (Huff, 1979). However, storks cannot be assumed to have a causal connection with human fertility: larger families tend to be associated with larger houses, and these often have more chimney pots, where the storks tend to set up their nests. Suffice to say that it is not difficult to draw false inferences from statistics, and great caution is required. This topic will not be explored further here, save to refer the reader to Irvine and Evans (1979) and Spirer *et al.* (1998).

29.4. Models as part of tunnel fire safety decision-making
29.4.1 Models need the *potential* to be of value

For the use of a fire model to be acceptable as part of the tunnel fire safety decision-making process, it is necessary for the model to have the ability to assist in bringing about a better understanding than otherwise in a particular case; that is, a model that is used must have the ability or *potential* to be of value in the tunnel fire safety context. Further, whether or not a particular model is valuable in a given case depends upon the case and how the model is used and how the results are interpreted. A model that has the potential to be valuable may not be valuable in a specific application because it may not be appropriate for that particular use. Further, even if a model may be regarded as acceptable for use in a particular application, whether or not it is really of value rather than misleading depends upon how it is used. Sensistivity studies are essential. The necessity of seeing a computer-based model in a supportive role only is provided, together with a discussion of limitations of models, by Beard (1992).

Many of the current fire models that have been applied to tunnels would be expected to have the ability to be of value. This does not mean, though, that such a model will necessarily produce results that are quantitatively reliable, even if a model has been employed for a suitable case in a way that is acceptable. Having said this, the qualitative trends predicted by a model may be valuable although actual calculated numbers may be inaccurate. As an example, a model may predict the direction of change of a particular variable, such as radiation or temperature, with change in the value of a particular design parameter, for example, whether temperature goes up or down as the parameter varies. It becomes necessary to decide whether a predicted qualitative trend exists in the real world or not. Assessments of models that are as independent as possible are required as part of the process of trying to determine this. In this sense, in order to be independent, an assessment of a model must be conducted by a person or people who do not have an interest in seeing a model depicted in either a favourable or disfavourable way. Assessments that are totally independent could not be achieved, but every effort should be made to try to ensure that an assessment of a model is as independent as it possibly can be.

Overall, we need to consider changes that would be required in order to produce a structure within which use of fire models may become generally acceptable as part of tunnel fire safety decision-making. As part of this, model assessment groups (MAGs) should be formed that would assess given models in as comprehensive a way as possible: both qualitative and quantitative aspects. Assessments of this kind should be carried out continually and be iterative: an assessment of a model may change in the light of new knowledge or data. However, assessments cannot represent absolute knowledge, and should not be regarded in that way. With these kinds of warnings, though, the assessments produced by MAGs should prove to be of great assistance. The results of assessments by model assessment groups should be made available to possible users and all those concerned with tunnel fire safety, including the general public.

Some sources have produced guidance concerning this problem: for example, the Model Evaluation Protocol produced by the European Union (EC, 1994); a document produced by the International Standards Organization (BSI, 1999); guidelines from the Society of Fire Protection Engineers (SFPE, 2009); and a paper from the American Society for Testing and Materials (ASTM, 2005), and that to emerge from a workshop at the Building Research Establishment, UK (Kumar, 2004). The general problem remains, though, of creating a procedure for producing comprehensive, iterative, assessments of fire models, in as 'independent' a way as possible, and incorporating these into a framework that is acceptable to all parties concerned. Beard (1997) provides an outline of a procedure for assessing a deterministic fire model. It is not definitive but is put forward in order to help to generate discussion in this area. In the longer term, there would be different procedures for different types of models.

Brannigan and Meeks (1995) have also discussed the need to create an acceptable regulatory framework within which fire models may be employed.

29.4.2   Fire models and 'validation'

The word 'validation' has become widely used in relation to fire models, especially computer-based fire models. However, this is unfortunate because the word is misleading, especially to those outside the subject area of fire safety engineering. To say that a model has been 'validated' implies that it has been 'proven correct'. However, this can never be the case, and, indeed, it is directly contrary to the scientific method. The word itself carries a partiality, and a more neutral word, such as 'testing', is preferable. One often hears that a particular model has been 'validated' for a particular kind of use, and this carries the implication, especially to those outside the field, that the results will be accurate (in the sense of corresponding to the real world) if applied to that case. It can never, though, be guaranteed that results are accurate, even if the results from the model for a similar case have been found to be reasonably close to experiment. Often, the word 'validation' is used as, effectively, equivalent to 'comparison with experiment', for example in the study carried out by Khoudja (1988). (See, also, the way the words 'validation' and 'verification' have been used in Chapter 15.) Other people have used the word to mean that a computer program has been shown to be consistent with its 'formal specification' (Ashley, 1985). This leads to the question of how a fire model may be 'formally specified', if at all. Further, the word 'validation' has been used in a much wider sense, to mean that a piece of software 'functions properly in a total system environment' (Glass, 1979). It would be better for the word not to be employed, but for people to say explicitly and clearly what they mean. For example, if a model user wishes to refer to 'comparison with experiment', then it would be preferable to use that phrase rather than to refer to 'validation'. At the very least, if the word 'validation' is to be used, then a precise, explicit and comprehensive definition of the way in which the word is being used by a writer in the particular case under consideration should be given.

29.5.   Illustrative case

In order to illustrate the kinds of errors that may be present and some of the issues that would need to be considered by a model assessment group, the case of predicting temperature using a deterministic model is looked at. Ideally, it is required that the predicted temperature corresponds to the real temperature, where the 'real temperature' is taken to mean the actual temperature in a fire that a modeller is trying to simulate.

29.5.1   Sources of error when using theoretical models

While the following is couched in terms of deterministic modelling, the broad categories of error apply to probabilistic modelling as well. Possible sources of error are summarized in Figure 29.2, which shows an analogy based on a set of lenses. The vertical lines correspond to, for example, the temperature at a given position or an average temperature over a given volume, at a given time, of the hot gases in a smoke layer resulting from a fire. The 'real' temperature is denoted by the point T on the left-hand side, and the temperature calculated in a simulation, T5, is on the far right-hand side. Each of the sources of error will be considered. Before that, though, it is important to make two points.

Figure 29.2 Categories of error (represented via a lens analogy)

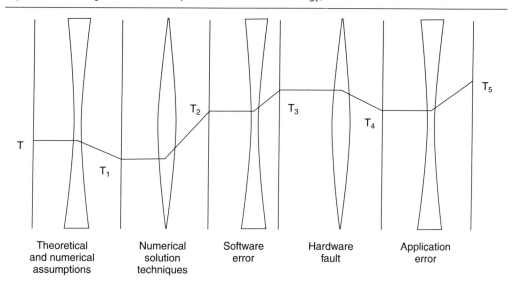

Theoretical and numerical assumptions	Numerical solution techniques	Software error	Hardware fault	Application error

1. The type of model being used for a simulation may influence the state variable being used for a comparison between theory and experiment. For example, a control volume model may assume a uniform temperature in a control volume, and simulate its value. If comparing calculated temperatures with experiment, therefore, it becomes necessary to estimate an average temperature over a volume equivalent to that control volume, inferred from individual thermocouple readings. On the other hand, a field model (i.e. CFD based) predicts, effectively, the temperature at a point because a cell in a CFD-based simulation will, in general, be much smaller than a control volume of a control volume model, even though a space in a field model calculation would be divided into a number of cells and uniformity assumed for the conditions within each cell. The relatively large number of cells used in a CFD model means that the temperature at the position of a single thermocouple can effectively be simulated. The general point is that the variable chosen may be conditioned by the nature of the model.
2. Calculating an average value of a variable from the raw data of an experiment may be done in more than one way; for example, calculating the depth of an upper smoke layer may be done using the method of Mitler (1984) or Cooper (Cooper, 1982; Cooper *et al.*, 1982).

The main categories of error are described briefly below (for further discussion, see Beard (1997)).

29.5.1.1 Lack of reality of the theoretical and numerical assumptions in the model

The first 'lens' in the analogy represented in Figure 29.2 corresponds to the fact that the conceptual and numerical assumptions in a model are only an approximation to the real world. For example, a control volume model may assume one of a number of different models for entrainment into the fire plume (e.g. see Beyler, 1986). A CFD model may assume, *inter alia*, one of a number of turbulence models (Woodburn and Britter, 1994). Also, for example, Gao *et al.* (2004) have used CFD to compare large eddy simulation (LES) and k–ε submodels, and found significant differences in estimated back-layering. (See also Miloua *et al.* (2011) and references within for considerations on using different combustion and turbulence models in CFD.) As well as variability in the qualitative assumptions made, there will often be latitude to put in different numerical values for parameters because of uncertainty in a particular application and still be able to claim that the numbers used are 'reasonable' or 'plausible'. For example, there would usually be scope for using different possible values for parameters related to turbulence or convective heat transfer. If a specific conclusion is desired, it may not be difficult to devise a set of 'plausible' input assumptions that imply that conclusion. Thus, models may be inappropriately used in this way.

29.5.1.2 Lack of fidelity of the numerical solution procedures

The second lens in the analogy (see Figure 29.2) represents error resulting from the numerical solution procedures used to solve the equations that make up the model. Systems of equations in models need to be solved numerically rather than analytically, except in the case of a simple model with relatively simple equations. 'Analytical' refers to a theoretical, general, solution being found to a problem, without using a computer, leading to exact results. Different numerical solution techniques may be employed to conduct a specific mathematical procedure, and they may well produce different results. An example is provided by the grids used in CFD-based models to form the cells. Results usually depend upon the resolution of the grid, as illustrated in the studies by Galea and Markatos (1988) and Rhodes (1998). The differences between the results of using a coarse grid (or 'mesh') and those for a fine grid may be significant. However, Coyle and Novozhilov (2007), using the FDS, have found that specifying a finer grid may significantly worsen predictions. They also question the claims made for the FDS concerning temperature, and say that sometimes a zone model can give better predictions than CFD. Further, Cheong (2008), using the FDS, found that a coarser grid gave a better fit to results from the Runehamar test series than a finer grid. Cong (2008–09), using the FDS, found a finer grid is not necessarily better, and considered both uniform and non-uniform grids. Also, results from a CFD model will in general depend on the boundary conditions assumed (Woodburn and Britter, 1994) as well as the time step (Rhodes, 1998), in addition to other factors. (For further considerations of deterministic models, see Chapters 15 and 16.)

29.5.1.3 Direct mistakes in software

The third lens in the metaphor of Figure 29.2 corresponds to software mistakes. In addition to the error sources that have already been considered, it is possible that the software will not be an accurate representation of the model and numerical solution procedures. According to one estimate, there may be around eight errors per thousand lines of computer source code, and, even for safety-critical applications, there could be around four errors per thousand lines of computer code (Jackson, 1993). Errors may result from mistakes in producing the software that is intended to represent the model and its numerical solution techniques. There is also the possibility that the physical system that is being modelled may go into a condition that the software is not suitable for; this may be related to lack of realism of the theoretical and numerical assumptions made in the model.

Examples of software error are very hard to find because of the intrinsic nature of the error and because of a lack of willingness to be open about such matters by those who have specific commercial or other interests in a case. Some examples, though, have come into the public domain. One specific case relates to a computer code that was intended for a commercial nuclear reactor in

Canada. The nuclear regulatory body demanded independent examination, and a team from McMaster University found some serious errors (Channel 4, 1992). Other examples are given by Beard (1997).

It is very important that the source codes for computer programs used as part of tunnel fire safety decision-making be open to examination by the public in general and the scientific community in particular; commercial considerations should not be allowed to stand in the way of this. This point has been discussed further by Beard (1992a). As a general rule, it must be assumed that there will be errors in any complex piece of software. Procedures need to be produced in order to try to reduce the likelihood of software errors existing in models that may be used as part of tunnel fire safety decision-making. Examination of software error should be part of the independent assessment of a model.

29.5.1.4 Faults in the computer hardware
It is usually assumed that computer hardware is very reliable, and the possibility of a computer making a mistake because of its hardware has been generally ignored, at least by those who are not in the field of computer science. It is the case, though, that a fault may exist in hardware as a result of (1) mistakes in the design of micro-processors, (2) faults in the manufacture of micro-processors or (3) a combination of (1) and (2).

A specific case is seen in the fault in the Pentium processor, which came to be realised during 1994. In this case, the manufacturer 'made no public announcement when it first detected the error' (Arthur, 1994). It was effectively left up to users to find the fault, by chance, and attempt to contact other users.

29.5.1.5 Mistakes in application
A fire model user may make an error whilst putting numerical or other input into a model or in the analysis of output due to (1) misunderstanding of the model or its numerical solution techniques, (2) misunderstanding of the design of the software or (3) a slip in inserting input or reading output.

Specific information is hard to find. One study, though, has suggested that, in very general terms, an error exists of the order of 1 in 1000 for relatively simple mistakes such as misreading a number or pressing a wrong computer key (HSE, 1991).

29.5.2   Inadequate documentation
In addition to these categories, there may be 'effective errors' in using a model because of poor documentation. For example, an ability may be implied for a model that it does not possess. This is a crucial point: model documentation must state clearly and explicitly the conditions for which the software is suitable or unsuitable. In one particular case, for example, a field model was, in reality, suitable only for smooth walls. However, the documentation did not mention this, implying that the model could also be applied in the case of rough walls.

29.5.3   Comparison between theoretical prediction and the results of experiments
A theoretical prediction for, say, a gas temperature cannot be directly compared with such a temperature in a fire. Likewise for other variables. This means that it is necessary to compare with experimental results. This is more difficult than might be thought, although there is a humorous saying: 'Nobody believes a theory except the theoretician; everybody believes an

Figure 29.3 Ostensibly 'identical' tests; results from two experiments intended to be the same. Curves smoothed for clarity

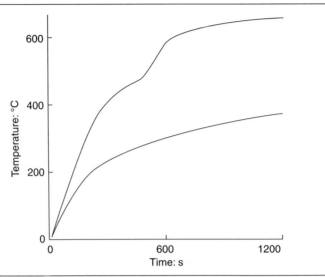

experiment, except the experimenter'. However, it is necessary to adopt an inquiring attitude to experimental results as well as to theoretical predictions. False inferences drawn from the results of an experiment may be associated with the following.

- Lack of control of the conditions of an experiment (e.g. ambient humidities may vary from day to day or month to month, leading to different temperatures for ostensibly 'identical' experiments (Figure 29.3)).
- Design of the experiment (e.g. different sets of thermocouple positions will lead to different sets of 'raw data').
- Direct error in measurement (e.g. error associated with a thermocouple reading; amongst other things, difficulty in controlling tip emissivity may affect measured gas temperatures).
- Raw data processing algorithms, i.e. obtaining the final 'results' from an experiment from raw data (e.g. assumptions associated with finding an average temperature from a set of raw data).

These sources of flexibility and error have been considered by Beard (1992a). It may be specifically noted here that experimental heat release rate (HRR) measurements depend upon how they are done (measurement location, method, assumptions, etc.): for example, see the different HRR estimates for the Eureka HGV test (French, 1994; Sorlie and Mathisen, 1994) and Table 14.2. Also, ostensibly 'identical tests' may produce quite different results (e.g. see Drysdale, 1999) with regard to the CBUF (furniture) project. Also, Figure 29.3 is based upon the temperatures measured in two tests that were intended to be identical. The particular case is not relevant; however, the results of Figure 29.3 were not a product of 'bad science' – they were the result of work in a highly regarded laboratory conducted by respected scientists. Variability of this kind does exist in well-conducted experimental work. It shows that there is a need for experimental tests to be repeated in as identical conditions as possible (i.e. tests should be replicated) and for distributions of experimental results to be produced, for each given case. It has been estimated, for example, that the test that reached 202 MW in the Runehamar series was very approximately at the 59th percentile (Beard, 2009). This raises problems: replication of experimental tests is expensive, and there is a lack of willingness to carry out such replication because of this. It is very important that it be done, though, and it strongly suggests the need for collaboration at an international level that is aimed at producing acceptable data sets and distributions of results from replicated tests. Results from a single experimental test may well not be at the mean of a distribution of replicated test results for the same case. Further, there is a need for large-scale experimental tests in addition to smaller-scale tests (Johnson, 1991; Beard, 1992b). While important work has been conducted at large scale, for example, the EUREKA test series, the Memorial Tunnel Fire Ventilation Test Program and the test series carried out in the Runehamar Tunnel, Norway, there is a great need for more large-scale testing in order to be able to understand fire and smoke dynamics much better.

In comparing any theoretical prediction with experimental results there are three generic types of comparisons; brief descriptions are given here, and further discussion is given by Beard (2000). The types are as follows.

1. An *a priori* comparison in which the modeller has, effectively, not 'seen' or used any results from an experiment being used for comparison.
2. A *blind* comparison in which the modeller has, effectively, not 'seen' all the results from an experiment being used for comparison but some limited data from that experiment have been used as input, for example, heat release rate or mass loss rate over time.
3. An *open* comparison in which data from an experiment being used for comparison have been seen and possibly used. The modeller is free to adjust input after initial comparisons.

In comparison types (1) and (2) above, the modeller is not free to adjust the input after the comparison. Most of the comparisons in the literature are of type (3). Very few are of type (2): in the tunnel context, mention may be made of Bettis *et al.* (1994) and Miloua *et al.* (2011) (see also Chapter 17). Extremely few indeed are of type (1): examples are given by Beard (1990, 1992c) and Rein *et al.* (2009). Comparisons of all three types are needed. (See also the CFD study carried out as part of the Benelux tests (Directorate-General for Public Works and Water, 2002), which concluded that, quantitatively, 'substantial deviations' can occur.) When reporting a comparison, it should be made explicit which type it is, and full details given. In using models for real-world decision-making, we are effectively in the realm of type (1): we cannot have the results from a real fire in a tunnel which is yet to happen, if it does happen. While there is a place for *open* comparisons, far more *a priori* and *blind* comparisons need to be carried out and reported in the public scientific literature.

29.6. The *potential* of a specific model in tunnel fire safety decision-making

A summary list of important points concerning the potentiality of a fire model to be valuable as part of fire safety decision-making has been given by Beard (1997). Here, it will simply be stressed that.

■ Models (especially computer-based models) should be used in a supportive role only and should not play a sole part in decision-making. They should be used in conjunction with other sources of knowledge/experience/intuition.

- In comparing prediction with experiment for a deterministic model, it should be made clear and explicit whether this is on the basis of *a priori*, *blind* or *open* comparison. Full details should be given, to enable a reader, in principle, to repeat the simulation.
- Sensitivity studies are vital: conceptual and numerical input assumptions need to be varied to see the effect on output.
- Uncertainty of input leaves much scope for different values to be used as input and for this still to be claimed to be 'plausible'. That is, it is often not known whether a particular value used for an input parameter is realistic or not.

29.7. An acceptable 'methodology of use'

Even with a model that is accepted as being able to be of value in assessing a design option it is necessary to have an appropriate *methodology of use*, with the aim of trying to ensure that the model is used in a way that is acceptable to all concerned, including the general public. A methodology of this kind should encourage a user to be open and *explicit* about the assumptions being made in a particular use of the model. It should also stimulate the carrying out of sensitivity studies. In all, it should encourage a user to be clear about what has been done and to be conscious of the need to be direct and explicit to other people and in reports produced.

The situation that exists in the fire safety engineering field today is similar to that which existed in structural engineering in relation to finite element models which has led to the development of a 'methodology of use' for that subject area. That methodology is known as the safety-critical structural analysis (SAFESA) methodology (Knowles and Maguire, 1995; NAFEMS, 1995). An essential part of the process consists of estimating the sensitivity to different conceptual and numerical assumptions. An outline of a generic 'methodology of use' for a deterministic fire model is given by Beard (1997): it is intended to help to generate discussion and is not put forward as a definitive methodology. Eventually, there should be different methodologies for different types of models.

Even with a suitable methodology there is no guarantee that a model will necessarily be applied in an acceptable way. An appropriate methodology can only help and encourage a model user to be systematically comprehensive and explicit; and to be as conscientious as possible in the use of the model. Using an acceptable methodology of use may still result in a 'poor' and unacceptable application of a model in a specific 'real-world' application. That is, there is a need for competent and acceptable application of an 'acceptable' model, using a suitable 'methodology of use', and this leads to the concept of a 'knowledgeable user'.

29.8. A 'knowledgeable user'

It is essential that a user apply a methodology of use in a suitable way to a model in a given application. It is not enough to simply have an 'acceptable' model and an 'acceptable' methodology of use: it is also vital to have a 'knowledgeable user'. For example, a methodology of use may stress that how a real-world case is represented, or 'idealised', within the model must be made explicit. Even then, any specific idealisation would still be devised by the user. Further, how to carry out a sensitivity analysis (and to what extent), and interpret results, would be decided by the user. This means that the user must be knowledgeable about both tunnel fire science and the limitations and conditions of applicability of the model. There is a need for education and experience in these matters. In addition, drawing upon the fire knowledge and experience of people other than the person who is conducting the study would generally be necessary. It is essential that results be interpreted in the light of other knowledge and experience in order for a model to be used in a justifiable way in a real-world application. Overall, even with an 'acceptable' model and an 'acceptable' methodology of use, the possibility would still exist that the model, or the methodology of use, will be misused or results misinterpreted (or both).

29.8.1 Sensitivity studies

These considerations illustrate how a 'close' correspondence between a single simulation using a theoretical model and the results of any given experiment does not necessarily mean that the model is a 'good' one. Further, a 'bad' correspondence between the results of a model and the results of a single experiment does not necessarily mean that the model is a poor one. Sensitivity studies need to be conducted when using a model in a real-world application; that is, changing the input (conceptual and numerical) to see the effect on output for a given case. Only in this way can the uncertainty in the input even start to be addressed.

However, carrying out comprehensive sensitivity studies for a model being used in a particular case is not easy. An exhaustive sensitivity study for a typical control volume model would, in principle, involve carrying out many thousands of runs, and even a restricted study would involve hundreds of simulations (Khoudja, 1988). For example, a model with 16 parameters to be varied, and considering only two values for each parameter, would require 2^{16} (i.e. 65 536) simulations for a 'full factorial' analysis (Khoudja, 1988). For a restricted study, the choice of which factors to vary becomes crucial. A significant sensitivity may not come to light. While a very restricted sensitivity study might involve tens of simulations, carrying out such a study may still be of great value, depending upon how simulations are selected and results interpreted. Inappropriate interpretation of results is a very real danger.

29.8.2 Interpretation of results from a theoretical model

The issue of the inappropriate interpretation of results from a theoretical model is an extremely important one. This concerns the knowledge and experience of the user, and has been discussed to some degree by Beard (1992a); it will not be pursued further here. It relates to both the concepts of an acceptable *methodology of use* and of a *knowledgeable user*.

29.8.3 Quantitative and qualitative results

While it may not be possible to assume that results from a model are quantitatively accurate, results may display a measure of qualitative accuracy and be capable of assisting in decision-making. This has already been mentioned in Section 29.4.1 and by Beard (1992a). It must, however, be realised that the qualitative inferences drawn from using a model in a given case may be erroneous.

29.9. Evacuation modelling

Modelling of human behaviour and evacuation has taken place in relation to fires in buildings for many years. In recent times, such models have been applied to evacuation in tunnel fire emergency situations: see, for example, the proceedings of conferences in the series *Safety in Road and Rail Tunnels* or in the series *Tunnel Fires and Escape from Tunnels*. However, it should not be assumed that an evacuation model that is applicable for building fires will also be applicable for tunnel fires. Even in relation to building fires, evacuation models are under scrutiny in relation to their basic assumptions and whether or not these are justifiable. The plain truth is that, at the current time, we know extremely little about how human beings behave during tunnel fire emergencies. Basic empirical work needs to be carried out. (For more on this theme, see Chapters 19–21 on human behaviour.)

29.10. Conclusion

Theoretical models, especially computer-based models, are capable of being very valuable in assisting tunnel fire safety decision-making. However, models are also capable of being misleading and causing decisions to be made that should not be accepted. Models should only ever be used in a supportive role, in the light of other fire knowledge and experience. Further, in order to be able to use fire models acceptably, several conditions need to exist.

1. A model must have the *potential to be valuable* and help a user to gain a better understanding than otherwise in a particular real-world case. This implies that a procedure for independent assessment should be established. This should lead to a way of accepting, or not accepting, each model for specific uses in specific ways. It would be appropriate for MAGs to be created as part of this (Beard, 1997). This requires that each MAG would be made up of people who are as 'independent' as possible of the model being examined: they should have no interest in promoting or discouraging the use of any particular model that is being looked at.
2. A model must be *used in an acceptable way*. For this purpose, a *methodology of use* should be constructed that attempts to help a user to apply a model in an acceptable way. Essentially, it would encourage a user, via the stages and demands of the methodology, to be conscious about the process and open and explicit about what was being done. This is crucially important: full details should be out in the open. Overall, the results from a model must be seen in relation to the limitations and conditions of applicability of any model being used. Sensitivity studies are vital in any application to a real-world case, and this should be emphasised as part of an acceptable methodology of use.
3. *Users of models must have adequate knowledge* about both the model and fire behaviour. In principle, it would be possible for an acceptable methodology of use to be applied to a model that has the potential to be of value in a particular case, but in a way that is not suitable. A methodology of use would provide a framework that encourages and helps a user to apply a model in an appropriate way in a given case. However, a user must be knowledgeable about the model and fire behaviour in order to try to reduce the likelihood of misinterpretation of theoretical results. Theoretical results must always be interpreted in the context of other knowledge and experience. It is worth mentioning, at this point, the existence of a proposal for a 'core curriculum' in fire safety engineering (Magnusson *et al.*, 1995). Such a core curriculum should take explicit account of the general limitations and conditions of applicability of using models, especially computer-based models, in relation to fire safety decision-making. General requirements for the acceptable use of fire models, of all kinds, need to be included. The currently proposed core curriculum could be improved in this regard. Beyond educational courses that would provide a 'broad-brush' knowledge of fire safety engineering and modelling, there also need to be specialist courses in specific models that would include specific limitations and conditions of applicability (Beard, 2005). It should be realised, though, that knowledgeable use of a model depends upon experience as well as 'education'. Given this, while a user must have knowledge of fire dynamics, it is also essential to draw upon the knowledge and experience of others in specific cases.

The 'fire model triangle' of Figure 29.4 shows these three essential requirements. Models are able to be of considerable help in decision-making in relation to tunnel fire safety. However, it is also possible that their use may lead to inappropriate decisions being taken, and this could well be dangerous. The creation of a coherent context should help to make fire models valuable and

Figure 29.4 Fire model triangle

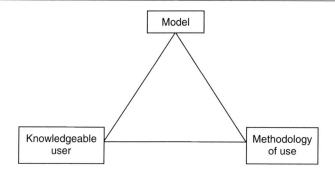

acceptable rather than problematic in the future. In addition to fire development models, computer-based evacuation models are also starting to be applied to tunnel fires. At the present time, however, too little is known about human behaviour during tunnel fire emergency situations to be able to realistically model evacuation.

REFERENCES

Arthur C (1994) Flawed chips bug angry users. *New Scientist*, 10 December 1994: p. 18.

Ashley N (1985) Program validation. In: Anderson T (ed.), *Software: Requirements, Specification and Testing*. Blackwell, Oxford.

ASTM (2005) *Standard Guide for Evaluating the Predictive Capability of Deterministic Fire Models*, E1355-97. American Society for Testing and Materials, West Conshohocken, PA.

Beard AN (1984/1985) A stochastic model for the number of deaths in a fire: further considerations. *Fire Safety Journal* **8**: 201–226.

Beard AN (1990) *Evaluation of Fire Models*. Ten reports for the Fire Research and Development Group of the UK Home Office. Unit of Fire Safety Engineering, University of Edinburgh, Edinburgh. (These reports were used as a basis for guidance for UK fire brigades.)

Beard AN (1992a) Limitations of computer models. *Fire Safety Journal* **18**: 375–391.

Beard AN (1992b) On comparison between theory and experiment. *Fire Safety Journal* **19**: 307–308.

Beard AN (1992c) Evaluation of deterministic fire models; part 1 – introduction. *Fire Safety Journal* **19**: 295–306.

Beard AN (1997) Fire models and design. *Fire Safety Journal* **28:** 117–138.

Beard AN (2000) On *a priori*, blind and open comparisons between theory and experiment. *Fire Safety Journal* **35**: 63–66.

Beard AN (2004) Risk assessment assumptions. *Civil Engineering and Environmental Systems* **21**: 19–31.

Beard AN (2005) Requirements for acceptable model use. *Fire Safety Journal* **40**: 477–484.

Beard AN (2009) HGV fires in tunnels: fire size and spread. *2nd International Tunnel Safety Forum for Road and Rail, Lyon*, 20–22 April 2009. TMI, Tenbury Wells, pp. 103–111.

Beck VR (1994) The development of a risk–cost assessment model for the evaluation of fire safety in buildings. *4th International Symposium on Fire Safety Science, Ottawa*, 13–17 July 1994. International Association for Fire Safety Science, London, pp. 817–828.

Bettis RJ, Jagger SF, Lea CJ, Jones IP, Lennon S and Guilbert PW (1994) The use of physical and mathematical modelling to assess the hazards of tunnel fires. *8th International*

Symposium on Aerodynamics and Ventilation of Vehicle Tunnels, Liverpool, 6–8 July 1994. BHR Group, London.

Beyler C (1986) Fire plumes and ceiling jets. *Fire Safety Journal* **11**: 53–75.

Brannigan V and Meeks C (1995) Computerized fire risk assessment models: a regulatory effectiveness approach. *Journal of Fire Sciences* **13**: 177–196.

BSI (1999) *Fire Safety Engineering – Part 3: Assessment and Verification of Mathematical Fire Models*, BS ISO TR 13387-3:1999. British Standards Institution, London.

Carvel RO, Beard AN, Jowitt PW and Drysdale DD (2001) Variation of heat release rate with forced longitudinal ventilation for vehicle fires in tunnels. *Fire Safety Journal* **36**: 569–596.

Channel 4 (1992) *Dispatches*, 4 March 1992. Channel 4 Television, London.

Cheong MK (2008) Assessment of credible vehicle fires in road tunnel. PhD thesis, University of Canterbury, New Zealand.

Cong BH, Liao GX, Yao B, Qin J and Chow WK (2008–09) Numerical studies on extinguishing solid fires by water mist. *Journal of Applied Fire Science* **18**: 241–270.

Cooper LY (1982) A mathematical model for estimating available safe egress time in fires. *Fire and Materials* **6**: 135–144.

Cooper LY, Harkleroad M, Quintiere J and Rinkinen W (1982) An experimental study of upper hot layer stratification in full-scale multi-room fire scenarios. *Journal of Heat Transfer* **104**: 741–749.

Cox G, Chitty R and Kumar S (1989) Fire modelling and the King's Cross fire investigation. *Fire Safety Journal* **15**: 103–106.

Coyle P and Novozhilov V (2007) Further validation of fire dynamics simulator using smoke management studies. *International Journal on Engineering Performance-Based Fire Codes* **9**: 7–30.

Directorate-General for Public Works and Water (2002) *Project 'Safety Test' – Report on Fire Tests*. Directorate-General for Public Works and Water, Civil Engineering Division, Utrecht. (2nd Benelux tests.)

Drysdale D (1999) *Introduction to Fire Dynamics*, 2nd edn. Wiley, Chichester.

Drysdale DD and MacMillan AJR (1992) The King's Cross fire: experimental verification of the trench effect. *Fire Safety Journal* **18**: 75–82.

EC (1994) *Model Evaluation Protocol*. European Communities, Directorate-General 12, Scientific Research and Development, Brussels.

Fennel D (1988) Investigation into the King's Cross Underground Fire. HMSO, London.

French SE (1994) Eureka 499 – HGV fire test (Nov 1992). *International Conference on Fires in Tunnels, Borås*, 10–11 October 1994, pp. 63–85.

Galea ER and Markatos NC (1988) Modelling of aircraft cabin fires. In: *Second International Symposium on Fire Safety Science, Tokyo*, 13–17 June 1988.

Gao PZ, Liua SL, Chow WK and Fong NK (2004) Large eddy simulations for studying tunnel smoke ventilation. *Tunnelling and Underground Space Technology* **19**: 577–586.

Geyer TAW, Morris MI and Hacquart RY (1995) Channel Tunnel Safety Case: quantitative risk analysis methodology. *2nd International Conference on Safety in Road and Rail Tunnels, Granada*, 3–5 April 1995. ITC, Tenbury Wells.

Glass RL (1979) *Software Reliability Guidebook*. Prentice-Hall, Englewood Cliffs, NJ.

Hawker CR (1995) Offshore safety cases – BG E&P's experience. *Conference on 'Safety Cases – Can We Prove the Next Ten Years will be Safer than the Last Ten?' London*. Organised by IBC Technical Services Ltd.

Hostikka S and Keski-Rahkonen O (1998) *Results of CIB W14 Round Robin for Code Assessment*. Technical Research Centre of Finland, Espoo.

HSE (1991) *Human Reliability Assessment – A Critical Overview (Second Report)*. Advisory Committee on Safety of Nuclear Installations, Study Group on Human Factors, Health and Safety Executive, Bootle.

Huff D (1979) *How to Lie with Statistics*. Penguin Books, Harmondsworth.

Irvine J and Evans J (1979) *Demystifying Social Statistics*. Pluto Press, London.

Jackson D (1993) New developments in quality management as a pre-requisite to safety. *First Safety-Critical Systems Symposium, Bristol*, 9–11 February 1993. (Quoted during the presentation not in the proceedings.)

Johnson P (1991) Letter to the Editor. *Fire Safety Journal* **17**: 415–416.

Khoudja N (1988) *Procedures for Quantitative Sensitivity and Performance Validation Studies of a Deterministic Fire Model*, NBS-GCR-88-544. National Institute of Standards and Technology, Gaithersburg, MD.

Knoflacher H (2001) The accident at the Tauern Tunnel on 29th May 1999. *Conference on 'How Can Tunnels be Safe Enough from Fire?', London*, 6 June 2001. (Organised by the Institution of Fire Engineers.)

Knowles NC and Maguire JR (1995) On the qualification of safety-critical structures: the SAFESA approach. *3rd Safety-Critical Systems Symposium, Brighton*, 7–9 February 1995.

Kumar S (2004) *Computer Modelling Performance Assessment Scheme for Performance-based Fire Safety Engineering*. Building Research Establishment, Garston, Watford.

Magnusson SE, Drysdale DD, Fitzgerald RW, Mowrer F, Quintiere J, Williamson RB and Zalosh RG (1995) A proposal for a model curriculum in fire safety engineering. *Fire Safety Journal* **25**: special issue.

Miloua H, Azzi A and Wang HY (2011) Evaluation of different numerical approaches for a ventilated tunnel fire. *Journal of Fire Sciences* **29**: 403–429.

Mitler HE (1984) Zone modelling of forced ventilation enclosures. *Combustion Science and Technology* **39**: 83–106.

NAFEMS (1995) *SAFESA Technical Manual*. NAFEMS, Glasgow.

Powell S and Grubits S (1999) Tunnel design with TRAFFIC – tunnel risk assessment for fire incidents and catastrophes. *International Conference on Tunnel Fires and Escape from Tunnels, Lyon*, 5–7 May 1999. ITC, Tenbury Wells.

Rein G, Empis C and Carvel R (eds) (2007) *The Dalmarnoick Fire Tests: Experiments and Modelling*. School of Engineering, Edinburgh University, Edinburgh.

Rein G, Torero JL, Jahn W, Stern-Gottfried J, Ryder NL, Desanghere S, Lázaro M, Mowrer F, Coles A, Joyeux D, Alvear D, Capote JA, Jowsey A, Abecassis-Empis C and Reszka P (2009) Round-robin study of 'a priori' modelling predictions of the Dalmarnock Fire Test One. *Fire Safety Journal* **44**: 590–602.

Rhodes N (1998) The accuracy of CFD modelling techniques for fire prediction. *3rd International Conference on Safety in Road and Rail Tunnels, Nice*, 9–11 March 1998. ITC, Tenbury Wells, pp. 109–115.

Russel B (1979) *History of Western Philosophy*. Unwin, London.

Simcox S, Wilkes NS and Jones IP (1988) *Fire at King's Cross Underground Station, 18th November 1987: Numerical Simulation of the Buoyant Flow and Heat Transfer*. Report AERE-G 4677. UK Atomic Energy Authority, Harwell.

Society of Fire Protection Engineers (2009) *Guidelines for Substantiating a Fire Model for a Given Application*. Draft for Comments. Society of Fire Protection Engineers, Bethesda, MD.

Sorlie R and Mathisen HM (1994) Measurements and calculations of the heat release rate in tunnel fires: fire test heavy goods vehicle. *International Conference on Fires in Tunnels, Borås*, 10–11 October 1994, pp. 104–116.

Spirer HF, Spirer L and Jaffe AJ (1998) *Misused Statistics*. Dekker, New York.

The Times (2003) Old faithful makes light of growing old. *The Times*, 11 March 2003.

Tuovinen H, Holmstedt G and Bengtson S (1996) Sensitivity calculations of tunnel fires using CFD. *Fire Technology* **32:** 99–119.

Woodburn PJ and Britter RE (1994) *Comparison of Numerical and Experimental Results for a Fire in a Tunnel*. Health and Safety Executive, Bootle.

Handbook of Tunnel Fire Safety, 2nd edition
ISBN: 978-0-7277-4153-0

ICE Publishing: All rights reserved
doi: 10.1680/htfs.41530.635

Chapter 30
Decision-making and risk assessment

Alan Beard
Civil Engineering Section, Heriot-Watt University, Edinburgh, UK

30.1. Introduction

The general shift away from prescriptive to performance-based decision-making with regard to tunnel fire safety is a double-edged sword. In some ways it is a very desirable shift, but in other ways it may backfire. Whatever else it implies, it means that there is a need to assess the risk in some way. Prescriptive regulations, including 'best practice' codes and guides, have played a vital role in society, and should continue to do so.

The key objectives of tunnel fire safety decision-making may be seen as to prevent (1) fatality and injury, (2) property loss and (3) disruption of operation. In the real world, these cannot be achieved in absolute terms, and the aim tends to become to achieve acceptable ranges for the risks associated with these various forms of harm. (See also Chapters 5 and 29 and SAFE-T (2006).)

30.2. Prescriptive and risk-based approaches

With a purely prescriptive approach, tunnel designers, operators and users are effectively unaware of what the risks are with regard to the three categories above, in explicit terms at least. Historical statistics give us some idea of the risk implicit in a particular system; however, there is a crucial problem with simply looking at statistics, and that is this: the system changes over time. Simply considering historical statistics with regard to a particular tunnel over a long period, say 20 years, may be very misleading because it is certain that the system as it exists at one point will be different to the system that exists 20 years later; or even 5 or 10 years later. To consider just one factor alone, increasing traffic volume probably means that the systems associated with most road tunnels have changed dramatically in recent years. While a prescriptive approach would not recognise this (at least explicitly), it would be recognised in a 'risk-based' approach; or at least it should be. That is, a risk-based approach has the *potential* to be very valuable in helping us to cope with decision-making in an increasingly complex and ever-changing world. Having said that, there are significant problematic issues associated with the use of risk assessment techniques, as will be mentioned later. (See also a report produced for the European Parliament on tunnel safety and risk assessment (Beard and Cope, 2008). This report makes 25 recommendations.)

It may be noted that the design of the Channel Tunnel involved the use of a significant amount of 'risk assessment'. This has not, however, prevented extremely serious fires taking place in the tunnel within a relatively short time period.

Both prescriptive and risk-based approaches have their positive and negative aspects: while prescriptive codes do not allow us to understand the risk explicitly, they often represent a rich seam of knowledge and experience grounded in the real world. Conversely, while a risk-based approach does, in principle, allow us to appreciate what the risk is, there are considerable problems associated with assessing risk and being able to use that modelling as part of tunnel fire safety decision-making in a reliable, effective and acceptable way.

With regard to tunnel safety, specific pressures spurring a move towards an approach that includes risk assessment are being provided by the European Union (EU) via the 2004 Directive on road tunnels (EU, 2004) and, significantly, by the global insurance industry via a code of practice for tunnel construction (discussed by Dix, 2004a).

30.3. Approach to design

In principle, a new tunnel may be designed, or an existing tunnel upgraded, to produce an acceptable risk by the use of (1) prescriptive regulations, (2) qualitative risk assessment or (3) quantitative risk assessment; or a combination of these. It should be noted that quantitative risk assessment may mean deterministic risk assessment (i.e. using deterministic models) or non-deterministic risk assessment (i.e. using probabilistic methods or points schemes or statistical techniques); or a combination of these. Physical models and tests may also be employed as part of the approach.

A fundamental question arises: *what is a healthy mixture of prescriptive requirements, qualitative risk assessment and quantitative risk assessment?*

Beyond that, what is to be regarded as acceptable in relation to each of these three elements?

Some countries, most notably Japan, do not use formal risk assessment (i.e. quantitative risk assessment, at least) other than, possibly, in very rare cases. The essence of the system for road tunnels in Japan seems to be to assign a new or existing tunnel to one of five categories based on tunnel length and traffic volume. Measures to be used basically derive from that categorisation (Ota, 2003). Also, risk assessment may be applied in a purely qualitative way, as in Rodriquez *et al.* (2011).

If it is to be accepted that risk assessment forms part of an approach, then it needs to be decided how this is to be done and what the essential elements of a risk assessment are. The essence of a risk assessment may be regarded as centring on the following.

1. Hazard identification. That is, attempting to find answers to the basic question: what could possibly happen within this system which may lead to harm? The word 'harm' may be seen in relation to the key objectives of fire safety identified above.
2. Risk analysis. This centres on attempting to identify and describe possible consequences.
3. Risk evaluation. This focuses on deciding what, if anything, should be done to change the system.

The above may be regarded as applying to both an existing tunnel system, for example, upgrading, or to a tunnel system that exists on the drawing board, i.e. in the design stage. In fact, the three primary categories above could be regarded as applying to many types of risk, including, for example, financial or environmental. (See also Chapter 5.) It should be noted that there is a lack of uniformity in the use of terms in risk assessment: different people and organisations use terms in different ways (Beard and Cope, 2008).

In shifting to an approach that includes risk assessment (both qualitative and quantitative), several issues come to the fore.

- *Which methodology* should be used to address the primary categories above? What are to be the criteria for acceptance of a methodology?
- *Which models* should be used, and how, to assess the risk and make a decision? What are to be the criteria for acceptability of a model and how it is used? (See also Chapter 29.)
- *What are to be the criteria for acceptability of risk?* In essence, this is an ethical matter, not a technical one, and as such the criteria need to be generally acceptable to the public (see below).
- A *'knowledge base' is essential*: to underpin a risk assessment, a thorough knowledge and understanding of the system is required. This means that there is a need for both theoretical and experimental research and acquiring of information. Because the real world and its values continually change, it is vital that research continues indefinitely. The research is never 'finished'.

30.3.1 Measures of risk

Different indicators of life risk, including injury, may be used, and these may be expressed in deterministic or non-deterministic terms. Deterministic criteria may use, for example, temperature, toxic gas concentrations or heat fluxes. A probabilistic measure might be the probability of a person being killed per year. In general, two kinds of measures are used, relating to 'individual risk' (referring to a member of a specified group of people) and 'societal risk' (expressing risk to society as a whole). Different definitions of these types of risks are used. (For more on this subject, see Beard and Cope (2008), Trbojevic (2003) and Chapter 22.) In addition to measures of life risk, there would be measures for property risk and disruption of operation.

30.3.2 Lack of suitability of a model or technique for a specific case

A model may be used for a case for which it is not suitable. This is strongly linked to the assumptions made in the model being used. For example, a zone model may be employed for a case for which it is totally unsuitable because of the complex geometry of the case being considered. Further, with regard to non-deterministic risk assessment, Ferkl and Dix (2011) have described what they call the 'seven most deadly sins'; this also relates to the next section. (Additionally, see Chapter 29.)

30.3.3 Interpretation of results from models used in risk assessments

It is important not to misinterpret results from a model. For example, a probabilistic risk assessment was conducted in relation to the Channel Tunnel, and the safety case (Eurotunnel, 1994) concluded: 'The total estimated rate of fatalities amongst passengers and staff in the Channel Tunnel system is less than 6% of the risk associated with the equivalent size of the existing GB or French railway network.' This may be seen alongside the fact that there have been two very major fires in the tunnel since it opened in 1994, fortunately with no loss of life. While the occurrence of major fires does not conflict with the conclusions reached in the safety case, their existence does serve to highlight the need to be very circumspect and cautious in relation to risk assessment studies. A calculation of a relatively low risk does not mean that a major fire will not happen tomorrow: results from a risk assessment must not be misinterpreted. More generally, results from models, whether deterministic or non-deterministic, must always be seen in relation to the assumptions made and the nature of the techniques employed.

30.4. Lack of independence and assessment of risk

Bearing in mind the possibility of socio-economic, political or institutional pressures, it is important for a risk assessment to be conducted by a person who is as independent of the system under scrutiny as possible.

30.4.1 Flexibility and uncertainty

As a general rule, there will be many sources of flexibility in the way a situation may be represented in a model and uncertainty in conceptual and numerical assumptions. In such conditions, it is sensible to err on the cautious side. (See also Chapter 29.) Given the lack of specific knowledge about input to a model that often exists (i.e. uncertainty), and the flexibility that may exist in the way in which a specific case may be represented (i.e. 'idealised') in a model, there may well be considerable scope for an analyst to make conceptual or numerical assumptions that may be defended as 'plausible' and which present the option that a client may prefer (e.g. the cheapest) in a favourable light relative to other options. Upon deeper analysis, however, including sensitivity considerations, this may not correspond to a risk that the general public would regard as acceptable.

30.4.2 No person can be 'truly independent'

There is no such thing as a 'truly independent' person: everybody has a position in society and a set of skills and knowledge particular to them. A quote from Upton Sinclair, the novelist, makes a salutary point (Sinclair, 1994): 'it is difficult to get a man to understand something when his salary depends upon his not understanding it'.

It is important for an assessment to be carried out by a person who is as independent as possible of commercial, political, institutional or other pressures related to the particular case. Because every person has their own particular background and interests (including an 'independent' person conducting the analysis), an assessment for a specific tunnel, once conducted, should be examined by a second 'independent' person.

While it may be possible, in principle, for hazard identification and risk analysis to be carried out *relatively* free from socio-economic and other interests, the stage of risk evaluation reflects social values via criteria for acceptability, and will in general involve economic considerations. However, once the criteria for acceptability have been set down (as, for example, by the French and British governments in the case of the Channel Tunnel) then the interpretation and application of the criteria should be carried out in as independent a way as possible. The arguments above apply here also.

It is evident that the use of models as part of safety decision-making has a potentiality to lead to socially unacceptable design. The EU's Directive on tunnel safety (EU, 2004) says that risk analyses should be carried out by people who are 'functionally independent from the Tunnel Manager'. (See also Chapter 29.)

30.5. Acceptability of risk

While risk assessment should be carried out as 'independently' as possible by technically competent persons, it needs to be recognised that risk evaluation is essentially an ethical concern. As such, in essence, this means that deciding on criteria for acceptability should be carried out by the political representatives of the public. Such matters should not be effectively assigned to engineers as though they were purely technical considerations. Criteria for what is to be regarded as 'acceptable' need to be made clear and explicit.

Traditionally, there have been three key approaches to ethical decision-making.

1. The 'motivist' approach. This is associated with Immanuel Kant; whether or not an action is right or wrong depends upon the motive from which the act was done. This approach is often accepted in law: if a person kills another person deliberately, then that may be regarded as murder; if a person kills another person without intent to kill, then this may well not be regarded as murder.
2. The 'consequence' approach. According to this approach, whether or not an action is right or wrong depends entirely upon the subsequent effects of the action. One interpretation of this is associated with the Utilitarians in the 19th century (q.v. Jeremy Bentham and John Stuart Mill) who established the 'principle of utility'. This states that an action is right in so far as it tends to produce the greatest happiness for the greatest number. This has further been interpreted as: if an action produces an excess of beneficial effects over harmful ones, then it is right, otherwise not. The Utilitarian approach may be regarded as the genesis of cost–benefit analysis (CBA).
3. The 'deontological' approach. According to this, whether an action is right or wrong depends neither upon the motive from which it was done nor upon the consequences, but solely upon the kind of act it is. As an example, one of the Ten Commandments of the Old Testament of the Bible is 'Thou shalt not kill'. There is no conditionality regarding motive or consequence, or anything else. For example, it does not say 'Thou shalt not kill, unless you are being attacked'.

Different criteria have been put forward for acceptability of risk. Sometimes, criteria have been deterministic (specifying, for example, levels of temperature or gas concentrations) and sometimes non-deterministic (specifying, for example, frequencies or probabilities; e.g. see BSI, 2001; Beard and Cope, 2008). Non-deterministic criteria also include 'points' (see Chapter 18). The Norwegian government has adopted a 'vision zero' policy, whereby a long-term aim of no one being killed or permanently disabled as a result of road accidents has been established (Norwegian Public Roads Administration, 2006; Beard and Cope, 2008). Sweden and Denmark have also adopted 'vision zero' policies. It may be used as a guiding principle for gradual, continual, improvement. (See Beard and Cope (2008) and SAFE-T (2006) for more on acceptability of risk.)

30.5.1 Socio-economic, political and institutional pressures

Institutional pressure, often related to socio-economic or political influences, may be very strong. A dramatic example is afforded by the *Challenger* space shuttle disaster of 1986, where institutional pressure to go for a launch has been regarded as having won the day, despite evidence from engineers indicating a serious problem (Hayhurst, 2001). In the tunnel context, there may well be pressure to keep a tunnel operating normally, despite evidence that it should be closed or restrictions imposed. (See also Chapter 23.)

30.5.2 'Statistical death' and 'specific death'

As a general rule, decision-makers are unwilling to make decisions that will result in certain death for specific individuals, but they are willing to make decisions that effectively result in death for people whose identities are not known in advance. For example, in the UK, an average of approximately seven people are killed in road accidents each day. At the start of the day, it is known that about seven people will be dead by the end of the day, although their identities are not known. Such deaths may be thought of as 'statistical deaths'.

30.5.3 Secrecy

The withholding of information, i.e. secrecy, in relation to matters of public safety is not acceptable. Unfortunately, however, it is common. For example, many experimental tests are carried out by private companies, and full results are often not released into the public domain (see Chapters 1 and 8). Another example is seen in the case of the Channel Tunnel. Following the major fires that have occurred, it is intended to install a water mist fire protection system in the tunnel. The system will have a number of fire-fighting 'stations' (Wynne, 2010; FOGTEC, 2011; NCE, 2011). However, when the author of this chapter attempted to find out more details of the system from Eurotunnel, it was not made available on grounds of confidentiality with respect to subcontractors. The system, to the extent that information has been put into the public domain, has been the subject of critical comment, and 'encasing' the lattice rolling stock for heavy goods vehicles (HGVs) has been proposed as an alternative (Butcher, 2009, 2010). The question has been asked (Butcher, 2010; Wynne, 2010): if a fire is too large before a train reaches an extinguishing station, might the system be overwhelmed? Further, secrecy has been raised as an issue in an article that is critical of the Channel Tunnel Safety Authority (Eisner, 2000), especially in relation to its earlier decisions. This article also raises questions about the design of the tunnel system, including the semi-open HGV shuttle wagons. Yet another example of secrecy in relation to public safety information is seen in the fact that the computer source codes for models that may be used as part of fire safety decision-making are often kept secret, or a large amount of money may be charged for all or parts of a code – often with limited access restrictions. (See Chapter 29.)

30.5.4 A general duty of care: the Kaprun case

It has been argued (Beier, 2008) that in the case of the Kaprun funicular railway fire in Austria in 2000 the 'fundamental problem was that the designers complied with the regulations for a 'seilbahn' (cable car) ... no one thought about a fire nor did regulators ask anyone to think about it'. However, it is argued by Beier (2008) that there needs to be a general 'obligation to take care' and 'not merely comply with the law'. (See also Chapter 25.)

30.5.5 'Safe' systems and 'degree of safety'

To have a 'safe' system implies that the risk is zero, and this is impossible for real-world tunnel systems. The word 'safe' is similar to the word 'reliable', in that it is impossible to have a 'reliable system' in the sense that it would imply that the probability of failure is zero. In practical terms, it is necessary to think and speak in terms of 'degree of safety', just as it is necessary to think and talk about 'degree of reliability', rather than using the terms 'safe' and 'reliable' *per se*.

30.5.6 Prevention or protection?

It has been proposed that the control of pesticide risk within the EU be directly based on 'hazard' (BBC, 2008). As an example, it was stated that if a pesticide has been shown to be carcinogenic, then it should be banned. This may be seen as eliminating the 'crucial event', i.e. exposure to the pesticide, rather than controlling risk by controlling possible consequences (see Chapter 5). The EU's Directive on tunnel safety (EU, 2004) refers to achieving a 'minimum level of safety' by 'prevention of critical events'. The term 'critical events' here may be regarded as equivalent to 'crucial events'. It seems, therefore, that controlling risk by primarily controlling hazard directly has official approval in the EU. This appears to be effectively endorsing 'prevention' rather than 'protection' as a principal motivation. (However, adequate protection would always be necessary in a tunnel safety context, as elimination of all crucial events is impossible.) Dix has made the point that the Directive provides a minimal provision for safety but that this may not be enough to discharge the legal responsibility of engineers (Dix, 2004a, b; Beard and Cope, 2008).

30.6. Cost–benefit analysis (CBA)

CBA may be used to assess different options for changing a design on the drawing board or for an existing system. In essence, for each option the 'cost' of putting that option into effect is calculated. This generally means the financial cost of putting that option into place. One option may be taken as the status quo, i.e. doing nothing, in which case the cost would be zero.

The 'loss' corresponds to the 'loss' associated with each option. For example, if the loss of life were the only consideration, then the loss might be taken to correspond to the expected number of deaths per 'person-km' or 'per vehicle-km' associated with a particular option (e.g. see Beard and Cope, 2008). This depends on the quantitative models used. (In principle, a CBA could be carried out on a very approximate qualitative basis, without the need for explicit quantitative values to be estimated. However, this would not be suitable for many cases.)

A given measure, for example the total 'cost + loss' or the 'benefit/cost ratio' associated with each option would need to be calculated. Such measures can only be calculated if a common measure is used for the loss and the cost. The measure generally taken is money. This means that, effectively, a monetary value needs to be assigned to a human life or what the decision-maker is prepared to pay to prevent a fatality. The Health and Safety Executive in the UK, for example, refers to the 'value of preventing a fatality' (HSE, 2001). (The Health and Safety Executive also describes the ALARP ('as low as reasonably practicable') principle (HSE, 2001; see also Beard and Cope, 2008).) Putting a monetary value on human life is a controversial idea, and often provokes strong reactions. If a quantitative CBA considering life loss is to be carried out, however, then it needs to be done, explicitly or implicitly. Such considerations should always be explicit and clear.

Having arrived at values of a chosen measure for different options, then a so-called 'optimum' level may be found by taking, for example, the minimum of the 'cost + loss' or the maximum of the benefit/cost ratio. (In general, these do not necessarily correspond to the same option.) In pure CBA terms, then, the corresponding option would be the one to be taken. In reality, things may be different: the option corresponding to this may not be chosen for other reasons, for example, social or political. Also, in principle, the 'optimum' option may not be chosen because of lack of confidence about the reliability of the calculations or the logic of the procedure, for example, concerns about what has and has not been included in the 'cost' and 'loss' and how these have been calculated.

30.6.1 Problems with cost–benefit analysis

A delegate at a conference on risk once volunteered the point of view that the first thing one should ask about a CBA is 'Who paid for it?' The whole issue of CBA is a controversial one, and there has been much critical comment (e.g. see Adams, 1995). Mooney (1977) says he regards CBA as a 'decision-aiding' tool rather than accepting the results of CBA in a strict 'decision-taking' sense. Given this context, some of the problems with CBA are given below.

- The approach is essentially 'one-dimensional'. That is, all costs and benefits must be expressed in terms of one dimension; generally money. How are qualitatively different costs and losses to be assigned monetary values? (Multi-dimensional decision-making approaches exist which do not reduce everything to one dimension.)

- A full CBA should include *all* costs and *all* benefits. Many costs or benefits may be outwith the immediate situation of concern, for example, there may be environmental costs to a particular option, which may not be included. Also, we may not know the effects of a particular option for some time.
- Adopting a CBA approach to decision-making, as described above, means that the actual risk level that is associated with an option that may be chosen is not consciously decided upon. For example, at the maximum of the benefit/cost ratio the risk level may actually be relatively high. That is, the risk level comes out as a 'side-product' of the analysis. Of course, having calculated an option to be chosen, using CBA, one may then estimate the risk associated with that option, and reject it if the risk is unacceptable. However, this latter part is not intrinsic to the CBA principle itself.

30.7. 'Cost-effectiveness' approaches

Cost-effectiveness approaches may be distinguished from CBA, and two such methods are summarized here.

- The 'fixed-money' approach. In essence, with this approach, it has been decided to spend £X on reducing risk. Different options may be assessed, and a risk level estimated for each. The option that, for example, saved the greatest number of lives for the money spent would be the one to be chosen, in principle.
- The 'target' approach. In essence, with this approach, a target risk level is decided upon, and different options that are capable of reducing the risk to the target level are assessed. If there is no good reason to choose otherwise, then the cheapest option would be taken. In the real world there may be good reasons why the cheapest option is not to be chosen.

30.8. Methodology for tunnel fire safety decision-making

A key issue relates to knowing what methodology to adopt when applying a risk-based approach. Methodologies range from a very 'hard' methodology, in which there is overwhelming agreement amongst the 'actors' or 'participants' or 'stakeholders' as to what the problem is and what is desirable, through to 'soft systems' methodologies. In a purely 'hard' methodology, a considerable knowledge and understanding of the system is necessary, with very little uncertainty and no iteration in the decision-making process. The method proceeds from 'problem' to 'solution' in a mechanical orderly manner (e.g. see The Open University, 1984). While such an approach may be suitable for some situations, for example, putting in a simple telephone system, it is not suitable for tunnel fire safety. At the other end of the spectrum are the 'soft systems' methodologies, for example the one by Checkland (1981). The essential features of a soft systems approach are the existence of different points of view amongst the people involved and affected, and lack of reliable knowledge about the system. There will usually be considerable uncertainty, and there may be differences of opinions as to what the 'problem' actually is. Classic soft systems problems are those associated with directly 'social' issues and, to some degree, with systems associated with areas such as health care.

30.8.1 Intermediate methodologies

Between the 'hard' and 'soft' ends of the spectrum of methodologies are the intermediate methodologies. It may be posited that an intermediate methodology would be appropriate for decision-making with respect to tunnel fire safety. Whatever methodology is employed, it would be appropriate for it to have the capacity to allow the features identified in Box 30.1 to be applied.

Box 30.1 Appropriate general features for incorporation into a tunnel fire safety decision-making methodology

---

1. Encourage a user to be explicit and, in particular, make all assumptions clear.
2. Be explicit about scenarios that have been thought of but which have not been taken account of in the design (in the sense of protecting against).
3. Allow for the fact that there will almost certainly be scenarios that have not been thought of (Beard, 2002, 2004). This implies that one should err on the 'cautious side', to try to allow for such scenarios. Also, the system changes (see Chapter 5).
4. Include iteration as part of the process.
5. Incorporate explicitly a capacity to learn from previous serious incidents.
6. Incorporate explicitly a capacity to learn from minor incidents or 'near misses' (e.g. see Koornneef, 2000; Skapinker, 2001; Bodart *et al.*, 2004). These incidents represent a great source of real-world knowledge.
7. Learn from experiences of operational staff (of different tunnels), emergency workers and tunnel users. Also, allow people to be open about mistakes, without incrimination.
8. Bring 'buried research' to the surface. Research relevant to a particular issue may have been carried out, but it is often not generally known and is effectively 'buried' (Beard, 2004).
9. Allow for different kinds of models to be employed: physical, qualitative and quantitative.
10. Give due consideration to the 'precautionary principle'. This may be regarded as being applicable where it is not totally clear whether there is a significant 'threat' from some aspect (e.g. the introduction of a new material). The United Nations Conference on the Environment and Development in 1992 stated that 'lack of full scientific certainty shall not be used as a reason for postponing cost effective measures'. There are different interpretations of the precautionary principle, from 'weak' to 'strong'. The EU has published a document on this principle (EC, 2000; Tait, 2001). By contrast, the 'cautionary principle' may be seen as applying where there is certainly known to be a significant 'threat' from an aspect. This may be regarded as similar to applying a 'safety factor'.
11. Include safety management. A safety management system (SMS) needs to be designed as part of a system being assessed on the drawing board. Further, the SMS needs to be as 'systemic' as possible. After the start of operation, a risk assessment needs to be conducted on a continual basis *as part of* the SMS. That is, effectively, the situation becomes reversed. Incorporate in the SMS an effective, explicit, capacity to learn from the experiences of staff, emergency workers and tunnel users.
12. Incorporate a realisation that the tunnel system will change over time (e.g. changing traffic patterns). The SMS being devised as part of the total tunnel system must include a capacity to learn from changes, of whatever kind, in the system as a whole.

Iteration would be an important part of the process. When being used as part of design, a first pass through a methodology that is acceptable may be solely, or largely, qualitative. In the course of later iterations, quantification may be introduced. Iterations would continue until it is realised that no new information is being introduced and no new insights are coming about.

A methodology that is intermediate but lies towards the hard end of the spectrum has been outlined by Charters (1992). The basic stages are (1) hazard identification, (2) frequency analysis/consequence analysis, (3) risk assessment, (4) risk acceptable/unacceptable (iteration loop to stage 1 through 'risk reduction' if necessary) and (5) end. While it contains an iteration loop (one characteristic of an intermediate methodology), the degree to which it is hard or not depends upon how much time and effort is put into each of the stages, for example the stage aimed at deciding whether or not the risk implicit in an option is acceptable.

Other intermediate methodologies, partial or implied, have been put forward, for example the approach of the Health and Safety Executive (UK) (2001), that described by Hammer and Miller (1998), that of Worm and Hoeskma (1998) and that of the British Standards Institution (2001).

An additional intermediate methodology is that which emerged from the SAFE-T project (Khoury et al., 2008). The key stages are (1) plan a safe tunnel, (2) (re)define safety features, (3) safety analysis, (4) evaluation (iteration loop to stage 2 if necessary), (5) safe and cost-justified tunnel system, (6) educate users, (7) future changes (iteration loop to stage 2 if necessary) and (8) service life of tunnel. This methodology draws out the category 'educate users' as an explicit part of the approach, although no other specific measures are identified as part of the methodology *per se*. It may be said, though, that while consideration of the education of users is certainly an important part of any decision-making process, the behaviour of drivers and other users occurs within the context of an entire tunnel system. The behaviour of drivers is affected by the infrastructure and operation as well as by 'education'. The need to account for the education of users should not result in a failure to take full account of other aspects of the system.

Another intermediate methodology is that constructed by the author and referred to in Beard and Cope (2008) (Figure 30.1). This aims to provide a generic methodology for design, including risk assessment. In using such a methodology, it is essential to apply the features of Box 30.1 at each stage, as appropriate. It is intended that much more time be spent in the earlier stages than in many methodologies, and an iteration loop is included after every stage. Iteration is an important part of the process, and it is essential that it be conducted in a 'genuine sense' and not purely as a formality. Iteration forms a part of methodologies in other fields (e.g. see Nuclear Energy Agency, 2004). A tendency to 'rush ahead to the next stage' should be avoided, and plenty of time needs to be spent in the earlier stages. While all the features of Box 30.1 should be applied, particular mention may be made of learning from 'near misses'. 'Near misses' represent a very great source of information and knowledge about the behaviour of real-world systems, and we should tap this source much more than we do at the present time. Further, in general, there needs to be a willingness to learn from those 'on the spot' and from workers and staff involved in the entire system (including the emergency services) as well as tunnel users. As part of this, there should be a willingness to allow people to be open about mistakes, without incrimination. Also, once lessons have been learnt, they need to be acted upon effectively. For example, after the King's Cross (London) Underground station fire of 1987, a recommendation to resolve incompatibility between the radio system used by staff and that used by emergency services had, according to one report, still not been effectively acted upon by the time of the London bombings in 2005 (*Metro*, 2011). At Stage 4, the method of McQueen and Beard might be applied to help to consider an interim option (McQueen and Beard, 2004).

While the methodology of Figure 30.1 is intermediate, it leans more towards the softer end of the spectrum than does, for example, the methodology described by Charters.

Figure 30.1 A generic methodology for design, including risk assessment. In using such a methodology, the features of Box 30.1 should be applied, at each stage as appropriate

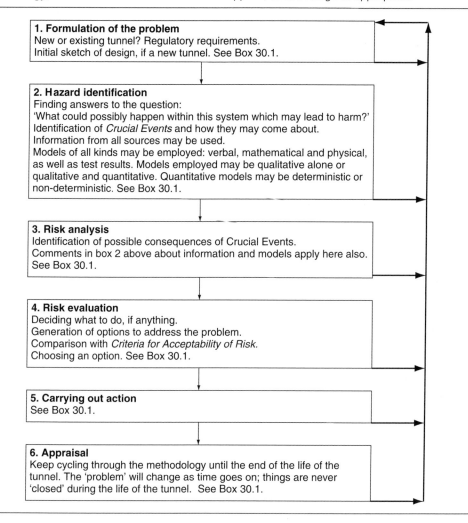

Having decided on an overall methodology, with a risk-based approach it becomes necessary to construct models in relation to tunnel fires, and the models constructed become ever more complex. There are fundamental problems associated with constructing and using models in a reliable and acceptable way. Every quantitative model makes conceptual assumptions, and these may be inadequate. There may be, for example, possible 'real-world' sequences that we simply do not know about and that, therefore, have not been considered in an analysis at all; this would be in addition to possibly unrealistic assumptions about sequences that have been included in an analysis. For example, a sequence involving a HGV on fire may be included in an analysis, but the assumptions about fire development and spread may be unrealistic. Considerations of this kind have been discussed further in Beard (2004). In addition to possible uncertainty or ignorance about conceptual assumptions, there is the problem of uncertainty about numerical assumptions. These difficulties mean that, even if a model has the potential to be valuable, acceptable use of a model is generally very problematic and requires a knowledgeable user employing an acceptable approach. As a general rule, the conditions do not yet exist for reliable and acceptable use of complex computer-based models as part of tunnel fire safety decision-making. These conditions need to be created. Given the complexity of the real world, it may be the case that some, perhaps many, models will be valuable in a qualitative sense only rather than quantitative. (See also Chapters 18, 25 and 29; Beard (2004, 2005); Beard and Cope (2008).)

REFERENCES

Adams J (1995) *Risk*. Routledge, London.

BBC (2008) Radio 4 news, 3 July 2008.

Beard AN (2002) We don't know what we don't know. *7th International Symposium on Fire Safety Science, Worcester, USA*, 16–21 June 2002. Interscience Communications, London, pp. 765–776.

Beard AN (2004) Risk assessment assumptions. *Civil Engineering and Environmental Systems* **21**: 19–31.

Beard AN (2005) Requirements for acceptable model use. *Fire Safety Journal* **40**: 477–484.

Beard AN and Cope D (2008) Assessment of the Safety of Tunnels. Report to the European Parliament. Available on the website of the European Parliament under the rubric of Science & Technology Options Assessment (STOA).

Beier B (2008) Principles of the Single European Market: integrating fire safety and law. *Fire and Building Safety in the Single European Market, Edinburgh*, 12 November 2008.

Bodart X, Marlair G and Carvel R (2004) Fire in road tunnels and life safety: lessons to be learnt from minor incidents. *10th International Conference on Fire Science and Engineering (INTERFLAM), Edinburgh*, 5–7 July 2004.

BSI (2001) *Application of Fire Safety Engineering Principles to the Design of Buildings – Code of Practice*. BS 7974:2001. British Standards Institution, London.

Butcher W (2009) Passive fire protection in tunnels – the costs of history repeating itself. *Industrial Fire Journal* **77**: 14–16.

Butcher W (2010) Fires in tunnels – how well do we learn the lesson of history? *Rail Technology Magazine*, June/July 2010.

Charters D (1992) Fire risk assessment of rail tunnels. *1st International Conference on Safety in Road and Rail Tunnels, Basel*, 23–25 November 1992. ITC, Tenbury Wells.

Checkland P (1981) *Systems Thinking, Systems Practice*. Wiley, Chichester.

Dix A (2004a) Risk management takes on a key role. *Tunnel Management International* **7**: 29–32.

Dix A (2004b) Safety standards for road and rail tunnels – a comparative analysis. In: *International Conference on Tunnel Safety and Ventilation, Graz*, 19–21 April 2004.

EC (2000) *Communication from the Commission: On the Precautionary Principle*. COM (2000). European Commission, Brussels.

Eisner HS (2000) The Channel Tunnel Safety Authority. *Safety Science* **36**: 1–18.

EU (2004) Directive 2004/54/EC of the European Parliament and of the Council of 29 April 2004 on minimum safety requirements for tunnels in the Trans-European Road Network. *Official Journal of the European Communities* L 167.

Eurotunnel (1994) *The Channel Tunnel: A Safety Case*. Eurotunnel, Folkestone.

Ferkl L and Dix A (2011) Risk analysis – from the Garden of Eden to its seven most deadly sins. *14th International Symposium on Aerodynamics and Ventilation of Tunnels, Dundee*, 11–13 May 2011. BHR Group, Cranfield.

FOGTEC (2011) Available at: http://www.fogtec-international.com (accessed 22 June 2011).

Hammer RW and Miller DW (1998) Railway safety management: building on the new Southern Railway experience. *3rd International Conference on Safety in Road and Rail Tunnels, Nice*, 9–11 March 1998. ITC, Tenbury Wells, pp. 747–760.

Hayhurst M (2001) I knew what was about to happen. Special report on space exploration. *The Guardian*, 23 January 2001.

HSE (2001) *Reducing Risks, Protecting People*. Health and Safety Executive, London.

Khoury G, van den Horn B, Molag M, Kalstrom H and Trijssenaar I (2008) *Global Approach to Tunnel Safety*. Report WP7.1 of Work Package 7: Harmonized European Guidelines for Tunnel Safety, 2006. European Union, Brussels. Available at: http://www.uptun.net (Work-package 5) (accessed 22 June 2011).

Koornneef F (2000) *Organized Learning from Small-Scale Incidents*. Delft University Press, Delft.

McQueen A and Beard A (2004) A Creative Approach to Assessing Proposals. *5th International Conference on Creative Thinking*, University of Malta, Msida, Malta, 21–22 June 2004. Published by the Malta University Press, 2007, under the title 'Creative Thinking: Designing Future Possibilities'. Edited by SM Dingli.

Metro (2011) 7/7 problems 'were pointed out in 1987'. *Metro* (UK newspaper), 9 February 2011.

Mooney GH (1977) *The Valuation of Human Life*. Macmillan Press, London.

NCE (2011) Eurotunnel trials anti-fire system. Online News, *New Civil Engineer*, 2 February 2011.

Norwegian Public Roads Administration (2006) *Vision, Strategy and Targets for Road Traffic Safety in Norway*. Norwegian Public Roads Administration, Oslo.

Nuclear Energy Agency (2004) *Stepwise Approach to Decision Making for Long-term Radioactive Waste Management*. Nuclear Energy Agency, OECD, Geneva.

Ota Y (2003) Japanese guide-lines for road tunnel safety. *2nd International Conference on Traffic and Safety in Road Tunnels, Hamburg*, 19–21 May 2003.

Rodriquez R *et al.* (2011) The New Pajares Railway Link, tunnel safety design: risk analysis, civil works and provided facilities. *3rd International Tunnel Safety Forum for Road and Rail, Nice*, 4–6 April 2011. TMI, Tenbury Wells, pp. 57–74.

SAFE-T (2006) Work package 5: Harmonised Risk Assessment. Framework Programme 5, funded by the European Union, 2006. Available at: http://www.mep.tno.nl/SAFET/Work-packages.

Sinclair U (1994) *I, Candidate for Governor: And How I got Licked*. University of California, Berkeley.

Skapinker M (2001) Deadly bureaucracy. *Financial Times*, 1 August 2001.

Tait J (2001) More Faust than Frankenstein: the European debate about the precautionary principle and risk regulation for genetically modified crops. *Journal of Risk Research* **4**: 175–189.

The Open University (1984) *The Hard Systems Approach*. The Open University, Milton Keynes.

Trbojevic M (2003) Development of risk criteria for a road tunnel. *5th International Conference on Safety in Road and Rail Tunnels, Marseille*, 6–10 October 2003. TMI, Tenbury Wells, pp. 159–168.

Worm EW and Hoeskma J (1998) The Westerschelde Tunnel: development and application of an integrated safety philosophy. *3rd International Conference on Safety in Road and Rail Tunnels, Nice*, 9–11 March 1998. ITC, Tenbury Wells, pp. 761–772.

Wynne A (2010) Eurotunnel unveils secret fire suppression system. *New Civil Engineer*, 25 November 2010.

Part VII

Specific topics

Handbook of Tunnel Fire Safety, 2nd edition
ISBN: 978-0-7277-4153-0

ICE Publishing: All rights reserved
doi: 10.1680/htfs.41530.651

Chapter 31
The UPTUN project: a brief summary

Kees Both
Efectis Nederland BV, Rijswijk, The Netherlands

(Author's note: it is impossible to summarise all the findings of the UPTUN project, and to give adequate credit to all people who have contributed to the project in a short chapter. Further information is freely available at http://www.uptun.net, where all reports can be downloaded.)

31.1. Introduction

The acronym UPTUN stands for *Cost-effective Sustainable and Innovative Upgrading Methods for Fire Safety in Existing Tunnels*. In response to the series of tunnel fire accidents around the turn of this century, the European Commission launched various calls for proposals as part of the Fifth and Sixth Framework Programmes, to strengthen European innovation and research.

A large number of stakeholders combined in the early 2000s, and seven large research projects, all focused on tunnel fire safety, were funded. The projects were complementary, and had several joint objectives. These overlaps were most evident in the two successful and well-attended international 'Safe and Reliable Tunnels' symposia in Prague (February 2004) and Lausanne (May 2006). The seven projects were as follows.

- *DARTS* (Durable and Reliable Tunnel Structures; http://www.dartsproject.net; 2001–2004). The initiative included eight European partners, and was focused on optimising tunnel design for long life and reduced cost.
- *FIT* (Fire in Tunnels; http://www.etnfit.net; 2001–2005). This 'thematic network' included 33 partners from 12 European countries. The project gathered information from all over Europe and around the world about existing research results and general experiences with regard to fire prevention and mitigation in transport tunnel facilities.
- *UPTUN* (http://www.uptun.net; 2002–2006). This project was funded by the European Community under the 'Competitive and Sustainable Growth' programme (1998–2002), contract G1RD-CT-2002-0766, project GRD1-2001-40739. UPTUN involved 42 partners from 18 European countries, and is described in this chapter.
- *Safe Tunnel* (Safety in Road Tunnels; http://www.crfproject-eu.org; 2001–2004). This project involved nine partners. The main focus was to reduce the extent and number of fire accidents in road tunnels.
- *SIRTAKI* (Safety Improvement in Road and Rail Tunnels using Advanced Information Technologies and Knowledge Intensive Decision Support Models; http://www.sirtakiproject.com; 2001–2004). The initiative was shared by 12 European partners. The main focus of the project was to reform operational concepts with regard to safety and emergency management.

- *Virtual Fires* (Virtual Real Time Emergency Simulator; http://www.virtualfires.org; 2001–2004). This involved eight partners from five European countries. The objective was to develop a suitable and practical simulator to train fire-fighters in confining and fighting fires in tunnels. A computer model was used to create virtual simulations of fires in tunnel situations.
- *Safe-T* (Safety in Tunnels; http://www.safetunnel.net; 2003–2007). This was a thematic network aimed at harmonising the European requirements regarding tunnel safety. Experiences gathered at the national level were compiled and assessed. Given the background of the negative experience with the former operative concept of the Mont Blanc Tunnel, special emphasis was on cross-border operative concepts. The experience of regional authorities, fire-fighters and emergency rescue services were of special importance for this project.
- *L-SURF* (Large Scale Underground Research Facility; http://www.L-Surf.org). This was a project to assess the need for and to develop an underground facility for testing tunnel safety systems. It resulted in the creation of the L-SURF Foundation.

31.2. UPTUN objectives and work programme

A large proportion of existing transport tunnels, built several decades ago, was designed on the basis of estimated growth rates for traffic. However, traffic volumes have grown at a faster rate than anticipated, and have also changed in composition (more combustible and flammable goods). Consequently, the safety level in existing tunnels throughout Europe has tended to decrease.

In addition to this, accidents in recent years have drawn widespread attention to the risks of fires in tunnels. This has had two major consequences. First, the fires themselves have resulted in fatalities, casualties, economic damage and, in several cases, lengthy shut-downs of the tunnels themselves. Second, the perceived risk of fire is also likely to have discouraged tunnel usage in some cases (tunnels might become an unwanted impediment for trade). Both of these consequences will have added to congestion, and hence to noise, particulate and air-borne pollution, with negative environmental and health implications. As a result of the accidents, and the media attention they have received, the public acceptance level of fires causing major losses has also decreased.

The main problem statement is as follows: unsafe, or supposedly unsafe, tunnels hamper the use and the development of sustainable transport systems, required for a healthy European economy. Upgrading the safety level in tunnels is, with existing technology and within the legislation and guideline frameworks of the member states, in most cases, either nearly impossible or too costly.

Moreover, fire safety has rarely been approached in an integrated fashion, considering all aspects (probability of incidents, consequences of fires, human response, structural response, emergency response teams and tunnel operators) in a similar manner. This may have resulted in an adverse interaction between preventive mitigating measures or non-optimal safety investments.

The main objectives of the UPTUN project were as follows.

- Development of innovative technologies for tunnel applications, with comparison to and assessment of existing technologies. The focus was on technologies in the areas of detection and monitoring, mitigating measures, influencing human response and protection against structural damage. The main output was a set of innovative, cost-effective technologies.

■ Development, demonstration and promotion of procedures for safety-level evaluation, including decision support models and knowledge transfer. The main output was a risk-based evaluating and upgrading model.

These objectives have been realised, as evidenced by the following.

■ New technologies that have been developed: water mist, water curtains, smoke compartmentation, CCTV detection techniques, training tools and programmes, repair mortars and software assessment tools, including local and regional cost-effective models.
■ Safety-level assessment criteria that have been drawn up and integrated in manual and automatic upgrade procedures/models. Knowledge transfer through dissemination at various levels: papers and presentations, international symposia (Prague, Lausanne), dedicated workshops throughout Europe and beyond (Australia, China, the USA), the establishment of a manual for good practice, and an UPTUN summer course.

The desired spin-offs of the UPTUN project were

■ the restoration of faith in tunnels as 'safe' parts of the transport system
■ the levelling out of trade barriers imposed by supposedly unsafe tunnels
■ an increased awareness by stakeholders of the need to develop initiatives to link all relevant research.

The work was divided into seven technical work packages (WPs), which are summarised in this chapter

    WP 1: Prevention, detection and monitoring
    WP 2: Fire development and mitigation measures
    WP 3: Human response
    WP 4: Fire effect and tunnel performance; system structural response
    WP 5: Evaluation of safety levels and upgrading of existing tunnels
    WP 6: Full-scale experimental proof – demonstration
    WP 7: Promotion, dissemination, education/training and socio-economic impact.

The first four WPs were to increase insight and to develop new measures to reduce the probability of, and to mitigate the consequences of, fires in tunnels. The fifth and sixth WPs were mainly focused on the development of an innovative integral upgrading approach. The last WP was to promote and disseminate the results and to address the socio-economic impact.

Furthermore, since not all aspects could be foreseen from the start of the project, nor could all problems be solved within UPTUN, strong links were established with relevant research projects at national and international levels (e.g. the European projects DARTS, FIT and Safe-T).

31.3. WP 1: Prevention, detection and monitoring

A wide range of detection and monitoring devices are currently installed in tunnels all over Europe, including fire detectors, smoke detectors, gas detectors and video-based detection systems. However, data on the reliability of existing detection and monitoring systems in tunnel applications are scarce.

Tunnel fire incidents generally start with a stopped vehicle (because of rear-end collision or breakdown) before the fire is detectable. New technologies aimed at incident detection should be significantly faster than conventional fire-detection systems.

Video detection provides (1) wide area detection, (2) direct incident detection and (3) real-time analysis of camera images, providing operators with visual information to enable rapid response.

Within the UPTUN project, a new fire detection image-processing algorithm for CCTV systems was developed. Future research should be focused on sensor or technology fusion where two technologies or functionalities are combined to result in an even more reliable system.

31.4. WP 2: Fire development and mitigation measures

The focus of this WP was on existing and innovative mitigating systems. In support of this objective, advanced fire simulations were used, and experimental work carried out where appropriate.

To provide suitable mitigation tools, the fire development and the fire hazards have to be fully explored and understood. Thus, this WP project documented this knowledge by

- carrying out data analysis of historical accidents
- evaluating the potential fire hazard from today's traffic
- using the most advanced mathematical fire simulation tools
- carrying out experiments where data were lacking, to increase knowledge about tunnel fires and to investigate tools to mitigate these fires.

The findings of WP 2 on fire hazards were implemented in a set of design scenarios, acceptance criteria and in the development of several fire mitigation systems. Engineering guidance with respect to the design of mitigation systems for tunnel fires, as well as the capability of advanced fire simulation tools, has been produced.

Mitigation measures, including current technology and innovative cost efficient tools, have been evaluated. All proposed tools have been verified as far as possible within the budget and time-scale of UPTUN. The effects of the mitigation of tunnel fires have been evaluated for different ventilation regimes in terms of corresponding changes in heat release rate, temperatures, toxicity and visibility.

To provide more knowledge about fire scenarios, several fire scenarios were defined based on normal cargo for heavy goods vehicles (HGV). These were tested in the Runehamar tunnel (Norway), and the resulting fires were the largest ever tested under experimental conditions. These results have been used around the world to reconsider design fires used for the fire safety design of tunnels. These tests are described in Chapter 1.

Criteria and methods to measure the quality of fire mitigation are vital for system optimisation and to achieve an acceptable level of safety. The acceptance criteria in WP 2 were derived from information about tenability, and the exposure of humans and fire-fighters in a hazardous fire environment, as well as from information on target criteria for constructions and mitigation.

31.4.1 Acceptance criteria
31.4.1.1 Escape from tunnels
It is concluded that criteria for asphyxiating gases must be based on doses rather than on concentration. An acceptable evacuation time must be the basis, and the doses must thus be calculated

from an appropriate model. Therefore, if the concept of fractional effective dose (FED) is used, the target criteria used in buildings should also apply in tunnels.

31.4.1.2 Fire brigades
Target criteria for fire brigade action in tunnels are poorly documented in the literature. There has been some basic work carried out in Sweden and in France, and proposals for target criteria are available. These requirements are prescriptive and not performance based. It is, however, not possible to apply a performance-based approach at the moment, since there are no models available.

31.4.1.3 Constructions
The verification of different functions in fire, such as the load-bearing or separating function, requires that the fire resistance is sufficient. Available design methods can be divided into different categories, of which the main categories can be described as design using either a deterministic or a probabilistic approach. Although the focus is generally on the deterministic approach and an equivalent level of safety, an attempt was made to describe and categorise all available methods in order to be able to discuss and evaluate their applicability for tunnels.

Statistics show that human injury is rarely caused by structural collapse. This may not be the case with fire service intervention and people involved in restoration work following a fire. A special hazard for fire-fighters is the risk of spalling of concrete and the total loss of tunnel integrity.

31.4.1.4 Mitigation
To be suitable for consideration for use in tunnels, a suppression system must be able to control fire growth within acceptable parameters and to prevent spread to adjacent vehicles. It should limit the growth of a fire by pre-wetting adjacent combustibles, controlling ceiling gas temperatures to prevent structural damage and providing the possibility of manual fire-fighting.

Investigations were performed into the mitigation performance and capacity of

- a water curtain system provided by APT Engineering, Italy
- a sprinkler system
- water mist systems using nozzles provided by VID, Denmark and Fogtec, Germany.

The results from these tests gave indications on the mitigation quality and performance of each system. All systems tested showed some mitigation capacity. However, each system has its own strengths and weaknesses. The performance and capabilities of each system have to be carefully assessed against each particular application.

An engineering guidance document was produced for the installation of water-spraying mitigation systems, given by the deliverable UPTUN WP2 D251, *Water Based Fire Safety Systems* (*Summary of Water Based Fire Safety Systems in Road Tunnels and Sub Surface Facilities*). This document presented general performance details (based on the testing carried out as part of the UPTUN project) and recommendations with respect to where such a system may be particularly valuable. (For further, and more detailed, information on the UPTUN testing, refer to UPTUN reports R230, *Evaluation of Current Mitigation Technologies in Existing Tunnels, Task 2.3: Final Report*, and D241, *New Innovative Technologies, Task 2.4: Final Report*.)

UPTUN has proposed design fires, related to fire growths, for different fire scenarios to be considered for tunnels.

31.5. WP 3: Human response

The objectives of WP 3 were as follows.

- To ascertain the current knowledge about human behaviour in tunnels (users, operators and emergency response teams): how do people respond to accidents and incidents in tunnels, what can be learned from real accidents, what studies have already investigated the response of operators to accidents or from rescue services, and what knowledge still needs to be gathered?
- To investigate the evacuation behaviour of tunnel users and the information that will lead to actual evacuations in order to decrease pre-movement time.
- To enhance way-finding of escaping tunnel users by using innovative means, taking into account human behaviour, smoke and temperature.
- To evaluate the operator task in the case of special events in order to arrive at a definition of critical elements in their tasks and suggestions for solving the bottle-necks.
- To evaluate the operation of crisis management teams dealing with tunnel calamities, with an emphasis on the overload of (correct or incorrect) information, the various communication channels and training aspects.

Car and truck simulators were used to observe and find ways to improve driver behaviour. The study also focused on the effects of the EU tunnel leaflet on a driver's perception of the situation in terms of risk and coping (for this, the subjects studied were truck drivers).

Important issues that were focused on in this study were as follows.

- How do truck drivers perceive, cope and behave in a tunnel fire incident?
- Is there a difference in driver behaviour and response to fire incidents between informed drivers (who have read the EU leaflet) and non-informed drivers?

Figure 31.1 shows the driving simulator that was used for this study and what people saw. While driving the simulated truck, participants were confronted with the scenario shown in Figure 31.1. The data for truck driver behaviour for different groups is shown in Figure 31.2.

Figure 31.1 (a) The truck driving simulator and (b) the fire accident that participants saw (http://www.uptun.net)

Figure 31.2 Behaviour of drivers in a fire incident for different groups (http://www.uptun.net)

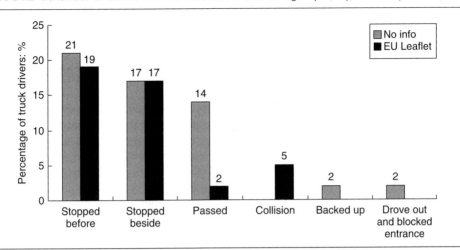

Various egress exercises were carried out using vehicles in a real tunnel. The drivers were invited to take part in a driving study, and were unexpectedly placed in an 'emergency' egress situation. By performing different evacuation tests, data for updating evacuation models have been produced. Various different way-finding and signage methods were also used in different tests. The idea was to test and develop systems that could help people evacuate under adverse conditions (e.g. low visibility). The research has concentrated on active techniques of way-finding and guidance through smoke. The system design emphasis, apart from its effectiveness in smoke, has been to ensure a high retro-fit potential.

31.6. WP 4: Fire effect and tunnel performance: system structural response

The main objective of WP 4 was to define, develop, assess and promote new methodologies and innovative materials/technologies to eliminate and/or at least reduce the risk of unacceptable behaviour of the structural systems of a tunnel.

WP 4 aimed at obtaining

- an assessment of the performance of the integral tunnel structure in all fire phases, from ignition, through growth, to the fully developed stage and the decay period
- the optimisation of the thermal and structural behaviour of all tunnel components designed for active and passive safety
- an increase in the load-bearing capacity under accidental conditions
- a reduction in and the limitation of non-operational time and repair retro-fitting work
- an evaluation of existing technology, with the main emphasis on cost–benefit (including maintenance)
- the establishment of methods for safer design and evaluation and recommendations for optimal tunnel systems.

The critical elements were identified in a two-step process

- first, a systematic investigation was carried out covering incidents in tunnels and specific tunnel fire tests

■ second, a brief in-depth analysis was made of these studies, in order to identify the most critical areas worth exploring further for improvements and innovations.

The new results of the WP included a check-up, comparison and 'validation' of all design tools and methodologies used by the partners for the design of tunnel structures and the prediction of their behaviour during fire, creating recommendations for numerical modelling.

Check-up, comparison and validation of all design tools and methodologies used by the partners for the design of tunnel structures and prediction of their behaviour during fire was obtained.

In addition to Eurocode EN 1992-1-2, there is a requirement for a data set for concrete characteristics during fire (variation in temperature). These new data were defined, checked and validated. Given that these data are available, a higher reliability in the estimation of fire perfor- mance should be achievable. The data are particularly important for thermal field analyses (thermal simulation tools seem to be quite reliable).

An experimental large-scale fire test was carried out on a suspended ceiling, which was identified as a critical structural component, based on a review of actual tunnel fires.

The study allowed the proposal of an optimised design of the ceiling. Recommendations for improving the design of the suspended ceiling were clearly defined. The tests themselves high- lighted the necessity of standards in experimental procedures.

In order to define a methodology for a low-cost quick refurbishment of a tunnel after fire, the specific objectives were to define, qualify and propose a quick, low-cost, in situ damage assessment method and to develop and qualify a repair shotcrete, proposing repair/recovery procedures. Particular effort was given to the definition, production application and verification of new concrete materials and to quick economic in situ damage-assessment methods.

For the assessment of the thermal damage of concrete, digital camera colorimetry and/or drilling resistance tests were used as innovative tools. Using digital camera colorimetry, the concrete colour after temperature loading could be assessed, and a clear relation between colour and temperature load level was found (pink or red, 300–600°C; whitish grey, 600–900°C; buff, 900–1000°C).

In the drilling resistance test, the measured resistance was compared with a reference from known virgin material. It was necessary to monitor the power, bit rotation and hole depth, in order to obtain a relation between resistance and damage (significant resistance changes occur above a temperature load of 400°C).

As an important innovative result, an anti-spalling concrete material was defined, developed and produced. New materials were developed for the resistance of fires with less damage and with less repair.

31.7.   WP 5: Evaluation of safety levels and upgrading of existing tunnels

The main objective was to determine and develop a pilot version of both a manual and an auto- matic upgrade (software-based) tool. In the tool, for illustration purposes, advanced, though limited, models were included (and further developed) with respect to smoke dispersion and evacuation. The key innovative element in the upgrading software was the incorporation of micro- and macro-socio-economic consequence modelling.

An example of the upgrade procedure is presented. Evacuation scenarios were calculated for four *safety upgrade options*.

1. No side exit and no detection/alarm: users can evacuate only by tunnel portals. Each user starts to evacuate following their own perception of tenability conditions (e.g. a drop in visibility conditions).
2. No side exit, but detection/alarm (3 minutes after ignition): users start evacuation at the same time due to an alarm signal.
3. Side exits and no detection/alarm: users can evacuate by tunnel portals and by safety exits integrated in the side of the tunnel every 200 m.
4. Side exits and detection/alarm.

Results show, for example, that, for the first upgrade option, 'self-evacuation from portals without detection/alarm or side exits', *there are nine fatalities* downstream of the fire. The results with side exits (third upgrade option) show that the number of fatalities is considerably reduced, to two. However, the second and fourth upgrade options, with detection and alarm, show that *nobody dies*. It is also interesting to highlight that people exit from the portals by preference, even when side exits are provided in the fourth upgrade option.

31.8.   WP 6: Full-scale experimental proof – demonstration

In WPs 1–4, the individual safety features were investigated, and in some cases tested singly, but not in combination. WP 5 brought together the innovative features in a theoretical evaluation and upgrading model, but its predictions required experimental verification.

The purpose of this WP was, therefore, to bring together the key innovative fire safety features developed in the previous WPs and to demonstrate their effectiveness experimentally in combination, both in terms of safety for people and of safety of the structure. This is best undertaken at full scale (both equipment and the tunnel) under realistic representative fire conditions.

The main objectives of WP 6 were

■ to demonstrate experimentally the effectiveness of the innovative fire safety features in combination
■ to provide feedback to WPs 1–5 in terms of the interaction of their individual features with the features developed in other WPs
■ to validate the theoretical model developed in WP 5
■ to make recommendations for upgrading based on real testing (nozzles provided by Fogtec, Germany).

The validation of safety features and monitoring devices allowed the correct installation of the systems identified and studied, and the upgrade of the Virgolo Tunnel in Italy for the fire simulation.

Full-scale experiments provide valuable knowledge for comprehensive and foresighted planning in many areas of tunnel safety. The demonstration was organised in the week between 14 and 18

February 2005 in different phases, in order to demonstrate the function and efficacy of the fire safety features, as well as the interactions of each system developed, both for road and for rail tunnels.

To achieve all the objectives, four fire demonstrations were organised. The fuel used in each case was diesel contained in stainless steel pools, each with a surface area of $2\,m^2$, and each able to release about 5 MW of energy. A specially designed compartment (a mini-tunnel) was built, on whose walls and roof concretes, mortars and anchorages were applied and exposed to extremely high temperatures. More than 300 temperature sensors were distributed, not only at different heights in the tunnel but also at different depths in the walls.

The systems and elements installed and demonstrated in the tunnel were 100 m of pipes for the water mist system for fire suppression; two water shields for smoke containment; two air plugs for smoke containment; six optical signalling systems; several stationary and mobile samplers for air quality evaluation during fire; 80 m of rail tracks; an $18\,m^3$ compartment (mini-tunnel) for material and anchorage tests; six innovative fire-resistant concretes and mortars applied to the tunnel walls; one fire-resistant paint applied to the tunnel walls; anchorage systems for illumination and ventilation systems; several fire detection and monitoring systems; five video cameras, including two stereoscopic cameras and one infrared camera; and hundreds of sensors.

Contributions to the real-scale fire demonstration were also offered by local partners (highway police, professional and voluntary fire services, an emergency unit, universities (ten institutes) and external partners. All experiments and results of the real-scale tests are given in the progress report *Test Report – Virgl/Virgolo Tunnel: Real Scale Tunnel Fire Tests* (182 pp., available in English, German and Italian).

The water mist system proved to be an efficient fire suppression technique. A second temperature increase after activation of the system was a result of the combination between burning oil and water.

The temperature separation effect (air/gas) before and after activating the water shields can be considered to be significant. With regard to the measured gradient before activating the mitigation system, the water shield gave an additional effect of 20–35°C, on average. The opacity measurement outside the separated fire section did not register any significant improvement or change in visibility. Many other conclusions were made: consult the full report for details.

31.9. WP 7: Promotion, dissemination, education/training and socio-economic impact

One of the key objectives of this WP was to promote the idea of a European tunnel safety board, comprising adequate representation of stakeholders and the balanced participation of (tunnel) member states, with operational links to ongoing research projects in the field, and supported by an independent organisation continuously providing relevant information obtained from studies of incidents and accidents. Such a tunnel safety board could interact with the European Commission as a discussion, monitoring and dissemination platform, prioritising, initiating and promoting (new) research, especially in the aftermath of incidents and accidents.

After a systematic investigation of different types of organisation, this initiative finally resulted in the foundation of the Committee on Operational Safety of Underground Facilities (COSUF) within the International Tunnelling Association (ITA).

31.10. Conclusions

UPTUN was a huge project, comprising seven WPs, the average size of each being equivalent to a Fifth Framework research project. All deliverables have been produced, and all milestones met.

Mutually beneficial linkages have been formed between the different WPs and with other tunnel safety projects within the Fifth and Sixth Framework Programmes and beyond, thus creating higher efficiencies and positive synergies. Given its size, its scope of work, the R&D and its innovative features, UPTUN has become an essential structural element in the Fifth Framework Programme, and is helpful to the development of Sixth Framework Programme Integrated Projects and also to the EU Directive. UPTUN members have taken initiatives beyond the call of duty, thus bringing into UPTUN additional resources and technologies and many unscheduled products (e.g. papers, reports, experiments, innovations), indicating that UPTUN members are proactive and enthusiastic.

UPTUN has created the platforms and tools for successful development and, above all, the opportunities and tools for the much needed cost-effective upgrading of existing tunnels.

31.11. Recommendations for further research

Continuation of some specific research items is highly recommended in the following technical areas

- detection (moving, developing/growing fires, etc.)
- upgrading, specifically dedicated to rail and metro tunnels
- smoke compartmentation
- incident investigation and evaluation (also near misses)
- (test protocols for) water-based suppression systems
- interaction between (and optimisation of) measures
- fire scenarios for new vehicles and rolling stock
- extensive development of the 'upgrading tool', providing owners/designers with a strategic tool to select appropriate upgrade options, and also allowing them to take account of local and more global economic effects of (not) taking measures
- establishing and/or connecting large-/full-scale test and demonstration sites (L-SURF).

Acknowledgements

Sincere thanks to all my colleagues in the UPTUN project, and to those in the European Commission overseeing and giving feedback. Special thanks to the Steering Board members and Work Package and Task Leaders, as well as to the Advisory Board, for their valuable comments and supportive criticism.

Handbook of Tunnel Fire Safety, 2nd edition
ISBN: 978-0-7277-4153-0

ICE Publishing: All rights reserved
doi: 10.1680/htfs.41530.663

Chapter 32
The River Tyne road tunnels and fixed fire suppression

Peter Hedley
Tyne Tunnels, Wallsend, UK

32.1.   Introduction

This chapter describes work regarding the Tyne Tunnels to provide an additional tube and refurbishment of the 1960s road tunnel. It includes a brief account of the first installation in a road tunnel in the UK of a fixed fire suppression system (FFSS). The chapter is not intended to be an exhaustive illustration, nor is it intended to be academic work. However, what follows is, hopefully, an example of good practice that contributes to the debate which can arise during road tunnel construction and/or refurbishment that raises the question: 'Should fire protection systems be active, passive or a mixture of both?'

Note: for the purpose of this chapter, active fire protection is defined as an installed fire protection system that detects and/or suppresses fires. Typical systems include fire sprinklers that are either automatically or manually operated, and fire detection systems such as smoke detectors or systems that detect smoke, heat or radiated energy.

Passive fire protection is defined here as an integral component of structural fire protection and fire safety in a building. Passive fire protection attempts to contain fires or slow the spread, through the use of fire-resistant materials that protect the structure and thus the integrity of the building.

The Tyne Tunnel facilities are located in the north-east corner of the UK, near to the city of Newcastle-upon-Tyne.

32.2.   Organising the change to the system

The original facilities, before November 2007, consisted of three tunnels crossing the River Tyne between Wallsend and Jarrow.

- Pedestrian and cyclist tunnels – two tunnels, with the pedestrian tunnel being 3.3 m in diameter and 274.5 m long and the cyclist tunnel being 3.7 m in diameter, and also 274.5 m long. The tunnels were built to serve the 'industrial communities' along the River Tyne and to feed the heavy industries of the time. Construction occurred between 1946 and the opening in 1951.
- The Tyne vehicle tunnel – a single-tube two-lane bi-directional vehicle tunnel, 1650 m long with a 5% gradient, passing in the region of 38 000 vehicles every 24 hours. The traffic profile is 90% car and light van, 8% large goods vehicles (above 3500 kg in weight), 0.8%

motorcycles and 1.2% disabled motorists, whose cars are exempt from vehicle excise licensing requirements because they receive the highest mobility allowance. Furthermore, 90% of the customers are regular local users. The Tyne vehicle tunnel was constructed between 1961 and its opening in 1967.

The New Tyne Crossing (NTC) project is promoted by the Tyne and Wear Integrated Transport Authority – the 'grantor'. The project is to build an additional 'immersed tube' vehicle tunnel, and, once this new tunnel is open and operational, to refurbish the 1967 existing road tunnel. In addition to the construction and refurbishment works, the project includes a 30-year concession requirement to run the four tunnels (two road tunnels, one pedestrian tunnel and one cyclist tunnel). The construction company is Bouygues Travaux Publics, and the concession-aire is TT2 (Tyne Tunnel 2). The key dates for the project are as follows.

- PFI feasibility study: 1996
- application under the Transport and Works Act: 2005
- public inquiry: February 2006 (6 weeks' duration)
- Secretary of State's decision: 2007
- preferred bidder: 12 December 2007
- financial close: 23 November 2007
- transfer of tunnels and personnel to concessionaire: 1 February 2008
- new tunnel expected to open: 2011
- older tunnel expected to be upgraded: by the end of 2011.

It is intended that the new tunnel will be operated as a bi-directional tunnel until the refurbished tunnel opens. The UK design standard for road tunnels (Highways Agency, 1999) requires that a body is set up called the Tunnel Design and Safety Consultation Group (TDSCG). This group brings together all the stakeholders for the road tunnel and allows coordinated input and evaluation of structure, infra-structure, plant, systems, equipment and procedures for the construction of the new tunnel or the refurbishment of an existing tunnel. The stakeholders on the NTC TDSCG are listed below.

- The grantor: the Tyne and Wear Integrated Transport Authority (which was originally the Tyne and Wear Passenger Transport Authority)
- The concessionaire (chair of the TDSCG): TT2
- The construction contractor: Bouygues Travaux Publics
- The technical approval authority: the UK Highways Agency
- The Emergency Services:
  - Tyne and Wear Fire and Rescue Service
  - Northumbria Police
  - Northumbria Ambulance Service
  - Tyne and Wear Emergency Planning Unit
- the Health and Safety Executive
- the NTC construction design and management (CDM) coordinator
- construction company designers
- the Port of Tyne
- the UK Environment Agency.

The NTC TDSCG was initiated in the year 2000 by the grantor's professional advisor: the consultancy firm ARUP. During the project's planning phase, ARUP had not only prepared a 'reference design' in conjunction with other specialist consultants (such as Royal Haskoning), they had, in addition, produced an 'outline fire safety strategy'.

The terms of reference for the TDSCG are as outlined below.

1. To review the basic configurations of the tunnels and their service buildings and the functionality of plant and equipment to be installed at the tunnels, their service buildings and remote monitoring and control sites as submitted by the preferred bidder (the contractor), in advance of the start of the detailed design phase of the scheme.
2. To establish agreed written procedures for the safe operation in all practical configurations and for maintenance of the tunnels, approach roads and service buildings. These written procedures will be embodied in the TDSCG reports for both the new and refurbished existing vehicle tunnels, and shall clearly establish the duties and responsibilities of all parties involved.
3. To arrange and carry out staged emergency incidents, involving a full-scale fire test, as defined in Appendix D of the UK design standard for road tunnels (Highways Agency, 1999), just prior to the opening of the tunnels to traffic.
4. To review the written procedures and the TDSCG report in light of the experience gained from the staged emergency exercises, and to also agree modifications to the written procedures where necessary.

32.3. Deciding on a new system

The considerations of the TDSCG from 2000 to August 2010 often centred around dealing with fires within the tunnel bore. Since the opening of the vehicle tunnel in October 1967, the emergency response at the Tyne tunnels has been delivered via an in-house 'first-response' fire-fighting/safety force, using Land Rovers equipped with aqueous film-forming foam (AFFF) as their fire-fighting medium. The operational/emergency philosophy contained the following.

■ Each Land Rover rapid response vehicle (RRV) has two people.
■ 14 seconds after stopping, the staff and the RRV could produce 6% AFFF for a period of about 12 minutes.
■ The staff deployed on the RRV benefited from:
  – a 2-year training plan with six core skills (breathing apparatus, fire-fighting, security duties, first aid, vehicle breakdown recovery and vehicle examination)
  – two hot fire drills per year
  – one full operational exercise every 4 years
  – one table top exercise per year.
■ The in-house staff also had two heavy recovery vehicles that could lift 13.5 tonnes and tow 200 tonnes.
■ An operational shift had a minimum of two first-response teams, with further support available from their colleagues on shift.

However, during the TDSCG debate regarding the existing Tyne Tunnel emergency response and the possibilities for a future fire-fighting response in the NTC and the refurbished existing Tyne Tunnel, consensus could not be achieved amongst all stakeholders. In short, the Fire Brigade was highly in favour of fitting a FFSS in both the new and the existing road tunnels, whilst the existing Tyne Tunnel Operations believed that their in-house emergency response was the best answer, and they were in fact regarded as equivalent to an 'intelligent sprinkler system'. To achieve consensus, the chair of the TDSCG tabled a proposal to resolve the matter: this was for all protagonists to agree to an independent cost–benefit analysis – with regard to the installation of a FFSS for the new (southbound) tunnel and also for the refurbished (northbound) tunnel. This cost–benefit analysis took the form of the 'Fire Suppression Requirement Appraisal – Independent Experts Report', and it was carried out by Dr Fathi Tarada of the consultants Halcrow. His report was issued to the TDSCG in October 2007 (Tarada, 2007; see also Chan and Tarada, 2009).

In the cost–benefit analysis report, Tarada stated that since the values of the risk parameters were themselves uncertain, Monte Carlo simulations had been undertaken using assumed ranges of the relevant parameters. The result of the simulations included the likely benefits of the fire suppression system including reductions in injuries and fatalities, minimising traffic delays, reduced attendance costs by the emergency services and reductions in infrastructure repair costs. These benefits were then set against the capital and maintenance costs of the fire suppression systems.

The findings indicated a 62% probability that the installation of a fire suppression system would generate a positive return to society, with the majority of the benefit (71%) accruing from reductions in traffic delays. The calculated mean benefit-to-cost ratio was 1.27, with a standard deviation of 0.58, based on a mean whole-life cost for a fire suppression system of £9.7 million (over 60 years).

Therefore, Tarada completed his summary by stating: 'Considering the net societal benefits calculated from the quantified risk assessment presented in the report, the consultancy (Halcrow) recommended that a fixed fire suppression system is installed in the New Tyne Crossing' (Tarada, 2007). Tarada acknowledged that: 'This recommendation is subject to review by the stakeholders in the Tunnel Design and Safety Consultation Group, acceptance by the Tyne and Wear Transport Authority' (the body that is now the Tyne and Wear Integrated Transport Authority), 'and approval by the Highways Agency, acting as the Technical Approval Authority'.

In line with the prior agreement, all parties accepted the recommendation to install a FFSS in the new southbound tunnel and the refurbished northbound tunnel. This was a monumental decision for the UK road tunnel operating industry, as the Tyne Tunnels would be the first road tunnel in the UK to benefit from the installation of an FFSS.

The operator's considerations, most of which had been input into the cost–benefit analysis, raise interesting points.

32.4. Comparison of a FFSS with an in-house First-Response approach

The emergency services assessed the installation of a deluge water mist system as being the most important of all options, as they believed the equipment would stop any fire growing, stop fire spreading by keeping adjacent vehicles cool and prevent an escalation of the incident until fire-fighters could approach from the safety of a positive-pressure evacuation corridor that enjoyed 2-hour fire protection. From an operator's viewpoint, TT2 was also comfortable with this due to the FFSS equipment allowing the company to reduce the fire hazard to people by

■ reducing the fire hazard by cooling the source of ignition

- automating the fire-fighting response, thereby removing people, as far as is reasonable and practicable, away from the hazard – this includes both the in-house first responders and the emergency services
- providing a tenable environment within which motorists can be evacuated.

Both the operator, TT2, and the emergency services, Tyne and Wear Fire and Rescue Service, acknowledge that although some, perhaps many, vehicles are becoming safer, this is not necessarily the case with all. Concern has been expressed about, for example, larger fuel tanks being used in HGVs (Beard and Cope, 2008). There is also concern about the possible passage of hydrogen-powered vehicles through tunnels (Beard and Cope, 2008). The fire hazard arising from vehicles cannot be removed from the road tunnel, and the most effective step to ameliorate the hazard is to mechanise the response with an effective control measure. Whilst it was also acknowledged by the operator and the fire brigade that the deluge water mist (sprinkler system) was primarily to protect the structure and the infrastructure, the massive benefit of 'people protection' and being a 'tool to allow evacuation safely from vehicle to escape corridor' cannot be diminished. It is intended that the FFSS replaces the 'first-response team' approach described above. Tyne and Wear Fire and Rescue Service practices its response relating to the Tyne Tunnels regularly, and quotes an average response time of 8 minutes, with an agreed response of two fire appliances to either end of the road tunnel.

32.5. The FFSS selected
The *Fogtec (2010) General Description, High Pressure Water Mist Fire Suppression System* has been developed for road tunnel applications, and uses pure water as a fire-fighting medium (see also Chapters 5 and 8). Specially designed nozzles atomise the water, at a minimum pressure of 50 bar, into very fine droplets of mist with high kinetic energy. The mist floats in the air like a gas. The immense surface area of water produced in this way means that better use can be made of the cooling effect of the water; in addition, the water is able to vaporise quickly. During vaporisation, the volume of the water is increased, displacing some of the oxygen at the fire source. These two effects combine to control the fire.

The cooling effect of the water brings about a rapid drop in temperature to a lower level, to enable people to escape or carry out further fire-fighting measures. At the same time, radiated heat to objects in the vicinity is reduced and the chance of fire spreading is lessened.

The very small amount of water usage leads to minimal water damage, and considerably less water needs to be stored in comparison with conventional sprinkler/deluge systems.

Since the FFSS uses no chemicals, the environment is not damaged and there is no health risk to people present during the discharge of the system.

Comment should also be made that with regards to the installation of a FFSS, it makes a great deal of sense to install the system whilst one tube is being built and the other is under a massive 'strip out' and refurbishment. Because the site remains operational, the 'strip out' and refurbishment of the existing tube will not occur until the new tube has been completed and has become operational, allowing the existing tunnel to be closed.

32.6. The safety culture underpinning the operating philosophy
The safety culture that underpins the operating philosophy within which the FFSS sits is very important. Quite simply, the safety culture that exists at the Tyne Tunnels is 'early detection and early intervention'. The following factors impinge on this.

- As an operator, TT2 recognises that free-flowing traffic passing through a road tunnel without let or hindrance is the safest operating condition.
- The Tyne Tunnel has an onsite control room that is open and staffed 24 hours a day, 365 days per year. The control room staff are highly trained, and cannot staff this area alone unless they have had some $3\frac{1}{4}$ years' training.
- The organisation has an on-site first-response section that is equipped with RRVs and is trained in six core skills (fire-fighting, breathing apparatus, vehicle examination, vehicle recovery, security and 'blue light' driving).
- TT2's concession agreement sets performance targets and has penalties should the targets not be met. Included in these performance targets are:
  - the concessionaire shall respond to an emergency in the tunnel within 2 minutes
  - removal of a broken-down car from the tunnel within 10 minutes
  - the minimum level of emergency equipment and or emergency systems shall be operational 100% of the time.
- The CCTV system has automatic vehicle incident detection (AVID) installed, complete with smoke algorithms. This, when linked to the 24/7 staffed control room, with a duty controller monitoring the site and mobile patrols touring the site, satisfies the 'early detection' requirement.
- The fire safety system is controlled via a master control room, and an indication panel is sited in the control room. A communication link is provided that allows data displayed on the master control and indication panel to be repeated on the supervisory control and data acquisition (SCADA) system display.

Upon activation of a fire system alarm (smoke), the Tyne Tunnel operator is alerted by an audible alert signal and a flashing light that is provided via the SCADA interface (see also Chapter 5). On confirmation of the incident by the duty controller, the system will automatically select pre-set modes for ventilation, traffic management and for the public address system (which will broadcast evacuation instructions). Also, on confirmation by the controller, the water mist should commence immediately.

The duty controller is able to change the pre-set ventilation mode or delay activation if necessary. The operator can adjust, i.e. switch, the zones to be activated; each fire zone is approximately 25 m long. Also, the operator can cancel the activation when, for example, the incident turns out to be a 'blown injector' or 'stuck exhaust brake', etc. – i.e. lots of smoke but no 'fire'. Initially, the fire suppression zones applicable to the incident area will be displayed and made ready for activation. Should confirmation by the operator not occur, a 'dead-man's' safety function will then engage and trigger the appropriate FFSS response.

32.7. Use of the FFSS and ventilation at the same time

It is important to control smoke, and that implies having an 'emergency ventilation velocity' that is able to do this. There will be a longitudinal ventilation system only in the tunnels. Importantly, operators understand the difficulty of evacuating people from the comfort zone of their own motor-cars, etc. In reality, one does not get 100% evacuation, which is why achieving the emergency ventilation velocity is so important: this velocity prevents back-layering of the smoke, thus maintaining a tenable environment for the motorists trapped behind the incident and, possibly, refusing to leave their vehicles. At this point, a question must be raised: 'How could an operator of a road tunnel, with a FFSS installed and operational, not operate that FFSS to protect those drivers unwilling or, indeed, unable to evacuate the road tunnel by means of the evacuation corridor?' TT2's answer to this important question is: 'As an operator with a FFSS we could not defend not using the FFSS, and therefore we would use it at the earliest possible opportunity'. The TT2 position has been comprehensively debated both within and outside the TDSCG, and all stakeholders are comfortable with this. The FFSS manufacturer (and installer) has tested the system at the Tyne Tunnel's critical ventilation velocity – the FFSS worked well and no negative factors were detected. This did not surprise the Tyne Tunnels manager, as he had previously had discussions with another FFSS manufacturer that had carried out similar water mist tests under ventilation conditions and had achieved successful mist operation with ventilation velocities over 4 m/s, and the only observed effect was a 'concave' bowing on the leading edge of the fire zone.

32.8. The positive return to society

Tarada (2007) in his cost–benefit analysis alluded to the need to protect the transport asset to ensure it remains available on the Tyne and Wear Strategic Highway Network. TT2 as an operator understood this point, as its past experience showed that a 30-minute complex HGV breakdown at 07:30 hours (during the morning peak traffic time) resulted in a loss of transit of about 2500 vehicles. However, this loss of vehicle transits was not the whole story: when the local highway network was analysed, at the time of breakdown about 6500 vehicles were standing in the congestion area, and, if about 2500 vehicle transits had been lost to this crossing, it appears that the vehicles did not use the other crossings of the River Tyne; instead, they dissipated. This dissipating traffic could be viewed as people returning home and not attending work (injurious to the regional economy). Furthermore, should the tunnel be lost to the transport network, 40 000 vehicles per day would be forced to use the other five crossings of the Tyne, all of which are at capacity, causing massive congestion throughout Tyne and Wear. Goodwin (2005) has assessed the cost to the UK per year arising from the congestion caused by the work of utility companies (water, gas, electricity, etc.) as £4.3 billion. Other commentators put the total annual cost for traffic congestion in the UK at £20 billion. Therefore, societal and economic risk relating to the loss of a transport asset is important.

Whilst protecting the structure and infrastructure has a positive return to society, and it was the motivation in the 2007 cost–benefit analysis for the installation of a FFSS within the new and the refurbished tunnels, the 'spin off' of superior protection of people and reduction of life risk cannot be trivialised or dismissed. Furthermore, the British Chamber of Commerce estimates that a death arising from an accident costs the British economy as a whole £1.5 million. Thus, stakeholders could see the benefits arising from the installation of a FFSS in the Tyne Tunnels despite an estimated whole-life cost over 60 years of £9.7 million.

32.9. Observations on decisions regarding the New Tyne Crossing

The following observations may be of interest.

- When the decision was made to install an active fire protection system in both the new southbound tunnel and the refurbished existing tunnel, the specification for the passive fire protection was not reduced in any way. Therefore, the increase in safety was many-fold: from protecting people to maintaining the availability of the highway network by ensuring a fire incident will not cause prolonged closure of the tunnel.
- During discussions within the TDSCG, the Tyne and Wear Fire and Rescue Service was consistent and insistent that the only fire safety solution acceptable to it was a solution that included the installation of a FFSS.

- As an operator, TT2 believes that it must use the FFSS at the earliest opportunity to protect life and to protect the asset. This has a good strategic fit with the organisation's culture, which demands 'early detection, early intervention'. Furthermore, this clear position with regard to the operation of the FFSS enjoys complete support from all stakeholders within the TDSCG.
- TT2 accepts that a FFSS will cool smoke and prevent stratification of the smoke during fire incidents; however, TT2 is also confident that the FFSS can be used at the same time as the ventilation system is delivering an 'emergency ventilation velocity' (the air speed that is recognised to prevent back-layering of the smoke). Therefore, it is a responsible, considered response to use the FFSS at the same time as the longitudinal ventilation system.

REFERENCES

Beard AN and Cope D (2008) *Assessment of the Safety of Tunnels*. Report to the European Parliament. Available on the website of the European Parliament, under the rubric of Science & Technology Options Assessment (STOA).

Chan E and Tarada F (2009) Crossing points. *Fire Risk Management* March: 34–37.

Goodwin P (2005) *Utilities Street Work and the Cost of Traffic Congestion*. University of the West of England, Bristol.

Highways Agency (1999) *Design Manual for Roads and Bridges*, vol. 2. *Highways Structures Design (Sub-structures and Special Structures)*, section 2. *Special Structures*, part 9, BD78/99. *Design of Road Tunnels*. Stationery Office, London.

Tarada F (2007) Fire Suppression Requirement Appraisal – Independent Experts Report. Unpublished.

Handbook of Tunnel Fire Safety, 2nd edition
ISBN: 978-0-7277-4153-0

ICE Publishing: All rights reserved
doi: 10.1680/htfs.41530.671

Chapter 33
Hydrogen-powered cars and tunnel safety

Yajue Wu
University of Sheffield, UK

33.1. Introduction

Hydrogen is a very clean fuel and ideal for transport systems, either for direct use in internal combustion (IC) engines or for fuel cells to power electrically driven vehicles. Combustion of hydrogen can produce very low emissions of nitrogen oxide together with some water vapour. In fuel cells, it only produces water vapour and, so far as is known, there are no fuel-related health effects; therefore, hydrogen cars have the potential to bring significant benefits to the environment. Under normal, i.e. non-emergency, conditions, hydrogen cars may be considered as very clean vehicles in road tunnels, and would be expected to bring benefits for air quality control in tunnels. From the risk assessment point of view, however, since hydrogen poses a high risk of fire and explosion, the impact of hydrogen cars on tunnel safety is a complex issue. Although hydrogen cars might be considered as relatively safe in the open air (i.e. relative to other vehicles), there is certainly a concern regarding the safety of hydrogen cars in a confined space with restricted ventilation, such as underground parking or road tunnels.

This chapter summarises an initial assessment of the characteristics of fuel release and fire hazards from hydrogen-powered cars in tunnels, and possible mitigation methods. To simplify the hazard analysis, a hydrogen car is being treated as having two components: the car body and the fuel on board. It is reasonable to assume a hydrogen car body poses similar hazards to those of any other conventional car of similar size. Since hydrogen disperses in gaseous form, and most conventional transport fuels are liquid, hydrogen-gas-related hazards become a unique feature for hydrogen car risk assessment. In this chapter, the discussion is only about hydrogen-fuel-related hazards in tunnels; with particular regard to cars.

33.2. Fuel release scenarios and fuel release rate

The characteristics of the fuel release are largely determined by the method of hydrogen fuel storage on board. The most common methods in use are hydrogen compressed gas for fuel cell vehicles and liquefied hydrogen for IC engines. Hydrogen has a very light molecule, and to make storage of hydrogen in gaseous form efficient, hydrogen gas needs to be compressed to a very high pressure in order to reduce storage size. Currently, hydrogen stored on board a vehicle is mainly in high-pressure compressed gas form, with pressures of 250, 350 and 450 bar. The common storage capacity of hydrogen on board a vehicle is approximately 3–5 kg: this could be higher if high storage pressure is used.

Under normal operational conditions, hydrogen fuel should be very well contained, and the tunnel safety risk might be expected to be the same as for any conventional car. However, for

671

risk assessment, it is necessary to consider possible accident scenarios, which are likely to result when vehicles are involved in extreme situations and result in loss of fuel. One of the worst fuel release scenarios from such a high-pressure 'bottle' is a high-velocity jet release. The gas velocity could easily reach from 200 m/s up to sonic velocity at 1294 m/s for hydrogen. A few hydrogen jet release demonstration tests have been conducted, and the results showed that it took only a few minutes to release the whole contents of hydrogen fuel. Controlled hydrogen fuel release tests showed that the 5 kg of fuel was emptied in 2.5 minutes at a mass loss rate of 0.033 kg/s. If the fuel calorific value is used to assist in describing the hazard posed by the fuel release, this would be expected to result in a 4 MW hydrogen fire if the hydrogen jet were to be ignited. It is reasonable to expect that hydrogen fuel release from a failure of valves in a storage system would result in a hydrogen release rate of about 0.03–0.05 kg/s, being equivalent to about a 4–6 MW heat release rate if the fuel jet were ignited. The heat release rate could be higher if a catastrophic failure mode in the storage tank occurred and resulted in loss of fuel in a very short period.

33.3.   The hazards of hydrogen release in tunnels

Hydrogen has quite distinctive physical and chemical kinetic properties. It has a very wide flammability range and a very high burning velocity. At atmospheric pressure and ambient temperature, the flammability limits for a hydrogen–air mixture are 4.1–74.8%. The burning velocity varies from 270 to 350 cm/s. The range for hydrogen detonation in a hydrogen–air mixture is 18.3–59%. The flame temperature for 19.6 vol% in air is 2321 K. The range of reported auto-ignition temperatures for a stoichiometric hydrogen–air mixture is 773–850 K. The minimum spark energy for the ignition of hydrogen in air is 0.017 mJ at atmospheric pressure. The minimum spark energy required for the ignition of hydrogen in air is considerably less than that for methane (0.29 mJ) or petrol (0.24 mJ). The difference between the hydrocarbon fuels and hydrogen clearly indicates that hydrogen can be ignited much more easily. The low ignition energy indicates that a hydrogen jet has a very high probability of ignition, resulting in a hydrogen jet flame.

A demonstration of a hydrogen car fire (Swain, 2001) and a petrol car in the open air showed that the hydrogen release generated a vertical high velocity jet flame with a flame length of approximately 5 m. Although the temperature of a hydrogen flame is high, the flame has relatively low thermal radiation due to the low emissivity. During a release in the open air, the leak from a petrol car formed a pool fire, the flame engulfed the body of the car and substantial smoke was generated from the burning car. In contrast to the petrol car, the body of the hydrogen car remained clear of the fire, and the fire remained as a jet flame throughout the test. The test led to the conclusion that a hydrogen car may, perhaps, be regarded as less hazardous than a petrol car *in the open air*. However, if those burning tests had been carried out inside a typical road tunnel, the safety concerns of having such a long jet flame in a confined space, with a very restricted ceiling height, becomes apparent.

The low ignition energy and the very wide flammability range of a hydrogen–air mixture are of particular importance for the hazard assessment of hydrogen cars in tunnels. The low ignition energy indicates that the hydrogen fuel would be likely to be ignited as soon as it is released. Prompt ignition (i.e. soon after the start of release) can prevent a hydrogen cloud building up inside the tunnel, and therefore prevent explosions inside the tunnel. However, in that case there would be the problem of a possible jet flame. If the ignition is not prompt, then the gas cloud may build up, and there would be a very significant explosion risk. The jet flame scenarios discussed in Wu (2008) are based on the assumption of immediate ignition. Computational fluid dynamics (CFD) simulations of hydrogen jet flames in an arbitrary tunnel (102 m long and with a 5 m by 5 m square cross-section) have been carried out for 'normal' hydrogen release rate (6 MW fire) and 'high' fuel release rate (30 MW fire) scenarios. It was indicated that a 2.5 m/s ventilation velocity has the potential, according to CFD modelling, to completely eliminate the back-layering in the case of a normal hydrogen release rate and to keep the back-layering under control in the case of a high release rate. This would need to be tested experimentally.

A jet flame hazard may be unique for hydrogen cars, compared with other fuels. For a high release rate, the flame inside a tunnel might be in a state of oxygen deficit. This would be expected to result in impingement of a hydrogen jet flame on the tunnel ceiling, and to produce a high temperature ceiling flow reaching a substantial distance and causing damage to tunnel infrastructure. It is indicated that there would be intense heat downstream, and this would need to be dealt with.

Although hydrogen has a low ignition energy and can be ignited easily, there is always a possibility that ignition may not occur within a very short period after the start of release. Substantial hydrogen release into a tunnel without immediate ignition following the onset of release would be expected to result in the formation of a hydrogen cloud inside the tunnel, and to pose explosion hazards. The consequences of a hydrogen explosion inside a tunnel are much more difficult to predict.

Given the highly hazardous nature of hydrogen, it is essential that vehicles be of adequate design. However, research by Molkov (2010) raises serious concerns about pressure relief devices (PRDs). It was concluded that 'current PRDs and on-board storage tanks are unacceptable and should be re-designed'. Also, concerns have arisen about test methods. Lonnermark (2010) has raised questions about several new fuels, not just hydrogen: for example, dimethyl ether (DME) and ethanol. General questions include the following.

- What scenarios do the methods represent?
- How well defined is the fire source?
- What does the temperature measurement represent?
- Are jet flames represented in the test methods?

33.4. Mitigation measures

Mitigation measures which might be considered are: (1) dilution of fuel for un-ignited hydrogen and (2) heat removal from ignited hydrogen. This section includes some very preliminary ideas on possible methods of dealing with hydrogen leaks and mitigating the effects of hydrogen fires. It is evident that there needs to be far more work on this. The ideas below may be regarded as intended for consideration and discussion.

33.4.1 Hydrogen dilution using ventilation

If there is an un-ignited hydrogen release inside a tunnel, then ventilation could be used to dilute the hydrogen fuel. For hydrogen dilution, the ventilation should aim at diluting the hydrogen concentration to a level less than the lower flammability limit of a hydrogen–air mixture, which is 4.1%. During the dilution process, and after, measures should be taken to eliminate possible ignition sources, such as sparks or sources of friction. In particular, for a hydrogen release, it would be necessary to stop the traffic to avoid the ignition of hydrogen via running engines or moving vehicles.

33.4.2   Flame cooling by ventilation, and a water-based fire protection system

When hydrogen is ignited during a release, intense heat is released over a very short period of time, and there is a build up of a high-temperature zone downstream, which could damage the tunnel facilities and structure and act as an ignition source to the surroundings. The priority in dealing with a hydrogen flame inside a tunnel is to mitigate the heat in the immediate area affected by the hydrogen flame.

In the case of a hydrogen flame inside a tunnel, tunnel ventilation could, in theory, be used to deliver cool air to disperse the heat and to reduce the temperature. However, in reality, the ventilation could extend the hazardous high-temperature zone downstream. The cooling effect from the ventilation is influenced by the tunnel geometry and type of ventilation system. It might be possible, in principle, that the ventilation could provide sufficient cooling in large and short tunnels which have a few metres of clear head-space above the hydrogen vehicles. However, for a long tunnel with a low ceiling, the ventilation may not be effective and may produce more problems downstream. It is evident that the possible use of forced ventilation in trying to deal with a hydrogen flame inside a tunnel is problematic.

Tests on traditional deluge and water mist systems in tunnel fires have indicated that water-based fire protection has the potentiality to remove the heat effectively and provide almost instant cooling inside a tunnel: see Chapter 8 on fire suppression in tunnels. Among the water-based fire protection systems, water mist systems utilise very fine water droplets, and are regarded as being effective in extinguishing gaseous flames. Water mist systems could, perhaps, be used to remove the intense heat generated by hydrogen flames. Once the heat has been removed, the ventilation could be used in a 'normal' way to clear the tunnel atmosphere.

33.4.3   Other scenarios and hydrogen residues

For enclosed spaces, such as car garages, hydrogen accumulation and residues (i.e. other than resulting from an accident or other incident) could be an issue, and continuous ventilation is recommended to prevent hydrogen concentration building up. In tunnels, the ventilation already exists in the form of natural ventilation or via forced-ventilation systems. Hydrogen accumulation, other than in the case of an accident or other unwanted event, is unlikely to be a matter of significant concern in tunnels. Hydrogen is very active and can bond with other molecules easily. Therefore, any hydrogen residues are not expected to stay very long inside a tunnel. This is an advantage for hydrogen compared with other hydrocarbon fuels. That is, hydrocarbon fuel vapour may accumulate inside a tunnel, but hydrogen may diffuse quickly or bond with other molecules. It is unlikely that hydrogen would accumulate in 'routine' operation, so normal ventilation routines should, in principle, be sufficient to keep a tunnel clear of hydrogen residues. If ignition were to occur, however, then forced ventilation may increase the high-temperature zone downstream (see above).

33.5.   Outlook

An initial assessment and discussion of the fire hazards and fire scenarios associated with allowing hydrogen cars to use existing tunnels has been given. It is clear that there are significant uncertainties regarding how to deal with the potential hazards related to hydrogen release in tunnels. In a report to the European Parliament (Beard and Cope, 2008), it was recommended that hydrogen-powered vehicles be prevented from passing through road tunnels until a comprehensive and exhaustive independent assessment has been conducted. It was concluded that, even after such an assessment, it would not be wise to allow hydrogen-powered vehicles to pass through tunnels.

It is clear that problematic issues regarding vehicle design and how to deal with ignited and un-ignited hydrogen inside a tunnel must be addressed satisfactorily if tunnels are to be regarded as having acceptable safety measures and procedures for accidents and other incidents involving hydrogen-powered cars.

The ideas put forward in this chapter are tentative, and much discussion and consideration is required, as the issue is a complex one. Thorough experimental testing, carried out independently, is essential in the consideration of any possible measures. As indicated above, after substantial investigation and consideration, it may come to be regarded as prudent not to allow hydrogen-powered vehicles to pass through tunnels. This chapter has concerned itself with cars. Other hydrogen-powered vehicles, for example buses, should, of course, be examined as well.

REFERENCES

Beard AN and Cope D (2008) *Assessment of the Safety of Tunnels*. Report to the European Parliament. Available on the website of the European Parliament, under the rubric of Science & Technology Options Assessment (STOA).

Lonnermark A (2010) Special fire risks associated with new energy carriers. *International Conference: FIVE – Fires in Vehicles, Gothenburg*, 29–30 September 2010. SP, Borås, Sweden.

Molkov V (2010) Safety issues of hydrogen-powered vehicles. *International Conference: FIVE – Fires in Vehicles, Gothenburg, Sweden*, 29–30 September 2010. SP, Borås, Sweden.

Swain MR (2001) Fuel leak simulation. *Proceedings of the 2001 Hydrogen Program Review*, NREL/CP-570–30535. Department of Energy, Washington, DC.

Wu Y (2008) Assessment of the impact of jet flame hazard from hydrogen cars in road tunnels. *Transportation Research, Part C* **16**: 246–254.

Handbook of Tunnel Fire Safety, 2nd edition
ISBN: 978-0-7277-4153-0

ICE Publishing: All rights reserved
doi: 10.1680/htfs.41530.677

Index

Page numbers in *italics* refer to figures separate from the corresponding text.